PHYTOTHERAPIES

PHYTOTHERAPIES

Efficacy, Safety, and Regulation

Edited by

IQBAL RAMZAN

WILEY

Published by John Wiley & Sons, Inc., Hoboken, New Jersey
Published simultaneously in Canada

Library of Congress Cataloging-in-Publication Data:

Phytotherapies : efficacy, safety, and regulation / edited by Iqbal Ramzan.
p. ; cm.
Includes bibliographical references and index.
ISBN 978-1-118-26806-3 (cloth)
I. Ramzan, Iqbal, 1951–, editor.
[DNLM: 1. Phytotherapy–methods. 2. Phytotherapy–standards.
3. Quality Assurance, Health Care. WB 925]
RS164
615.321–dc23

2014049520

Set in 10/12pt Times LT Std by SPi Global, Pondicherry, India

10 9 8 7 6 5 4 3 2 1

1 2015

CONTENTS

LIST OF CONTRIBUTORS

Mohi Iqbal Mohammed Abdul, College of Pharmacy, Taibah University, Madina, Kingdom of Saudi Arabia

Bolanle A. Adeniyi, Department of Pharmaceutical Microbiology, University of Ibadan, Ibadan, Nigeria

Maureen V. Boost, Faculty of Health and Social Sciences, The Hong Kong Polytechnic University, Hong Kong

Hak-Kim Chan, Faculty of Pharmacy, The University of Sydney, Sydney, New South Wales, Australia

Kelvin Chan, Faculty of Pharmacy, The University of Sydney; and National Institute for Complementary Medicine, University of Western Sydney, Sydney, New South Wales, Australia

Meiwan Chen, State Key Laboratory of Quality Research in Chinese Medicine, Institute of Chinese Medical Sciences, University of Macau, Taipa, Macau, China

Jun-Lae Cho, Faculty of Pharmacy, The University of Sydney, Sydney, New South Wales, Australia

Sigrun Chrubasik-Hausmann, Institute of Forensic Medicine, University of Freiburg, Freiburg, Germany

Stephen J. Clarke, Sydney School of Medicine and Northern Clinical School, Kolling Institute of Medical Research, Royal North Shore Hospital, The University of Sydney, Sydney, New South Wales, Australia

Alain Cuerrier, Institut de recherche en biologie végétale, l'Université de Montréal, Montréal, Canada

Pei H. Cui, Faculty of Pharmacy, The University of Sydney, Sydney, New South Wales, Australia

Colin C. Duke, Faculty of Pharmacy, The University of Sydney, Sydney, New South Wales, Australia

Gabrielle Escalante, Department of Pharmacy Practice, College of Pharmacy, PAHO/WHO Collaborating Center for Traditional Medicine, University of Illinois at Chicago, Chicago, Illinois, USA

Sophia Yui Kau Fong, School of Pharmacy, The Chinese University of Hong Kong, Shatin, Hong Kong

Dong Fu, Faculty of Pharmacy, The University of Sydney, Sydney, New South Wales, Australia

Qiong Gao, School of Pharmacy, The Chinese University of Hong Kong, Shatin, Hong Kong

Cheong Hian Goh, Audit and Licensing Division, Health Products Regulation Group, Health Sciences Authority, Singapore, Republic of Singapore

Paul W. Groundwater, Faculty of Pharmacy, The University of Sydney, Sydney, New South Wales, Australia

David E. Hibbs, Faculty of Pharmacy, The University of Sydney, Sydney, New South Wales, Australia

Tom Hsun-Wei Huang, Faculty of Medicine, The University of Sydney, Sydney, New South Wales, Australia

Fangming Jin, School of Pharmacy, Shaanxi University of Chinese Medicine, Xi'an, Shaanxi, China; and Global Therapeutics Pty Ltd, Byron Bay, New South Wales, Australia

Moon-Sun Kim, Faculty of Pharmacy, The University of Sydney, Sydney, New South Wales, Australia

Hwee-Ling Koh, Department of Pharmacy, Faculty of Science, National University of Singapore, Singapore, Republic of Singapore

Temitope O. Lawal, Department of Pharmaceutical Microbiology, University of Ibadan, Ibadan, Nigeria

George Q. Li, Faculty of Pharmacy, The University of Sydney, Sydney, New South Wales, Australia

Vivian Wan Yu Liao, Faculty of Pharmacy, The University of Sydney, Sydney, New South Wales, Australia

Ernest V. Linek, Banner & Witcoff, Ltd., Boston, Massachusetts, USA

Hao Liu, Faculty of Health and Social Sciences, The Hong Kong Polytechnic University, Hong Kong

Gail B. Mahady, Department of Pharmacy Practice, College of Pharmacy, PAHO/WHO Collaborating Center for Traditional Medicine, University of Illinois at Chicago, Chicago, Illinois, USA

Laura J. Mahady, The Barrow Neurological Institute and Arizona State University, Phoenix, Arizona, USA

Andrew J. McLachlan, Faculty of Pharmacy and Centre for Education and Research on Ageing, The University of Sydney and Concord Hospital, Sydney, New South Wales, Australia

Pooja Mikkilineni, Department of Pharmacy Practice, College of Pharmacy, PAHO/WHO Collaborating Center for Traditional Medicine, University of Illinois at Chicago, Chicago, Illinois, USA

Rosina Yau Mok, School of Pharmacy, The Chinese University of Hong Kong, Shatin, Hong Kong

Christina L. Nance, Department of Pediatrics, Baylor College of Medicine, Immunology, Allergy and Rheumatology, Texas Children's Hospital, Houston, Texas, USA

Rajeshwar Narlawar, School of Chemistry, The University of Sydney, Sydney, New South Wales, Australia

Jennifer A. Ong, Faculty of Pharmacy, The University of Sydney, Sydney, New South Wales, Australia

Iqbal Ramzan, Faculty of Pharmacy, The University of Sydney, Sydney, New South Wales, Australia

Michel Rapinski, Institut de recherche en biologie végétale, l'Université de Montréal, Montréal, Canada

Michael Rieder, Departments of Paediatrics, Physiology and Pharmacology, and Medicine, Schulich School of Medicine & Dentistry; and Robarts Research Institute, Western University, London, Ontario, Canada

Bashar Saad, Qasemi Research Center-Al-Qasemi Academic College, Baga Algharbiya, Israel; and Faculty of Arts and Sciences, Arab American University Jenin, Jenin, Palestine

Stephen Sagar, Departments of Oncology and Medicine, McMaster University, Hamilton, Ontario, Canada

Susan J. Semple, Sansom Institute for Health Research, School of Pharmacy and Medical Sciences, University of South Australia, Adelaide, South Australia, Australia

Gint Silins, Cullens Patent & Trade Mark Attorneys, Brisbane, Queensland, Australia

Bradley S. Simpson, Flinders Centre for Innovation in Cancer, School of Medicine, Flinders University, Bedford Park, South Australia, Australia

Huanxing Su, State Key Laboratory of Quality Research in Chinese Medicine, Institute of Chinese Medical Sciences, University of Macau, Taipa, Macau, China

Daniel M.-Y. Sze, School of Medical Sciences and Health Innovations Research Institute (HiRi), RMIT University, Australia

Jennifer Tan, E-TQCM Consultants Limited, Tsuen Wan, Hong Kong

Jian-Bo Wan, State Key Laboratory of Quality Research in Chinese Medicine, Institute of Chinese Medical Sciences, University of Macau, Taipa, Macau, China

Shengpeng Wang, State Key Laboratory of Quality Research in Chinese Medicine, Institute of Chinese Medical Sciences, University of Macau, Taipa, Macau, China

Yi-Tao Wang, State Key Laboratory of Quality Research in Chinese Medicine, Institute of Chinese Medical Sciences, University of Macau, Taipa, Macau, China

Lynn Weekes, NPS MedicineWise, Surry Hills, New South Wales, Australia

Elizabeth M. Williamson, The School of Pharmacy, Whiteknights, Reading, Berkshire, United Kingdom

Raimond Wong, Departments of Oncology and Medicine, McMaster University, Hamilton, Ontario, Canada

Yin Cheong Wong, School of Pharmacy, The Chinese University of Hong Kong, Shatin, Hong Kong

Wai-Ping Yau, Department of Pharmacy, Faculty of Science, National University of Singapore, Singapore, Republic of Singapore

Xiao-Jing Zhang, State Key Laboratory of Quality Research in Chinese Medicine, Institute of Chinese Medical Sciences, University of Macau, Taipa, Macau, China

Qi (Tony) Zhou, Faculty of Pharmacy, The University of Sydney, Sydney, New South Wales, Australia

Zhong Zuo, School of Pharmacy, The Chinese University of Hong Kong, Shatin, Hong Kong

PREFACE

This book focuses on many facets of the use of Phytotherapies in preventing or treating illness and diseasc internationally. Phytotherapies are variously defined by practitioners and scientists but include herbal therapies, therapies used by indigenous peoples around the world and alternative medicines as opposed to classical western medicines.

I had not considered editing a book on Phytotherapies as the focus of my scientific career has not been on Phytotherapies. However, in the last 10 years I have investigated some of the pharmacological and more specifically, the hepatotoxic effects of kava. This interest has arisen from my background in that I was born and educated in Fiji where kava is widely used for social and ceremonial functions for centuries.

I was very surprised to receive an invitation from Jonathan Rose at Wiley some 2 years ago to ask if I had an interest in editing such a book. Not having edited an entire book previously, I naively agreed, of course, not realizing the magnitude of the task ahead. I realize now that editing an entire book is exponentially more challenging than contributing to a Book Chapter or even publishing numerous journal articles.

The topic interested me for several other reasons. The Faculty of Pharmacy at the University of Sydney had for many years hosted a Herbal Medicines Education and Research Centre (HMREC) and the Faculty also offered a Masters degree in Herbal Medicines. This program was moderately successful financially and the Centre was closed following an external review that I instituted as Dean of the Faculty. However, I do believe that it is important to examine in a scientific manner, the various forms of Phytotherapies used around the globe as use of such therapies continues to increase.

Phytotherapies are at the heart of disease management in countries such as China and India where they are used instead of and alongside Western medicines.

In the west, the use of Phytotherapies continues to grow at a phenomenal rate. Whether this reflects the dissatisfaction with modern western medicine or the perception that Phytotherapies are natural and thus free of any adverse effects is open to conjecture. However, there is certainly a belief especially among younger people that Phytotherapies are promising alternatives to modern drugs not only in promoting well-being and preventing disease but also in managing some conditions.

Identifying suitable Chapter authors was very challenging due to the diverse and varied nature of the field. I believed it was important to identify suitable scientists with the research and scientific credentials to bring reputational credit to such a book and to ensure balanced and erudite debate. This was confounded by language and cultural sensitivities relating to Phytotherapy use and the evidence base for use in different cultural and ethnic contexts.

Having succeeded in identifying potential Chapter authors the other interesting observation was that while these authors have individually made a strong contribution to the evidence base for the use of Phytotherapies some were also philosophically committed to clinical paradigms that promote the use of Phytotherapies. Separating this attachment to the adoption of Phytotherapies from the scientific evidence for their use was an additional challenge that I had not anticipated in accepting to edit such a book.

If you, the reader, like this book and find it informative and useful in either your practice, for your students or indeed as a resource in your scientific library, then I hope I have been able to objectively separate out the evidence base and summarise some of the science in this vast field of Phytotherapies. The other issue which I wanted to come across in the book is the rightful acknowledgment of the breath of the topic and the variety of the evidence base that is available for the use of Phytotherapies.

I want to thank Angela Teklic for her tireless effort in assisting me with the formatting of the Chapters and making sure that the Book complied with the Wiley template. Both of us underestimated this mammoth task but the attention to detail displayed by Angela made my life more bearable. Eleanor Luntao was very valuable in making sure the contributor agreements were in place and that permissions to reproduce published material were obtained. Eleanor's dedication during the proof-reading stage was also exemplary. Finally, I would like to thank my wife, Dr Lynn Weekes AM, who had to spend many hours alone while I spent days editing this book.

Professor IQBAL RAMZAN

The University of Sydney
February 2015

1

PHYTOTHERAPIES—PAST, PRESENT, AND FUTURE

IQBAL RAMZAN AND GEORGE Q. LI
Faculty of Pharmacy, The University of Sydney, Sydney, New South Wales, Australia

1.1 OVERVIEW OF PHYTOTHERAPY

1.1.1 Definition

Phytotherapy, or the use of herbal medicines to prevent or treat a disease, is a traditional medical practice based on medicinal plants. It is a branch of complementary and alternative medicine (CAM) or traditional medicine, which refers to traditional medicine systems and various forms of Indigenous medicine [1]. Many different cultures have developed herbal medicine systems, for example, Western herbal medicines, Chinese herbal medicines, Ayurvedic and Unani medicines, and Australian Indigenous medicines [2]. Phytotherapy is the basis of modern pharmaceutical science, with about 25% of the drugs prescribed today, such as digoxin, aspirin, and paclitaxel being derived from plants [3].

Western herbal medicine and orthodox medicine share to a large degree a common physiologic and diagnostic system, but they are different in many important ways as well. Herbs are complex mixtures of chemicals, which may have several distinct and concurrent pharmacological activities, while pharmaceutical drugs are mostly single chemical entities. Modern herbal medicines are becoming part of integrative clinical management in medical textbooks as exemplified in *Natural Standard Herbal Pharmacotherapy* [4].

Traditional Chinese Medicine (TCM) is another popular traditional medical system in China and worldwide. It includes various practices including Chinese

Phytotherapies: Efficacy, Safety, and Regulation, First Edition. Edited by Iqbal Ramzan.

herbal medicine, acupuncture, and massage, sharing a fundamental principle that the human body is part of the whole universe. The treatment goals are harmonization and balance using a holistic approach. The basic theories of TCM are Yin and Yang theory, Five-Element theory, Zang Fu (viscera and bowels) theory, Meridian, Qi, Blood and Fluid theory, Syndrome Differentiation, and Treatment theory. Detailed information on TCM can be found in textbooks on Chinese medicine [5–8]. For example, the blockage by Phlegm is closely related to excessive fat retention in metabolic syndrome and the management with herbal formulations and other modalities is to eliminate the Phlegm [9]. Treatment of diabetes with TCM focuses on nourishing Yin, clearing Heat, producing Body Fluid, and moistening Dryness using herbal formulae composed of herbs such as Rehmannia (*Rehmannia glutinosa*) and yam (*Dioscorea opposita*) [10].

Modernization of TCM and integration with orthodox medicine and science is a model accepted in China, covering education, clinical practice, and research. Modern pharmacologic and clinical studies have been used to examine claims of traditional practice; chemistry and chemical analysis are used for quality control of Chinese herbal medicines. Pharmacological and chemical studies have revealed connections between nature of herbal medicines and pharmacological activities, herbal tastes, and chemical components. For example, ephedra is warm as it contains ephedrine, a sympathomimetic amine; pungent herbs contain essential oils; sour herbs contain acid and tannins; sweet herbs contain sugars, proteins, and amino acids; bitter herbs contain alkaloids and glycosides; and salty herbs contain inorganic salts. Pharmacokinetic studies demonstrate a link between the tissue distribution of active chemical constituents and the attributive meridians of Chinese herbal medicines.

The World Health Organization (WHO) has a long-term interest in promoting traditional medicines and has produced a series of publications on global atlas [11], good agricultural practices [12], and monographs on selected medicinal plants [13], providing scientific information on the safety, efficacy, and quality control of widely used medicinal plants. The latest version of *WHO Traditional Medicine Strategy* (2014–2023) was developed to support Member States in harnessing the potential contribution of traditional medicine to health, wellness, and health care; and promoting the safe and effective use of traditional medicines by regulating, researching, and integrating traditional medicine products, practitioners, and practice into health systems [14].

1.1.2 International Trend in the Usage of Complementary Medicines

Complementary medicines have maintained their popularity in all regions of the world. The global market for herbal medicines is significant and growing rapidly. In China, traditional herbal preparations account for approximately 40% of the total health care delivered [1]. In the United States, over 42% of the population have used complementary or alternative medicine at least once. Total out-of-pocket expenditure relating to alternative therapies in 1997 was conservatively estimated at $27.0 billion, which is comparable with the projected 1997 out-of-pocket expenditure for all US physician services [15]. In the United Kingdom, estimate of annual out-of-pocket expenditure on practitioner visits in 1998 was £450 million [16].

In Australia, it has been reported that in 2000, 52% of the population used at least one nonmedically prescribed complementary medicine [17]. The estimated expense on complementary medicines was nearly twice the patient expenditure on pharmaceutical medicines during 1992–1993 [17]. The expenditure on alternative therapies in 2000 was $AUD 2.3 billion [18]. In 2005, the annual out-of-pocket expenditure was estimated to be $AUD 4.13 billion [19]. More recent studies have indicated that complementary medicines are finding a growing preference amongst patients with chronic or serious diseases who are looking for natural options to assist in the ongoing management of these conditions. For instance, St. John's wort preparations have low rates of side effects and good compliance, comparatively low cost, making it worthy of consideration in the management of mild-to-moderate depression [20]. An overview of complementary medicines use and regulation in Australia is available in the Australian government's commissioned report, *Complementary Medicines in the Australian Health System* [21].

1.2 PRECLINICAL RESEARCH ON PHYTOTHERAPIES

1.2.1 Pharmacognosy and Quality Standardization of Phytotherapies

Pharmacognosy is the study of medicinal materials, mainly plants, using theory and methods of modern sciences such as botany, zoology, chemistry, pharmacology, and traditional medicines to study the origin, production, harvesting and processing, identification and evaluation, chemical components, physical and chemical properties, resource development, pharmacology, toxicology, and therapeutic application of herbal medicines to ensure the quality of herbal materials and to develop new herbal resources. Its main focus is on the study of authentication and quality control of herbs [22].

Plant descriptions are used in the identification of herbal materials. They are first classified by the plant parts of origin, such as roots and rhizomes, stems, leaves, flowers, fruits, or whole herbs. Then the macroscopic and microscopic descriptions are included in each monograph. Some microscopic features reflect the secondary metabolites, starch granules, resin ducts, and oil cells. The macroscopic features are still very useful for authentication; for example, the colors of herbs such as yellow coptis, brown rhubarb, and black valerian are related to their alkaloid, anthraquinone, and iridoid contents, respectively.

Pharmacognosy, particularly correct identification and high quality of the herb, is the foundation of safety, clinical efficacy, and research on phytotherapy. It is a subject most relevant to professionals in testing laboratories, herbal dispensing, and regulatory bodies. Pharmacognosy is the principal discipline employed in national and international pharmacopeia in the form of the following topics: species identification using plant taxonomy, macroscopic identification using morphology, microscopic identification using anatomy, and quality control with analytical methods. The WHO monographs are examples of such comprehensive monographs [13], while *British Pharmacopoeia* used as statutory standards in Europe and Australia focuses on chemical analysis for quality control [23].

Bioequivalence is a useful concept in the quality standardization of herbal medicines. *European Guideline on the Investigation of Bioequivalence* defined bioequivalence as same active substances and similar bioavailability that results in similar clinical effectiveness and safety [24]. To approve two products to be bioequivalent, the following studies need to be carried out: pharmaceutical equivalence (quality standardization), pharmacokinetic equivalence (same bioavailability and time-to-peak concentration), pharmacodynamic equivalence (*in vivo* and *in vitro*), and therapeutic equivalence (clinical study). For example, a study found that the bioavailability of ginkgolide A, ginkgolide B, and bilobalide of two different *Ginkgo biloba* commercial brands were clearly different and did not demonstrate bioequivalence of test and reference products. The slow *in vitro* dissolution of the test product resulted in a large decrease in bioavailability [25]. The bioequivalence concept implies the need for a comprehensive platform for evaluation of herbal products [22].

Kudzu root is an example of a herb requiring a comprehensive quality control platform. Kudzu is one of the most commonly used Chinese herbal medicines for the treatment of diabetes, cardiovascular disease, and many other conditions. It includes two closely related species, *Pueraria lobata* and *Pueraria thomsonii*, which are not well-differentiated in pharmacopoeias. Isoflavonone puerarin is currently used as a marker for quality control of the species [26]. Recent studies indicate that ultra-performance liquid chromatography combined with partial least square discriminant analysis (PLS-DA) was more effective than using puerarin alone in differentiating the two species [27]. HPTLC coupled with multivariate classification analyses has also been used effectively to differentiate the two species [28].

Similarly, multiple markers have been used in the quality control of propolis. High-performance liquid chromatography with UV detection has been used to simultaneously quantify the eight major bioactive phenolic compounds in Chinese propolis [29] and a rapid thin-layer chromatography combined with chemometric fingerprinting has also been used to distinguish Chinese propolis from poplar tree gum [30].

1.2.2 Pharmacological Studies and Identification of Bioactive Compounds

Herbal pharmacology is the study of the function and mechanism of action of herbal medicines in biological systems and the pharmacokinetics of herbal compounds with modern scientific methods to understand the underlying nature of the likely clinical application. Herbal medicines are unique in that they contain multiple components and can act on multiple pharmacologic targets. The major types of herbal pharmacology research are *in vitro* studies at the cellular or tissue level to uncover the mechanism of action of the herbal components at the molecular level, for example, cytotoxicity in cancer cell lines; whole animal models to test preclinical properties of herbal medicines and to determine the pharmacokinetic properties, for example, streptozotocin-induced diabetic rats and human clinical studies to confirm the efficacy and safety of the herbal medicines. For instance, preclinical and limited clinical evidence have shown pentacyclic triterpenoids including the oleanane, ursane, and lupane groups have multiple biological activities and may contribute to their use in traditional medicine for the treatment of diabetes and diabetic complications [31]. Increasing evidence also has shown common chemical components such as gallic acid, a common phenolic compound, playing some role in the potential health benefits of food and nutraceuticals [32, 33]. Quercetin is clinically used as a nutraceutical for cardiovascular disease [34], and berberine has been used for the management of diabetes [35].

St. John's wort is an example of a herb with a huge body of research on the chemistry, analysis, and pharmacological actions. The active compounds may include naphthodianthrones (e.g., hypericins), flavonoids (rutin, quercetin), and phloroglucinols (hyperforin) individually or in combination. St. John's wort extracts have been found to interact with a number of neurotransmitter systems implicated in depression and in psychiatric illness generally, such as uptake of serotonin, noradrenaline, and dopamine and to interact with γ-aminobutyric acid (GABA) receptors, monoamine oxidases, catechol-*O*-methyltransferase, and dopamine-beta hydroxylase [36]. However, the exact active compound(s) and mechanism(s) are still to be fully defined.

Lavender flower (*Lavandula officinalis*) is another example of a herb having multiple actions. This herb is used for anxiety, insomnia, antimicrobial activity, dyspepsia, wounds, and sores, and pharmacological studies have focused on anxiety, but cover other actions. Lavender oil showed significant dose-dependent anxiolytic activity in rats and mice, comparable to that of the standard anxiolytic agent lorazepam and also increased pentobarbital-induced sleeping time [37]; lavender oil also lowered the mean heart rate in dogs [38]. Mechanistic studies revealed it inhibited voltage-dependent calcium channels in synaptosomes, primary hippocampal neurons [39], and increased the dopamine D3 receptor subtype in the olfactory bulb of mice [40]. The lavender essential oils are dominated by oxygenated monoterpenes including linalyl acetate, linalool, 1,8-cineole, fenchone, camphor, nerol, and borneol. However, the exact compositions are dependent on the varieties and steaming process [38, 41, 42], which can impact the biological and clinical outcomes. While the current actions are mostly based on the total effects of the essential oils, identification of active ingredients should help future quality standardization of the extracts.

Overall, for most herbal medicines, the mechanism of action and the nature of active constituents are still not well-defined. Furthermore, most research involving herbal medicines concentrates on establishing biological activities of purified single compounds, or crude extracts without a defined fingerprint of the extract or formulation. New research platforms need to be multidisciplinary in nature to cover the research from single constituent activity to multiple biological activities linking to various standardized extracts.

1.2.3 Application of Proteomics and Metabolomics in Phytotherapy Research

To address the multi-ingredient and multitarget nature of herbal medicines and TCM formulae, network pharmacology or systems biology approach has been used in phytotherapy research in the past few years [43, 44]. Protein–protein interaction network and topological attributes related to the biological targets of the ingredients were integrated to identify active ingredients in herbal medicines [45].

Progress in analytical techniques, such as matrix-assisted laser desorption/ionization time-of-flight mass spectrometry (MALDI-TOF-MS) combined with bioinformatics, proteomics, and metabolomics have attracted increasing attention. The use of metabolomics has led to the discovery of the metabolite 2-aminoadipic acid as a marker of diabetes risk in humans [46] and two differentiating urinary metabolites involved in key metabolic pathways of sugar have been identified in high-fat-diet-induced type 2 diabetic rats [47].

Proteins and small metabolites are more responsive to disease, environment, and drug treatment and may be more relevant to the holistic approach in traditional medicines [48, 49]. Omic studies may provide answers on epigenetic effects on gene expression and polymorphisms of cytochrome P450 liver enzymes or P-glycoprotein [50]. Herbal medicines have elicited changes in proteins in wound healing in rats [51] and liver HepG2 cells [52]. Treatment with berberine of patients with type 2 diabetes and dyslipidemia led to a highly significant decrease in the concentrations of 13 fatty acids, suggesting that berberine might play a pivotal role in the treatment of type 2 diabetes by downregulating the high level of free fatty acids [35]. In rats, epimedium herb was shown to reverse perturbations in plasma levels of phenylalanine, tryptophan, cholic acid, and other metabolites regulating oxidant–antioxidant balance, amino acid, lipid, and energy metabolism, respectively, and gut microflora [53].

1.3 CLINICAL RESEARCH ON PHYTOTHERAPIES

1.3.1 Efficacy of Popular Phytotherapies

Clinical evidence on herbal medicine comes as case reports and/or clinical data. In the past 5 years, there have been over two hundred systematic reviews on herbal medicines and traditional Chinese medicines published in the Cochrane Library, including reviews on the most popular herbs, such as ginkgo (*Ginkgo biloba*), St. John's wort (*Hypericum perforatum*), ginseng (*Panax ginseng*), valerian (*Valeriana officinalis*), hawthorn (*Crataegus monogyna*), echinacea (*Echinacea* species), milk thistle (*Silybum marianum*), bitter melon (*Momordica charantia*), and black cohosh (*Cimicifuga* species).

Although a large number of trials report positive outcomes, the reviews reveal no conclusive evidence on the efficacy of the popular herbs, including ginkgo for cognitive impairment and dementia [54], ginseng for cognition [55], echinacea for preventing and treating the common cold [56], milk thistle for alcohol or nonalcohol hepatitis and other liver diseases [57], bitter melon for type 2 diabetes mellitus [58], and black cohosh and phytoestrogens for menopausal symptoms [59, 60]. A positive conclusion has been drawn for St. John's wort for major depression, as available evidence suggests that the hypericum extracts tested in the relevant trials are superior to placebo in patients with major depression; and are as effective as standard antidepressants [61]. In addition, hawthorn extract is beneficial in symptom control as an adjunct for chronic heart failure treatment [62].

Lavender is another popular phytotherapy with positive effects for the management of anxiety. In a randomized, double-blind, double-dummy trial, 539 adults with generalized anxiety disorder received 160 or 80 mg lavender preparation, Silexan®, 20 mg paroxetine, or placebo once daily for 10 weeks. Silexan was more efficacious than placebo [63]. A systematic review of seven trials concluded Silexan was significantly superior to placebo in patients with subsyndromal anxiety and was comparable to lorazepam [64].

1.3.2 Chinese Herbal Medicines

In TCM, herbal medicines are normally used in formulae that are based on classic prescriptions and subtypes of clinical syndrome. Therefore, clinical trials often involve different formulae, making meta-analysis of trials impossible or difficult.

In the past 5 years, a large proportion of published systematic reviews in the Cochrane library are on TCM, covering common chronic conditions such as osteoporosis [65], hypertriglyceridemia [9], fatty liver [66], acute bronchitis [67], severe acute respiratory syndrome and irritable bowel syndrome [68], premenstrual syndrome [69], and type 2 diabetes mellitus [70]. Many small and less-rigorous trials report positive findings. However, they should be interpreted with caution due to inappropriate methodology, small sample size, and lack of confirmatory data. Similar findings are noted for single-herb preparations, including Danshen (*Salvia miltiorrhiza*) preparations for acute myocardial infarction [71], puerarin injection (*Pueraria lobata*) for unstable angina pectoris [72], and Sanchi (*Panax notoginseng*) for acute ischemic stroke [73]. There is insufficient evidence to support their claims and high-quality trials are needed to support their clinical use.

1.3.3 Food Nutrition and Translational Research

There is no clear border between phytotherapies and foods since many phytotherapies are also used as foods and many foods contain phytochemicals. Recent studies on the impact of nutrition on health and life span have shed some new light on the understanding and the management of metabolic syndrome and cardiovascular disease. In a study with mice fed one of 25 diets *ad libitum*, longevity and health were optimal when protein was replaced with carbohydrate. High-protein diet intakes were associated with hepatic mammalian target of rapamycin (mTOR) activation and circulating branched-chain amino acids and glucose [74]. A cross-sectional study of 1015 Chinese patients who underwent coronary angiography indicated that high animal-protein diet was positively associated with hyperhomocysteinemia, whereas high plant-protein diet was inversely associated with total homocysteine concentrations [75]. In a prospective study of 1003 patients who underwent coronary angiography, higher concentrations of plasma *S*-Adenosyl-L-homocysteine are independently associated with an increased risk of cardiovascular events [76].

At the same time, great interest has been placed on the function of micronutritions in food. One such example is the anthocyanins in rice and fruits. Anthocyanins may play an important role in atherosclerosis prevention, by suppressing oxidative stress-induced endothelial injury in endothelial cells [77, 78], mouse peritoneal macrophages [79], apolipoprotein E-deficient mice [80], and dyslipidemic subjects [81]. The consumption of bayberry juice containing polyphenols for a period of 4 weeks protects against nonalcoholic fatty liver disease in young adults by antioxidant and anti-inflammatory effects [82]. Since food and nutrition are consumed by the public on a daily basis, this research finding will directly impact on an individual's lifestyle.

1.4 SAFETY OF PHYTOTHERAPIES

Phytotherapies are generally shown to be well-tolerated in clinical studies. According to Cochrane reviews, clinical studies often report no additional side effects compared with placebo, as shown with echinacea [56], ginkgo [54], St. Johns wort [61], and Chinese herbal medicines [70]. However, some herbal medicines exhibit toxicity and serious adverse effects. For instance, ephedra causes hypertension, heart attacks, and strokes due to the alkaloid ephedrine, and aristolochia leads to kidney toxicity [83].

Kava (*Piper methysticum*) has been shown to be more effective in a placebo-controlled trial in the treatment of generalized anxiety disorder (GAD) [84]. However, it is associated with over 100 reports of spontaneous adverse hepatic effects. The unexpected toxicity may be related to pharmacokinetic interactions between kavalactones and coadministered drugs or alcohol involving cytochrome P450 enzyme system [85], or inflammation [86] and involvement of liver macrophages [87]. Some authors propose that contaminant hepatotoxins including molds might have caused rare kava hepatotoxicity in humans [88]. Understanding the underlying mechanisms and quality standardization will help to reduce or prevent future toxicity.

While there are many reports and studies on the toxicity of herbal medicines, the standard and ranking criteria of toxicity used for scheduling of herbal medicines remain unclear. Scientific evidence on toxicity comes from systematic reviews, randomized clinical trials, case reports, animal studies, cellular studies, and chemical studies. A scheduling platform has been proposed based on analysis of all available data. Herbs with high toxicity leading to injury or death, for example, aristolochia should be prohibited for medicinal use, while some toxic herbs should be restricted for medicinal use prescribed by qualified practitioners [83]. This will improve regulation and scheduling of Chinese herbal medicines internationally.

Drug–herb interactions pose major concerns for health practitioners. While many interactions are theoretically possible and predicted from preclinical studies, the interactions of St. John's wort with pharmaceuticals have been confirmed in clinical studies. Combining St. John's wort with other antidepressants is strongly discouraged due to potentiation of pharmacodynamic effects. Because St. John's wort can induce CYP3A4 and 2C19, its concurrent use with conventional drugs can decrease the blood concentration of antidepressants, anticonvulsants, antineoplastic drugs, cyclosporin, digoxin, oral contraceptives, and warfarin [89–91]. The popular Chinese herb rhubarb may either induce or inhibit activities of CYP1A2, CYP2C6, CYP2E1, and CYP3A1 and modify the metabolism of antidiabetic drug saxagliptin in rats [92]. As drug–herb combinations are common practice in China, clinical evaluation of safety and efficacy of drug–Chinese herb interactions are required.

1.5 PROFILE OF RESEARCH IN COMPLEMENTARY MEDICINE

1.5.1 International Profile

To obtain an overall profile of phytotherapy research internationally, we have searched major databases for publication counts up to May 2014 using "phytotherapy" as the key word in All Fields in Scopus and retrieved 87,636 publication counts. The publication number increased 10 times from 326 to 3779 from 1994 to 2004, but increased less than 3 times from 2004 to 9698 in 2013 and remained steady in the last 3 years (Fig. 1.1). Although the data did not include publications not using phytotherapy as a key word, this trend indicates that recent progress on phytotherapy research has been slower in quantity. The top countries with highest publication counts during the past two decades in Scopus were India, the United States, and China (Fig. 1.2). The publication counts reflect not only research output but also public interest and scope of phytotherapy industry in these countries. Phytotherapies are part of traditional medicine systems in India and China, and are widely accepted in the United States. They are regulated as dietary supplements in the United States, but as medicines in China.

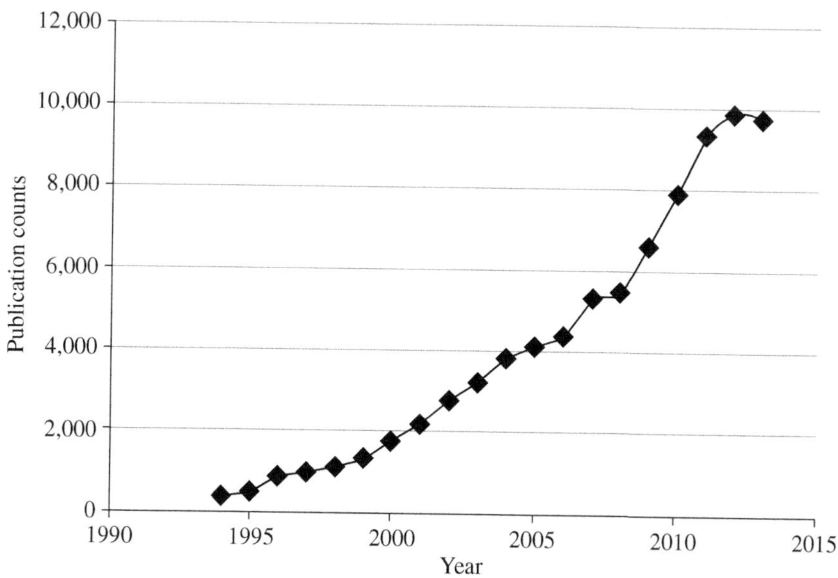

FIGURE 1.1 Publication counts during the past two decades in Scopus.

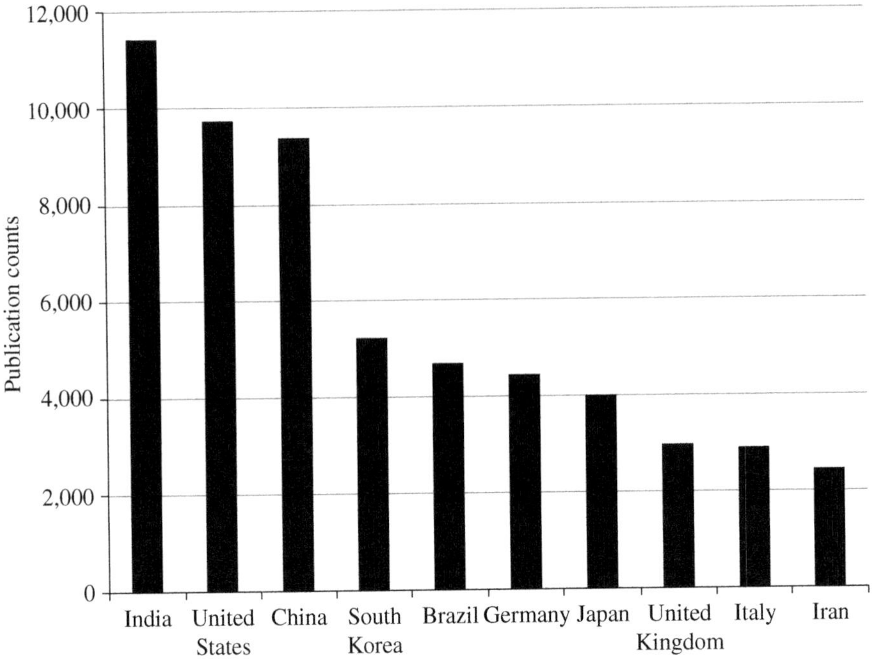

FIGURE 1.2 Top 10 countries with highest publication counts during the past two decades in Scopus.

1.5.2 Australian Profile of Research in Complementary Medicines

Australia plays a leading role in regulation, education, and research on phytotherapies, particularly TCM. Complementary medicine is a listed medicine that needs to meet the requirements of safety and good manufacturing practice standards under the regulation of Therapeutic Goods Administration. Many universities and private colleges offer diploma and/or undergraduate or even postgraduate degrees in herbal medicines and TCM. While herbalists are regulated by professional associations, TCM practitioners are regulated nationally under Australian Health Practitioner Regulation Agency together with other health professions since 2012. Lectures, workshops, and practicals have been introduced into the undergraduate Bachelor of Pharmacy and postgraduate (Master of Pharmacy) courses at the University of Sydney to equip students with the knowledge and skills to provide clinical advice on herbal products available in pharmacy and supermarkets.

In 2007, the Australian Federal government provided a grant to establish the National Institute of Complementary Medicine (NICM) at the University of Western Sydney and approximately AUD $2 million was used to support three NICM collaborative centers: University of Sydney NICM Collaborative Centre for Traditional Chinese Medicine, University of Queensland NICM Collaborative Centre for Transitional Preclinical and Clinical Research in Nutraceuticals and Herbal

Medicine, and Swinburne University of Technology NICM Collaborative Centre for the Study of Natural Medicines and Neurocognition in Health and Disease. A further $5.3 million for 13 projects was provided by the National Health and Medical Research Council (NHMRC). The projects covered different disciplines of CAM and included the following research areas on herbal medicines [93]:

1. Alternative medicines from medicinal plants of Aboriginal people of northern NSW
2. Novel probiotics and naturally sourced extracts as treatment strategies for chemotherapy-induced intestinal mucositis
3. Clinical and physiological evaluation of Chinese herbal medicine for constipation predominant irritable bowel syndrome
4. Complementary medicines based on propolis produced by honeybees from Australian flora
5. A randomized placebo-controlled trial of a herbal preparation in functional dyspepsia: cost-effectiveness and mechanisms

Complementary and alternative medicine was a field of research in the 2012 round of Excellence in Research for Australia (ERA), which was carried out to evaluate research excellence in Australia by the Australian Federal Government. The profile of CAM based on the data during 2008–2010 is listed in Table 1.1 [94]. Overall, CAM (four-digit Field of Research Code, FoR 1104) is a very small component of medical and health sciences (FoR 11). In comparison with Pharmacology and Pharmaceutical Sciences (FoR 1115), CAM had 4, 6, 6% in research income, unit of

TABLE 1.1 Profile of Complementary and Alternative Medicine (CAM) Research from Excellence in Research for Australia (ERA) 2012

	Complementary and Alternative Medicine (FoR 1104)	Pharmacology and Pharmaceutical Sciences (FoR 1115)	Percentage 1104/1115 (%)
Researcher EFT[a]	110.7	444.8	25
Research outputs	814.5	5,210.3	16
Research income	$5,883,895	$139,406,197	4
Unit of evaluation assessed[b]	1	17	6
Esteem count[c]	2.3	38.3	6
Patents	0	23.5	—
Research commercialization income	—	$15,280,047	—

[a]Researchers EFT, researchers full-time equivalent.
[b]Unit of evaluation assessed, number of university included in ranking.
[c]Esteem includes fellowship of learned academy, recipient of a nationally competitive research fellowship, membership of a statutory committee.

evaluation assessed and esteem count, respectively, and no patents and research commercialization income. However, CAM had 25% of researchers and 16% of the publications output in comparison with Pharmacology and Pharmaceutical Sciences, which were very substantial. The data indicate many researchers are publishing in CAM, but they have not attracted similar government funding as Pharmaceutical Sciences. One possible way forward is for governments to establish international joint research centers to bring together different research teams with different areas of technical competency and expertise to form new research platforms, which will lead to multidisciplinary strategies to address the complexity of phytotherapies.

1.6 SUMMARY AND FUTURE DIRECTIONS

Phytotherapy, or the use of herbal medicines to prevent or treat a disease, is a modality of complementary and alternative medicine, or traditional medicine. Its popularity is maintained not only in developing countries but also in Western countries. The long-term traditional knowledge and rich source of medicinal plants have attracted enormous modern scientific research, providing an evidence base for the rationale of traditional practice, and pharmaceutical development and integration into medical practice. In this chapter, we have introduced topics and issues involved in preclinical and clinical disciplines in phytotherapy and further critical reviews may be found in the following chapters of this book.

In the modern and developing economies, chronic diseases such as obesity, diabetes, metabolic syndrome, mental illness, and cancer are leading causes of preventable deaths and are not successfully managed by current clinical and public health measures. There is a clear mandate for the identification of novel approaches, including the development of phytotherapies with respect to both clinical treatment and prevention. There is also a demand for researchers in phytotherapy to be more competitive and have higher profile and impact. Translational research and latest technology such as systems biology, proteomics, and metabolomics are some of the promising approaches to providing a stronger evidence base for traditional medicines, and also to develop new products and formulations for the prevention and treatment of life-threatening chronic conditions for the benefit and well-being of humankind.

REFERENCES

[1] WHO (2002) *WHO Traditional Medicines Strategy 2002–2005*. Geneva: WHO.

[2] Heinrich M, Barnes J, Gibbons S, Willamson EM (2012) *Fundamentals of Pharmacognosy and Phytotherapy*. Edinburgh: Churchill Livingstone, Elsevier.

[3] Wachtel-Galor S, Benzie IFF (2011) Herbal medicine: an introduction to its history, usage, regulation, current trends, and research needs. In: Benzie IFF, Wachtel-Galor S, editors. *Herbal Medicine: Biomolecular and Clinical Aspects*. Boca Raton: CRC Press.

[4] Ulbricht U, Seamon S (2010) *Natural Standard Herbal Pharmacotherapy: An Evidence-Based Approach*. St. Louis: Mosby/Elsevier.

[5] Maciocia G (2005) *The foundations of Chinese medicine*. Philadelphia: Elsevier Churchill Livingstone.

[6] Bensky D (1993) *Chinese Herbal Medicine: Materia Medica*. Seattle: Eastland Press.

[7] Bensky D (1990) *Chinese Herbal Medicine: Formulas & Strategies*. Seattle: Eastland Press.

[8] Hicks J (2013) *Principles of Chinese Herbal Medicine*. London: Jessica Kingsley Publishers.

[9] Liu ZL, Li GQ, Bensoussan A, Kiat H, Chan K, et al. (2013) Chinese herbal medicines for hypertriglyceridaemia. *Cochrane Database Syst Rev* 6: CD009560.

[10] Li GQ, Kam A, Wong KH, Zhou X, Omar EA, et al. (2012) Herbal medicines for the management of diabetes. *Adv Exp Med Biol* 771: 396–413.

[11] WHO (2005) *WHO Global Atlas of Traditional, Complementary and Alternative Medicine*. Kobe: World Health Organization Centre for Health Development.

[12] WHO (2003) *WHO Guidelines on Good Agricultural and Collection Practices (GACP) for Medicinal Plants*. Geneva: World Health Organization.

[13] WHO (1999) *WHO Monographs on Selected Medicinal Plants*, Vol 1. Geneva: WHO.

[14] WHO (2014) *WHO Traditional Medicines Strategy 2014–2023*. Geneva: WHO.

[15] Eisenberg DM, Davis RB, Ettner SL, Appel S, Wilkey S, et al. (1998) Trends in alternative medicine use in the United States, 1990–1997: results of a follow-up national survey. *JAMA* 280: 1569–1575.

[16] Thomas KJ, Nicholl JP, Coleman P (2001) Use and expenditure on complementary medicine in England: a population based survey. *Complement Ther Med* 9: 2–11.

[17] MacLennan AH, Wilson DH, Taylor AW (1996) Prevalence and cost of alternative medicine in Australia. *Lancet* 347: 569–573.

[18] MacLennan AH, Wilson DH, Taylor AW (2002) The escalating cost and prevalence of alternative medicine. *Prev Med* 35: 166–173.

[19] Xue CC, Zhang AL, Lin V, Da Costa C, Story DF (2007) Complementary and alternative medicine use in Australia: a national population-based survey. *J Altern Complement Med* 13: 643–650.

[20] Solomon D, Ford E, Adams J, Graves N (2011) Potential of St John's Wort for the treatment of depression: the economic perspective. *Aust N Z J Psychiatry* 45: 123–130.

[21] Expert Committee on Complementary Medicines in the Health System (2003) *Complementary Medicines in the Australian Health System, Report to the Parliamentary Secretary to the Minister of Health and Ageing*. Canberra: Commonwealth of Australia.

[22] Razmovski-Naumovski V, Tongkao-on W, Kimble B, Qiao VL, Lin B-L, et al. (2010) Multiple chromatographic and chemometric methods for quality standardisation of Chinese herbal medicines. *World Sci Technol* 12: 99–106.

[23] British Pharmacopoeia Commission (2013) *British Pharmacopoeia*. London: Stationery Office.

[24] Committee for Medicinal Products for Human Use (2010) *Guidance on the Investigation of Bioequivalence*. London: European Medicines Agency.

[25] Kressmann S, Biber A, Wonnemann M, Schug B, Blume HH, et al. (2002) Influence of pharmaceutical quality on the bioavailability of active components from *Ginkgo biloba* preparations. *J Pharm Pharmacol* 54: 1507–1514.

[26] Wong KH, Li GQ, Li KM, Razmovski-Naumovski V, Chan K (2011) Kudzu root: traditional uses and potential medicinal benefits in diabetes and cardiovascular diseases. *J Ethnopharmacol* 134: 584–607.

[27] Wong KH, Razmovski-Naumovski V, Li KM, Li GQ, Chan K (2013) Differentiation of *Pueraria lobata* and *Pueraria thomsonii* using partial least square discriminant analysis (PLS-DA). *J Pharm Biomed Anal* 84: 5–13.

[28] Wong KH, Razmovski-Naumovski V, Li KM, Li GQ, Chan K (2014) Differentiating Puerariae Lobatae Radix and Puerariae thomsonii Radix using HPTLC coupled with multivariate classification analyses. *J Pharm Biomed Anal* 95: 11–19.

[29] Sha N, Huang HL, Zhang JQ, Chen GT, Tao SJ, et al. (2009) Simultaneous quantification of eight major bioactive phenolic compounds in Chinese propolis by high-performance liquid chromatography. *Nat Prod Commun* 4: 813–818.

[30] Tang TX, Guo WY, Xu Y, Zhang SM, Xu XJ, et al. (2014) Thin-layer chromatographic identification of Chinese propolis using chemometric fingerprinting. *Phytochem Anal* 25: 266–272.

[31] Alqahtani A, Hamid K, Kam A, Wong KH, Abdelhak Z, et al. (2012) The pentacyclic triterpenoids in herbal medicines and their pharmacological activities in diabetes and diabetic complications. *Curr Med Chem* 20: 908–931.

[32] Huang TH, Peng G, Kota BP, Li GQ, Yamahara J, et al. (2005) Anti-diabetic action of *Punica granatum* flower extract: activation of PPAR-gamma and identification of an active component. *Toxicol Appl Pharmacol* 207: 160–169.

[33] Kam A, Li KM, Razmovski-Naumovski V, Nammi S, Chan K, et al. (2014) Gallic acid protects against endothelial injury by restoring the depletion of DNA methyltransferase 1 and inhibiting proteasome activities. *Int J Cardiol* 171: 231–242.

[34] Egert S, Bosy-Westphal A, Seiberl J, Kurbitz C, Settler U, et al. (2009) Quercetin reduces systolic blood pressure and plasma oxidised low-density lipoprotein concentrations in overweight subjects with a high-cardiovascular disease risk phenotype: a double-blinded, placebo-controlled cross-over study. *Br J Nutr* 102: 1065–1074.

[35] Gu Y, Zhang Y, Shi X, Li X, Hong J, et al. (2010) Effect of traditional Chinese medicine berberine on type 2 diabetes based on comprehensive metabonomics. *Talanta* 81: 766–772.

[36] Nahrstedt A, Butterweck V (2010) Lessons learned from herbal medicinal products: the example of St. John's Wort (perpendicular). *J Nat Prod* 73: 1015–1021.

[37] Kumar V (2013) Characterization of anxiolytic and neuropharmacological activities of Silexan. *Wien Med Wochenschr* 163: 89–94.

[38] Komiya M, Sugiyama A, Tanabe K, Uchino T, Takeuchi T (2009) Evaluation of the effect of topical application of lavender oil on autonomic nerve activity in dogs. *Am J Vet Res* 70: 764–769.

[39] Schuwald AM, Noldner M, Wilmes T, Klugbauer N, Leuner K, et al. (2013) Lavender oil-potent anxiolytic properties via modulating voltage dependent calcium channels. *PLoS One* 8: e59998.

[40] Kim Y, Kim M, Kim H, Kim K (2009) Effect of lavender oil on motor function and dopamine receptor expression in the olfactory bulb of mice. *J Ethnopharmacol* 125: 31–35.

[41] Zuzarte M, Goncalves MJ, Cavaleiro C, Dinis AM, Canhoto JM, et al. (2009) Chemical composition and antifungal activity of the essential oils of *Lavandula pedunculata* (Miller) Cav. *Chem Biodivers* 6: 1283–1292.

[42] Perino-Issartier S, Ginies C, Cravotto G, Chemat F (2013) A comparison of essential oils obtained from lavandin via different extraction processes: ultrasound, microwave, turbohydrodistillation, steam and hydrodistillation. *J Chromatogr A* 1305: 41–47.

[43] Zhou W, Wang Y (2014) A network-based analysis of the types of coronary artery disease from traditional Chinese medicine perspective: potential for therapeutics and drug discovery. *J Ethnopharmacol* 151: 66–77.

[44] Shi SH, Cai YP, Cai XJ, Zheng XY, Cao DS, et al. (2014) A network pharmacology approach to understanding the mechanisms of action of traditional medicine: bushen-huoxue formula for treatment of chronic kidney disease. *PLoS One* 9: e89123.

[45] Wang L, Li Z, Shao Q, Li X, Ai N, et al. (2014) Dissecting active ingredients of Chinese medicine by content-weighted ingredient-target network. *Mol Biosyst* 10: 1905–1911.

[46] Wang TJ, Ngo D, Psychogios N, Dejam A, Larson MG, et al. (2013) 2-Aminoadipic acid is a biomarker for diabetes risk. *J Clin Invest* 123: 4309–4317.

[47] Sun H, Zhang S, Zhang A, Yan G, Wu X, et al. (2014) Metabolomic analysis of diet-induced type 2 diabetes using UPLC/MS integrated with pattern recognition approach. *PLoS One* 9: e93384.

[48] Li SS (2007) Commentary—the proteomics: a new tool for Chinese medicine research. *Am J Chin Med* 35: 923–928.

[49] Xin GZ, Qi LW, Shi ZQ, Li P, Hao HP, et al. (2011) Strategies for integral metabolism profile of multiple compounds in herbal medicines: pharmacokinetics, metabolites characterization and metabolic interactions. *Curr Drug Metab* 12: 809–817.

[50] Sarris J, Ng CH, Schweitzer I (2012) "Omic" genetic technologies for herbal medicines in psychiatry. *Phytother Res* 26: 522–527.

[51] Hsiao CY, Tsai TH, Chak KF (2012) The molecular basis of wound healing processes induced by lithospermi radix: a proteomics and biochemical analysis. *Evid Based Complement Alternat Med* 2012: 508972.

[52] Cheng ZX, Liu BR, Qian XP, Ding YT, Hu WJ, et al. (2008) Proteomic analysis of anti-tumor effects by Rhizoma Paridis total saponin treatment in HepG2 cells. *J Ethnopharmacol* 120: 129–137.

[53] Huang D, Yang J, Lu X, Deng Y, Xiong Z, et al. (2013) An integrated plasma and urinary metabonomic study using UHPLC-MS: intervention effects of *Epimedium koreanum* on "Kidney-Yang Deficiency syndrome" rats. *J Pharm Biomed Anal* 76: 200–206.

[54] Birks J, Grimley Evans J (2009) *Ginkgo biloba* for cognitive impairment and dementia. *Cochrane Database Syst Rev* (2): CD003120.

[55] Geng J, Dong J, Ni H, Lee MS, Wu T, et al. (2010) Ginseng for cognition. *Cochrane Database Syst Rev* (12): CD007769.

[56] Karsch-Volk M, Barrett B, Kiefer D, Bauer R, Ardjomand-Woelkart K, et al. (2014) Echinacea for preventing and treating the common cold. *Cochrane Database Syst Rev* (2): CD000530.

[57] Rambaldi A, Jacobs BP, Gluud C (2007) Milk thistle for alcoholic and/or hepatitis B or C virus liver diseases. *Cochrane Database Syst Rev* (4): CD003620.

[58] Ooi CP, Yassin Z, Hamid TA (2012) *Momordica charantia* for type 2 diabetes mellitus. *Cochrane Database Syst Rev* 8: CD007845.

[59] Leach MJ, Moore V (2012) Black cohosh (*Cimicifuga* spp.) for menopausal symptoms. *Cochrane Database Syst Rev* 9: CD007244.

[60] Lethaby A, Marjoribanks J, Kronenberg F, Roberts H, Eden J, et al. (2013) Phytoestrogens for menopausal vasomotor symptoms. *Cochrane Database Syst Rev* 12: CD001395.

[61] Linde K, Berner MM, Kriston L (2008) St John's wort for major depression. *Cochrane Database Syst Rev* (4): CD000448.

[62] Pittler MH, Guo R, Ernst E (2008) Hawthorn extract for treating chronic heart failure. *Cochrane Database Syst Rev* (1): CD005312.

[63] Kasper S, Gastpar M, Muller WE, Volz HP, Moller HJ, et al. (2014) Lavender oil preparation Silexan is effective in generalized anxiety disorder—a randomized, double-blind comparison to placebo and paroxetine. *Int J Neuropsychopharmacol* 17: 859–869.

[64] Kasper S (2013) An orally administered lavandula oil preparation (Silexan) for anxiety disorder and related conditions: an evidence based review. *Int J Psychiatry Clin Pract* 17 (Suppl 1): 15–22.

[65] Liu Y, Liu JP, Xia Y (2014) Chinese herbal medicines for treating osteoporosis. *Cochrane Database Syst Rev* 3: CD005467.

[66] Liu ZL, Xie LZ, Zhu J, Li GQ, Grant SJ, et al. (2013) Herbal medicines for fatty liver diseases. *Cochrane Database Syst Rev* 8: CD009059.

[67] Jiang L, Li K, Wu T (2012) Chinese medicinal herbs for acute bronchitis. *Cochrane Database Syst Rev* 2: CD004560.

[68] Liu JP, Yang M, Liu YX, Wei M, Grimsgaard S (2006) Herbal medicines for treatment of irritable bowel syndrome. *Cochrane Database Syst Rev* (1): CD004116.

[69] Jing Z, Yang X, Ismail KM, Chen X, Wu T (2009) Chinese herbal medicine for premenstrual syndrome. *Cochrane Database Syst Rev* (1): CD006414.

[70] Liu JP, Zhang M, Wang WY, Grimsgaard S (2004) Chinese herbal medicines for type 2 diabetes mellitus. *Cochrane Database Syst Rev* (3): CD003642.

[71] Wu T, Ni J, Wu J (2008) Danshen (Chinese medicinal herb) preparations for acute myocardial infarction. *Cochrane Database Syst Rev* (2): CD004465.

[72] Wang Q, Wu T, Chen X, Ni J, Duan X, et al. (2006) Puerarin injection for unstable angina pectoris. *Cochrane Database Syst Rev* (3): CD004196.

[73] Chen X, Zhou M, Li Q, Yang J, Zhang Y, et al. (2008) Sanchi for acute ischaemic stroke. *Cochrane Database Syst Rev* (4): CD006305.

[74] Solon-Biet SM, McMahon AC, Ballard JW, Ruohonen K, Wu LE, et al. (2014) The ratio of macronutrients, not caloric intake, dictates cardiometabolic health, aging, and longevity in ad libitum-fed mice. *Cell Metab* 19: 418–430.

[75] Xiao Y, Zhang Y, Wang M, Li X, Xia M, et al. (2013) Dietary protein and plasma total homocysteine, cysteine concentrations in coronary angiographic subjects. *Nutr J* 12: 144.

[76] Xiao Y, Zhang Y, Wang M, Li X, Su D, et al. (2013) Plasma S-adenosylhomocysteine is associated with the risk of cardiovascular events in patients undergoing coronary angiography: a cohort study. *Am J Clin Nutr* 98: 1162–1169.

[77] Yi L, Chen CY, Jin X, Mi MT, Yu B, et al. (2010) Structural requirements of anthocyanins in relation to inhibition of endothelial injury induced by oxidized low-density lipoprotein and correlation with radical scavenging activity. *FEBS Lett* 584: 583–590.

[78] Xia M, Ling W, Zhu H, Wang Q, Ma J, et al. (2007) Anthocyanin prevents CD40-activated proinflammatory signaling in endothelial cells by regulating cholesterol distribution. *Arterioscler Thromb Vasc Biol* 27: 519–524.

[79] Xia M, Hou M, Zhu H, Ma J, Tang Z, et al. (2005) Anthocyanins induce cholesterol efflux from mouse peritoneal macrophages: the role of the peroxisome proliferator-activated receptor {gamma}-liver X receptor {alpha}-ABCA1 pathway. *J Biol Chem* 280: 36792–36801.

[80] Li D, Wang D, Wang Y, Ling W, Feng X, et al. (2010) Adenosine monophosphate-activated protein kinase induces cholesterol efflux from macrophage-derived foam cells

and alleviates atherosclerosis in apolipoprotein E-deficient mice. *J Biol Chem* 285: 33499–33509.

[81] Qin Y, Xia M, Ma J, Hao Y, Liu J, et al. (2009) Anthocyanin supplementation improves serum LDL- and HDL-cholesterol concentrations associated with the inhibition of cholesteryl ester transfer protein in dyslipidemic subjects. *Am J Clin Nutr* 90: 485–492.

[82] Guo H, Zhong R, Liu Y, Jiang X, Tang X, et al. (2014) Effects of bayberry juice on inflammatory and apoptotic markers in young adults with features of non-alcoholic fatty liver disease. *Nutrition* 30: 198–203.

[83] Kim EJ, Chen Y, Huang JQ, Li KM, Razmovski-Naumovski V, et al. (2013) Evidence-based toxicity evaluation and scheduling of Chinese herbal medicines. *J Ethnopharmacol* 146: 40–61.

[84] Sarris J, Stough C, Bousman CA, Wahid ZT, Murray G, et al. (2013) Kava in the treatment of generalized anxiety disorder: a double-blind, randomized, placebo-controlled study. *J Clin Psychopharmacol* 33: 643–648.

[85] Li XZ, Ramzan I (2010) Role of ethanol in kava hepatotoxicity. *Phytother Res* 24: 475–480.

[86] Zhang LY, Rowe A, Ramzan I (2011) Does inflammation play a role in kava hepatotoxicity? *Phytother Res* 25: 629–630.

[87] Zhang L, Rowe A, Braet F, Ramzan I (2012) Macrophage depletion ameliorates kavalactone damage in the isolated perfused rat liver. *J Toxicol Sci* 37: 447–453.

[88] Teschke R, Sarris J, Lebot V (2013) Contaminant hepatotoxins as culprits for kava hepatotoxicity—fact or fiction? *Phytother Res* 27: 472–474.

[89] Barnes J, Anderson LA, Phillipson D (2007) *Herbal Medicines*. London: Pharmaceutical Press.

[90] Braun L, Cohen M (2010) *Herbs and Natural Supplements: An Evidence-Based Guide*. Chatswood: Elsevier Australia.

[91] Brinker FJ (2001) *Herb Contraindications and Drug Interactions: With Extensive Appendices Addressing Specific Conditions, Herb Effects, Critical Medications, and Nutritional Supplements*. Sandy: Eclectic Medical Publications.

[92] Gao J, Shi Z, Zhu S, Li GQ, Yan R, et al. (2013) Influences of processed rhubarbs on the activities of four CYP isozymes and the metabolism of saxagliptin in rats based on probe cocktail and pharmacokinetics approaches. *J Ethnopharmacol* 145: 566–572.

[93] National Health and Medical Research Council (NHMRC) (2008) *Complementary and Alternative Medicine Research—Special Call for Research Applications*. Canberra: NHMRC. http://www.nhmrc.gov.au/grants/outcomes-funding-rounds/historical-information-capacity-building-grants/complementary-and-alt. Accessed on December 5, 2014.

[94] Australian Research Council (2013) *Excellence in Research for Australia 2012, National Report*. Canberra: Commonwealth of Australia.

2

QUALITY CONTROL AND QUALITY ASSURANCE OF PHYTOMEDICINES: KEY CONSIDERATIONS, METHODS, AND ANALYTICAL CHALLENGES

WAI-PING YAU[1], CHEONG HIAN GOH[2], AND HWEE-LING KOH[1]

[1] *Department of Pharmacy, Faculty of Science, National University of Singapore, Singapore, Republic of Singapore*
[2] *Audit and Licensing Division, Health Products Regulation Group, Health Sciences Authority, Singapore, Republic of Singapore*

2.1 INTRODUCTION

Phytomedicine is gaining popularity globally. For example, herbs like *Panax ginseng* have continued to be widely used for enhancing general health and well being [1]. In some Asian and African countries, it is estimated that as high as 80% of the population depends on traditional medicines for primary health care, with herbal medicines identified as the most lucrative form of traditional medicines [2]. Apart from its traditional use, there is also a paradigm shift to use phytomedicines for treating diseases, and plants have been important sources of drugs [3], for example, the antimalarial drug artemisinin and the anticancer drug taxol.

According to the definition by the World Health Organization (WHO), phytomedicines, or herbal medicines, include crude herbs, herbal materials, herbal preparations, and finished herbal products, which contain as active ingredients parts of plants, or other plant materials, or in combinations [4]. They may be subject to local

Phytotherapies: Efficacy, Safety, and Regulation, First Edition. Edited by Iqbal Ramzan.

processes such as steaming or roasting, more rigorous processes (including extraction, fractionation, purification, or concentration, other physical or biological processes) and/or further incorporated with excipients as a final product [4]. Not many countries have national policies for phytomedicines and disparity in regulations of such products can arise globally from the variations in definitions and classification of herbal medicines by each country as a food, a dietary supplement, or a herbal medicine [5].

The constituents of phytomedicines can be very complex and may contain different chemical constituents that contribute to their therapeutic effects [6]. Furthermore, active principles in herbal medicines may not be clearly understood. Unlike conventional pharmaceutical products, which are generally produced from synthesized compounds by means of reproducible manufacturing techniques and procedures, herbal medicines are often prepared from plants harvested from different geographical locations or at different seasons. As plant materials are chemically and naturally variable, it may not always be possible to assure the quality and consistency of the crude plant materials. Procedures and techniques used in processing and manufacturing can further influence the properties and composition of the herbal medicines. In this regard, quality assurance (QA) and quality control (QC) in the processing and manufacture of phytomedicines under Good Manufacturing Practices (GMPs) are essential in ensuring the standards for product safety and quality. Currently, manufacturers of herbal medicinal products in Europe and botanical dietary supplements in the United States are required to conform to current GMPs (cGMPs) [7, 8].

QA can be described as a high-level process-oriented approach that focuses on defect prevention (or "first-time right" principle) and ensuring its intended process output ("fit for its purpose" principle), while QC is product-oriented and focuses more specifically on the process outputs including defect identification. By definition, QA encompasses all aspects of the engineering or quality-related activities that can influence product quality. This includes the control and storage of starting materials, GMP during production and processing, product design and development, as well as the analytical assessments and other arrangements made in the product life cycle to ensure product quality and its intended use [4].

QC is the part of GMP relating to sampling, specifications, and testing. The primary QC activities of proper documentation and release procedures ensure that the necessary and relevant specified tests are actually carried out and that materials are only released for use, or released for sale or supply, following a satisfactory quality assessment. Indeed, QC is not only confined to laboratory operations but can be involved in all decisions that pertain to the quality of the product [4].

As part of QA and QC, different aspects of analytical assessments are typically encompassed in the processing and manufacturing of herbal medicines. In light of the complexity and diverse nature of the ingredients, this chapter will examine the key considerations for quality assessments, provide an overview of the analytical tools used, and highlight the key challenges faced to ensure consistent quality for phytomedicines.

2.2 KEY CONSIDERATIONS IN QC/QA OF PHYTOMEDICINES

With the inherent complexity of naturally grown medicinal plants/herbal substances and the limitations in the analytical techniques for characterization, quality assurance of phytomedicines will require control of starting source materials, storage, and production processes. Specifications for the phytomedicines produced should be established as a control strategy to ensure product quality and consistency. Specifications are generally established based on scientific data and acceptable compendial monographs and describe the tests established for quality assessments, the references to the analytical and biological procedures, and the acceptance criteria for the analytical procedures [9].

Characteristics that define the quality of the herbal substance/preparation and herbal medicinal product should be identified and included in the specifications. Key considerations that are relevant for the quality assessments of phytomedicines include its identification, aspects of contamination, issues of adulteration and substitution, contents such as the active principles and characteristic constituents and their standardization, product stability, and processing (Table 2.1).

2.2.1 Identification and Good Agricultural and Collection Practices (GACP)

To ensure consistent quality of phytomedicines, herbal starting materials should first be adequately controlled and defined through an established identification process. Characterization of these materials includes an evaluation of the botanical, morphological, and phytochemical aspects of the plant that will be used for the manufacture of the preparation or herbal medicinal product.

Botanical identification essentially verifies the plant materials to be used in the phytomedicines. Specifications that are established for quality assessment include the accepted scientific name (with name of genus, species/subspecies, variety and family of the plant), the synonyms and associated common names, the parts of the plant used for each preparation, and the geographical source and the conditions under which the herbal substance is obtained [10]. Preferably, the site of field collection, the time of harvesting and stage of growth, the pesticides used during growth, and drying and storage conditions should be assessed [11].

For the morphological assessments, macroscopic and microscopic methods from the pharmacopoeial standards may be used and included as part of the approved quality specifications of the herbal starting materials. Reference samples of the herbal substances should be available for use in these comparative tests [11]. For herbal substances that have constituents of known therapeutic activity, an assay of the constituent contents should be conducted for the phytochemical assessments, in accordance with test procedures in the starting material specifications. The established range of the constituent content (or also known as the active marker) shall describe the acceptance criteria that assure reproducibility of the quality of the herbal medicinal product [11]. In the case of herbal substances where constituents of claimed therapeutic activity are not known, assays of marker substances may be carried out but the choice of the markers in these test procedures should be justified [11].

TABLE 2.1 Key Considerations for Quality Assessments of Phytomedicines

Key Considerations	Specific Aspects
1. Identification of herbal starting materials (including Good Agricultural and Collection Practices (GACP))	Botanical • Scientific name (genus, species/subspecies, variety, family of the plant) • Synonyms and associated common names • Parts of the plant used • Cultivation, harvesting and postharvesting conditions (e.g., geographical location, cultivation conditions, site of field collection, conditions under which the herbal substance is harvested) Morphological • Macro- and microscopic characteristics Phytochemical • Constituent or marker content
2. Contamination	Microorganisms • Bacteria and Molds (Fungi and Yeasts) Toxic heavy metals • Arsenic, cadmium, copper, iron, lead, mercury, nickel, and zinc Radiation • Activity concentration and types of radioactive contamination Pesticides • Amount present in herbal materials and tolerable intake (TI) Mycotoxins and endotoxins • Aflatoxins, ochratoxin A Residual solvents
3. Substitution	Herbal substitution
4. Adulteration	Detecting undeclared compounds/ingredients
5. Contents and standardization	Targeted constituent contents for adjustment
7. Stability	Concentration of active constituents and other substances and chromatographic fingerprints of phytomedicines
8. Processing and Good Manufacturing Practices (GMP)	Many methods of extraction and processing, with in-process control tests

To date, several guidelines have been established, such as the Good Agricultural and Collection Practices (GACP) Guidelines by WHO [10], European Medicines Agency [12], and China State Food and Drug Administration [13]. These guidelines generally focus on cultivation conditions, with respect to the expectations in cross-contamination control and traceability of plant materials [14]. Considerations of drying and postharvesting processes that can impact the moisture control and promote microbial growth are also described [10]. Such controls can alleviate the problem of herb misidentification and ensure that quality of the starting herbal ingredients is reproducible through good practices for cultivation, harvesting, and postharvesting processes [15]. GACP can also control the aspects of contamination and product safety, which will be discussed in the following sections.

2.2.2 Contamination

Contamination of phytomedicines can arise from two sources: intrinsic factors during cultivation or extrinsic factors such as the mitigating processes during manufacture and/or storage. As such, herbal substances should be monitored for potential contaminants with microorganisms, toxic heavy metals, residues of pesticides and fumigation agents, mycotoxins (aflatoxins, ochratoxin A), endotoxins, and also residual solvents during processing [11]. Approved specifications for the phytomedicines should establish the permissible limits, including the tolerable intake (TI) that can assure safe consumption of these herbal materials [16]. Additionally, these established limits should also meet the requirements that are stipulated by the appropriate governmental regulatory agencies for product approval [17].

2.2.2.1 ***Test for Microorganisms and Toxic Heavy Metals*** Microorganisms and heavy metals from the soil can inadvertently be present in the harvested herbal materials. These contaminants should be monitored and controlled. Indeed, in a study on some 334 crude herb samples that were collected throughout China, it was reported that at least one toxic heavy metal was detected in all the samples (100%) and 115 samples (34%) had detectable levels of arsenic, cadmium, chromium, lead, and mercury [18]. Currently, there are different regulatory controls to the microbial and heavy metals that can be present in phytomedicines. Table 2.2 provides a summary of the legal permissible toxic heavy metal limits and microbial limits applicable to the different finished herbal products in Singapore [17]. Acceptance limits for the herbal substances or herbal preparations will need to be established in the specifications and justified by the manufacturer. Generally, the total viable aerobic count method is commonly preferred for bacteria and fungi determination [16]. In some herbal materials, the inherently high contents of tannins, essential oils, or other antimicrobial substances can pose challenges to the microbial determination. Antimicrobial properties in these test specimens will need to be first removed, using methods such as filtration, neutralization, or serial dilutions [16]. Toxic heavy metals can be determined quantitatively by atomic absorption spectrometry (AAS). The specific toxic heavy metals that are typically assayed include arsenic, cadmium, copper, iron, lead, mercury, nickel, and zinc.

2.2.2.2 ***Test for Pesticides and Fumigation Agents*** Pesticides and sometimes fumigating agents are used during the different stages of cultivation, storage, transport, distribution, and processing against pests and unwanted deterioration [16]. In this way, contaminants arising from residual pesticides or illegal use of unapproved pesticides during cultivation can also pose potential health risks and need regular

TABLE 2.2 Toxic Heavy Metal and Microbial Limit Requirement in Singapore[a]

Parameter	Permissible Limits
1. Heavy metals	
Arsenic	5 ppm
Copper	150 ppm
Lead	20 ppm
Mercury	0.5 ppm
2. Microbial limits for:	
Oral herbal medicinal products	
Total aerobic microbial count	Not more than 10^5 CFU/g or ml
Yeast and mold count	Not more than 5×10^2 CFU/g or ml
Escherichia coli	
Salmonellae	Absent in 1 g or ml
Staphyloccocus aureus	
Topical herbal medicinal products	
Total aerobic microbial count	Not more than 10^4 CFU/g or ml
Yeast and mold count	Not more than 5×10^2 CFU/g or ml
Pseudomonas aeruginosa	
Staphyloccocus aureus	Absent in 1 g or ml
3. Specific naturally occurring substances	
Ephedra alkaloids	<1%
Lovastatin	<1%
Sodium borate	<5%
Lobelia alkaloids	<0.1%
Aconite alkaloids	Dosing of <60 mcg/day

CFU, Colony Forming Unit; ppm, parts per million.
[a]From Ref. [17].

monitoring. This was observed in a study that examined 36 types of crude herbal products procured from large herbal drug wholesale markets in South Korea. Seven imported (*Paeoniae Radix*, *Cuscuta Semen*, *Atractylodes alba Rhizoma*, *Zingiberis Rhizoma*, *Atractylodes Rhizoma*, *Polygalae Radix*, and *Myristicae Semen*) and two domestic herbal materials (*Platycodi Radix* and *Atractylodes Rhizoma*) were found to be contaminated with eight pesticides (0.034–0.579 mg/kg) such as benzene hexachloride (BHC), procymidone, and endosulfan [19]. In a separate study, harvested wild plants were found to contain significantly higher contaminant levels of pesticides than cultivated samples, thereby suggesting that local sources of industrial or agricultural pollution could be important factors contributing to contamination of phytomedicines [18].

Sulfur fumigation is commonly claimed to prevent pest infestation and microbial contamination, together with claimed advantages that it facilitates drying of herbs and helps to preserve the herbs' freshness and color [20]. Several reviews have discussed in detail the toxicities of sulfur dioxide that arise from sulfur fumigation, the harmful effects of associated sulfiting agents, as well as other potential chemical and pharmacokinetic modifications and their impacts on the bioactive ingredients in the fumigated herbs [20, 21]. The degradation of peoniflorin and formation of peoniflorin sulfonate observed in sulfur-fumigated *Paeoniae alba Radix* (*Paeonia lactiflora* Pall., Paeoniaceae) and similar observations of reduction in various ginsenosides, with formation of ginsenoside sulfonate derivatives in sulfur-fumigated *Ginseng Radix et Rhizoma* (*Panax ginseng* C.A. Meyer, Araliaceae) are some good examples [20]. In this regard, the SFDA of China has restricted the permissible limits of residual sulfur dioxide to 150 mg/kg for most crude and prepared slices of crude herbs, except for some herbs that should not exceed 400 mg/kg [21]. The Chinese Pharmacopeia Commission listed the following 10 herbs that should not exceed 400 mg/kg of residual sulfur dioxide: *Dioscoreae Rhizoma* (*Dioscorea opposita* Thumb., Dioscoreaceae), *Achyranthis bidentatae Radix* (*Achyranthes bidentata* Bl., Amaranthaceae), *Puerariae thomsonii Radix* (*Pueraria thomsonii* Benth., Leguminosae), *Asparagi Radix* (*Asparagus cochinchinensis* [Lour.] Merr., Liliaceae), *Gastrodiae Rhizoma* (*Gastrodia elata* Bl., Orchidaceae), *Trichosanthis Radix* (*Trichosanthes kirilowii* Maxim., Cucurbitaceae), *Bletillae Rhizoma* (*Bletilla striata* (Thunb.) Reichb. f., Orchidaceae), *Paeoniae alba Radix* (*Paeonia lactiflora* Pall., Paeoniaceae), *Atractylodis macrocephalae Rhizoma* (*Atractylodes macrocephala* Koidz., Asteraceae), and *Codonopsis Radix* (*Codonopsis pilosula* (Franch.) Nannf., Campanulaceae) [22].

Column or gas chromatography (GC) coupled with mass spectrometry (MS) is generally the preferred method for the determination of pesticide residues. The TI of pesticides and fumigating agents as established in the specifications of phytomedicines will be indicative of the estimated amounts that can be consumed by humans without significant health risk [16].

2.2.2.3 ***Test for Mycotoxins and Endotoxins*** Accidental consumption of mycotoxin- or endotoxin-contaminated products can lead to serious health problems. High-performance liquid chromatography (HPLC) coupled with fluorescence detection remains the mainstay for determination of mycotoxins such as aflatoxins, for example, B1, B2, G1, and G2 [16]. The current practice adopts a risk assessment approach that does not require mandatory routine testing for mycotoxins for all the herbal substances [23]. However, routine mycotoxin testing will be necessary for herbal sources known to be at risk of mycotoxin formation and possible contamination of the herbal substance has been reported in literature [23]. The test for alcohol-soluble toxins (e.g., aflatoxins and ochratoxin A) should still be considered even when the herbal preparation has undergone an extraction process [23]. A screening study of 84 sampled herbs in Spain concluded that medicinal and aromatic herbs were prone to mycotoxin multicontamination, in particular *Salvia officinalis* (sage leaves), *Matricaria chamomilla* (chamomile flower), *Valeriana officinalis* (valerian root), *Cassia angustifolia* (senna leaves), and *Rheum officinalis* (rhubarb stem) [24]. In the same study, mycotoxin contamination of other therapeutic herbs, such as *Peumus boldus* (boldus leaves), Arctium sp. (burdock leaves), *Taraxacum officinale* (dandelion plant), *Rhamnus frangula* (frangula bark), and *Ginkgo biloba* (ginkgo leaves), have also been reported [24].

2.2.2.4 ***Test for Radiation*** Herbal substances are only tested for radioactive contamination in cases of nuclear accidents or when there are concerns of suspected airborne radioactive contamination [23]. Such analytical considerations arise as herbs can be potentially subject to prolonged radiation exposure or uptake of radionuclides by the plants from contaminated soil. In one report, the radioactivity levels of 18 different crude herbs from the Serbian market were evaluated by a gamma spectroscopy system to monitor the safe use of these herbs [25]. In this study, the activity concentrations of determined radionuclides were not at dangerous levels hazardous to public health. However, much higher 137Cs activity concentration (82.5 Bq/kg) was observed for the imported *Rhamnus frangula* L. (or buckthorn) than the other herbs examined (0.3–8.9 Bq/kg) and this was possibly related to the geographic location of cultivation [25]. In general, the activity concentrations and the types of radioactive contamination can be monitored, but a generalized method of radiation measurement for the different types of facility is currently not available [16].

2.2.2.5 Test for Residual Solvents Organic solvents that are used or produced during manufacturing and processing may remain as trace amounts in the phytomedicines. Based on the potential risks of these solvents, the approved specifications for the phytomedicines should include appropriate QC tests that exclude, for example, class 1 solvents such as benzene in the herbal preparations/products [16].

2.2.3 Substitution

Herbal substitution is common, whereby one plant material may be replaced by another due to cost consideration or its unavailability. Such practice can potentially compromise product quality and safety, which QC needs to monitor and deter. Expensive herbs are usually subject to substitution. For example, the root of *Hydrastis canadensis* has been reported to be substituted with *Rumex crispus*, *Mahonia nervosa*, *Coptis* spp., *Xanthorhiza simplicissima*, and *Berberis vulgaris*, which are nontoxic [26]. In another example, however, the Chinese herb *Stephania tetrandra* is said to be substituted with *Aristolochia fangchi*, a known renal toxin and carcinogen [26]. Findings from a molecular epidemiologic study in Taiwan have also suggested that aristolochic acid (and possibly the related consumption of aristolochic acid-containing herbs) may be the primary cause of upper urinary tract urothelial carcinomas in Taiwan [27].

2.2.4 Adulteration

Adulteration is common in herbal medicines and this can have negative impact on product efficacy, consumer safety, and possible legal problems for the pharmaceutical industries. Reports of phytomedicines being adulterated with undeclared synthesized chemical drug compounds to enhance/provide the therapeutic claims of phytomedicines are not uncommon. Examples include phytomedicines marketed for sexual performance enhancement being adulterated with phosphodiesterase type 5 enzyme inhibitors and their analogues [28–31]. Consumption of undeclared compounds/ingredients in adulterated phytomedicines may result in adverse effects and toxicity, contraindications, and drug–drug interactions in unsuspecting consumers.

2.2.5 Contents and Standardization

The contents of the herbal preparation affect the safety and efficacy of the product. According to WHO, herbal medicines with constituents of known therapeutic activity are often standardized (i.e., adjusted to a defined content of such constituents) [4]. It may also be possible to standardize a herbal product with respect to some marker compounds (both active and nonactive constituents). According to the European Medicines Agency, standardization means adjusting the herbal substance/preparation to a defined content of a constituent or a group of constituents with known therapeutic activity respectively either by adding excipients or by blending batches of the herbal substance and/or herbal preparation (e.g., standardized extracts) [9]. Approved specifications for the phytomedicines should therefore include the acceptable tolerance (with both upper and lower limits) set for the defined active constituents with known therapeutic activity or for other marker substances, in which the standardized herbal preparation is to be adjusted [9]. For the QC tests conducted, the equivalent quantity of the actual herbal substance should be documented. The various analytical methods will be discussed in Section 2.3.

2.2.6 Stability

Stability is an important aspect in ensuring QA of phytomedicines. Stability profiles should be considered in totality, on both the active constituents and other substances present in the phytomedicines. In general, stability can be demonstrated if the proportional content of the active constituents is comparable with the initial fingerprint chromatogram. During the proposed shelf life, any variation in the active constituent content or surrogate marker contents used should not exceed ±5 or ±10% of the respective declared assay values [11]. Various factors, including those arising from processing steps, can potentially influence the reactivity of herbal constituents in hydrolytic or oxidative reactions and should be monitored to ensure stability of herbal extracts. These factors include pH, light, moisture/humidity, temperature, air/oxidation, or secondary factors like endogenous enzymatic degradation, presence of metal ions, residual solvents, and other concomitant compounds that are present [32, 33].

2.2.7 Processing

The processes in making herbal preparations are critical to the quality of phytomedicines. Various processing methods include washing, drying, grinding, boiling, steaming, frying, and different extraction methods (e.g., maceration, reflux, percolation, Soxhlet extraction, ultrasonication, microwave-assisted extraction, pressurized liquid extraction, steam distillation, and supercritical fluid extraction) [33, 34]. In particular, the different processing conditions can result in associated degradation and affect water content, thereby impacting the active substance, the overall quality and composition of the phytomedicines produced. For example, postharvest drying and processing caused significant decrease in the main constituent contents (senkyunolide A, z-ligustilide, and coniferylferulate) of *Ligusticum chuanxiong rhizome* but increased the contents of other bioactive constituents (senkyunolides I and H, riligustilide, levistolide A, and ferulic acid) [35]. In this case, variations to the chemical profile and contents of the main constituent component were attributed to process-induced hydroxylation, dimerization, and hydrolysis reactions, which could also potentially alter the overall therapeutic effect [35]. Hence, to monitor and assure consistency in product quality, in-process control tests are routinely conducted as part of GMP.

2.3 METHODS FOR QC/QA OF PHYTOMEDICINES

A number of analytical tools are used for macroscopic and microscopic evaluation and fingerprinting analysis of phytomedicines for the purposes of QC and QA (Table 2.3). These include traditional techniques such as chromatography and spectroscopy and the newer "omics" technologies, which are discussed as follows.

2.3.1 Macroscopic Evaluation

Consistent quality for phytomedicines can only be assured if the starting crude plant materials are appropriately controlled [9]. Morphological and organoleptic evaluation of a herbal sample based on shape, size, color, texture, surface and fracture characteristics, appearance of the cut surface, odor, and taste is conducted as a first step to establish the botanical identity and species authentication of the sample. Authentic specimens and/or pharmacopoeial monograph descriptions for the plant material serve as references for comparison [16]. Other than authentication of raw herbal materials, macroscopic investigation has also been used to identify Chinese Materia Medica (CMM) from their decoction dregs (which are the residues after decoction) [25, 26]. Studies have demonstrated that four pairs of commonly misidentified CMM, namely Baizhu (*Atractylodis macrocephalae Rhizoma*) and Cangzhu (*Atractylodis Rhizoma*), Baishao (*Paeoniae alba Radix*) and Chishao (*Paeoniae rubra Radix*), Fenge (*Puerariae thomsonii Radix*) and Shanyao (*Dioscoreae Rhizoma*), Dihuang (*Rehmanniae Radix*) and Huangjing (*Polygonati Rhizoma*), could be distinguished and authenticated based on differences in gross morphological and microscopic characteristics of the decoction dregs of the pairs of CMM [36, 37]. For example, in terms of the macroscopic differences in the decoction dregs of Baizhu (*Atractylodis macrocephalae Rhizoma*) and Cangzhu (*Atractylodis Rhizoma*), the former are irregularly flaky in shape and have a weak odor whereas the latter are occasionally branched and irregularly moniliform or nodular–cylindrical in shape and have a strong odor [36].

2.3.2 Microscopic Evaluation

In addition to macroscopic examination, microscopic evaluation of the histological characteristics of a crude herb is also performed and compared with reference materials [16]. This is particularly essential for the preliminary identification and

TABLE 2.3 Methods for Quality Assessment of Phytomedicines

	Methods	Specific Aspects
1.	Macroscopic evaluation (including organoleptic evaluation)	Physical characteristics • Shape, size, color, texture, surface, and fracture characteristics • Appearance of cut surface of crude plant materials, • Odor and taste
2.	Microscopic evaluation	Histological characteristics
3.	Physicochemical analysis	Physicochemical properties • Moisture content • Specific gravity • Optical rotation, refractive index • Melting point • Viscosity, solubility in different solvents • Ash values (total ash, sulfated ash, acid-insoluble ash, water-soluble ash) • Acid value • Saponification value • Contents of essential oils
4.	Chemical fingerprinting	Chromatography • TLC • GC • HPLC Electrophoresis • CE Spectroscopy • Vibrational spectroscopy (IR, NIR, and Raman spectroscopies) • AAS • NMR • MS Hyphenated techniques • GC-MS • LC-MS • Others (e.g., HPLC-NMR, CE-MS, TLC-MS, TLC-HPLC-MS, HPLC-NMR- MS)
5.	DNA fingerprinting	DNA fingerprinting techniques • RAPD analysis • ISSR analysis • AFLP analysis • Microarray-based DNA profiling
6.	"Omics" technology	Types of "omics" techniques • Genomics and transcriptomics • Proteomics • Metabolomics (chromatography-based methods, NMR-based methods, and MS-based hyphenated methods)

authentication of powdered herbs and small fragments of crude herbs, which cannot be determined by macroscopic evaluation [38, 39]. The use of both light and fluorescence microscopy for microscopic investigation of the mechanical tissues (including stone cells and fibers), conducting tissues (including vessels), and ergastic substances (including calcium oxalate crystals and secretions) in powdered samples of 16 Chinese herbal medicines has been reported [40]. Microscopic analysis is also potentially valuable as an initial screen for adulterants and contaminants in herbal drugs. Mineral arsenicals, including orpiment (mainly containing As_2S_3), realgar (mainly containing As_4S_4), arsenolite, and arsenic trioxide have been successfully identified and differentiated by using light microscopy coupled with polarized microscopy [41].

2.3.3 Physicochemical Analysis

To aid in the identification of specific active constituents and detection of adulterants and contaminants present in phytomedicines, analytical tests to evaluate physicochemical properties are used. Physicochemical analyses include the determination of moisture content, specific gravity, optical rotation, refractive index, melting point, viscosity, solubility in different solvents, ash values (total ash, sulfated ash, acid-insoluble ash, water-soluble ash), acid value, saponification value, essential oils, etc., [16, 38, 39, 42]. General test methods of some of these analyses for phytomedicines are detailed in the WHO guidelines for the QC of herbal materials [16], while specific test procedures for individual phytomedicines are provided in pharmacopoeial monographs, such as in the British Pharmacopoeia, American Herbal Pharmacopoeia, and European Pharmacopoeia.

2.3.4 Chemical Fingerprinting

Quality assessment of phytomedicines is often based on the analysis of marker compounds. Nonetheless, this approach has its limitations. First, marker identification is a challenge due to the lack of unique chemical constituents in a number of herbal species. Second, the complex composition of phytomedicines cannot be adequately reflected based on a few chemical markers. In light of these, chemical fingerprint analysis is employed as a comprehensive approach to provide characteristic profiles that reflect the complex chemical composition of different phytomedicines. Chemical fingerprint analysis utilizes chromatographic, electrophoretic, spectroscopic, or hyphenated techniques for the identification, as well as quality evaluation and control of medicinal herbs. Of these, chromatographic fingerprinting is one of the QC tests for phytomedicines as specified by both the US Food and Drug Administration [8] and the European Medicines Agency [9]. The following sections give an overview of various methods for chemical fingerprinting.

2.3.4.1 Chromatography

Thin Layer Chromatography (TLC) TLC is a simple, rapid, inexpensive, and versatile planar chromatographic fingerprint analysis technique that is capable of simultaneous qualitative analysis of multiple different samples in parallel [43–45]. It has long been used for the preliminary characterization of crude plant materials and herbal drug preparations such as tinctures and powdered extracts [45]. In fact, TLC is one of the key identity tests specified in most pharmacopoeial monographs that are used by industry as a basis for meeting cGMP standards and QC requirements [26, 44].

With the demands for increased reproducibility, accuracy, and high-throughput screening, instrumental high-performance TLC (HPTLC) has been introduced and is currently employed for the purposes of QA and QC of phytomedicines. HPTLC fingerprinting is used in the selection of appropriate plants for cultivation as raw material for the production of phytomedicines, confirming and establishing the identity of crude herbal materials or preparations, detection of substitutions, screening for adulterants and contaminants, and establishing batch-to-batch consistency and stability testing [26]. Validated HPTLC methods with high accuracy, precision, reliability, and reproducibility have also been reported and proposed for quantitative monitoring of phytomarkers, such as betulinic acid in *Nelumbo nucifera* Gaertn. Rhizomes [46], taraxerol in *Clitoria ternatea* L. (Shankapushpi) [47], 6-gingerol in *Zingiber officinale* rhizomes (ginger) [48], which can be used for routine quality testing of phytomedicinal extracts. In addition to the high degree of reproducibility and consistency essential for cGMP compliance, HPTLC has the capability of real-time documentation and electronic storage of results which provides analytical documentation as part of cGMP requirements [26, 45].

Gas Chromatography (GC) GC analysis of volatile and thermostable compounds in phytomedicines provides characteristic GC fingerprints based on the composition and relative concentrations of the volatile compounds in specific herbs [44]. This is useful for the identification of medicinal herbs. In addition, GC is one of the principal techniques recommended by WHO for the detection of pesticide residues for the purpose of QC of phytomedicines [16].

With technological advances, comprehensive two-dimensional (2D) GC (GC × GC) has improved separation efficiency via a double separation process using two columns with different stationary phases, making it a powerful technique for analysis of complex mixtures. GC × GC is being applied as a promising separation tool, combined with other techniques, most commonly with MS, for characterization of chemical composition of volatile compounds in phytomedicines [49]. This will be further discussed in Section 3.4.4 on GC-MS.

High-Performance Liquid Chromatography HPLC, in particular HPLC with diode array detector (DAD), is one of the most important and most widely used techniques in the analysis and QC of phytomedicines [44]. Some of its many applications include the following:

- Analysis of specific compounds in herbal materials (e.g., identification and quantification of flavonoids and ellagic acid derivatives in *Drosera anglica*, *D. intermedia*, *D. madagascariensis*, *D. rotundifolia*, and *D. peltata*) [50, 51]
- Evaluation of effects of processing on the chemical constituents and chromatographic fingerprints (e.g., the effect of steaming on the chromatograms

and saponin concentrations of steamed *Panax notoginseng* compared with raw *P. notoginseng*) [52–54]

- Fingerprinting analysis of herbs for botanical identification (e.g., *Isatis indigotica* root samples compared with authentic reference standard) [55]
- Fingerprinting analysis of herbs for detection of substitutes (e.g., *Stephania tetrandra* being substituted by *Aristolochia fangchi*) [56]
- Screening of herbal preparations for adulterants (e.g., phosphodiesterase-5 inhibitors, codeine) [57–59]

Over the years, advancement in HPLC technology has brought about the introduction of ultra-performance liquid chromatography (UPLC), which has increased efficiency and speed of separation, and the introduction of 2D-HPLC, which has improved resolution and separation efficiency [60]. UPLC [61–63], 2D-HPLC [64, 65], and even HPLC × UPLC [66] have been applied to the analysis of a variety of phytomedicines to date. Most of these are coupled with MS as hyphenated techniques that will be discussed subsequently.

2.3.4.2 Capillary Electrophoresis (CE) CE is another analytical tool that has been applied to the fingerprinting of herbal medicines and the determination of various specific compounds present in phytomedicines, such as alkaloids, flavonoids, terpenes, and organic acids [67]. Its advantages over conventional chromatographic techniques, which include higher separation efficiency, more rapid analysis time, lower solvent and sample consumption per run, and faster method development, make it an attractive analytical technique for the QC of phytomedicines [67].

In addition, CE is versatile as it allows a range of separation modes to be employed depending on the compounds that need to be analyzed. The most common mode of CE is capillary zone electrophoresis (CZE), with ultraviolet (UV) detection. One example of the application of CZE is the detection and quantification of total β-escin (a complex mixture of saponins) in dry, hydroalcoholic and hydroglycolic extracts of *Aesculus hippocastanum* L. (horse chestnut) [68]. While CZE-UV may be limited by the sensitivity of UV detection, online preconcentration strategies, such as field enhancement sample stacking (FESS), have been developed to overcome this limitation. It has been demonstrated that a 16- to 73-fold sensitivity enhancement could be achieved by a FESS-CZE method with UV detection for the analysis of organic acids (e.g., linolenic acid, lauric acid, ascorbic acid) in three traditional Chinese medicines, namely *Portulaca oleracea* L., *Crataegus pinnatifida*, and *Aloe vera* L. [69].

Other than UV detection, electrochemical detection has also been coupled with CZE. The use of CZE with a capacitively coupled contactless conductivity detector (C4D) has been applied to selectively screen for seven anorexic and antidepressant drug adulterants (amfepramone, sibutramine, fenproporex, fluoxetine, paroxetine, sertraline, and bupropion) in 106 herbal weight loss preparations [70] and also for diuretics (furosemide, hydrochlorothiazide, chlorthalidone, and amiloride) and laxatives (phenolphthalein) in addition to anorexics (amfepramone) and antidepressants (fluoxetine and paroxetine) in 26 commercial herbal formulations for weight loss [71].

Microemulsion electrokinetic capillary chromatography (MEEKC), which is another mode of CE, has also been applied to the analysis of phytomedicines. One study incorporated the use of a novel additive of cationic surfactant-coated multi-walled carbon nanotubes for MEEKC separation and quantification of eight phenolic compounds, four glycosides, and one phenanthraquinone in a complex herbal matrix preparation (Compound Xueshuantong capsule) [72]. This was shown to improve the separation of the solutes as compared with conventional MEEKC [72].

2.3.4.3 Spectroscopy

Vibrational Spectroscopy Vibrational spectroscopic methods, which include infrared (IR), near-IR (NIR), and Raman spectroscopies, have also been applied in the QC of phytomedicines. Compared with chromatographic methods, these techniques are simple, fast, and require minimal sample prepreparation. Hence, these have been used as rapid techniques for the identification and manufacturing process management of phytomedicines [60]. NIR spectroscopy is often combined with chemometrics for spectral data preprocessing and calibration. It has been utilized to discriminate between raw and processed herbs, for example, 40 batches of raw *Dipsacus asperoides* (a common traditional Chinese medicine) and their corresponding processed samples treated by stir-baking with salt solution [73], as well as to quantify the content of herbal products, for example, the content of oren powder extract (an ethanol-soluble component of *Coptis chinensis* used in Kampo medicine) in suppositories [74]. The use of Raman spectroscopy has been reported for the analysis of the chemical composition of *Paeoniae alba Radix* (white peony root) decoctions [75] and for the differentiation of ginseng samples from those of *Anthriscus sylvestris*, *Radix glehniae*, and balloonflower root [76].

Like the chromatographic technologies, 2D vibrational spectroscopy (2D-IR, 2D-NIR, and 2D-Raman spectroscopies) with enhanced spectral resolution has also been introduced and used to discriminate herbs of different species, from different geographical regions, or those that have undergone different processing methods [60]. The use of multistep IR macro-fingerprinting (consisting of Fourier transform-IR, second derivative IR, and 2D-IR correlation spectroscopy) combined with soft independent modeling of class analogy (a statistical pattern recognition method) has been reported for rapid analysis and discrimination of 179 samples of four *Cistanche* species, namely *Cistanche deserticola*, *C. tubulosa*, *C. salsa*, and *C. sinensis*, collected from different regions in China [77]. In addition, this technique enabled high-throughput classification of the 179 samples and examination of the quality consistency among the samples of each species within a short period of time [77].

Atomic Absorption Spectroscopy AAS is an essential technique for the determination and quantitation of heavy metals, such as arsenic, cadmium, copper, iron, lead, mercury, nickel, and zinc, in phytomedicines [9, 16]. Due to the potential toxic effects of heavy and trace metals, the quantification of metals in phytomedicines as part of QC is important. The use of AAS has been reported in the determination of the content of heavy metals (specifically arsenic, lead, and mercury) and trace metals (specifically copper, chromium, manganese, iron, and nickel) in materials of eight different plant species, including leaf of *Aloe vera*, leaf of *Centella asiatica*, flower of *Calendula officinalis*, fresh fruit of *Cucumis sativus*, leaf of *Camellia sinensis*, leaf of *Clitoria ternatea*, leaf of Piper betel, and flower of *Tagetes erecta*, which are commonly used medicinal herbs in Ayurveda [78].

Nuclear Magnetic Resonance (NMR) NMR spectroscopy is one of the principal techniques for structural analysis [79]. Over the years, NMR spectroscopy has also developed into one of the important tools for analysis of phytomedicines due to its simple sample preparation, high sample throughput, high reproducibility, and nondestructive nature. Not only is NMR spectroscopy used for qualitative fingerprint analysis, quantitative NMR (qNMR) is also applied for quantitation of chemical constituents in phytomedicines for QC and standardization of herbal medicinal preparations [60, 80]. Compared with quantitative analysis by HPLC that requires pure authentic standards of the compounds to be quantified, quantitation by qNMR can be conducted without the need for reference marker standards, which is an advantage when pure certified reference materials may not be available [80]. It was also demonstrated that qNMR technique is as reliable as HPLC for standardization and quantitative analysis of extracts from three Indian medicinal plants, namely *Eugenia jambolana*, *Withania somnifera*, and *Aegle marmelos*, and their herbal products [80].

In addition, 2D diffusion-ordered spectroscopy (DOSY), 1H-NMR, and three-dimensional (3D) DOSY-correlation spectroscopy (COSY), 1H-NMR, are powerful emerging techniques applied for structural elucidation and multivariate fingerprinting of complex phytomedicines for the identification and detection of adulterants [81, 82]. In one study, 13 of 20 herbal medicines and dietary supplements marketed as natural slimming products were analyzed by these techniques and were found to be adulterated with sibutramine alone or in combination with phenolphthalein [62]. In another study, 8 of 17 herbal dietary supplements marketed as natural substances for the enhancement of sexual function were analyzed and found to contain compounds related to the synthetic phosphodiesterase-5 inhibitors, namely sildenafil, tadalafil, vardenafil, hydroxyhomosildenafil, thiosildenafil, and thiomethisosildenafil [63].

Mass Spectrometry (MS) With its high selectivity, sensitivity, and resolution, MS is a mainstay for structural elucidation and is usually combined with chromatographic separation techniques. In fact, MS-based hyphenated techniques, in particular LC-MS and GC-MS, are among the most prominent and powerful combination techniques for the analysis of phytomedicines and have been applied to fingerprinting analysis, standardization, identification of active compounds, and screening for adulterants in phytomedicines [83]. These are discussed in the next section.

2.3.4.4 Hyphenated Techniques

GC-MS GC-MS is the first successful hyphenated technique that was introduced in the 1960s. Coupling the capability of GC to produce high-quality chromatographic fingerprints due to its good separation ability and the strength of MS to provide structural information on the analytes, GC-MS is the method of choice for the analysis of volatile and thermally stable constituents in phytomedicines, including the identification of both known and unknown chemical compounds [44]. In addition, an optimized selective pressurized liquid extraction and GC-MS/MS method was applied for the simultaneous determination of 52 pesticide residues in four medicine and food dual-purpose herbs, including the root of *Pueraria thomsonii* Benth., *Pogostemon cablin* Benth., *Houttuynia cordata* Thunb., and *Disoscorea opposita* Thunb [84].

GC×GC in tandem with high-resolution time-of-flight (TOF) MS has also been applied in the separation and identification of 167 volatile components (50 monoterpenes, 36 sesquiterpenes, 31 esters and acids, 9 aldehydes and ketones, 6 alcohols, 3 ethers, 12 phenyl compounds, and 20 other components) in the volatile oil of *Pericarpium Citri reticulatae* (Chenpi) [49]. Compared with previous studies, this is a significantly larger number of Chenpi volatile components identified, which was achieved by the use of GC×GC that provides high-resolution separation of the components based on volatility difference on the first-dimension nonpolar column and then based on polarity on the second medium polar column [49]. This significantly reduced the problem of co-elution and incomplete separation of Chenpi volatile constituents by one-dimensional GC or GC-MS in previous studies [49].

LC-MS LC-MS-based methods, including HPLC-MS [50, 56, 57, 59, 85–89], UPLC-MS [61–63], 2D-HPLC-MS [64], and HPLC×UPLC-MS [66], are highly sensitive and selective hyphenated techniques for the analysis of complex samples. The ability to analyze a wide range of chemical compounds, including thermolabile compounds, renders LC-MS-based methods to be among the most popular, important, and powerful techniques for fingerprinting, identification, quantitation, and structural elucidation of phytomedicines containing multiple components. For example, HPLC/electrospray ionization (ESI)-MS/MS has been applied to the determination of four indoxyl derivatives (indican, isatin, indirubin, and indigotin) in the roots and leaves of *Isatis indigotica* (Chinese woad), named "Ban-Lan-Gen" and "Da-Qing-Ye," respectively [87]. UPLC-MS was used for quantitative analysis of four ginsenosides Rb1, Rb3, Rd, and F2 in *Gynostemma pentaphyllum*, so as to differentiate the sweet and bitter taste variants of this herb which have different clinical applications in Chinese medicine and are otherwise difficult to differentiate by taste or morphological characteristics [63].

An example of the application of 2D-HPLC-MS is the analysis of *Schisandra chinensis* (a famous traditional Chinese medicine known as Beiwuweizi in Chinese) using 2D-HPLC-DAD-ESI-MS, as well as chemometric analyses (specifically principal component analysis (PCA) and hierarchical clustering analysis (HCA)) to process the 3D chromatographic fingerprints obtained [64]. More than 40 constituents in *Schisandra chinensis* were separated by the 2D-HPLC system and among them, 14 constituents were identified by UV and mass spectral data [64].

UPLC has also been used as the second dimension in 2D liquid chromatographic analysis. HPLC×UPLC-TOF-MS has been utilized for the analysis of Qingkailing injection (a complex extract of Traditional Chinese Medicine) that consists of several components, including bile acids, amino acids, flavonoids, organic acids, nucleotides, iridoid glycoside, pigments, volatile compounds, inorganic compounds, etc.

[66]. From the four-dimensional data generated by HPLC×UPLC-TOF-MS, as many as 398 components of Qingkailing injection were separated, which was sevenfold more than the 54 components separated using HPLC×HPLC-MS [66]. In addition, analysis time based on HPLC×UPLC-TOF-MS (250 min) could be reduced by 75% as compared with HPLC×HPLC-MS (1110 min) [66].

Other Hyphenated Techniques Other than GC-MS and LC-MS, a range of other MS-based hyphenated techniques have also been applied to the analysis of phytomedicines. These include: CE-MS, which has been used to analyze alkaloids in different Chinese medicinal herbs such as *Rhizoma Coptidis* (Huanglian) [90], aconite root *Radix Aconiti praeparata* (Wutou), and the seed of *Strychnos pierrian* (Maqianzi) [91], as well as TLC-MS and TLC-HPLC-MS, which have been applied to the fingerprinting of phenolic acid fractions selectively extracted from *Salvia lavandulifolia* [92].

HPLC-NMR is another powerful technique for the rapid and detailed structural characterization and identification of constituents in phytomedicines, for example, flavonoids and ellagic acid derivatives in *Drosera* species [50, 51], stilbenes, phloroglucinols, and flavonoids from *Lysidice brevicalyx* extracts [88] and cardiac glycosides from *Periploca forrestii* extracts [89]. Another hyphenated technique, HPLC-NMR-MS, has also been utilized for structural identification of phytoconstituents, for example, furanocoumarins (byakangelicol, oxypeucedanin, imperatorin, phellopterin, and isoimperatorin) in the extracts of *Angelica dahurica* [93].

2.3.5 DNA Fingerprinting

While the identification, QC, and standardization of phytomedicines rely mainly on macroscopic and microscopic evaluation and chemical fingerprinting, these approaches may be limited by the botanical expertise of the evaluator and extrinsic factors such as growth or cultivation conditions, storage conditions, and processing methods that affect the chemical composition of the phytomedicines [94]. By contrast, the genetic makeup of each herbal species is unique and not affected by external factors, except for geographical origin. DNA analysis is one of the most reliable methods of herbal identification [60]. Hence, to complement other methods of identification, DNA fingerprinting techniques have been applied in species characterization and authentication, detection of adulterants, standardization, identification of the geographical origin of phytomedicines, and functional characterization of important genes for the determination of herbal quality at the DNA level.

Randomly amplified polymorphic DNA (RAPD) analysis has been applied to evaluate the genetic diversity among 11 accessions of *Stevia rebaudiana* (cultivated and wild varieties) from different geographical regions of India [95]. Population genetic diversity among six natural populations of *Limonium sinense* in China was also elucidated using RAPD, as well as inter-simple sample repeat (ISSR) and amplified fragment length polymorphism (AFLP) analyses [96]. Microarray-based DNA profiling has also been applied for fingerprinting medicinal herbs [97], for example, Salvia species, which is useful for species differentiation [98].

2.3.6 "Omics" Technology

The field of "omics" technology, which includes genomics, transcriptomics, proteomics, and metabolomics, has evolved rapidly in this post-genome era, with emerging applications in the quality evaluation of phytomedicines. Ongoing large-scale projects, including the Medicinal Plant Consortium (http://medicinalplantgenomics.msu.edu/), the PhytoMetaSyn Project (http://www.phytometasyn.com/), and the 1KP Project (http://www.onekp.com/index.html), are generating "omics" datasets on various species of plants, including medicinal plants, which serve as useful reference resources.

2.3.6.1 Genomics and Transcriptomics Advances in DNA sequencing techniques brought about the advent of high-throughput next-generation sequencing (HT-NGS) technology, which was first commercially available around 2005 [99, 100]. HT-NGS is now emerging as a potential tool to complement or even replace traditional DNA fingerprinting techniques for whole genome and transcriptome profiling of phytomedicines for purposes of quality assessment [101, 102]. In one example, deep transcriptome technology (also known as RNA sequencing, which provides whole transcriptome expression profiles of selected plant tissues or cells) was applied to the transcriptome profiling of *Ophiorrhiza pumila* hairy roots and cell suspension cultures using the Illumina platform, so as to further the understanding of the biosynthesis of camptothecin (an anticancer alkaloid) and anthraquinones [103].

2.3.6.2 Proteomics Advances in ionization technology for MS, in particular ESI and matrix-assisted laser desorption/ionization (MALDI) that enable the ionization of large, polar, and thermolabile biomolecules (including proteins and peptides) have contributed to significant progress in proteomics [104]. In addition to ESI-MS and MALDI-TOF-MS, 2D gel electrophoresis has also been used in peptide mass fingerprinting to facilitate authentication and identification of phytomedicines for quality evaluation and standardization [105]. An example is the application of 2D gel electrophoresis-based proteomic approach to differentiate between *Panax ginseng* (Oriental ginseng) and *Panax quinquefolius* (American ginseng), among different plant parts (main root, lateral roots, rhizome head, and skin) of *P. ginseng* and between wild-grown *P. ginseng* main root and *P. ginseng* culture cells [106]. In addition, proteome analysis has also been applied to investigate the molecular mechanisms underlying the effects of phytomedicines, for example, the hepatoprotective effects of a traditional Chinese medicine, Yin-Chen-Hao Tang (consisting of *Artemisia annua* L., *Gardenia jasminoides* Ellis, and *Rheum palmatum* L.), was studied by a combination of 2D gel electrophoresis and MALDI-TOF-MS [107].

2.3.6.3 Metabolomics Metabolic fingerprint analysis of phytomedicines has been made possible with the advent of metabolomics (i.e., a comprehensive and quantitative analysis of all metabolites in a system) (also used interchangeably with the term, metabonomics) [108]. Various analytical techniques are used in metabolic profiling, which include chromatography-based methods (e.g., HPTLC [95], HPLC-DAD

[109]), NMR-based methods (e.g., 1H-NMR [110–113], 2D-NMR [110–112]), and MS-based hyphenated methods (e.g., HPLC-MS [114, 115], UPLC-quadrupole TOF (qTOF)-MS [61, 62, 116], GC-MS [117]). In addition, chemometrics-derived pattern recognition analysis (e.g., PCA, HCA, partial least-squares-discriminant analysis (PLS-DA), partial least-squares projections to latent structures (PLS), and orthogonal projections to latent-structures-discriminant analysis (OPLS-DA)) is used as the statistical approach to analyze the wealth of metabolomics data generated [118]. The metabolomics approach has been used in the quality assessment of phytomedicines [34] and some examples of its applications include:

- Discrimination of different plant parts, varieties, species, geographic origins, and suppliers (e.g., investigation of the metabolite compositional differences between the leaves and flowers of *Tussilago farfara* L. (Coltsfoot) by 1H-NMR and 2D-NMR with PCA and PLS-DA [110]; differentiation of *Magnolia officinalis* and *M. officinalis* var. *biloba* by 1H-NMR and 2D-NMR with PCA, HCA, and PLS-DA [112]; discrimination of species and geographical origin of dry root samples of *Angelica acutiloba*, *A. acutiloba* Kitagawa var. *sugiyamae* Hikino, and *A. sinensis* cultivated in different regions in Japan and China, by GC-MS with PCA and OPLS-DA [117]; discrimination between *Tanacetum parthenium* (L.) Schultz Bip (Compositae) (feverfew) samples from different suppliers by 1H-NMR and PCA [113])
- Determination of cultivation age (e.g., differentiation of the roots of *Panax ginseng* C.A. Meyer (Araliaceae) according to cultivation ages by 1H-NMR and 2D-NMR with PCA, PLS-DA, HCA, and PLS [111])
- Evaluation of the effects of storage period and processing methods (e.g., discrimination of dried Citrus peels (Chenpi) with different storage periods through a targeted metabolomic approach by HPLC-DAD and PCA of 11 phenolic compounds of regular (1 year) and long-term (3 year) stored Chenpi samples [109]; time-dependent metabolite profiling and discrimination of raw and steamed *Panax notoginseng* (Burk.) F.H. Chen (commonly known as Tianqi or Sanqi) by UPLC-qTOF-MS with chemometric data analysis [61, 62])

2.4 CHALLENGES

Quality evaluation of phytomedicines has made much progress over the years with the aid of technological advances in analytical techniques. The use of different analytical techniques has enabled species authentication, detection of adulterants, evaluation of contaminants, quantitation, and standardization of active ingredients in phytomedicines. An example of such applications in a QA program covering the acquisition, botanical authentication, contaminant determination, standardization, and GMP production of a 20-herb Chinese herbal formula has been reported [119]. Nonetheless, analytical challenges still exist as specifics relating to the quality assessment of most phytomedicines are generally not clear or not established.

To begin with, analysis of phytomedicines is often confounded by the complex composition and associated interactions between phytochemicals that contribute to the holistic therapeutic effect of the individual herbs or preparations. Standardization of phytomedicines based on specific selected chemical constituents is the current approach to ensure that the phytomedicines contain concentrations of these chemical constituents within specified range believed to provide a certain level of biological or therapeutic activity. However, these constituents may or may not be the active components. To date, not all active constituents of phytomedicines have been isolated, characterized, or quantified [120]. In addition, active phytocomponents such as glycosides, alkaloids, and organic acids, have been reported to undergo metabolism or degradation [121]. As explained in the earlier sections, treatment by sulfur fumigation or processing conditions that affect stability can generate new derivatives and chemical entities. When these structurally complex active constituents undergo extensive or unexpected biological metabolism, metabolite identification and quantification may be somewhat complicated and difficult [121]. Although instrumental analytical tools are available to separate the individual components in phytomedicines for qualitative and quantitative analyses and for subsequent biological activity evaluation, it is not clear what all the chemical constituents and their respective concentrations/proportions are, to yield an appropriate balance between efficacy and safety in most phytomedicines. It has been suggested that the markers for QC should be correlated with their efficacy and safety [122]. Nevertheless, given the complexity of the composition of phytomedicines and the multitude of interactions among the various known and unknown chemical constituents that are not yet clearly understood, the identification of individual or group(s) of phytochemical markers that correlate with the efficacy and safety of the phytomedicines for quality assessment remains a challenge [122, 123].

In addition, there is a lack of internationally recognized standards for quality, including standard QC profiles, for individual raw herbs and polyherbal preparations that may have undergone different processing methods [124, 125]. Harmonization and establishment of internationally accepted standards is needed to ensure consistency in product quality across the industry [124]. The current approach is to ensure batch-to-batch consistency of the chemical compositions or fingerprint chromatograms of phytomedicines, or to ensure equivalence of herbal extracts based on the same starting materials, extraction process and conditions, solvent system and native extract ratio [126]. Nevertheless, the ideal approach is to also ensure phytoequivalence, that is, to ensure equivalence between the effects of herbal preparations where the efficacy of one preparation has been previously established and used as the basis for comparison with other preparations. However, demonstration of phytoequivalence in phytomedicines is still a challenge today. It has been suggested that the application of "omics" technology (genomics, transcriptomics, proteomics, and metabolomics) may advance understanding of the efficacy of phytomedicines [127]. Integrated "omics" analysis as a systems biology approach to elucidate and compare efficacy profiles of phytomedicines holistically may hold promise in assessing phytoequivalence for QA/QC of phytomedicines.

Being readily available over the counter and of natural origin, phytomedicines are widely perceived as safe by consumers. However, adverse effects due to toxicity, contraindications, and herb–drug interactions have been reported with the use of phytomedicines. Aside from the inherent toxicity of some herbs, adverse effects of phytomedicines may also be attributed to extrinsic factors as a result of QC lapses. One of these extrinsic factors is the misidentification or substitution of the herbal materials. The use of a wrong herb leads to a lack of efficacy, or more adversely, could lead to toxic effects in consumers if the herb is toxic. QC of the herbs is therefore crucial to ensure that the correct herbal material is used. Although methods for the quality assessment of starting herbal materials are in place, the lack of authenticated herbs or botanical reference materials, in many cases, against which test samples can be compared, poses a challenge for QA/QC. Furthermore, while certain countries have pharmacopoeial monograph descriptions and specifications for physicochemical analysis of plant materials (e.g., in the British Pharmacopoeia, American Herbal Pharmacopoeia, European Pharmacopoeia, and Chinese Pharmacopoeia), many other countries and regions have not established a clear framework for quality assessment of phytomedicines. This poses an analytical challenge for quality evaluation of phytomedicines that are unique to these other countries.

Adulteration with undeclared synthetic chemical drug compounds to enhance the claimed therapeutic effects of phytomedicines could also result in adverse effects in consumers. While advances in analytical techniques have enabled the detection of adulterants, not all regulatory authorities or analytical laboratories may have access to appropriate or advanced analytical instruments, hence limiting post-marketing QA surveillance of phytomedicines. In addition, although general adulterant screening methods using instrumental analytical techniques are available [58], an ideal method for routine screening of the unlimited possibilities of adulterants and their structurally diverse analogues is lacking. Identification of analogues of known drugs is particularly challenging, unlike targeted analysis for known compounds or adulterants. This problem is exacerbated by the lack of reference standards or chemical reference materials for comparison and confirmation [79]. Furthermore, reports of tadalafil being concealed in the gelatin matrices of capsule shells of phytomedicines [28, 29] highlight an additional challenge of novel method of adulteration. Awareness of the potential presence of adulterants in matrices other than the capsule contents is critical for the detection of such adulterants.

In addition, a proper sample preparation approach prior to instrumental analysis is crucial to ensure repeatability and accuracy of analysis of phytochemicals for the quality assessment of phytomedicines [122]. However, the complex composition of phytomedicines and differences in solubility and extraction efficiency of the various phytoconstituents pose inherent challenges in the choice of extraction solvents and extraction methods. There is no single ideal method for routine extraction of the different components in phytomedicines for subsequent qualitative or quantitative analysis. Appropriate sample preparation approaches for determination and quantitation of the components in different phytomedicines need to be developed and optimized.

2.5 CONCLUSIONS

The analytical components in QA/QC form an essential basis for quality assessment of phytomedicines. The employment of new, improved, and evolving technologies has enhanced quality assessment of phytomedicines by enabling species identification, detection of substitutes and contaminants, screening for adulterants, standardization of specified chemical constituents, and stability testing. Moving forward, concerted efforts should be made to address the critical challenges underlying analyses of phytomedicines. These include, in particular, the establishment of quality specifications, making available chemical reference materials and authentic herbs, harmonization of quality assessment framework, as well as the development and improvement of integrative analytical strategies. Despite the many challenges, as quality, safety, and efficacy are closely interrelated, the QA/QC of phytomedicines are essential. By strengthening the analytical capacity, regulators, manufacturers, and researchers can work together toward achieving the successful QA/QC of phytomedicines.

REFERENCES

[1] Jia L, Zhao Y (2009) Current evaluation of the millennium phytomedicine--ginseng (I): etymology, pharmacognosy, phytochemistry, market and regulations. *Curr Med Chem* 16: 2475–2484.

[2] WHO (2002) *Traditional Medicine Strategy 2002–2005*. World Health Organization, Geneva.

[3] Pilcher H (2004) Herbal medicine spawns antimalarial chemical. *Nature*. DOI:10.1038/news040816-7.

[4] WHO (2007) *Quality Assurance of Pharmaceuticals: A Compendium of Guidelines and Related Materials*. Vol. 2, 2nd updated edition, Good manufacturing practices and inspection. World Health Organization, Geneva.

[5] WHO (2005) *National Policy on Traditional Medicine and Regulation of Herbal Medicines: Report of a WHO Global Survey*. World Health Organization, Geneva.

[6] Xue TH, Roy R (2003) Studying traditional Chinese herbal medicine. *Science* 300: 740–741.

[7] EudraLex (2006) *The Rules Governing Medicinal Products in the European Union. Volume 4: EU Guidelines to Good Manufacturing Practice—Medicinal Products for Human and Veterinary Use. Annex 7: Manufacture of Herbal Medicinal Products*. European Commission, Brussels.

[8] FDA (2004) *Guidance for Industry: Botanical Drug Products*. U.S. Department of Health and Human Services, Food and Drug Administration, Center for Drug Evaluation and Research (CDER), Rockville.

[9] EMA (2011) *Guideline on Specifications: Test Procedures and Acceptance Criteria for Herbal Substances, Herbal Preparations and Herbal Medicinal Products/Traditional Herbal Medicinal Products*. European Medicines Agency, London.

[10] WHO (2003) *WHO Guidelines on Good Agricultural and Collection Practices (GACP) for Medicinal Plants*. World Health Organization, Geneva.

[11] EMA (2011) *Guideline on Quality of Herbal Medicinal Products/Traditional Herbal Medicinal Products*. European Medicines Agency, London.

[12] EMEA (2006) *Guideline on Good Agricultural and Collection Practice (GACP) for Starting Materials of Herbal Origin*. European Medicines Agency, London.

[13] SFDA (2002) *Good Agricultural Practice for Chinese Crude Drugs (Interim)*. State Food and Drug Administration, P.R. China

[14] Andre P, Su HB, Gao WY (2011) Ensuring the compliance of raw herbal materials stemmed from China with European Good Agricultural and Collection Practice. *Chin Herb Med* 3: 251–256.

[15] Chan LW, Sia CH, Tan KY (2012) Current scientific status and regulatory control of traditional/herbal medicinal products: globalization challenges. *Pharm Eng* 32: 1–15.

[16] WHO (2011) *Quality Control Methods for Herbal Materials*. World Health Organization, Geneva.

[17] HSA (2014) *Health Products Regulation: Requirements of Test Reports*. Health Sciences Authority, Singapore.

[18] Harris ES, Cao S, Littlefield BA, Craycroft JA, Scholten R, et al. (2011) Heavy metal and pesticide content in commonly prescribed individual raw Chinese Herbal Medicines. *Sci Total Environ* 409: 4297–4305.

[19] Oh CH (2007) Multi residual pesticide monitoring in commercial herbal crude drug materials in South Korea. *Bull Environ Contam Toxicol* 78: 314–318.

[20] Kan WL, Ma B, Lin G (2011) Sulfur fumigation processing of traditional Chinese medicinal herbs: beneficial or detrimental? *Front Pharmacol* 2: 84.

[21] Jiang X, Huang LF, Zheng SH, Chen SL (2013) Sulfur fumigation, a better or worse choice in preservation of Traditional Chinese Medicine? *Phytomedicine* 20: 97–105.

[22] SFDA (2013) *Commentary on 2010 Chinese Pharmacopoeia Second Supplement Revision: Residue Limit of Sulfur Dioxide in TCMs and their Formulations*. State Food and Drug Administration, P.R. China.

[23] EMA (2013) *Questions & Answers on Quality of Herbal Medicinal Products/Traditional Herbal Medicinal Products*. European Medicines Agency, London.

[24] Santos L, Marin S, Sanchis V, Ramos AJ (2009) Screening of mycotoxin multicontamination in medicinal and aromatic herbs sampled in Spain. *J Sci Food Agric* 89: 1802–1807.

[25] Jevremovic M, Lazarevic N, Pavlovic S, Orlic M (2011) Radionuclide concentrations in samples of medicinal herbs and effective dose from ingestion of 137Cs and natural radionuclides in herbal tea products from Serbian market. *Isotopes Environ Health Stud* 47: 87–92.

[26] Upton RT (2010) Use of high-performance thin layer chromatography by the American Herbal Pharmacopoeia. *J AOAC Int* 93: 1349–1354.

[27] Chen CH, Dickman KG, Moriya M, Zavadil J, Sidorenko VS, et al. (2012) Aristolochic acid-associated urothelial cancer in Taiwan. *Proc Natl Acad Sci USA* 109: 8241–8246.

[28] Medsafe (2010) New Zealand Medicines and Medical Devices Safety Authority. Adulterated erectile dysfunction products: on the rise in New Zealand. Prescriber Update 31: 20.

[29] Venhuis BJ, Tan J, Vredenbregt MJ, Ge X, Low MY, et al. (2012) Capsule shells adulterated with tadalafil. *Forensic Sci Int* 214: e20–e22.

[30] Patel DN, Low WL, Tan LL, Tan MM, Zhang Q, et al. (2012) Adverse events associated with the use of complementary medicine and health supplements: an analysis of reports in the Singapore Pharmacovigilance database from 1998 to 2009. *Clin Toxicol (Phila)* 50: 481–489.

[31] Low MY, Zeng Y, Li L, Ge XW, Lee R, et al. (2009) Safety and quality assessment of 175 illegal sexual enhancement products seized in red-light districts in Singapore. *Drug Saf* 32: 1141–1146.

[32] Gafner S, Bergeron C (2005) The challenges of chemical stability testing of herbal extracts in finished products using state-of-the-art analytical methodologies. *Curr Pharm Anal* 1: 203–215.

[33] Mukherjee PK (2002) *Quality Control of Herbal Drugs – an Approach to Evaluation of Botanicals*. Business Horizons, New Delhi.

[34] Ning Z, Lu C, Zhang Y, Zhao S, Liu B, et al. (2013) Application of plant metabonomics in quality assessment for large-scale production of traditional Chinese medicine. *Planta Med* 79: 897–908.

[35] Li SL, Yan R, Tam YK, Lin G (2007) Post-harvest alteration of the main chemical ingredients in *Ligusticum chuanxiong* Hort. (Rhizoma Chuanxiong). *Chem Pharm Bull (Tokyo)* 55: 140–144.

[36] Wong L, Liang Z, Chen H, Zhao Z (2011) Authentication of Chinese Materia Medica decoction dregs, part 1: comparison of morphological and microscopic features of four Chinese Materia Medica before and after decoction. *Microsc Res Tech* 74: 320–328.

[37] Wong L, Liang Z, Chen H, Zhao Z (2012) Authentication of Chinese Materia Medica decoction dregs. Part II: comparison before and after decoction of four Chinese Materia Medica that mainly comprise storage tissue. *Microsc Res Tech* 75: 164–175.

[38] Silva Júnior JOC, Costa MR, Teixeira FM, Barbosa WLR (2011) Processing and quality control of herbal drugs and their derivatives. In: Shoyama Y, editor. *Quality Control of Herbal Medicines and Related Areas*. InTech, Croatia, pp. 195–222.

[39] Surekha D, Thiruvengada Rajan VS, Amruth Kumar N, Angala Parameswari S, Madhusudhana Chetty C (2011) A review on role of quality control and quality assurance system in regulation of herbal drugs. *Int J Rev Life Sci* 1: 97–105.

[40] Wang YQ, Liang ZT, Li Q, Yang H, Chen HB, et al. (2011) Identification of powdered Chinese herbal medicines by fluorescence microscopy, part 1: fluorescent characteristics of mechanical tissues, conducting tissues, and ergastic substances. *Microsc Res Tech* 74: 269–280.

[41] Li Q, Chu C, Wang YQ, Chen HB, Li P, et al. (2011) Authentication of the 31 species of toxic and potent Chinese Materia Medica by microscopic technique assisted by ICP-MS analysis, part 4: four kinds of toxic and potent mineral arsenical CMMs. *Microsc Res Tech* 74: 1–8.

[42] Kamboj A (2012) Analytical evaluation of herbal drugs. In: Vallisuta O, editor. *Drug Discovery Research in Pharmacognosy*. InTech, Croatia, pp. 23–60.

[43] Zhang Y, Sun S, Dai J, Wang W, Cao H, et al. (2011) Quality control method for herbal medicine - chemical fingerprint analysis. In: Shoyama Y, editor. *Quality Control of Herbal Medicines and Related Areas*. InTech, Croatia, pp. 171–194.

[44] Liang YZ, Xie P, Chan K (2004) Quality control of herbal medicines. *J Chromatogr B Analyt Technol Biomed Life Sci* 812: 53–70.

[45] Meier B, Spriano D (2010) Modern HPTLC—a perfect tool for quality control of herbals and their preparations. *J AOAC Int* 93: 1399–1409.

[46] Mukherjee D, Kumar NS, Khatua T, Mukherjee PK (2010) Rapid validated HPTLC method for estimation of betulinic acid in *Nelumbo nucifera* (Nymphaeaceae) rhizome extract. *Phytochem Anal* 21: 556–560.

[47] Kumar V, Mukherjee K, Kumar S, Mal M, Mukherjee PK (2008) Validation of HPTLC method for the analysis of taraxerol in *Clitoria ternatea. Phytochem Anal* 19: 244–250.

[48] Rai S, Mukherjee K, Mal M, Wahile A, Saha BP, et al. (2006) Determination of 6-gingerol in ginger (*Zingiber officinale*) using high-performance thin-layer chromatography. *J Sep Sci* 29: 2292–2295.

[49] Qin K, Zheng L, Cai H, Cao G, Lou Y, et al. (2013) Characterization of chemical composition of Pericarpium Citri reticulatae volatile oil by comprehensive two-dimensional gas chromatography with high-resolution time-of-flight mass spectrometry. *Evid Based Complement Alternat Med* 2013: 237541.

[50] Zehl M, Braunberger C, Conrad J, Crnogorac M, Krasteva S, et al. (2011) Identification and quantification of flavonoids and ellagic acid derivatives in therapeutically important Drosera species by LC-DAD, LC-NMR, NMR, and LC-MS. *Anal Bioanal Chem* 400: 2565–2576.

[51] Braunberger C, Zehl M, Conrad J, Fischer S, Adhami HR, et al. (2013) LC-NMR, NMR, and LC-MS identification and LC-DAD quantification of flavonoids and ellagic acid derivatives in *Drosera peltata. J Chromatogr B Analyt Technol Biomed Life Sci* 932C: 111–116.

[52] Lau AJ, Seo BH, Woo SO, Koh HL (2004) High-performance liquid chromatographic method with quantitative comparisons of whole chromatograms of raw and steamed *Panax notoginseng. J Chromatogr A* 1057: 141–149.

[53] Lau AJ, Woo SO, Koh HL (2003) Analysis of saponins in raw and steamed *Panax notoginseng* using high-performance liquid chromatography with diode array detection. *J Chromatogr A* 1011: 77–87.

[54] Lau AJ, Koh HL (2006) Quality control of herb: principles and procedures, using Panax as an example. In: Leung PC, Fong HHS, Xue CC, editors. *Current Review of Chinese Medicine: Quality Control of Herbs and Herbal Material.* World Scientific Publishing Co. Pte. Ltd., Singapore, pp. 87–115.

[55] Zou P, Hong Y, Koh HL (2005) Chemical fingerprinting of *Isatis indigotica* root by RP-HPLC and hierarchical clustering analysis. *J Pharm Biomed Anal* 38: 514–520.

[56] Koh HL, Wang H, Zhou S, Chan E, Woo SO (2006) Detection of aristolochic acid I, tetrandrine and fangchinoline in medicinal plants by high performance liquid chromatography and liquid chromatography/mass spectrometry. *J Pharm Biomed Anal* 40: 653–661.

[57] Zou P, Oh SS, Hou P, Low MY, Koh HL (2006) Simultaneous determination of synthetic phosphodiesterase-5 inhibitors found in a dietary supplement and pre-mixed bulk powders for dietary supplements using high-performance liquid chromatography with diode array detection and liquid chromatography-electrospray ionization tandem mass spectrometry. *J Chromatogr A* 1104: 113–122.

[58] Liu SY, Woo SO, Koh HL (2001) HPLC and GC-MS screening of Chinese proprietary medicine for undeclared therapeutic substances. *J Pharm Biomed Anal* 24: 983–992.

[59] Liu SY, Woo SO, Holmes MJ, Koh HL (2000) LC and LC-MS-MS analyses of undeclared codeine in antiasthmatic Chinese proprietary medicine. *J Pharm Biomed Anal* 22: 481–486.

[60] Jiang Y, David B, Tu P, Barbin Y (2010) Recent analytical approaches in quality control of traditional Chinese medicines—a review. *Anal Chim Acta* 657: 9–18.

[61] Toh DF, New LS, Koh HL, Chan EC (2010) Ultra-high performance liquid chromatography/time-of-flight mass spectrometry (UHPLC/TOFMS) for time-dependent profiling of raw and steamed *Panax notoginseng*. *J Pharm Biomed Anal* 52: 43–50.

[62] Chan EC, Yap SL, Lau AJ, Leow PC, Toh DF, et al. (2007) Ultra-performance liquid chromatography/time-of-flight mass spectrometry based metabolomics of raw and steamed *Panax notoginseng*. *Rapid Commun Mass Spectrom* 21: 519–528.

[63] Lu JG, Zhu L, Lo KY, Leung AK, Ho AH, et al. (2013) Chemical differentiation of two taste variants of *Gynostemma pentaphyllum* by using UPLC-Q-TOF-MS and HPLC-ELSD. *J Agric Food Chem* 61: 90–97.

[64] Wang S, Wang C, Zhao X, Mao S, Wu Y, et al. (2012) Comprehensive two-dimensional high performance liquid chromatography system with immobilized liposome chromatography column and monolithic column for separation of the traditional Chinese medicine *Schisandra chinensis*. *Anal Chim Acta* 713: 121–129.

[65] Ni Y, Liu Y, Kokot S (2011) Two-dimensional fingerprinting approach for comparison of complex substances analysed by HPLC-UV and fluorescence detection. *Analyst* 136: 550–559.

[66] Ma S, Liang Q, Jiang Z, Wang Y, Luo G (2012) A simple way to configure on-line two-dimensional liquid chromatography for complex sample analysis: acquisition of four-dimensional data. *Talanta* 97: 150–156.

[67] Ganzera M (2008) Quality control of herbal medicines by capillary electrophoresis: potential, requirements and applications. *Electrophoresis* 29: 3489–3503.

[68] Dutra LS, Leite MN, Brandao MA, de Almeida PA, Vaz FA, et al. (2013) A rapid method for total beta-escin analysis in dry, hydroalcoholic and hydroglycolic extracts of *Aesculus hippocastanum* L. by capillary zone electrophoresis. *Phytochem Anal.* DOI:10.1002/pca.2425.

[69] Zhu Q, Xu X, Huang Y, Xu L, Chen G (2012) Field enhancement sample stacking for analysis of organic acids in traditional Chinese medicine by capillary electrophoresis. *J Chromatogr A* 1246: 35–39.

[70] De Carvalho LM, Cohen PA, Silva CV, Moreira AP, Falcao TM, et al. (2012) A new approach to determining pharmacologic adulteration of herbal weight loss products. *Food Addit Contam Part A Chem Anal Control Expo Risk Assess* 29: 1661–1667.

[71] Moreira AP, Motta MJ, Dal Molin TR, Viana C, de Carvalho LM (2013) Determination of diuretics and laxatives as adulterants in herbal formulations for weight loss. *Food Addit Contam Part A Chem Anal Control Expo Risk Assess* 30: 1230–1237.

[72] Cao J, Li P, Chen J, Tan T, Dai HB (2013) Enhanced separation of Compound Xueshuantong capsule using functionalized carbon nanotubes with cationic surfactant solutions in MEEKC. *Electrophoresis* 34: 324–330.

[73] Xin N, Gu XF, Wu H, Hu YZ, Yang ZL (2012) Discrimination of raw and processed *Dipsacus asperoides* by near infrared spectroscopy combined with least squares-support vector machine and random forests. *Spectrochim Acta A Mol Biomol Spectrosc* 89: 18–24.

[74] Teraoka R, Abe H, Sugama T, Ito K, Aburada M, et al. (2012) Nondestructive prediction of oren extract powder, a herbal medicine, in suppositories by chemometric near-infrared spectroscopy. *J Nat Med* 66: 329–332.

[75] Huang H, Chen WW, Yu Y, Li ZY, Lin D, et al. (2012) Raman spectroscopic analysis of *Paeoniae Radix* Alba decoction based on Raman technology. *Zhongguo Zhong Yao Za Zhi* 37: 3569–3572.

[76] Wan QE, Liu HP, Zhang HM, Liu SH (2012) Identification of Ginseng and its counterfeit by laser Raman spectroscopy. *Guang Pu Xue Yu Guang Pu Fen Xi* 32: 989–992.

[77] Xu C, Jia X, Xu R, Wang Y, Zhou Q, et al. (2013) Rapid discrimination of Herba Cistanches by multi-step infrared macro-fingerprinting combined with soft independent modeling of class analogy (SIMCA). *Spectrochim Acta A Mol Biomol Spectrosc* 114: 421–431.

[78] Nema NK, Maity N, Sarkar BK, Mukherjee PK (2012) Determination of trace and heavy metals in some commonly used medicinal herbs in Ayurveda. *Toxicol Ind Health.* DOI:10.1177/0748233712468015.

[79] Patel DN, Li L, Kee CL, Ge X, Low MY, et al. (2013) Screening of synthetic PDE-5 inhibitors and their analogues as adulterants: analytical techniques and challenges. *J Pharm Biomed Anal.* DOI:10.1016/j.jpba.2013.04.037.

[80] Chauthe SK, Sharma RJ, Aqil F, Gupta RC, Singh IP (2012) Quantitative NMR: an applicable method for quantitative analysis of medicinal plant extracts and herbal products. *Phytochem Anal* 23: 689–696.

[81] Vaysse J, Balayssac S, Gilard V, Desoubdzanne D, Malet-Martino M, et al. (2010) Analysis of adulterated herbal medicines and dietary supplements marketed for weight loss by DOSY 1H-NMR. *Food Addit Contam Part A Chem Anal Control Expo Risk Assess* 27: 903–916.

[82] Balayssac S, Trefi S, Gilard V, Malet-Martino M, Martino R, et al. (2009) 2D and 3D DOSY 1H NMR, a useful tool for analysis of complex mixtures: application to herbal drugs or dietary supplements for erectile dysfunction. *J Pharm Biomed Anal* 50: 602–612.

[83] Sarker SD, Nahar L (2012) Hyphenated techniques and their applications in natural products analysis. *Methods Mol Biol* 864: 301–340.

[84] Du G, Xiao Y, Yang HR, Wang L, Song YL, et al. (2012) Rapid determination of pesticide residues in herbs using selective pressurized liquid extraction and fast gas chromatography coupled with mass spectrometry. *J Sep Sci* 35: 1922–1932.

[85] Koh HL, Lau AJ, Chan EC (2005) Hydrophilic interaction liquid chromatography with tandem mass spectrometry for the determination of underivatized dencichine (beta-N-oxalyl-L-alpha,beta-diaminopropionic acid) in Panax medicinal plant species. *Rapid Commun Mass Spectrom* 19: 1237–1244.

[86] Lau AJ, Holmes MJ, Woo SO, Koh HL (2003) Analysis of adulterants in a traditional herbal medicinal product using liquid chromatography-mass spectrometry-mass spectrometry. *J Pharm Biomed Anal* 31: 401–406.

[87] Zou P, Koh HL (2007) Determination of indican, isatin, indirubin and indigotin in *Isatis indigotica* by liquid chromatography/electrospray ionization tandem mass spectrometry. *Rapid Commun Mass Spectrom* 21: 1239–1246.

[88] Qu J, Hu YC, Li JB, Wang YH, Zhang JL, et al. (2008) Structural characterization of constituents with molecular diversity in fractions from *Lysidice brevicalyx* by liquid chromatography/diode-array detection/electrospray ionization tandem mass spectrometry and liquid chromatography/nuclear magnetic resonance. *Rapid Commun Mass Spectrom* 22: 755–765.

[89] Li Y, Wu X, Li J, Wang Y, Yu S, et al. (2010) Identification of cardiac glycosides in fractions from *Periploca forrestii* by high-performance liquid chromatography/diode-array detection/electrospray ionization multi-stage tandem mass spectrometry and liquid chromatography/nuclear magnetic resonance. *J Chromatogr B Analyt Technol Biomed Life Sci* 878: 381–390.

[90] Chen J, Zhao H, Wang X, Lee FS, Yang H, et al. (2008) Analysis of major alkaloids in *Rhizoma Coptidis* by capillary electrophoresis-electrospray-time of flight mass spectrometry with different background electrolytes. *Electrophoresis* 29: 2135–2147.

[91] Feng HT, Yuan LL, Li SF (2003) Analysis of Chinese medicine preparations by capillary electrophoresis-mass spectrometry. *J Chromatogr A* 1014: 83–91.

[92] Sajewicz M, Staszek D, Natic M, Waksmundzka-Hajnos M, Kowalska T (2011) TLC-MS versus TLC-LC-MS fingerprints of herbal extracts. Part III. Application of the reversed-phase liquid chromatography systems with C18 stationary phase. *J Chromatogr Sci* 49: 560–567.

[93] Kang SW, Kim CY, Song DG, Pan CH, Cha KH, et al. (2010) Rapid identification of furanocoumarins in *Angelica dahurica* using the online LC-MMR-MS and their nitric oxide inhibitory activity in RAW 264.7 cells. *Phytochem Anal* 21: 322–327.

[94] Heubl G (2010) New aspects of DNA-based authentication of Chinese medicinal plants by molecular biological techniques. *Planta Med* 76: 1963–1974.

[95] Chester K, Tamboli ET, Parveen R, Ahmad S (2013) Genetic and metabolic diversity in *Stevia rebaudiana* using RAPD and HPTLC analysis. *Pharm Biol* 51: 771–777.

[96] Ding G, Zhang D, Yu Y, Zhao L, Zhang B (2013) Analysis of genetic variability and population structure of the endemic medicinal *Limonium sinense* using molecular markers. *Gene* 520: 189–193.

[97] Niu L, Mantri N, Li CG, Xue C, Pang E (2011) Array-based techniques for fingerprinting medicinal herbs. *Chin Med* 6: 18.

[98] Olarte A, Mantri N, Nugent G, Wohlmuth H, Li CG, et al. (2013) A gDNA microarray for genotyping Salvia species. *Mol Biotechnol* 54: 770–783.

[99] Pareek CS, Smoczynski R, Tretyn A (2011) Sequencing technologies and genome sequencing. *J Appl Genet* 52: 413–435.

[100] Egan AN, Schlueter J, Spooner DM (2012) Applications of next-generation sequencing in plant biology. *Am J Bot* 99: 175–185.

[101] Sucher NJ, Hennell JR, Carles MC (2012) DNA fingerprinting, DNA barcoding, and next generation sequencing technology in plants. *Methods Mol Biol* 862: 13–22.

[102] Sucher NJ, Hennell JR, Carles MC (2013) Genomic and transcriptomic profiling: tools for the quality production of plant-based medicines. In: Chandra S, Lata H, Varma A, editors. *Biotechnology for Medicinal Plants*. Springer, Berlin/New York, pp. 439–455.

[103] Yamazaki M, Mochida K, Asano T, Nakabayashi R, Chiba M, et al. (2013) Coupling deep transcriptome analysis with untargeted metabolic profiling in *Ophiorrhiza pumila* to further the understanding of the biosynthesis of the anti-cancer alkaloid camptothecin and anthraquinones. *Plant Cell Physiol* 54: 686–696.

[104] Koh HL, Yau WP, Ong PS, Hegde A (2003) Current trends in modern pharmaceutical analysis for drug discovery. *Drug Discov Today* 8: 889–897.

[105] Cho WC (2007) Application of proteomics in Chinese medicine research. *Am J Chin Med* 35: 911–922.

[106] Lum JH, Fung KL, Cheung PY, Wong MS, Lee CH, et al. (2002) Proteome of Oriental ginseng *Panax ginseng* C.A. Meyer and the potential to use it as an identification tool. *Proteomics* 2: 1123–1130.

[107] Sun H, Zhang A, Yan G, Han Y, Sun W, et al. (2013) Proteomics study on the hepatoprotective effects of traditional Chinese medicine formulae Yin-Chen-Hao-Tang by a

combination of two-dimensional polyacrylamide gel electrophoresis and matrix-assisted laser desorption/ionization-time of flight mass spectrometry. *J Pharm Biomed Anal* 75: 173–179.

[108] Lindon JC, Nicholson JK, Holmes E (2007) *The Handbook of Metabonomics and Metabolomics*. Elsevier B.V, Amsterdam/Oxford.

[109] Choi MY, Chai C, Park JH, Lim J, Lee J, et al. (2011) Effects of storage period and heat treatment on phenolic compound composition in dried Citrus peels (Chenpi) and discrimination of Chenpi with different storage periods through targeted metabolomic study using HPLC-DAD analysis. *J Pharm Biomed Anal* 54: 638–645.

[110] Zhi HJ, Qin XM, Sun HF, Zhang LZ, Guo XQ, et al. (2012) Metabolic fingerprinting of *Tussilago farfara* L. using (1)H-NMR spectroscopy and multivariate data analysis. *Phytochem Anal* 23: 492–501.

[111] Yang SO, Shin YS, Hyun SH, Cho S, Bang KH, et al. (2012) NMR-based metabolic profiling and differentiation of ginseng roots according to cultivation ages. *J Pharm Biomed Anal* 58: 19–26.

[112] Jiang Y, Vaysse J, Gilard V, Balayssac S, Dejean S, et al. (2012) Quality assessment of commercial *Magnoliae officinalis* cortex by (1)H-NMR-based metabolomics and HPLC methods. *Phytochem Anal* 23: 387–395.

[113] Bailey NJ, Sampson J, Hylands PJ, Nicholson JK, Holmes E (2002) Multi-component metabolic classification of commercial feverfew preparations via high-field 1H-NMR spectroscopy and chemometrics. *Planta Med* 68: 734–738.

[114] Wang Z, Hu H, Chen F, Zou L, Yang M, et al. (2012) Metabolic profiling assisted quality assessment of Rhodiola rosea extracts by high-performance liquid chromatography. *Planta Med* 78: 740–746.

[115] Li Y, Wang Y, Su L, Li L, Zhang Y (2013) Exploring potential chemical markers by metabolomics method for studying the processing mechanism of traditional Chinese medicine using RPLC-Q-TOF/MS: a case study of *Radix Aconiti*. *Chem Cent J* 7: 36.

[116] Farag MA, Wessjohann LA (2012) Metabolome classification of commercial Hypericum perforatum (St. John's Wort) preparations via UPLC-qTOF-MS and chemometrics. *Planta Med* 78: 488–496.

[117] Kobayashi S, Putri SP, Yamamoto Y, Donghyo K, Bamba T, et al. (2012) Gas chromatography-mass spectrometry based metabolic profiling for the identification of discrimination markers of Angelicae Radix and its application to gas chromatography-flame ionization detector system. *J Biosci Bioeng* 114: 232–236.

[118] Gad HA, El-Ahmady SH, Abou-Shoer MI, Al-Azizi MM (2013) Application of chemometrics in authentication of herbal medicines: a review. *Phytochem Anal* 24: 1–24.

[119] Ip SP, Zhao M, Xian Y, Chen M, Zong Y, et al. (2010) Quality assurance for Chinese herbal formulae: standardization of IBS-20, a 20-herb preparation. *Chin Med* 5: 8.

[120] Xie PS, Wong E (2005) Direction of TCM modernisation and quality control mode. In: Leung PC, Xue CCL, editors. *Chinese Medicine: Modern Practice*. World Scientific Publishing Co. Pte. Ltd., Singapore, pp. 99–105.

[121] Wu L, Hao H, Wang G (2012) LC/MS based tools and strategies on qualitative and quantitative analysis of herbal components in complex matrixes. *Curr Drug Metab* 13: 1251–1265.

[122] Li SP, Zhao J, Yang B (2011) Strategies for quality control of Chinese medicines. *J Pharm Biomed Anal* 55: 802–809.

[123] Jordan SA, Cunningham DG, Marles RJ (2010) Assessment of herbal medicinal products: challenges, and opportunities to increase the knowledge base for safety assessment. *Toxicol Appl Pharmacol* 243: 198–216.

[124] Li CG, Moyle K, Xue CC (2003) Problems and challenges of Chinese herbal medicine. In: Leung PC, Xue CC, Cheng YC, editors. *A Comprehensive Guide to Chinese Medicine*. World Scientific Publishing Co. Pte. Ltd., Singapore, pp. 85–112.

[125] Shinde VM, Dhalwal K, Potdar M, Mahadik KR (2009) Application of quality control principles to herbal drugs. *Int J Phytomed* 1: 4–8.

[126] TGA (2011) *Guidance on Equivalence of Herbal Extracts in Complementary Medicines*. Therapeutic Goods Administration, Canberra.

[127] Sarris J, Ng CH, Schweitzer I (2012) "Omic" genetic technologies for herbal medicines in psychiatry. *Phytother Res* 26: 522–527.

3

PRECLINICAL (*IN VIVO*) AND LABORATORY (*IN VITRO*) EVIDENCE OF PHYTOMEDICINE EFFICACY

MOHI IQBAL MOHAMMED ABDUL[1] AND TOM HSUN-WEI HUANG[2]

[1] *College of Pharmacy, Taibah University, Madina, Kingdom of Saudi Arabia*
[2] *Faculty of Medicine, The University of Sydney, Sydney, New South Wales, Australia*

3.1 INTRODUCTION TO DEVELOPMENT OF DRUGS FROM NATURE

Since antiquity, herb-derived drugs have been used as medicines for the treatment of a range of diseases. Medicinal plants have played a crucial role in the world's health and fathered the evolution of modern medicine. Medicinal plants are distributed worldwide, but they are most abundant in tropical countries. Over the last 20 years, interest in drugs derived from plants, particularly, of phytotherapeutic origin has increased intensely. Approximately 25% of all modern medicines are directly or indirectly derived from plants [1–4]. For example, for antitumor and antimicrobial drugs, about 50–60% of the medicines currently available or in the late stages of clinical trials are derived from natural origin [3].

Natural products have stimulated many developments in organic chemistry leading to advances in synthetic methodologies and to the possibility of delivering various lead compounds with improved properties. Traditional development of a new drug entity is generally marked by a number of preclinical and clinical investigations that follow a regimented sequence with regard to contents and logic [5]. Conversely, for medications of phyto-origin, the "reverse pharmacology approach" is adopted in which the subjects (i.e., patients) experience a reverse path from "clinics to bench" rather than the classical "bench to clinics." The concept of "reverse pharmacology"

Phytotherapies: Efficacy, Safety, and Regulation, First Edition. Edited by Iqbal Ramzan.

was created in India to develop pharmaceuticals from Ayurvedic medicines [6]. However, one of the major challenges with "reverse pharmacology" is to accurately perform a scientific study with a plant extract when not knowing which part of the extract is responsible for the principle pharmacologic activity [7]. Typically, an extract possesses multiple mixtures of active, partially active, and inactive compounds. Such rate-limiting factors affect the reproducibility of studies, compounded by the fact that batch variations of active lead(s) may occur especially the conditions of how the raw material(s) is grown that may differ with each yield. Furthermore, lacking stringent quality control over how extract was prepared further hinders reproducibility.

Therefore, adopting the "bench to clinic" philosophy of conducting pharmacological experiments ensures accurate employment of critical reasoning and well-crafted experimental design. An ideal place to start botanical extract screening is to begin with *in vitro* assays, which are designed with the intention to better understand the mechanism(s) of action. The dilemma with such an approach is the tendency to attribute a pharmacological effect to almost every plant extract and create science headlines that sound too good to be true. If every health claim of phytotherapeutics were true, the cure for diseases like cardiovascular, diabetes, and oncology-related disorders would be a reality. Of course, sadly, this is not the case.

In the last decade, drug discovery scientists are strongly encouraged to reconfirm the findings of *in vitro* with *in vivo* data before results are published. This also assists in establishing the pharmacokinetic signature and toxicity related to the active lead compound(s). Animal models are crucial to drug discovery regardless of whether the actives are of herbal or synthetic origin. Relevant animal models with relevant correlation to human diseases are crucial for active compound evaluation and comparative analysis [8]. However, translating finding from "proof of concept" to phase 1 clinical trials remains a critical and at times insurmountable challenge [8]. If successful, post a "positive hit" through *in vitro* screening, the fraction or active compound(s) should be further studied using *in vivo* models until a single active compound is identified. This lead molecule can then be further investigated with regard to its mechanism of action utilizing either molecular pharmacological or biochemical assays. Furthermore, by adopting computational techniques, the natural ligand can be used as a structural template allowing generations of new synthetic molecules guided by ligand–receptor-based simulation modeling. Mother Nature holds the key for curing all diseases, and it is up to us as scientists to adopt not just the right technique(s) but the right mind frame in embarking on the journey of drug discovery.

3.2 USE OF *IN VITRO* AND *IN VIVO* MODELS IN HERB DRUG RESEARCH: LEARNING THUS FAR

3.2.1 *In Vitro* Assays

In botanical research, *in vivo* and *in vitro* bioassays are commonly used to screen new extracts and subsequent lead compounds to better understand their mechanism(s) of action. These assays are usually affordable and easy to perform. Traditionally, a suitable cell line has been chosen and grown in culture wells or flasks. After optimal cell passages are identified, several doses of the test material(s) are applied to generate a dose–response relationship, normally in comparison to control and to a positive control. The problem faced with botanical extracts is that these are multicomponent mixtures and potentially contain multiple leads. Hence, to optimize *in vitro* screening, the extract fractions are normally prepared based on their polarities segregating compounds that dissolve in media spanning from polar to nonpolar.

Many herbal extracts are not water soluble; therefore, a solvent that is not toxic to cell lines is selected for use as control. Often, dimethylsulfoxide (DMSO) is used as a solvent for extracts. The disadvantage is that DMSO might also elicit an effect on the cells, especially if cell lines are exposed to a DMSO concentration greater than 0.1% [9]. Hence, one should always have a control setting where the cells are treated with the preparation medium, without the herbal extract.

Another issue is solubility; testing herbal extracts using *in vitro* assays is challenging as they are considered as solid particles that do not behave in the same way as soluble molecules, thus the difficulties in defining an appropriate dose level [10]. Larger microparticles in extracts generally sediment rapidly and come into contact with the cells, thus inducing unnecessary oxidative stress and/or toxicity. A possible solution to address this is to centrifuge the extract before testing and to use only the clear supernatant in the *in vitro* assay. However, a phytochemical analysis to determine the composition of the supernatant is essential. *In vitro* testing can continue if there are no significant changes in the phytochemical composition of the extract after centrifugation. If the composition and/or phytochemical or other properties are affected by particulate materials in the extract, then this issue needs to be addressed, and analytical or computational assessment of the cellular components *in vitro* will be required [10].

Commonly, an extract or a pure compound that exhibits favorable activity in *in vitro* bioassay does not necessarily give positive *in vivo* results. Thus, many of the "hits" generated by traditional *in vitro* screening of herbal extracts turn out to be invalid once tested in animals, resulting in a waste of resources and time. Pharmacokinetic and toxicologic failures are the two main reasons for such failure. The effects found in tissue cultures are quite often not typical of an intact organism response. Therefore, researchers/scientists should consider these aspects before overinterpretation of data from *in vitro* assays. As the old saying goes, "*In vitro simplicitas, in vivo veritas*" (*in vitro* simplicity, *in vivo* the truth) [11].

3.2.2 *In Vivo* Assays

In vivo experiments for modern day drug discovery is a necessity before conducting any human clinical trial. Fundamentally, this is crucial as it helps to accurately translate the drug dosage from animal studies to humans. Unfortunately, there is a general misunderstanding and confusion about how to appropriately translate a drug dose from animals to human. Often, the animal dose is extrapolated to a human's equivalent dose simply by conversion based on body weight.

Research conducted by Reagan-Shaw et al. [12] and Baur et al. [13] clearly demonstrated the relationship of dose translation from animal to human studies and the fatal misinterpretation of data based on such an assumption. In this instance, a dose of 22.4 mg/kg body weight of resveratrol in mice was found to provide protection against aging [13]. The media reported that a 60 kg human would have to consume 1344 mg of resveratrol (22.4 × 60) per day in order to achieve health benefits. Based on this calculation, if a bottle of red wine contains approximately 2 mg of resveratrol, a person would have to drink 672 bottles of wine [12]. This obviously is not humanly possible. For a more accurate conversion of drug doses, the Food and Drug Administration [14] suggested that the extrapolation of animal to human doses (mg/kg) is correctly performed via normalization to the body surface area (mg/m^2). Using this equation to convert the dose in a mouse to a dose based on the surface area for humans, the equivalent resveratrol dose would be 1.82 mg/kg, which would correspond to 109 mg for a 60 kg person. However, this means a person will still need to consume 55 bottles of wine in order to get any health benefits. Hence, maybe, the beneficial effects of wine as reported initially should be critically qualified, since if indeed a person was to consume such a quantity, it would lead to a potential health risk.

Another aspect to consider in *in vivo* experiments is the route by which the herbal extract should be administered to animals. Almost all botanical preparations traditionally have been used in the form of infusions, decoctions, or as macerates for oral consumption; extract and isolated compounds should use the same route of administration [15]. Unfortunately, some extracts have been administered intraperitoneally or even intravenously, thus moving away from the traditional ethnomedical mode of application [15]. In general, the route in which a medication is administered has a major influence on the pharmacokinetics of a compound. The simplest method to administer a substance is to mix it with food or drinking water [16]. However, this method is not scientifically accurate nor is it practicable with substances that are insoluble or chemically unstable. Mixing compounds with food must be carried out carefully and accurately, especially when extracts are incorporated, since they represent multicomponent mixtures that might contain compounds that are sensitive to heat or pressure to which they are exposed during the preparation process. Thus, it will be necessary to determine release and recovery rates from food pellets in order to estimate the amount of compound that will be taken up by the animals. Further, the daily food and water intake of the animals must be known before starting the experiment so that the amount of substance to be mixed with food or drinking water can be accurately calculated [16]. Oral administration by diet or drinking water is not suitable when the exact amount of substance intake is required since food and water wastage is very common in rodents. Therefore, it is impossible to determine the exact amount of diet or water intake of rodents without placing them into metabolic cages [16]. There are many examples in the literature where these basic considerations were not taken into account and the consequences of carelessly using different routes of administration are potentially faulty since they can lead to inappropriate conclusions and serious misinterpretations. Hence, animal models pertinent to specific mechanism(s) of action may need to be developed to validate mechanistic hypothesis in preclinical phases, and therefore, translation of these animal disease models to progressive human diseases remains a scientific challenge.

3.3 CARDIOVASCULAR- AND STROKE-RELATED DISEASES: *IN VITRO* AND *IN VIVO* FOCUS

There are numerous therapeutic areas where phytomedicines have been researched and numerous books have been written on these respective areas. In a quest to provide the updated information on the therapeutic areas of current interest, we will present here the data from the past 5 years (2010–2014) on the preclinical evidence of phytomedicines in the diseases known to be of greatest impact worldwide. The World Health Organization fact sheet [17] states the 10 leading causes of death as mentioned in Figure 3.1, of which the first 8 can be clearly attributed to disease states for which treatment can be sought. However, for the purpose of this book chapter, the examples specific to drug discovery will be cardiovascular and stroke focused, as these two diseases are currently the leading causes of deaths globally [17].

3.3.1 Cardiovascular Diseases

Cardiovascular disease has been one of the leading causes of mortality for many decades. Despite a long list of the drugs and extensive ongoing research for finding better treatments for cardiovascular complications, there is still unmet medical need in this area. For this reason, hope of phytomedicines use for supportive treatment or for a breakthrough treatment is always present.

3.3.1.1 Drug Development Methods for Cardiac Function Conventional methods used to assess the cardiac functions of drugs can also be applied to phytomedicines that comprise *in vitro* and *in vivo* methods. In either conventional drug development approach or reverse pharmacology drug development approach, *in vitro* and *in vivo* methods hold their own status. These methods give more insight into the mechanism of action/site of action of the drug and provide more freedom to try different approaches in contrast with the clinical studies in humans.

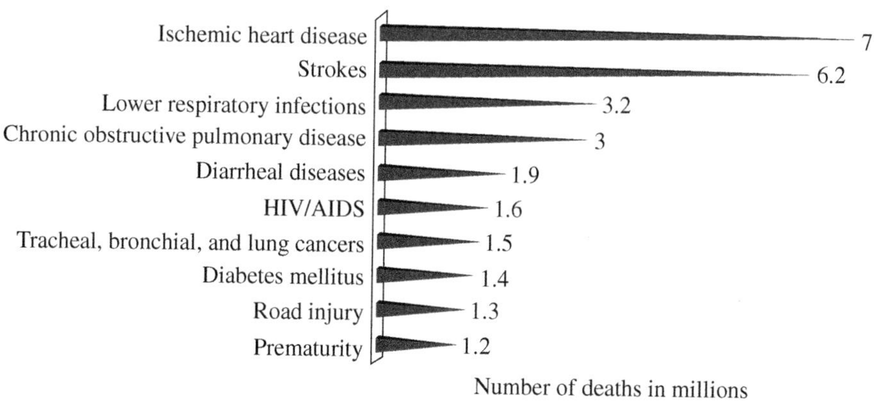

FIGURE 3.1 Leading causes of death worldwide in the year 2011. Adopted from the World Health Organization fact sheet [17].

TABLE 3.1 *In Vivo* Models for Studying Cardiac Effects of Drugs or Phytomedicines[a]

Parameter Studied	*In Vivo* Method
Cardiac output	Direct method: By using a flow probe • Electromagnetic • Ultrasound Indirect method: • Dye dilution technique
Diastolic pressure	By inserting a catheter in the right ventricle
Electrical activity	ECG electrodes can be inserted at various points in the heart
Induced arrhythmias	Chemical[b] • Glycosides for guinea pig • Chloroform for mice • Aconitine for rats • Catecholamines for dogs Electrical • By subjecting animals to electrical impulses through electrodes inserted at different locations in the heart Occlusion of the coronary artery and diminishing oxygen supply[c] • Suitable with rats, pigs, rabbits, and primates

[a] From Ref. [18].
[b] The chemical method to induce arrhythmias used for small animals and the choice of the individual chemical differ with the animal used.
[c] This is regarded as more clinically relevant. Myocardial infarction development is determined by the duration of occlusion and size of infarct. Surgically removed infarct is then visualized by the use of dyes to distinguish normal tissue from the infarct.

For cardiac function assessment, two main measurements are desired [18]:

1. *Measurement of mechanical function of the heart*, which basically is assessing the cardiac output and blood pressure for *in vivo* studies or ventricular pressure for *in vitro* studies. This can be studied in the presence or absence of a drug candidate; in our case, this can be an extract from a phytomedicine or a constituent or mixture of constituents derived from a phytomedicine.
2. *Measurement of electrical activity of the heart*, which basically is assessing the effects of the phytomedicine on the electrical impulse generation and transmission. *In vivo*, this uses electrocardiograms (ECG), and *in vitro*, numerous other approaches are used including the use of high-throughput screening methods to screen the effects of numerous drug molecules on the activity of cardiac ion channels (Na^+, K^+, and Ca^{2+} ion channels) in a short period of time.

For *in vitro* or *in vivo* studies, induction of cardiac condition to evaluate the efficacy of phytomedicine can be achieved by using different techniques (Tables 3.1 and 3.2).

Phytomedicines and Cardiac Activity: Usually, 10 years is the time frame for developing a drug, but with the advent of technology and with the use of new approaches, this time span maybe reduced to approximately 4.5 years [19].

TABLE 3.2 *In Vitro* Models for Studying Cardiac Effects of Drugs or Phytomedicines[a]

Parameter Studied	*In Vitro* Method
Actions on ion channels	Patch clamp techniques Used where whole-cell recording is desired
Pharmacological efficacy by studying mechanical, electrical, and biochemical changes in heart tissues	Isolated tissues: • Isolated atria • Isolated ventricle • Isolated papillary muscle • Coronary perfused right ventricular wall Isolated perfused whole heart • Langendorff heart preparation

[a]From Ref. [18].

Phytomedicines studied earlier than this time period potentially should have transformed into a clinical drug candidate; otherwise, this is regarded as a nonfeasible approach. For this reason, in this section, we will limit ourselves to discussing the phytomedicines recently studied for their cardiac activity in the past 5 years (2010–2014). Table 3.3 summarizes the evidence on phytomedicines or their constituents, respectively, and the study outcomes and methodology utilized in these studies. The following criteria were used in selecting studies to be included in Table 3.3:

1. Studies published during 2010–2014.
2. Studies aimed to study phytomedicines or their constituents including combinations of phytomedicines.
3. Studies with the primary aim to study the cardiovascular efficacy of phytomedicines.
4. Studies using only preclinical approaches to assess the efficacy of phytomedicines, namely, *in vitro*, *in vivo*, or *ex vivo* methods, were included.

3.3.2 Stroke

Stroke is a central nervous system complication with ischemia being the main cause; 88% of cases are ischemia related. Strokes are referred to as encephalic vascular accident due to complications in the encephalon resulting from vascular disturbances. It ranges from mild to serious and consequences can be temporary or permanent, which are acute or chronic in nature. Acute stroke usually involves necrosis of neural cells due to ischemia—depletion of oxygen and nutrient supply. Chronic strokes lead to lifelong diseases like amyotrophic lateral sclerosis, Huntington's disease, Alzheimer's disease, and Parkinson's disease. Molecular evidence suggests that neurodegenerative diseases are mediated by neuroinflammation, oxidative stress, and apoptotic cell death. Preclinical studies for the evaluation of newer agents including phytomedicines usually involve studying the protective nature of an intervention in comparison to the control of neuroinflammation, oxidative stress, and apoptotic cell death of neural cells [35, 36].

TABLE 3.3 ***In Vitro* and *In Vivo* Evidence of Phytomedicines or Their Constituents for Efficacy in Cardiovascular Diseases**

Phytomedicine/Phytoconstituent	Study Outcome	Method
Cynodon dactylon [20]	Increased heart rate	*In vivo* study in zebra fish
Sida acuta [20]	Ethanolic extract decreased heart rate	*In vivo* study in zebra fish
[6]-Gingerol (from ginger) [21]	Gingerol was reported to inhibit angiotensin II type 1 receptor and hence has the potential to lower blood pressure and strengthen the heart	*In vitro* analysis using cell-based calcium metabolism assay
Flowers, leaves, and stem of *Polygonum orientale* [22]	Polygonum plant parts showed protective effects on H9C2 myocardial cells. Potential for use in antimyocardial ischemia	*In vitro* analysis performed on the myocardial cells injured by oxidative stress by hydrogen peroxide
Luteolin, a flavonoid found in fruits, vegetables, and herbs [23]	Reported to possess anti-inflammatory, antioxidant, and anticarcinogenic properties. Potential cardioprotective agent	Mostly *in vivo* data from rats
Paeonol and danshensu in combination [24]	Cardioprotective effects. Potential to be used in myocardial infarction	Protective against isoproterenol-induced myocardial injury in rats
Hydroxymethanolic leaf exact of *Mimosa pigra* [25]	Phytomedicine showed antioxidant actions *in vitro*. The herb's ability as a vasorelaxant was also demonstrated *in vivo*	Antioxidant properties studied *in vitro* using 1,1-diphenyl-2-picrylhydrazyl radical scavenging and the oxygen radical absorbance capacity. Vasorelaxant effects studied *in vivo* in hypoxic pulmonary hypertensive rats
Genistein (phytoestrogen from soy) [26]	Genistein successively reversed monocrotaline-induced pulmonary hypotension in male rats. Demonstrated *in vitro* activity likely to be due to inhibition of human pulmonary artery smooth muscle cell proliferation probably via estrogen receptor-β	Genistein given for 10 days to male rats with prior administration of monocrotaline for 21 days reversed pulmonary hypertension induced by monocrotaline in contrast to control rats
Resveratrol-rich grape extract [27]	Resveratrol demonstrated to prevent atherosclerotic lesions in pigs fed with atherogenic diet	Three groups of pigs fed with atherogenic diet; the first group fed on resveratrol-rich grape extract, the second group fed on grape extract lacking resveratrol, and the third group with resveratrol only. Prevention of atherosclerotic lesions was prominent in pigs fed with resveratrol-rich grape extract

Delphinidin-3-glucoside (an anthocyanin) [28]	Delphinidin-3-glucoside inhibits platelet activation and attenuates thrombus growth at both arterial and venous shear stresses, which likely contributes to its protective role against thrombosis and cardiovascular diseases	Experiments performed in human and murine platelet-rich plasma and purified platelets. Thrombus growth studies done in human and murine blood in perfusion chambers
Hawthorn exact [29]	Demonstrated vasoprotective actions of the phytomedicine extract	Mechanics of the endothelial surface layer in *ex vivo* murine aortae and sodium permeability of endothelial cells *in vitro*
Paeoniflorin (a Chinese herb) [30]	Phytomedicine exhibited potential to be used for its antiarrhythmic effects with less proarrhythmic potential	Whole-cell patch clamp technique was used to measure ion channel currents
Quercetin [31]	A potential phytoconstituent for use in atherosclerosis	Human CRP transgenic mice and ApoE*3Leiden transgenic mice were used to study anti-inflammatory and antiatherosclerotic properties of quercetin
Volatile oil of Radix *Angelica sinensis* (VOAS) and its active constituent, n-butylidenephthalide (BP) [32]	Both the volatile oils of the phytomedicine and its active constituent exhibited antiangiogenic properties *in vitro* and *in vivo*. Potential source of compounds that can be used in cardiovascular diseases	Several experiments performed *in vitro* (human umbilical vein endothelial cell), *ex vivo* (mouse aortic ring model), and *in vivo* (zebra fish) models
Roots of *Paeonia lactiflora* and *P. suffruticosa* [33]	Numerous constituents of the phytomedicine demonstrated anticoagulant and/or platelet aggregation inhibition properties	*In vitro* assessment of platelet aggregation and anticoagulation properties
Herbal formulation of combination of *Zingiber officinale*, *Terminalia belerica*, and *Orchis mascula* (ZTO) [34]	ZTO demonstrated antidyslipidemic, antihypertensive, and endothelial modulatory properties	Different animal models including tyloxapol and high-fat diet-induced dyslipidemia and spontaneously hypertensive rats (SHR) were used

3.3.2.1 In Vivo Models for Studying Strokes Mice and rats are most commonly used for stroke research, but rabbits, dogs, swine, and primates can also be used for studying strokes (Table 3.4). Selection of the right species is important in order to extrapolate the research findings to humans. For instance, dogs are not suitable to study strokes due to ischemia because of the difference in blood supply in dog encephalon to human encephalon, whereas monkeys closely resemble humans for such studies [35].

Phytomedicines and Strokes: To list the evidence of *in vitro* and *in vivo* efficacy of phytomedicines for strokes, the following criteria were implemented in selecting studies. The evidence is presented in Table 3.5:

1. Studies published during 2010–2014.
2. Studies aimed to study phytomedicines or their constituents including combinations of phytomedicines.

TABLE 3.4 Methods Used To Induce Strokes in Different Animals

Stroke Induction Method	Procedure	Species Used
Embolism	Blood clot is either injected or formation of clot is stimulated by photocoagulation in the arteries supplying the encephalon or by injecting sodium laurate into the internal carotid artery	Rats [37, 38], mice [39, 40], rabbits [41, 42]
Middle cerebral artery occlusion. Occlusion of arteries	Physical obstruction of blood flow in arteries. Targeted arteries: • Middle cerebral artery • Perforating artery occlusion • Bilateral carotid artery	Rats [43, 44], mice [45]
Endothelin-1 (ET-1) injection	Use of ET-1 to create infarcts in the white matter of the internal capsule underlying the sensorimotor cortex	Rats [46, 47]
Genetic model of cerebral autosomal dominant arteriopathy with subcortical infarcts and leukoencephalopathy (CADASIL) syndrome	Transgenic mice to study CADASIL	Mice [48]
Immunization with HL60 promyelocytic cell differentiation factor	Rabbits immunized with synthetic HLDF differentiation factor develop hemorrhagic stroke with thrombosis of small cerebral vessels. More serum antibodies lead to more severe hemorrhage	Rabbit [49]

TABLE 3.5 ***In Vitro* and *In Vivo* Evidence On Phytomedicines or Their Constituents for Their Efficacy in Strokes**

Phytomedicine/Phytoconstituent	Study Outcome	Method
Danshen and gegen in combination [50]	Combination of herbs showed additive vasodilator effects and hence may be useful in cerebrovascular diseases. Vasodilator effects are thought to be due to activation of potassium channels coupled with inhibition of calcium channels	*In vitro* studies on basilar artery rings obtained from rats. Basilar artery is a posterior cerebral and vertebral artery, which is a major encephalic artery in the animal
Huang-Lian-Lie-Du decoction [51]	Herb reported to reduce cerebral injury by ameliorating disturbances in energy metabolism, membrane and mitochondrial metabolism, and neurotransmitter and amino acid metabolism and alleviating the oxidative stress from reactive oxygen species and the inflammatory damage and recovery from the destructive osmoregulation	Stroke parameters were studied in rats fed with the herb in contrast to positive control and no treatment. Middle cerebral artery occlusion (MCAO) is used to induce ischemic stroke
Traditional Chinese medicine, a mixture of *Radix Rehmanniae Preparata*, *Radix Polygoni Multiflori*, Colla Cornus Cervi, *Eucommia ulmoides* Oliv, Epimedium brevicornum Maxim, and *Loranthus parasiticus* [52]	Mixture studied was reported to suppress expression of Nogo-A, a myelin-associated inhibitor with a strong axon growth inhibitory activity, and also suppress the expression of $p75^{(NTR)}$, a coreceptor for Nogo-A	Nogo-A and $p75^{(NTR)}$ expression was studied in rats divided into four groups: control sham-operated, placebo, and herb-treated groups, respectively. MCAO model was used to induced ischemia
EGb761, a standard extract of Ginkgo biloba [53]	EGb761 had significant therapeutic effects on ischemic stroke	Herb extract significantly increased the apparent diffusion coefficient (ADC) and average diffusion coefficient (DCavg) both in the peripheral and central zone, improved behavior scores, as well as enhanced the phosphorylation of AKT and CREB and the expression of BDNF in the brain

3. Studies with the primary aim to study the efficacy of phytomedicines in strokes.
4. Studies using only preclinical approaches to assess the efficacy of phytomedicines, namely, *in vitro*, *in vivo*, or *ex vivo* methods, were included.

3.4 CONCLUSIONS

The authors recommend that when conducting preclinical testing of botanicals, a sequential approach should be adopted. This begins with the phytochemical analysis of the plant extract in order to determine a possible pattern of constituents. Next, the extract should be examined *in vivo* in relevant animal models after oral administration in appropriate doses to substantiate the ethnopharmacological use. Once the extract displays positive activity *in vivo*, a bioguided fractionation process using an adequate *in vitro* screening model in relevant doses should follow. However, one should be aware that coeffectors may be present if the activity decreases or even disappears during purification [54]. Attention should also be given to determine the right solvent used and the IC50. A fraction that yields a positive hit should be further investigated in the selected *in vivo* model until a single active compound is identified. Once a single active compound is detected, a pharmacokinetic profile should be determined followed by final *in vivo* testing of the isolated pure compound. Molecular pharmacological or biochemical assays *in vitro* can then be carried out as a very last step to evaluate the mechanism(s) of action. These assays only provide useful data with pure isolated compounds when information regarding their *in vivo* activity and bioavailability are available. This is to prevent trivial findings especially when extracts are applied in *in vitro* molecular pharmacological assays or if isolated compounds are tested on a molecular level before information regarding their bioavailability is available. Considerations should also be given, prior to testing, to such issues as the physicochemical properties of the tested material, choice of realistic doses, adequate test models, appropriate routes of administration and suitable statistical evaluations that will enable proper data interpretation.

The medicinal use of herbal drugs in the last 25 years has seen a tremendous growth both in the discovery of new ligands derived from plants and maturation of the techniques used in drug discovery. The growth of the botanical market has attracted much interest on the part of pharmaceutical companies, which has in turn stimulated strong interest in preclinical pharmacological testing and well-designed randomized clinical trials to prove their safety and efficacy. Furthermore, abundant scientific journals have dedicated significant efforts to publishing both basic and clinical scientific studies on herbal medicines, which in turn has aided both traditional and orthodox physicians to better understand herbal drugs. However, further advancement can be made to provide an accurate assessment of the quality, efficacy, and safety of most herbal medicines. What is important to note is that there has been two paradigm shifts in modern medicine in the last 10 years, that is, the gradual renunciation of the long-standing reliance on single-agent therapy in favor of a multidrug therapy and the transition to a new kind of multitarget therapy, through which drugs with protective, repair, and/or immunomodulatory mechanisms are more effective rather than use of a drug targeting a single disease-causing mechanism. Phytomedicine research has a good chance of contributing to these new strategies through the development of new and better drugs for an evidence-based and rational phytotherapy. One major challenge will be to investigate the multivalent and multi-target actions of plant constituents and standardized extracts, with the aim of rationalizing the therapeutic superiority of many plant extracts over single isolated constituents. Phytomedicine and chemosynthetic pharmaceutical research find themselves in a race to develop new medicines, with fewer or no side effects, for preventive and therapeutic application in illnesses for which causality-based treatments are nonexistent or imperfect.

REFERENCES

[1] Farnsworth NR, Morris RW (1976) Higher plants—the sleeping giant of drug development. *Am J Pharm Educ* 148: 46–52.

[2] De Smet PAGM (1997) The role of plant-derived drugs and herbal medicines in healthcare. *Drugs* 54: 801–840.

[3] Cragg GM, Newman DJ, Snader KM (1997) Natural products in drug discovery and development. *J Nat Prod* 60: 52–60.

[4] Shu YZ (1998) Recent natural products based drug development: a pharmaceutical industry perspective. *J Nat Prod* 61: 1053–1071.

[5] Patwardhan B, Mashelkar RA (2009) Traditional medicine-inspired approaches to drug discovery: can Ayurveda show the way forward? *Drug Discov Today* 14(15–16): 804–811.

[6] Olejniczak K, Guenzel P, Bass R (2001) Pre-clinical testing strategies. *Drug Inf J* 35: 321–336.

[7] Calixto JB (2000) Efficacy, safety, quality control, marketing and regulatory guidelines for herbal medicines (phytotherapeutic agents). *Braz J Med Biol Res* 33: 179–189.

[8] Khanna I (2012) Drug discovery in pharmaceutical industry: productivity challenges and trends. *Drug Discov Today* 17: 1088–1102.

[9] Rodríguez-Burford C, Oelschlager DK, Talley LI, Barnes MN, Partridge EE, et al. (2003) The use of dimethylsulfoxide as a vehicle in cell culture experiments using ovarian carcinoma cell lines. *Biotech Histochem* 78: 17–21.

[10] Lison D, Huaux F (2011) *In vitro* studies: ups and downs of cellular uptake. *Nat Nanotechnol* 6: 332–333.

[11] Vogel HG, Vogel W (2002) *Drug Discovery and Evaluation*. New York: Springer.

[12] Reagan-Shaw S, Nihal M, Ahmad N (2008) Dose translation from animal to human studies revisited. *FASEB J* 22: 659–661.

[13] Baur JA, Pearson KJ, Price NL, Jamieson HA, Lerin C, et al. (2006) Resveratrol improves health and survival of mice on a high-calorie diet. *Nature* 444: 337–342.

[14] Food and Drug Administration (2002) *Estimating the Safe and Starting dose in Clinical Trials for Therapeutics in Adult Healthy Volunteers*. Rockville, MD: Center for Drug Evaluation and Research/Center for Biological and Research/US Food and Drug Administration.

[15] ICH Harmonized Tripartite Guideline (2000). Safety Pharmacology Studies for Human Pharmaceuticals S7A. Available at http://www.ich.org/fileadmin/Public_Web_Site/ICH_Products/Guidelines/Safety/S7A/Step4/S7A_Guideline.pdf. Accessed January 2014

[16] Nebendahl K, Hauff P (2011) Drug administration. In: Kiessling F, Pichler BJ, Hauff P, editors. *Small Animal Imaging*. Heidelberg: Springer; 93–119.

[17] WHO (2014). The Top 10 Causes of Death. Fact Sheets 2013; The 10 Leading Causes of Death in the World. Available at http://who.int/mediacentre/factsheets/fs310/en/. Accessed January 2014

[18] Pugsley MK (2003) *Cardiac Drug Development Guide. Methods in Pharmacology and Toxicology*. Totowa: Humana Press.

[19] Gupta SK (2009) *Drug Screening Methods—Pre-clinical Evaluation of New Drugs*. New Delhi: Jaypee Brothers Medical Publishers (P) Ltd.

[20] Kannan RR, Vincent SG (2012) *Cynodon dactylon* and *Sida acuta* extracts impact on the function of the cardiovascular system in zebrafish embryos. *J Biomed Res* 26: 90–97.

[21] Liu Q, Liu J, Guo H, Sun S, Wang S, et al. (2013) [6]-gingerol: a novel AT(1) antagonist for the treatment of cardiovascular disease. *Planta Med* 79: 322–326.

[22] Zheng L, Li J, Chen H, Wang YL, Wang AM, et al. (2012) Study on fingerprint-pharmacology correlation of protective effect of *Polygonum orientale* on myocardial cell oxidative injury induced by H202. *Zhongguo Zhong Yao Za Zhi* 37: 2585–2588.

[23] Xu T, Li D, Jiang D (2012) Targeting cell signaling and apoptotic pathways by luteolin: cardioprotective role in rat cardiomyocytes following ischemia/reperfusion. *Nutrients* 4: 2008–2019.

[24] Li H, Xie YH, Yang Q, Wang SW, Zhang BL, et al. (2012) Cardioprotective effect of paeonol and danshensu combination on isoproterenol-induced myocardial injury in rats. *PLoS One* 7: e48872.

[25] Rakotomalala G, Agard C, Tonnerre P, Tesse A, Derbré S, et al. Extract from *Mimosa pigra* attenuates chronic experimental pulmonary hypertension. *J Ethnopharmacol* 48: 106–116.

[26] Matori H, Umar S, Nadadur RD, Sharma S, Partow-Navid R, et al. (2012) A soy phytoestrogen, reverses severe pulmonary hypertension and prevents right heart failure in rats. *Hypertension* 60: 425–430.

[27] Azorín-Ortuño M, Yañéz-Gascón MJ, Pallarés FJ, Rivera J, González-Sarrías A, et al. (2012) A dietary resveratrol-rich grape extract prevents the developing of atherosclerotic lesions in the aorta of pigs fed an atherogenic diet. *J Agric Food Chem* 60: 5609–5620.

[28] Yang Y, Shi Z, Reheman A, Jin JW, Li C, et al. (2012) Plant food delphinidin-3-glucoside significantly inhibits platelet activation and thrombosis: novel protective roles against cardiovascular diseases. *PLoS One* 7: e37.

[29] Peters W, Drueppel V, Kusche-Vihrog K, Schubert C, Oberleithner H (2012) Nanomechanics and sodium permeability of endothelial surface layer modulated by hawthorn extract WS 1442. *PLoS One* 7: e29972.

[30] Wang RR, Li N, Zhang YH, Ran YQ, Pu JL (2011) The effects of paeoniflorin monomer of a Chinese herb on cardiac ion channels. *Chin Med J (Engl)* 124: 3105–3111.

[31] Kleemann R, Verschuren L, Morrison M, Zadelaar S, van Erk MJ (2011) Anti-inflammatory, anti-proliferative and anti-atherosclerotic effects of quercetin in human *in vitro* and *in vivo* models. *Atherosclerosis* 218: 44–52.

[32] Yeh JC, Cindrova-Davies T, Belleri M, Morbidelli L, Miller N, et al. (2011) The natural compound n-butylidenephthalide derived from the volatile oil of Radix *Angelica sinensis* inhibits angiogenesis *in vitro* and *in vivo*. *Angiogenesis* 14: 187–197.

[33] Koo YK, Kim JM, Koo JY, Kang SS, Bae K, et al. (2010) Platelet anti-aggregatory and blood anti-coagulant effects of compounds isolated from *Paeonia lactiflora* and *Paeonia suffruticosa*. *Pharmazie* 65: 624–628.

[34] Aziz N, Mehmood MH, Gilani AH (2013) Studies on two polyherbal formulations (ZPTO and ZTO) for comparison of their antidyslipidemic, antihypertensive and endothelial modulating activities. *BMC Complement Altern Med* 13: 371.

[35] Chik SC, Or TC, Luo D, Yang CL, Lau AS (2013) Pharmacological effects of active compounds on neurodegenerative disease with gastrodia and uncaria decoction, a commonly used poststroke decoction. *Scientific World Journal* 2013: 896873.

[36] Casals JB, Pieri NC, Feitosa ML, Ercolin AC, Roballo KC, et al. (2011) The use of animal models for stroke research: a review. *Comp Med* 61: 305–313.

[37] Wang CX, Todd KG, Yang Y, Gordon T, Shuaib A (2001) Patency of cerebral microvessels after focal embolic stroke in the rat. *J Cereb Blood Flow Metab* 21: 413–421.

[38] Toshima Y, Satoh S, Ikegaki I, Asano T (2000) A new model of cerebral microthrombosis in rats and the neuroprotective effect of a Rho-kinase inhibitor. *Stroke* 31: 2245–2250.

[39] Atochin DN, Murciano JC, Gürsoy-Ozdemir Y, Krasik T, Noda F, et al. (2004) Mouse model of microembolic stroke and reperfusion. *Stroke* 35: 2177–2182.

[40] Lozano JD, Abulafia DP, Danton GH, Watson BD, Dietrich WD (2007) Characterization of a thromboembolic photochemical model of repeated stroke in mice. *J Neurosci Methods* 162: 244–254.

[41] Lapchak PA (2010) Translational stroke research using a rabbit embolic stroke model: a correlative analysis hypothesis for novel therapy development. *Transl Stroke Res* 1: 96–107.

[42] Brown AT, Skinner RD, Flores R, Hennings L, Borrelli MJ (2010) Stroke location and brain function in an embolic rabbit stroke model. *J Vasc Interv Radiol* 21: 903–909.

[43] Gharbawie OA, Auer RN, Whishaw IQ (2006) Subcortical middle cerebral artery ischemia abolishes the digit flexion and closing used for grasping in rat skilled reaching. *Neuroscience* 137: 1107–1118.

[44] Lindner MD, Gribkoff VK, Donlan NA, Jones TA (2003) Long-lasting functional disabilities in middle-aged rats with small cerebral infarcts. *J Neurosci* 23: 10913–10922.

[45] Shibata M, Ohtani R, Ihara M, Tomimoto H (2004) White matter lesions and glial activation in a novel mouse model of chronic cerebral hypoperfusion. *Stroke* 35: 2598–2603.

[46] Frost SB, Barbay S, Mumert ML, Stowe AM, Nudo RJ (2006) An animal model of capsular infarct: endothelin-1 injections in the rat. *Behav Brain Res* 169: 206–211.

[47] Hughes PM, Anthony DC, Ruddin M, Botham MS, Rankine EL, et al. (2003) Focal lesions in the rat central nervous system induced by endothelin-1. *J Neuropathol Exp Neurol* 62: 1276–1286.

[48] Ruchoux MM, Domenga V, Brulin P, Maciazek J, Limol S, et al. (2003) Transgenic mice expressing mutant Notch3 develop vascular alterations characteristic of cerebral autosomal dominant arteriopathy with subcortical infarcts and leukoencephalopathy. *Am J Pathol* 162: 329–342.

[49] Gapon MV, Dranitsyna SM, Minkevich NI, Gruden' MA, Babichenko II, et al. (2006) Experimental model of hemorrhagic stroke: rabbit immunization with HL-60 promyelocytic cell differentiation factor. *Bull Exp Biol Med* 141: 272–274.

[50] Deng Y, Ng ES, Kwan YW, Lau CB, Cheung DW, et al. (2013) Cerebral vasodilator properties of Danshen and Gegen: a study of their combined efficacy and mechanisms of actions. *Phytomedicine* 21: 391–399.

[51] Wang PR, Wang JS, Yang MH, Kong LY (2014) Neuroprotective effects of Huang-Lian-Jie-Du-Decoction on ischemic stroke rats revealed by (1) H NMR metabolomics approach. *J Pharm Biomed Anal* 88: 106–116.

[52] Yan Y, Li T, Liu L, Zhou H (2012) Effect of tonifying liver and kidney-essence herbs on expression of Nogo-A and p75(NTR) in cerebral ischemic stroke rats model. *J Tradit Chin Med* 32: 664–668.

[53] Zhang Z, Peng D, Zhu H, Wang X (2012) Experimental evidence of Ginkgo biloba extract EGB as a neuroprotective agent in ischemia stroke rats. *Brain Res Bull* 87: 193–198.

[54] Nahrstedt A, Butterweck V (2010) Lessons learned from herbal medicinal products: the example of St. John's Wort (perpendicular). *J Nat Prod* 73: 1015–1021.

4

CLINICAL EFFICACY TRIALS WITH NATURAL PRODUCTS AND HERBAL MEDICINES

CHRISTINA L. NANCE

Department of Pediatrics, Baylor College of Medicine, Immunology, Allergy and Rheumatology, Texas Children's Hospital, Houston, Texas, USA

4.1 INTRODUCTION

The idea of plants as source of drugs turned into practice in human medicine over millennia and is still viable today. In recent decades, more than half of newly registered drugs have close relationships to secondary metabolites [1, 2]. However, it is clear that although several botanicals have been characterized and used for hundreds of years (Traditional Chinese Medicine (TCM), Ayurveda, Siddha, Unani) [3, 4], there have been several challenges and limitations toward progress in this field. To promote interdisciplinary study of botanicals, knowledge and understanding must be in the context of rigorous science with research activities ranging from plant characterization to preclinical and early phase clinical studies.

Clinical trials, prospective biomedical or behavioral research study of human subjects that is designed to answer specific questions about biomedical or behavioral interventions, are used to determine whether new biomedical interventions are safe, efficacious, and effective.

Natural compounds with exceptionally rich ethnopharmacological tradition, like active components of herbals used in traditional Chinese or Ayurvedic medicine, often performed poorly in Western clinical trials, indicating serious problems either with pharmacokinetic (PK) parameters and consequently in bioavailability or with overall efficacy.

Phytotherapies: Efficacy, Safety, and Regulation, First Edition. Edited by Iqbal Ramzan.

There are multiple factors inherently incorporated into any clinical trial. These parameters must be teased out at some point and weighted appropriately in an unbiased manner. Such factors that give strength and validity are the clinical sites, depth and breadth of study populations, power of the study, statistical analysis, randomization, and blinding status. The hallmark of rigorous testing is the properly designed and implemented randomized control trial (RCT). The slow pace of growth of possible therapeutics may be attributed to the lack of translational scientists engaged in conducting well-designed trials. As well as that, clinical trials, properly performed, are very laborious and time/money consuming.

Past trials of preventive agents offer important lessons that can form the basis of design and conduct of future clinical trials such as the need for more preclinical and early phase work before undertaking phase III trials [5, 6]. Other lesson that may apply to the issue is that safety can be improved in iterative generations of agents and trials [7, 8]. Taken directly from the aspirin study in adenoma prevention trials the highlighting of the benefit of preclinical and phase II testing, as well as the imperative for broad, sensitive toxicological, and human safety assessments was paramount [9, 10]. The Sulindac® combination trial demonstrated the importance of synergy between agents that can lead to lower doses, improved efficacy, and fewer or less severe toxicities.

Therefore, our primary goals are to systematically accelerate the design, to conduct successful clinical trials to evaluate botanicals/biologics for the prevention and treatment, and to achieve efficient translation of these discoveries into the standards for clinical practice that will ultimately impact disease morbidity and mortality [11].

4.2 TRIALS IN VARIOUS DISEASE STATES

The spectrum of translation of natural products to the clinical setting is far flung. With the breadth and depth of the field these days, it would be an impossible feat to list a full compendium of all natural product human clinical trials. What follows is a very broad overview of translation efforts scanning selected examples of multiple disease states and types of natural products.

4.2.1 Profile: RCT of Natural Product in Rheumatoid Arthritis (RA)

A meta-analysis of terpenoids in *Tripterygium wilfordii* Hook F (TwHF), commonly known as thunder god vine on RA, is presented. Most of the clinical information comes from uncontrolled clinical trials, retrospective reports, and a few multicenter trials. In this meta-analysis, 105 potential clinical trials were identified and reviewed with a resultant 10, only, as valid randomized clinical trials, regardless of blinding parameters [12]. Of the original 105, exclusion was due to either, nonrandomization, nonhuman subjects, failure to meet inclusion criteria, no reportable data, and duplicated trials or nonrelevance. Limitations of the meta-analysis incorporated the issue of selection bias with nine of the ten RCT being Chinese populations, implication of a high-risk parameter. RCT were of poor quality regarding the designs, reporting, and methodologies. Only one study applied adequate randomization, double-blinding, and allocation concealment [13]. The quality of an RCT is compromised if the investigators, participants, and outcome assessors are not blinded, as knowledge of group assignment can influence responses to an intervention [14]. The inadequate allocation concealment can result in exaggerated estimates of treatment effect [15]. Heterogeneity, as evidenced by quality of the reports, intervention methods, doses, and durations of treatment, along with different efficacy, applicability, and toxicity both among RCT and within subgroups, brings variability. Therefore, considering the low methodological quality of these RCT, more RCTs are needed before a recommendation of efficacy of TwHF can be made.

Another meta-analysis of three RCTs of RA involving 287 patients with a median treatment period of 3 months with standardized rosehip powder was performed [16]. The findings gave evidence that rosehip powder consistently reduced pain scores. Since the patenting of standardized rosehip powder, there have been a number of clinical trials exploring the efficacy of this preparation (Hypen Vital®) in conditions such as osteoarthritis, RA, and inflammatory bowel disease (IBD). Clinical research includes open label and RCT with a duration of 6 months. These studies found rosehip to be extremely safe, with occasional mild allergic reactions or gastrointestinal complaints but no serious adverse events (AsE) [17].

4.2.2 Asthma

There have been multiple clinical studies in the area of asthma. One detailed example is the work of Kelly-Pieper and colleagues [18]. The objective of this particular study was to evaluate the safety and hematological tolerability of nonsteroidal Chinese herbal formula antiasthma herbal medicine intervention (ASHMI) in adult subjects with allergic asthma. ASHMI comprises three herbs: Ling-Zhi (*Ganoderma lucidum*), Ku-Shen (*Sophora flavescens*), and Gan-Cao (*Glycyrrhiza uralensis*) and was derived from a TCM 14-herb formula. The trial design was a randomized, double-blind, placebo-controlled, dose-escalation, phase I trial aimed at developing a botanical drug under the United States Food and Drug Administration (FDA) Investigational New Drug (IND) title. Subjects received one of three doses of ASHMI or placebo: 600 mg (two capsules), 1200 mg (four capsules), or 1800 mg (six capsules) twice daily for 1 week. Four ASHMI and two placebo subjects were treated at each dose level. Subjects continued to use their conventional asthma medications for the duration of the study. The outcome measures were vital signs, physical examination, laboratory data, and electrocardiogram data. These were monitored throughout the study to assess occurrence of AE. Immunomodulatory studies were performed to evaluate the effect of ASHMI on cytokine, chemokine, and growth factor levels. A total of 20 nonsmoking, allergic subjects with asthma were included in the study. Eight subjects (four ASHMI and four placebo) reported mild gastrointestinal (GI) symptoms. No grade 3 AEs were observed during the study period. Vital signs, electrocardiogram findings, and laboratory results obtained pre- and post-treatment remained within normal range. No abnormal immunological alterations were detected. These results showed that ASHMI was safe and well tolerated in allergic asthmatic patients. A larger phase II trial is now being initiated to assess the clinical effectiveness of ASHMI [18].

4.2.3 Cancer

The predominant generation of information from clinical trials of natural products is in the field of cancer research, thus providing a fair body of the accumulated wealth of information here. In summary, several new derivatives of known anticancer agents are undergoing clinical trials [19]. Many plant-derived chemopreventive agents (botanicals) have been shown to suppress cancer [4, 20]. Examples of these studies are presented later in this chapter.

Some of the leading plant-derived compounds have entered human clinical trials for specific cancers such as curcumin in phase I for colorectal cancer and phase II for pancreatic cancer and geinstein in two separate phase II studies for pancreatic cancer and prostate cancer, respectively [21–24]. Other botanicals moving through cancer clinical trials are resveratrol in phase I for colon cancer, silymarin in phase II for acute promyelocytic leukemia and liver cancer and green tea in phase I for solid tumors [25–27].

As an example, the natural product, indole-3-carbinol (13C), has been shown to possess cancer-preventive properties. Rosen and colleagues [28] reported a long-term clinical study using 13C for the treatment of recurrent respiratory papillomatois. Clinical trials with 13C have focused on hormone-responsive cancers, including cervical dysplasia, breast cancer, vulvar intraepithelial neoplasia, and prostate cancer [29, 30]. Phase I and II double-blinded RCTs of 13C in women with cervical intraepithelial neoplasia found significant reduction in symptoms, lesion size, and severity, as well as significant improvement in estrogen metabolism [31–33].

4.2.4 Cardiovascular Disease

In the field of cardiovascular disease, there have been a plethora of clinical trials with natural products. One such systematic review performed by Shang et al. [34] evaluated thc current evidence for the benefit and side effects of oral *Panax notoginseng* for coronary heart disease (CHD). Seventeen randomized placebo-controlled clinical trials (1747 participants) were reviewed. The analysis showed no significant effect on reduction of cardiovascular events;but it significantly alleviated angina pectoris, these trials were deemed to be of low quality and with bias.

A single clinical trial by Leung et al. conducted three randomized, double-blind placebo-controlled clinical trials on different target populations with a two-Chinese herb combination of Salvia Miltiorrhizae Radix et Rhizoma (Danshen) and Puerariae Lobatae Radix (Gegen) [35]. This formula has been studied extensively on cardiovascular biological platforms. In the laboratory, the formula was found to have the biological effects of anti-inflammation, antioxidation, and vasodilation. Clinical trials using ultrasonic carotid intima thickness as a surrogate marker showed highly significant benefits and no significant AEs [35].

4.2.5 Diabetes

In the field of diabetes, there is a wealth of emerging clinical trials with natural products. Therefore, what follows are in-depth reports and analysis of clinical trials of natural and herbal medicines in this field. Our first look is at a meta-analysis of RCT evaluating the efficacy and safety of TCM on diabetic peripheral neuropathy (DPN) performed by Hao et al. [36]. The primary outcome measures were the absolute values or changing of motor or sensory nerve conduction velocity (NCV), and the secondary outcome measurements were clinical symptom improvements and AEs. The methodological quality was assessed by a Jadad scale and the 12 criteria recommended by the Cochrane Back Review Group. A total of 163 studies claimed using RCT design. And 10 studies with 653 individuals were selected as eligible to fit the criteria of the meta-analysis and further authenticated based on the Jadad score ≥3. These 10 studies were all of high methodological quality with a low risk of bias. Meta-analysis showed the effects of NCV favoring TCM when compared with Western conventional medicines ($P=0.05$ or $P=0.01$). There was a significant difference in the total efficacy rate between the two groups ($P=0.001$). Adverse effects were reported in all of the 10 studies included and were well tolerated in all patients with DPN. Despite the apparently positive findings and low risk of bias, it is premature to conclude the efficacy of TCM for the treatment of DPN because of the high clinical heterogeneity and small sample sizes of the relevant studies. However, TCM therapy was safe for DPN. Further standardized preparation, large sample size, and rigorously designed RCT are required [36].

Turning next to a double-blind randomized placebo-controlled trial investigating the efficacy of a Chinese herbal formula, *Jiangtang Xiaozhi*, in managing impaired glucose control and insulin resistance in individuals with pre-diabetes and controlled diabetes is reviewed [37]. A total of 71 patients with pre-diabetes or "controlled" diabetes were randomized to receive three capsules of *Jiangtang Xiaozhi* ($n=39$) or placebo ($n=32$) three times daily for 16 weeks with a follow-up 8 weeks later. The primary outcome was change in glycemic control as evidenced by fasting blood glucose (FBG), postprandial plasma glucose, and glycosylated hemoglobin (HbA1c). Other measures included change in fasting insulin, insulin resistance and sensitivity, lipids, C-reactive protein, body mass index, waist girth, blood pressure, and health-related quality of life and safety. In the current study, the 16-week *Jiangtang Xiaozhi* treatment did not lower fasting blood glucose, but it improved serum insulin and HDL cholesterol in a Western population with pre-diabetes or controlled diabetes. This trial may have been underpowered. Dosage needs to be considered before commencing a longer adequately powered trial [37].

To evaluate the efficacy and safety of radix astragali and its prescriptions for diabetic retinopathy, a computer-based online and manual search was conducted for RCTs addressing radix astragali and its prescriptions for diabetic retinopathy [38]. A total of 16 RCT carried out in China, involving 977 Chinese subjects, were identified from an initial group of 147 studies. Meta-analysis indicated that the effect of radix astragali and its prescriptions in improving visual acuity and fundus manifestations, lowering FBG, plasma viscosity, was superior to that of the control group. By contrast, the efficacy of radix astragali and its prescriptions was not superior to those of the control group in decreasing HbA1C. Analysis suggested that the studies were of low methodological quality. Radix astragali and its prescriptions were superior to other treatments for diabetic retinopathy in terms of improving visual acuity and fundus manifestations, reducing FBG and plasma viscosity. The evaluated studies were of low methodological quality [38].

4.2.6 Dermatology

In dermatology, several natural products, such as honey, green tea, and vitamin C, have been used as topical treatments for a variety of diseases. A systematic review was performed to explore the cutaneous effects of each of these three products. The unique antibacterial, anti-inflammatory, and antioxidant properties of honey were shown to contribute to wound healing, especially in ulcers and burns. Green tea, among many health benefits, demonstrated protection from ultraviolet-induced events, such as photo-immunosuppression and skin cancer growth. Vitamin C, known for its antioxidant properties and key role in collagen production, has been shown to produce positive effects on skin hyperpigmentation and aging. Future large well-designed clinical trials are needed in order to further investigate the potential of these agents as dermatological therapies [39].

4.2.7 Gastroenterology

The field of gastroenterology also has been an area of intense investigation of natural and herbal medicines. There is a broad base of investigation to review. To begin, we take a look at a previously licensed natural product. To assess the safety, pharmacology, and clinical effects of tormentil extracts (TEs) in patients with active ulcerative colitis (UC), TE was given as a commercially available ethanolic dry extract from rhizome of *Tormentilla erecta*, which contains 200 mg/capsule (RatioGast®, Ratiopharm, Ulm, Germany) and licensed in Germany for the treatment of unspecific diarrhea [40]. TE contains 15–22% tannins, which are considered the active compound and also triterpenes and flavonoids [41]. A total of 16 patients with active UC received TE in escalating doses of 1200, 1800, 2400, and 3000 mg/day for 3 weeks each. Each treatment phase was followed by a 4-week washout phase. The outcome parameters were side effects, clinical activity index, C-reactive protein, and tannin levels in patient sera. TE appeared safe up to 3000 mg/day [40].

Meta-analysis review of the induction of clinical response and remission of IBD by the use of herbal medicines was performed [42]. Initially, 41 trials were examined with a resulting analysis of seven controlled trials involving 474 patients demonstrating that herbal medicines may safely induce clinical response and remission in patients with IBD without significant effects on endoscopic and histological outcomes. The results of sub-analyses based on plant type demonstrated that induction of clinical remission was obtained only by *Artemisia absinthium* and *Boswellia serrate*. Induction of clinical response was gained by only *Aloe vera* and *Triticum aestivum*. *Boswellia serrata*, in one study evaluating recurrence rate, did not cause prevention of relapse. Induction of AEs by any of these agents was significant compared to that of placebo. The results indicated that herbal medicines may induce clinical remission and clinical response in patients with IBD. Endoscopic efficacy was investigated in two studies, both in patients with UC. Herbal medicines did not demonstrate any significant effect on induction of endoscopic remission and endoscopic response. Overall, the results show that herbal medicines may induce clinical efficacy in patients with IBD, but the evidence is too limited to make any confident conclusions. Meta-analysis of clinical trials that have compared efficacy of herbal medicines with that of conventional drugs such as aminosalicylates may be helpful. This is being carried out by the author of this chapter. Further high-quality, large controlled trials using standardized preparations are warranted to better elucidate the effects of these natural products in IBD [42].

A few well-designed clinical studies have been performed on a number of such herbal supplements; but in general, the current knowledge remains limited to make a definitive decision about their effectiveness in treating IBD symptoms. The use of peppermint extracts has been studied in a number of clinical trials that evaluated the administration of enteric-coated peppermint oil capsules to IBD patients. The duration of the trials ranged from 4 to 8 weeks and was not divided into specific IBD subtypes. The trials showed a significant reduction in abdominal pain and severity compared to placebo after 4 weeks and a significant increase in quality of life, although the effects did not last once peppermint was discontinued [43–45]. Turmeric was evaluated in IBD patients and induced decreases in IBD symptoms and increased quality of life if given in two different doses of 72 and 144 mg/day over 8 weeks [46]. However, this study lacks a double-blind and placebo-controlled design that reduces the strength of the obtained data.

Another double-blind, placebo-controlled study compared curcuma extract from which turmeric is derived with a placebo and a fumitory extract [47]. Both curcuma and fumitory extracts did not show any significant improvements in abdominal pain and distension compared to the placebo group.

The first combination product that has received some interest from patients and healthcare providers alike is Iberogast®, a mixture of nine herbal plant extracts that was originally used mainly for functional dyspepsia in Germany [48, 49]. The product has been on the market for more than 30 years. Iberogast for use in IBD symptoms was investigated in 208 patients with various IBD subtypes in the United States [50]. This study adheres to the clinical trial standards by utilizing a randomized, double-blind, placebo-controlled study protocol over a 4-week period. The extract consists of liquid extracts from chamomile flowers, bitter candytuft, angelica root, caraway fruits, milk thistle, lemon balm leaves, greater celandine, licorice root, and peppermint leaves [51]. The study indicates that Iberogast significantly improves quality of life and reduces abdominal pain in IBD patients [52]. This appears to be mediated through effects on serotonin, acetylcholine, and opioid receptors in the GI tract [51].

TCM has provided a range of different treatment approaches over centuries, among these a number of TCM herbal mixtures that are specifically formulated based on the patient's symptoms [53]. Such individualized medicine is not uncommon but has the obvious limitation of resisting standardization and adopting the rigorous clinical trial design that is used as a major determinant of effectiveness in Western medicine. *Tong Xie Yao Fang* (TXYF) is one such TCM that has been studied in several clinical trials, but often adjustments were made to the herbal combination based on the predominant IBD symptom presentation [54–56]. A review of 12 studies with modified TXYF preparations resulted in the general conclusion that the extracts improved IBD symptoms, in particular abdominal pain, distension, flatulence, and diarrhea [55]. However, the study design and end points were diverse among these studies complicating a direct comparison of outcomes. A more streamlined and standardized approach to TXYF clinical trials is warranted.

4.2.8 Viral Infections

There are multiple natural products with antiviral properties. Two specific viral infections will be reviewed here. First, to assess the safety and efficacy of *Pleurotus ostreatus* in patients with human immunodeficiency virus (HIV) and antiretroviral therapy (ART)-induced hyperlipidemia, a single-arm, open-label, proof-of-concept study of 8 weeks' duration with a target enrollment of 20 subjects was conducted [57]. Study patients with ART-induced elevated non-HDL cholesterol levels (>160 mg/dl) were enrolled. Participants received packets of freeze-dried *P. ostreatus* (15 g/day) to be taken orally for the 8-week trial period. Lipid levels were quantified every 2 weeks to assess efficacy. A total of 126 patients were screened to enroll 25, of which 20 completed the 8-week study. Subjects had a mean 13.7 years of HIV infection. *P. ostreatus* as administered in this experiment did not lower non-HDL cholesterol in HIV patients with ART-induced hypercholesterolemia. Small changes in HDL and triglycerides were not of a clinical magnitude to warrant further study [57].

A study of hepatitis C virus (HCV)-positive patients was carried out as a single-center, single-arm, and open-label phase II design. The study agent, a mixture of seven botanicals, is based on a traditional Chinese Medicine herbal formula (xiao-chai-hu-tang in Chinese or sho-sai-ko-to in Japanese, abbreviated as, SST) [58]. It is available as a granular powder from the water extract of Bupleurum root (Bupleurum Chinese, Order: Apiales, Family; Apiaceae) 29% (w/w), Pinellia tuber (Pinellia ternate, Order: Alismatales, Family: Araceae) 21%, Baikal skullcap root (*Scutellaria baicalensis*, Order: Lamiales, Family: Lamiaceae) 12.5%, Ginseng root (*Panax ginseng*, Order: Apiales, Family: Araliaceae) 12.5%, Jujube fruit (*Ziziphus Jujuba*, Order: Rosales, Family: Rhamnaceae) 12.5%, Licorice root (*Glycyrrhiza uralensis*, Order: Fabales, Family: Fabaceae) 8%, and Ginger rhizome (Zingiber officinale, Order: Zingiberales, Family: Zingiberaceae) 4%. The agent used in this trial is manufactured by Honso Pharmaceutical Co. Ltd., Nagoya, Japan, as a standard commercial pharmaceutical product according to Current Good Manufacturing Practice (cGMP). Each unit dose contains 2.5 g of extract granules individually wrapped in water and air tight plastic packets. Each 7.5 g daily dose should have 24.7–46.0 mg of Glycyrrhizin, 110.6–205.6 mg of Baicalin, and 6.5–19.7 mg of Saikosaponin. Quality assurance data were submitted to the US FDA to obtain IND status. A total of 24 chronic HCV patients received SST, 2.5 g (p.o.) three times daily for 12 months. Liver function, HCV viral load, and liver biopsy histology were assessed before and after the intervention. SST may improve liver pathology in selected HCV patients who are not candidates for interferon-based treatment. Larger, controlled studies of this botanical formulation are warranted [58].

4.3 NATURAL PRODUCT: GREEN TEA

Following the introductory survey of natural products on disease states, this section undertakes a more critical review of the translation of a single type of natural product from bench-to-bedside. The focus here is on examples of the specific compounds of plant origin that are phenolic by structure and chemical characteristic, the polyphenols. Phenolic compounds of plant origin, specifically green tea catechins, belong to reactive chemical species and share certain characteristics, like antioxidant activity, which is considered beneficial for human health. They are seldom pharmacologically inert and known to share certain metabolic pathways, by which they are conjugated and excreted, like other xenobiotics, from the human body [59, 60]. Plant phenolics are well suited to the concept of reverse pharmacology, describing activities in which the time progresses from the state of proven safety and efficacy of an agent toward study of its molecular mechanism of action. What is probably most important is that a vast knowledge (chemical and biological) on individual members of this class has been accumulated already [61].

There is a long history of tea consumption in the world. Epidemiological studies suggest that the consumption of tea may reduce cancer and cardiovascular disease risk, resulting from its polyphenol components [62–68]. Green, black, and oolong teas are produced from the tea plant *Camellia sinensis* by different manufacturing processes [69, 70]. As a consequence of these differences, only green tea maintains its original composition of polyphenols [69]. Polyphenolic compounds in green tea include flavanols, flavandiols, flavanoids, and phenolic acids accounting for 30% of the dry weight of green tea leaves [69]. Some of these exhibit a broad spectrum of physiological and pharmacological properties such as antioxidant, radical-scavenging, anti-inflammatory, antiviral and antibacterial, antiatherogenic, and anticarcinogenic activities [61, 71]. It is reported that human intake of all flavonoids is a few hundred milligrams to 650 mg/day in our Western diet. Their entry into the body takes place via the gastrointestinal tract; therefore, this organ and especially the epithelial lining cells are exposed to fairly high concentrations of flavonoids [72].

4.3.1 Green Tea Catechin, Epigallocatechin Gallate (EGCG)

Most of the polyphenols in green tea are flavanols, commonly known as catechins. These mainly consist of four compounds: (–)-epicatechin (EC), (–)-epigallocatechin (EGC), (–) epicatechin, gallate (ECG), and epigallocatechin gallate (EGCG) (Fig. 4.1) [61], which may be present at concentrations of up to 1 mg/ml in a cup of tea [73]. The concentration of EGCG and other catechins in green tea can vary slightly, depending upon the variety of tea plant and the harvesting season [70]. EGCG is not found in other plants and is the major catechin in tea [74]. EGCG (molecular weight, 458.4) is the largest of the simple isoflavanoids found in tea but

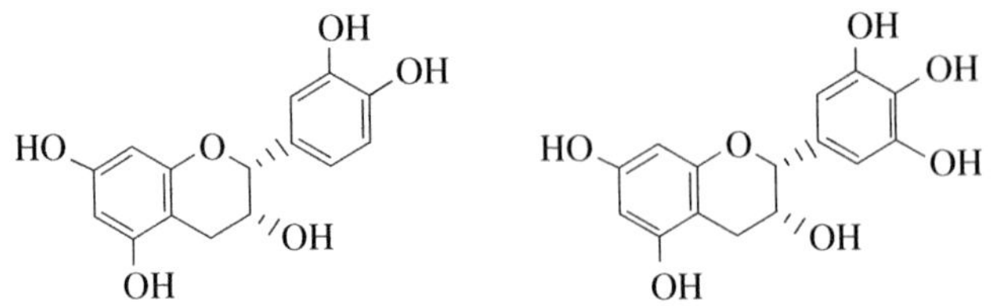

(–) Epicatechin (EC) (–) Epigallocatechin (EGC)

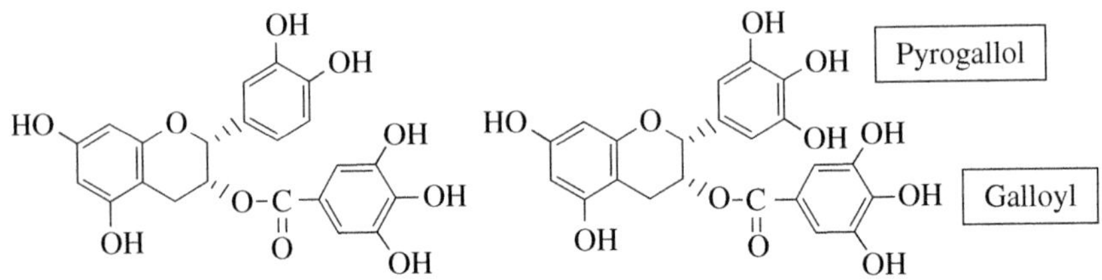

(–) Epicatechin gallate (ECG) (–) Epigallocatechin gallate (EGCG)

FIGURE 4.1 Structures of the polyphenolic catechins found in green tea.

is still a small molecule and, along with the other tea polyphenols, appears to lack toxicity following human consumption [74]. In fact, tea, and therefore EGCG, is on the US FDA's list of compounds generally recognized as safe and approved for human consumption [75, 76].

EGCG accounts for approximately 65% of the total amount of catechins in green tea. EGCG binds strongly to many biological molecules and affects a variety of enzyme activities and signal transduction pathways at micromolar or nanomolar levels attributable to its pyrogallol and galloyl moieties [77, 78]. It is this specific component of green tea that is thought to be responsible for the vast array of claimed health benefits. Properties ascribed to EGCG include neuroprotective, antitumorigenic, anti-inflammatory, antioxidative, antiproliferative, antibacterial, and antiviral effects [68, 77, 79–81].

There have been no reports of clinical toxicity when green tea is consumed as a beverage throughout the day. Consumption of up to 20 cups of green tea per day is not uncommon in certain populations [82]. A crucial aspect of translating the observed effects of EGCG to clinically relevant strategies is the necessity of achieving physiologically relevant concentrations. As tea catechins have poor bioavailability, most of the ingested EGCG does not reach the blood as a significant fraction is eliminated pre-systemically [83]. Phase I clinical trials involving pharmacokinetic studies of EGCG have shown that only a small percentage of the orally administered catechin appeared in the blood. Drinking the equivalent of two cups of green tea resulted in a mean peak plasma EGCG level of 170 nM after 1.5 h [25, 78, 84].

4.4 EGCG CLINICAL TRIALS

4.4.1 Polyphenon E

Formulations of EGCG have become critical as purified EGCG is considered to be too costly for clinical use and is poorly absorbed [85–87]. Investigators have used supplemental and/or synthetic forms of EGCG and other tea catechins to help characterize dose–response relationships, other biological effects and to promote the design and development of chemopreventive drug candidates based on EGCG. Therefore, a decaffeinated supplemental preparation of tea catechins, Polyphenon E, was developed by the Division of Cancer Prevention (DCP), Chemoprevention Branch at the National Cancer Institute (NCI) and in conjunction with Mitsui Norin Co., Ltd., contains approximately 65% EGCG and 30% other catechins [88–90].

Single-compound green tea extracts (GTs) and EGCG have been studied in the laboratory for decades, but the well-defined makeup, standardization, and cheaper cost over nonstandardized compounds make Polyphenon E a prime candidate for human clinical trials [91].

Polyphenon E is stable under normal storage and can be readily packaged into capsules, which are also stable, according to the complete stability tests supported by DCP/NCI. Mitsui Norin prepared a voluminous Drug Master File showing the Chemistry, Manufacturing and Controls of the manufacturing facility (filed with the US FDA) and was approved as a cGMP facility for the manufacturing of Polyphenon E [76]. To date, there have been over 25 human clinical studies with Polyphenon E (registered with the NIH clinicaltrials.gov) primarily in cancer.

The results of a phase I/II trial looking at the safety, tolerability, and efficacy of Polyphenon E and its effectiveness in treating patients with stage 0, I, or II chronic lymphocytic leukemia (CLL) led to the announcement in July 2008 by the US FDA of an orphan drug designation for the treatment of CLL with Polyphenon E [91].

4.4.2 Safety, Toxicity, and Pharmacokinetics

In a phase I pharmacokinetic study of Polyphenon E, 20 human subjects were given single doses of either 200, 400, 600, or 800 mg EGCG, with each capsule containing 200 mg of EGCG and 68 mg of other catechins. Peak plasma EGCG levels of 200–400 ng/ml (0.4–0.8 μM) could be achieved after the administration of these formulations at doses equivalent to the EGCG content of 8–16 cups of green tea [83].

A subsequent in-depth study of the safety and plasma kinetics of multiple-dose administration of purified EGCG and Polyphenon E was performed. This study examined once-daily and twice-daily dosing regimens of EGCG and Polyphenon E over a 4-week period in 40 healthy volunteers [82]. In this study, standardized, defined, and decaffeinated green tea polyphenol oral products in amounts similar to the EGCG content in 16 Japanese-style cups of green tea were consumed once daily or in divided doses twice daily (4 capsules/day) for 4 weeks. EGCG intake at doses of 400 mg twice daily and 800 mg once daily established peak serum concentrations averaging 150–290 ng/ml. Peak concentrations were reached between 2.4 and 4.2 h.

Once-daily dosing of 800 mg resulted in approximately a 60% increase in the systemic exposure of free EGCG after chronic Polyphenon E administration. Purified EGCG and Polyphenon E were administered with food in these studies. All AEs during the 4-week period rated as mild and overall were similar to placebo. Abdominal discomfort (19%) was the most frequent AE. Other side effects included headache (6%), excess gas (6%), and dizziness (6%) [82].

On the basis of the reported AEs and clinical laboratory data in this trial, the study agents and dosing schedules have been found to be safe and well tolerated by the study subjects for at least 1 month. The reported AEs were rated as mild events. The more common events include headache, stomach ache, abdominal pain, and nausea, which have been reported in subjects receiving green tea polyphenol treatment, as well as in subjects receiving placebo. There were no significant changes in blood counts and blood chemistry profiles after 4 weeks of green tea polyphenol treatment [82].

4.4.3 Metabolism

After oral intake, ECGC undergoes extensive hepatic first-pass metabolism, including glucuronidation, sulfation, and methylation. In humans, EGCG is present mainly in the free form in plasma [83]. EGCG is degallated to EGC. EGCG is not detected in the urine after ingestion [83, 92]. These data suggest that EGCG is not excreted in the urine but transported to the liver for excretion into bile and feces.

Drug interactions may result for either pharmacokinetic or pharmacodynamic reasons. Pharmacokinetic interactions can affect absorption, transport, distribution, metabolism, or elimination of a drug. Pharmacodynamic interactions alter the pharmacological response to a drug but may not affect the drug's body concentrations. Risk of drug interactions with EGCG may be decreased due to minimal interaction with major cytochrome P450 (CYP) isozymes. Two studies showed no significant alteration in the disposition of medications primarily dependent on the CYP2D6 or CYP3A4 pathways of metabolism [93, 94]. Four weeks of Polyphenon E 800 mg once daily did not change the activities of CYP1A2, CYP2D6, and CYP2C9, but inhibited CYP3A4 activity by approximately 20% compared to baseline ($p=0.01$) [95]. These results suggest that EGCG administration may be less likely to affect the pharmacokinetics of commonly used pharmaceutical drugs.

4.4.4 Clinical Studies

Standardized GTE has been examined in a randomized double-blind placebo-controlled trial carried out for 3 months in obese, hypertensive patients. Both groups had extensive multiparameter baseline characteristics, which facilitated efficacy of the intervention. GTE supplementation was demonstrated to exert positive effect on blood pressure, carbohydrate metabolism, and lipid profile [96].

A randomized, double-blind, two-group parallel clinical trial of 197 subjects was conducted to compare the effects of green tea catechins against placebo for the prevention of influenza infection for 5 months in a cohort in Japan. The results indicated statistically significant preventive effect on clinically defined influenza infection (OR, 0.25: 95% CI, 0.07–0.76, $P=0.022$) and was well tolerated [97, 98].

4.4.5 Cancer Studies

Tea catechins are the only flavonoids investigated in clinical trials for oral cancer. The first clinical trial using green tea for treatment of an oral pre-malignant lesion was a double-blind, placebo-controlled, randomized, phase II trial that resulted, over 6 months of intervention, in a significant reduction (37.9%) in size of the oral lesion [99].

Another randomized, placebo-controlled, phase II trial evaluated the chemopreventive potential of GTE in oral cancer. Efficacy of GTE on reduction of oral lesion was not significantly demonstrated [100].

Over the last 8 years, researchers have explored the ability of Polyphenon E to increase glutathione S-transferase enzymes, which has the potential to render cancerous chemicals harmless; its efficacy in preventing bladder cancers; and its effect on growth and activation of the epidermal growth factor signaling pathways in human colon cancer cells [91].

A phase I clinical trial of EGCG at doses of 600 mg daily for 1 year resulted in a 90% reduction in the rate of high-grade prostate intraepithelial neoplasia-positive men developing prostate cancer. No significant side effects were observed [101].

Daily oral EGCG in the Polyphenon E preparation was well tolerated at doses up to 2000 mg twice per day for up to 6 months in patients with asymptomatic Rai stage 0 to II CLL. Decline in CLL and lymphadenopathy during Polyphenon E therapy were observed in the majority of patients. A phase II trial evaluating the efficacy of Polyphenon E (2000 mg twice per day) in patients with asymptomatic Rai stage 0 to II CLL was initiated in November 2007 [102].

The largest randomized, placebo-controlled trial to evaluate the potential of oral Polyphenon E intervention in HPV-related cervical disease in 98 women with persistent high-risk HPV infection and concomitant low-grade cervical intraepithelial neoplasia (CIN1) as cervical cancer prevention was undertaken by Garcia et al. [103] in Phoenix, Arizona, and Southern Pines, North Carolina, USA. Definitive results show that Polyphenon E is acceptable, safe, and well tolerated, but without resolution of persistent high-risk HPV and related CIN1 [103].

Meta-analysis of clinical studies of tea on gynecological cancers reports an inverse association for green tea intake and risk of ovarian cancer (OR=0.66, 95% CI: 0.654, 0.80) and for green tea and risk of endometrial cancer (OR=0.78, 95% CI: 0.62, 0.98). There was no association for black tea and ovarian cancer risk (OR=0.94, 95% CI: 0.87, 1.02) and a positive association with endometrial cancer risk (OR=1.20, 95% CI: 1.05, 1.38) [104]. All were case-control studies based in China or Japan. The meta-analyses may have been confounded by age, body size, coffee intake, or alcohol use because these factors were not consistently adjusted for in the individual studies [104].

Meta-analysis of four randomized, double-blind, placebo-controlled trials testing the efficacy of green tea catechins, Polyphenon E ointment formulation, for the treatment of cervical lesions and external genital warts (EGW) was identified. EGW clearance was statistically significant [104].

The following is an overview of a successful registration of a tea extract product with the US FDA. The initial fact-finding trial was successfully performed in a clinical setting at the Bejing Cancer Center Hospital in 1990 in China. Application of Polyphenon E ointment on genital warts (*Condyloma acuminata*) eliminated the warts effectively at the Bejing Cancer Center Hospital in 1990 and in subsequent phase I clinical trials [105]. This study demonstrated the therapeutic effect of tea catechins on benign tumors.

On this basis, a German pharmaceutical company, MediGene AG, spent time and money on phase II and phase III clinical trials, internationally. In 2006, the US FDA approved the marketing of the ointment as a prescribed Botanical Drug in the United States [106].

Polyphenon E (MediGene AG, Martinsried, Germany) is a standardized extract of green tea leaves from *Camellia sinensis* containing mainly tea polyphenols, mostly consisting of catechins (>85%). The main catechin in Polyphenon E ointment is EGCG, which has the highest biological activity within the catechins with immuno-stimulatory, antiproliferative, and antitumor properties [107–111]. Polyphenon E ointment is a self-administered product based on natural ingredients and has promising therapeutic properties. Two controlled, randomized phase III trials have shown efficacy and safety in the clearance of EGW [112, 113].

Several case reports have implicated concentrated extracts of over the counter Chinese green tea in causing hepatotoxicity, usually with a mixed hepatocellular-cholestatic picture [114, 115]. Cumulative amount of extract consumed ranged from 5.9 to 240 g. Most patients were younger than 40 years of age (men, 8 of 11 cases) and (women, 9 of 11). Ten of eleven patients recovered after discontinuing the implicated substance. One patient required liver transplantation. Additionally, while EGCG alone is active in suppressing cancer, while EGCG has shown synergistic effects with other anticancer drugs [116–120].

The mechanisms underlying EGCG anticancer effects include antioxidant, modification of carcinogen metabolism, prevention of DNA damage, induction of cell cycle arrest and apoptosis, inhibition of metastasis, proteasome inhibition, and modulation of multiple signal transduction pathways [121]. The inconsistent results among the various studies were probably due to various confounding factors. To minimize these, more potent tools were used in later well-designed clinical intervention studies, including a defined GTE, pharmacologically formulated EGCG-enriched fractions such as Polyphenon E.

4.5 HUMAN CLINICAL STUDY: EGCG AND HIV-1 INFECTION

4.5.1 Translational Medicine: EGCG: Bench-to-Bedside

On the basis of the literature, the concentration range of EGCG likely in the plasma after consuming the equivalent of 2–3 cups of green tea is within the range of 0.1–0.6 μmol/l and with greater consumption of green tea (7–9 cups) at the level of 1–2 μmol/l [68, 78, 87, 92, 122]. Therefore, interference of gp120 binding to $CD4^+$ T cells was assessed at the physiologically relevant levels of 0.2–2.0 μmol/l [123]. These studies have demonstrated evidence of high affinity binding of EGCG to the CD4 molecule with a K_d of 10 nM with subsequent inhibition of gp120 binding to human $CD4^+$ T cells. EGCG binds in the same molecular pocket on CD4 as does HIV-1-gp120. EGCG at concentrations equivalent to those achieved by the consumption of green tea is able to reduce the attachment of gp120 to CD4 (when present at physiological concentrations) by a factor of between 10- and 20-fold [123].

4.5.2 Phase I Clinical Trial: Polyphenon E in HIV-1 Infection

Currently, translation of the green tea catechin, EGCG, from bench-to-bedside in the area of HIV is in the clinical trial stage. A placebo-controlled, dose-blind, dose-escalation study to evaluate the safety, tolerability, pharmacokinetics, and antiviral activity of EGCG as monotherapy for 14 days in ARV-naive and ARV-experienced, HIV-1-infected subjects was designed and initiated. The purpose of this study is to determine the safety, toxicity, dosing, and antiviral effects of EGCG in capsule form (Polyphenon E), administered orally twice daily at three different doses in HIV-1-infected clinically stable, treatment-naive and treatment-experienced adults not on concomitant ARV therapy. Several potential doses of EGCG administered as Polyphenon E have been selected as likely to achieve the concentrations previously shown *in vitro* to be necessary for anti-HIV-1 activity (0.2–5 μM = 92–2292 ng/ml).

The Polyphenon E capsules administered in the trial are manufactured, stored, distributed, and evaluated for stability under contract to NCI, DCP using cGMPs as outlined in the US Code of Federal Regulations. Placebo capsules contain the inactive excipients. Dark green, opaque, size 0 hard gelatin capsules are packaged in high-density polyethylene containers with child-resistant closures and stored under ambient conditions. EGCG is dispensed in capsules containing 200 mg of active drug, Polyphenon E.

The primary objectives of this clinical trial are to evaluate the safety and tolerability of Polyphenon E for 14 days in HIV-1-infected participants and to evaluate the change in viral load from baseline in the HIV-1-infected participants treated with Polyphenon E for 14 days. The secondary objectives are to evaluate $CD4^+$ T lymphocyte counts and evaluate pharmacokinetic characteristics in the HIV-1-infected participants treated with Polyphenon E for 14 days.

HIV-1-infected clinically stable, treatment-naive and treatment-experienced adults with no AIDS-defining events during the 12 weeks prior to screening will be enrolled. There will be three treatment arms, each consisting of eight evaluable participants. Two participants in each study arm will be randomized to receive placebo. Dosing will be escalated sequentially contingent on the safety profile of previous doses. To qualify for enrollment, subject plasma HIV-1-RNA levels must be greater than 1000 copies/ml at screening, $CD4^+$ T lymphocyte counts must be stable and greater than or equal to 250 cells/mm^3 at screening.

Furthermore, catechins are cheap natural compounds. EGCG (purity ≥98%), the most popular catechin containing a galloyl moiety, is less than $5 each gram on the Chinese market. After EGCG is mixed into Polyphenon E (EGCG purity ≥60%), the cost is just $0.06, which is nearly 700 times cheaper than raltegravir. Thus, catechins that contain a galloyl moiety may reduce the prescription costs for HIV-1 patients and may be advantageous among low-income populations [124]. EGCG represents a potential low-cost inhibitor of HIV infection that could be combined with current anti-HIV therapy.

4.6 CONCLUSION

Translation of natural and herbal medicines from bench-to-bedside can be a very arduous path. There are inherent difficulties by the nature of the compound, itself. Along with that, there are the rigid and stringent details of executing a successful RCT that is paramount in achieving the validation and statistical significance required to ultimately bring the natural product drug to the clinic. Suitable pharmaceutical formulation must be refined, to supply therapeutic dose in a patient-friendly manner, rendering it bioavailable and efficacious. Successful completion of this should be evidenced by convincing results of clinical trials. Of particular note is the result of negative outcomes of the natural product tested in the clinical trial. Lessons learned regarding the overall design of clinical trials underscore the importance of the randomized, placebo-controlled design and the need for long-term follow-up and monitoring to meet FDA requirements and promote acceptance in the marketplace. Applying these and other lessons to the design of future chemoprevention trials should facilitate the translation of novel preventive agents to the clinic.

REFERENCES

[1] Liang X, Fang W-S (2006) *Medicinal Chemistry of Bioactive Natural Products*. Hoboken: John Wiley & Sons, Inc..

[2] Buss AD, Butler MS (2010) *Natural Product Chemistry for Drug Discovery*. Cambridge: Royal Society of Chemistry.

[3] Gogtay N, Bhatt H, Dalvi S, Kshirsagar N (2002) The use and safety of non-allopathic Indian medicines. *Drug Safety* 25: 1005–1019.

[4] Jia L (2012) Cancer complementary and alternative medicine research at the US National Cancer Institute. *Chinese Journal of Integrative Medicine* 18: 325–332.

[5] The Alpha-Tocopherol, Beta Carotene Cancer Prevention Study Group (1994) The effect of vitamin E and beta carotene on the incidence of lung cancer and other cancers in male smokers. The Alpha-Tocopherol, Beta Carotene Cancer Prevention Study Group. *New England Journal of Medicine* 330: 1029–1035.

[6] Omenn GS, Goodman GE, Thornquist MD, Balmes J, Cullen MR, et al. (1996) Risk factors for lung cancer and for intervention effects in CARET, the Beta-Carotene and Retinol Efficacy Trial. *Journal of the National Cancer Institute* 88: 1550–1559.

[7] Fisher B, Costantino JP, Wickerham DL, Redmond CK, Kavanah M, et al. (1998) Tamoxifen for prevention of breast cancer: report of the National Surgical Adjuvant Breast and Bowel Project P-1 Study. *Journal of the National Cancer Institute* 90: 1371–1388.

[8] Vogel VG, Costantino JP, Wickerham DL, Cronin WM, Cecchini RS, et al. (2006) Effects of tamoxifen vs raloxifene on the risk of developing invasive breast cancer and other disease outcomes: the NSABP Study of Tamoxifen and Raloxifene (STAR) P-2 trial. *JAMA* 295: 2727–2741.

[9] Baron JA, Cole BF, Sandler RS, Haile RW, Ahnen D, et al. (2003) A randomized trial of aspirin to prevent colorectal adenomas. *New England Journal of Medicine* 348: 891–899.

[10] Sandler RS, Halabi S, Baron JA, Budinger S, Paskett E, et al. (2003) A randomized trial of aspirin to prevent colorectal adenomas in patients with previous colorectal cancer. *New England Journal of Medicine* 348: 883–890.

[11] Meyskens FL, McLaren CE, Pelot D, Fujikawa-Brooks S, Carpenter PM, et al. (2008) Difluoromethylornithine plus sulindac for the prevention of sporadic colorectal adenomas: a randomized placebo-controlled, double-blind trial. *Cancer Prevention Research* 1: 32–38.

[12] Liu Y, Tu S, Gao W, Wang Y, Liu P, et al. (2013) Extracts of *Tripterygium wilfordii* Hook F in the treatment of rheumatoid arthritis: a systemic review and meta-analysis of randomized controlled trials. *Evidence-Based Complementary and Alternative Medicine* 2013: 1–11.

[13] Goldbach-Mansky R, Wilson M, Fleischmann R, Olsen N, Silverfield J, et al. (2009) Comparison of *Tripterygium wilfordii* Hook F versus sulfasalazine in the treatment of rheumatoid arthritis: a randomized trial. *Annals of Internal Medicine* 151: 229–240.

[14] Schulz KF, Grimes DA (2002) Blinding in randomised trials: hiding who got what. *The Lancet* 359: 696–700.

[15] Schulz KF, Grimes DA (2002) Allocation concealment in randomised trials: defending against deciphering. *The Lancet* 359: 614–618.

[16] Cohen M (2012) Rosehip: an evidence based herbal medicine for inflammation and arthritis. *Australian Family Physician* 41: 495.

[17] Chrubasik S, Chrubasik C, Neumann E, Müller-Ladner U (2009) The anti-inflammatory efficacy of powdered rose hip-a review. *Zeitschrift für Phytotherapie* 30: 227–231.

[18] Kelly-Pieper K, Patil SP, Busse P, Yang N, Sampson H, et al. (2009) Safety and Tolerability of an Antiasthma Herbal Formula (ASHMI™) in adult subjects with asthma: a randomized, double-blinded, placebo-controlled, dose-escalation Phase I study. *Journal of Alternative and Complementary Medicine* 15: 735–743.

[19] WHO (2011) Who Global Forum: Addressing the Challenge of Non-communicable Diseases. The Russian Federation, Moscow, April 27, 2011.

[20] Institute of Medicine (2010) Extending the Spectrum of Precompetitive Collaboration in Oncology Research—Workshop Summary. Institute of Medicine, Washington, DC, July 22, 2010.

[21] Sharma RA, Euden SA, Platton SL, Cooke DN, Shafayat A, et al. (2004) Phase I clinical trial of oral curcumin biomarkers of systemic activity and compliance. *Clinical Cancer Research* 10: 6847–6854.

[22] Dhillon N, Aggarwal BB, Newman RA, Wolff RA, Kunnumakkara AB, et al. (2008) Phase II trial of curcumin in patients with advanced pancreatic cancer. *Clinical Cancer Research* 14: 4491–4499.

[23] NIH (2010) Genistein in Treating Patients with Pancreatic Cancer that can be Removed by Surgery. Updated June 2, 2010. http://www.clinicaltrials.gov/ct2/show/study/NCT00689195?term=Genisteinandrank=2 (Accessed November 14, 2014).

[24] Taylor CK, Levy RM, Elliott JC, Burnett BP (2009) The effect of genistein aglycone on cancer and cancer risk: a review of in vitro, preclinical, and clinical studies. *Nutrition Reviews* 67: 398–415.

[25] Pisters KM, Newman RA, Coldman B, Shin DM, Khuri FR, et al. (2001) Phase I trial of oral green tea extract in adult patients with solid tumors. *Journal of Clinical Oncology* 19: 1830–1838.

[26] NIH (2009) Resveratrol for Patients with Colon Cancer. Updated January 2, 2009. http://www.clinicaltrials.gov/ct2/show/NCT00256334?term=Resveratrolandrank=11 (Accessed November 14, 2014).

[27] NIH (2010) Silymarin (Milk Thistle Extract) in Treating Patients with Acute Lymphoblastic Leukemia who are Receiving Chemotherapy. Updated February 25, 2010. http://www.clinicaltrials.gov/ct2/show/NCT00055718?term=silymarinandrank=9 (Accessed November 14, 2014).

[28] Rosen CA, Bryson PC (2004) Indole-3-carbinol for recurrent respiratory papillomatosis: long-term results. *Journal of Voice* 18: 248–253.

[29] Michnovicz JJ, Adlercreutz H, Bradlow HL (1997) Changes in levels of urinary estrogen metabolites after oral indole-3-carbinol treatment in humans. *Journal of the National Cancer Institute* 89: 718–723.

[30] Michnovicz JJ, Bradlow HL (1991) Altered estrogen metabolism and excretion in humans following consumption of indole-3-carbinol. *Nutrition and Cancer* 16: 59–66.

[31] Naik R, Nixon S, Lopes A, Godfrey K, Hatem M, et al. (2006) A randomized phase II trial of indole-3-carbinol in the treatment of vulvar intraepithelial neoplasia. *International Journal of Gynecological Cancer* 16: 786–790.

[32] Del Priore G, Gudipudi DK, Montemarano N, Restivo AM, Malanowska-Stega J, et al. (2010) Oral diindolylmethane (DIM): pilot evaluation of a nonsurgical treatment for cervical dysplasia. *Gynecologic Oncology* 116: 464–467.

[33] Bell MC, Crowley-Nowick P, Bradlow HL, Sepkovic DW, Schmidt-Grimminger D, et al. (2000) Placebo-controlled trial of indole-3-carbinol in the treatment of CIN. *Gynecologic Oncology* 78: 123–129.

[34] Shang Q, Xu H, Liu Z, Chen K, Liu J (2013) Oral *Panax notoginseng* preparation for coronary heart disease: a systematic review of randomized controlled trials. *Evidence-Based Complementary and Alternative Medicine* 2013: 12.

[35] Leung P-C, Koon C-M, Lau CBS, Chook P, Cheng WKF, et al. (2013) Ten years 2019; research on a cardiovascular tonic: a comprehensive approach 2014; from quality control and mechanisms of action to clinical trial. *Evidence-Based Complementary and Alternative Medicine* 2013: 1–6.

[36] Hao C-Z, Wu F, Lu L, Wang J, Guo Y, et al. (2013) Chinese herbal medicine for diabetic peripheral neuropathy: an updated meta-analysis of 10 high-quality randomized controlled studies. *PLoS One* 8: e76113.

[37] Grant SJ, Bensoussan A, Chang D, Kiat H, Klupp NL, et al. (2009) Chinese herbal medicines for people with impaired glucose tolerance or impaired fasting blood glucose. *Cochrane Database of Systematic Reviews* 4. DOI: 10.1002/14651858.CD006690.

[38] Duan J (2013) Systematic review and meta-analysis of 16 randomized clinical trials of radix astragali and its prescriptions for diabetic retinopathy. *Evidence-Based Complementary and Alternative Medicine* 2013: 1–13.

[39] Barbosa NS, Kalaaji AN (2014) CAM use in dermatology. Is there a potential role for honey, green tea, and vitamin C? *Complementary Therapies in Clinical Practice* 20: 11–15.

[40] Huber R, Ditfurth AV, Amann F, Güthlin C, Rostock M, et al. (2007) Tormentil for active ulcerative colitis: an open-label, dose-escalating study. *Journal of Clinical Gastroenterology* 41: 834–838.

[41] Sticher O, Steinegger E, Hänsel R (1999) *Pharmakognosie-Phytopharmazie*. Berlin, Heidelberg: Springer, pp. 870–890.

[42] Rahimi R, Nikfar S, Abdollahi M (2013) Induction of clinical response and remission of inflammatory bowel disease by use of herbal medicines: a meta-analysis. *World Journal of Gastroenterology* 19: 5738.

[43] Alam M, Roy P, Miah A, Mollick S, Khan M, et al. (2013) Efficacy of Peppermint oil in diarrhoea predominant IBS-a double blind randomized placebo-controlled study. *Mymensingh Medical Journal* 22: 27–30.

[44] Merat S, Khalili S, Mostajabi P, Ghorbani A, Ansari R, et al. (2010) The effect of enteric-coated, delayed-release peppermint oil on irritable bowel syndrome. *Digestive Diseases and Sciences* 55: 1385–1390.

[45] Cappello G, Spezzaferro M, Grossi L, Manzoli L, Marzio L (2007) Peppermint oil (Mintoil®) in the treatment of irritable bowel syndrome: a prospective double blind placebo-controlled randomized trial. *Digestive and Liver Disease* 39: 530–536.

[46] Bundy R, Walker AF, Middleton RW, Booth J (2004) Turmeric extract may improve irritable bowel syndrome symptomology in otherwise healthy adults: a pilot study. *Journal of Alternative and Complementary Medicine* 10: 1015–1018.

[47] Brinkhaus B, Hentschel C, Keudell CV, Schindler G, Lindner M, et al. (2005) Herbal medicine with curcuma and fumitory in the treatment of irritable bowel syndrome: a randomized, placebo-controlled, double-blind clinical trial. *Scandinavian Journal of Gastroenterology* 40: 936–943.

[48] von Arnim U, Peitz U, Vinson B, Gundermann K-J, Malfertheiner P (2007) STW 5, a phytopharmacon for patients with functional dyspepsia: results of a multicenter, placebo-controlled double-blind study. *The American Journal of Gastroenterology* 102: 1268–1275.

[49] Rösch W, Vinson B, Sassin I (2002) A randomised clinical trial comparing the efficacy of a herbal preparation STW 5 with the prokinetic drug cisapride in patients with dysmotility type of functional dyspepsia. *Zeitschrift für Gastroenterologie* 40: 401–408.

[50] Madisch A, Holtmann G, Plein K, Hotz J (2004) Treatment of irritable bowel syndrome with herbal preparations: results of a double-blind, randomized, placebo-controlled, multi-centre trial. *Alimentary Pharmacology and Therapeutics* 19: 271–279.

[51] Simmen U, Kelber O, Okpanyi S, Jaeggi R, Bueter B, et al. (2006) Binding of STW 5 (Iberogast) and its components to intestinal 5-HT, muscarinic M3, and opioid receptors. *Phytomedicine* 13: 51–55.

[52] Ottillinger B, Storr M, Malfertheiner P, Allescher H-D (2013) STW 5 (Iberogast®)—a safe and effective standard in the treatment of functional gastrointestinal disorders. *Wiener Medizinische Wochenschrift* 163: 65–72.

[53] Wang Z-J, Li H-X, Wang J-H, Zhang F (2008) Effect of Shugan Jianpi Granule () on gut mucosal serotonin-positive cells in patients with irritable bowel syndrome of stagnated Gan-qi attacking Pi syndrome type. *Chinese Journal of Integrative Medicine* 14: 185–189.

[54] Leung WK, Wu JC, Liang S, Chan L, Chan FK, et al. (2006) Treatment of diarrhoea-predominant irritable bowel syndrome with traditional Chinese herbal medicine: a randomized placebo-controlled trial. *The American Journal of Gastroenterology* 101: 1574–1580.

[55] Bian Z, Wu T, Liu L, Miao J, Wong H, et al. (2006) Effectiveness of the Chinese herbal formula TongXieYaoFang for irritable bowel syndrome: a systematic review. *Journal of Alternative and Complementary Medicine* 12: 401–407.

[56] Pan F, Zhang T, Zhang Y-H, Xu J-J, Chen F-M (2009) Effect of Tongxie Yaofang granule in treating diarrhoea-predominate irritable bowel syndrome. *Chinese Journal of Integrative Medicine* 15: 216–219.

[57] Abrams DI, Couey P, Shade SB, Kelly ME, Kamanu-Elias N, et al. (2011) Antihyperlipidemic effects of *Pleurotus ostreatus* (oyster mushrooms) in HIV-infected individuals taking antiretroviral therapy. *BMC Complementary and Alternative Medicine* 11: 60–68.

[58] Deng G, Kurtz RC, Vickers A, Lau N, Yeung KS, et al. (2011) A single arm phase II study of a Far-Eastern traditional herbal formulation (sho-sai-ko-to or xiao-chai-hu-tang) in chronic hepatitis C patients. *Journal of Ethnopharmacology* 136: 83–87.

[59] Quideau S, Deffieux D, Douat-Casassus C, Pouységu L (2011) Plant polyphenols: chemical properties, biological activities, and synthesis. *Angewandte Chemie International Edition* 50: 586–621.

[60] Crozier A, Jaganath IB, Clifford MN (2009) Dietary phenolics: chemistry, bioavailability and effects on health. *Natural Product Reports* 26: 1001–1043.

[61] Nance C, Shearer W (2003) Is green tea good for HIV-1 infection? *Journal of Allergy and Clinical Immunology* 112: 851–853.

[62] Dufresne C, Farnworth E (2001) A review of latest research findings on the health promotion properties of tea. *Journal of Nutritional Biochemistry* 12: 404–421.

[63] Deana R, Turetta L, Donella-Deana A, Dona M, Brunati AM, et al. (2003) Green tea epigallocatechin-3-gallate inhibits platelet signalling pathways triggered by both proteolytic and non-proteolytic agonists. *Thrombosis and Haemostasis* 89: 866–874.

[64] Ludwig A, Lorenz M, Grimbo N, Steinle F, Meiners S, et al. (2004) The tea flavonoid epigallocatechin-3-gallate reduces cytokine-induced VCAM-1 expression and monocyte adhesion to endothelial cells. *Biochemical and Biophysical Research Communications* 316: 659–665.

[65] Sakata R, Ueno T, Nakamura T, Sakamoto M, Torimura T, et al. (2004) Green tea polyphenol epigallocatechin-3-gallate inhibits platelet-derived growth factor-induced proliferation of human hepatic stellate cell line LI90. *Journal of Hepatology* 40: 52–59.

[66] Nakachi K, Matsuyama S, Miyake S, Suganuma M, Imai K (2000) Preventive effects of drinking green tea on cancer and cardiovascular disease: epidemiological evidence for multiple targeting prevention. *Biofactors* 13: 49–54.

[67] Yang C, Landau J (2000) Effects of tea consumption on nutrition and health. *Journal of Nutrition* 130: 2409–2412.

[68] NCI (1996) Clinical development plan: tea extracts. Green tea polyphenols. Epigallocatechin gallate. *Journal of Cellular Biochemistry. Supplement* 26: 236–257.

[69] Mukhtar H, Ahmad N (2000) Tea polyphenols: prevention of cancer and optimizing health. *American Journal of Clinical Nutrition* 71: 1698s–1702s.

[70] Chu D, Juneja L (1997) General chemical composition of green tea and its infusion. In: Yamamoto L, Juneja D, Chu C, Kim M, editors. *Chemistry and Applications of Green Tea*. New York: CRC Press.

[71] Yang CS, Wang X, Lu G, Picinich SC (2009) Cancer prevention by tea: animal studies, molecular mechanisms and human relevance. *Nature Reviews Cancer* 9: 429–439.

[72] Hoensch H, Richling E, Kruis W, Kirch W (2010) Colorectal cancer prevention by flavonoids. *Medizinische Klinik (Munich, Germany)* 105: 554–559.

[73] Kanaka S, Kim M, Taniguchi M, Yamamoto T (1989) Antibacterial substances in japanese green tea extract against streptococcus mutans, a cariogenic bacterium. *Agricultural and Biological Chemistry* 53: 2307–2231.

[74] Hamilton-Miller JM (1995) Antimicrobial properties of tea (*Camellia sinensis* L.). *Antimicrobial Agents and Chemotherapy* 39: 2375–2377.

[75] Paterson I, Anderson E (2005) The renaissance of natural products as drug candidates. *Science* 310: 451–453.

[76] Yukihiko H (2011) Tea catechins and their applications as supplements and pharmaceutics. *Pharmacological Research* 64: 100–104.

[77] Zaveri NT (2001) Synthesis of a 3,4,5-trimethoxybenzoyl ester analogue of epigallocatechin-3-gallate (EGCG): a potential route to the natural product green tea catechin, EGCG. *Organic Letters* 3: 843–846.

[78] Lee MJ, Maliakal P, Chen L, Meng X, Bondoc FY, et al. (2002) Pharmacokinetics of tea catechins after ingestion of green tea and (–)-epigallocatechin-3-gallate by humans: formation of different metabolites and individual variability. *Cancer Epidemiology, Biomarkers and Prevention* 11: 1025–1032.

[79] Sartippour MR, Shao ZM, Heber D, Beatty P, Zhang L, et al. (2002) Green tea inhibits vascular endothelial growth factor (VEGF) induction in human breast cancer cells. *Journal of Nutrition* 132: 2307–2311.

[80] Maeda-Yamamoto M, Kawahara H, Tahara N, Tsuji K, Hara Y, et al. (1999) Effects of tea polyphenols on the invasion and matrix metalloproteinases activities of human fibrosarcoma HT1080 cells. *Journal of Agricultural and Food Chemistry* 47: 2350–2354.

[81] Yamaguchi K, Honda M, Ikigai H, Hara Y, Shimamura T (2002) Inhibitory effects of (–)-epigallocatechin gallate on the life cycle of human immunodeficiency virus type 1 (HIV-1). *Antiviral Research* 53: 19–34.

[82] Chow HH, Cai Y, Hakim IA, Crowell JA, Shahi F, et al. (2003) Pharmacokinetics and safety of green tea polyphenols after multiple-dose administration of epigallocatechin gallate and Polyphenon E in healthy individuals. *Clinical Cancer Research* 9: 3312–3319.

[83] Chow H, Cai Y, Alberts D, Hakim I, Dorrl R, et al. (2001) Phase I pharmacokinetic study of tea polyphenols following single-dose administration of epigallocatechin gallate and Polyphenon E. *Cancer Epidemiology, Biomarkers and Prevention* 10: 53–58.

[84] Asres K, Seyoum A, Veersham C, Bucar F, Gibbons S (2005) Naturally derived anti-HIV agents. *Phytotherapy Research* 19: 557–581.

[85] Fujiki H, Suganuma M, Imai K, Nakachi K (2002) Green tea: cancer preventive beverage and/or drug. *Cancer Letters* 188: 9–13.

[86] Fujiki H, Suganuma M, Okabe S, Sueoka E, Sueoka N, et al. (2001) Cancer prevention with green tea and monitoring by a new biomarker, hnRNP B1. *Mutation Research* 480–481: 299–304.

[87] Lambert JD, Yang CS (2003) Mechanisms of cancer prevention by tea constituents. *Journal of Nutrition* 133: 3262S–3267S.

[88] Clark J, You M (2006) Chemoprevention of lung cancer by tea. *Molecular Nutrition & Food Research* 50: 144–151.

[89] McLarty J, Bigelow R, Smith M, Elmajian D, Ankem M, et al. (2009) Tea polyphenols decrease serum levels of prostate-specific antigen, hepatocyte growth factor, and vascular endothelial growth factor in prostate cancer patients and inhibit production of hepatocyte growth factor and vascular endothelial growth factor *in vitro*. *Cancer Prevention Research* 2: 673–682.

[90] Chang P, Mirsalis J, Riccio E, Bakke J, Lee P, et al. (2003) Genotoxicity and toxicity of the potential cancer-preventive agent Polyphenon E. *Environmental and Molecular Mutagenesis* 41: 43–54.

[91] Graff S (2009) Natural Products: Research probes anticancer mechanisms of Polyphenon E. *Journal of the National Cancer Institute* 101: 627–628.

[92] Lee M, Wang Z, Li H, Chen L, Sun Y, et al. (1995) Analysis of plasma and urinary tea Polyphenols in human subjects. *Cancer Epidemiology, Biomarkers and Prevention* 4: 393–399.

[93] Chow HH, Hakim I, Vining DR, Crowell J, Cordova CA, et al. (2006) Effects of repeated green tea catechin administration on human cytochrome P450 activity. *Cancer Epidemiology, Biomarkers and Prevention* 15: 2473–2476.

[94] Donovan J, Chavin K, DeVane C, Taylor R, Wang J, et al. (2004) Green tea (*Camellia sinensis*) extract does not alter cytochrome P450 3A4 or 2D6 activity in healthy volunteers. *Drug Metabolism and Disposition* 32: 906–908.

[95] Chow HH, Hakim I, Vining DR, Crowell JA, Ranger-Moore J, et al. (2005) Effects of dosing condition on the oral bioavailability of green tea catechins after single-dose administration of Polyphenon E in healthy individuals. *Clinical Cancer Research* 11: 4627–4633.

[96] Bogdanski P, Suliburska J, Szulinska M, Stepien M, Pupek-Musialik D, et al. (2012) Green tea extract reduces blood pressure, inflammatory biomarkers, and oxidative stress and improves parameters associated with insulin resistance in obese, hypertensive patients. *Nutrition Research* 32: 421–427.

[97] Matsumoto K, Yamada H, Takuma N, Nino H, Sagesaka Y (2011) Effects of green tea catechins and theanine on preventing influenza infection among healthcare workers: a randomized controlled trial. *BMC Complementary and Alternative Medicine* 11: 1–7.

[98] Rowe CA, Nantz MP, Bukowski JF, Percival SS (2007) Specific formulation of *Camellia sinensis* prevents cold and flu symptoms and enhances $\gamma\delta$ T cell function: a randomized, double-blind, placebo-controlled study. *Journal of the American College of Nutrition* 26: 445–452.

[99] Li N, Sun Z, Han C, Chen J (1999) The chemopreventive effects of tea on human oral precancerous mucosa lesions. *Experimental Biology and Medicine* 220: 218–224.

[100] Tsao AS, Liu D, Martin J, Tang X-M, Lee JJ, et al. (2009) Phase II randomized, placebo-controlled trial of green tea extract in patients with high-risk oral premalignant lesions. *Cancer Prevention Research* 2: 931–941.

[101] Bettuzzi S, Brausi M, Rizzi F, Castagnetti G, Peracchia G, et al. (2006) Chemoprevention of human prostate cancer by oral administration of green tea catechins in volunteers with high-grade prostate intraepithelial neoplasia: a preliminary report from a one-year proof-of-principle study. *Cancer Research* 66: 1234–1240.

[102] Shanafelt T, Call T, Zent C, LaPlant B, Bowen D, et al. (2009) Phase I trial of daily oral Polyphenon E in patients with asymptomatic rai stage 0 to II chronic lymphocytic leukemia. *Journal of Clinical Oncology* 27: 3808–3814.

[103] Garcia FA, Cornelison T, Nuño T, Greenspan DL, Byron JW, et al. (2014) Results of a phase II randomized, double-blind, placebo-controlled trial of Polyphenon E in women with persistent high-risk HPV infection and low-grade cervical intraepithelial neoplasia. *Gynecologic Oncology* 132: 377–382.

[104] Butler LM, Wu AH (2011) Green and black tea in relation to gynecologic cancers. *Molecular Nutrition & Food Research* 55: 931–940.

[105] Cheng SJ, Wang DC, Hara Y (1998) Composition for Treating Condiloma Acuinta. Google Patents, US Patent 5,911.

[106] MedieGene AG (2006) *MedieGene AG Obtains Approval of Polyphenon E Ointment in the USA*. Medigene AG.

[107] Yang GY, Liao J, Kim K, Yurkow EJ, Yang CS (1998) Inhibition of growth and induction of apoptosis in human cancer cell lines by tea polyphenols. *Carcinogenesis* 19: 611–616.

[108] Rosl F, Das BC, Lengert M, Geletneky K, Zur Hausen H (1997) Antioxidant-induced changes of the AP-1 transcription complex are paralleled by a selective suppression of human papillomavirus transcription. *Journal of Virology* 71: 362–370.

[109] Ahmad N, Gupta S, Mukhtar H (2000) Green tea polyphenol epigallocatechin-3-gallate differentially modulates nuclear factor kB in cancer cells versus normal cells. *Archives of Biochemistry and Biophysics* 376: 338–346.

[110] Beltz L, Bayer D, Moss A, Simet I (2006) Mechanisms of cancer prevention by green and black tea polyphenols. *Anti-Cancer Agents in Medicinal Chemistry* 6: 389–406.

[111] Khan N, Afaq F, Saleem M, Ahmad N, Mukhtar H (2006) Targeting multiple signaling pathways by green tea polyphenol (–)-epigallocatechin-3-gallate. *Cancer Research* 66: 2500–2505.

[112] Tatti S, Swinehart JM, Thielert C, Tawfik H, Mescheder A, et al. (2008) Sinecatechins, a defined green tea extract, in the treatment of external anogenital warts: a randomized controlled trial. *Obstetrics and Gynecology* 111: 1371–1379.

[113] Stockfleth E, Beti H, Orasan R, Grigorian F, Mescheder A, et al. (2008) Topical Polyphenon E in the treatment of external genital and perianal warts: a randomized controlled trial. *British Journal of Dermatology* 158: 1329–1338.

[114] Gloro R, Hourmand-Ollivier I, Mosquet B, Mosquet L, Rousselot R, et al. (2005) Fulminant hepatitis during self-medication with hydroalcoholic extract of green tea. *European Journal of Gastroenterology and Hepatology* 17: 1135–1137.

[115] Bonkovsky HL (2006) Hepatotoxicity associated with supplements containing Chinese green tea (*Camellia sinensis*). *Annals of Internal Medicine* 144: 68–71.

[116] Zhang X, Zhang H, Tighiouart M, Lee JE, Shin HJ, et al. (2008) Synergistic inhibition of head and neck tumor growth by green tea (–)-epigallocatechin-3-gallate and EGFR tyrosine kinase inhibitor. *International Journal of Cancer* 123: 1005–1014.

[117] Stearns ME, Wang M (2011) Synergistic effects of the green tea extract epigallocatechin-3-gallate and taxane in eradication of malignant human prostate tumors. *Translational Oncology* 4: 147.

[118] Sartippour MR, Pietras R, Marquez-Garban DC, Chen H-W, Heber D, et al. (2006) The combination of green tea and tamoxifen is effective against breast cancer. *Carcinogenesis* 27: 2424–2433.

[119] Sun X, Huo X, Luo T, Li M, Yin Y, et al. (2011) The anticancer flavonoid chrysin induces the unfolded protein response in hepatoma cells. *Journal of Cellular and Molecular Medicine* 15: 2389–2398.

[120] Somers-Edgar TJ, Scandlyn MJ, Stuart EC, Le Nedelec MJ, Valentine SP, et al. (2008) The combination of epigallocatechin gallate and curcumin suppresses ERα-breast cancer cell growth *in vitro* and *in vivo*. *International Journal of Cancer* 122: 1966–1971.

[121] Khan N, Mukhtar H (2008) Multitargeted therapy of cancer by green tea polyphenols. *Cancer Letters* 269: 269–280.

[122] Behr M, Small P (1997) Inhibition of carcinogenesis by tea. *Nature* 389: 134–135.

[123] Williamson MP, McCormick TG, Nance CL, Shearer WT (2006) Epigallocatechin gallate, the main polyphenol in green tea, binds to the T-cell receptor, CD4: potential for HIV-1 therapy. *Journal of Allergy and Clinical Immunology* 118: 1369–1374.

[124] Jiang F, Chen W, Yi K, Wu Z, Si Y, et al. (2010) The evaluation of catechins that contain a galloyl moiety as potential HIV-1 integrase inhibitors. *Clinical Immunology* 137: 347–356.

5

NOVEL FORMULATIONS AND DRUG DELIVERY SYSTEMS FOR PHYTOTHERAPIES

Shengpeng Wang[1], Meiwan Chen[1], Qi (Tony) Zhou[2], and Hak-Kim Chan[2]

[1] *State Key Laboratory of Quality Research in Chinese Medicine, Institute of Chinese Medical Sciences, University of Macau, Taipa, Macau, China*
[2] *Faculty of Pharmacy, The University of Sydney, Sydney, New South Wales, Australia*

5.1 LIMITATIONS OF CONVENTIONAL FORMULATIONS FOR HERBAL MEDICINES

5.1.1 Barriers in Physicochemical and Biological Properties

The development path of herbal medicines from natural sources to clinical application faces numerous barriers: poor solubility, low permeability, extensive metabolism, and reduced target tissue uptake. Though many medicinal herbs are usually administered orally at relatively high doses, their plasma and tissue concentrations can be lower than the effective therapeutic doses due to poor bioavailability, resulting in only limited efficacy. Herbal medicines may contain various constituents of which the hydrophilic compounds cannot easily penetrate the intestinal membrane, whereas the hydrophobic compounds may not be soluble in the intestinal lumen. Although some compounds can be rapidly absorbed, their bioavailability may be compromised by the extensive metabolism in the intestine and liver (so-called gut and/or liver first-pass effect, respectively). For example, only 0.1% of epicatechin-3-gallate (EGCG) is bioavailable after intragastric administration of green tea [1] because EGCG is extensively glucuronidated [2].

Phytotherapies: Efficacy, Safety, and Regulation, First Edition. Edited by Iqbal Ramzan.

Chemical stability of certain herbal medicines, especially the volatile compounds, can also make the formulation design of phytotherapies challenging. For example, essential oils extracted from herbal medicines are a rich source of active ingredients (e.g., terpenoids and phenolic acids) [3]. However, when exposed to atmospheric pressure, light, heat, or oxygen, volatile compounds will easily evaporate and/or decompose.

5.1.2 Challenges in Quality and Safety Assurance

Quality and safety assurance remain a critical challenge for the development of herbal medicines. Current quality assurance criteria for herbal medicines depend mainly on herbal verification, chemical identification, and quantitative analysis of selected markers [4]. However, content of each active constituent could vary by the herb origin and batch. In many cases, the selected marker components may even be biologically inactive [5, 6]. Recently, DNA fingerprinting and bar-coding have been employed for precise authentication of herbs, but the data from these molecular approaches may not reflect their pharmacological or clinical efficacy [7]. The situation is more complicated in the multiherb formulae. In addition, there are external quality problems including contamination (e.g., heavy metals, pesticides, microbes, and mycotoxins), misidentification, and adulteration that also jeopardize the quality and safety of herbal medicines [8].

Generally, herbal medicines are deemed as "gentle medicines" because of their minor adverse effects. In China, development of herbal injections is a significant milestone for the modernization of traditional Chinese medicines (TCMs). TCM injection has become a common clinical practice in China and 136 TCM injectable products have been approved [9]. However, serious adverse events have frequently occurred in the last decade (e.g., aristolochic acid nephropathy) and have caused safety concerns [10]. Incomplete guidance of clinical preparation, incompatibility with other injections, and inappropriate quality control methods are considered to be the main reasons for the severe adverse effects [9, 11]. To address these side effect issues, current standards for quality control are unquestionably inadequate [12].

Due to the limited knowledge of chemical composition and bioactivity of each ingredient in herbs, reliable quality evaluation systems are highly in demand [7]. Chromatographic fingerprint technique is becoming an effective approach to assess the quality of TCM. It is widely accepted by the various regulatory authorities including the European Medicines Agency (EMA), US Food and Drug Administration (FDA), and World Health Organization (WHO) for evaluating chemical consistency between different batches of herbal medicines [13]. Recently, metabolomics has also attracted interests as a potential tool for quality control, metabolic phenotyping, and pharmacological and toxicological evaluation of herbal medicines [12].

5.1.3 Conventional Formulations Limit the Therapeutic Efficacy of Herbal Medicines

The route of administration for herbal medicines can significantly affect their therapeutic efficacy. Classically, herb drugs are administrated via oral (including tablets, powders, pellets, and decoction forms) and transdermal routes. However, even without considering the limited bioavailability, concentrations of some active ingredients in herbs can be low and thereby highly concentrated extracts are used with relatively high doses (up to 20–30 g). On the other hand, the active herbal constituents (e.g., cantharidin and celastrol) may also cause toxicity due to their unwanted nonselective effects on normal tissues. Therefore, it is difficult for the traditional formulations of herbal medicines to achieve the optimal efficacy within the therapeutic doses [14]. Furthermore, herbal medicines are more commonly used to treat chronic disorders (or for routine health care) than acute diseases [14]. Safety issues are more prominent for these long-term therapies.

Although herbal injections have been used in clinics and represent a revolutionary breakthrough for the administration of herbal medicines, the safety and dose rationale remain controversial. For example, germacrone is used as a marker for the quality control of the Zedoary turmeric oil injection. However, severe adverse effects have been frequently reported and are associated with the high concentration of Tween 80 (as solubilizer) and the poor stability of final preparation [15]. It is obvious that conventional delivery methods for herbal medicines are insufficient. Designing novel formulations by multidisciplinary approaches has become necessary to maximize the therapeutic efficacy of herbal medicines.

5.2 CRUCIAL ISSUES OF DEVELOPING NOVEL DELIVERY SYSTEMS FOR HERBAL MEDICINES

5.2.1 How Novel Delivery Systems Follow the Tradition?

Traditional herbal medicines exhibit multilevel, multitarget, and coordinated intervention effects [10]. While at present the delivery systems of herbal medicines mainly focus on single active ingredient, the reports on the total extracts of herbal medicines are scarce [16]. This is due to the extraordinarily complicated chemical composition of medicinal herbs. It is challenging to efficiently formulate these constituents with varying properties and retain the integrity of a herb or prescription [17].

Although challenging, it is very important to resolve the limitations of novel herbal nanomedicines via the oral route as it is the most popular delivery method for herbal medicines. For example, nanocarriers assembled using degradable polymers should avoid drug leakage in the gastrointestinal tract [18]. Meanwhile, the particle size determines the diffusion of nanoparticles through the mucus layer and the elimination by mucociliary clearance [19].

Novel formulations of herbal extracts have received numerous attention from pharmaceutical industry. Currently, there are several companies such as Cosmetochem International AG that specialize in formulating liposomal herbal extracts. Indena has developed a series of herbal products via Phytosome® technology. Absorption of active constituents in these novel formulations is enhanced when administrated transdermally and systemic bioavailability is improved for the oral dosage forms [20]. As an example, absorption of standardized, decaffeinated green tea catechins (GTC) extract in healthy human subjects was compared with the GCT produced by Phytosome. Volunteers were given a single dose of GTC (400 mg) for 6 h and the epigallocatechin-3-gallate (EGCG) levels in plasma were measured. The Phytosome formulation achieved EGCG peak plasma concentration of approximately 4.0 mM, which was significantly higher than that of the standardized extract (around 2.0 mM) [21, 22]. Meriva®, a patented complex of curcuminoids formulated with soy phosphatidylcholine, demonstrated superior bioavailability and stability [21, 23]. These findings shed new light on the formulation of herbal abstracts and warrant further investigation.

The purpose of using guiding herbs in TCM is to direct the primary components to the target sites [17]. This traditional concept coincides with the modern pharmaceutical theory of targeting delivery. Guiding herbs in TCM may be further developed for targeting delivery systems by the following strategies: (i) surface modification of nanocarriers by active constituents (e.g., glycyrrhizin) of guiding herbs as a "magic bullet" to achieve tissue targeting and (ii) co-delivering active constituents (e.g., curcumin) of guiding herbs along with other chemotherapeutic agents for reversing multidrug resistance. Combining the passive targeting by nanoparticles and active targeting by guiding herbs may achieve superior therapeutic efficacy.

5.2.2 Pharmacokinetic Research on Delivery Systems for Herbal Medicines

Novel delivery systems including nanomedicines could enhance the therapeutic efficacy of herbal medicines by altering their administration route, *in vivo* release and pharmacokinetic profile. Pharmacokinetic characteristics, including biodistribution and metabolic pathway, are crucial for the evaluation of therapeutic and toxic responses of nanodrugs [24]. Due to the unique pharmacokinetic features of nanoparticles and the complicated chemical composition of medicinal herbs, there are some factors that need to be considered to achieve a better understanding of their pharmacokinetic characteristics: (i) as herbal medicines are chemically and biologically complex system, the selection of a marker is a prerequisite issue. In this case, systems biology approaches (i.e., metabolomics) have been applied in identifying the therapeutic mechanisms of herbs and understanding the metabolic changes [25, 26]; (ii) molecular pharmacokinetics/pharmacodynamics (PK/PD) and physiologically based pharmacokinetics (PBPK) are also useful to measure the pharmacokinetic profile of drug delivery systems [27]; and (iii) the transport, metabolism, excretion, and immune response mechanisms of nanoparticles may differ significantly from those of traditional forms [28]. In future, advanced analytical methods along with the aid of mathematical and computational techniques may answer the question of "when, where, and what should be delivered," by assessing the *in vivo* pharmacokinetic parameters of active ingredients in herbal medicines and clarifying the drug–drug interactions within a herb or prescription.

5.2.3 Safety Considerations on Delivery Systems for Herbal Medicines

Human exposure to nanoparticles is becoming more frequent as a consequence of the rapid expansion of nanomedicines [29, 30]. The use of nanoparticles has raised safety concerns. Therefore, a new subdiscipline of nanotechnology termed "nanotoxicology," has emerged in recent years [31]. Herbal medicines are generally considered as safe based on the long-term observation in human subjects, but it is necessary to verify whether novel nanoparticle systems would enhance or reduce the toxicity of herbal medicines. For example, Doxil® (doxorubicin HCL liposome injection) can reduce cardiotoxic risk from doxorubicin. However, Palmar–Plantar skin reactions and mucositis become more frequent [32]. Inhalation of ultrafine inorganic particles (<100 nm) could also induce inflammation, oxidative stress, and distal organ involvement [33]. Moreover, the doses of traditional oral formulations are generally high due to poor bioavailability, thus the clinical doses of more efficient nanoparticle delivery systems would require recalibration. With the expanding usage of nanomedicines, toxicological studies are necessary to transition herbal nanomedicines into the clinical arena.

5.3 NOVEL DELIVERY SYSTEMS OF HERBAL MEDICINES

5.3.1 Pulmonary Delivery of Herbal Medicines

Inhaled phytotherapies have been used for many years for the treatments of respiratory diseases (i.e., asthma, respiratory infections, and tracheitis). Local delivery of herbal medicines may provide more rapid onset of drug action and higher bioavailability compared to traditional oral administration. Delivering herbal medicines directly to target sites in airways could offer a high local drug concentration without the first-pass effect, minimize systemic exposure, and reducing adverse effects [34]. Pulmonary drug delivery systems may also be employed for systemic herbal drugs because the large surface area, extensive vascularization, and thin alveolar epithelium of lungs can facilitate drug absorption into the blood circulation.

The first report of inhaled phytotherapy can be traced back to 2000–3000 B.C. when Indians inhaled the smoke of *Atropa belladonna* leaves for medical use. Nowadays, traditional inhalation phytotherapies by means of smoke, vapor, and steam are still popular in Africa, India, China, and some other Asian countries [35]. However, there is an urgent need to formulate these traditional inhaled therapies into reliable pharmaceutical products via modern formulation design so as to improve quality, safety, delivery efficiency, and patient adherence. Nebulization of injectable TCM solutions, including Shuang-Huang-Lian, Yu-Xing-Cao, and Qing-Kai-Ling, is now becoming common clinical practice in China for the treatments of respiratory infections and asthma but no product is specifically developed for inhalation. The use of injectable solution for inhalation therapy may cause toxicity because lung cytotoxicity and dose factors for the pulmonary administration are not appropriately evaluated. Han et al. demonstrated at that Shuang-Huang-Lian solutions at high concentrations could cause lung inflammation (>6 mg/kg) and alveolar fusion (>12 mg/kg) [36], while the clinical doses of reported nebulization therapies were 60–100 mg/kg [37]. Therefore, development of nebulized TCM products is necessary, instead of simply adopting the injection formulations and doses.

Pressurized metered dose inhalers (pMDIs) are currently the only approved inhalation dosage form for TCM in China, including Shuang-Huang-Lian MDI for respiratory infections and Ying-Huang-Ping-Chuan and Zhi-Chuan-Ling MDIs for asthma. In a pharmacokinetic study, the absolute bioavailability of Shuang-Huang-Lian MDI was determined to be 89%, and there were no significant differences in serum AUC_{0-6h}, C_{max}, and T_{max} values between the pMDI and intravenous formulations at a dose of 500 mg [38]. Clinical studies have shown the Shuang-Huang-Lian pMDI was more efficacious than the oral dosage form for the treatment of acute respiratory tract infections caused by *Staphylococcus aureus* [39]. However, because of the global banning of chlorofluorocarbon (CFC) use, the traditional pMDI propellant CFC has been replaced by the more environmental-friendly hydrofluoroalkanes (HFAs). It is a challenge to reformulate pMDIs of TCM since the compatibility between herbal active ingredients and HFA could be a significant issue with regard to formulation performance and stability. Dry powder inhalers (DPIs) may be an alternative dosage form to nebulization and pMDI.

DPIs consist of a powder formulation and an inhaler device. Dry powders of drugs is generally more stable than solutions or suspensions. In passive DPIs, powder aerosols are generated by patient's normal inspiratory action without using any propellant [40]. DPI devices are in portable sizes and convenient to carry and use. No DPI of TCM has yet been approved but there is increasing interest to develop TCM DPIs. Jet-milling is a common approach to micronize drugs into inhalable sizes, but the jet-milled powder is notoriously cohesive due to high particle surface energy [41]. Particle engineering via spray drying [42] or surface coating [43] has been utilized to reduce powder cohesion and improve aerosolization. Inhalable powders of Shuang-Huang-Lian were prepared by spray drying and formulated to carrier-based formulations [44]. The FPF total (<4.4 μm) of the Shuang-Huang-Lian powder through a Cyclohaler was significantly improved from 14.7 to 32.4% when the particle surface roughness was increased. Inhalable TCM particles were also produced for notoginseng [45] and *Trollius chinensis* [46] by spray drying. However, both chemical and physical stability of the spray-dried TCM powders have not been evaluated in these studies. As most spray-dried powders of small-molecule drugs are amorphous and could be physically unstable [47], stability information is critical to examine the suitability of spray drying as an approach to generate inhalable TCM particles.

Nanoparticles of TCM could be formulated into inhalation formulations. Besides the enhanced aqueous solubility and dissolution rate, nanoparticles could provide rapid drug absorption and reduced clearance from the lung [48]. There is a concern that nanoparticles may be exhaled during expiration due to the extremely small mass. A novel strategy of forming nano-aggregate microparticles may solve the problem [49]. Kwok et al. developed inhalable mannitol microparticles containing cyclosporine nanoparticles with enhanced dissolution [49]. This strategy can also be applied for TCM nanoparticles.

5.3.2 Nanocarriers of Herbal Medicines for Drug/Gene Delivery

Polysaccharides, which consist of repeating monosaccharide components, can be found in the natural sources of alginates, plants (e.g., pectin), microbes (e.g., xanthan gum), and animals (e.g., chitosan) [50]. Due to their favorable properties of biocompatibility, low cytotoxicity, ease of chemical modification, and nonimmunogenicity, polysaccharides are extensively studied for drug delivery and medical applications [51]. *Angelica sinensis* (Oliv.) Diels has been used for treating various gynecological diseases in China [52]. In a recent study, *Angelica sinensis* polysaccharide was chemically modified with a polyethylenimine (PEI, 1200 Da) as a gene vector for the delivery of plasmid encoding transforming growth factor beta 1 (TGF-β1) to the mesenchymal stem cells. The cationized polysaccharide from *Angelica sinensis* showed higher transfection efficiency and lower toxicity than Lipofectamine 2000 and PEI (25 kDa), indicating its potential for nonviral gene delivery [53]. Similar study was conducted on *Ganoderma lucidum* polysaccharide (GLP), a β-glucan from the Ganoderma lucidum (Reishi or Lingzhi), for antitumor effect [54].

Sanguis Draxonis, also known as Dragon's Blood or Resina Draconis, is a red resin processed from the extract of *Dracaena cochinchinensis* (Lour.) S.C. Chen (Agavaceae) [55]. Sanguis Draxonis has been used to improve blood circulation and also has potential for treating diabetes by enhancing insulin secretion. By trapping insulin into Sanguis Draxonis nanoparticles through a deposition method, the degradation of insulin by proteolytic enzymes in the gastrointestinal tract after oral administration was effectively minimized [56]. A hypoglycemic effect was also observed in rats with streptozotocin-induced diabetes [56]. With the proven safety of many herbs, herbal ingredients may be a promising source of nanocarriers in developing novel drug delivery systems.

5.3.3 Surface Modification of Nanocarriers by Herbal Medicines

Attaching active constituents of herbal medicines to the surface of nanoparticles could improve therapeutic efficacy by modifying nanoparticle properties [57]. For example, mannose-coated polymeric nanoparticles that can be specifically recognized by the mannose receptors were developed for targeting the immune system [58]. Modification of surface charge for polyisohexylcyanoacrylate nanoparticles by positively charged chitosan enables the attachment of negatively charged small interfering RNA (siRNA), which is promising for cancer therapy [59].

Some small-molecular-weight compounds from herbal medicines have showed potential in specific organ-targeted delivery. Glycyrrhetinic acid (GA) and glycyrrizin (GL) are the major active ingredients from *Glycyrrhiza uralensis* Fisch. GA and GL are used as ligands for liver targeting based on the specific binding sites/receptors of hepatocyte membrane [60, 61]. Several nanocarriers based on the GA-modified sulfated chitosan were synthesized [62]. The doxorubicin-targeted micelles produced with these modified chitosan exhibited significantly higher *in vitro* liver cytotoxicity against HepG2 cells than the unmodified formulation. A targeting effect of the modified micelles on liver cancer cells was also demonstrated [62].

5.3.4 Herbal Medicines as Photosensitizers for Photodynamic Therapy

Photodynamic therapy has been explored to improve cancer diagnosis and treatment [63]. Clinical studies revealed that photodynamic therapy is effective in several types of cancers, particularly in early stage cancers [64]. Traditional photosensitizers have several limitations such as poor selectivity, which drive researchers' attention to the naturally derived materials. Hypericin, exacted from *Hypericum perforatum*, has been examined recently as a natural photosensitizer [65]. Hypericin-loaded polylactic acid

(PLA) nanoparticles showed a higher photoactivity than the free hypericin on NuTu-19 ovarian cancer cells, which demonstrated the potential of using hypericin-loaded nanoparticles in photodynamic therapy [66].

5.4 SUMMARY

Formulation and drug delivery of herbal medicines is challenging since various components with different properties are present in a herb or herbal prescription. The active constituents from a herbal source generally have poor solubility, low permeability, poor pharmacokinetic parameters, and potential toxicity. Traditional formulations that are unable to overcome these problems will result in low therapeutic efficacy and unwanted cytotoxicity. Novel drug delivery systems including nanomedicines and local delivery can be leveraged to maximize the clinical outcomes and minimize the adverse effects of phytotherapies by targeting the specific sites or cells (Fig. 5.1). Further development of these novel herbal drug delivery systems is essential to progress them from laboratory to bedside.

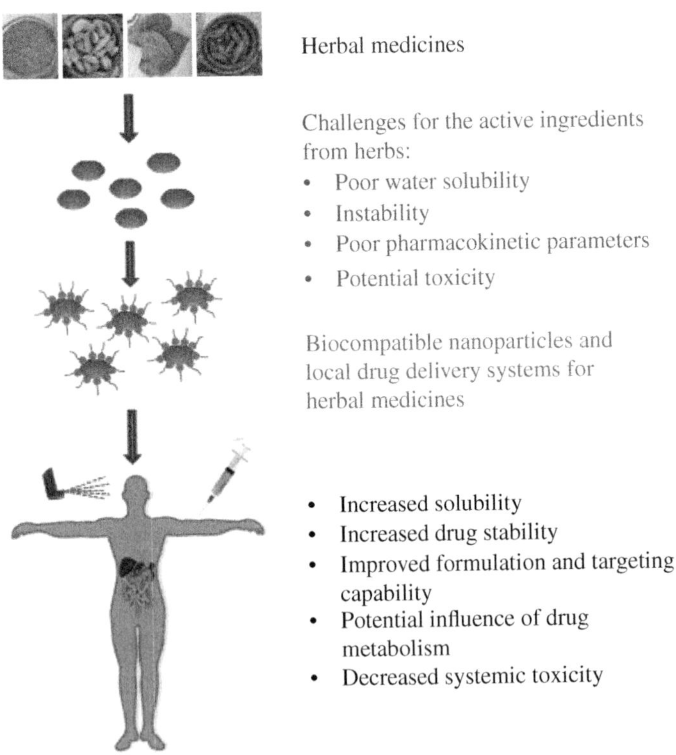

FIGURE 5.1 Developing novel delivery systems for herbal medicines (some illustrations used in this figure are courtesy of vectorolie, cooldesign, and digitalart, kindly provided and permitted by FreeDigitalPhotos.net).

REFERENCES

[1] Chen L, Lee MJ, Li H, Yang CS (1997) Absorption, distribution, elimination of tea polyphenols in rats. *Drug Metab Dispos* 25: 1045–1050.

[2] Lambert JD, Lee MJ, Lu H, Meng X, Hong JJ, et al. (2003) Epigallocatechin-3-gallate is absorbed but extensively glucuronidated following oral administration to mice. *J Nutr* 133: 4172–4177.

[3] Hosseini SF, Zandi M, Rezaei M, Farahmandghavi F (2013) Two-step method for encapsulation of oregano essential oil in chitosan nanoparticles: preparation, characterization and in vitro release study. *Carbohydr Polym* 95: 50–56.

[4] Ip SP, Zhao M, Xian Y, Chen M, Zong Y, et al. (2010) Quality assurance for Chinese herbal formulae: standardization of IBS-20, a 20-herb preparation. *Chin Med* 5: 8.

[5] Pallavi K, Kumaraswamy G, Shruthi (2011) Pharmacognostic investigation and antibacterial activity of Triticum aestivum. *J Pharm Res* 4: 3355–3359.

[6] Chen H, Li F, Jia JG, Diao YP, Li ZX, et al. (2010) Effects of traditional Chinese medicine on intestinal mucosal permeability in early phase of severe acute pancreatitis. *Chin Med J* 123: 1537–1542.

[7] Che CT, Wang ZJ, Chow MS, Lam CW (2013) Herb-herb combination for therapeutic enhancement and advancement: theory, practice and future perspectives. *Molecules* 18: 5125–5141.

[8] Zhang JH, Wider B, Shang HC, Li XM, Ernst E (2012) Quality of herbal medicines: challenges and solutions. *Complement Ther Med* 20: 100–106.

[9] Bian Z, Shang H, Cheng C, Wu T, Li Y, et al. (2010) Review of adverse reactions to injections of Chinese materia medica. *J Evid Based Med* 3: 88–94.

[10] Wang SP, Wu X, Tan M, Gong J, Tan W, et al. (2012) Fighting fire with fire: poisonous Chinese herbal medicine for cancer therapy. *J Ethnopharmacol* 140: 33–45.

[11] Cornelius VR, Perrio MJ, Shakir SA, Smith LA (2009) Systematic reviews of adverse effects of drug interventions: a survey of their conduct and reporting quality. *Pharmacoepidemiol Drug Saf* 18: 1223–1231.

[12] Wang L, Cui X, Cheng L, Yuan Q, Li T, et al. (2010) Adverse events to Houttuynia injection: a systematic review. *J Evid Based Med* 3: 168–176.

[13] Li X, Chen H, Jia W, Xie G (2013) A metabolomics-based strategy for the quality control of traditional Chinese medicine: shengmai injection as a case study. *Evid Based Complement Alternat Med* 2013: 836179.

[14] Kumar K, Rai AK (2012) Miraculous therapeutic effects of herbal drugs using novel drug delivery systems. *Int Res J Pharm* 3: 27–30.

[15] Zhao Y, Wang CG, Chow AHL, Ren K, Gong T, et al. (2010) Self-nanoemulsifying drug delivery system (SNEDDS) for oral delivery of Zedoary essential oil: formulation and bioavailability studies. *Int J Pharm* 383: 170–177.

[16] Li DC, Zhong XK, Zeng ZP, Jiang JG, Li L, et al. (2009) Application of targeted drug delivery system in Chinese medicine. *J Control Release* 138: 103–112.

[17] Gong J, Chen MW, Zheng Y, Wang SP, Wang YT (2012) Polymeric micelles drug delivery system in oncology. *J Control Release* 159: 312–323.

[18] Bernkop-Schnurch A (2013) Nanocarrier systems for oral drug delivery: do we really need them? *Eur J Pharm Sci* 49: 272–277.

[19] Primard C, Rochereau N, Luciani E, Genin C, Delair T, et al. (2010) Traffic of poly(lactic acid) nanoparticulate vaccine vehicle from intestinal mucus to sub-epithelial immune competent cells. *Biomaterials* 31: 6060–6068.

[20] Ajazuddin, Saraf S (2010) Applications of novel drug delivery system for herbal formulations. *Fitoterapia* 81: 680–689.

[21] Kidd PM (2009) Bioavailability and activity of phytosome complexes from botanical polyphenols: the silymarin, curcumin, green tea and grape seed extracts. *Altern Med Rev* 14: 226–246.

[22] Pietta P, Simonetti P, Gardana C, Brusamolino A, Morazzoni P, et al. (1998) Relationship between rate and extent of catechin absorption and plasma antioxidant status. *Biochem Mol Biol Int* 46: 895–903.

[23] Marczylo TH, Verschoyle RD, Cooke DN, Morazzoni P, Steward WP, et al. (2007) Comparison of systemic availability of curcumin with that of curcumin formulated with phosphatidylcholine. *Cancer Chemother Pharmacol* 60: 171–177.

[24] Liu CX, Si DY, Xiao XF, He X, Li YZ (2012) Drug metabolism and pharmacokinetics of nanodrugs from Chinese medicines and natural products. *Curr Drug Metab* 13: 659–666.

[25] Wang XJ, Sun H, Zhang AH, Sun WJ, Wang P, et al. (2011) Potential role of metabolomics approaches in the area of traditional Chinese medicine: as pillars of the bridge between Chinese and Western medicine. *J Pharm Biomed Anal* 55: 859–868.

[26] Wang SP, Hu YY, Tan W, Wu X, Chen RE, et al. (2012) Compatibility art of traditional Chinese medicine: from the perspective of herb pairs. *J Ethnopharmacol* 143: 412–423.

[27] Yamashita F, Hashida M (2013) Pharmacokinetic considerations for targeted drug delivery. *Adv Drug Deliv Rev* 65: 139–147.

[28] Li MG, Al-Jamal KT, Kostarelos K, Reineke J (2010) Physiologically based pharmacokinetic modeling of nanoparticles. *ACS Nano* 4: 6303–6317.

[29] Botelho MA, Martins JG, Ruela RS, Queiroz DB, Ruela WS (2010) Nanotechnology in ligature-induced periodontitis: protective effect of a doxycycline gel with nanopartícules. *J Appl Oral Sci* 18: 335–342.

[30] Gonzaga LW, Botelho MA, Queiroz DB, Fechine P, Freire R, et al. (2012) Nanotechnology in hormone replacement therapy: safe and efficacy of transdermal estriol and estradiol nanoparticles after 5 Years Follow-Up Study. *Lat Am J Pharm* 31: 442–450.

[31] Lewinski N, Colvin V, Drezek R (2008) Cytotoxicity of nanoparticles. *Small* 4: 26–49.

[32] Wang AZ, Langer R, Farokhzad OC (2012) Nanoparticle delivery of cancer drugs. *Annu Rev Med* 63: 185–198.

[33] Nel A, Xia T, Madler L, Li N (2006) Toxic potential of materials at the nanolevel. *Science* 311: 622–627.

[34] Zhou Q, Morton DAV, Yu HH, Jacob J, Wang JP, et al. (2013) Colistin powders with high aerosolisation efficiency for respiratory infection: preparation and *in vitro* evaluation. *J Pharm Sci* 102: 3736–3747.

[35] Mohagheghzadeh A, Faridi P, Shams-Ardakani M, Ghasemi Y (2006) Medicinal smokes. *J Ethnopharmacol* 108: 161–184.

[36] Han R, Ye J-X, Quan L-H, Liu C-Y, Yang M, et al. (2011) Evaluating pulmonary toxicity of Shuang–Huang–Lian *in vitro* and *in vivo*. *J Ethnopharmacol* 135: 522–529.

[37] Wang Q, Li AQ (2002) Comparative study on the clinical treatment of bronchiolitis using nebulized or intravenously delivered Shuang-Huang-Lian. *Herald Med* 21: 306–307.

[38] Xu K-J, Sun K-X, Lu Y-C (1992) Study on human bioavailability of the injection and the aerosol of Shuang Huang Lian. *Chin J Hosp Pharm* 12: 484–486.

[39] Wang Y-H, Xu K-J, Jiang W-S (1995) Experimental and clinical study of Shuanghuanglian aerosol in treating acute respiratory tract infection. *Chin J Integr Tradit West Med* 6: 347–350.

[40] de Boer AH, Chan HK, Price R (2012) A critical view on lactose-based drug formulation and device studies for dry powder inhalation: which are relevant and what interactions to expect? *Adv Drug Deliv Rev* 64: 257–274.

[41] Zhou Q, Morton DAV (2012) Drug-lactose binding aspects in adhesive mixtures: controlling performance in dry powder inhaler formulations by altering lactose carrier surfaces. *Adv Drug Deliv Rev* 64: 275–284.

[42] Chew NYK, Tang P, Chan HK, Raper JA (2005) How much particle surface corrugation is sufficient to improve aerosol performance of powders? *Pharm Res* 22: 148–152.

[43] Zhou QT, Qu L, Larson I, Stewart PJ, Morton DAV (2010) Improving aerosolization of drug powders by reducing powder intrinsic cohesion via a mechanical dry coating approach. *Int J Pharm* 394: 50–59.

[44] Yang JJ, Liu CY, Quan LH, Liao YH (2012) Preparation and in vitro aerosol performance of spray-dried Shuang-Huang-Lian corrugated particles in carrier-based dry powder inhalers. *AAPS PharmSciTech* 13: 816–825.

[45] Wang H-M, Ting-Ming F, Li-Wei G (2013) Preparation of *Panax notoginseng* saponins-tanshinone a composite method for pulmonary delivery with spray-drying mothod and its characterization. *Zhongguo Zhong Yao Za Zhi* 38: 559–563.

[46] Fan T, Zhu Y-F, Qing-Ri C, Cui J-H (2013) Preparation, *in vitro* evaluation of excipient-free dry powder inhalation of extraction of *Trollius chinensis*. *Zhongguo Zhong Yao Za Zhi* 38: 2096–2100.

[47] Vehring R (2008) Pharmaceutical particle engineering via spray drying. *Pharm Res* 25: 999–1022.

[48] Zhang J, Wu L, Chan H-K, Watanabe W (2011) Formation, characterization and fate of inhaled drug nanoparticles. *Adv Drug Deliv Rev* 63: 441–455.

[49] Yamasaki K, Kwok PC, Fukushige K, Prud'homme RK, Chan H-K (2011) Enhanced dissolution of inhalable cyclosporine nano-matrix particles with mannitol as matrix former. *Int J Pharm* 420: 34–42.

[50] Sinha VR, Kumria R (2001) Polysaccharides in colon-specific drug delivery. *Int J Pharm* 224: 19–38.

[51] Raemdonck K, Martens TF, Braeckmans K, Demeester J, De Smedt SC (2013) Polysaccharide-based nucleic acid nanoformulations. *Adv Drug Deliv Rev* 65: 1123–1147.

[52] Yi LZ, Liang YZ, Wu H, Yuan DL (2009) The analysis of radix angelicae Sinensis (Danggui). *J Chromatogr A* 1216: 1991–2001.

[53] Deng W, Fu M, Cao Y, Cao X, Wang M, et al. (2013) *Angelica sinensis* polysaccharide nanoparticles as novel non-viral carriers for gene delivery to mesenchymal stem cells. *Nanomedicine* 9: 1181–1191.

[54] Li N, Hu YL, He CX, Hu CJ, Zhou J, et al. (2010) Preparation, characterisation and anti-tumour activity of Ganoderma lucidum polysaccharide nanoparticles. *J Pharm Pharmacol* 62: 139–144.

[55] Xin N, Li YJ, Li Y, Dai RJ, Meng WW, et al. (2011) Dragon's Blood extract has anti-thrombotic properties, affecting platelet aggregation functions and anticoagulation activities. *J Ethnopharmacol* 135: 510–514.

[56] Hou ZQ, Zhang ZX, Zhang CX, Huang M (2004) Use of natural plant exudates (Sanguis Draxonis) for sustained oral insulin delivery with dramatic reduction of glycemic effects in diabetic rats. *J Control Release* 97: 467–475.

[57] Hirsjarvi S, Dufort S, Bastiat G, Saulnier P, Passirani C, et al. (2013) Surface modification of lipid nanocapsules with polysaccharides: from physicochemical characteristics to *in vivo* aspects. *Acta Biomater* 9: 6686–6693.

[58] Rieger J, Freichels H, Imberty A, Putaux JL, Delair T, et al. (2009) Polyester nanoparticles presenting mannose residues: toward the development of new vaccine delivery systems combining biodegradability and targeting properties. *Biomacromolecules* 10: 651–657.

[59] Pille JY, Li H, Blot E, Bertrand JR, Pritchard LL, et al. (2006) Intravenous delivery of anti-RhoA small interfering RNA loaded in nanoparticles of chitosan in mice: safety and efficacy in xenografted aggressive breast cancer. *Hum Gene Ther* 17: 1019–1026.

[60] Lin AH, Liu YM, Huang Y, Sun JB, Wu ZF, et al. (2008) Glycyrrhizin surface-modified chitosan nanoparticles for hepatocyte-targeted delivery. *Int J Pharm* 359: 247–253.

[61] Zhang CN, Wang W, Liu T, Wu YK, Guo H, et al. (2012) Doxorubicin-loaded glycyrrhetinic acid-modified alginate nanoparticles for liver tumor chemotherapy. *Biomaterials* 33: 2187–2196.

[62] Tian Q, Wang XH, Wang W, Zhang CN, Wang P, et al. (2012) Self-assembly and liver targeting of sulfated chitosan nanoparticles functionalized with glycyrrhetinic acid. *Nanomedicine* 8: 870–879.

[63] Dolmans DEJGJ, Fukumura D, Jain RK (2003) Photodynamic therapy for cancer. *Nat Rev Cancer* 3: 380–387.

[64] Agostinis P, Berg K, Cengel KA, Foster TH, Girotti AW, et al. (2011) Photodynamic therapy of cancer: an update. *CA-Cancer J Clin* 61: 250–281.

[65] Saw CLL, Heng PWS, Olivo M (2007) Potentiation of the photodynamic action of hypericin. *J Environ Pathol Toxicol Oncol* 27: 23–33.

[66] Zeisser-Labouebe M, Lange N, Gurny R, Delie F (2006) Hypericin-loaded nanoparticles for the photodynamic treatment of ovarian cancer. *Int J Pharm* 326: 174–181.

6

PHYTOTHERAPIES USED BY INDIGENOUS POPULATIONS

BRADLEY S. SIMPSON[1] AND SUSAN J. SEMPLE[2]

[1] *Flinders Centre for Innovation in Cancer, School of Medicine, Flinders University, Bedford Park, South Australia, Australia*
[2] *Sansom Institute for Health Research, School of Pharmacy and Medical Sciences, University of South Australia, Adelaide, South Australia, Australia*

6.1 INTRODUCTION

The advancements in modern Western medicine have provided innovative means of treating disease; however, there remain populations across the globe, particularly in developing countries and remote regions where plants are still relied on as sources of primary and/or complementary treatments for curing illnesses [1]. In this chapter, the use of plants in Indigenous medicine systems will be examined, in particular the medicinal use of plants by the Australian Aboriginal peoples.

Culturally, plants are inextricably linked with ancient (traditional) civilizations where they have served as symbolic icons in myth and religion, practically as tools and most importantly for medicinal use. The significance of plants as sources of medicines provides the basis for their continuity in maintaining the health and well-being of Indigenous populations. Not surprisingly, various cultures have developed independent philosophies pertaining to the use of plants as therapeutic interventions, yet in many instances, fundamental aspects of these medicinal systems may resonate with one another, for example, the combination of constituents from within the plant or plant part acting holistically in providing the efficacy of that medicine.

Ancient cultures including those of the hunter/gatherer have presumably suffered from both disease (including infectious diseases) and injury for thousands of years [2].

Phytotherapies: Efficacy, Safety, and Regulation, First Edition. Edited by Iqbal Ramzan.

It was therefore necessary that these populations developed systems that would enable them to treat and heal a variety of illnesses. A common means of achieving this appears to be via the concomitant use of plant, spiritual and psychological therapies. Traditional medicine is the more common term, used to collectively describe these "systems."

The World Health Organization defines traditional medicine as

> the sum total of the knowledge, skills and practices based on the theories, beliefs and experiences Indigenous to different cultures, whether explicable or not, used in the maintenance of health, as well as in the prevention, diagnosis, improvement or treatment of physical and mental illnesses [3].

It seems each major ancient civilization has developed a traditional medicine practice of some form or another including traditional Chinese medicine (TCM, China), Ayurvedic medicine (India), and Unani medicine (originating from Greece but practiced predominantly in South Asia and Middle Eastern countries). Furthermore, Indigenous peoples throughout the world have developed systems of medicine incorporating plant species native to their area, including the use of specific parts of these plants and methods of preparation.

According to the World Health Organization, traditional medicine use (including plants) by Indigenous populations in developing countries such as Africa ranges between 60 and 80% in order to meet primary healthcare needs [1]. For Indigenous communities in many regions of the world, there is still a significant dependence on plants and the associated medicinal knowledge for community survival. The Bolivian Andes are home to the Apillapampa people who belong to a large community living in a country where 30% of the population does not receive modern medical care. In a study of the medicinal plant use of these people, there was a strong, positive correlation between distance from the nearest village and the number of plants used for medicinal purposes [4]. Additionally, despite living only a short distance from a larger urban area, the neighboring Asháninka Native Community Bajo Quimiriki, District Pichanaki, Junín, Peru, is reliant on traditional plant knowledge and use in daily life (D. Claudie, 2006, Traditional use of Northern Kaanju Kuuku I'yu medicinal plants. Chairman—Chuulangun Aboriginal Corporation, personal communication).

Balick and Cox [5] analyzed the types of ailments for which Indigenous traditional medicine was used in 15 geographical areas. They found the greatest emphasis upon the treatment of easily detectable symptoms, including gastrointestinal complaints, inflammation, fever, skin ailments, and gynecological disorders. Less emphasis was placed on the treatment of illness with poorly defined symptoms such as cardiovascular disease and cancer.

By virtue of geographical location, some Indigenous communities are at greater risk of contracting particular diseases. In Bangladesh, mosquito-infested forests are a breeding ground for malaria, with tribes inhabiting the Khagrachari, Bandarban, Tangail, and Mymensingh districts relying heavily on plants for treating the disease [6].

Patterns also appear where specific plant families are commonly used throughout medicinal systems of different Indigenous populations. Interestingly, some recent analyses conducted in geographically separated regions that have distinct flora including New Zealand, South Africa, and Nepal have indicated that traditional use of plants is often clustered toward certain taxonomic groups [7, 8] despite little chance for cross-cultural exchange, suggesting the likelihood of independent discovery. Several commonly used plant families have been identified, which include Asteraceae (daisy flower family), Convolvulaceae (the "morning glory" family), Cucurbitaceae (gourd family), Euphorbiaceae (spurge family), Lamiaceae (mint family), Malvaceae (hibiscus or mallow family), and Solanaceae (nightshade family) [7]. In more recent times, the adoption of sedentary and nontraditional lifestyles in some Indigenous communities is influencing the prevalence of some diseases. For example, in Australia, New Zealand, and North America, type 2 diabetes and related conditions have been identified as priority areas for research and health services among Indigenous populations [9]. In the Cree Nation of Eeyou Istchee (Quebec, Canada) [10], collaborative research between Cree people and researchers from the University of Montreal are tackling this issue by investigating plants traditionally used for treating symptoms of diabetes (e.g., *Larix laricina*) as a means for developing alternative therapeutic approaches and integrating them into health clinics to treat obesity and diabetes among these people [11].

While it is well accepted that plants have served as primary sources of traditional medicinal therapies, the question is, why have plants been so popular and successful in being able to treat a wide range of ailments? The answer to this may vary depending on which perspective is taken. However, from a biochemical viewpoint, plants contain the same molecular building blocks found in all life forms (i.e., nucleic acids, amino acids, carbohydrates, lipids). The metabolism of these gives rise to chemicals known as secondary metabolites. Some of these compounds are either structurally similar to biomolecules found within the human body or interact with proteins in the human body that are structural analogous but functionally distinct to plant biosynthetic enzymes [12]. A comparative example to support this is phytosterols and sterols (e.g., cholesterol) found in humans [13]. Both contain a similar sterol backbone, with differences in the decoration of functional groups found on these molecules. Similarly linked to this concept is a plant's ability to defend itself against infectious pathogens (bacteria, viruses, fungus). Plants produce specific secondary metabolites, known as phytoalexins, which are synthesized on demand in defense against microorganism invasion [14]. Such compounds might be appropriately exploited for use in controlling similar microbial pathogens to which humans are susceptible. While phytotherapies may only form one aspect of a traditional medicine system, Indigenous populations have developed intricate means and ways for intelligently exploiting plants for the purpose of treating disease and injury.

6.2 PHYTOTHERAPIES OF INDIGENOUS AUSTRALIANS

6.2.1 Introduction

There is a wealth of information published in the literature that covers systems of TCM, Ayurveda, and the like. In contrast, the published content and knowledge of traditional Australian Aboriginal medicine plant use is less complete and rather fragmented, being scattered across scientific journals, government reports, reference books, and so on. It has been suggested that both J.H. Maiden's book entitled *The Useful Native Plants of Australia and More Recently Traditional Aboriginal Medicines in the Northern Territory of Australia* and a related publication by the Aboriginal Communities of the Northern Territory contain the most integrated collection of work on this subject [15]. More recently, an initiative called Customary Medicinal Knowledgebase (CMKb) operated by Macquarie University is aiding in defragmenting and creating a centralized repository of Australian Aboriginal medicinal plant knowledge [16].

In keeping with the theme of this book, the remainder of the chapter is reserved for exploring phytotherapies used by Australian Indigenous peoples. Key aspects of traditional Australian Aboriginal medicine philosophy, medicine preparation, genera of plants that are particularly prized in Australian Indigenous medicine, efficacy, and safety will be highlighted. The authors acknowledge that the information may be limited in its representation of specific clans of Australian Aboriginal people and should therefore be taken as a general overview.

6.2.2 Philosophy and Knowledge Transmission

Australian Indigenous peoples form one of, if not the oldest continuous surviving cultures in the world, having migrated from Southeast Asia some 40,000–50,000 years ago [17]. During this time, each community (population) has developed its own unique knowledge related to the medicinal flora growing on local homelands, meaning there is no one Aboriginal pharmacopoeia [18]. Development of such knowledge has been achieved empirically through observation, interpretation, trial and error, serendipity, and experimentation. The discovery and development of such traditional medicine knowledge incorporate similar mechanisms as used in contemporary medicine research [19]:

> ...you look at the animals to see what plants they eat. If you see the animal get sick, then you know you that plant is no good. (D. Claudie, 2012, Traditional use of Northern Kaanju Kuuku I'yu medicinal plants. Simpson B, personal communication)

With the breakup of traditional societies, knowledge about the medicinal use of plants is one aspect of culture that is lost [20]. The arrival of Europeans in Australia resulted in the disruption of Aboriginal society and traditional ways of life. In some areas, Indigenous people were forbidden to carry out traditional practices or to collect traditional foods and medicinal plants [21, 22], resulting in the loss of invaluable knowledge [23]. Traditional medical practices, however, still play an important role in many areas of Australia today.

Australian Indigenous people do not have a traditional written language, meaning that intergenerational transfer of traditional knowledge is preserved through oral means, including stories, songs, and dances [24]. The extensive knowledge that the Australian Aboriginal peoples had of their physical environment, including its plants and animals, is reflected in the large number of very specific terms for these in the various Aboriginal languages [25]. Commonly used Aboriginal terms for plants may differentiate types of plants at a level of sophistication similar to a botanist or biologist using post-Linnean classification of European languages [25, 26].

Similar to some other Indigenous cultures [27], the oral tradition on which Australian Aboriginal medicine is based presents a great challenge for Australian Indigenous populations as many of the holders of the knowledge are elderly and often pass on before the wealth of information can be captured. While written material recorded since European settlement can preserve some aspects of knowledge and culture, often these can fail to capture the complexity and context of the knowledge and the depth of meaning [28]. However, many Aboriginal people agree that it is important for the information to be recorded while the holders of knowledge are alive to avoid further decline of information relating to "old-time" plants [23]. To circumvent this, involvement of children from a very young age can play a pivotal role in ensuring they become connected with and accustomed to the medicinal plant knowledge and cultural practices of their community [19]. In fact, children are exposed to phytotherapies from essentially the moment they are born. In a unique act of preventative medicine practice, infants, just born into the world, are initiated in a smoking ceremony. The smoking ceremony for new babies, involving women only (mothers and grandmothers), holds great ritual significance in many Aboriginal communities as a means of strengthening and protecting the mother and newborn both in the short term and over their lifetime. It is also believed that the act can stimulate breast milk, reduce bleeding, and fight off infection [29]. The nature of the ritual involves digging a fire pit (hole) and establishing a hot fire (using hardwood) to produce hot coals [26]. The leaves from several species including *Acacia aneura*, *Acacia lysiphloia*, *Eremophila longifolia*, *Eulalia aurea*, *Exocarpos latifolius*, and *Dodonaea viscosa* may be used for burning to produce a continuous flow of smoke [19]. One might speculate that species high in essential oils might be used, as many essential oils are unsaturated (contain carbon–carbon double bonds) and would undergo incomplete combustion to produce soot. A recent study examining the medicinal species *E. longifolia* found that partially pyrolyzed leaf oils (produced under similar conditions to burning the leaves in a fire) showed enhanced antimicrobial and antioxidant activities compared to nonpyrolyzed oils [30]. Interestingly, not all species used in smoking ceremonies are particularly rich in essential oils. These species do however contain other components including di- and triterpenes, tannins, and saponins, which may contribute to smoke production [19]. Once an appropriate amount of smoke is being liberated, the baby is held over the smoke for several minutes and the mothers sit or lie over the pit allowing the smoke to smother their body. This ritual is repeated over the course of several days (Fig. 6.1) [19].

Pearn's [19] description of children living in some of Australia's Aboriginal communities suggests they are exposed to other aspects of traditional medicine practices as philosophy of traditional healings is not simply restricted to the use of plants. The spiritual health of individuals is recognized as a major contributor to the overall health of the individual. In certain parts of Australia, children are appointed as apprentice healers, where old folk healers (known as nungungi in some Central Australian groups) educate them in the role of becoming a traditional community

FIGURE 6.1 *Acacia aneura*. Reproduced with permission from Martin O'Leary.

healer, acquiring knowledge that is distinct from ethnobotanical knowledge. This recognition is a special honor as few children may display the attributes and powers required to become senior healers [19]. It is considered a specialized role in learning precious knowledge, which is different from that of the general knowledge, held by all members. These individuals will become holders of information and come to learn who they can and cannot pass such knowledge to. From an ethnobotanical perspective, they go on to hold a high position in the community where only they have the authority to collect plant material and tread sacred land, which is closely linked with the spiritual and cosmic beliefs of those people. In some Central Australian communities, children may also play an important role in both the collection of plant material and the capture of food [31].

6.2.3 Ailments Treated with Medicinal Plants

Traditional Australian Aboriginal concepts about disease have varied between groups. In general terms, there was a belief that less serious ailments, with external symptoms such as colds (running nose, sore throat), skin infections, eye infections, joint pain, and fever, were able to be treated with medicines. In contrast, more serious internal illnesses were considered to be of spiritual or supernatural origin [26, 32–35]. Hence, for the treatment of internal illness, spiritual matters were first attended to, and then relief of symptoms could be obtained by the use of traditional medicines such as plant remedies [26].

In many communities, being sick (unhealthy) does not simply refer to an imbalance in physiological well-being. For example, in the Wik-Mungkan community of Cape York Peninsula, Queensland, Australia, sickness encapsulates the disparity in an individual's body, spirit (soul), environment, and society [36]. Hence, a holistic approach is required in order to bring about a cure. This contrasts with the Western biomedical model that incorporates a reductionist approach, focusing on defined causes, treating specific tissues using logically derived treatments without addressing an individual's psychosocial state and how it relates to their illness [18].

Plants have been used by the Australian Aboriginal peoples for a variety of conditions. Records of Aboriginal plant use show that a high proportion of plant remedies have been used for the management of gastrointestinal complaints, respiratory illness (including the common cold), headache, fever, joint and muscle pain, skin ailments, wounds, eye and ear infections, and childbirth [15, 26, 35, 37, 38].

Hunter/gatherer civilizations were potentially susceptible to different injury types due to their dynamic lifestyle, more likely suffering burns, insect/animal bites, bruises, broken bones, and rashes as opposed to highly infectious diseases and diet-related disorders, which are akin to the modern era [39]. Therefore, different treatment regimes are apparent when comparing ancient and modern civilizations. Of course, with the influence of Western lifestyle on Indigenous populations of today, Indigenous people are no longer able to entirely rely on traditional medicine alone (discussed later).

The physiological healing aspects of Indigenous peoples' phytotherapies form only one aspect of their mechanism of action. The psychological and spiritual belief that a plant "works" appears to be as powerful as the constituents that might contribute to the efficacy. To demonstrate this point, a certain plant used by Northern Kaanju Kuuk I'yu people (Cape York Peninsula) with whom the authors have worked is said that it

> will not work without the psychological part. (D. Claudie, 2008, Traditional use of Northern Kaanju medicinal plants. Semple S, personal communication)

Therefore, one must believe that the treatment they are receiving will help them if it is to be of any benefit.

6.2.4 How Plant Medicines Have Been Used

6.2.4.1 Internal versus External Use It has been estimated that less than 10% of traditional Australian Aboriginal medicines used in Central Australia have been taken internally [26]. This may reflect the difficulty of accurate dosing without measuring devices [38]. An internal medicine requires more accurate dosing than an externally applied remedy in order to produce the desired effect without unwanted toxicity. Additionally, there may be quite large variations in the amounts of biologically active secondary compounds in the plant material. Plants growing in different areas, under different growing conditions, or different forms of the same plant may produce different quantities of certain secondary compounds.

Although a number of Aboriginal remedies have been used as applications to the skin, these remedies could still produce systemic effects. Some compounds have the ability to traverse the layers of skin and enter blood vessels and then the general circulation. Transdermal drug delivery is used to administer a number of drugs used in Western medicine [40]. These include steroid hormones, alkaloids such as nicotine and scopolamine, and nitrate compounds such as nitroglycerin.

In traditional Australian Aboriginal medicine, crude aqueous extracts of plant material have been applied as washes to large areas of the skin. Hydration of the skin and the presence of detergent-type compounds are known to facilitate the absorption of some compounds from the skin [41]. Some Aboriginal medicinal plants have been used as applications to the head [26, 34, 37]. Areas of the head and neck are known to have a high rate of transdermal absorption of drugs [40].

6.2.4.2 Use of Single versus Multiple Plant Ingredients In contrast to traditional phytotherapies from some other systems like TCM where multiple plants (and plant parts) are commonly used together in the same medicine, Australian Indigenous medicine generally relies on the use of one plant at any one time. In rare circumstances, multiple parts of the same plant may be used together (e.g., roots and leaves) [19]. There are some examples in the literature of the use of combination plant therapies. The species *Scaevola spinescens* (Goodeniaceae) has been used medicinally by some groups of Australian Aboriginal people for a variety of ailments [15, 42]. It may be used in combination with another plant, *Codonocarpus cotinifolius* (Gyrostemonaceae), as a reputed cure for cancer [43, 44].

6.2.4.3 Seasonal and Other Factors Resulting in Variation in Plant Use In Aboriginal folklore, the use of a specific plant to treat a particular illness can often be restricted by the season. Aboriginal people have developed intricate seasonal calendars that do not coincide with seasons to which most Westerners are familiar with (i.e., summer, autumn, winter, and spring) [45]. The surrounding environment provides cues that tell them which part of the temporal cycle they are in. The appearance and behavior of particular animal species and the flowering pattern a certain plant displays contribute to providing signals for seasonal events and hence when a plant may be considered to be "pharmacologically active" or "inert." Further to this, the use of some plants may be dictated by the age or even the sex of the plant (for dioecious species). For example, the all-purpose medicinal plant *Brachychiton diversifolius* (Northern Kurrajong), which is used when there is no specific remedy for a diagnosed illness, requires strict use of leaves from juvenile plants as opposed to mature plants to provide its claimed benefits [19].

6.2.4.4 Other Cultural Uses Australian Indigenous phytotherapies are not necessarily restricted to the purpose of curing a physical ailment or disease. The use of the term "phytotherapy" may extend to the adoption of certain plant species used in the courtship of the opposite sex. Selected plants have a unique "perfume" or "scent" to them. Individuals would leave the plant material near the shelters of the person they are attracted to as a means of increasing the libido of that person. A *Pittosporum* species is used in this way by the rainforest people in the Bloomfield area, Queensland [46]. An alternative approach to this is using plants that are secretly placed on a fire to produce smoke, which, when breathed in by people within the vicinity of the fire, puts them to sleep (like a sleeping tablet), allowing the man and woman to go off together.

6.2.4.5 Food as Medicine It is now widely recognized that traditional foods play an important role in providing health benefits to Indigenous communities around the world [47]. While plants play an important role in Aboriginal medicine systems, the collection and use of traditional plant-based bush foods (mainly native fruits and vegetables) can also play an equally important role in the overall well-being on community members [48].

6.2.4.6 Use of Common Plants In some areas of Australia, a nomadic or seminomadic lifestyle was necessary due to limited food and water resources. This meant that many of the remedies were common plants that could be easily found when required, rather than having to be carried while travelling [38]. This was also reflected by multipurpose use of single botanical sources to treat a wide range of conditions. For example, the roots and leaves of the small shrub *Grewia retusifolia* (emu berry) have been used to treat a variety of disorders including diarrhea, headache, fatigue, boils, infected sores, and scabies. A decoction of the plant has also been known for the treatment of uveitis (eye inflammation) [19]. While on the move, Indigenous people learned to adapt and substitute one plant for another for the treatment of similar conditions, as they would often be on unfamiliar territory where commonly used medicinal plants may not grow. Most remedies were chiefly made from scratch using raw material with very few medicines prepared and stored for later use. An exception to this includes some highly regarded medicinal plant species, for example, *Eremophila alternifolia*, which were dried, kept, and carried for future use [37, 38].

6.2.5 Methods of Plant Preparation

Depending on the intended application, plants for medicinal purposes may be prepared (extracted) using a variety of methods. These include infusion, decoction, maceration, direct application (which may require mechanical crushing), roasting, and emollient preparation (emu fat used as an excipient or carrier vehicle) [19]. One or more methods may be used in conjunction with one another to produce a medicinal extract. The following provides a snapshot of methods described in the literature [26, 37] as well as our own observations in working with traditional healers:

Infusion—Similar to the method when preparing a cup of tea, plant material (leaves, roots, bark, flowers, and fruits) is allowed to infuse through in hot water.

Decoction—The above method may be modified for a more complete extraction whereby plant material is boiled for a short period in water to give a decoction. The boiling aids in the breaking down of plant fibers and damages cell walls, releasing an array of pharmacological agents different from that during an infusion process. Some decoctions are used as internal mixtures or as body washes. Steam from a hot water extract may also be inhaled for respiratory conditions (Fig. 6.2) [37].

Maceration—This is a gentle technique where plant material seeps into cold water for an extended period of time, preserving labile constituents such as fragrances and complex molecules like tannins.

Direct crushing (application)—As the term suggests, botanical material is crushed in the absence of solvent (i.e., water) to liberate active chemicals. This is typical of a plant that is used in the oral cavity, with crushing performed by teeth and applied to the affected area. Additionally, substances like sap or saponin-rich lathers from plant material may be directly applied to the skin. Leaves of some plant species may be crushed and the vapors inhaled directly [15, 37, 38]. Crushed plant material may also be inserted into the nasal cavity as an inhalation or dried material inhaled like a snuff [22, 26, 37, 38].

Roasting—This involves "cooking" of plant material, for example, bark, which is ground to a powder and incorporated with an excipient (emu fat) and applied to the affected area (e.g., skin burns).

Emollient—Incorporating the above five methods of extraction, animal fat may be used as an excipient (particularly for topical use on the skin), in a similar manner that creams and gels are used for conventional medicines as a means of delivering (for absorption of) active constituents.

Smoking—As described earlier, smoking ceremony in which plant material is burned to release fragrant smoke is another important method for medicinal plant preparation [26].

Traditional medicine knowledge and practice is not static, but appears to evolve as is the case with modern-day scientific knowledge. That is, the more advanced one becomes, the more refined is the technology. In "old times," equipment used for preparing water extracts would have come in the form of wooden bowls or large shells, perhaps using hot stones to heat water [22, 23, 38]. Since European settlement, metal containers have replaced wooden bowls, which can be used to boil water on hot coals. More recently, some Indigenous groups have incorporated their traditional medicines into different formulations such as soaps and creams, which may be sold as part of small-scale business enterprises [49]. These examples illustrate the evolving use of new technology in Australian Aboriginal medicinal practices.

It should come as no surprise that the constituents extracted using the methods described earlier are predominantly water soluble. Alcohol or other organic solvents have never played a role in traditional Aboriginal medicine preparations. It is well understood that many potentially active components may be more lipophilic (fat soluble), and therefore, a limitation of water-based extraction methods is that

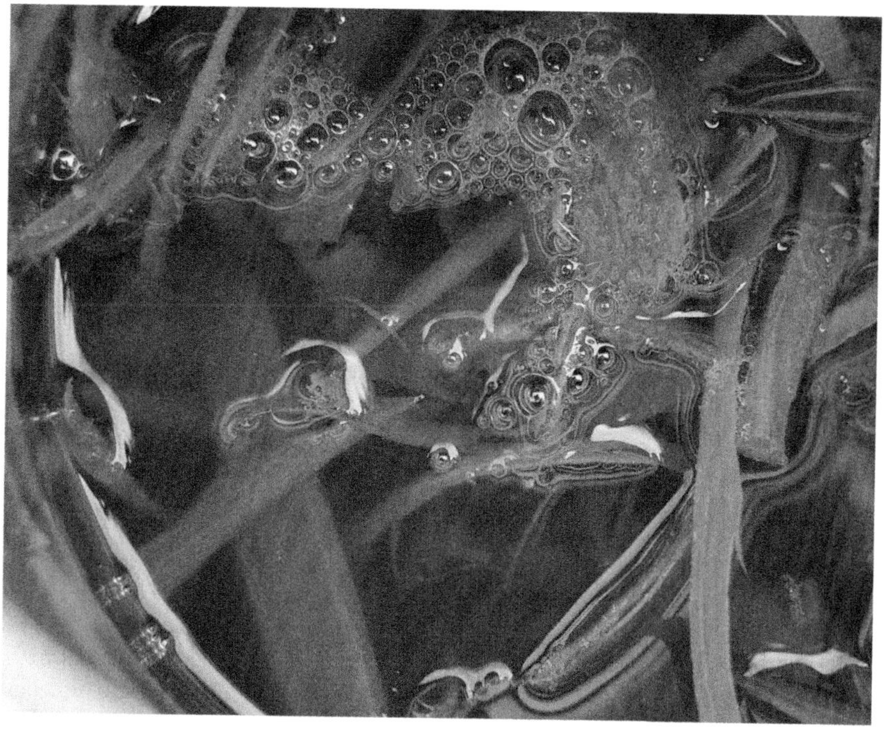

FIGURE 6.2 Boiling a decoction. Reproduced with permission from Chuulangun Aboriginal Corporation.

these compounds may never find their way into the preparation. However, other constituents in the plant may enhance the capacity of more fat-soluble components to dissolve in water.

In contrast, many plants produce resin, latex (a milklike substance), and sap. These contain constituents that may be fat soluble. These raw plant materials are applied directly to the skin and mucosal surfaces to promote absorption of these substances.

6.2.6 Prized and Commonly Used Plants in Australian Indigenous Medicine

So far, this chapter has focused on the philosophy and mechanics of Australian Indigenous phytotherapies, highlighting some plant species to exemplify the points raised. A number of literature sourced, although reasonably fragmented, describe a spectrum of plant species and their uses in traditional Aboriginal medicine. To avoid reiterating the freely available information, this section includes a discussion of some plant families and genera that are considered highly valuable by particular Indigenous communities or that are widely used and documented in a number of ethnobotanical reports. These illustrate some of the common ways that plants are used in Australian Indigenous medicine.

6.2.6.1 Eremophila Arguably, one of the most highly prized genera used by Australian Indigenous peoples is the genus *Eremophila* (Scrophulariaceae). The genus consists of more than 200 species, all of which are endemic to Australia and a number of which have recorded traditional uses. The origin of the name is from the Greek words *eremos* (desert) and *phileo* (love), that is, "desert loving." Species may be found growing over the vast majority of the Australian continent but predominantly in the semiarid to arid regions [50]. The leaves of this genus are highly resinous and are the dominant part of the plant used for medicinal purposes including colds, flu, internal pain, congestion, sores, cuts, scabies, toothache, headaches, diarrhea, promotion of sleep, and general well-being [51]. A number of chemical and pharmacological studies support the interesting phytochemistry and medicinal activities of plant species in this genus [52]. Aside from their significant uses as medicinal plants, some *Eremophila* species are present as a component in ceremonial rituals including initiations (Fig. 6.3) [50].

6.2.6.2 Acacia There are over 1000 species of the Australian acacias or "wattles" that are members of the subfamily Mimosoideae within the family Leguminosae. They range from small shrubs to trees and are one of the dominant tree types in the Australian landscape. It is recorded that at least 30 of these species have uses in traditional Australian Aboriginal medicine [53]. Different parts of the plants may be used including the leaves (or phyllodes), stems, bark, and roots. *Acacia* species have been used as aqueous extracts, smoke therapies, and direct applications. Aqueous decoctions of *Acacia* species such as *A. ligulata* and *A. lysiphloia* and *A. kempeana* have been used as cough medicines and as a body wash for colds and chest infections

FIGURE 6.3 *Eremophila duttonii.*

[26, 37, 54, 55]. Some *Acacia* species have played an important role in smoking ceremonies around childbirth. For example, the phyllodes of *A. lysiphloia* have been used to produce a fortifying smoke treatment for the mother and newborn [37]. *Acacia* species have also been used as direct applications. Phyllodes that are shaped like sharp spines, such those of *A. tetragonophylla* ("dead finish"), have been inserted directly into warts on the skin to assist in wart removal [37].

6.2.6.3 Eucalyptus and Melaleuca Another dominant plant family in the Australian landscape and in traditional medicine is Myrtaceae. This large family includes the genera *Eucalyptus* and *Melaleuca*. A number of species in these genera are medicinally important.

Melaleuca species have volatile oils in their leaves, and a number have been used as inhalation therapies for respiratory conditions [37]. The leaves may be crushed or smoked in a fire and the vapors inhaled. The bark of some *Melaleuca* species is also a valued medicine. *Melaleuca leucadendra*, commonly known as the weeping paperbark, has soft papery bark. Traditionally, an aqueous extract of the bark has been used as a body wash and as an internal mixture for fever and other associated symptoms [23, 37]

The leaves of *Eucalyptus* species are also rich in volatile oils, which may be used for inhalation or other therapy. The river red gum *Eucalyptus camaldulensis*, for example, has a variety of recorded uses [22, 23, 54]. A decoction of leaves or bark has been used as a wash for fever, headache, chest infections, aching joints, and cold and influenza symptoms, while it has been taken internally for coughs, nasal congestion, and sore throat. It also has recorded use as a smoke and steam bath therapy for general sickness, colds, influenza, and fever. *Eucalyptus* species are also a source of medicinal kinos. Kino is a deeply colored exudate rich in polyphenolic (tannin) compounds. *Eucalyptus* species exude kino from the trunk and branches. Due to their polyphenolic nature, the kinos of various *Eucalyptus* species have astringent properties and a variety of uses in traditional medicine including as an internal remedy for diarrhea and external remedy for sores, burns, and other skin afflictions [56].

6.2.6.4 Pterocaulon Certain genera in the "daisy *plant*" family Asteraceae are recorded as important medicinal plants for particular Indigenous groups. An example is the genus *Pterocaulon*.

Pterocaulon serrulatum is a favored medicinal plant, especially for colds [23, 37, 38, 54, 55]. A decoction of the strongly scented leaves has been used as a wash for skin sores and the symptoms of colds, chest infections, fever, and influenza and as a drink for headache, colds, and fever. A plug of crushed leaves inserted into the nasal cavity may be used for managing the symptoms of head colds and sore throat [23, 37, 38, 54, 55]. *Pterocaulon sphacelatum* is a similarly used species and a valued medicine that may be dried and stored for future use [23, 26, 37, 38].

6.2.6.5 Solanaceae Including* Nicotiana *and* Duboisia *Species The leaves from species belonging to the genus *Nicotiana*, including *N. benthamiana*, *N. cavicola*, *N. excelsior*, *N. gossei*, and *N. rosulata* subsp. *ingulba*, have been used throughout Australia, but particularly Central Australia as a narcotic agent [57]. These plants contain the pharmacologically active and addictive alkaloid nicotine, often coexisting with the structurally related analogue nornicotine. After mixing leaves with ash of burned wood, the subsequent preparation is chewed. The highly vascularized nature of the oral epithelium means nicotine enters the bloodstream without undergoing first-pass metabolism in the liver. This ethnomedicine commonly became known as "pituri." This term is also used to describe the plant *Duboisia hopwoodii*, for which a similar use is found [58]. Both *Nicotiana* spp. and *D. hopwoodii* were and continue to be of immense value in Aboriginal culture, for its psychological effects and social integration and as a commodity of trade between different Indigenous groups.

6.3 CHALLENGES OF A CHANGING ENVIRONMENT

Post-European settlement in Australia has resulted in many native plant species becoming rare and endangered, predominantly as a result of introduced pest animal and exotic plant species. Additionally, many Indigenous people have adopted a more Western-centric lifestyle, which has altered their plant-use practices, with these factors having consequences for the environment and intergenerational transfer of knowledge as a whole [45]. Innovative approaches to help prevent further medicinal plant knowledge from being irrevocably lost may include collaborative initiatives between Indigenous communities and academia [59, 60]. Such projects might be recognized as strategies for improving cultural, social, and potentially economical outcomes for Australian Indigenous people and their communities. It is well recognized that Australian Indigenous peoples' knowledge of natural materials is commercially attractive and valuable, with exploitation of such materials including bush foods, tools, and artwork being noteworthy examples. This is perhaps as a result of a softer regulatory framework surrounding the protection of Indigenous knowledge in the past. Inappropriate exploitation of medicinal products based on Indigenous knowledge continues to be of grave concern for Australian Aboriginal communities. Many Western-driven research and development schemes (e.g., pharmaceutical and cosmetic companies) continue to attempt to develop commercial products based on Australian native medicinal and food plants. A historical example includes the commercial development of a product derived from the Australian native—the "cork-wood" plant, *Duboisia myoporoides* (Solanaceae), traditionally used as a narcotic for the relief of stomach pain [57] from which the product Buscopan® is based on. The product contains the semisynthetic butyl bromide derivative of the active ingredient hyoscine (scopolamine), a tropane alkaloid. Cultivations of this plant in both Australia and South America serve as source of hyoscine [61]. A more recent example concerns the fruit of the Australian plant Kakadu plum (*Terminalia ferdinandiana*), which has been traditionally used by Northern Australian Indigenous communities for both food and medicine. The fruit is known to have an exceptionally high content of vitamin C. The US-based cosmetic company, Mary Kay, attempted to patent compositions containing extracts of the fruit for cosmetic purposes in a 2007 patent application. This caused considerable anxiety and anger among Indigenous communities, some of whom are involved in sustainable wild-harvest industries based on this plant [62].

Currently, changes to legislation in response to international initiatives including the Nagoya Protocol [63] are trying to establish greater certainty for industry and Indigenous communities around issues of access to resources that derives from such sharing. However, currently, most collaborative partnerships between Indigenous communities and Western scientific researchers rely heavily on investing time in building appropriate research partnerships and are left for small-scale academic operations to undertake. A challenge for academic-based research programs is that there are not always the resources to take a product through the entire product development pipeline. In particular, demonstrating the safety of a product, regardless of its historical use, is an area where a collaborating commercial industrial partner could contribute to a greater extent. Perhaps appropriate collaborative partnerships with industry where time is invested in face-to-face dialogue and appropriately understanding a plant's traditional use may help to overcome such issues.

6.3.1 Safety of Australian Phytotherapies

Australian Aboriginal plant therapies are predominantly used for external application followed by inhalation or use in the oral cavity, with only a minority being administered internally. For example, an all-over body ache might be treated using mashed-up leaves in water and the resulting liquid used as a body wash or the same plant may be used in a smoke treatment in which the entire body is fumigated by slow burning leaves. Species belonging to the genus *Eremophila* are commonly used in this way. Indigenous Australians have developed a profound detailed knowledge base of native Australian plants. Contained within this web of knowledge is an understanding of plant species that are safe to use and those that are not. Aboriginal people hold an immense toxicological understanding of their botanical diversity, and it forms a fundamental part of traditional knowledge. Toxicology is as important a factor as is the efficacy of a substance.

Knowing that many plants can be highly toxic may explain why internal use is often avoided in Australian Indigenous medicine. Under most circumstances, the actual "dose" or "quantity" of material used (e.g., infusion of leaves) may be arbitrary, where measurements might be described as "about a handful," "a few leaves," or "a branch will do" [23]. This also holds true for the volume of water that might be used during preparation of such a medicine; hence, the concentration also appears vague and subjective. The understanding of dose–response properties by Indigenous people ensures that highly toxic species are administered in appropriate doses to avoid unwarranted side effects [64].

Observing the interrelationship between animal behavior and plants has provided native Australians with clues to plants that are acceptable or those to avoid for use as food or medicines. Contrary to this, there are plants that exist within the pharmacopoeias of Australian Indigenous people that have selective toxicity. That is to say, animals may be safeguarded from consuming certain fruits and other plant parts from specific species; however, the same plants and plant parts (e.g., fruit) are toxic to humans. For example, in the Pilbara region of Western Australia, the red fruits of the snake vine plant (*Mukia maderaspatana*) are said to only be safe for emus to eat, while the whole plant can be boiled in water and liquid used to relieve sore eyes (note: *Tinospora smilacina* is also known as snake vine and used in a similar manner) [65]. In addition, the flannelbush (*Solanum lasiophyllum*) contains a round tomato-like fruit but is inedible and considered "kangaroo food" only [65].

Generally, frequency of treatment is not a strict requirement and is used quite loosely with a therapy being applied until symptoms subside. The terminal stem of the leaves from the medicinal plant *Dodonaea polyandra* is administered into the oral cavity as a treatment for toothache using a dosage frequency similar to that of paracetamol (i.e., fresh material replaced approximately every 4 h).

In the teaching of children about plants that are used for medicinal and cultural purposes, an emphasis is placed on learning how to distinguish plants that might look similar but have different properties [65]. This is to ensure that an error is not made of selecting the wrong plant, which could lead to negative consequences. While many medicinal plants might be considered safe to consume, there are circumstances where excessive use of particular phytotherapies is said to elicit undesirable effects. In the case of berries from the native plum plant *Psydrax latifolia*, overconsumption causes a burning sensation in the mouth [65].

6.3.2 Development and Regulation of Australian Indigenous Medicines

There is considerable potential for plants used by Indigenous communities to be developed as herbal or complementary medicine products for use in the wider community. However, the lack of a long written history as a source of evidence for efficacy has limited the development of Australian plants as medicinal products. The Australian medicine regulatory authority, the Therapeutic Goods Administration (TGA), regulates both complementary and conventional medicines. The Australian Regulatory Guidelines for Complementary Medicines [66] are currently undergoing reform to provide better transparency and confidence for the consumer of complementary medicine-based products. The reform is also looking to provide a more workable definition of a history of traditional use for sponsors looking to engage in the development of Australian Aboriginal medicinal plants. This involves recognition that oral histories concerning plant use may be used as a substitute to written historical records in order to support the claims of health benefits made for a product. This is provided that "evidence is obtained independently from multiple practitioners or members of Indigenous group(s) who maintain such a history" [67] and "collected by an ethnographic professional" [67]. Adoption of oral histories as evidence would overcome one of the major barriers to Indigenous medicines being able to be listed as therapeutic products in Australia. However, scientific evidence of safety will need to be provided in conjunction with oral evidence from traditional use, recognizing the issues around frequency of dosing and detection of delayed effects that may occur with traditional use. This evidence may be acquired in many ways with one such option being via collaborative research partnerships between Indigenous communities and academia [60]. This type of initiative can provide a platform and opportunity for Indigenous people to be at the forefront of development of their own resources into medicines within both Indigenous and Western contexts.

Importantly, such an approach may help to provide a vehicle for ensuring greater recognition of Australian Aboriginal medicine knowledge.

6.3.3 Integration of Traditional and Western Medicine in Indigenous Populations

In this contemporary era, Western medicine practice is the predominant form of primary healthcare for Australian Aboriginal people, although traditional systems of Aboriginal medicine still operate within communities today. The survival of this complex medicine system is a testament of the power within such an approach, which is entrenched in the culture of Aboriginal Australia. The philosophy of Australian Aboriginal medicine was covered earlier in this chapter. Extending on this in a modern-day context, Aboriginal people may choose to use Western medicine as a means of treating the majority of their medical conditions; however, explanation of the causes of sickness continues to be described in terms of traditional beliefs. This is evident in Aboriginal health clinics in Australia where patients have a range of health services on offer [18]. This includes Western biomedical services but also more traditional forms of healing (including spiritual forms). For many Indigenous people, the choice of treatment may be based on the cause of the disease. For example, if the health issues are of "natural" origin, they may opt for a conventional medication at a clinic or plant medicine remedy. Importantly, traditional bush medicines (phytotherapics) are offered in clinics, in particular those situated in remote locations [68]. In fact, there is a tendency for patients to feel more comfortable and at ease if a traditional phytotherapy treatment is available and used. In contrast, if the origin of illness is sorcery, no medicine whether modern or traditional will suffice. In such cases, the services of a traditional healer (known as ngangkari in the Northern Territory) will be requested and utilized [69]. Integrating traditional phytotherapies with Western biomedicine into health clinics is important in providing culturally appropriate healthcare to Indigenous Australian populations [18]. It ensures that their rights to access traditional forms of healing are respected and services synonymous with their culture are respected and upheld in the manner intended by their ancestors.

6.4 CONCLUSIONS

Plants have historically and still remain an important source of medication for many Indigenous populations around the world. Culturally, there appears to be a link between people and plants, regardless of origin, with different methods of medicine preparation having been empirically designed and adopted depending on the intended use. The survival of such populations, cultures, and traditions is a testament to the value and role plants have played throughout history. The modern-day paradox to this is that intergenerational transfer of cultural traditional knowledge is on the decline, which may ultimately affect the survival of fragile populations. This is disconcerting for cultures such as those akin to Australian Aboriginal peoples where transmission of knowledge is purely by oral means. This chapter has highlighted some of the most prized plants used as medicines by Australian Indigenous people and their methods of preparation. There still remains a huge scope for Western science to learn from Indigenous medicine. This presents a unique opportunity to work together with Indigenous communities to explore the potential of their plants for development into efficacious and safe medicines. One challenge to overcome here is working around the lack of written history as a source of evidence for efficacy. In Australia, suggested changes to the regulatory framework, in which oral history relating to plant use is more clearly defined and recognized, may facilitate novel medicinal products appearing on the market that are based on Australian Aboriginal traditional medicines.

REFERENCES

[1] World Health Organization (2002) *WHO Traditional Medicine Strategy 2002–2005*. Geneva: WHO.

[2] Wolfe ND, Dunavan CP, Diamond J (2007) Origins of major human infectious diseases. *Nature* 447: 279–283.

[3] World Health Organization (2000) *General Guidelines for Methodologies on Research and Evaluation of Traditional Medicine*. Geneva: WHO.

[4] Vandebroek I, Calewaert JB, De Jonckheere S, Sanca S, Semo L, Van Damme P, et al. (2004) Use of medicinal plants and pharmaceuticals by Indigenous communities in the Bolivian Andes and Amazon. *Bulletin of the World Health Organization* 82: 243–250.

[5] Balick MJ, Cox PA (1996) *Plants, People and Culture, the Science of Ethnobotany*. New York: Scientific American Library.

[6] Rahmatullah M, Hossan S, Khatun A, Seraj S, Jahan R (2012) Medicinal plants used by various tribes of Bangladesh for treatment of malaria. *Malaria Research and Treatment* 2012: 5.

[7] Saslis-Lagoudakis CH, Williamson EM, Savolainen V, Hawkins JA (2011) Cross-cultural comparison of three medicinal floras and implications for bioprospecting strategies. *Journal of Ethnopharmacology* 135: 476–487.

[8] Saslis-Lagoudakis C, Savolainen V, Williamson E, Forest F, Wagstaff S, Baral S, et al. (2012) Phylogenies reveal predictive power of traditional medicine in bioprospecting. *PNAS* 109: 15835–15840.

[9] McNamara BJ, Sanson-Fisher R, D'Este C, Eades S (2011) Type 2 diabetes in Indigenous populations: quality of intervention research over 20 years. *Preventive Medicine* 52: 3–9.

[10] Leduc C, Coonishish J, Haddad P, Cuerrier A (2006) Plants used by the Cree Nation of Eeyou Istchee (Quebec, Canada) for the treatment of diabetes: a novel approach in quantitative ethnobotany. *Journal of Ethnopharmacology* 105: 55–63.

[11] Harbilas D, Vallerand D, Brault A, Saleem A, Arnason J, Musallam L, et al. (2012) Larix laricina, an Antidiabetic alternative treatment from the Cree of Northern Quebec pharmacopoeia, decreases glycemia and improves insulin sensitivity *in vivo*. *Evidence-Based Complementary and Alternative Medicine* 2012: 10.

[12] Kellenberger E, Hofmann A, Quinn RJ (2011). Similar interactions of natural products with biosynthetic enzymes and therapeutic targets could explain why nature produces such a large proportion of existing drugs. *Natural Product Reports* 28: 1483–1492.

[13] Moghadasian MH (2000) Pharmacological properties of plant sterols: *in vivo* and *in vitro* observations. *Life Sciences* 67: 605–615.

[14] Ahuja I, Kissen R, Bones AM (2012) Phytoalexins in defense against pathogens. *Trends in Plant Science* 17: 73–90.

[15] Lassak EV, McCarthy T (1993). *Australian Medicinal Plants*. North Ryde: Methuen Australia.

[16] Gaikwad J, Khanna V, Vemulpad S, Jamie J, Kohen J, Ranganathan S (2008) CMKb: a web-based prototype for integrating Australian Aboriginal customary medicinal plant knowledge. *BMC Bioinformatics* 9(Suppl 12): S25.

[17] Rasmussen M, Guo X, Wang Y, Lohmueller KE, Rasmussen S, Albrechtsen A, et al. (2011) An Aboriginal Australian genome reveals separate human dispersals into Asia. *Science* 334: 94–98.

[18] Ralph-Flint J (2001) Cultural borrowing and sharing: Aboriginal bush medicine in practice. *Australian Journal of Holistic Nursing* 8: 43–46.

[19] Pearn J (2005) The world's longest surviving paediatric practices: some themes of Aboriginal medical ethnobotany in Australia. *Journal of Paediatrics and Child Health* 41: 284–290.

[20] Farnsworth NR (1966) Biological and phytochemical screening of plants. *Journal of Pharmaceutical Sciences* 55: 225–276.

[21] Burden J (1994) Health: a holistic approach. In: Bourke C, Bourke E, Edwards B, editors. *Aboriginal Australia*. St Lucia: University of Queensland Press, pp. 157–178.

[22] Zola N, Gott B (1992) *Koorie Plants Koorie People: Traditional Aboriginal Food, Fibre and Healing Plants of Victoria*. Melbourne: Koorie Heritage Trust.

[23] Smith N (1991) Ethnobotanical field notes from the Northern Territory, Australia. *Journal of the Adelaide Botanical Gardens* 14: 1–65.

[24] Locher C, Semple SJ, Simpson BS (2013) Traditional Australian Aboriginal medicinal plants: an untapped resource for novel therapeutic compounds? *Future Medicinal Chemistry* 5: 733–736.

[25] Thieberger N, McGregor W (1994) *Macquarie Aboriginal Words: A Dictionary of Words from Australian Aboriginal and Torres Strait Islander Languages*. Macquarie University: Macquarie Library.

[26] Latz P (1995) *Bushfires and Bushtucker: Aboriginal Plant Use in Central Australia*. Alice Springs: IAD Press.

[27] Stevens A (2008) A different way of knowing: tools and strategies for managing Indigenous knowledge. *Anglais* 58: 25–33.

[28] Berndt RM, Berndt CH (1988) *World of the First Australians: Aboriginal Traditional Life—Past and Present*. Canberra: Aboriginal Studies Place.

[29] Kildea S, Wardaguga M (2009) Childbirth in Australia: Aboriginal and Torres Strait Islander women. In: Selin H, editor. *Childbirth Across Cultures*. Dordrecht: Springer Netherlands, pp. 275–286.

[30] Sadgrove N, Jones G (2013) A possible role of partially pyrolysed essential oils in Australian Aboriginal traditional ceremonial and medicinal smoking applications of *Eremophila longifolia* (R. Br.) F. Muell (*Scrophulariaceae*). *Journal of Ethnopharmacology* 147: 638–644.

[31] Latz P (1999) *Pocket Buchtucker: A Field Guide to the Plants of Central Australia and their Traditional Uses*. Alice Springs: IAD Press.

[32] Cawte J (1996) *Healers of Arnhem Land*. Sydney: University of New South Wales Press.

[33] Cawte JE (1975) Australian Aboriginal medicine before European contact. *Annals of Internal Medicine* 82: 422–423.

[34] Peile AR (1997) *Body and Soul: An Aboriginal View*. Carlisle/Rossmoyne: Hesperian Press/Pallottines in Australia.

[35] Clarke P (2008) Aboriginal healing practices and Australian bush medicine. *Journal of the Anthropological Society of South Australia* 33: 3–38.

[36] Edwards S, Heinrich M (2005) Integrating Wik and Kugu (Australian Aboriginal) phytotherapeutic knowledge with Western science. *Revista de Fitoterapia* 5: 159–162.

[37] Barr A, Chapman J, Smith N, Wightman G, Knight T, Mills L, et al. (1993) *Traditional Aboriginal Medicines in the Northern Territory of Australia*. Darwin: Conservation Commission of the Northern Territory.

[38] Low T (1990) *Bush Medicine: A Pharmacopoeia of Natural Remedies*. North Ryde: Angus and Robertson.

[39] Milton K (2000) Hunter-gatherer diets—a different perspective. *The American Journal of Clinical Nutrition* 71: 665–667.

[40] Margetts L, Sawyer R (2007) Transdermal drug delivery: principles and opioid therapy. *Continuing Education in Anaesthesia, Critical Care & Pain* 7: 171–176.

[41] Prausnitz MR, Langer R (2008) Transdermal drug delivery. *Nature Biotechnology* 26: 1261–1268.

[42] Gott B (1992) *SAUSE Database. South Australian Plants Used by Aborigines*. Monash University, Melbourne: Department of Ecology and Evolutionary Biology.

[43] Reid EJ, Betts TJ (1979) Records of Western Australian plants used by Aboriginals as medicinal agents. *Planta Medica* 36: 164–173.

[44] Snow D. (1981) The sad saga of *Scaevola spinescens*. In: Snow D, Western Australia. Public Health Department, editors. *The Progress of Public Health in Western Australia 1829–1977*. Perth: Public Health Department, pp. 129–136.

[45] Clarke PA (2003) Australian ethnobotany: an overview. *Australian Aboriginal Studies* 2003(2): 21–38.

[46] Clarke PA (2007) *Aboriginal People and their Plants*. Sydney: Rosenberg Publishing Pty. Ltd.

[47] Pieroni A, Price L (2006) *Eating and Healing: Traditional Food as Medicine*. New York: Food Products Press.

[48] O'Dea K (1991) Traditional diet and food preferences of Australian Aboriginal hunter-gatherers. *Philosophical Transactions of the Royal Society B: Biological Sciences* 334: 233–241.

[49] Tjurma Bush Products Pty Ltd (2013) Tjurma Bush Products: Traditional and Contemporary health products. Semaphore, SA: Tjurma Bush Products. Available online at http://www.bushrub.com/ (Accessed November 17, 2014).

[50] Richmond GS (1993) A review of the use of *Eremophila* (Myoporaceae) by Australian Aborigines. *Journal of the Adelaide Botanical Gardens* 15: 101–107.

[51] Ghisalberti EL (1994) The ethnopharmacology and phytochemistry of *Eremophila* species (Myoporaceae). *Journal of Ethnopharmacology* 44: 1–9.

[52] Singab AN, Youssef FS, Ashour ML, Wink M (2013) The genus *Eremophila* (*Scrophulariaceae*): an ethnobotanical, biological and phytochemical review. *Journal of Pharmacy and Pharmacology* 65: 1239–1279.

[53] Wickens K, Pennacchio M (2002) A search for novel biologically active compounds in the phyllodes of *Acacia* species. *Conservation Science Western Australia* 4: 139–144.

[54] O'Connell JF, Latz PK, Barnett P (1983) Traditional and modern plant use among the Alyawara of central Australia. *Economic Botany* 37: 80–109.

[55] Webb LJ (1969) The use of plant medicines and poisons by Australian Aborigines. *Mankind* 7: 137–146.

[56] Locher C, Currie L (2010) Revisiting kinos—an Australian perspective. *Journal of Ethnopharmacology* 128: 259–267.

[57] Lassak EV, McCarthy T (2011) *Australian Medicinal Plants*. Chatswood: New Holland Publishers.

[58] Ratsch A, Steadman KJ, Bogossian F (2010) The pituri story: a review of the historical literature surrounding traditional Australian Aboriginal use of nicotine in Central Australia. *Journal of Ethnobiology and Ethnomedicine* 6: 1746–4269.

[59] Packer J, Brouwer N, Harrington D, Gaikwad J, Heron R, Yaegl Community Elders, et al. (2012) An ethnobotanical study of medicinal plants used by the Yaegl Aboriginal community in northern New South Wales, Australia. *Journal of Ethnopharmacology* 139: 244–255.

[60] Claudie DJ, Semple SJ, Smith NM, Simpson BS (2012) Ancient but new. Developing locally-driven enterprises based on traditional medicines in "Kuuku I'yu" (Northern Kaanju homelands, Cape York, Queensland, Australia). In: Drahos P, Frankel S, editors. *Indigenous Peoples' Innovation: IP Pathways to Development*. Canberra: ANU E Press, pp. 29–55.

[61] Foley P (2006) *Duboisia myoporoides*: the medical career of a native Australian plant. *Historical Records of Australian Science* 17: 31–69.

[62] Holcombe S, Janke T (2012) Patenting the Kakadu plum and the Marjarla tree: biodiscovery, intellectual property and Indigenous knowledge. In: Rimmer M, McLennan A, editors. *Intellectual Property and Emerging Technologies: The New Biology*. Cheltenham/Northampton: Edward Elgar, pp. 293–319.

[63] Convention on Biological Diversity (2010) *Nagoya Protocol on Access to Genetic Resources and the Fair and Equitable Sharing of Benefits Arising from their Utilization (ABS) to the Convention on Biological Diversity*. Canada: Secretariat of the Convention on Biological Diversity.

[64] Hegarty M, Hegarty E (2001) Food safety of Australian plant bushfoods. Publication no. 01/28. Rural Industries Research and Development Corporation, Australia.

[65] Young L (2007) *Lola Young Medicine Woman and Teacher*. Fremantle: Fremantle Arts Centre Press.

[66] Therapeutics Goods Administration (2005) Part 1 Registration of Complementary Medicines in Australian Regulatory Guidelines for Complementary Medicines. Australia: Department of Health and Aging, pp. 1–113.

[67] Therapeutic Goods Administration (2012) Consultation: evidence required to support indications for listed medicines (excluding sunscreens and disinfectants). Version 2.0, Draft report. Canberra: Australian Government Department of Health and Ageing. Available online at http://www.tga.gov.au/newsroom/consult-cm-evidence-listed-medicines-120827.htm (Accessed July 14, 2013).

[68] Saethre EJ (2007) Conflicting traditions, concurrent treatment: medical pluralism in remote Aboriginal Australia. *Oceania* 77: 95–110.

[69] Traditional Healers of Central Australia (2013) *Ngangkari/Ngaanyatjarra Pitjantjatjara Yankunytjatjara Women's Council Aboriginal Corporation*. Broome: Magabala Books Aboriginal Corporation.

7

PHYTOTHERAPIES FROM TRADITIONAL CHINESE MEDICINE

MICHAEL RIEDER[1,2]

[1] *Departments of Paediatrics, Physiology and Pharmacology, and Medicine, Schulich School of Medicine & Dentistry, London, Ontario, Canada*
[2] *Robarts Research Institute, Western University, London, Ontario, Canada*

7.1 TRADITIONAL CHINESE MEDICINE

China has one of the oldest continuous civilizations in the world, originating in the Yellow River region with the first written records dating back to the Shang Dynasty (1700–1046 B.C.). As would be expected, China has also enjoyed a long history of medical care, which has included the development of a system and philosophy of traditional medicine that has and continues to make considerable use of therapies based on plants and plant extracts, frequently in combination [1]. The first written description of the use of herbal medicines date to approximately 1100 B.C., with the fundamental doctrinal resource for Traditional Chinese Medicine being the Yellow Emperor's Inner Canon (*Huangdi Neijing*; 黃帝內經), often referred to as the Textbook of the Yellow Emperor [2]. The exact date of composition is unclear and appears to be between the period of the Warring States and the Han Dynasty (between 200 and 200 B.C.). An important event in Chinese history that enabled intellectuals to be able to collate material from earlier scholars was the establishment of the empire, centralizing government in China and ending five centuries of conflict between the Warring States. The end of conflict was marked by a significant increase in intellectual exchange across the territory of the previous Warring States. Thus, the authors of the book were able to take into account earlier written accounts to produce a comprehensive basis for Traditional Chinese Medicine. The text itself consists of 81 chapters

Phytotherapies: Efficacy, Safety, and Regulation, First Edition. Edited by Iqbal Ramzan.

written in the form of a discourse, being a record of questions and answers between the legendary Yellow Emperor and six of his ministers; as an example, at the start of the book the emperor asks Qi Bo, the Imperial Family physician, why it is that in the earlier days people often lived for 100 years while today people often only live for 50 years. Qi Bo responds that previously people had lived in balance, but now lived under stress, which depleted their Qi. Following this, there are a series of chapters on yin and yang and on Qi and the spirit followed by a number of chapters on diagnosis, specific diseases, and therapy [2]. This book was also the first to provide a mature synthesis of the conceptual framework of the Five Phases and yin and yang.

The first formulary that linked yin and yang with specific prescriptions was the Treatise on Cold Damage Disorders (*Shang Han Lun*; 傷寒論), written approximately 220 A.D. during the past years of the Han Dynasty [3]. This book has 12 chapters and divides diseases into 6 different groups or "channels," for which 112 herbal prescriptions are provided. Over the subsequent, centuries a number of additional prescriptions were developed and characterized; one of the activities that new dynasties would engage in was supporting the writing of new texts that collected herbal prescriptions, all of which were written following the basic conceptual framework developed in the *Huangdi Neijing*. While the basic synthesis underlying the *Huangdi Neijing* was not challenged by later authors, new observations were incorporated into subsequent works to help clarify difficult points and to support classical theories of disease with more recent developments. As an example, the Classic of Difficult Problems (*Nan Ching* 黃帝八十一難經) was written at the end of the first century B.C. to address 81 selected passages in the *Huangdi Neijing* that was considered to be especially enigmatic [4]. There were many more texts written after this time, but interestingly the fundamental concepts—the role of yin and yang and the importance of the flow of Qi—are not challenged in these more recent works. Many of these more recent texts covered specific topics, including infectious disease, acupuncture, and special populations such as pregnant woman and children [5]. The first official pharmacopeia, the *Xin Xiu Ben Cao*, was written in 659 A.D. and included 850 medications [5]. The importance of this book and similar works was demonstrated by the fact that this was obligatory reading for all medical students, which on average required a year of study. This book also was recommended reading for Japanese physicians.

The widespread use of herbal medicine should not be surprising given that China is home to more than 30,000 species of flowering plants, of which more than 5,000 have been evaluated for use in medical therapy. The spread of knowledge with respect to Chinese herbal medicine was greatly enhanced by the invention of the printing press with movable type by Pi Sheng in 1040 A.D. Traditional Chinese medicine was firmly established throughout the Empire, a position of pre-eminence that was held until the twentieth century. By the eighteenth century, a period of relative stagnation occurred, during which the emphasis was on classical writings and new discoveries were discouraged. This was also the time when foreign influences came to enter China, often through missionaries such as Jesuits. The herbal drugs that the missionaries brought, such as quinine, were readily incorporated into Chinese formularies while missionaries and foreign doctors often impressed Chinese patients with their surgical abilities, surgery not being emphasized in classical Chinese medical teaching [5].

During the early twentieth century, a number of foreign doctors and increasing numbers of Chinese doctors with foreign training introduced Western medicine to China, a movement that accelerated rapidly during World War II and as a consequence of the Therapeutic Revolution [6]. The discovery of specific therapy, which began with the use of sulfonamides to treat bacterial infection, transformed the face of Western medicine as well as the impact of Western medicine on health care in China. A number of Western-style medical schools were established, such as Peking Union Medical College (1906) and the University of Hong Kong's Faculty of Medicine (1911). Despite the increasing interest in and use of Western medicine, Traditional Chinese Medicine has continued to be an important part of health-care delivery in China. Currently the Chinese government allows (and regulates) both systems of care independently.

The increasing global interest in complementary and alternative medicine has triggered an accompanying interest in Traditional Chinese Medicine outside of China. This has resulted in a number of interesting developments, for example, the creation of regulatory agencies to certify Traditional Chinese Medicine practitioners in Canada and Australia, the development of companies both in China and beyond who export Chinese medicines worldwide, and the creation of unique networks such as the Consortium for Globalization of Chinese Medicine, which is an international organization made up of universities and companies from across the world with a shared interest in Traditional Chinese Medicine (http://www.tcmedicine.org). A part of these efforts has included systems to preserve and protect traditional and classical concepts underlying Traditional Chinese Medicine [7].

7.2 KEY CONCEPTS IN TRADITIONAL CHINESE MEDICINE

As noted earlier, a remarkably consistent fact with respect to Traditional Chinese Medicine has been the preservation of the core conceptual framework over more than two millennia [5]. The roots of this framework lie in ancient Taoist philosophy, in which all things are believed to derive from a single origin with natural energy or life force Qi being made up of two opposing parts, yin and yang, that blend to create a harmonious whole (Fig. 7.1) [8]. Yin the dark part containing a bright circle is considered to be slow, cold, wet, and yielding and is associated with femininity, water, the moon, and night, whereas yang, the bright part containing a dark circle, is considered to be fast, warm, dry, and solid and is associated with masculinity, fire, the sun, and daytime. A key concept is that yin and yang cannot be considered in isolation; that is to say, the optimal condition is achieved when both yin and yang are in balance. Qi cannot be created or destroyed but rather exists in a dynamic balance in which it changes its manifestation. An increase or decrease in one manifestation of Qi is accompanied by a decrease or increase in the other, creating a vacuity. As an example, night sweats, insomnia, dry mouth, and rapid pulse are signs and symptoms of yin vacuity, whereas cold limbs, diarrhea, pale tongue, and slow pulse are signs and symptoms of yang vacuity.

An important difference between Traditional Chinese Medicine and Western medicine lies in the fundamental approach, in that while Western medicine tends to

FIGURE 7.1 Yin and yang. The central duality underpinning Traditional Chinese Medicine; yin is the dark area ("shady place" or "north slope") and yang is the light area ("sunny place" or "south slope").

embrace a pathophysiological model in which understanding disease pathogenesis at a more and more focused level is an important driver. Traditional Chinese Medicine is much more phenomenological. This is not to say that underlying mechanism is unimportant in Traditional Chinese Medicine, but highly detailed observations of disease manifestations are much more important to the practitioner and scholar involved in Traditional Chinese Medicine than are detailed investigations of causation [8]. This also drives the importance of diagnosis in Traditional Chinese Medicine as well as the considerable effort taken, again both by practitioners and scholars, to accurately describe and characterize various disease patterns. The classical four steps to make a diagnosis include observation of the patient, carefully listening to the complaint, directed questioning, and physical examination including pulse analysis. While these four steps are similar to those employed by careful clinicians using Western medicine, Traditional Chinese Medicine places much more importance and devotes much more detail to analysis of the pulse and pulse diagnosis. The importance of diagnosis is also a key element in the selection of therapy, as the diagnosis is not merely of a symptom complex but also of the personality type of the patient as expressed in the five phases.

The Five Phase Theory of Systemic Correspondence is a conceptual framework to understand how natural phenomena are expressed, known as the Wu Xing, (五行) (Fig. 7.2) [9]. The Five Phases consist of wood, fire, earth, metal, and water, while the Five Organ Networks are liver, heart, spleen, lung, and kidney (Table 7.1) [10]. The classical view is that the five phases also describe five distinct personality types, with unique emotional and psychological characteristics that have been described as

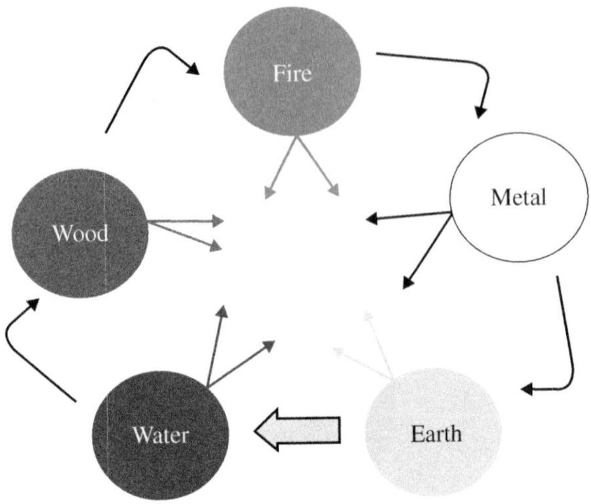

FIGURE 7.2 The five phases: the arrows on the outer circle represent generation or creation while the arrows in the inner circle represent destruction [10].

TABLE 7.1 The Five Phases: Associated Colors, Directions, Seasons, and Organs[a]

Phase	Wood	Fire	Earth	Metal	Water
Color	Green	Red	Yellow	White	Blue
Direction	East	South	Center	West	South
Season	Spring	Summer	Late summer	Autumn	Winter
Yin organ	Liver	Heart	Spleen	Lung	Kidney
Yang organ	Gall bladder	Small intestine	Stomach	Large intestine	Urinary bladder
Emotion	Anger	Happiness	Love	Grief	Fear
Sensory organ	Eyes	Tongue	Mouth	Nose	Ears
Life	Birth	Youth	Adulthood	Old age	Death
Taste	Sour	Bitter	Sweet	Spicy	Salty

[a]From Ref. 10.

the pioneer, wizard, peacemaker, alchemist, and philosopher [10]. The relevance of this to Traditional Chinese Medicine and phytotherapy is that the disorders and remedies, which develop among people of the various archetypes, are assessed and treated with approaches unique to that archetype. Thus, prescribing for a health problem involves the assessment of not only the health problem but also the patient, in what could be considered a very early example of personalized medicine!

7.3 HERBAL MEDICINE AND TRADITIONAL CHINESE MEDICINE

Accurate diagnosis is at the core of Traditional Chinese Medicine and sets the stage for therapy, with treatment typically including a variety of modalities such as exercise, diet, acupuncture, and massage and herbal medication [11]. Although animal and mineral elements are also used in Traditional Chinese Medicine, herb-derived therapies are by far the most commonly used medications in Traditional Chinese Medicine. There are several aspects of the use of herbal medications in the context of Traditional Chinese Medicine that are worth considering.

The first relates to the fundamental conceptual framework of Traditional Chinese Medicine, that being the importance of balance. As opposed to other theoretical frameworks of health and disease, the goal is not to address a specific pathophysiological process but to restore balance. Thus, many herbal therapies used in Traditional Chinese Medicine have been developed for long-term rather than for acute use. Additionally, as noted earlier, assessment of the five phases is a key part of development of a therapeutic plan designed for the individual patient, as patients with different makeup would be likely to be prescribed different therapies for a disorder that appears to be superficially similar.

A second issue related to phytotherapeutics is that many of the herbal therapies used are extracts, which are typically aqueous. Thus, teas and soups are relatively common delivery vehicles for Traditional Chinese Medicine herbal formulations. This is in contrast to other systems of traditional or herbal medicine, for example, German herbal medicine, where the use of alcohol extracts is much more common.

The final consideration relates to how the formulations are developed and prepared. In contrast to conventional Western post-Therapeutic Revolution small molecules, which are typically single agent therapies, the majority of Traditional Chinese Medicine herbal prescriptions are for herbal combinations. The use of herbal products in combination is not uncommon among practitioners of complementary medicine. As an example, the original formulation that William Withering studied for the therapy of dropsy was a herbal preparation made by a folk medicine practitioner in Shropshire that contained more than 20 herbs. Withering's training in botany and medicine, provided him with a novel skillset that enabled him to identify one herb, Foxglove or *Digitalis purpurea*, as the most likely source of a medicinal product to treat heart failure. His subsequent studies demonstrated that an extract from Foxglove, the cardiac glycoside digitalis, was responsible for the beneficial therapeutic effects that had been observed [12]. The principle of single agent therapy has been emphasized for many years, a primary hypothesis that has been challenged more recently by data demonstrating improved outcomes with combination therapy for complex problems as diverse as cancer, drug-resistant infection, and chronic pain [13–16]. In these cases, combination therapy has been designed to take advantage of synergies in drug action or disease susceptibility. This is not to say that Traditional Chinese Medicine does not in some cases rely on a single herb. For example, Ren Shen (Ginseng) is used to strengthen Qi in conditions characterized by a weak pulse, shortness of breath, and cold limbs, as an example after blood loss. However, while single herbs have been used, combination therapy is more common. In the case of herbal preparations for Traditional Chinese Medicine, the combinations characteristic of the prescriptions recorded in texts such as the *Xin Xiu Ben Cao* are the result of many decades of observation and thoughtfully considered clinical experience.

These prescriptions were developed to enhance efficacy with synergistic interactions while reducing the dose of any single herb to provide safer formulations. Interestingly, the mutual interactions between herbs are classically described in terms of government officials, each herb in the prescription is designated as either an "emperor," "minister," "assistant," or "messenger" herb (Table 7.2) [17]. A Chinese folk saying with regards to this is that "prescribing medication is like commanding an army" in which the various elements of the army, infantry, cavalry, or light troops, have specific, well-defined, and often complementary roles. Each specific role in a Chinese Traditional Medicine prescription may be undertaken by a single herb or by multiple herbs. Many prescriptions include as many as 15 herbs, each chosen to fulfill a specific purpose. The emperor herb is key to the herbal combination in that the emperor has the most potent effect on the symptoms or imbalances being addressed. The minister herb acts to enhance the activity of the emperor herb and may also target secondary symptoms. The assistant herb may have a number of roles, and classically three types of assistant herbs are described: the "helpful assistant" acts to enhance the effects of the emperor herb, the "corrective assistant" acts to reduce the adverse effects produced by the emperor and/or minister herbs, and the "opposing assistant" acts to diminish the effects of the emperor herb. In the latter case, this is typically in complex mixtures where the effect of combinations of emperor and minister herbs may be too potent. The messenger herb acts to direct the actions of the other herbs on a particular part of the body and may have other effects such as altering the taste of the prescription. The herbs chosen for the various roles are carefully selected based on their characteristics (Table 7.3).

The prescription of Traditional Chinese Medicine is very dependent on accurate diagnosis taking into account the five phases. Diagnosis must take into account not only the signs and symptoms of the disease but also the underlying makeup of the person. Once the practitioner determines this, the prescription can be prepared.

To illustrate how this works in practice, an example would be useful. Gui Zhi Fu Zi Tang (桂枝附子汤) is a classical herbal prescription used to reduce cold and to expel wind and dampness, characterized by symptoms of fever and chills, nasal congestion, stiff neck, and dry heaves [18]. It is a decoction made up of a number of herbs whose action is intended to reduce pathogenic effects in the muscles and regulate protective Qi. The components of the prescription include Gui Zhi (cinnamon,

TABLE 7.2 Characteristics of the Four Types of Herbs Used in Prescriptions of Classical Chinese Medicine

Characteristic	Emperor	Minister	Assistant	Messenger
Alternate name	King or monarch	Deputy	Clerical, adjunct, or auxiliary	Convoy, guide, dispatcher, or emissary
Role	Directly addresses the disorder in question	Enhances the efficacy of the emperor herb	Minimize adverse effects of the other herbs and deal with secondary symptoms	Directs the activity of the other herbs; may alter the taste of the formulation

TABLE 7.3 Characteristics of Herbs Used in Traditional Chinese Medicine[a]

Characteristic	Effect
Herbs that release the exterior	These herbs are intended to prevent disorders from penetrating into the body. They include spicy warm herbs that induce perspiration and warm the body as well as spicy cold herbs that expel pathogens associated with heat.
Herbs that clear heat	These herbs reduce heat generated internally; these include herbs that deal with herbs from toxicity and from damp heat.
Downward draining herbs	These herbs include herbs that are purgatives and laxatives.
Herbs that dispel wind dampness	These herbs treat painful obstructions characterized typically by arthritis and act to reduce inflammation and swelling.
Herbs that drain dampness	These herbs are used to treat dampness such as edema as well as with disorders associated with urination.
Aromatic herbs that drain dampness	These herbs are used to treat nausea, vomiting, and diarrhea.
Herbs that transform phlegm	These herbs are used to treat disorders associated with thick phlegm, which include not only respiratory disorders but also goiter and nausea and vomiting.
Herbs that relieve food stagnation	There herbs are intended to treat abdominal distension and treat abdominal masses.
Herbs that regulate Qi	These herbs are intended to treat blockages of the digestive system as well as liver and lung disease.
Herbs that regulate blood	These herbs can be either herbs that stop bleeding or herbs that increase the circulation. The herbs intended to stop bleeding are ideally used with other herbs that address the reason for the bleeding.
Herbs that warm the interior	These herbs are intended to increase metabolism and enhance digestion.
Herbs that calm the spirit	These herbs are calming and are used to treat problems such as insomnia and anxiety.
Herbs that open the heart	These herbs are used for conditions such as angina.
Herbs to clear internal wind and tremors	These herbs are used to treat muscle spasms and involuntary movements.
Tonic herbs	These are preventative herbs intended to address imbalances in yin and yang in order to prevent disease.
Herbs that stabilize and bind (astringent herbs)	These herbs are intended to reduce excessive discharges such as diarrhea or excessive urination and are ideally taken with other herbs that address the reason for the excessive discharge.
Herbs to treat parasites	These herbs are intended to remove parasites.
Herbs for external uses	These are herbs intended to be applied topically for skin disorders or problems such as sprains.

[a] Modified from Ref. 10.

Cinnamomi ramulus), Bai Shao (Paeonia, *Paeoniae radix*), Da Zao (Chinese date, *Ziziphus jujuba*), Sheng Jiang (ginger, *Zingiber officinale*), and Zhi Gan Cao (licorice, *Glycyrrhizae radix*, prepared in honey). The classical formula incorporates Gui Zhi as the emperor herb, Bai Shao as the minister herb, Da Zao and Sheng Jiang as assistant herbs, and Zhi Gan Cao as the messenger herb. The rationale behind this mixture is that the emperor herb Gui Zhi acts to release cold from the muscle layer, with Bai Shao helping to preserve yin and nutritive Qi. In the case of the minister herb, this prescription is indicated for patients who are unable to relieve cold by sweating, indicating a weakness in Qi. Thus, the action of the minister herb is to enhance Qi to assist the emperor herb in driving out the cold. Of the two assistant herbs, Da Zao assists the minister herb in developing nutritive Qi while Sheng Jiang acts to both release the exterior and to warm the center. The messenger herb, Zhi Gan Cao, acts to harmonize the effects of the other herbs.

Thus, it can be seen that the development of therapies in the classical mode is complex and requires not only a very careful assessment of the patient but also a comprehensive knowledge of the effects, interactions, and adverse events of a wide range of herbs [19]. To perform this competently requires considerable study and careful attention to detail. Many of the classical references needed to master the complex field of Traditional Chinese Medicine have until quite recently been only available in Cantonese or Mandarin, limiting the opportunity to study these works by scholars and practitioners from other parts of the world.

7.4 ISSUES IN THE DEVELOPMENT OF PHYTOTHERAPY FROM TRADITIONAL CHINESE MEDICINE

Given the extensive experience with herbal formulations in Traditional Chinese Medicine, one might have expected that there would have been considerable development of phytotherapies based on herbal extracts. In fact, despite the common use of herbal medicine as a mainstay of treatment in much of China, there was until recently very little development of these approaches outside of China. There are several reasons for this. While Imperial China had been outward looking under the Ming Dynasty, the successors of the Ming Dynasty, the Qing Dynasty, were increasingly isolationist. As well, the highly centralized imperial government discouraged innovation to the point of stagnation, such that by the time European powers began to seriously intervene in Chinese affairs after the mid-nineteenth century, the Chinese military found itself seriously out-classed. The ease with which British forces in the First Opium War (1839–1842) were able to destroy Chinese naval squadrons and capture fortifications such as Amoy was a testament to the fact that, although brave, Chinese forces were fighting with weapons better suited to 1640 than 1840. The Chinese Imperial government was never able to strike a balance between modernization and maintaining the Mandate of Heaven, and thus Western influences, as noted earlier, came to increasingly dominate areas such as science and medicine. As noted earlier, the fact that many primary information sources were only available in Chinese has been until recently been a limitation for scholars outside of China to explore possibilities in Chinese Traditional Medicine. As well, contact with the West was limited initially following the establishment of the People's Republic of China, although over the past two decades China has taken great strides to increase the presence and activity of China in the world.

There are also several practical issues that impact as well. The patients on whom the large classical work is written are largely Han Chinese, a reasonably homogenous population. It is now appreciated that there may be genetically determined differences between populations that can impact on both drug efficacy and drug safety [20–22]. A major problem historically has been the standardization of herbs used; herbal activity and purity is known to vary widely when grown in different regions [23]. For example, studies of products prepared from a species of ginseng (*Panax ginseng*) have between 40- and 200-fold difference in ginsenoside content between preparations from different areas [24]. More recently, a problem with herbal medications has been contamination and adulteration, which is a problem that is not unique to herbs of Chinese origin [23].

Over the past two decades the increasing interest in Traditional Chinese Medicine, the development of techniques for standardization such as chromatography and DNA fingerprinting, and the formation of research consortia between Chinese universities and universities elsewhere has resulted in a substantial increase in the amount of research in and the development of unique phytotherapies derived from Traditional Chinese Medicine [23, 25].

7.5 PHYTOTHERAPIES DEVELOPED FROM TRADITIONAL CHINESE MEDICINE

As noted earlier, many of the factors limiting the development of phytotherapies from Traditional Chinese Medicine have changed considerably over the past two decades, and this has been accompanied by a sharp increase in the use of and research into therapeutic applications of Traditional Chinese Medicine well beyond the borders of China [23, 26]. While there are a large number of herbs used in Traditional Chinese Medicine, there are some that are much more commonly used than others (Table 7.4).

With the increasing interest in Chinese herbal medicine, there has been an increase in both the amount of research in this area as well as the rigor and range of techniques used [23, 26]. This has included the collection and translation of literature from classical and traditional sources in forms that are indexed and can be searched using approaches such as systemic reviews that can be accomplished by a two-step approach, first reviewing large compendia that can be cross-referenced to identify and characterize key citations [27]. One interesting development has been the study of how to modify the diagnostic approach from Traditional Chinese Medicine, which as noted earlier, is a crucial element in the decision as to which prescription to use in randomized clinical trials [28]. A systematic review and analysis has demonstrated that while there are issues in standardizing and validating syndrome patterns, there does appear to be ways to use syndrome patterns for the design of randomized-controlled clinical trials [29]. Given that evidence-based medicine has become a central element

TABLE 7.4 Commonly Used Chinese Herbs[a]

Herb	Scientific Name	Common Name
Ginseng	Genus *Panax*	*Panax ginseng* (Ren Shen or Chinese ginseng)
	Panax notoginseng	Xi Yang Shen or American ginseng
	Panax bipinnatifidus	Red ginseng (Hangul); ginseng heated and dried
	Panax ginseng	White ginseng; ginseng dried without heating
	Panax japonicus	Sun ginseng; prepared by steam treatment of white ginseng
	Panax quinquefolius	
	Panax vietnamensis	
	Panax wangianus	
	Panax zingiberensis	
	Panax trifolius	
	Series Panax	
Siberian Ginseng	*Eleutherococcus senticosus*	Ci Wu Jia
Ginger	*Zingiber officinale*	Sheng Jiang (fresh ginger rhizome)
		Gan Jiang (dried ginger rhizome)
Cinnamon	*Cinnamomum cassia*	Gui Zhi (cinnamon twig)
Licorice	*Glycyrrhiza glabra*	Gan Cao
Ephedra	*Ephedra sinica*	Ma Huang
	Ephedra intermedia	

[a]Modified from Ref. 10.

in decision analysis for many treatments and is increasingly being used to make regulatory or reimbursement decisions, this is a very important development [30].

The development of phytotherapies from Traditional Chinese Medicine and the application to other populations also need to take into consideration the fact that the variability in response to therapy in many countries may be larger than that in China, where greater than 90% of the population are Han Chinese. Genetic and anthropological sources of variation may lead to differences in efficacy and safety in different populations and thus the design and conduct of clinical trials should take this into account. As well, responses in vulnerable groups, such as, young children or the extreme elderly, may vary from those of more typical adults and thus studies should be conducted among these groups if a phytotherapy derived from Traditional Chinese Medicine is to be used for these types of patients.

The development of phytotherapies based on Traditional Chinese Medicine has focused primarily on chronic disorders and is often investigated in the light of prevention or for amelioration of symptoms or to address adverse effects associated with treatment with Western medicine. An example is diabetes. Diabetes, notably type II diabetes, is a more and more common problem worldwide. A number of Chinese herbs have shown promise for use as oral hypoglycemic agents and further research is ongoing [31]. Similar work is being undertaken for rheumatoid arthritis [32].

One commonality between the various research streams investigating the potential for phytotherapies derived from Traditional Chinese Medicine is the goal to produce interventions that can be used orally for chronic therapy, thus placing a very high expectation of safety. As well, it is not surprising that many disorders being studied are diseases for which Western medicine approaches are not optimal, often with less than desired efficacy or with more than desired toxicity. The safety concern with respect to the development of herbal therapies is notable in that many of the herbal components used in Traditional Chinese Medicine have potent effects and can have potent adverse effects [33]. We have demonstrated that extracts of the *Ganoderma lucidum* (Lingzhi) can produce concentration-dependent toxicity in not only cultured cells but also in peripheral blood mononuclear cells isolated from adults and children [33]. Indeed, some herbs are sufficiently toxic to be considered for use as chemotherapeutic agents [34].

While some therapies have been directed at specific disorders, others have targeted disease mechanisms such as inflammation. Given the increasing recognition of the importance of inflammation in processes as varied as heart disease to arthritis, this suggests that there is potentially a wide range of uses for phytotherapies derived from Traditional Chinese Medicine [35, 36]. The use of complementary and alternative medicine approaches for inflammatory disorders is of obvious importance to patients given the many adverse effects associated with conventional Western anti-inflammatory therapies. Extracts of *Tripterygium wilfordii* (Lei Gong Teng) have been proven to have robust anti-inflammatory activity albeit with potential for adverse effects such as liver injury, gastrointestinal distress, and leucopenia [37]. Similar considerations apply to the study of herbal agents as part of and/or derived from Traditional Chinese Medicine in the therapy of chronic problems such as neuropathic pain [38].

One of the areas where there has been the most interest for development of phytotherapies derived from Traditional Chinese Medicine has been in the area of cancer care. In this context, while there has been interest in the use of toxic herbs as chemotherapeutic agents, the major interest has been in the use of herbal therapy as adjunctive or symptomatic therapy when used in addition to conventional Western therapy [34, 39, 40]. This was clearly demonstrated in a recent review in which nearly 3000 Chinese studies, including more than 250,000 patients, on the use of Traditional Chinese Medicine for cancer care were analyzed [40]. Of these studies, 90% used herbal medication and more than 70% were conducted with Chinese and Western medicine approaches combined [40]. The improvement of clinical symptoms was the most commonly reported outcome.

The large number of studies conducted demonstrate both that there is potential and need for the most commonly studied tumors, lung cancer, and liver tumors or other cancers that are often associated with a poor outcome, requiring optimal therapy to be defined. As an example, there are many studies evaluating the potential for herbs derived from Traditional Chinese Medicine to relieve cachexia associated with cancer therapy [41].

The number of potential phytotherapies that may be derived from Traditional Chinese Medicine is numerous and the diseases for which they could be used are many as well, ranging from the ones described earlier to asthma and dysmenorrhea [31, 32, 37, 38, 42–44]. While it would take a very large textbook to describe in detail all of these studies, one recent example is worth examining in detail as it provides an excellent insight into the issues and possibilities with respect to the development of phytotherapy from Traditional Chinese Medicine.

7.6 HUANG QIN TANG AND THE DEVELOPMENT OF PHY906

Huang Qin Tang (黃芩湯) is a Traditional Chinese Medicine prescription also known as *Scutellaria* decoction that was first described in 220 A.D. in *Shang Han Lun* [3, 45]. This is a classical herbal preparation used for the treatment of gastrointestinal problems such as nausea, vomiting, and diarrhea. It comprises four herbs, Huang Qin (黃芩) or baical skullcap root (*Scutellaria baicalensis Georgi*), Da Zao (大棗) or Chinese date, Bai Shao (白芍) (*Ziziphus Jujuba Mill*), white peony root (*Paeonia lactiflora Pall*), and Zhi Gan Cao (炙甘草) or licorice prepared with honey (*Glycyrrhizae uralensis Fisch*). The potential for the use of Huang Qin Tang or derivatives was explored by a team of investigators at Yale led by Dr. Y.C. Cheng, the Henry Bronson Professor of Pharmacology in collaboration with industrial partners. The first step of this research was to explore a number of derivatives of Huang Qin Tang in a stepwise fashion, including careful analytical studies on the identity of key components, exploration of gene expression, and study of the biological activity of key components in a relevant animal model [46]. The first step in the process was to develop a standardized hot water extraction technique for the 18 different batches of Huang Qin Tan obtained from a number of different companies. The components were then identified with liquid chromatography/mass spectroscopy (LS/MS) [47]. The extracts were then incubated with extensively validated and well-known cell culture systems, in this case Jurkat, KB, and HepG2 cells. The expression of key genes was then studied with the validation of expression of key genes by real-time polymerase chain reaction. The extracts that were felt to be of interest after screening were evaluated to determine the effect of incubation of these drugs with the cancer chemotherapy drug CPT-11 when administered to 4–6-week-old female BDF-1 mice who had been implanted with murine Colon 38 colorectal cancer cells [46]. This process involved a large amount of experimental work, for example, PHY-906, the extract that eventually emerged as the most likely candidate, had 64 peaks identified on LC/MS analysis of which 39 were felt to be worthy of more detailed study. The genetic expression studies looking at the expression of 1800 genes found a unique signature for PHY-906 involving 20 genes [46]. Batch-to-batch equivalence of PHY-906 was determined using comparison-chemical and comparison-bioresponse fingerprinting [46]. This careful quantification and validation enabled the investigators to prepare a sufficient quality of standardized PHY-906 to conduct clinical trials of the efficacy of this formulation in patients with colon cancer. This study demonstrated a significant improvement in nausea and vomiting among patients treated with chemotherapy for rectal or colon cancer when treated concurrently with PHY-906 [48–50]. It should be noted that care was also taken during this study to determine if there was toxicity associated with PHY-906 and that PHY-906 therapy did not alter the pharmacokinetics of the chemotherapeutic drugs [48–50].

This finding triggered interest in other effects of PHY-906. A study using increasing doses of PHY-906 in an animal model of colon cancer demonstrated concentration-dependent increase in anti-tumor effect and preservation of body weight while also promoting recovery of intestinal injury associated with chemotherapy [51]. This may be related to the observation that PHY906 promotes the expression of intestinal progenitor or stem cell markers following treatment with chemotherapy and inhibiting inflammation that is associated with chemotherapy [51]. More detailed studies of the effect of PHY906 on the tumor microenvironment demonstrated that the combination of PHY906 and chemotherapy, in the case of this study, CPT-111, altered pro-apoptotic and pro-inflammatory pathways in a manner that would direct the immune system toward tumor rejection, thus potentially enhancing the effect of chemotherapy by enlisting the immune system [52]. This appeared to only occur when PHY-906 and chemotherapy were used together, providing a contemporary validation of some of the concepts of combination therapy first described in Traditional Chinese Medicine literature 1800 years ago. The efficacy of this combination has had dramatic results in animal models (Fig. 7.3).

This finding and the positive results of the Phase I trial have led to a number of other studies using PHY-906. A Phase II trial has demonstrated that combination therapy with chemotherapy plus PHY-906 may allow for much higher chemotherapy doses among patients with advanced pancreatic cancer [53]. A murine study has demonstrated that PHY-906 appeared to reduce radiation-related toxicity while not protecting tumors, suggesting that there may be a role for PHY-906 in reducing adverse events associated with radiation therapy for cancer [54]. Pre-clinical studies have been conducted to determine

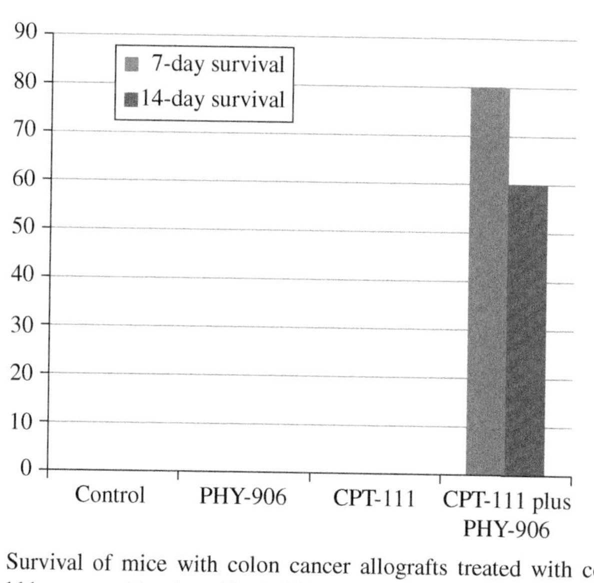

FIGURE 7.3 Survival of mice with colon cancer allografts treated with control vehicle PHY-906, CPT-111, or a combination of both CPT-111 and PHY-906 [28]. There was survival at 7 and 14 days post-grafting only in mice treated with the combination.

the potential role for combination therapy with PHY-906 in a number of tumors including colorectal cancer, liver cancer, lung cancer, leukemia, breast cancer, ovarian cancer, renal cancer, and pancreatic cancer [55]. PHY-906 has been used in combination with drugs such as CPT-111, VP-16, gemcitabine, capecitabine, oxaliplatin, and sorafenib [55]. Of interest, removal of any of the four herbs from PHY-906 results in a decrease in either the prevention of loss of weight, anti-tumor effect, or effect on mortality that can influence all three parameters [55].

PHY-906 appears to be a success story in developing phytotherapy from Traditional Chinese Medicine and indeed is probably one of the most-studied herbal medications. There are some key lessons that can be learned from the development of PHY-906 that is informative to those interested in this area.

The first is the importance of leadership and a strong team. Dr. Y.C. Cheng, the leader of the project, is a very well trained classical pharmacologist with a long track record of research in viral and cancer chemotherapy. In developing this project, Dr. Cheng sought out and worked closely with a number of collaborators, including Dr. Edward Chu, a medical oncologist at Yale as well as with a number of colleagues both in industry and in Traditional Chinese Medicine. These industrial collaborations were important as they provided the team with a number of preparations of Huang Qin Tan, which were studied extensively and from which PHY-906 was developed. It should be noted that collaboration has been greatly facilitated by the creation of the Consortium for Globalization of Chinese Medicine, which as noted earlier is an international consortium of universities, research institutes, and companies with the common goal of providing Traditional Chinese Medicine to the global community (http://www.tcmedicine.org). The working groups in the consortium include groups on Herbal Resources and Quality Control, groups in which academic investigators form close relationships with traditional practitioners and industry. In the case of PHY-9906, these relationships with industry enabled the team to secure a robust and validated supply of the herbs needed to make PHY-906, a supply that over the course of the studies was carefully evaluated and re-evaluated to ensure consistency.

The importance of the rigor with which the team approached this project cannot be over-emphasized. The careful characterization of key components, biological validation, and study in animal models were crucial to finding a formulation that was both effective and safe. This same rigor applied to the clinical studies, in which safety and efficacy were carefully evaluated but also, an issue central to oncology, the potential effect of the intervention in response to chemotherapy and tumor growth. The story of PHY-906, which continues to evolve, is an excellent case study for those interested in developing phytotherapies from Traditional Chinese Medicine.

7.7 GINSENG

While the example of PHY-906 appears to be a clear success story, there are lessons that can be taken from the medical use of ginseng, one of the most popular herbs used in Traditional Chinese Medicine. Ginseng has been commonly used in Traditional Chinese Medicine for more than 1800 years [56]. Ginseng in its classical sense is 1 of 11 species of slow-growing perennial plants of the family Araliaceae, genus *Panax*. Ginseng is a slow-growing plant that is only seen in cooler climates, growing only in the Northern hemisphere and growing best in forests. Ginseng in the wild is only found in North America, northeast China, eastern Siberia, Bhutan, and Korea with *Panax vietnamensis*, the species endemic to Vietnam, being the most southern-growing species.

Initially, the most commonly used type of ginseng, the type cited in most classical textbooks of Traditional Chinese Medicine, was wild ginseng [57]. However, as demand for ginseng expanded and the amount of land not under cultivation in northeast China declined, a shortage of ginseng developed. During the exploration of North America, Jesuit missionaries confirmed the taxonomy of North American ginseng, *Panax quinquefolius*, in relation to the Asian ginseng, *Panax ginseng*. The demand for ginseng led to industrial scale wild harvesting of ginseng; as an example, in 1841 nearly 300,000 kg of ginseng was harvested in North America and shipped to Asia [58]. This in turn led to a scarcity of wild ginseng in North America and resulted in attempts to cultivate ginseng. Cultivation was frequently unsuccessful, given the relatively slow growth of ginseng and the climatic restrictions on growth, ginseng growing best in cool hardwood forests. There were frequent crop failures and the quality of cultivated ginseng was often not the same as that of wild ginseng [58]. Ginseng cultivation has continued on a limited scale in Wisconsin in the United States and in Canada in Ontario and British Columbia. There is a price differential for cultivated versus wild ginseng such that there continues to be harvesting of wild ginseng.

The use of ginseng has continued to grow, notably over the past several decades. There has been considerable interest and investigation directed at the mechanism of action of ginseng and on the use of ginseng as an immunomodulator in a wide range of disorders ranging from cardiovascular and neurological diseases to diabetes [59–64].

Studies with ginseng have clearly demonstrated the importance of selection of the original plant and the extraction method used. An issue with ginseng that is likely to be of greater importance is supply, notably for wild ginseng. Cultivated ginseng tends not to have the gnarled roots that are most valued by Asian buyers, and hence there continues to be harvesting of wild ginseng. The issue of supply of wild ginseng relates to three facts, the first being that ginseng is relatively slow growing, the second that there is an active harvest, and the third being competition with deer. The demand for wild ginseng has increased as has the price commensurate with the increasing affluence of the middle class in Asia. Wild ginseng used to command \$200 a pound in Appalachia but currently is being sold for between \$500 and \$600 a pound. This has fueled a brisk harvest. Current conservation practices in Canada and the United States have resulted in a large North American deer population; with an estimate being that the current deer population is twice what it was prior to European settlement. Deer are active competitors for ginseng. It is estimated that, without addressing these issues, much of the wild ginseng in North America is at risk of extinction; in 1975, American ginseng was listed in Appendix II of the Convention on International Trade in Endangered Species of Wild Fauna and Flora (CITES) that indicates a threat of extinction unless trade controls are enacted [58].

7.8 MOVING FORWARD

Traditional Chinese Medicine is built on a framework of two thousand years of experience and offers a wide breadth of possibilities for the development of plant-based therapies. The worldwide interest in alternative sources of medications coupled with a global spirit of cooperation and collaboration suggest that there are many opportunities for these developments.

To take full advantage of these opportunities, investigators need to be open-minded, collaborative, and innovative. Being open-minded includes considering efficacy from a wide perspective as well as being careful to avoid harm and to ensure that any therapies developed do not adversely interact with other treatments [65, 66]. Current technology and new approaches to syndrome patterning and clinical trial design provide investigators with previously unavailable capacity to robustly study and develop therapies that can be used globally to pursue the shared goal of better health for everyone.

REFERENCES

[1] Kaptchuk TJ (2000) *The Web That Has No Weaver: Understanding Chinese Medicine*, 2nd ed. New York, NY: McGraw-Hill.

[2] Zhu M, trans (2001) *The Medical Classic of the Yellow Emperor*. Beijing, China: Foreign Language Press.

[3] Hinrichs TJ, Barnes LL (2013) *Chinese Medicine and Healing: An Illustrated History*. Cambridge, MA: Belknap Press.

[4] Unschuld PU, trans (1986) *Nan-ching: The Classic of Difficult Disease*. Berkley, CA: University of California Press.

[5] Hong FF (2004) History of Medicine in China: when medicine took an alternative path. *McGill J Med* 8: 79–84.

[6] Weinshilboum RM (1987) The therapeutic revolution. *Clin Pharmacol Ther* 42: 481–484.

[7] Liu C, Gu M (2011) Protecting traditional knowledge of Chinese medicine: concepts and proposals. *Front Med* 5: 212–218.

[8] Xutian S, Dongyi D, Wozniak J, Junion J, Boisvert J (2004) Comprehension of the unique characteristics of traditional Chinese medicine. *Am J Chin Med* 40: 231–244.

[9] Sivin N (1987) *Traditional Medicine in Contemporary China: A Partial Translation of Revised Outline of Chinese Medicine (1972) with an Introductory Study on Change in Present-day and Early Medicine*. Ann Arbor, MI: University of Michigan, Center for Chinese Studies.

[10] Beinfield H, Korngold E (2007) *Between Heaven and Earth: A Guide to Chinese Medicine*. San Francisco, CA: Chinese Medicine Works.

[11] O'Brien KA, Xue CC (2003) The theoretical framework of Chinese medicine. In: *A Comprehensive Guide to Chinese Medicine*. Leung PC, Xue CC, Eds. Singapore: World Scientific, pp. 47–84.

[12] Goldthorp WO (2009) An account of the foxglove and some of its medicinal uses. *Br Med J* 338: 1393.

[13] Chaparro LE, Wiffine PI, Moore RA, Gilron I (2012) Combination pharmacotherapy for the treatment of neuropathic pain in adults. *Cochrane Database Syst Rev* 7: DOI:10.1002/14651858.CD008943.pub2.

[14] Burke SL, Rose WE (2014) New pharmacological treatments for methicillin-resistant *Staphylococcus aureus* infection. *Expert Opin Pharmacother* 15: 483–491.

[15] Penings PS (2013) HIV drug resistance: problems and perspectives. *Infect Dis Rep*: 5 (Suppl 1): e5.

[16] Locatelli F, Moretta F, Rutella S (2013) Management of relapsed acute lymphoblastic lymphocytic leukemia in childhood with conventional and innovative approaches. *Curr Opin Oncol* 25: 707–715.

[17] Che CT, Wang ZJ, Chow MSS, Lam CWK (2013) Herb-Herb combination for therapeutic enhancement and advancement: theory, practice and future perspectives. *Molecules* 18: 5125–5141.

[18] Chen CY, Kuo TL, Sheu SY, Kuo TF (2010) Preventive effects of Chinese herb chai-hu-gui-zhi-tang extract on water immersion restraint stress-induced acute gastric ulceration in rats. *J Vet Med Sci* 72: 679–685.

[19] Chen JK Chen TT (2004) *Chinese Medical Herbology and Pharmacology*. City of Industry, CA: Art of Medicine Press.

[20] Chung WH, Hung SI, Hong HS, Hsih MS, Yang LC, et al. (2004) Medical genetics: a marker for Stevens-Johnson syndrome. *Nature* 428: 486.

[21] McCormack M, Alfirevic A, Bourgeois S, Farrell JJ, Kasperavičiūtė D, et al. (2011) HLA-A*3101 and carbamazepine-induced hypersensitivity reactions in Europeans. *N Engl J Med* 364: 1134–1143.

[22] Rieder M (2012) New ways to detect adverse drug reactions in children. *Pediatr Clin North Am*: 59: 1071–1092.

[23] Yaw L, Siow YL, Yuewen Gong Y, Kathy KW, Au-Yeung KKW, et al. (2005) Emerging issues in traditional Chinese medicine. *Can J Physiol Pharmacol* 83: 321–334.

[24] Harkey MR, Henderson GL, Gershwin ME, Stern JS, Hackman RM (2001) Variability in commercial ginseng products: an analysis of 25 preparations. *Am J Clin Nutr* 73: 1101–1106.

[25] Liu Y, Liu J, Yin P, Gao M, Deng C, et al. (2011) High throughput identification of components from traditional Chinese medicine herbs by utilizing graphene or graphene oxide as MALDI-TOF-MS matrix. *J Mass Spectrom* 46: 804–815.

[26] Raskin I, Ribnicky DM, Komarnytsky S, Ilic N, Poulev A, et al. (2002) Plants and human health in the twenty-first century. *Trends Biotechnol* 10: 522–531.

[27] May BH, Lu C, Xue CCI (2012) Collections of traditional Chinese medical literature as resources for systematic searches. *J Altern Complement Med* 18: 1101–1107.

[28] Jiang M, Lu C, Zhang C, Yang J, Tan Y, et al. (2012) Syndrome differentiation in modern research of traditional Chinese medicine. *J Ethnopharmacol* 140: 634–642.

[29] Shin BC, Kim S, Cho YH (2013) Syndrome pattern and its application in parallel randomized controlled trials. *Chin J Integr Med* 19: 163–171.

[30] Oxman AD, Sackett DL, Guyatt GH (1993) Users guides to the medical literature I. How to get started. The Evidence Based Medicine Working Group. *JAMA* 270: 2093–2095.

[31] Yang LX, Liu TH, Huang ZT, Li JE, Wu LI (2011) Research progress on the mechanism of single-Chinese medicinal herbs in treating diabetes mellitus. *Chin J Integr Med* 17: 235–240.

[32] Keisuke I, Bian BL, Xiang-Dong L, Takashi S, Akira I (2011) Action mechanisms of complementary and alternative medicine therapies for rheumatoid arthritis. *Chin J Integr Med* 17: 723–730.

[33] Gill SK, Rieder MJ (2008) Toxicity of a traditional Chinese medicine, *Ganoderma lucidum*, in children with cancer. *Can J Clin Pharmacol* 15: e275–e285.

[34] Liu X, Wang Q, Song G, Zhang G, Ye Z, et al. (2014) The classification and application toxic Chinese material medica. *Phytother Res* 28: 334–347.

[35] Medzhitov R (2008) Origin and physiological roles of inflammation. *Nature* 454: 428–435.

[36] Wang Q, Kuang H, Su Y, Sun Y, Feng J, et al. (2013) Naturally derived anti-inflammatory compounds from Chinese medicinal plants. *J Ethnopharmacol* 146: 9–39.

[37] Hoyle GW, Hoyle CI, Chen J, Chang W, Williams RW, et al. (2010) Identification of triptolide, a natural diterpenoid compound, as an inhibitor of lung inflammation. *Am J Physiol Lung Cell Mol Physiol* 298: 830–836.

[38] Schröder S, Beckmann K, Franconi G, Meyer-Hamme G, Friedemann T, et al. (2013) Can medical herbs stimulate regeneration or neuroprotection and treat neuropathic pain in chemotherapy-induced peripheral neuropathy? *Evid Based Complement Alternat Med* 2013: 1–18.

[39] World Health Organization (2009) *Global Health Risks: Mortality and Burden of Disease Attributable to Selected Major Risk*. Geneva, Switzerland: World Health Organization.

[40] Li X, Yang G, Li X, Zhang Y, Yang J, et al. (2013) Traditional Chinese medicine in cancer care: a review of controlled clinical studies published in Chinese. *PLoS One* 8: e60338.

[41] Cheng KC, Li YX, Cheng JT (2012) The use of herbal medicine in cancer-related anorexia/cachexia treatment around the world. *Curr Pharm Des* 18: 4819–4826.

[42] Tan YH, Zhang AL, Chen DC, Xue CC, Lenon GB (2013) Chinese herbal medicine for atopic dermatitis: a systematic review. *J Am Acad Dermatol* 69: 295–304.

[43] Hook IL (2014) Danggui to *Angelica sinensis* root: are potential benefits to European women lost in translation? A review. *J Ethnopharmacol* 152: 1–13.

[44] Hong MI, Song Y, Li XM (2011) Effects and mechanisms of Chinese medicines for asthma. *Chin J Integr Med* 17: 483–491.

[45] Qi F, Li A, Inagaki Y, Gao J, Li J, et al. (2010) Chinese herbal medicines as adjuvant treatment during chemo or radio-therapy for cancer. *Biosci Trends* 4: 297–307.

[46] Tilton R, Paiva AA, Guan JQ, Marathe R, Jiang Z, et al. (2010) A comprehensive platform for quality control of botanical drugs (PhytomicsQC): a case study of Huangqin Tang (HQT) and PHY906. *Chin Med* 5: 30.

[47] Ye M, Liu SH, Jiang Z, Lee Y, Tilton R, et al. (2007) Liquid chromatography/mass spectrometry analysis of PHY906, a Chinese medicine formulation for cancer therapy. *Rapid Commun Mass Spectrom* 21: 3593–3607.

[48] Farrell MP, Kummar S (2003) Phase I/IIA randomized study of PHY906, a novel herbal agent, as a modulator of chemotherapy in patients with advanced colorectal cancer. *Clin Colorectal Cancer* 2: 253–256.

[49] Kummar S, Copur MS, Rose M, Wadler S, Stephenson J, et al. (2011) A phase I study of the Chinese herbal medicine PHY906 as a modulator of irinotecan-based chemotherapy in patients with advanced colorectal cancer. *Clin Colorectal Cancer* 2: 85–96.

[50] Liu SH, Jiang Z, Su TM, Gao WY, Leung CH, et al. (2004) Developing PHY906 as a broad-spectrum modulator of chemotherapeutic agents in cancer therapy. *Proc Am Assoc Cancer Res* 45: 128.

[51] Lam W, Bussom S, Guan F, Jiang Z, Zhang W, et al. (2010) The four-herb Chinese medicine PHY906 reduces chemotherapy-induced gastrointestinal toxicity. *Sci Transl Med* 2: 45ra59.

[52] Wang E, Bussom S, Chen J, Quinn C, Bedognetti D, et al. (2011) Interaction of a traditional Chinese medicine (PHY906) and CPT-11 on the inflammatory process in the tumor microenvironment. *BMC Med Genomics* 4: 38.

[53] Saif MW, Ji J, Lamb L, Kaley K, Elligers K, et al. (2014) First-in-human phase II trial of the botanical formulation PHY906 with capecitabine as second-line therapy in patients with advanced pancreatic cancer. *Cancer Chemother Pharmacol* 73: 373–380.

[54] Rockwell S, Grove TA, Liu Y, Cheng YC, Higgins SA, et al. (2013) Preclinical studies of the Chinese Herbal Medicine formulation PHY906 (KD018) as a potential adjunct to radiation therapy. *Int J Radiat Biol* 89: 16–25.

[55] Liu SH, Cheng YC (2012) Old formula, new Rx: the journey of PHY906 as cancer adjuvant therapy. *J Ethnopharmacol* 140: 614–623.

[56] Yun TK (2001) Brief introduction of *Panax ginseng* C.A. Meyer. *J Korean Med Sci*: 16 (Suppl): S3–S5.

[57] Park HJ, Kim DH, Park SJ, Kim JM, Ryu JH (2012) Ginseng in traditional herbal prescriptions. *J Ginseng Res* 36: 225–241.

[58] McGraw JB, Lubbers AE, Van der Voort M, Mooney EH, Furedi MA, et al. (2013) Ecology and conservation of ginseng (*Panax quinquefolius*) in a changing world. *Ann N Y Acad Sci* 1286: 69–91.

[59] Baek SH, Base ON, Park JH (2012) Recent methodology in ginseng analysis. *J Ginseng Res* 36: 119–134.

[60] Nah SY (2014) Ginseng ginsenoside pharmacology in the central nervous system: involvement in the regulation of ion channels and receptors. *Front Physiol* 5: 1–13.

[61] Liu J, Wang Y, Qui L, Yu Y, Wang C (2014) Saponins of *Panax notoginseng*: chemistry, cellular targets and therapeutic opportunities in cardiovascular diseases. *Expert Opin Investig Drugs* 23: 523–529.

[62] Kang S, Min H (2012) Ginseng the "immunity boost": the effects of *Panax ginseng* on immune system. *J Ginseng Res* 36: 354–368.

[63] Cho IH (2012) Effects of Panax ginseng in neurodegenerative diseases. *J Ginseng Res* 36: 354–368.

[64] Im DS, Nah SY (2013) Yin and Yang of ginseng pharmacology: ginsenosides vs gintonin. *Acta Pharmacol Sin* 34: 1367–1373.

[65] Posadzki P, Watson LK, Ernst E (2013) Adverse effects of herbal medicines: an overview of systematic reviews. *Clin Med* 13: 7–12.

[66] Haefeli WE, Carls A (2014) Drug interactions with phytotherapeutics in oncology. *Expert Opin Drug Metab Toxicol* 10: 359–377.

8

INTEGRATING TRADITIONAL GRECO-ARAB AND ISLAMIC DIET AND HERBAL MEDICINES IN RESEARCH AND CLINICAL PRACTICE

Bashar Saad

Qasemi Research Center- Al-Qasemi Academic College, Baga Algharbiya, Israel
Faculty of Arts and Sciences, Arab American University Jenin, Jenin, Palestine

8.1 INTRODUCTION

Traditional Arab-Islamic herbal-based medicines are increasingly used, mainly in chronic liver disease, diabetes, obesity, infertility, impotence, psychosomatic ailments, and many other diseases. The current use of these medicines has historical roots in Arab-Islamic medicine. The founder of this medicine is the Prophet Mohammad (570–632) (PBUH) himself, as a significant number of statements (Hadith) concerning medicines are attributed to him. He stated that "Allah (God) has sent down both the disease and the cure and He has appointed a cure for every disease, so treat yourselves medically," "There is no disease that Allah has created, except that He also has created its treatment," "Make use of medical treatment, for Allah has not made a disease without appointing a remedy for it, with the exception of one disease, namely old age," and "For every disease, Allah has given a cure." These Hadiths encouraged early Muslims to engage in seeking out a medicine for every illness known to them and hence initiated the foundations of Arab-Islamic medicine. Early Muslims utilized many plants and animal products mentioned in the Holy Quran and in the Hadith of the Prophet for health promotion, for example, dates, black seeds, olive leaf and olive oil, honey, and camel milk. Later on, these

Phytotherapies: Efficacy, Safety, and Regulation, First Edition. Edited by Iqbal Ramzan.

products formed the basis for the Prophet's medicine (Al-Tibb al-Nabawi), which includes medical treatments, prescriptions of diseases, prevention, health promotion, and spiritual aspects that were recommended by the Prophet (PBUH) to his companions.

Arab-Islamic medicine laid down the principles for the development of Greco-Arab and Islamic medicine that was developed in the golden age of the Arab-Islamic civilization (seventh to fifteenth century). This civilization, which extended from Spain to Central Asia and India, became the wellspring of prosperity and brilliant medical developments and innovations, as well as great achievements not only in medicine but also in chemistry, natural sciences, mathematics, astronomy, architecture, and philosophy. Arab-Islamic medicine was initially based on traditional medical knowledge developed in Arabia, Mesopotamia, Persia, Greece, Egypt, Rome, and India. As a result of this translation activity, a great portion of Greco-Roman scientific heritage has been preserved. There is no doubt that Arab-Islamic medicine was a venue for innovation and change and not simply a continuation of Greek and Persian achievements. As will be seen in the course of this chapter, Arab and Muslim physicians and scholars developed a large and complex evidence-based medical literature exploring and synthesizing theory and practical medicine (Fig. 8.1).

The establishment of evidence-based medicine and pharmacy in the Arab-Islamic world laid down the principles of clinical investigation and, later on, for the development of modern Western medicine and pharmacy. Arabs and Muslims contributed many insights of their own to the development of medicine while acknowledging the knowledge they received from other civilizations. This synthesis resulted in a richer and universal medical system, based on scientific roles and experimentation. The Arab and Muslim scholars Al Tabbari (838–870), Al Razi (Rhazes, 846–930), Al Zahrawi (Albucasis 936–1013), Al-Biruni (973–1050), Ibn Sina (Avicenna, 980–1037), Ibn al Haitham (960–1040), Ibn al Nafees (1213–1288), Ibn Khaldun (1332–1395), Ibn al Baitar (1197–1248), and Ibn Zuhr (Avenzoar, 1091–1161) are regarded as among the great medical authorities of the medieval world, physicians whose textbooks were used in European universities up to the sixteen century. They made accurate diagnoses of diabetes, gout, cancer, plague, diphtheria, leprosy, rabies, and epilepsy. Avicenna's and Rhazes's works on infectious diseases led to the introduction of quarantine as a means of limiting the spread of these diseases. Other physicians laid down the principles of clinical investigation and drug trials, animal tests, and they uncovered the secret of sight. They mastered operations for hernia and cataract, filled teeth with gold leaf, and prescribed spectacles for defective eyesight. And they passed on rules of health, diet, and hygiene that are still largely valid today. While, as mentioned earlier, Greco-Arab and Islamic medicine laid the foundation of modern medicine, some of the currently practiced therapies may seem irrelevant to the modern world. These include magical procedures and folkloric practices of local tradition (Fig. 8.2).

The development of the independent, evidence-based, academically oriented status of pharmacy as a profession charged with ensuring effective and safe use of medicines started in Baghdad in the ninth century. Pharmacists, or *saydalaneh* in

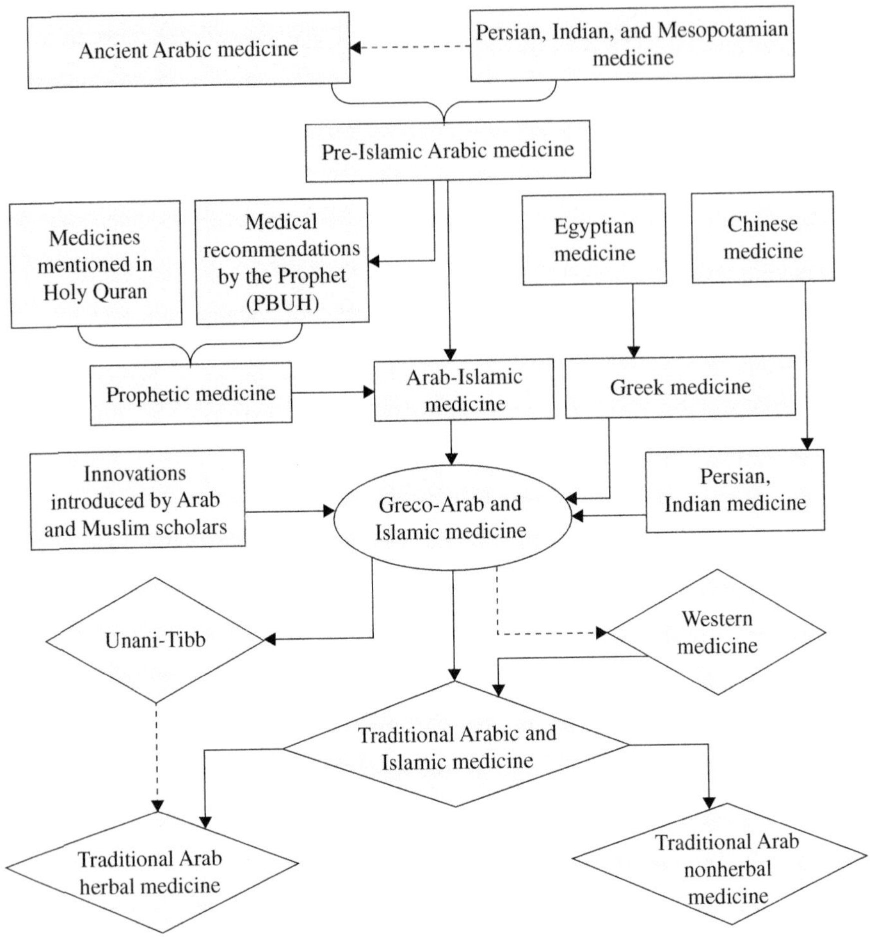

FIGURE 8.1 Development of Greco-Arab and Islamic medicine.

Arabic, translated and interpreted accumulated data on drugs, searched for potential natural products as sources for new drugs, and they even started to elucidate physicochemical properties of these products. Medicines were classified according to their effects on the human body, for example, stimulants, tonics, diuretics, expectorants, topical antiseptic cleansers, analgesics and anesthetics, digestive aids, and oral health agents. Saydalaneh were able to develop a large number of new medicines, including musk, myrrh, cassia, tamarind, nutmeg, senna, camphor, sandalwood, cloves, aconite, ambergris, and mercury. They also developed syrups and juleps and pleasant solvents such as rose water and orange-blossom water as a means of administering drugs. The first pharmacy shop was apparently in Baghdad and drugs were produced and distributed commercially, then dispensed by physicians and pharmacists in a variety of forms: ointments, pills, elixirs, confections,

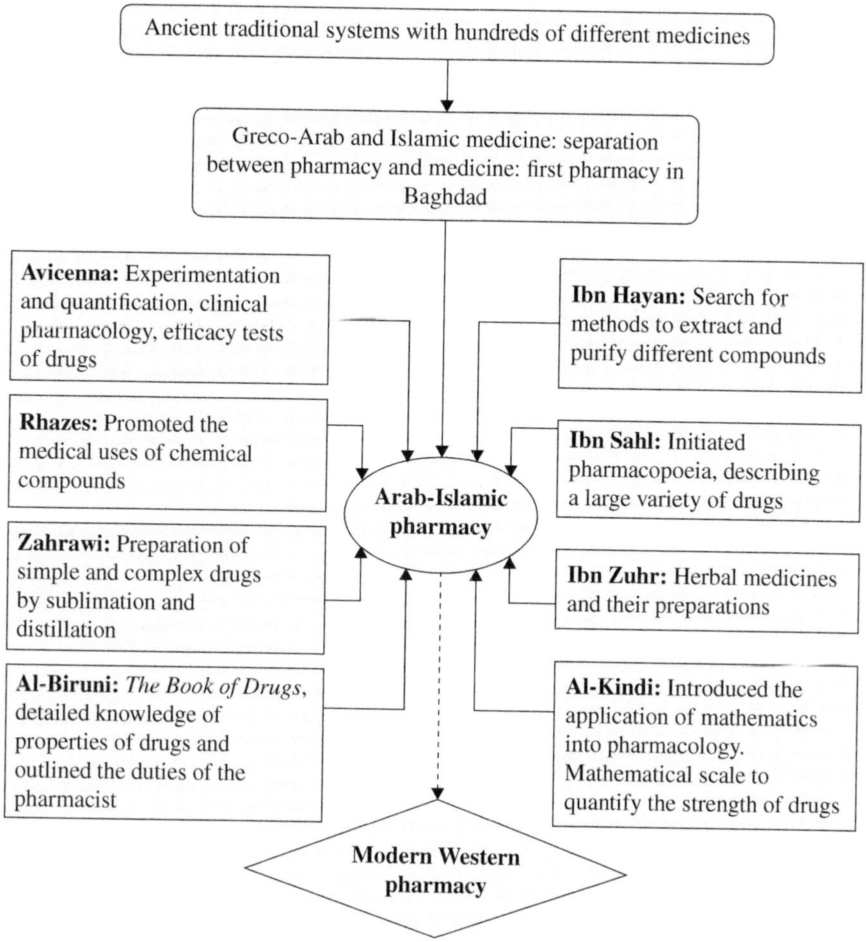

FIGURE 8.2 Scholars of the Greco-Arab and Islamic medicine and their contributions.

tinctures, suppositories, and inhalants. Saydalaneh were required to pass examinations and be licensed and were then monitored by the state [1–18]. The selection of potential natural products as a source of new medicines was based on traditional knowledge developed in the pre-Islamic era based on a long history of trial and error and then by theoretical and practical knowledge introduced by Islam. These include natural products mentioned in the Holy Quran or in the Hadith of the Prophet Mohammad (PBUH), for example, honey, milk, dates, black seeds, olive leaf, and olive oil. In addition, theoretical and practical knowledge developed in other medical systems, which became available to Arab-Islamic scholars after the translation of Greek and Persian scripts, played a central role in developing new medicines. Arab-Muslim physicians developed hundreds of new natural product-based remedies. They were not guided by a long history of trial and error, but mainly by scientific methods, which led to the production of evidence-based medication. Avicenna discussed in his second book, on simple drugs (*materia medica*), the nature and quality of drugs and the way that compounding them influences their effectiveness. He stated "You can tell the potency of drugs in two ways, by analogy (qiyas) and by experiment (tajribah). We say experimenting leads to knowledge of the potency of a medicine with certainty after taking into consideration certain conditions."

According to the school of Greco-Arab and Islamic medicine, the body should be treated as a whole and not just as a series of organs and tissues. Physicians noted that there are individual differences in the severity of disease symptoms and in the individual's ability to cope with disease and healing. Hippocrates thus laid the foundations of the modern theory that thoughts, ideas, and feelings, which he proposed to originate in the brain, can influence health and the process of disease. Rhazes supported this concept by his recommendations. He said: "when the disease is stronger than the natural resistance of the patient, medicine is of no use. When the patient's resistance is stronger than the disease, the physician is of no use. When the disease and the patient's resistance are equally balanced, the physician is needed to help tilt the balance in the patient's favour." In another statement he said "The physician, even though he has his doubts, must always make the patient believe that he will recover, for state of the body is linked to the state of the mind." Later on, Avicenna who defined medicine as "the science from which we learn the states of the human body with respect to what is healthy and what is not; in order to preserve good health when it exists and restore it when it is lacking" supported the views of Rhazes. He stated that "We have to understand that the best and most effective remedy for the treatment of patients should be through the improvement of the power of the human body in order to increase its immune system, which is based on the beauty of the surroundings and letting him listen to the best music and allowing his best friends to be with him." It is now clear that the mind and the body interact, influence, and regulate each other. The perception of stress can lead to the production of "stress hormones," as well as products of the immune system. These "stress hormones" act in a feedback pathway to regulate their own production and the production of certain immune products. These immune products act on the brain to modify behavior and the ability to perceive and to respond to stressful challenges by inducing lethargy, fever, and nausea.

Based on the recommendations of Rhazes and Avicenna, Greco-Arab and Islamic medicine treated patients through a scheme starting with physiotherapy and diet, and if this failed, drugs were used. Rhazes's treatment scheme started with diet therapy, he noted that "if the physician is able to treat with foodstuffs, not medication, then he has succeeded. If, however, he must use medications, then it should be simple remedies and not compound ones." Drugs were divided into two groups: simple and compound drugs. Physicians were aware of the interaction between drugs, thus they used simple drugs first. If these failed, compound drugs, consisting of two or more compounds, were used. If these conservative measures failed, surgery was undertaken (Fig. 8.3).

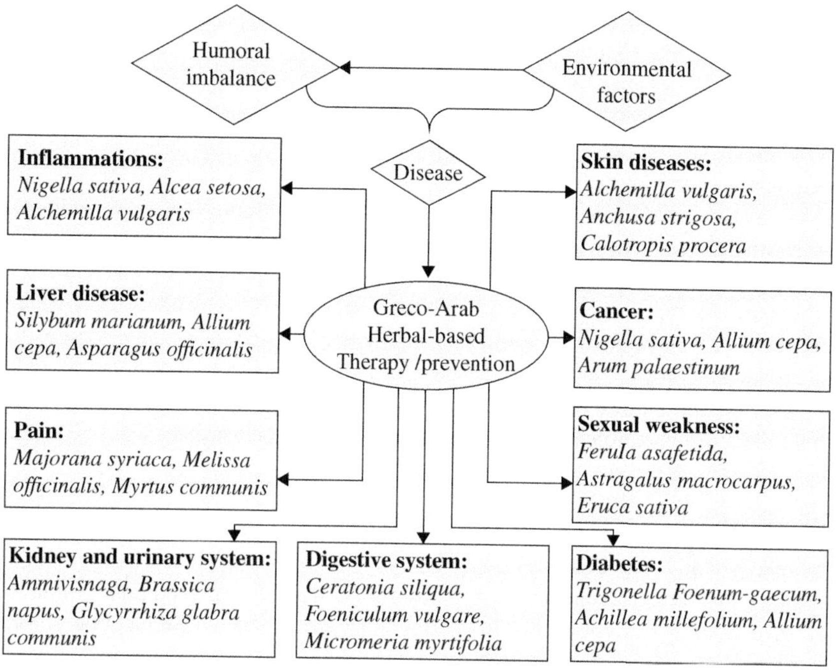

FIGURE 8.3 Greco-Arab and Islamic medicine-based herbal medicines.

8.2 FOOD THERAPY IN GRECO-ARAB AND ISLAMIC MEDICINE

Populations living in the Mediterranean region benefit from a lower incidence rate of chronic diseases and a longer life expectancy than North Americans or Northern Europeans. In general, traditional Mediterranean diet includes a significantly large amount and variety of plant foods, for example, fruits, vegetables, wild edible plants, breads, seeds, nuts, and olive oil. Therefore, it guarantees an adequate intake of carotenoids, vitamins, tocopherols, α-linolenic acid, various important minerals, and several possibly beneficial non-nutrient substances such as polyphenols and anthocyanins and dietary fiber.

Diet is a matter of faith in Arab-Islamic culture and plays an important role in maintaining a healthy body, soul, and spirit. The Prophet Mohammad (PBUH) stated "The stomach is the central basin of the body and the veins are connected to it. When the stomach is healthy, it passes on its condition to veins and in turn the veins will circulate the same and when the stomach is putrescence, the veins will absorb such putrescence and issue the same." Indeed, the Prophet used to prefer food for ailments even more than herbs or medicines. He used everything from barley soup to honey to camel's milk to heal his followers and advised them to eat certain foods to prevent or cure other diseases. For instance, the Prophet (PBUH) mentioned figs and then stated, "If I had to mention a fruit that descended from paradise I would say this is it because the paradisiacal fruits do not have pits … eat from these fruits for they prevent hemorrhoids, prevent piles and help gout." Figs are a high source of fiber, as well as potassium and vitamin B6. Fiber results in bulkier stools, which lessen the incidence of constipation, hemorrhoids, and colon cancer. Fiber also lowers cholesterol and the risk of heart disease. Dates are mentioned in 20 places in the Quran. The Prophet is reported to have said: "if anyone of you is fasting, let him break his fast with dates. In case he does not have them, then with water. Verily water is a purifier." Melon is one of the best recommendations for health the Prophet (PBUH) has given us. Melon is one of the few fruits and vegetables rich in vitamin C, beta-carotene, and potassium. Concerning olive oil, the Prophet (PBUH) said "Eat olive oil and massage it over your bodies since it is a holy tree." Black seeds were regarded as a medicine for that cures all types of diseases. The Prophet once stated, "The black seed can heal every disease, except death." Later on, Rhazes said, "As long as you can heal with food, do not heal with medication." Therefore, in the Greco-Arab medical school, the patients were treated through a scheme starting with diet, exercises, and water baths; if this failed, drugs were used and at last surgery would be used [1–3, 19–33].

8.2.1 Honey

The honeybee makes honey from countless varieties of plant flowers, and it is logical to assume that honey contains many substances of food and medicinal value that modern medicine has yet to discover. Beyond carbohydrates (about 75%), honey contains numerous compounds such as organic acids, proteins, amino acids, minerals, polyphenols, vitamins, and aroma compounds. It should be noted that the composition of honey depends greatly on the botanical origin [33–39].

Honey has been known since thousands of years for its food and therapeutic values. Honey's magic healing effects is praised in the Old and New Testaments, the Holy Quran, the sacred books of India, China, Persia, and Egypt [34–36]. In Arab-Islamic medical schools, as in other medical schools, including Ayurvedic, Chinese, and Roman traditions, honey is considered as a healthy drink and prescribed in the treatment of wounds. The potential therapeutic value of honey is mentioned in the Holy Quran: "And thy Lord has inspired the Bees, to build their hives in hills, on trees and in man's habitations, from within their bodies comes a drink of varying colours, wherein is healing for mankind, Verily in this is a Sign, for those who give thought." AlBasri, a tenth-century Arab philosopher, mentioned uncooked honey for swollen intestine, whereas cooked honey was good for inducing vomiting when poisonous drug was ingested. For that purpose, he recommended mixing one pound of sesame oil with one-third pound of cooked honey. Rhazes (864–932) prescribed honey ointment made of flour, honey vinegar for skin disease and nerve injuries, and honey water for bladder wounds. His book, Al Hawi (*Encyclopedia of Medicine*), a comprehensive medical textbook of medicine, which was translated from Arabic to Latin in the thirteenth century and became a standard textbook of medicine up to the 1700s, stated: "Honey is the best treatment for the gums. To keep the teeth healthy mix honey with vinegar and use as mouth wash daily. If you rub the teeth with such a preparation, it will whiten the teeth. Honey does not spoil and could also be used to preserve cadavers."

Likewise, Avicenna (980–1037) wrote in his Canon: "Honey is good for prolonging life, preserve activity in old age. If you want to keep your youth, take honey. If you are above the age of 45, eat honey regularly, especially mixed with chestnut powder. Honey and flour could be used as dressing for wounds. For lung disease, early stage of tuberculosis, use a combination of honey and shredded rose petals. Honey can be used for insomnia on occasions." Therefore, honey has to be recommended as part of an overall holistic approach to health and should be incorporated into one's everyday diet [12, 34–36].

As mentioned earlier, honey has been used in Arab-Islamic medicine in the treatment of wounds. Over the past five decades, scientific reports affirmed the effectiveness of honey in treating wounds, burns, and serious infections. Many studies have demonstrated that honey has antimicrobial activities *in vitro* and a small number of clinical case studies have shown that application of honey to severely infected skin wounds is capable of clearing infection from the wound and enhancing tissue repair. Honey works differently from antibiotics, which attack the bacteria's cell wall or inhibit intracellular metabolic pathways. Honey has been reported to have an inhibitory effect to around 60 species of bacteria including aerobes and anaerobes, gram-positives and gram-negatives. An antifungal action has also been reported for some yeasts and species of *Aspergillus* and *Penicillium* as well as all the common dermatophytes. Honey is hygroscopic, meaning it draws moisture out of the environment and thus dehydrates bacteria. Its high sugar content hinders the growth of microbes, but the sugar content alone is not the sole reason for honey's antimicrobial effects. When honey is diluted with water, reducing its high sugar content, it still inhibits the growth of various bacterial species that cause wound infections. Over 100 substances are candidates for the particular antibacterial property of the honey, but the active ingredient has not yet been identified.

Honey has been found to contain significant anti-oxidant activity including glucose oxidase, catalase, ascorbic acid, flavonoids, phenolic acids, carotenoid derivatives, organic acids, and amino acids. The antioxidative activity of honey polyphenols can be measured *in vitro* by comparing the oxygen radical absorbance capacity (ORAC) with the total phenolics concentration. There is a significant correlation between the antioxidant activity, the phenolic content of honey, and the inhibition of the *in vitro* lipoprotein oxidation of human serum.

Anti-inflammatory effects of honey in humans were studied in a recent study [40]. The mean plasma concentration of thromboxane and PGE(2) were significantly reduced after ingestion of 70 g honey. At day 15, plasma concentrations of thromboxane, PGE, and PGF decreased by 48, 63, and 50%, respectively. The ingestion of honey decreased inflammation in an experimental model of inflammatory bowel disease in rats. Honey administration is as effective as prednisolone treatment in an inflammatory model of colitis. The postulated mechanism of action is by preventing the formation of free radicals released from the inflamed tissues. The reduction of inflammation could be due to the antibacterial and antioxidant effects of honey or to a direct anti-inflammatory effect. The latter hypothesis was supported in animal studies, where anti-inflammatory effects of honey were observed in wounds with no bacterial infection [39–43].

8.2.2 Olive Oil

In the Arab-Islamic world, olive oil has been commonly used in cooking, cosmetics, pharmaceuticals, and soaps and as a fuel for traditional oil lamps. The Prophet said, "Eat olive oil and massage it over your bodies since it is a holy tree." He also stated that olive oil cures 70 diseases. The Quran also mentions olives as a holy (mubarak) plant: "By the fig and the olive and the Mount of Sinai and this secure city (Mecca)." Olive oil is mentioned in the Quranic verse: "God is the light of heavens (paradise) and earth. An example of His light is like a lantern inside which there is a torch, the torch is in a glass bulb, the glass bulb is like a bright planet lit by a blessed olive tree, neither Eastern nor Western, its oil almost glow, even without fire touching it, light upon light."

As mentioned earlier, the Mediterranean diet, in which olive oil is the main source of fat, has been associated with a low cardiovascular and cancer mortality. The beneficial effects of olive oil on coronary heart disease (CHD) risk factors are now attributed to the high mono-unsaturated fatty acid (MUFA) content and other minor compounds found in the olive oil. Evidences from epidemiological studies suggest that a higher proportion of MUFA in the diet is linked with a low risk of coronary heart disease. There is a large body of clinical data that show that consumption of olive oil can provide cardiac health benefits, such as favorable effects on cholesterol regulation and inhibition of LDL oxidation. It also exerts anti-inflammatory, antithrombotic, antihypertensive, as well as vasodilatory effects both in animals and in humans. Olive oil, however, besides having a high MUFA content, contains oleic acid, which has multiple pharmacologically active components. Olive oil phenolics have shown to have antioxidant properties, higher than that of vitamin E, on lipids and DNA oxidation. They prevent endothelial dysfunction by decreasing the expression of cell adhesion molecules and increasing nitric oxide (NO) production and inducible NO synthesis by quenching vascular endothelium intracellular free radicals. In addition, olive oil phenolic compounds inhibited platelet-induced aggregation and have been reported to enhance the expression of the gene of the antioxidant enzyme glutathione peroxidase. Other potential properties include anti-inflammatory and chemopreventive activity. In animal models, olive oil-derived phenolics retained their antioxidant properties *in vivo* and delayed the progression of atherosclerosis. So far, most of the cardioprotective effects of olive oil in the context of the Mediterranean diet have been attributed to its high MUFA content. It must be noted; however, that oleic acid is one of the predominant fatty acids in widely consumed animal foods in Western diets, such as poultry and pork. A direct association of meat intake with the plasma oleic acid concentration was observed in a Swedish female population. In this population, oleic acid plasma concentrations were higher than those of females of Granada in Spain, without differences in polyunsaturated fatty acid (PUFA) levels. Thus, perhaps a high oleic acid intake is not the sole primary agent responsible for the healthy properties of olive oil. In spite of the promising results displayed in experimental studies, evidence concerning the consumption of phenolic compounds in olive oil is still under investigation. If the beneficial properties of olive oil in humans can be attributed solely to its MUFA content, any type of olive oil, rapeseed/canola oil, or MUFA-enriched fat would provide the same health benefits. Thus, public health implications are involved in order to specifically recommend olive oil and which type of olive oil (i.e., virgin olive oil rich in phenolic compounds), as individualized nutritional strategies for coronary heart disease (CHD) prevention. Taken collectively, one key conclusion is that olive oil is more than a MUFA fat.

The phenolic content of an olive oil can account for greater benefits on blood lipids and oxidative damage than those provided by the MUFA content of the olive oil. Therefore, it is beneficial to use olive oil rich in phenolic compounds in order to achieve additional cardiovascular benefits against cardiovascular risk factors. In addition, olive oils with high phenolic content are in general more bitter and greener than those with low phenolic content and for some individuals the taste may be too strong. Olive oil must be taken as a part of a healthy and pleasant dietary pattern [44–46].

8.2.3 Dates

Dates were among the favored food of the Prophet who said, "Whoever takes seven 'Ajwa dates in the morning will not be affected by magic or poison on that day." Furthermore, Prophet Mohammad (PBUH) is reported to have said: "if anyone of you is fasting, let him break his fast with dates. In case he does not have them, then with water. Verily water is a purifier." Therefore, in Islamic countries, dates and yogurt or milk is a first meal when the sun sets during the month of Ramadan.

Dates contain a high percentage of fructose and glucose. Ten minerals were reported, the major being selenium, copper, potassium, and magnesium. The consumption of 100 g of dates can provide over 15% of the recommended daily requirement of these minerals. Dates contain vitamin C and vitamins B1 (Thiamine), B2 riboflavin, nicotinic acid (niacin), and vitamin A. Vitamins B-complex and C are the major vitamins in dates. In addition, dates contain high levels of water insoluble dietary fiber, which are important for the health of the digestive tract. Many studies recommend the consumption of adequate amounts of dietary fiber from a variety of plant foods. Dates are a good source of antioxidants, mainly carotenoids and phenolics [9, 26–29]. The average contents of phenolics ranged from 193.7 mg/100 g for fresh dates, to 239.5 mg/100 g for dried dates. In general, drying is regarded as unfavorable due to the possibility of inducing oxidative decomposition either enzymatically by polyphenol oxidase and glycosidase or by thermal degradation of phenolic compounds. The oxygen radical absorbance capacity (ORAC) values (a measure for total antioxidant content) are about 1656 µmol/100 g in fresh dates and reduced after drying on average to 1025 µmol/100 g. The antioxidant content of other dried fruits ranged between 340 µmol/100 g for apricot and 3383 µmol/100 g for figs. Thus, in comparison with these fruits, dates are a good source of antioxidants. This finding is supported by other *in vitro* studies published on date antioxidants. For instance, a recent *in vitro* study measured antioxidant and antimutagenic properties of date extract. There was a dose-dependent inhibition of superoxide and hydroxyl radicals by an aqueous extract of date fruit. The amount of fresh extract required to scavenge 50% of superoxide radicals was equivalent to 0.8 mg/ml of date fruit in the riboflavin photoreduction method. An extract of 2.2 mg/ml of date fruit was needed for 50% hydroxyl-radical-scavenging activity in the deoxyribose degradation method. Concentrations of 1.5 and 4.0 mg/ml completely inhibited superoxide and hydroxyl radicals, respectively. Aqueous date extract was also found to inhibit significantly the lipid peroxidation and protein oxidation in a dose-dependent manner. In an $Fe2^{+}$/ascorbate system, an extract of 1.9 mg/ml of date fruit was needed for 50%

inhibition of lipid peroxides. In a time-course inhibition study of lipid peroxide, at a 2.0 mg/ml concentration of date extract, there was a complete inhibition of TBARS formation in the early stages of the incubation period that increased during later stages of the incubation. Similarly, in the high $Fe2^{+}$/ascorbate induction system a concentration of 2.3 mg/ml inhibited carbonyl formation measured by DNPH reaction by 50%. Moreover, a concentration of 4.0 mg/ml completely inhibited lipid peroxide and protein carbonyl formation.

Date fruit extract also produced a dose-dependent inhibition of benzo(α)pyrene-induced mutagenicity on *Salmonella* tester strains TA-98 and TA-100 with metabolic activation. Extract from 3.6 mg/plate and 4.3 mg/plate was required for 50% inhibition of His+ revertant formation in TA-98 and TA-100, respectively. These results indicate that date fruit has quite potent antioxidant and antimutagenic activity and indicates the presence of compounds with potent free-radical-scavenging activity [47–52].

8.2.4 Carob (*Ceratonia siliqua*)

The carob tree is native to the eastern Mediterranean region, where it is known as *kharob*. The main constituents of carob are large carbohydrates, which make carob gummy and able to act as a thickener to absorb water and help bind together watery stools. Carob contains up to 8% protein, vitamins A, B, B2, B3, and D. It is also high in calcium, phosphorus, potassium, and magnesium and contains iron, manganese, barium, copper, and nickel. Carob pulp preparation (carob fiber) is rich in tannins, insoluble dietary fiber, and polyphenols. The effects of dietary fiber consumption on body weight management may be related to gut hormones, which regulate satiety and energy intake. In humans, consumption of carob fiber was shown to have a high antioxidant capacity and to lower serum cholesterol and serum triglycerides. Furthermore, other studies showed that polyphenols might increase fat oxidation and energy turnover in humans and in mice. Consumption of an insoluble dietary fiber rich in polyphenols obtained from carob pulp reduces postprandial free fatty acids and triglyceride concentrations and affects substrate utilization toward lipid oxidation. Carob's tannins exhibit antimicrobial and antiviral activities. Furthermore, carob bean juice (CBJ) has been found to be a powerful adjunct to oral rehydration solution treatment in diarrhea. In children, the treatment of acute diarrhea with standard oral rehydration solution (ORS) provides effective rehydration but does not reduce the severity of diarrhea. As mentioned earlier, carob bean has been used in the Greco-Arab and Islamic medicine to treat diarrheal diseases. Clinical antidiarrheal effects of CBJ were tested in 80 children who were admitted to Tepecik Teaching Hospital in Izmir, Turkey, with acute diarrhea and mild or moderate dehydration. The children were randomly assigned to receive treatment with either standard WHO ORS alone or a combination of standard WHO ORS and CBJ. In the children receiving ORS+CBJ, the duration of diarrhea was shortened by 45%, stool output was reduced by 44%, and ORS requirement was decreased by 38% compared with children receiving ORS alone. Weight gain was similar in the two groups at 24 h after the initiation of the study. Hypernatremia was detected in three patients in the ORS group but in none of those in the ORS+CBJ group. The use of CBJ in combination with ORS did not lead to any clinical metabolic problem [53, 54].

8.2.5 Fig (*Ficus carica*)

The common fig fruit is used as food and for medicinal properties in Greco-Arab and Islamic medicine as well as in Ayurvedic and traditional Chinese medicine. The Prophet (PBUH) mentioned figs and then stated, "If I had to mention a fruit that descended from paradise I would say this is it because the paradisiacal fruits do not have pits … eat from these fruits for they prevent hemorrhoids, prevent piles and help gout." Figs are a high source of fiber, as well as potassium and vitamin B6. Fiber results in bulkier stools, which lessen the incidence of constipation, hemorrhoids, and colon cancer.

Pharmacological and chemical studies have demonstrated antineoplastic or anti-inflammatory effects of both the crude extract and pure compounds. Of particular interest and potential, due to their potent cytotoxic activity against a number of cancer cell lines, are the phenanthroindolizidine alkaloids and the triterpenoids with a C-18 carboxylic acid functional group. In fact, these alkaloids, which have also been found in a small number of other plant genera, are currently under active investigation as potential therapeutic leads. In addition to these cytotoxic compounds, several flavonoids, including anthocyanins, as well as other phenolic compounds, demonstrated antioxidant and anti-inflammatory activities. The sterols found in figs may also help bolster immunity, as well as inhibit inflammation and invasion while promoting apoptosis and differentiation. Coumarins, in many cases, are selectively cytotoxic to cancer cells and also have antioxidant activity and may interfere with the formation of the lipoxygenase product, 5-HETE, to suppress inflammation.

In addition to the potential anticancer and antioxidant properties, fig fruit products exhibit pleasant taste and extremely positive safety profile. The antioxidant action also translates to chronic anti-inflammatory action and decreased insulin resistance. Fig fruits also hold potential as functional foods aimed at normalizing metabolic syndrome and boosting wellness beyond the widely accepted role of figs in the diet for improving bowel performance and as a source of naturally sweet, readily available quick energy [55–59].

8.2.6 Pomegranate (*Punica granatum*)

Pomegranate is commonly known as Rumman in the Arab world and has long been used in traditional Arab-Islamic medicine to treat a variety of ailments, including sore throat, inflammation, and rheumatism. Additional traditional uses include the treatment of diarrhea and colic and to remove intestinal worms in children. The fruit is also used for treating bladder disturbances, strengthening gums, and soothing mouth ulcers. According to the Quran, pomegranates grow in the gardens of paradise. Pomegranates, along with dates and olives, are also mentioned in the following verse from the Holy Quran, which speaks of the dues that have to be paid upon each harvest, as well as the evil of wastefulness. "And it is He Who produces gardens trellised and untrellised and date-palms and crops of different shape and taste and olives and pomegranates, similar (in kind) and different (in taste). Eat of their fruits when they ripen, but pay the due thereof on the day of its harvest and waste not by extravagance. Verily, He likes not those who waste by extravagance."

Punicalagins are the most abundant polyphenols in pomegranate juice. These powerful antioxidants are absorbed into the human body. Other phytochemicals include beta-carotene and polyphenols catechins, gallocatechins, and anthocyanins such as prodelphinidins, delphinidin, cyanidin, and pelargonidin. The fruit also contains Vitamin C at 0.47 mg/100 g. The pharmacological uses of the pomegranate, as was seen with the two other plants of the Quran, dates and olives, are numerous. These include antioxidant, hormone replacement therapy, resolution of allergic symptoms, cardiovascular protection, oral hygiene, ophthalmic ointment, weight loss soap, and as an adjunct therapy to increase bioavailability of radioactive dyes during diagnostic imaging. Pomegranate-mediated antioxidant activity can be considered a means of lowering the threshold for inflammation. Antioxidant activity, as well as suppression of inflammation, may contribute to chemotherapeutic and chemo-preventive utility against cancer [1–3, 60].

8.2.7 Garlic (*Allium sativum*) and Onion (*Allium cepa*)

Garlic and onion are used both as a food and for medicinal applications. Garlic and onion are rich sources of several phytonutrients recognized as important elements of the Mediterranean diet, but are also used in the treatment and prevention of a number of diseases, including cancer, coronary heart disease, obesity, hypercholesterolemia, diabetes type 2, hypertension, cataract, and disturbances of the gastrointestinal tract (e.g., colic pain, flatulent colic, and dyspepsia). Garlic has a high concentration of sulfur-containing compounds. The thiosulphinates, including allicin, appear to be the active substances in garlic. Allicin is formed when alliin, a sulfur-containing amino acid, comes into contact with the enzyme alliinase when raw garlic is chopped, crushed, or chewed. Dried garlic preparations containing alliin and alliinase must be enteric coated to be effective because stomach acid inhibits alliinase. Because alliinase is also inactivated by heat, cooked garlic is less powerful medicinally. The antimicrobial, hypolipidemic, antioxidant, and antithrombotic effects that have been attributed to garlic are thought to be related to allicin and other breakdown products.

The pharmacological activities of garlic are related to the thiosulphinates, volatile sulfur compounds, which are also responsible for the pungency of garlic. Besides these low-molecular weight compounds, onion and garlic are characterized by more polar compounds of phenolic and steroidal origin, often glycosylated, showing interesting pharmacological properties. These latter compounds, compared to the more studied thiosulphinates, possess the advantage of not being pungent and more stable to cooking. Recently, there has been an increasing scientific attention on such compounds [1–3, 61, 62].

8.2.8 Edible Wild Plants

Traditional knowledge and experience are a golden source for the exploration and collection of wild plants. Compared with commonly cultivated vegetables, wild edible plants provide the diet with greater amounts of minerals. Additionally, several of these often so-called famine foods have proved to be important sources of high-quality protein and essential amino acids when compared with the WHO protein standard, as well as being rich in *n*-3 and *n*-6 essential fatty acids. Their antioxidant property, mainly from phytochemicals, was found to be two to three times higher than that of common vegetables. For these reasons, undomesticated greens are recognized as possessing a significant potential for widespread use and development, promoting global food security and nutrition.

Wild edible plants are appreciated for their health and medicinal properties and are commonly consumed in the eastern region of the Mediterranean, where a high percentage of individuals collect wild edible plants and consume them as part of traditional food. According to a survey by the group of Ali-Shtayeh at An-Najah University in Nablus, Palestine, there are 103 edible plants in the West Bank. Sixty-four of these plants are food plants that receive recognition as medicinal agents in the traditional Palestinian herbal medicine and represent a part of the Palestinian medicinal ethnoflora. The most significant species include *Majorana syriaca, Foeniculum vulgare, Malva sylvestris, Salvia fruticosa, Cyclamen persicum, Micromeria fruticosa, Arum palaestinum, Trigonella foenum-graecum, Gundelia tournefortii*, and *Matricaria aurea* [1–3, 63, 64].

8.2.8.1 ***Chicory* (Cichorium intybus)** Chicory is a well-known food and medicinal plant that has been known for its medicinal benefits since the first century. Chicory is cultivated widely throughout Europe for use in salads and it is used much like dandelion in European herbal medicine. That is, it is used in cleaning the body and supporting the liver and also in stimulating the eliminative processes via both the intestine and the kidneys. It is a tonifying plant and the fresh root is used traditionally in chest problems and cold conditions. Herbalists also use the plant as part of mixtures to treat dry coughs, chest pain, and bronchial problems. In the Mediterranean region, chicory is renowned for its digestive properties as a laxative and its blood properties in terms of treating anemia and strengthening blood.

Inulin and oligofructan, polysaccharides found in chicory, were reported in several studies to pass through the stomach and undergo fermentation in the colon. This leads to the selective stimulation of the healthy bifidobacteria population. The health consequences of this include the reduction of colonic diseases and diabetes. Inulin and oligofructan also have a significant effect on cholesterol levels, especially in reducing LDL cholesterol and increasing HDL cholesterol. In addition, other improvements in lipid metabolism, which may be signs of "blood purification" in the traditional herbal terminology, are induced by consuming chicory, along with a clearing out of body fat, bile, and cholesterol through fecal excretion. These changes may support general health and disease prevention. There is very little scientific evidence on the general health benefits of chicory. One study has demonstrated that elderly patients given chicory improve their hepatic function and rehabilitation. Traditional reports have described chicory as having antipyretic, anticolic, hypoglycemic, and hepatic properties.

Regarding its nutritional value, chicory is a rich source of folate, containing 110 mg of folate per half cup of chopped raw chicory. Therefore, an increased consumption of chicory might explain its reported properties in cases of folate-deficiency-related anemia in Greco-Arab and Islamic medicine. As for studies on its digestive properties, chicory is described to be bitter in taste and bitter plants have been used to treat digestive tract disturbances among various traditional systems, relieving gastrointestinal pains. Several sesquiterpene lactones found in chicory confer the bitter taste to the plant. The laxative effect of chicory can be explained by its high content of dietary fiber, having 3–6 g of dietary fiber per half cup of chopped raw chicory. Moreover, inulin, an indigestible carbohydrate, is found to some extent in the stalk of the plant. Inulin, like other dietary fibers, increases bowel movement and is thus responsible for the laxative and digestive-stimulant properties. Other reports of hepatoprotective activity and hypoglycemic effects of chicory are well supported by previous scientific literature [1–3, 64–66].

8.2.8.2 ***Palestinian Thyme*** **(Majorana syriaca)** Palestinian thyme possesses distinctive aroma with a slight warm pungent taste. It has served humans for thousands of years. Hippocrates prescribed it for bronchitis and pleurisy. Traditionally, *Majorana syriaca* has been used to remedy asthma, congestion, rheumatism, sore throats, wounds, ulcers, and tumors. Although *za'tar* is the word for *thyme* in the Arabic language, it is also a term that describes a combination of ground dried thyme leaves, salt, sesame seeds, and the fruits of the tree *Rhus coriaria* (Sumac), a very popular mixture that is used almost daily in the Middle East as food, additive in salads, and spice for pastry and meat.

Majorana syriaca contains Monoterpene hydrocarbons: α-Pinene, Myrcene, α-Terpinene, ρ-Cymene, and γ-Terpinene. Oxygenated monoterpenes: Linalool: Terpinen-4-ol, α-Terpineol, Thymol methyl ether, Carvacrol methyl ether, Thymol, and Carvacrol. The content of essential oil depends on soil, climate, and season. With its high content of volatile oils, the leaves are used in Greco-Arab and Islamic medicine as herbal tea to treat cold, flu, and cough. It has been reported that thyme in general possesses various medicinal benefits. For example, it has antibacterial and antifungal properties, and hence a solution of thyme with its most active ingredient, thymol, is used as over the counter antiseptic mouthwash product. Moreover, thyme extracts are frequently included in the popular cough syrups and prescribed to clear respiratory difficulties, including bronchial problems and coughs. The antimicrobial properties of thyme essential oils are mainly related to their high phenolic content. It is used as a powerful disinfectant in oral pharmaceutical preparations and flavoring agent for many food products [1–3, 67–69].

8.2.8.3 ***Fennel*** **(Foeniculum vulgare)** Fennel is known as Shumar in the eastern region of the Mediterranean and is a perennial herb. In Greco-Arab and Islamic medicine as well as in other different traditional medical systems, fennel is known for its laxative properties. It is also used as a muscle relaxant as well as to treat urinary disorders. In Arab countries, fennel is used for its therapeutic effects on the gastrointestinal system as a pain reliever as well as for its diuretic properties. Experimental as well as human studies demonstrated that fennel oil had antispasmodic and smooth muscle relaxing effects. This activity is due to the similarity found between the anethole, the major component in fennel oil, and the neurotransmitter dopamine. In animal studies, fennel was proved to have significant diuretic properties, which explains our informants' narratives [1–3, 63].

8.2.8.4 ***Gundelia*** **(Gundelia tournefortii)** Commonly known as *Akkoub* in the Arab world, gundelia is a nutritious food as well as medicinal plant. Nutrient analysis of raw Akkoub highlighted its abundance in calcium (642 mg/100 g) and iron (279 mg/100 g). It is recorded that the flowers, leaves, seeds, and stems of Akkoub are used as food sources. In the Middle East, the young and still undeveloped flower buds are sold in the local markets just like artichoke hearts; it is a highly sought item. In Arab-Islamic traditional medicine, *Akkoub* is known for its hypoglycemic and laxative properties [1–3, 63].

8.2.8.5 ***Purslane*** **(Portulaca oleracea)** Purslane is eaten as a salad and vegetable across the world and used traditionally in the treatment of a variety of conditions that include headache, painful urination, stomach ache, enteritis, lack of milk flow in nursing mothers, and in postpartum bleeding. Externally it is used to treat burns, earache, ulcers, itching skin, insect stings, inflammations, skin sores, eczema, and abscesses. These conditions are usually treated with the fresh herb used as a poultice or the expressed juice is also used.

Purslane was shown to have skeletal muscle relaxant effects both *in vitro* and *in vivo*. Water extracts of purslane were found to relax guinea pig gastric fundus, *teniae coli*, and rabbit jejunum as well as contracted the rabbit aorta and raised blood pressure. Topical application of the aqueous extract onto the skin was effective in relieving muscle spasms. Other effects include antibacterial and antifungal, wound healing, anti-inflammatory, uterine stimulant, and diuretic in rabbits. Although norepinephrine may account for some pharmacologic activities, the active principle for most of the biological activities and medicinal properties of purslane remain unidentified. Purslane contains large amounts of l-norepinephrine, a neurohormone that has vasopressor and antihypotensive activities and reduces hemorrhage at the tissue level. It also contains vitamins A, B1, B2, C, niacinamide, nicotinic acid, α-tocopherol, β-carotene, calcium oxalate, malic and citric acids, dopamine and dopa, coumarins, flavonoids, alkaloids, fatty acids—especially omega-3 acids whose concentration in purslane is the highest found in leafy vegetables—glutathione, glutamic acid, and aspartic acid [1–3, 63, 70].

8.2.8.6 ***High Mallow*** **(Malva sylvestris)** High mallow or *Malva* is traditionally used as a laxative and an anti-inflammatory agent. An isolated polysaccharide from malva leaves shows an anticomplement activity, thus modulating the inflammatory response. Moreover, nutrient analysis of a different species of malva eaten in the Arab world highlighted the plant as an important vegetable source of zinc, necessary for a healthy immune system. The laxative effects of malva could be attributed to its high mucilage content [1–3, 63].

8.3 MEDICINAL PLANTS

As mentioned earlier, Arab-Islamic physicians and scholars developed a large and complex medical literature exploring and synthesizing the theory and practice of herbal medicine. They appreciated and translated tens of thousands of Greco-Roman as well as Persian, Chinese, and Indian medical texts into Arabic. They introduced many new ideas about herbs and their efficacy and safety. Baghdad was an important center for Arab herbalism, as was Al-Andalus between 800 and 1400. Al-Dinawari (828–896) is considered to be the founder of Arabic botany for his *Book of Plants*, in which he described the phases of growth and the production of flowers and fruit of about 640 plants. Avicenna's *The Canon of Medicine* is considered the first pharmacopoeia and lists 800 tested plants and minerals. In particular, it introduced clinical trials and randomized controlled trials and efficacy tests. Al Zahrawi (Abulcasis, 936–1013) of Cordoba authored *The Book of Simples*, an important source for later European herbs. The experimental scientific methods were introduced into the field of *materia medica* in the thirteenth century by the Andalusian botanist Abu al-Abbas al-Nabati. Al-Nabati introduced empirical techniques in the testing, description, and identification of numerous *materia medica* and he separated unverified reports from those supported by actual tests and observations. This allowed the study of *materia medica* to evolve into the science of pharmacology. Later on Ibn al Baitar, who lived in Damascus, Syria (1197–1248), compiled *The Book on Drinks and Foods*, which is a collection of different drinks and foods. It is one of the most prestigious books in the Arabian pharmacopeia; it contains 260 references. Other pharmacopoeia books include that written by Abu-Rayhan Biruni in the eleventh century and Ibn Zuhr (Avenzoar) in the twelfth century. Daoud Al-Antaki used different herbs for treating patients and published a book on medicinal herbs summarizing the knowledge of his predecessors. Al Antaki described in his book 57 plants that were used as a source of simple drugs, or frequently as one ingredient in more complex herbal-based remedies. He described the plant as well as its preparations and form of administration. In addition, Al-Antaki mentioned non-indigenous plants, which were brought to the area specifically for their medicinal applications, such as Cornelian cherry, purging croton, and gardenia. He also described the pharmacological uses of typical agricultural crops, such as Caraway, carrot, wild coriander, pear, quince, sugar cane, and walnut (Fig. 8.4).

Medicines were produced in the medieval Arab-Islamic world in a variety of forms: ointments, pills, elixirs, confections, tinctures, suppositories, and inhalants. Herbal drugs were classified according to their effects on the human body, for example, diuretics (promote urination and thus expel toxins), expectorants (remove mucous accumulation), topical antiseptic cleansers, stimulants (prescribed to increase blood flow and raise energy level), tonics (general strength building and disease prevention), analgesics and anesthetics, digestive aids, and oral health [1–5].

The currently observed widespread use and popularity have also brought concerns and fears over the efficacy and safety of the "natural" products available on the market as well as the qualification of healers. It is well known that contamination, adulteration, inappropriate formulation, or lack of understanding of plant and drug interactions can lead to adverse reactions that are life-threatening or lethal to patients. Safety assessment of herbal-based preparations has often been neglected since traditional and prolonged use is usually considered evidence of its safety. Another important factor is the belief that these medicines are prepared according to the principles of the Greco-Arab tradition that forms the basis for the current conventional product. However, a history of traditional usage is not always a reliable guarantee of safety since it is difficult for

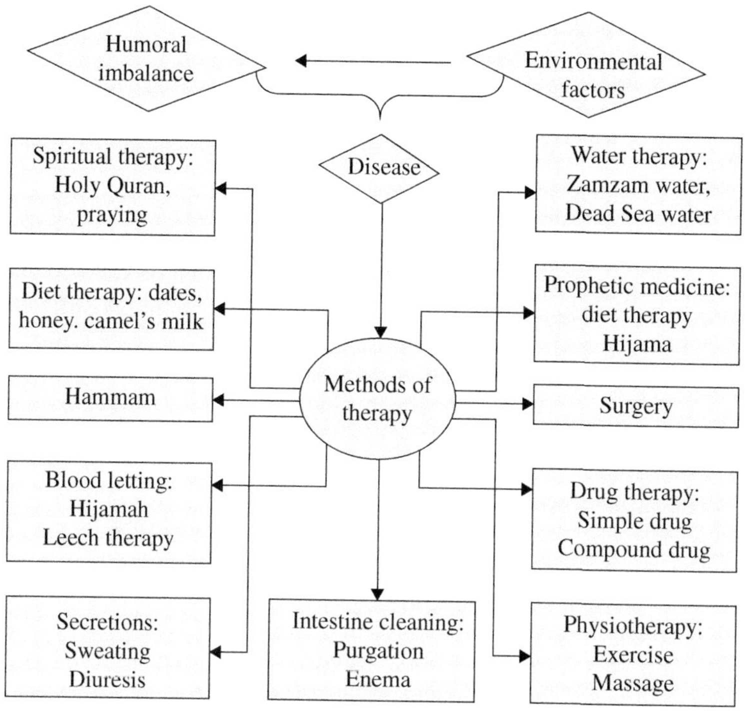

FIGURE 8.4 Methods of therapy used in Greco-Arab and Islamic medicine.

traditional practitioners to detect or monitor delayed effects (e.g., mutagenicity), rare adverse effects, and adverse effects arising from long-term use. Most reports concerning toxic effects of herbal medicines are associated with hepatotoxicity although reports of kidney, nervous system, blood, cardiovascular, dermatologic effects, mutagenicity, and carcinogenicity have also been published in the biomedical literature [1–5]. Recent publications have highlighted the severe side effects from certain herbal-based medicines. Several severe and even lethal side effects have been reported from the use of herbal preparations. These side effects may occur through several different mechanisms, including direct toxic effects of the herbs, effects of contaminants, and interactions with drugs or other herbs. Currently, it is easier to determine which herbal remedies might induce direct toxicity, because it is known which compounds they contain and which of these compounds can induce side effects in a significant proportion of users. Side effects may also occur due to contaminants in herbal-based preparations, as undeclared medicines are often illegally added to the herbs to produce a desired effect and heavy metals, including lead, mercury, or arsenic [1–5]. Furthermore, contamination with microorganisms, microbial toxins, and genetic factors might produce side effects and also affect the content of active constituents in the herbal product.

Most herbalists acknowledge that synthetic medicines are more effective in emergency situations where time is of the essence. However, they believe that over the long term medicinal plants are helpful in treatment as well as in the prevention of diseases and that in addition they provide nutritional and immunological support that synthetic drugs or purified herbal compounds lack. They argue that the different phytochemicals present in herbs will interact to potentiate therapeutic effects of the herb and reduce side effects. Potentiation can be defined as positive interactions that intensify the potency of a bioactive ingredient. Additive and synergistic effects are subsets of potentiation, where two or more compounds in a mixture interact to provide a combined effect that is equal to the sum of the effects of the single molecule (additive) or where combinations of bioactive substances exert effects that are greater than the sum of individual molecules (synergistic). Potentiation can exist between two components in a single plant extract, two components from two different plant extracts, or between a phytochemical and a synthetic drug. A good example of the multicomponent nature of botanicals is illustrated in the field of cancer research. Phytochemicals have been shown to affect various parts of signal transduction pathways including gene expression, cell cycle progression, proliferation, cell mortality, metabolism, and apoptosis. Combination chemotherapy has been the mainstay of cancer treatment for 40 years. It is therefore reasonable to assume that a mixture of compounds (phytochemical or synthetic) would have greater bioactivity than a single compound because a mixture of bioactive compounds has the ability to affect multiple targets. Studies have documented synergistic anticancer effects of phytochemicals including quercetin, catechins, resveratrol, and curcumin with various cancer drugs and/or other phytochemicals. In addition, natural products have been shown to overcome multiple drug resistance in tumors when used in combination with other natural products or drugs. Similar observations have been made in the field of antibiotic research. A number of plant extracts and natural products have been shown to work synergistically with existing antibiotics, restoring antibiotic activity against resistant strains of *Staphylococcus aureus* (methicillin resistant), *Escherichia coli*, and *Shigella* (Table 8.1).

According to recent surveys, there are about 670 medicinal plants in the Eastern region of the Mediterranean and in the coastal Mediterranean region in Egypt. The most commonly used medicinal plants in the Mediterranean region are now discussed briefly [1–4].

8.3.1 Black Seed (*Nigella sativa*)

Black seed is an annual herb commonly used in the Middle East, India, and is now gaining worldwide acceptance. Historical and traditional uses are extensively documented in ancient texts and historical documents. Black cumin seeds and oil are commonly used as a traditional tonic and remedy for many ailments as well as in confectionery and bakery. The seeds, known as black seeds, black cumin, or "Habatul-Barakah" in Arabic, have long been prescribed in Greco-Arab and Islamic medicine as well as in Indian and Chinese traditional medicine for the prevention and treatment of a wide range of diseases, including bronchial asthma, headache, dysentery, infections, obesity, back pain, hypertension, and gastrointestinal problems. Black seeds contain pharmacologically active compounds, namely, thymoquinone, dithymoquinone, thymohydroquinone, and thymol. These compounds are the main active compounds

TABLE 8.1 Commonly Used Cultivated Edible Plants in the Mediterranean Region and Their Evidence-Based Medical Properties

Medicinal herbs	Anti-oxi-dant	Anti-inflam-matory	Immuno-modu-latory	Hypo-tensive	Anti-diabetic	Anti-micro-bial	Anti-cancer	Anti-allergic	Hypo-lipid-emic	Vaso-constric-tive	Cardio-vascular	Sexual function	Hepato-pro-tective	Nervous system	Anti-obesity	Others
Olea europaea	+	+	+	+	+	+	+		+	+	+		+		+	+
Nigella sativa	+	+	+	+	+	+	+	+	+	+	+		+		+	+
Trigonella foenum-graecum		+			+				+			+	+			
Urtica dioica	+	+			+	+		+					+			
Ferula asafoetida						+			+			+	+			
Eruca sativa	+				+	+	+									
Melissa officinalis	+					+								+		+
Salvia fruticosa	+					+								+		
Portulaca oleracea		+			+	+										
Ammi visnaga						+				+						
Silybum marianum	+				+				+				+			
Cuminum cyminum						+	+		+						+	
Ruscus aculeatus										+						
Inula viscose		+				+										
Majorana syriaca	+					+										
Hypericum triquetrifolium	+	+												+		+
Cichorium intybus	+		+		+				+				+			

(*Continued*)

TABLE 8.1 *(Continued)*

Medicinal herbs	Anti-oxi-dant	Anti-inflam-matory	Immuno-modu-latory	Hypo-tensive	Anti-diabetic	Anti-micro-bial	Anti-cancer	Anti-allergic	Hypo-lipid-emic	Vaso-constric-tive	Cardio-vascular	Sexual function	Hepato-pro-tective	Nervous system	Anti-obesity	Others
Punica granatum	+	+	+				+	+			+				+	
Ruta chalepensi		+				+				+				+		
Conium maculatum														+		
Capparis spinosa	+	+						+	+	+		+	+			
Cyperus rotundus				+		+							+			
Sarcoporerium spinosum	+	+			+											
Atriplex halimus					+											
Origanum majorana	+			+		+								+		
Foeniculum vulgare	+	+				+			+							

responsible for the therapeutic effects of black seeds. Black seeds are rich in nutritional value; monosaccharide in the form of glucose, rhamnose, xylose, and arabinose are found in black seed. They contain a non-starch polysaccharide component, which is a useful source of dietary fiber. Black seeds are rich in fatty acids, particularly the unsaturated and essential fatty acids, for example, alpha-linoleic acid (omega 3) and linoleic acid (omega 6). In addition, seeds contain eight of the nine essential amino acids. Black seeds contain carotene, which is converted by the liver into vitamin A, the vitamin known for its anticancer activity. Black seed is also a source of calcium, iron, sodium, and potassium. Required only in small amounts by the body, these elements' main function is to act as essential cofactors for various enzymes [31, 71, 72].

Black seeds were used by ancient Egyptian and Greek physicians to treat nasal congestion, toothache, as a diuretic to promote menstruation, and to increase milk production. The Prophet Mohammad (PBUH) stated in one of his *hadiths* that "The black seed can heal every disease, except death." Avicenna referred to black seed in his Canon of Medicine, as the seed that stimulates the body's energy and helps recovery from fatigue and dispiritedness. In the Unani system of medicine, seeds are regarded as a valuable remedy for a number of diseases. The seed's oil has been used to treat skin conditions such as eczema and boils and to treat cold symptoms. The statement by Prophet Mohammad describing black seed, as "having a remedy for all illnesses" may not be as exaggerated as it appears. Recent research has provided evidence that black seed has the ability to significantly boost the human immune system—if taken over time. In the words of the Prophet, "hold onto the use of the seed," which also emphasizes consistent usage of the seed. Therefore, one important point is that black seed should be regarded as part of an overall holistic approach to health and should be incorporated into one's everyday diet. In this way, nutritional values and therapeutic properties contained in the black seed can assist in maintaining a healthy condition and supplying the immune system with the optimum resources it needs to help prevent and treat diseases. Therefore, in recognition of black seed's substantial nutritional components, as well as its specific medicinal properties, the body's ability to maintain health and promote the healing of a lasting nature is best increased through regular use of black seed.

Black seed is traditionally used in eastern Mediterranean for promoting milk production during breastfeeding. Black seed is an excellent source of added nutrition for both the mother (black seeds mixed with toasted flour, toasted sesame, and honey and prepared as cakes) and the growing child while its immune system boosting properties serve as a natural, safe way to build resistance against illness. In addition, as studies have shown, black seed helps increase milk production during breastfeeding [31, 71, 72].

Therapeutic potential and toxicological properties of the seeds have been extensively studied. A Medline and Google Scholar search using "*Nigella sativa*" and "medicine" reveals more than 1650 citations, including antioxidant, anti-inflammatory, antimicrobial, hypotensive, antinociceptive, choleretic, uricosuric, choleretic, antidiabetic and antihistaminic, immunomodulatory, anticancer, and antifertility effects. These effects are discussed here briefly [1, 2].

8.3.1.1 Antioxidant Activity One of the potential properties of *Nigella sativa* seeds is the ability of one or more of its constituents to reduce toxicity due to its antioxidant activities. *In vitro* investigations reveal that the seeds protect erythrocytes against lipid peroxidation and protein degradation, protect laryngeal carcinoma cells from LPS/cortisol-induced apoptosis, and inhibit the hemolytic activities of snake and scorpion venoms. Other *in vitro* studies indicate that the observed antitoxic effects of the seeds could be attributed to their antioxidant properties [2, 73]. For example, pretreatment of LPS-activated peritoneal macrophages with aqueous extract of the seeds caused a significant decrease in nitric oxide (NO) production and thymoquinone efficiently inhibited iron-dependent microsomal lipid peroxidation *in vitro* in a concentration-dependent manner.

Antioxidant activities of *Nigella sativa* seed oil were found in *in vivo* studies using different hepatic and kidney toxicity murine models [2, 74, 75]. For example, *Nigella sativa* seed oil protected against carbon tetrachloride (CCl_4)-induced hepatotoxicity coinciding with decreasing the elevated serum potassium and calcium levels; improving the serum lipid profile; elevating the reduced erythrocyte, leukocytes, and hemoglobin levels; decreasing the increased liver enzyme levels; and increasing the reduced antioxidant enzyme levels. Other studies indicate that treatment with the oil prevents CCl_4-induced liver fibrosis in rabbits and improves the antioxidant status.

Coupling the fact that black seeds have been widely used in Greco-Arab and Islamic medicine as well as in other traditional medicines with the aforementioned antitoxic properties, it is apparent that the seed's crude oil and its active components may reduce oxidative stress-mediated toxicity induced accidentally by environmental or infectious factors or by anticancer drugs. Since chemotherapy induces massive expansion of the immature granulocytes, which produce large amounts of NO, it might be feasible to follow chemotherapy with thymoquinone treatment that might alleviate the suppressive effects on the immune responses by chemotherapy-induced NO [2].

8.3.1.2 Anti-inflammatory Properties Inflammation is mediated by cytokines, eicosanoids, oxidants, and proteolytic enzymes secreted by macrophages and neutrophils. Nitric oxide (NO) initiates a wide range of toxic oxidative reactions causing tissue injury. In addition, inflammation is also mediated by two main enzymes: cyclooxygenase (COX) and lipoxygenase (LO). COX catalyzes the formation of prostaglandins (PGE) and thromboxane, while LO catalyzes the formation of leukotrienes (LT). Both PGE and LT function as the main mediators of allergy and inflammation [2, 76, 77].

Inhibitory effects of *Nigella sativa* oil and its active ingredients on the production of inflammatory mediators have been reported in several *in vitro* studies. For example, thymoquinone and the crude fixed oil inhibited both cyclooxygenase (COX) and lipoxygenase (LO) pathways of arachidonate metabolism in rat peritoneal leukocytes stimulated with calcium ionophore and inhibited non-enzymatic peroxidation in brain phospholipid liposomes. Furthermore, *in vitro* treatment of calcium- or ionophore-stimulated neutrophils with either crude extract or thymoquinone inhibited the formation of leukotrienes by LO. Thus, the inhibition of both COX and LO pathways are key factors mediating the anti-inflammatory effects of the crude oil of *Nigella sativa* and its active ingredients. These findings were confirmed in several *in vivo* studies. For instance, experimental allergic encephalomyelitis (EAE) is an autoimmune demyelinating disease of the central nervous system that is widely used as a test model for the human multiple sclerosis that is mediated by T lymphocytes. Oxidative stress plays a central role in the onset and progression of this disease [78]. Treatment of EAE animals with thymoquinone results in higher glutathione levels, no perivascular inflammation, or any disease symptoms. In addition, anti-inflammatory effects were seen in ulcerative colitis, an inflammatory disease characterized by cycles of acute inflammation, ulceration, and bleeding of the colonic mucosa. Several factors, such as eicosanoids, leukotrienes, platelet activating factor, and reactive oxygen species (ROS) have been implicated in the pathogenesis of this disease. Other investigations in rats showed that pretreatment of animals with thymoquinone led to complete protection against acetic acid-induced colitis with comparable or even higher effects than sulfasalazine, an anticolitis drug [76–82].

8.3.1.3 Anti-allergic Properties The antiallergic effect of *Nigella sativa* seed components could be attributed to its antihistaminic effects. *In vitro* studies support this notion. Aqueous extract of *Nigella sativa* has shown relaxant and antihistaminic effects on precontracted guinea pig tracheal chains. Moreover, thymoquinone caused a concentration-dependent decrease in the tension of the tracheal smooth muscle precontracted by carbachol and totally inhibited effects of histamine and serotonin on the guinea pig isolated tracheal and ileum smooth muscles. It is suggested that these effects of thymoquinone can be mediated, at least in part, by the inhibition of lipoxygenase products of arachidonic acid metabolism and possibly by nonselective blocking of the histamine and serotonin receptors. Preclinical and clinical studies have also shown antihistaminic effects for *Nigella sativa* seeds. Using gastric ulcer model induced by oral administration of ethanol, which caused a significant increase in mucosal histamine content, rat pretreated with *Nigella sativa* oil before induction of the ulcer showed a significant decrease in gastric mucosal histamine content. The effectiveness of *Nigella sativa* oil in the treatment of allergic rhinitis, bronchial asthma, and atopic eczema was confirmed in a clinical study. The oil decreased the IgE level, eosinophil count, and endogenous cortisol in plasma and urine. Dithymoquinone was found to suppress symptoms in the majority of patients suffering from bronchial asthma [2, 76, 77].

8.3.1.4 Antitumor Properties Antitumor affects were found in both *in vivo* and *in vitro* studies attempting to define the antitumor mechanisms of *Nigella sativa* seeds [2, 9, 20, 23–25]. The findings of these studies indicate the active ingredients of *Nigella sativa* oil, in particular thymoquinone, as a powerful chemopreventive agent against several experimental cancers, including fore-stomach, fibrosarcoma, colon, skin, and hepatic tumors. For example, treatment with aqueous and alcohol extracts completely inhibited the proliferation of cells from the breast cancer cell line (MCF-7). Other studies indicate that *Nigella sativa* extracts induced inhibition of the metastasis-induced factors, including type IV collagenase, metalloproteinase, and serine proteinase inhibitors, angiogenic protein-fibroblastic growth factor, tissue-type plasminogen activator, urokinase-type plasminogen activator, and plasminogen activator inhibitor type 1. It appears that antiangiogenic effects through inhibition of local tumor invasion and metastasis mediate the antitumor properties of *Nigella sativa* whole extract, thymoquinone, dithymoquinone, and other active ingredients. Thymoquinone and dithymoquinone were both found to be cytotoxic against different human tumor cells lines, including the pancreatic adenocarcinoma, human uterine sarcoma, and human leukemia, triggering their apoptosis through arresting the growth of these cells in the G1 phase of the cell cycle associated with increase in the gene and protein expression of p53 and inhibition of the antiapoptotic Bcl-2 protein. This indicates that the antineoplastic effect of thymoquinone is mediated by pro-apoptotic effects modulated by Bcl-2 protein and is linked to and dependent on p53.

Topical treatment with *Nigella sativa* inhibited two-stage initiation/promotion of skin carcinogenesis induced in mice by anthracene/croton oil, where the onset of papilloma formation was delayed and the mean number of papillomas was reduced. The active principle, fatty acids derived from *Nigella sativa*, completely inhibited the growth of Ehrlich ascites carcinoma and Dalton's lymphoma ascites cells. Moreover, oral feeding with *Nigella sativa* extract suppressed hepatic tumor in rats induced by diethylnitrosamine or by partial hepatectomy. Furthermore, the oil suppressed colon carcinogenesis induced by methylnitrosourea or by dimethylhydrazine. These antitumor effects of *Nigella sativa* oil might be attributed to the effect of thymoquinone, since administration of thymoquinone in drinking water resulted in significant suppression of forestomach tumor induced by benzo (α) pyrene. Using the same fibrosarcoma tumor model, administration of *Nigella sativa* extract 30 days after subcutaneous administration of methylcholanthrene restricted the fibrosarcoma tumor incidence to 33.3%, compared with 100% in control tumor-bearing mice, indicating a therapeutic potential. These observations demonstrate that thymoquinone, in addition to its prophylactic and therapeutic antitumor effects, can be a potential chemotherapeutic adjuvant to standard chemotherapy. This might lower the dose of standard chemotherapeutic drugs, while augmenting their antitumor efficacy. Suppression of immune cell function associated with chemotherapy, radiotherapy, and late stages in tumor-bearing hosts is mediated, at least in part, by NO produced by immature granulocytes that are massively generated under these conditions. Therefore, it is possible that the antitumor effects reported for *Nigella sativa* oil and thymoquinone are mediated by their ability to scavenge the NO produced by these cells. The impact of *Nigella sativa* ingredient, in particular thymoquinone, on these cells in the tumor-bearing hosts needs to be explored further. In addition, since chemotherapy induces massive expansion of the immature granulocytes, which produce large amounts of NO, it might be feasible to follow chemotherapy with thymoquinone treatment that might alleviate the suppressive effects on the immune responses by chemotherapy-induced NO. In addition to the possible antioxidant mediating antitumor effects of thymoquinone, it is also possible that its antitumor effect is mediated by the ability to suppress PEG and LT [2, 74–88].

8.3.1.5 Antidiabetic Effects In view of the traditional use of plant mixtures for treatment of diabetes, many scientific investigations have addressed the antidiabetic effects of plant mixtures containing *Nigella sativa*. These studies revealed that the blood glucose lowering effect was due to the inhibition of hepatic gluconeogenesis. For instance, an aqueous extract of a plant mixture containing *Nigella sativa* was found to lower the blood glucose level significantly after oral administration.

In addition, intraperitoneal administration of *Nigella sativa* seed oil produced a significant hypoglycemic effect in normal and alloxan-induced diabetic rabbits. Similar results were seen in rats treated with a mixture of *Nigella sativa* and other plant extracts.

Another study was designed to investigate the possible insulinotropic properties of *Nigella sativa* oil in streptozotocin and nicotinamide-induced diabetes mellitus in hamsters. After four weeks of treatment with *Nigella sativa* oil, significant decrease in blood glucose level together with significant increase in serum albumin level were observed, indicating that the hypoglycemic effect of *Nigella sativa* oil is, at least partially, mediated by a stimulation of beta cells coincident with an increase in serum insulin level and insulinotropic properties in type 2 diabetes model. In another study, the hypoglycemic effect of *Nigella sativa* was supposed to be mediated by extrapancreatic action rather than by stimulation of insulin release. A recent clinical study in human volunteers showed that 1 g of *Nigella sativa* twice daily caused a decrease in blood glucose level after 2 weeks of oral treatment [2, 83].

8.3.1.6 Antimicrobial Properties *Nigella sativa* seed oil and active ingredients have been found to exert antimicrobial activities, including antibacterial, antifungal, antiparasitic, and antiviral effects. Some of these antimicrobial effects have been attributed to the immunomodulatory properties of *Nigella sativa* seed components [2, 20–22]. For instance, *Nigella sativa* was found to exhibit antibacterial activity against several bacterial strains, such as *Escherichia coli, Bacillus subtilis, Streptococcus faecalis, Staphylococcus aureus*, and *Pseudomonas aeruginosa*, as well as against the pathogenic yeast *Candida albicans* and fungus. *In vivo* treatment with *Nigella sativa* oil induced a striking antiviral effect against *Murine cytomegalovirus* (MCMV) infection, indicating a promising therapeutic potential of *Nigella sativa* oil as an antiviral remedy. Both the nonspecific cells, including natural killer cells, and macrophages and specific cells, including T helper and T cytotoxic cells, control the immunity generated toward viral infection. Each cell population plays a central antiviral role at a certain time post-infection, where natural killer cells and macrophages are important during the early phase, while T cells are crucial for clearance of the virus at late stages. Leukocyte-derived cytokines, mainly IFN-γ, are seminal factors in mediating the antiviral response. Interestingly, the antiviral effect of the *Nigella sativa* oil was found to be associated with enhanced response of T helper and T cytotoxic cells and macrophages, augmenting their ability for IFN-γ production that is known to render mice more resistant to MCMV infection. It has been reported that viral infection induces apoptosis leading to lymphocyte depletion in the host and that antioxidant agents can inhibit virus-induced apoptosis as well as viral replication in target cells. Eventually, the antioxidant effect of the *Nigella sativa* oil may represent another mechanism that contributes to its antiviral activity [2, 83, 84].

8.3.2 Fenugreek (*Trigonella foenum-graecum*)

Fenugreek is extensively cultivated in the Mediterranean region. Defatted seeds of fenugreek, which are rich in fiber, saponins, and protein, have been described in early Greek and Latin pharmacopoeias as antihyperglycemic. In addition to the seed, other parts of the herb have also been investigated. Therapeutic effects include the delay of gastric emptying, slowing carbohydrate absorption and inhibition of glucose transport from the fiber content, as well as increased erythrocyte insulin receptors and modulation of peripheral glucose utilization. Fenugreek is another herb that was favored by the Prophet (PBUH) and herbalists for thousands of years.

Several clinical and animal studies show positive effect of the fenugreek seeds in the metabolism of glucose in the body. Fenugreek seeds contain a gel-like soluble fiber that combines with bile acid and lowers triglyceride and LDL cholesterol levels. To maximize their medicinal effect, fenugreek seeds are chopped finely and served as a flavorful preparation or soaked in water overnight. The nicotinic acid alkaloid, trigonelline and coumarin, proved to be the active ingredients responsible for its antidiabetic properties. In clinical trials, consumption of 100 g of defatted seed powder for 10 days improved fasting blood glucose (FBG) in diabetic subjects. Several active ingredients were purified from fenugreek seeds. Recent *in vitro* study has indicated that antidiabetic properties of *Trigonella foenum-graecum* extract are mediated, at least partially, through glucose transporter-4 GLUT4 translocation [89–91].

8.3.3 Sage (*Salvia officinalis*)

Sage has been used for centuries, especially by the Chinese to promote longevity and in Roman ceremonies as a sacred herb. The positive benefits of *Salvia officinalis* to health were reputedly known throughout Ancient Roman times and the Middle Ages. It is used to treat bronchial infections, colds, and coughs. Furthermore, *Salvia officinalis* is traditionally used to treat digestive disorders such as dyspepsia flatulence, poor digestion, and bloating, to reduce excessive perspiration, for example, during menopause. It is also used as a gargle or mouthwash to treat inflammations of the mouth or throat mucosa, such as pharyngitis, tonsillitis, stomatitis, gingivitis, and glossitis [89–91].

8.3.4 Khella (*Ammi visnaga*)

Khella has traditionally been used to treat respiratory diseases such as asthma, bronchitis, emphysema, and whooping cough, as well as cardiovascular disorders, premenstrual syndrome, liver and gall bladder disorders, and antispasmodic action on smaller bronchial muscles and coronary arteries. *Ammi visnaga* may vasodilate the coronary arteries, which increases the blood supply to the myocardium and as a result can be used to treat mild forms of angina. It is also used to treat spasms and constriction of the gallbladder and bile duct and facilitates the discharge of kidney stones and gallstones [89].

8.3.5 Milk Thistle (*Silybum marianum*)

Milk thistle is currently the most scientifically well-investigated medicinal plant in the treatment of liver disease. It has a long history of use in the Greco-Arab and Islamic medicine as well as in the European folkloric medicine as a liver tonic and in the treatment of chronic or acute liver disease, and protecting the liver against toxicity.

The active compounds of *Silybum marianum* are flavonolignans including silibin, silidianin, and silichristine, collectively known as silymarin. Silybin is the component with the highest biological activity and *Silybum marianum* extracts are usually standardized to contain 70–80% silybin. Silymarin is found in the entire plant but is concentrated in the fruit and seeds. Silymarin is not water-soluble and so cannot be taken as a tea. It is typically administered as an encapsulated standardized extract. The oral absorption is rather low; the peak plasma levels after an oral dose are achieved in 4–6 h in both animals and humans. Silymarin is cleared from the body predominantly via the bile and to a lesser extent the kidney. The elimination half-life is 6–8 h [89].

The terms milk thistle, flavonoids, silymarin, and silybin are generally used interchangeably; however, each of these compounds has specific characteristics and actions, with an intrinsic beneficial or toxic effect. Extracts of milk thistle, silymarin and silybin, are the most prescribed natural compounds, with different indications, but with no definitive results in terms of clinical efficacy. In the last 10 years, approximately 12,000 papers have been published on these substances, used as antioxidants or chemopreventives and anticancer agents and especially as hepatoprotectants. This volume of publications indicates that scientific interest in these molecules, or classes of molecules, is high worldwide. Numerous *in vitro* and *in vivo* studies confirm effects of silybin, its antifibrotic, anti-inflammatory, and antioxidant properties, as well as its metabolic effects, combined with this author's own knowledge of the literature. Results indicate that the bioavailability of silybin phytosome is higher than that of silymarin and is less influenced by liver damage; silybin does not show significant interactions with other drugs. Experimental studies have clearly demonstrated the antifibrotic, antioxidant and metabolic effects of silybin; previous human studies were insufficient for confirming the clinical efficacy in chronic liver disease, while ongoing clinical trials are promising. On the basis of literature data, silybin appears a promising drug for chronic liver disease [89, 92].

8.3.5.1 Detoxifying and Hepatoprotective Effects Many studies have demonstrated the beneficial hepatoprotective effects of treatment with silymarin. *Silybum marianum* and its derivatives have been used for centuries for the treatment of liver disease. Many scientific studies have been published pertaining to the potential use of *Silybum marianum* or its derivatives for the treatment of alcoholic liver disease. Several chemotherapeutic agents are metabolized by the liver and can exert hepatotoxicity, with the net result of drug withdrawal. Cancer patients taking these therapies often self-medicate with milk thistle because of its reputation as a liver protectant. Clinicians also prescribe it to cancer patients for the same purpose. The rationale behind milk thistle use is to provide support to the liver while it performs multiple functions, including responding to the increased metabolic demands caused by tumor growth, assisting in metabolizing products generated when a tumor is killed or reduced by chemotherapy and radiation and assisting in the processing of drugs prescribed to cancer patients. Silybin is also considered a potent inhibitor of human intestinal β-glucuronidase, blocking the release and reabsorption of free xenobiotics and their metabolites from their glucuronide conjugates. Liver being the primary organ that cleanses and detoxifies the blood, many detoxification pathways include a component of liver support or what is often called liver cleansing. For these reasons, silybin is commonly included in detoxification regimens. However, only one randomized, double-blind study has reported the effects of milk thistle in patients during cancer therapy; 50 children with acute lymphoblastic leukemia and grade 2 or higher hepatic toxicity were randomized to receive a milk thistle supplement (Siliphos®, Thorne Research, Dover, Idaho) (5.1 mg/kg/day) or placebo for 28 days. The authors reported significant reductions in AST levels ($P<0.05$) and a trend toward a significant reduction in ALT levels ($P<0.07$). A significantly larger number of children in the milk thistle group developed a greater than 50% reduction in total bilirubin at day 28 compared to placebo ($P<0.0069$) [89, 92, 93].

The most remarkable therapeutic properties of silymarin are its antitoxic effects in the treatment of *Amanita* mushroom poisoning. The *Amanita* genus is widespread in Europe and North America and mushroom collectors consider several species choice items. Unfortunately, this mushroom contains two extremely powerful hepatotoxins: amanitin (LD_{50} is 100 μg/kg body weight) and phalloidin. In mice, silymarin was 100% effective in preventing liver toxicity if given before or up to 10 min after Amanita toxin poisoning. Severe liver damage was avoided if silymarin was administered within 24 h. In a study with dogs, none of the dogs died when given silymarin 5–24 h after ingesting an LD_{50} dose of *Amanita phalloides* (85 mg/kg). In comparison, untreated dogs experienced a mortality rate of 33%. Liver enzyme studies and liver biopsies in the control and treated dogs demonstrated significant hepatoprotection for silymarin. The hepatoprotective effects of silymarin in humans after ingestion of *Amanita* toxins have been demonstrated repeatedly. In one series of 18 patients treated with silymarin, all patients survived except one particularly high-dose suicide. The authors concluded that administration of silymarin even up to 48 h after mushroom ingestion appears to be an effective measure to prevent severe liver damage in *Amanita phalloides* poisoning [89, 92, 93].

8.3.5.2 Alcoholic Liver Disease The metabolism of ethanol is primarily through conversion into acetaldehyde by three enzymes. These include catalase (CAT), alcohol dehydrogenases (ADH), and the microsomal ethanol oxidizing system (MEOS). Acetaldehyde is more hepatotoxic in its effects than ethanol. Acetaldehyde produces multiple effects in the body. Binding with proteins, glycoproteins, and membrane phospholipids results in cellular dysfunction such as swelling, impairment of the mitochondrial electron transport chain, and upregulation of protein kinase. Acetaldehyde also increases the production of cytokines interleukin (IL)-1a, IL-6, and tumor necrosis factor alpha (TNF-α) and promotes inflammatory responses via the activation of necrosis factor kappa beta (NF-kB). Furthermore, TNF-α promotes free radical production by mitochondria, activated neutrophils, and hepatic Kupffer cells. Numerous *in vitro* studies of Kupffer cells and other types of immune cells investigated the effect of *Silybum marianum* or its derivatives on the formation of the nitric oxide (NO), TNF-α, prostaglandin E2 (PGE2), and leukotriene B4 (LTB4) found beneficial effects of *Silybum marianum* its active compounds. For instance, controlled *in vitro* studies have demonstrated that silymarin inhibits NF-kB activation in a variety of cell lines. TNF-mediated NF-kB activation was inhibited in a dose-dependent manner. In addition, silymarin appeared to block the activation of NF-kB by phorbol ester, lipopolysaccharide (LPS), okadaic acid, and ceramide, partially inhibited NF-kB induction by hydrogen peroxide, and was found to inhibit NF-kB activation in all cell types studied [89, 92, 93].

8.3.5.3 Liver Regeneration Silymarin (100 mg/kg) was shown to enhance liver regeneration in hepatectomized rats, as shown by the increased weight of treated rats as compared with controls. Proliferative activity, as measured by the stathmokinetic test (counting numbers of mitotic cells in prepared slides of liver tissue from hepatectomized rats), was increased in treated animals as compared to controls. The rate of DNA synthesis in rats treated with silibin following partial hepatectomy was increased from 23 to 35% compared with controls. No change in DNA synthesis was seen in normal livers.

Clinical studies have varied greatly in quality, with the majority limited by inadequate sample size, lack of uniformity in the population treated, lack of standardization of preparations studied, variability in dosing regimens, inconsistent outcome measures, and a lack of information on concurrent use of alcohol during the treatment period. While *Silybum marianum* and its derivatives appear to be safe and the available evidence on the mechanisms of action appears promising, there is currently insufficient data from well-conducted clinical trials to recommend their use in patients with alcoholic liver disease [89, 92, 93].

Taking all this information, although silymarin appears safe and may have several properties that make it a potentially attractive therapy for alcoholic liver disease, such as effects on liver regeneration, lipid peroxidation, inflammation, and hepatic fibrogenesis, there are insufficient data from well-conducted clinical trials at present to routinely recommend its use for patients with alcoholic liver disease. The widespread availability for clinical trials of a standardized pure *Silybum*/silymarin/silibin product as proposed by the National Institutes of Health will be an important first step to the systematic study of whether this herbal compound may be an effective therapy for alcoholic and other liver diseases [89, 92, 93].

8.3.6 Marjoram (*Origanum majorana*)

Marjoram is an herbaceous and perennial plant native to southern Europe and the Mediterranean. Traditionally, it is used as a folkloric remedy against stiff joints and muscle spasms, including tics, excessive coughing, menstrual cramps, asthma, indigestion, headache, and rheumatism. Among the herbs of the Lamiaceae family, rosemary has been more extensively studied and its extracts are the first marketed natural antioxidants. *Origanum majorana*, which belongs to the same family, has gained the interest of many research groups as a potent antioxidant. A recent study has shown that *Origanum majorana* was effective in inhibiting the formation of advanced glycation end-products (AGEs). The antiglycation activities of *Origanum majorana* were attributed in part to their antioxidant activity and its abilities to scavenge reactive carbonyls. The ability of *Origanum majorana* to react with carbonyls was the major mechanism for inhibition of protein glycation. Furthermore, *Origanum majorana* alleviated oxidative stress under diabetic conditions through the inhibition of lipid peroxidation, preventing and/or delaying the onset of renal damage. These results suggested that *Origanum majorana* might prevent or improve AGE-associated chronic conditions [89].

8.3.7 Garlic (*Allium sativum*) and Onion (*Allium cepa*)

Garlic and onion are used both as a food and for medicinal applications. Garlic has been used for thousands of years for medicinal purposes. Sanskrit records mention its medicinal use about 5000 years ago and it has been used for at least 3000 years in Chinese medicine. The Egyptians, Babylonians, Greeks, and Romans used garlic for healing purposes. Garlic and onion are rich sources of several phytonutrients recognized as important elements of the Mediterranean diet but are also used in the treatment and prevention of a number of diseases, including cancer, coronary heart disease, obesity, hypercholesterolemia, diabetes type 2, hypertension, cataract, and disturbances of the gastrointestinal tract (e.g., colic pain, flatulent colic, and dyspepsia).

Garlic extracts and several of the isolated constituents have been tested in *in vitro* and *in vivo* studies for different biological activities. Garlic is one of the most popular traditional remedies universally used for its antidiabetic activities. These include lowering blood glucose level, inhibition of the formation of lipid peroxides, reactivation of the antioxidant enzymes, and restoring levels of GSH and metals (copper, zinc, iron, magnesium, and selenium) in STZ-diabetic rats. Several other studies investigated the effect of garlic on enzymes, biochemical parameters, and minerals in STZ/alloxan-diabetic mice/rats. In a randomized, single-blind, placebo-controlled clinical study, garlic tablets (300 mg twice daily for 2 weeks) significantly reduced serum total cholesterol and LDL cholesterol in type II diabetic patients [89, 94].

8.3.8 Tayun (*Inula viscose*)

Tayun has been regarded for centuries as one of the most effective medicinal plants in the Mediterranean region. *Inula viscosa* is traditionally used to treat infection, inflammation, fever, and external skin irritation. It is also effective in wound healing. The roots are used against cough and catarrh, as an antiseptic and expectorant, which loosens phlegm and supports mucus membranes [89].

8.3.9 Rocket (*Eruca sativa*)

Rocket is a member of the Brassicacae family. In recent years, it has gained greater importance as a salad vegetable and spice, especially among Middle Eastern populations and Europeans. *Eruca sativa* possess various therapeutic properties including inhibition of tumourigenesis, antiulcer, and hepatoprotective activities. *Eruca sativa*, known as Jarjeer in the Arab world, finds widespread use in Greco-Arab and Islamic medicine. These include antibacterial action (for eye infections) and increasing fertility and sperm production, as an aid to digestion and kidney function. The seeds and tender leaves are known in Arabian countries to increase sexual desire and are considered to be an aphrodisiac. It is also used as a carminative and to alleviate abdominal discomfort and improve digestion. Ibn Wahsiyya (ca. 900 A.D.) is quoted as stating that the ground seeds when mixed in a cream and spread on the face can be used for the treatment of acne. It has been reported that the rocket seed ethanolic extract possesses potent antioxidant and renal protective and diuretic activities. Phytochemical studies of rocket leaves and seeds have revealed the presence of glucosinolates [89, 95].

8.3.10 Nettle (*Urtica dioica*)

Nettle is the common name for any of the 30–45 species of flowering plants of the genus *Urtica*. The plant has a variety of uses in Western traditional medicine for genitourinary ailments (nocturia, frequency, dysuria, urinary retention, irritable bladder, and infections), kidney disorders, allergies, diabetes, internal bleeding (including uterine bleeding, epistaxis, and melena), anemia, GI tract ailments (diarrhea and dysentery and gastric hyperacidity), musculoskeletal aches, osteoarthritis, and alopecia. Nettle is widely used in Greco-Arab medicine to treat stomach ache, rheumatic pain, colds and cough, and liver insufficiency. It is also used as a hypotensive and anti-inflammatory agent. However, only a few of these uses have a scientific basis that supports their therapeutic uses. Nettle seems to have an antiproliferative effect on prostatic epithelial and stromal cells, which could be a potential mechanism of action in patients with benign prostatic hyperplasia (BPH). There is evidence that oral or topical use of nettle leaf extract might reduce pain in patients with osteoarthritis. Some clinicians use nettle leaf extract in combination with conventional nonsteroidal anti-inflammatory drugs (NSAIDs) or other analgesics. Evidence suggests that adding nettle might allow for a lower analgesic dose in some patients. Topically, stinging nettle leaf seems to relieve pain and disability in patients with osteoarthritis of the thumb, according to preliminary research. More evidence is needed to assess the potential of nettle for these uses. There is some data that nettle can lower blood glucose levels. Bnouham and colleagues [96] demonstrated that when administered 30 min before glucose loading aqueous extract of nettle (250 mg/kg) showed a pronounced glucose lowering effect ($33 \pm 3.4\%$ decrease compared to the control value 1 h after glucose loading). The same extract also exerted a hypotensive action in rats, although it causes vasoconstriction of the aorta via activation of alpha1-adrenergic receptors. In a recent study, nettle ethanol/water/extract almost doubled GLUT4 translocation in L6-GLUT4myc cells and increased about 1.6-fold in the insulin-stimulated state. There are few reports about the toxic effects of nettle; its root can cause gastrointestinal complaints, sweating, and allergic skin reactions and nettle juice can sometimes cause diarrhea. Topically, fresh nettle leaves can cause localized rash, itching, stinging, and tongue edema [89, 96–99].

8.3.11 Peppermint (*Mentha piperita*)

Peppermint has a long history of safe use in the Arab world, both in medicinal preparations and as a flavoring agent. Peppermint is used traditionally for several medicinal purposes for gastrointestinal tract ailments such as nausea, vomiting, diarrhea, cramps, flatulence, and dyspepsia. Other uses include treatment of common cold, inflammatory conditions of the mouth, pharynx, sinus, liver and gallbladder, and bowel. Peppermint alters the physiology of the gastrointestinal tract and has been used in clinical trials for the treatment of barium enema-related colonic spasm, dyspepsia, and irritable bowel syndrome. In nine studies, 269 healthy subjects/patients underwent exposure to peppermint either by topical intraluminal (stomach or colon) or oral administration. It was found that peppermint produces an inhibition of spontaneous peristaltic activity, reduces total gastrointestinal transit or gastric emptying, decreases the basal tone in the gastrointestinal tract, reduces the slow wave frequency in the esophagus, small intestine, which slows peristaltic movements and inhibits potassium depolarization-induced responses in the intestine. It was observed that peppermint relaxed the lower esophageal sphincter and that it was useful as an antispasmodic agent for double-contrast barium meal examination and in patients with dyspepsia.

Menthol, the active compound of peppermint, is a monoterpene that is widely used as a natural product in cosmetics, a flavoring agent and as an intermediate in the production of other compounds. Various extracts of peppermint contain menthol as a major active constituent and have been used for centuries as traditional medicines for a number of ailments including infections, insomnia, and irritable bowel syndrome as well as an insect repellent. Menthol's characteristic cooling sensation is due, in part, to the activation of sensory neurons generally termed transient receptor potential (TRP) channels. Aside from its cold-inducing sensation, menthol exhibits cytotoxic effects in cancer cells, induces reduction in malignant cell growth, and elicits synergistic excitation of GABA receptors and sodium ion channels resulting in analgesia. In addition, menthol was found to have antibacterial activity. Several toxic effects have been associated with the ingestion of peppermint oil such as heartburn, nausea, vomiting, allergic reactions, flushing, and headache [98, 100, 101].

8.3.12 Chamomile (*Chamomilla recutita*)

Chamomile is an annual herbaceous plant indigenous to Europe and Western Asia. Also known as German chamomile or wild chamomile, the plant is cultivated for its flower heads. Chamomile is one of the most popular single ingredient herbal teas, or tisanes. Infusions and essential oils from fresh or dried flower heads have aromatic, flavoring, and coloring properties. These are both used in a number of commercial products including soaps, detergents, perfumes, lotions, ointments, hair products, baked goods, confections, alcoholic beverages, and herbal teas. Chamomile tea, brewed from dried flower heads, has been used traditionally for medicinal purposes. The main constituents of the flowers include several phenolic compounds, primarily the flavonoids, apigenin, quercetin, patuletin, and luteolin. The principal components of the essential oil extracted from the flowers are the terpenoids α-bisabolol and its oxides and azulenes, including chamazulene. Chamomile has moderate antioxidant and antimicrobial activities and significant antiplatelet activity *in vitro*. Animal studies indicate potent anti-inflammatory action, some antimutagenic and cholesterol-lowering activities, as well as antispasmodic and anxiolytic effects. Clinical studies have shown that chamomile might also be effective for the treatment of dyspepsia and mucositis. Preliminary research suggests that it blocks slow wave activity in the small intestine, which may slow peristaltic movement. In a clinical trial of 98 patients receiving local radiation and systemic chemotherapy, chamomile oral rinse prevented mucositis secondary to radiation therapy and some by chemotherapeutic drugs. Kassi and colleagues [102] demonstrated that aqueous extracts of chamomile induce osteoblast differentiation and have anticancer effects on breast and uterine cancer cells *in vitro* (concentrations of 10–100 μg/ml). They concluded that chamomile extracts produce these effects by acting as a selective estrogen receptor modulator [98, 102].

8.3.13 Coriander (*Coriandrum sativum*)

Coriander is a small, aromatic, herbaceous, and perennial herb widely distributed in India, Pakistan, West Asia, the Mediterranean, and America. All parts of the plant are edible, but the fresh leaves and the dried seeds are commonly used in cooking. The main constituents of coriander seeds are essential oils, sugars (glucose, fructose, and sucrose), alkaloids, flavones, resins, tannins, anthraquinones, sterols (beta-sitosterol and beta-sitosteroline), and fixed oils. In the Greco-Arab traditional medicine, preparations containing coriander seed extract have been used as stimulants, carminatives, antispasmodics, diuretic, and antirheumatic. Coriander seeds have been reported to be potent antioxidants *in vitro* and have been shown to reduce serum cholesterol in experimental hyperlipidemia. The anti-inflammatory properties of coriander extract have been demonstrated in carrageenan-induced paw edema in animals [2, 103–105].

8.3.14 Anise (*Pimpinella anisum*)

Anise is a member of the Apiaceae family that includes fennel, caraway, cumin, cilantro, dill, and carrots. Anise is an annual grassy herb, 30–50 cm high with white flowers and small green to yellow seeds, which grows in the Eastern Mediterranean Region, West Asia, the Middle East, Mexico, Egypt, and Spain. It is commonly used to flavor candy, foods, and liqueurs. The seeds are used in traditional Arab medicine for a variety of conditions, particularly for their ability to produce a reduction in gas and bloating and to settle digestion-related problems. Seed-based remedies are commonly used in infants and children to induce relief from cases of colic; these remedies are also given to individuals of all ages to relieve symptoms associated with indigestion and nausea arising from a variety of causes. An additional therapeutic effect of the seeds is their antispasmodic properties, which are effective in reducing the symptoms of menstrual pain, the discomfort during asthma attacks, as well as in the treatment of whooping cough and other spasmodic coughs. Furthermore, remedies made from the seeds are also believed to increase breast milk production; these remedies may also be beneficial in the treatment of impotence and frigidity. The essential herbal oils derived from anise are also used in the treatment of similar complaints in patients. It is recommended that patients consume the essential oil while they are under health professional supervision. Pregnant women must also abstain from taking anise, with the exception of minute amounts, such as those normally used during cooking.

Various properties such as antimicrobial, antifungal, antiviral, antioxidant, and insecticidal effects have also been reported for the aniseeds. Findings also reveal that aniseeds can cause gastric protection, muscle relaxation, and affect the digestive system. In diabetic patients, it has hypoglycemic and hypo-lipidemic effects and reduces lipid peroxidation. Furthermore, aniseeds exhibited anticonvulsant effect, reduced morphine dependence, and induced conditioned place aversion in mice [105, 106].

8.3.15 Rosemary (*Rosmarinus officinalis*)

Rosemary is a woody shrub with fragrant evergreen needle-like leaves and blue flowers that last through spring and summer. Rosemary is a common household plant grown in many parts of the world. It is used for flavoring food, a beverage drink, as well as in cosmetics. The fresh and dried leaves are traditionally used throughout the Mediterranean region; they have a bitter, astringent taste and are highly aromatic, which complements a wide variety of foods. *Rosmarinus officinalis* is known for its muscle-relaxant effects, including the smooth muscles of the digestive tract and uterus. Thus, it is traditionally used to soothe an upset digestive system and relieve menstrual cramps. Several studies have indicated strong antioxidant and antimicrobial effects of *Rosmarinus officinalis*. The most important active constituents of rosemary are caffeic acid and its derivatives such as rosmarinic acid. These compounds have antioxidant effects. The phenolic compound, rosmarinic acid, derives one of its phenolic rings from phenylalanine via caffeic acid and the other from tyrosine via dihydroxyphenyl-lactic acid. Rosmarinic acid is well absorbed from the gastrointestinal tract and also from the skin. It increases the production of prostaglandin E2 and reduces the production of leukotriene B4 in human polymorphonuclear leucocytes and inhibits the complement system. It can be concluded that rosemary and its active compounds have a therapeutic potential in the treatment or prevention of bronchial asthma, spasmogenic disorders, peptic ulcer, hepatotoxicity, atherosclerosis, ischemic heart disease, cataract, inflammatory diseases, and cancer [105].

8.3.16 Devil's Dung (*Ferula asafetida*)

Devil's dung is a plant native to central Asia and it is held in high regard amongst Arab herbalists. *Zallouh* is the common name in the Middle East for the roots of the species *Ferula hermonis* growing on the slopes of Mount Hermon in the Syrian Golan Heights and has been used for centuries as a folk remedy to treat frigidity in women and erectile and sexual dysfunction in men. Greco-Arab and Islamic medicine supports its use as a sexual tonic to encourage potency. Al-Razi (Rhazes 841–926) reported that Indians used *Ferula asafoetida* L as the main botanical aphrodisiac, several centuries before his time. Ibn Sina (Avicenna) and Al-Antaki have also emphasized the aphrodisiac effect of *Ferula asafoetida* L [107, 108].

8.3.17 Ginger (*Zingiber officinale*)

Ginger has been used as a medicine in Chinese, Ayurvedic, and Greco-Arab and Islamic medicine since ancient times. In China, for example, the underground stem, or rhizome, of the plant ginger, has been used to aid digestion and treat stomach upset, diarrhea, and nausea for more than 2000 years. The rhizome of ginger has found widespread use in Greco-Arab and Islamic medicine. It is one of the plants that are mentioned in the Holy Quran as one of the drinks of Paradise: "And in it, their drink is mixed with ginger." Ginger has also been used to treat arthritis, colic, diarrhea, and heart conditions. In addition to these medicinal uses, ginger continues to be valued around the world as an important cooking spice and is believed to help manage the common cold, flu-like symptoms, headaches, and even painful menstrual periods. Currently, healthcare professionals recommend ginger for preventing or treating nausea and vomiting associated with motion sickness, pregnancy, and cancer chemotherapy. It is also used as a digestive aid for mild stomach upset, as adjunct therapy in inflammatory conditions such as arthritis, and may even be useful in heart disease or cancer [105].

REFERENCES

[1] Zaid H, Silbermann M, Ben-Aryeh E, Saad B (2012) Greco-Arab and Islamic herbal-derived anti-cancer modalities: from tradition to molecular mechanisms. *Evid Based Complement Alternat Med* 2012: 1–13.

[2] Saad B, Said O (2011) *Greco-Arab and Islamic Herbal Medicine: Traditional System, Ethics, Safety, Efficacy and Regulatory Issues*. Hoboken: John Wiley & Sons, Inc.

[3] Saad B, Zaid H, Said O (2013) *Tradition and perspectives of diabetes treatment in Greco-Arab and Islamic medicine*. In: Watson RR, Preedy VR (eds.) *Bioactive Food as Dietary Interventions for Diabetes*, pp. 319–326. San Diego: Academic Press.

[4] Saad B, Azaizeh H, Said O (2005) Tradition and perspectives of Arab herbal medicine: a review. *Evid Based Complement Alternat Med* 2: 475–479.

[5] Pormann PE, Savage-Smith E (2007) *Medieval Islamic Medicine*. Edinburgh: Edinburgh University Press.

[6] Michael Hamilton Morgan (2007) *Lost History the Enduring Legacy of Muslim Scientists, Thinkers and Artists*. Washington, DC: National Geographic Society.

[7] Bilal A, Jamal A (2007) Unani system of medicine. *Pharmacogn Rev* 1: 210–214.

[8] Hajar A, Bin AL, Ali H (2002) History of medicine. *Heart Views* 3: 10–17.

[9] Deuraseh N (2006) Health and medicine in the Islamic tradition based on the book of medicine (Kitab Al-Tibb) of Sahih Al-Bukhari. *JISHIM* 5: 2–14.

[10] Said O, Zaid H, Saad B (2009) *Greco-Arab and Islamic herbal medicine and cancer treatment/prevention*. In: Watson RR, Preedy VR (eds.) *Bioactive Foods and Extracts: Cancer Treatment and Prevention*, pp. 49–66. New York: Taylor & Francis Group.

[11] Saad B, Azaizeh H, Abu Hijleh G, Said O (2006) Safety of traditional Arab herbal medicine. *Evid Based Complement Alternat Med* 3: 433–439.

[12] Ibn Sena (Avicenna) (1994) *Al Qanun Fi al Teb (Arabic)*, pp. 77–78. Bairut, Lebanon: Dar Alfiker.

[13] Daud al-Antaki (1935) *Tadhkirat Uli l-al-Bab wa l-Jami li-L-'Ajab al- 'Ujab (Arabic)*. Cairo: Bulag.

[14] Ibn Albitar (1974) *Aljamea Limufradat Aladwiya Walaghdiya*. Cairo: Dar Bulaaq (manuscript from 12th century).

[15] Souayah N, Greenstein JI (2005) Insights into neurologic localization by Rhazes, a medieval Islamic physician. *Neurology* 65: 125–128.

[16] Ar-Razi (1956) *Kitab al-Hawi Fi Al-Tibb li-Muhammad Ibn Zakariyya ar Ra-Razi*, vol 1. Hyderabad: Al-Osmanya.

[17] Ar-Razi, Al-Mansuri Fi At-Tibb (1987) *The Book of Medicine for Mansur (Arabic). Edited by Hazim Al-Bakry Al-Siddiky*. Kuwait, Kuwait City: Institute of Arab Manuscripts, Arab League Educational Cultural and Scientific Organization.

[18] Abun-Nasr J (1987) *A History of the Maghrib in the Islamic Period*. Cambridge/New York: Cambridge University Press.

[19] Pak E, Esrason KT, Wu VH (2004) Hepatotoxicity of herbal remedies: an emerging dilemma. *Prog Transplant* 14: 91–96.

[20] Fugh-Berman A (2000) Herb-drug interactions. *Lancet* 355: 134–138.

[21] Rousseaux CG, Schachter H (2003) Regulatory issues concerning the safety, efficacy and quality of herbal remedies. *Birth Defects Res* 68: 505–510.

[22] Costa-Neto EM (2005) Animal-based medicines: biological prospection and the sustainable use of zootherapeutic resources. *An Acad Bras Cienc* 77: 33–43.

[23] Li JW, Vederas JC (2009) Drug discovery and natural products: end of an era or an endless frontier? *Science* 325: 161–165.

[24] Oumeish OY (1998) The philosophical, cultural and historical aspects of complementary, alternative, unconventional and integrative medicine in the Old World. *Arch Dermatol* 134: 1373–1386.

[25] Syed IB (2003) Spiritual medicine in the history of Islamic medicine. *JISHIM* 2: 45–50.

[26] Zimecki M, Kruzel ML (2007) Milk-derived proteins and peptides of potential therapeutic and nutritive value. *J Exp Ther Oncol* 6: 89–106.

[27] Saad B, Said O (eds.) (2011) *Contributions of Arab and Islamic scholars to modern pharmacology*. In: *Greco-Arab and Islamic Herbal Medicine*, pp. 87–100. Hoboken: John Wiley & Sons, Inc.

[28] Lev E (2003) Traditional healing with animals (zootherapy): medieval to present-day Levantine practice. *J Ethnopharmacol* 85: 107–118.

[29] Cragg GM, Newman DJ (2005) Biodiversity: a continuing source of novel drug leads. *Pure Appl Chem* 77: 7–24.

[30] Harvey AL (2008) Natural products in drug discovery. *Drug Discov Today* 13(19–20): 894–901.

[31] Koehn FE, Carter GT (2005) The evolving role of natural products in drug discovery. *Nat Rev Drug Discov* 4: 206–220.

[32] Salem ML (2005) Immunomodulatory and therapeutic properties of the *Nigella sativa* L. seed. *Int Immunopharmacol* 5: 1749–1770.

[33] Gali-Muhtasib H, Diab-Assaf M, Boltze C, Al-Hmaira J, Hartig R, et al. (2004) Thymoquinone extracted from black seed triggers apoptotic cell death in human colorectal cancer cells via a p53-dependent mechanism. *Int J Oncol* 25: 857–866.

[34] Simon A, Traynor K, Santos L, Blaser G, Bode U, et al. (2007) Medical honey for wound care—still the 'latest resort'? *Evid Based Complement Alternat Med* 6: 165–173.

[35] Bogdanov B, Jurendic T, Sieber R, Gallmann P (2008) Honey for nutrition and health: a review. *J Am Coll Nutr* 27: 677–689.

[36] Al-Quassemi R, Robinson RK (2003) Some special nutritional properties of honey—a brief review. *Nutr Food Sci* 33: 254–260.

[37] Jones R (2001) *Honey and healing through the ages*. In: Munn P, Jones R (eds.) *Honey and Healing*, pp. 1–4. Cardiff: International Bee Research Association IBRA.

[38] Albietz JM, Lenton LM (2006) Effect of antibacterial honey on the ocular flora in tear deficiency and meibomian gland disease. *Cornea* 25: 1012–1019.

[39] Irish J, Carter DA, Shokohi T, Blair SE (2006) Honey has an antifungal effect against *Candida* species. *Med Mycol* 44: 289–291.

[40] Al-Waili NS (2004) Investigating the antimicrobial activity of natural honey and its effects on the pathogenic bacterial infections of surgical wounds and conjunctiva. *J Med Food* 7: 210–222.

[41] Molan PC, Betts JA (2004) Clinical usage of honey as a wound dressing: an update. *J Wound Care* 13: 353–356.

[42] Molan PC (2006) The evidence supporting the use of honey as a wound dressing. *Int J Low Extrem Wounds* 5: 40–54.

[43] Al-Waili NS, Boni NS (2003) Natural honey lowers plasma prostaglandin concentrations in normal individuals. *J Med Food* 6: 129–133.

[44] Covas MI (2008) Bioactive effects of olive oil phenolic compounds in humans: reduction of heart disease factors and oxidative damage. *Inflammopharmacology* 16: 1–3.

[45] Fitó M, de la Torre R, Farré-Albaladejo M, Khymenetz O, Marrugat J, et al. (2007) Bioavailability and antioxidant effects of olive oil phenolic compounds in humans: a review. *Ann Ist Super Sanità* 43: 375–381.

[46] Goulas V, Exarchou V, Troganis AN, Psomiadou E, Fotsis T, et al. (2009) Phytochemicals in olive-leaf extracts and their antiproliferative activity against cancer and endothelial cells. *Mol Nutr Food Res* 53: 600–608.

[47] Al-Shahib W, Marshall RJ (2003) The fruit of the date palm: its possible use as the best food for the future? *Int J Food Sci Nutr* 54: 247–259.

[48] Al-Farsi MA, Lee CY (2008) Nutritional and functional properties of dates: a review. *Crit Rev Food Sci Nutr* 48: 877–887.

[49] Al-Rawahy F (2007) Compositional and functional characteristics of dates, syrups and their by-products. *Food Chem* 104: 943–947.

[50] Al-Farsi M, Alasalvar C, Morris A, Baron M, Shahidi F (2005) Comparison of antioxidant activity, anthocyanins, carotenoids and phenolics of three native fresh and sun-dried date (*Phoenix dactylifera* L.) varieties grown in Oman. *J Agric Food Chem* 53: 7592–7599.

[51] Almana HA, Mahmoud RM (1994) Palm date seeds as an alternative source of dietary fibre in Saudi bread. *Ecol Food Nutr* 32: 261–270.

[52] Vayalil PK (2002) Antioxidant and antimutagenic properties of aqueous extract of date fruit (*Phoenix dactylifera* L. Arecaceae). *J Agric Food Chem* 50: 610–617.

[53] Akşit S, Çağlayan S, Cukan K, Yaprak I (1998) Carob bean juice: a powerful adjunct to oral rehydration solution treatment in diarrhoea. *Paediatr Perinat Epidemiol* 12: 176–181.

[54] Slavin JL (2005) Dietary fibre and body weight. *Nutrition* 21: 411–418.

[55] Lansky EP, Paavilainen EM, Pawlus AD, Newman RA (2008) Ficus spp. (fig): ethnobotany and potential as anticancer and anti-inflammatory agents. *J Ethnopharmacol* 119: 195–213.

[56] Boukef K, Souissi HR, Balansard G (1982) Contribution to the study on plants used in traditional medicine in Tunisia. *Plant Med Phytother* 16: 260–279.

[57] Perez C, Canal JR, Torres MD (2003) Experimental diabetes treated with *Ficus carica* extract: effect on oxidative stress parameters. *Acta Diabetol* 40: 3–8.

[58] Saeed MA, Sabir AW (2002) Irritant potential of triterpenoids from *Ficus carica* leaves. *Fitoterapia* 73: 417–420.

[59] Salhi-Hannachi A, Chatt K, Saddoud O, Mars M, Rhouma A, et al. (2006) Genetic diversity of different Tunisian fig (*Ficus carica* L.) collections revealed by RAPD fingerprints. *Hereditas* 143: 15–22.

[60] Lansky EP, Newman RA (2007) *Punica granatum* (pomegranate) and its potential for prevention and treatment of inflammation and cancer. *J Ethnopharmacol* 109: 177–206.

[61] Tattelman E (2005) Health effects of garlic. *Am Fam Physician* 72: 103–106.

[62] Block E (1985) The chemistry of garlic and onions. *Sci Am* 252: 114–119.

[63] Jeambey Z, Johns T, Talhouk S, Batal M (2009) Perceived health and medicinal properties of six species of wild edible plants in north-east Lebanon. *Public Health Nutr* 23: 1–10.

[64] Ali-Shtayeh MS, Jamous R, Al-Shafie' J, Elgharabah W, Kherfan F, et al. (2008) Traditional knowledge of wild edible plants used in Palestine (Northern West Bank): a comparative study. *J Ethnobiol Ethnomed* 4: 13.

[65] James WP, Duthie GG, Wahle KW (1989) The Mediterranean diet: protective or simply non-toxic? *Eur J Clin Nutr* 43: 31–41.

[66] Kim M, Shin HK (1998) The water-soluble extract of chicory influences serum and lipid concentrations, cecal short-chain fatty acid concentrations and fecal lipid excretion in rats. *J Nutr* 128: 1731–1736.

[67] Ishaq J, El-Yousef S (1985) Isolation of essential oils of Palestinian thyme by GC/FID. *Bethlehem University J* 4: 162–175.

[68] Nam SH, Kang MY (2004) Antioxidant activity of 13 medicinal plants. *Pharm Biol* 42: 409–415.

[69] Cosentino S, Tuberoso CIG, Pisano B, Satta M, Arzedi E, et al. (1999) *In-vitro* antimicrobial activity and chemical composition of Sardinian Thymus essential oils. *Lett Appl Microbiol* 29: 130–135.

[70] Leung AY, Steven F (1996) *Encyclopedia of Common Natural Ingredients Used in Food, Drugs and Cosmetics*, 2nd edition. New York: John Wiley & Sons.

[71] Goreja WG (2003) *Black Seed: Nature's Miracle Remedy*. New York: Amazing Herbs Press.

[72] El-Dakhakhny M (1965) Studies on the Egyptian *Nigella sativa* L: IV. *Some pharmacological properties of the seeds' active principle in comparison to its dihydro compound and its polymer. Arzneimittelforschung* 15: 1227–1229.

[73] Mahmood MS, Gilani AH, Khwaja A, Rashid A, Ashfaq MK (2003) The *in vitro* effect of aqueous extract of *Nigella sativa* seeds on nitric oxide production. *Phytother Res* 17: 921–924.

[74] El-Dakhakhny M, Mady NI, Halim MA (2000) *Nigella sativa* L. oil protects against induced hepatotoxicity and improves serum lipid profile in rats. *Arzneimittelforschung* 50: 832–836.

[75] Nagi MN, Alam K, Badary OA, Al-Shabanah OA, Al-Sawaf HA, et al. (1999) Thymoquinone protects against carbon tetrachloride hepatotoxicity in mice via an antioxidant mechanism. *Biochem Mol Biol Int* 47: 153–159.

[76] El-Dakhakhny M, Madi NJ, Lembert N, Ammon HP (2002) *Nigella sativa* oil, nigellone and derived thymoquinone inhibit synthesis of 5-lipoxygenase products in polymorphonuclear leukocytes from rats. *J Ethnopharmacol* 81: 161–164.

[77] Mansour M, Tornhamre S (2004) Inhibition of 5-lipoxygenase and leukotriene C4 synthase in human blood cells by thymoquinone. *J Enzyme Inhib Med Chem* 19: 431–436.

[78] Al-Majed AA, Daba MH, Asiri YA, Al-Shabanah OA, Mostafa AA, et al. (2001) Thymoquinone-induced relaxation of guinea-pig isolated trachea. *Res Commun Mol Pathol Pharmacol* 110: 333–345.

[79] Islam SN, Begum P, Ahsan T, Huque S, Ahsan M (2004) Immunosuppressive and cytotoxic properties of *Nigella sativa. Phytother Res* 18: 395–398.

[80] Goldsby RA, Kind TJ, Osborne BA (2003) *Kuby Immunology*. New York: W.H. Freeman Company.

[81] Salem ML, Kadima AN, Cole DJ, Gillanders WE (2005) Defining the antigen-specific T cell response to vaccination and poly(I:C)/TLR3 signaling: evidence of enhanced

primary and memory CD8 T cell responses and anti-tumour immunity. *J Immunother*; 28: 220–228.

[82] Salem M, Kadima A, EL-Naggar S, Gillanders W, Cole D (2005) Cyclophosphamide preconditioning enhances the antigenspecific CD8 T cell response to peptide vaccination: evidence of enhanced innate immunity and induction of a beneficial cytokine milieu. 5th Annual Research Retreat Citadel, Charleston, SC, USA.

[83] Gilani AH, Jabeen Q, Khan M (2004) A review of medicinal uses and pharmacological activities of *Nigella sativa. Pak J Biol Sci* 7: 441–451.

[84] Agarwal R, Kharya MD, Shrivastava R (1979) Antimicrobial and anthelmintic activities of the essential oil of *Nigella sativa* Linn. *Indian J Exp Biol* 17: 1264–1265.

[85] Salem ML, Hossain MS (2000) Protective effect of black seed oil from *Nigella sativa* against *Murine cytomegalovirus* infection. *Int J Immunopharmacol* 22: 729–740.

[86] Worthen DR, Ghosheh OA, Crooks PA (1998) The *in vitro* antitumour activity of some crude and purified components of blackseed, *Nigella sativa* L. *Anticancer Res* 18: 1527–1532.

[87] Salomi MJ, Nair SC, Panikkar KR (1991) Inhibitory effects of *Nigella sativa* and saffron (*Crocus sativus*) on chemical carcinogenesis in mice. *Nutr Cancer* 16: 67–72.

[88] Salomi NJ, Nair SC, Jayawardhanan KK, Varghese CD, Panikkar KR (1992) Antitumour principles from *Nigella sativa* seeds. *Cancer Lett* 63: 41–46.

[89] Saad B, Said O (eds.) (2011) Commonly used herbal medicines in the mediterranean. In: *Greco-Arab and Islamic Herbal Medicine*, pp. 149–227. Hoboken: John Wiley & Sons, Inc.

[90] Zhou J, Chan L, Zhou S (2012) Trigonelline: a plant alkaloid with therapeutic potential for diabetes and central nervous system disease. *Curr Med Chem* 19: 3523–3531.

[91] Amin A, Alkaabi A, Al-Falasi S, Daoud S (2005) Chemopreventive activities of *Trigonella foenum graecum* (Fenugreek) against breast cancer. *Cell Biol Int* 29: 687–694.

[92] Ball KR, Kowdley KV (2005) A review of *Silybum marianum* (milk thistle) as a treatment for alcoholic liver disease. *J Clin Gastroenterol* 39: 520–528.

[93] Loguercio C, Festi D (2011) Silybin and the liver: from basic research to clinical practice. *World J Gastroenterol* 17: 2288–2301.

[94] Ashraf R, Aamir K, Shaikh AR, Ahmed T (2005) Effects of garlic on dyslipidemia in patients with type 2 diabetes mellitus. *J Ayub Med Coll Abbottabad* 17: 60–64.

[95] Alqasoumi S, Al-Sohaibani M, Al-Howiriny T, Al-Yahya M, Rafatullah S (2009) Rocket "*Eruca sativa*": a salad herb with potential gastric anti-ulcer activity. *World J Gastroenterol* 15: 1958–1965.

[96] Bnouham M, Merhfour FZ, Zyyat A, Mekhfi H, Aziz M, et al. (2003) Antihyperglycemic activity of the aqueous extract of *Urtica dioica. Fitoterapia* 74: 677–681.

[97] Federici E, Multari G, Gallo FR, Palazzino G (2005) Herbal drugs: from traditional use to regulation. *Ann Ist Super Sanita* 41: 49–54.

[98] Rodriguez-Fragosoa L, Reyes-Esparzaa J, Burchielb S, Herrera-Ruiza D, Torresc E (2008) Risks and benefits of commonly used herbal medicines in México. *Toxicol Appl Pharmacol* 15: 125–135.

[99] Heinrich M (2003) Ethnobotany and natural products: the search for new molecules, new treatments of old diseases or a better understanding of indigenous cultures. *Curr Top Med Chem* 3: 141–154.

[100] Melzer J, Rosch W, Reichling J, Brignoli R, Saller R (2004) Meta-analysis: phytotherapy of functional dyspepsia with the herbal drug preparation STW 5 (Iberogast). *Aliment Pharmacol Ther* 20: 1270–1287.

[101] Mizuno S, Kato K, Ono Y, Yano K, Kurosaka H, et al. (2006) Oral peppermint oil is a useful antispasmodic for double-contrast barium meal examination. *Gastroenterol Hepatol* 21: 1297–1301.

[102] Kassi E, Papoutsi Z, Fokialakis N, Messari I, Mitakou S, et al. (2004) Greek plant extracts exhibit selective estrogen receptor modulator (SERM)-like properties. *J Agric Food Chem* 52: 6956–6961.

[103] Goswami S, Singhai A, Pawar RS (2012) Phytochemical and pharmacological investigations on *Coriandrum sativum*: a review. *Asian J Pharm Edu Res* 1: 10–22.

[104] Wangensteen H, Samuelsen AB, Malterud KE (2004) Antioxidant activity in extracts from coriander. *Food Chem* 88: 293–297.

[105] Saad B, Said O (eds.) (2011) Herbal medicine. In: *Greco-Arab and Islamic Herbal Medicine*, pp. 47–69. Hoboken: John Wiley & Sons, Inc.

[106] Shojaii A, Fard MA (2012) Review of pharmacological properties and chemical constituents of *n*. *ISRN Pharm* 2012: 510795.

[107] Said O, Fulder S, Khalil K, Kassis E, Saad B (2009) Efficacy and safety assessments of *Ferula assa-foetida* L., traditionally used in Greco-Arab herbal medicine for enhancing male fertility, libido and erectile function. *Open Complement Med J* 1: 00–00 1.

[108] Saad B, Said O (eds.) (2011) *Arab medicinal plants: from traditional uses to scientific knowledge*. In: *Greco-Arab and Islamic Herbal Medicine*, pp. 303–338. Hoboken: John Wiley & Sons, Inc.

9

EVOLUTION OF HERBAL MEDICINES IN EUROPE AND ITS RELATIONSHIP WITH MODERN MEDICINE

ELIZABETH M. WILLIAMSON[1] AND KELVIN CHAN[2,3]

[1] *The School of Pharmacy, Whiteknights, Reading, Berkshire, United Kingdom*
[2] *Faculty of Pharmacy, The University of Sydney, Sydney, New South Wales, Australia*
[3] *National Institute for Complementary Medicine, University of Western Sydney, Sydney, New South Wales, Australia*

9.1 BACKGROUND

The histories of Western orthodox medicine (OM) and herbal medicine (HM) are inextricably intertwined because until the nineteenth century, with the advent of synthetic chemistry, all drugs were obtained from natural sources, and mostly from plants. Europe has always been well-connected to other parts of the world, via trade, colonization, and immigration, so its systems of medicine have evolved from the traditional medical practices of many ethnic groups and ancient civilizations from various parts of the world. These may differ widely in their philosophies and practical adaptations but often have herbal drugs in common: For example, ginger is used in almost all forms of traditional medicine, including European HM, despite being a tropical plant. European herbal medicine now uses plants from all over the world and especially China, India, South East Asia, and the Americas [1–4].

Many traditional practices have been preserved but also modified as time passes and knowledge progresses. From written records on tablets of clay, stone steles, papyrus, and vellum, there is evidence showing details of anatomical and pathological diseases, parasitic infections, bacterial diseases, bone-setting surgery, and the

Phytotherapies: Efficacy, Safety, and Regulation, First Edition. Edited by Iqbal Ramzan.

use of medicinal plants, as well as animal parts and minerals, to treat these ailments [1–4]. The development of currently accepted practices and chemical drugs in orthodox medicine has a comparatively short history: The remarkable medical evolution (or revolution) that brought the art of medieval medicine into a science-based discipline dates back less than 200 years. This rapid rate of success depended mainly on the advancement of physical sciences and technology, some of the most noticeable of which are shown in the timeline for conventional (orthodox) medicine in Table 9.1.

9.2 HISTORICAL PERSPECTIVE

The evidence we have for the use of medicinal plants comes from these surviving records, so this brief history of European herbal medicine is based on those that are (arguably) the most influential in describing the *materia medica* of the time and place. Medicine is about much more than the drugs it uses, but there are many excellent references on the history of medicine, which cover advances in surgery, microbiology, and diagnosis, so this short review is restricted to medicinal plant preparations or to ideas that have significantly affected the progression of European herbal medicine in some way. As the subject is so huge, the information is summarized in the form of a timeline in Table 9.2, highlighting important documents, mainly herbals, and the people responsible for introducing or promoting them. The geographical origins of the ideas and documents emphasize the global nature of the original sources that shaped European herbal medicine. Table 9.2 is a series of snapshots, not a deep analysis, of the constantly changing patterns of influence that have woven the complex cultural fabric of herbal medicine use in and around Europe. There are gaps in the resources available, in some cases because written records did not survive, for example, in wet tropical areas where bones, wood, and paper decompose easily. Plants, fortunately, do not respect geographical and political boundaries, so the use of herbs as medicines, spices, and aroma ingredients (and sometimes all three) can be common in adjacent and distant, related and unrelated cultures, and thus relevant to all.

As early as 3000 B.C., the Babylonian, Chinese, and Egyptian cultures all had records of their medical practices and *materia medica*. The Ebers papyrus contains a list of remedies including castor oil seeds. The earliest records of Chinese herbal medicine (around 2800 B.C.) were found in the *Shen Nong Ben Cao Jing* (The Herbal Classic of the Divine Plowman) [4]. The Greek inheritance of medicine at around 500 B.C. was derived from the Babylonian and Egyptian medicine, with possible influences from the East (Chinese medicine, and Indian Ayurvedic medicine and Hindu surgery, which existed around the second millennium). Medicine was harsh: emetics, purgatives, enemas, sneezing-powders, leeching, cupping, and bleeding were widely practiced. Greek physicians handed down this medical knowledge to the Romans. A period of magic and supernatural beliefs dominated medicine until Greek philosophers such as Pythagoras and Empedocles introduced the concept of science separating medicine from magic. During the time of Hippocrates (460–375 B.C.), medical opinion had largely discarded the concepts of magic and religion.

TABLE 9.1 Evolution of Orthodox Medicine in Europe

Origin	*Ancient Traditional Medical Practices of Various Ethnic Groups*
3000 B.C.	• Babylonian, Egyptian, and Chinese cultural and medical records gave initial contributions.
500 B.C.	• In Europe and surrounding regions, the Greek inheritance of Babylonian and Egyptian medicine, with influence from the East (Ayurvedic and Chinese medicine) led the progress.
510 B.C. to 476 A.D.	• The Roman Empire brought prosperity, law, the Latin language, and Christianity to Europe and the Mediterranean region, but adapted medical practice from the Greek physicians.
476 A.D. to Fourteenth Century	• Muslim influence on translation and compilation of Greek medical works into Arabic, forming a source of knowledge for orthodox medicine from the Dark Age and surviving through the Middle Age (fourth to eighth century) to modern times.
Fourteenth to Seventeenth Century	• Influence of the Renaissance started in Italy to France, increasing knowledge of anatomy and physiology of human body. • Translation of Arabic medical literature to Latin. • Influence of the evolution of physical sciences, mainly due to Galileo Galilee (1564–1642) and Isaac Newton (1642–1727).
Late Eighteenth Century to Nineteenth Century	• Influence of the Industrial Revolution in Britain (1830), then in France and Belgium (after 1870), and in Russia (after 1900). • Modernization of industry, social, and economic issues. • Development of natural sciences, such as chemistry, mathematics, and physics, laid the groundwork for medical developments.
Late Eighteenth Century to Nineteenth Century	• This had a profound influence on the progress of modern medicine from medieval practices (which was based mostly on mainly religious and superstitious beliefs) into a twentieth century science-based discipline, in only 100 years. • Isolation and identification of active ingredients from medicinal plants that were used over centuries and discovery of endogenous compounds with medicinal uses from animal sources, leading to the first therapeutic revolution of the twentieth century.
Growth	*Period of Rapid Medical Advances*
Late Nineteenth to Twentieth Century	Some key points are listed down: • Development of public health legislation and education. • Development of standardized and vaccination for preventative measures. • Development of anesthetics and anesthesia (William Morton, 1846) helping surgical progress. • Development of microscope helping bacteriology and sterile techniques that helped indirectly the progress of medical surgery (Robert Koch, Louis Pasteur, and Joseph Lister). • Discoveries of chemotherapeutic agents stimulated the development of rational therapeutics—emerging pharmaceutical industry.
1905	• Salvarsan's selective toxicity against syphilis (1905) led the introduction of sulphonamides.

(*Continued*)

TABLE 9.1 *(Continued)*

Early 1900s	• Prontosil's control of streptococcal infections led to the launch of the more effective, less toxic, sulphapyridine in 1938; other antimicrobials included penicillin (1928), streptomycin (1943), and chloramphenicol (1949). • Development of rational therapeutics continued under the influence of bioassays and receptor assays using isolated tissues for endogenous neurotransmitters, hormones, and so on. • Discovery of autonomic functions (due to the discovery of pharmacological actions of acetylcholine by Henry Dale in 1940) led to the introduction of muscle relaxants and cardiovascular drugs.
Early 1900s	• Discovery of insulin (action in 1921 by Banting and Best; structure clarified in 1955) standardized by blood sugar concentrations in mice and isolated rat diaphragm. • Discovery of neuroleptics for the treatment of schizophrenia (chlorpromazine in 1950). • Discovery of tricyclic antidepressants for treatment of depression (imipramine in 1952). • Introduction of benzodiazepine as tranquillizers and hypnotics (1960s), many other modern synthetic analogues for almost all diagnosed diseases.
Growth	*Period of Rapid Medical Advances*
1961	• The incidence of thalidomide-induced malformation of newborns nearly halted all new drug development due to teratogenic effects. Better quality assurance was demanded for testing toxicity, mutagenicity, carcinogenicity, and teratogenicity.
1970s	• Various government interventions on standards and quality assurances in laboratory testing and practice had a great influence on all industries. Good laboratory practice (GLP) guidelines were set to promote and coordinate experiments for the purpose of bringing about safety and quality control.
1970–1990s	• Setting up legislation by various governments of developed countries to implement good manufacturing practice (GMP) and good clinical trial practice (GCTP) for R&D of new drugs and biotechnological products with medicinal uses.
Twentieth Century Onward	• Further development of biotechnology in gene therapy, diagnosis, and biotechnological pharmaceuticals. • Development of vaccines against parasites of medical concern. • Preventive medicine, reducing the cost of primary healthcare. • Improvement of quality of life. • Acceptance of quality-assured herbal medicines in orthodox medical practice.

TABLE 9.2 A Timeline of Major Influences On European Herbal Medicines

Era	Author and/or Date Where Known	Title/Description of Major Work(s)	Region	Brief Notes on Significance and Content of Published Works
Antiquity	Unknown c. 1550 BCE	Papyrus Ebers [1, 5]	Egypt	One of the oldest and greatest medical texts, found in a tomb in Thebes (Luxor), purchased by Georg Ebers in 1872, contains nearly 700 recipes and magic formulae, some of which were possibly effective (e.g., pomegranate bark for roundworms) but many of which were not, and simply unpleasant (e.g., excrement and animal parts).
	Unknown c. 660 BCE	Assyrian Herbal [6, 7]	Mesopotamia/ Iraq	Said to be copied from a Sumerian tablet from 2201 to 2177 BCE, it describes similar herbs to the Papyrus Ebers and many plants named are still used today, e.g., saffron, colocynth, turmeric, myrrh, mandrake, poppy, and sesame.
Greek and Roman Period	Hippocrates 460–370 BCE	Corpus Hippocraticum [1, 8]	Greece	Hippocrates believed illness was caused by environmental agents, not supernatural causes. He advocated rest, hygiene, and the "healing power of nature" and although he did not write a specific herbal, his work influenced medicine for centuries.
	Aristotle 384–322 BCE	Corpus Aristotelicum [1, 9]	Greece	Probably the earliest "natural historian," Aristotle attempted to classify living things until Linnaeus devised his binomial system. He wrote extensively, and "De Plantis" (On Plants), often ascribed to Aristotle, may be the work of Nicolaus of Damascus.
	Theophrastus (Tyrtamus) 372–287 BCE	Historia Plantarum; De Causis Plantarum [1, 3, 10, 11]	Greece	Theophrastus established the careful observation and description now associated with the science of botany. He describes, e.g., "Scythian root" (probably liquorice), hellebore and all-heal, as well as middle eastern and oriental herbs such as frankincense and myrrh, cinnamon, and cassia.

(*Continued*)

TABLE 9.2 *(Continued)*

Era	Author and/or Date Where Known	Title/Description of Major Work(s)	Region	Brief Notes on Significance and Content of Published Works
	Celsus (Aulus Cornelius Celsus) 25–50 BCE	De Medicina [12, 13]	Rome (or Roman Gaul)	Celsus gave instructions on the preparation of poultices, pessaries, ointments, liniments, pills, and others. He used essential oils such as thyme, cedar, and cypress on wounds, and tannins to stem bleeding (e.g., acacia). He described southernwood, elecampane, violet, saffron, cinnamon, ammoniacum, and linseed, still used today.
	Dioscorides (Pedanius Dioscorides) 40–90	De Materia Medica [1–3, 12, 14]	Turkey	Contains about 1000 substances (plant, mineral, animal); it documents medicines used by Greek and Romans and their foreign territories of the era, and is considered to be the basis of modern pharmacopoeias. It includes common European and exotic species such as nigella, garlic, ginger, and opium, many previously unknown.
	Pliny the Elder (Gaius Plinius secundus) 23–79	Naturalis Historia [1, 12, 15]	Italy	The work consists of 37 books: nos 20–27 cover medicinal plants. Describes opium, aconite, wormwood, mandrake, hemlock, erigeron, ivy, juniper, and many common herbal drugs still in use. He died in the eruption of Mount Vesuvius in 79 CE.
	Soranus of Ephesus 98–138	Gynaecia [12, 16]	Turkey, Egypt, and Rome	A Roman gynecologist who described herbal formulae for use as fertility regulators and in childbirth, Soranus described the use of pomegranate, rue, myrtle, and "Cyreniac" juice (from a species of Ferula, called silphium). In the absence of proven pregnancy, contraception (including abortifacients) was widely accepted.
	Galen (Aelius/Claudius Galenus) 131–200	De Simplicibus medicamentorum facultatibus [1–3, 12, 17]	Turkey parents and Rome	Galen believed in Hippocrates humoral theory and influenced Western medicine for >1300 years. His mixtures gave rise to the term "galenical" and he introduced the universal antidote and panacea "Theriac," containing wine and herbs including saffron, squill, opium, as well as viper's flesh.
The Middle Ages and the Arab Influence	Mesue the Elder (Masawaiyh) 777–857	Opera Medicinalia [6, 7]	Persia/Iran	Combines knowledge on herbs used by the Greeks, Persians, Arabs, Indians, and Babylonians, as monographs containing directions for the preparation of medicinal formulae such as syrups and confections, with notes on their medicinal uses.

Rhazes (Ibn Zakaryia al-Razi) 865–952	Kitâb al-Hawi fi al-Tibb [6, 7, 18, 19]	Persia/Iran	Rhazes criticized Galen for his humoral theory and advocated hygiene, restricted diets, and good nutrition. He used peppermint as a digestive, prunes for constipation, and cumin and coriander to stimulate the appetite. He used "soregan," Colchicum speciosum, which contains colchicine, to treat gout (still used for this purpose).
Avicenna (IbnSina) 980–1037	Al-Qanun fi al-Tibb [1, 2, 6, 7, 20]	Persia/Iran	The "Canon of Medicine" followed Galen and Hippocrates and was used in Europe and the Islamic world until the eighteenth century. Book 5 describes the preparation of medicines and includes squill, absinthe, agrimony, barberry, myrtle, thyme, scammony, elecampane, rose, poppy, cinnamon, pomegranate, and others still used.
Ibn Al-Baitar 1197–1248	Kitâb al-Jami fi al-Adwiya al-Mufrada [6, 7]	Muslim Spain	The *Book of Simple Drugs and Food* details about 1400 medicines, mainly plants, of which about 200 had not been previously described, including some from Persia and India. It describes the production of syrups and waters from aromatic plants such as cinnamon, and oils from olives and sesame.
Bald (reputed author) Cild (scribe) 900–950	The Leech Book of Bald [1, 21]	England	This was the manual of a Saxon doctor (or "leech") in the time of Alfred the Great. It was written in the vernacular, not Latin, and shows a remarkable knowledge of native plants. The leech book contains some Roman and Arab herbs as well as chamomile, nettle, plantain, knotweed, and periwinkle.
Hildegard von Bingen 1098–1179	Physica; Causae et curae [22]	Germany	A nun, visionary, saint, musician, composer, and healer, St Hildegard wrote extensively on food and diet, and her medical works influenced Brufels and Fuchs (q.v.). She used nettle, calendula, vervain, opium poppy, mullein, fennel, and yarrow among others.
Unknown Twelfth century	Kitâb al-Diryâq [23]	Mesopotamia/ Iraq	The *Book of Theriac* documents the transmission of Hellenic medicine into the Arab world. It includes recipes for Theriac from many physicians including Galen. Liquorice, cardamom, valerian, opium, gentian, sesame, black pepper, and sweet flag are herbal ingredients added to the traditional viper's flesh.

(*Continued*)

TABLE 9.2 (*Continued*)

Era	Author and/or Date Where Known	Title/Description of Major Work(s)	Region	Brief Notes on Significance and Content of Published Works
	Platearius (Matthaeus Platearius) 1130–1161	Liber de Simplici Medicina [1, 12]	Italy	Matthaeus was the son of Trotula Platearius, one of the first female doctors recorded in history. Also known as "Circa Instans," the book is one of the first herbals from the new printing processes. It is a re-work of earlier Arab texts and Dioscorides, and describes similar herbs such as sundew and the castor oil plant.
	Bartholomew Anglicus 1203–1272	De Proprietatibus Rerum [24–26]	Britain	The only original treatise on herbs written by an Englishman during the Middle Ages, first printed in 1470, the book ran to 25 editions. It describes "Celidonia" (probably greater celandine), "Genesta" (probably broom), and mandrake, among many other edible herbs.
	Pandectarius (Matteo Sylvatico) 1280–1342	Opus Pandectarum Medicinae [27]	Italy	The "pandects" contain many errors; however, they contributed greatly to the accurate description of the properties and uses of plants. They include Artemisia species, Aloe, Carduus, Carthamus, and many herbs cited by Dioscorides, Avicenna, Serapion, and Galen.
	Serapion the Younger Probably twelfth century	Liber de Simplicibus Medicamentis [1, 12]	Arabia (or Greece)	At least three medieval physicians were named Serapion and are often confused. Serapion the Younger described Croton tigliumas a powerful purgative to be used only "with great care and precaution," and the use of pennyroyal, savin, and iris as menstrual regulators.
The Renaissance and Early Modern Period	Unknown Ca. 1404–1438	Voynich manuscript [28]	Italy/New World	Discovered in Italy in 1912 by Wilfrid Voynich, the manuscript is written in an obscure language or code. It has been considered a hoax, but the origin of some of the plants suggests it may be meso-American. It covers 303 plants, of which 37 have been identified to date, and include Passiflora, Viola, Valeriana, Urtica, Actaea, and Opuntia species.
	Garcia de Orta 1490–1570	Colóquios dos simples [29]	Portugal	"Colloquies on the Simples and Drugs of India" is a highly original work published in Goa by a Portuguese pioneer of tropical medicine. Benzoin, camphor, cannabis, cardamom, cinnamon, cloves, ginger, neem, opium, rhubarb, senna, and many other herbs, spices, and medicinal plants still in wide usage today, are described.

Jacob Meydenbach (printer) 1491	Ortus (or Hortus) sanitatis [1, 3]	Germany	The "Garden of Health" quotes Arabian (Rhazes, Avicenna), Greek (Dioscorides, Galen), and Roman (Pliny, Cato) physicians; printed a year before Columbus sailed to America, it is the last herbal to deal exclusively with Old World medicines. Contains familiar plants such as hops and mandrake but illustrations are highly stylized and often unrecognizable.
Otto Brunfels 1488–1534	Kräuterbuch; Herbarum Vivae Eicones [3, 30, 31]	Germany	Brunfels introduced information about German plants not found in Dioscorides and advocated the use of local, rather than imported exotic, medicine plants. He describes the use of marjoram, heartsease, valerian, and others still used.
William Turner 1508–1568	Libellus de re Herbaria Novus; A New Herball [24, 26, 30]	Britain	The "Libellus" used woodcut illustrations from Fuchs but also original contributions. The "Herball" was written in English and included species such as bog myrtle, mandrake, mulberry, larkspur, wild pansy, and Solomon's seal, with at least 200 native to England.
William Turner 1508–1568	Libellus de re Herbaria Novus; A New Herball [24, 26, 30]	Britain	The "Libellus" used woodcut illustrations from Fuchs but also original contributions. The "Herball" was written in English and included species such as bog myrtle, mandrake, mulberry, larkspur, wild pansy and Solomon's seal, with at least 200 native to England.
Paracelsus (Philippus Theophrastus Bombastus von Hohenheim) 1493–1541	Das BuchParagranum; Opus Paramirum [1, 3, 4]	Switzerland	Paracelsus advocated chemical medicine, stating that only the dose differentiates a medicine from a poison, but he also insisted that active principles could be extracted from plants. He believed in the Doctrine of Signatures and describes St John's wort and comfrey in those terms. He is said to have invented laudanum (opium tincture).
Pietro Andrea Mattioli 1501–1577	Commentarii in Libros sex Pedacii Dioscoridis [1, 12]	Italy	Mattioli, physician to the Italian aristocracy, first translated Dioscorides' works and added to them, producing a series of "Commentaries," which became best-sellers. He introduced species such Rhodiola rosea and Lactuca for headache, and first described the tomato.
Carolus Clusius (Charles de l'Escluse) 1526–1609	Rariorum Plantarum Historia [3, 30, 32]	Netherlands	Clusius studied the flora of Austria, Spain, and Hungary and described >600 new species. He studied botany in its own right (rather than as a branch of medicine) and helped to establish the Hortus Academicus at Leiden university, thus starting the Dutch tulip industry.

(*Continued*)

TABLE 9.2 *(Continued)*

Era	Author and/or Date Where Known	Title/Description of Major Work(s)	Region	Brief Notes on Significance and Content of Published Works
	Leonhard Fuchs 1542	De HistoriaStirpium [1, 3, 30]	Germany	His "Notable commentaries on the history of plants" provided accurate botanical details for identification of medicinal plants, with >400 native to Germany and Austria, and the first reports of maize and chili peppers. Fuchs described the use of Digitalis for "dropsy and internal swellings," picked up by Withering in 1780.
	Pietro Castelli 1629; 1653	Discorso della differenzatrag lisemplici freschi et I secchi. Cinchona Pamphlet [33]	Italy	Castelli wrote of the difference between the use of fresh and dried herbs, and also >150 pamphlets, among which was one where he describes the use of cinchona for malaria. This is the first mention of cinchona in Europe; in 1667 it was incorporated into the London Pharmacopoeia.
Twentieth Century and Beyond	Nicholas Culpeper 1616–1664	The English Physitian; The Compleat Herbal [1, 3, 30]	Britain	Culpeper's books, which are still in print, had a huge influence on medicine in England and even in the North American colonies. He identified each disease and herb with an astrological house. He described many common herbs such as burdock, cranesbill, garlic, hawthorn, mugwort, pennyroyal, milk thistle, and also foxglove for heart disease.
	Franz Köhler (publisher) 1887	Köhler's Medizinal-Pflanzen [30]	Germany	Notable for its ~300 chromolithography plates, it includes many medicinal plants from all over the world, such as aconite, boldo, celandine, cinchona, cinnamon, coca, coltsfoot, comfrey, foxglove, goldenseal, guarana, Indian hemp, liquorice, deadly nightshade, opium poppy, pyrethrum, quillaja, rue, senna, star-anise, strychnine tree, valerian, and wintergreen.
	Potter's Herbals 1907–2003	Potter's Herbal Cyclopedia [34]	Britain	Potter's Cyclopaedia of Botanical Drugs and Preparations first edition; many editions have been published since then, up to 2003. The later versions contain scientific references and some Ayurvedic and Chinese herbs have been added.
	Mrs Maud Grieve 1931	A Modern Herbal [35]	Britain	A contemporary herbal that echoes the herbals of the past, this book is still in print and available online. It contains numerous medicinal plants from all over the world.
	British Herbal Medicine Association 1983	British Herbal Pharmacopoeia [36]	Britain	The first pharmacopoeia specifically devoted to herbal medicines. Contains monographs of 215 widely used herbs in the UK, of which 128 are European, 14 Asian, 7 African, and 66 from the Americas.

Bundesinstitut für Arzneimittel und Medizinprodukte 1984–1994	Commission E Monographs: Therapeutic Guide to Herbal Medicines [37]	Germany	These comprise 380 monographs evaluating the safety and efficacy of herbs, originally for licensed medical prescribing in Germany. They were re-published in English by the American Botanical Council and are now used mainly for medico-legal purposes, e.g., for registering a herbal product under the Traditional Herbal Medicinal Products Directive.
Jean Valnet 1920–1995	Docteur Nature L'aromathérapie La Phytotherapie [38]	France	A herbalist and army doctor Valnet, along with Rene Gattefosse, was a pioneer of aromatherapy and used essential oils to treat wounded soldiers in the Second World War (in the absence of orthodox medicines), with considerable success.
Maurice Mességué 1921–present	Des Hommes et des Plantes [39]	France	A highly successful herbalist, in 1945 Mességué was prosecuted for practicing medicine illegally although none of his patients would testify against him. In "Of People and Plants: The Autobiography of Europe's Most Celebrated Herbal Healer" he claims to have treated Winston Churchill, Chancellor Adenauer and Pope John XXIII.
Modern scientific herbal medicine books, e.g., 1960–present	Principles and Practice of Phytotherapy, 1999 [40]; Lehrbuch der Phytotherapie 1960 [41]; Rationale Phytotherapie, 1996 [42]	UK Germany	These books (and others published in Europe) introduced the use of scientific references to support their recommendations. This evidence includes chemical, pharmacological, and clinical data. The date of the first edition is given, but the books are updated regularly and the reference is to the latest edition.
European Parliament 2004	The Traditional Herbal Medicinal Products Directive (THMPD) [43]	European Union	Introduction of the THMPD, which came into full force on 2011, controls the quality of herbal medicines for sale. Traditional Herbal Registration (THR) requires evidence on traditional use for 30 years (15 of which may be outside the EU) and a dossier documenting quality control procedures.

NB All dates are approximates and are given as an indication of chronological progress. Sources and further reading citations are, where possible, reputable, Web-based, free access resources, some from university websites, and listed in the references. Documents are included on the basis of their relevance to the use of medicinal plants, rather than for advancements in specialties such as surgery or diagnosis.

Hippocrates' teachings and writings, including the stress on diagnosis, knowledge of the course of observed diseases, and prognosis, have made such an impact on modern medicine that his legacy has been seen as the Code of Practice for over 2000 years and continues presently. The extent of the Roman Empire at the height of the power encircled the Mediterranean, reached north to the Rhine, the Danube, and central Scotland and spread into Armenia and Mesopotamia. Interestingly, the empire adopted much medical works from the Greek physicians. Among them, Galen (130–201 A.D.) was the authority on anatomy and experimental physiology that identified the connection of blood, heart, and arteries. This medical knowledge had been spread afar throughout Europe and North Africa under the influence of the Roman Empire [1–4, 12].

During the period of the Middle Ages (after the fall of the Roman Empire from 476 A.D. to the late fifteenth century), Greek medical manuscripts were preserved and translated into Arabic. The Moslem influence from 622 A.D., spreading from Persia to North Africa and Spain in the west and Afghanistan, India, Indonesia, Malaysia, Pakistan, Turkey, and some of the Balkans, produced many great physicians. Al-Razi recorded his findings for distinguishing smallpox from measles in the ninth century AD. Avicenna (980–1037) wrote the Canon of Medicine, a text in use in France until the seventeenth century. Abu al-Qasim (936–1013) of Cordoba advanced the status of surgery by his careful illustrated text. These Muslim physicians of that era added to the preserved Greek medicine their own experience and discoveries. The result was a compiled version of Greek and Arabic medicine that became the source of knowledge for Western medicine throughout the Middle Ages right up to early modern times [1–4, 6, 7, 12, 44]. The "Hortus (or 'Ortus') sanitatis," from 1491, depicts an imaginary discussion between the most eminent medical doctor scholars of that era, as shown in Figure 9.1.

9.3 EUROPEAN HERBAL MEDICINE: RELATIONSHIP WITH MODERN MEDICINE

During that same period, the Benedictines collated and translated the books of Hippocrates and Galen, with the formation of the first secular medical school in Europe at Salerno in Southern Italy. Subsequently other medical schools were set up at Montpellier, Bologna, and Padua. Many of the Greek medical works were translated from Arabic into Latin. Thus, modern medicine has its origin from this period onward when physical sciences and later biological sciences began to develop rapidly from the Industrial Revolution in the eighteenth century. The most important events that also relate to herbal medicines, as documented in surviving sources, are shown in Table 9.2, which needs no further explanation except a reminder that it is a selection and not a comprehensive list.

9.4 SUMMARY

In summary, the history of European herbal medicine follows the evolution of conventional medicine, at least until the advent of chemical medicine, and it is impossible to separate it completely from that of other regions of the world. First the Arab influence,

FIGURE 9.1 Frontispiece of "Hortus sanitatis," 1491. The woodcut represents an imaginary conversation between classical and medieval European and Arabian physicians: Aristotle (center), Hippocrates (right), Rhazes (left), and behind them, from left to right, is Galen, Avicenna, Serapion, Dioscorides, Pandectarius, and Platearius (see Table 9.2).

then the colonization of countries in other continents by European nations and reciprocal immigration patterns, led to an influx of ideas in science and medicine, from the East and the Americas especially, but it has led to a rich and varied tradition that includes plants from all over the world. European herbal medicine continues to survive and even flourish, and this is being recognized with initiatives to regulate it more closely. Herbal products for sale, intended for self-medication, are expected to comply with quality standards and are now regulated under the European Directive "European Traditional Herbal Medicinal Products Directive (THMPD) 2004/24/EC" [43]. There is an increasing awareness of the potential for herbal medicines to interact with conventional drugs, and this data is now being made available to health-care professionals who may need to advise the public (e.g., [45]). The drive toward evidenced-based medicine in all types of therapeutic modalities may eventually mean that phytotherapy can be a safe and beneficial part of the integrated medicine of the future.

REFERENCES

NB: References cited are, as far as possible, general and well-known, easily and often freely available. Most websites are from university, museum, government departments, or other reputable sources. Website dates show the year (December) they were accessed.

[1] Griggs B (1997) *New Green Pharmacy. The Story of Western Herbal Medicine*. 2nd Edition. Random House, London.

[2] Heinrich M, Barnes J, Gibbons S, Williamson EM (2012) *Fundamentals of Pharmacognosy and Phytotherapy*. 2nd Edition. Elsevier, Edinburgh.

[3] Pavord A (2005) *The Naming of Names: The Search for Order in the World of Plants*. Bloomsbury Press, New York.

[4] Chan K, Lee H (2001) *The Way Forward for Chinese Medicine*. CRC Press, London.

[5] Bryan CP (1930) Papyrus Ebers. Geoffrey Bles, UK. Appleton and Co, New York. Available at http://oilib.uchicago.edu/books/bryan_the_papyrus_ebers_1930.pdf. February 2015.

[6] Anon (2013) Botany, Herbals and Healing in Islamic Science and Medicine II. 4. Botanical Contributions from the Western Islamic Tradition. Available at http://muslimheritage.com/topics/default.cfm?ArticleID=1164. November 17, 2014.

[7] Anon (2013) Islamic Encyclopedia. Available at http://islamicencyclopedia.org/public/index/index. November 17, 2014.

[8] Anon (2013) Hippocrates (460 BC–377 BC) eBooks@Adelaide. Available at http://ebooks.adelaide.edu.au. November 17, 2014.

[9] Anon (2013) Aristotle's Natural Philosophy. Stanford Encyclopedia of Philosophy. Available at http://plato.stanford.edu/entries/aristotle-natphil/. November 17, 2014.

[10] Anon (2013) Theophrastus Biography. Available at http:/www.egs.edu/library/theophrastus/biography. November 17, 2014.

[11] Anon (2013) Theophrastus: Full Text of 'Enquiry into Plants'. Available at http://archive.org/stream/enquiryintoplant02theouoft/enquiryintoplant02theouoft_djvu.txt. November 17, 2014.

[12] Riddle JM (1994) *Contraception and Abortion from the Ancient World to the Renaissance*. Harvard University Press, Cambridge.

[13] Anon (2013) Celsus: 'On Medicine' Book 5. Available from http://penelope.uchicago.edu/Thayer/E/Roman/Texts/Celsus/5*.html. November 17, 2014.

[14] Osbaldeston TA (2000) Dioscorides: De Materia Medica: Being an Herbal with Many Other Medicinal Materials. Translated by Tess Anne Osbaldestan (2000). Ibidis Press, Johannesburg. Available at http://www.cancerlynx.com/dioscorides.html. February 2015.

[15] Thayer WB (1914) Pliny the Elder: The Natural History. Available at http://peneople.uchicago.edu/Thayer/E/Roman/texts/Pliny_the_Elder/home.html. February 2015.

[16] Dunn P (1995) Soranus of Ephesus (circa AD 98–138) and perinatal care in Roman times. *Arch. Dis. Child. Fetal Neonatal Ed.* 73: F51–52.

[17] Anon (2013) Galen. http://www.greekmedicine.net/whos_who/Galen.html. November 17, 2014.

[18] Nikaein F, Zargaran A, Mehdizadeh A (2012) Rhazes' concepts and manuscripts on nutrition in treatment and health care. *Anc. Sci. Life.* 31(4): 160–163. Available at http://www.ncbi.nlm.nih.gov/pmc/articles/PMC3644752/. November 17, 2014.

[19] Changizi Ashtiyani S, Golestanpour A, Shamsi M Tabatabaei SM, Ramazani M (2012) Rhazes' prescriptions in treatment of gout. *Iran Red Crescent Med. J.* 14(2): 108–112. Available at http://www.ncbi.nlm.nih.gov/pmc/articles/PMC3372038/. November 17, 2014.

[20] Anon (2013) Avicenna: Canon of Medicine, Book 5. Available at http://ddc.aub.edu.lb/projects/saab/avicenna/book-five.html. November 17, 2014.

[21] Anon (2013) Bald's Leechbook. Available at http://britishlibrary.typepad.co.uk/digitisedmanuscripts/2013/10/anglo-saxon-medicine.html. November 17, 2014.

[22] Anon (2013) Hildegard of Bingen. Available at http://www.academia.edu/494644/You_Are_What_You_Eat_Hildegard_of_Bingens_Viriditas. November 17, 2014.

[23] Lindstrom A (2011) The Paris Kitâb al-Diryâq. *HerbalGram.* 91: 56-67.

[24] Rohde ES (2013) The Old English Herbals. Available at http://www.gutenberg.org/files/33654/33654-h/33654-h.htm. November 17, 2014.

[25] Barlow HM (1913) Old English Herbals 1525–1640. *Proc. R. Soc. Med.* 6 (Sect Hist Med): 108–149. Available at http://www.ncbi.nlm.nih.gov/pmc/articles/PMC2006232/pdf/procrsmed00871-0112.pdf. November 17, 2014.

[26] Anon (2013) Bartholomew Anglicus. Available at http://www.gutenberg.org/ebooks/6493. November 17, 2014.

[27] Anon (2013) Pandectarius. Available at http://www.wdl.org/en/item/10665/zoom/#contributors=Silvatico%2C+Matteo%2C+died+approximately+1342. November 17, 2014.

[28] Tucker AO, Talbot RH (2013) A Preliminary Analysis of the Botany, Zoology and Mineralogy of the Voynich manuscript. *HerbalGram.* 100: 70-85.

[29] Markham C (1913) Colloquies on the Simples and Drugs of India. Translation of Garcia da Orta 'Colloquios dos simples e drogas e cousas medicinaes da India' 1563. Available at https://archive.org/details/colloquiesonsimp00orta. February 2015.

[30] Missouri Botanic Garden Digital Library (2014) Botanicus. Available at http://www.botanicus.org/browse. November 17, 2014.

[31] Anon (2013) Otto Brunfels: Kräuterbuch. Available at http://www.botanicus.org/title/b12076028#Krauml;uterbuch (in German). November 17, 2014.

[32] Anon (2013) The Clusius Garden. Available at http://www.clusiusstichting.nl/Eng/garden.html. November 17, 2014.

[33] Lloyd Library and Museum (2013) Pietro Castelli, Available at http://www.lloydlibrary.org/meyer_castelli.html. November 17, 2014.

[34] Williamson E (2003) Potter's Herbal Cyclopedia. C W Daniels, Saffron Walden.

[35] Grieve M (1931) A Modern Herbal. Full text available online at: www.botanical.com/botanical/mgmh/mgmh.html. November 17, 2014.

[36] British Herbal Medicine Association (1983) *British Herbal Pharmacopoeia.* BHMA, London.

[37] Blumenthal M (2000) (ed) *The Complete German Commission E Monographs: Therapeutic Guide to Herbal Medicines*. American Botanic Council, Austin.

[38] Lis-Balchin M (2006) *Aromatherapy Science: A Guide for Healthcare Professionals*. Pharmaceutical Press, London/Chicago.

[39] Mességué M (1991) *Of People and Plants: The Autobiography of Europe's Most Celebrated Healer*. Inner Traditions/Bear and Co, Colchester.

[40] Bone K, Mills S (2013) *Principles and Practice of Phytotherapy: Modern Herbal Medicine*. 2nd Edition. Churchill Livingstone, Edinburgh/New York.

[41] Fintelmann V, Weiss RF (2009) *Lehrbuch der Phytotherapie (German Edition)*. Hippokrates-Verlag, Stuttgart.

[42] Schulz V, Hänsel R (2012) *Rationale Phytotherapie: Ratgeber für Ärzte und Apotheker (German Edition)*. Springer, Berlin.

[43] European Union (2004) European Traditional Herbal Medicinal Products Directive (THMPD) 2004/24/EC. Available at http://ec.europa.eu/health/human-use/herbal-medicines/index_en.htm. November 17, 2014.

[44] Francia S, Stobart A (ed) (2014) *Critical Approaches to the History of Western Herbal Medicine: From Classical Antiquity to the Early Modern Period*. Bloomsbury Press, London.

[45] Williamson EM, Driver S, Baxter K (ed) (2013) *Stockley's Herb-Drug Interactions*. 2nd Edition. Pharmaceutical Press, London.

10

CHEMICAL CLASSIFICATION AND CHEMISTRY OF PHYTOTHERAPEUTICS CONSTITUENTS

PEI H. CUI AND COLIN C. DUKE
Faculty of Pharmacy, The University of Sydney, Sydney, New South Wales, Australia

10.1 INTRODUCTION

Plants or plant extracts used for medicinal purposes, as in phytotherapy, are called phytomedicines. Knowledge of their chemical aspects is essential in the understanding of the processes required to manufacture, preserve, define the quality, and evaluate the efficacy of phytomedicines. Phytomedicines are regarded as very complex mixtures of "chemical entities" usually referred to as constituents, which may be defined as single chemical substances made up of molecules of a single structural type. Chemicals derived from plants may be referred to as phytochemicals. The chemistry of phytomedicines is extremely complex with regard to the number of chemical entities present and their complexity and diversity. The specific significance of the individual constitutes with regard to the medicinal use of the plant may not be known. Additional complexity may arise where combinations of constituents are necessary for medicinal action. The chemical complexity of plants and herbal products may be dealt with at various levels. Where the specific action of individual constituents is considered important, they may be described and defined individually. Sometimes, the medicinal actions of groups of related substances are believed to be

Phytotherapies: Efficacy, Safety, and Regulation, First Edition. Edited by Iqbal Ramzan.

similar, and the substances are considered as a group. The group by its name may define a narrow range of constituents with very similar structures or using another name for a broad range of substances with more diverse structures. This system of grouping of constituents makes it easier to deal with the chemical complexity of herbs. The appropriate term may be selected to describe the constituents at the level appropriate for the medicinal action of interest. For example, phytoestrogens signify the group of plant constituents or plant-derived substances acting like the mammalian steroid hormone estrogen.

The most important chemical constituents for medicinal action are described as "active constituents," which describe active constituents believed to have biological activity that is essential for or is an essential part of the medicinal effect. It may be considered that for some herbs there is a single active constituent or it may be considered that there are many activities. Biological activity (or bioactivity) relates to chemical structures of the natural constituents (natural products) and is the end result of metabolic processes in plants. The process of formation of natural metabolic products in plants (and other living organisms) is described as biosynthesis or biosynthetic processes. The natural relationships between constituents with regard to chemical structures and biological activities are most conveniently made through the biosynthetic pathway that has been determined in great detail. For the purpose of broad classification, an understanding of the origin and relationship of the major classes is necessary.

Natural materials, particularly from plants and microorganisms, have been and will continue to be rich sources of medicinal products. Natural plant constituents may be divided into two categories as primary and secondary metabolites. Primary metabolites are the essential biochemicals for the everyday survival of the plant, for example, for photosynthesis or respiration. Typical examples are sugars, starch, cellulose, protein, polynucleotides, and lipids. Secondary metabolites are the phytochemicals produced that are not required for everyday essential biochemical functions. The secondary metabolites are often important for the long-term defense and survival of the plant. Typical examples are lignins, tannins, alkaloids, terpenes, and essential oils. In nature, there are numerous examples of plants and microorganisms producing natural chemicals to protect themselves from attack from microorganisms, parasites, and animals. The stationary higher plants have developed a formidable array of "chemical weapons" to fend off attack from the full range of organisms and mammals, including humans.

Most substances described as natural products are regarded as secondary or end-products of the primary biochemical pathway. The primary biochemical pathways provide the means for the plant to derive a carbon source by way of photosynthesis to provide glucose. Glucose is the food material on which the whole biochemical pathway depends for the production of building materials (i.e., cellulose), as a precursor for primary metabolites leading to secondary metabolites and also as a stored energy source to be drawn upon in respiratory metabolism. The relationships between glucose, the primary metabolites, and secondary metabolites (including natural products) are illustrated in Figure 10.1 [1].

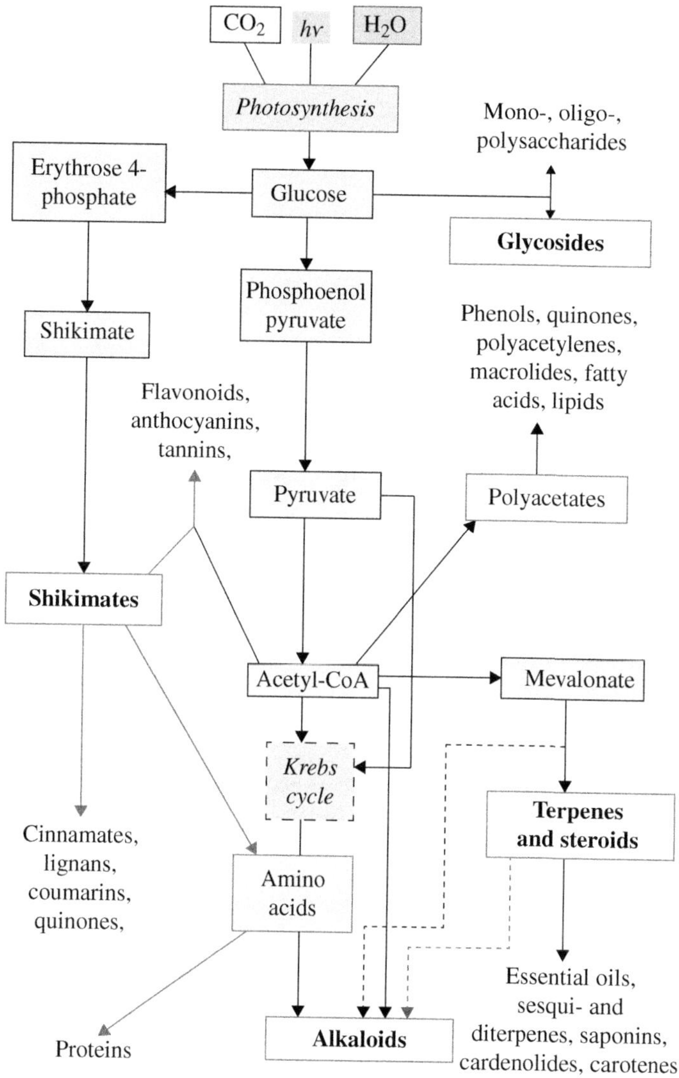

FIGURE 10.1 The relationships between glucose and the primary and secondary metabolites. Substrates and primary metabolites are shown in the gray boxes and the classes of secondary metabolites or natural products are shown in clear boxes with more specific examples listed nearby [1].

10.2 PHYTOCHEMICALS

10.2.1 Alkaloids

Typical alkaloids are alkaline organic compounds containing one or more nitrogen atoms, each connected to at least two carbon atoms within a heterocyclic ring system. However, vitamins and hormones, amino acids, peptides, and proteins are excluded

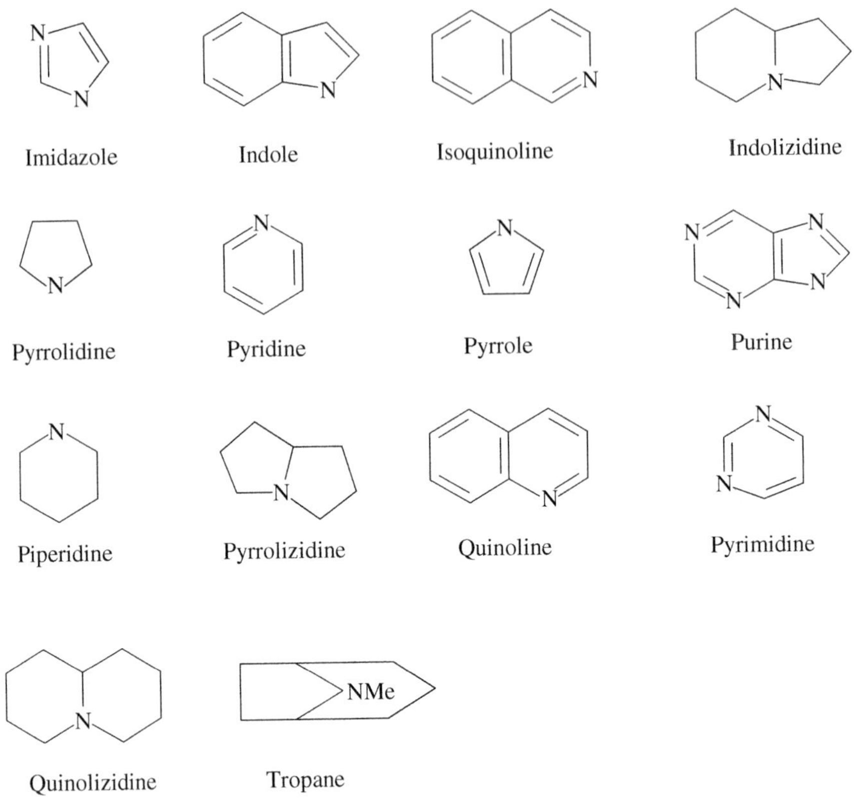

FIGURE 10.2 Structures of major classes of alkaloids.

even though they may structurally comply with this definition. Alkaloids are basic amines; they are not volatile and lack odor; and they have a bitter taste and are soluble in organic solvents and insoluble in water, although their salts are soluble. Many alkaloids, such as nicotine, coniine, sparteine, and lobeline, are colorless; and very few colored alkaloids exist, such as sanguinarine (red) and chelidonine (yellow).

Alkaloids are a very large and structurally diverse class of constituents with a wide range of pharmacological activities; it is one of the most important groups of naturally occurring substances of therapeutic interest. There are over 12,000 and more than 13 subclasses of alkaloids, purines, pyrrolidines, piperidines, pyrindines, and quinolones; the structures of major classes of alkaloids are presented in Figure 10.2, and examples of agents are characterized and listed in Table 10.1. Alkaloids are found in 15–30% of all flowering plants and are particularly common in certain families, such as *Fabaceae*, *Liliaceae*, *Ranunculaceae*, *Apocynaceae*, *Solanaceae*, and *Papaveraceae*. The most widely occurring alkaloids are caffeine and berberine. While the higher plants are the major source of alkaloids, they are also known to occur in lower plants and even in the organs of mammals [2].

Alkaloids vary in their structure and biological activities and are limited in distribution. They are regarded as secondary metabolites in the plant. Alkaloids have some

TABLE 10.1 Classifications of Alkaloids

Class	Examples
1. **Pyridine–piperidine alkaloids** These alkaloids have their nitrogen atoms in typical six-membered rings, the precursors are ornithine and nicotinic acid.	Coniine, nicotine, piperine, lobeline, and arecoline
2. **Pyrrolizidine alkaloids** These alkaloids contain two fused five-membered rings in which a nitrogen atom is common to both rings, the precursor is ornithine.	Echimidine, senecionine, symphytine, intermedidine, and symlandine
3. **Tropane alkaloids** These are bicyclic complex molecules containing pyrrolidine and piperidine ring structures, derived from precursors of ornithine and phenylalanine.	Hyoscamine, hyoscine, and atropine
4. **Quinoline alkaloids** These are bicyclic ring systems with the fusion of benzene and pyridine rings. Biosynthetically they are related to indole alkaloids, since both are derived from the same two precursors, tryptophan and loganin.	Quinine, quinidine, arborinine, and γ–fagarine.
5. **Isoquinoline alkaloids** These alkaloids have the basic structures of isoquinoline, benzylisoquinoline, or tetrahydroisoquinolines. They result from the condensation of a phenylethylamine derivative with phenylacetaldehyde derivatives, which are from phenylalanine or tyrosine. Sub-classifications: Morphinane alkaloids and benzylisoquinolines Protoberberines, protopines, and benzophenanthridine Ipecac alkaloids, and aporphine alkaloids.	Morphine, codeine, and thebaine Papaverine, oxyacanthine, and tubocurarine Cryptopine and protopine Emetine and cephaeline Berberine, hydrastine, and palmatine Chelidonine and sanguinarine Boldine
6. **Quinolizidine alkaloids** (**Norlupinane**) The quinolizidine structure consists of two carbon rings with a shared nitrogen atom. Their precursor is lysine. They are also referred to as "lupin" alkaloids.	Sparteine, cytisine, and myrine

(Continued)

TABLE 10.1 (*Continued*)

Class	Examples
7. **Indole** These alkaloids have basic structures containing a pyrrole ring fused to a benzene ring; their structures typically involve multiple ring systems, often complex in character. Biosynthesis of pure indoles involves the amino acid tryptophan as precursor.	Reserpine, rescinnamine, deserpidine, ajmaline, alstonine, alstonidine, vinblastine, vincristine, vincamine, majdine, majoridine, ergotamine, strychnine, and yohimbine
8. **Alkaloidal amines** These alkaloids do not contain heterocyclic nitrogen atoms. They are often referred to as "pseudo alkaloids." The precursors for alkaloidal amines are aromatic amino acids—phenylalanine, tyrosine, and tryptophan.	Ephedrine, pseudoephedrine, colchicine, muscarine, mescaline, psilocybin, and hordenine
9. **Purine alkaloids** The purines are derivatives of a heterocyclic nucleus consisting of the six-member pyrimidine ring fused to the five-membered imidazole ring.	Caffeine, theophylline, guaranine, and theobromine
10. **Steroidal alkaloids** These alkaloids are derived biogenetically from six isoprene units and could be classified as triterpenoids or steroids. They also contain nitrogen, which is not part of the steroidal nucleus.	Solasodine
11. **Terpenoid** They are derived from terpenoidal alkaloids; the nitrogen atom is inserted into the skeleton at a late biosynthetic stage.	Taxol and aconitine

TABLE 10.2 Common Alkaloids with Their Therapeutic Uses

Alkaloids	Plants	Major uses
Colchicines	*Colchicum autumnal* (Autumn crocus)	Gout [4]
Ephedrine	*Ephedra sinica* (Ma Huang)	Asthma [5]
Berberine	*Hydrastis Canadensis* (Goldenseal)	Central nervous system disorders [6] Anti-diabetic, anti-inflammatory, anti-depression, and anti-cancer [7]
Ergotamine	*Claviceps purpurea* (Ergot)	Migraine [8]
Reserpine	*Rauwolfia serpentina* (Snakeroot)	Hypertension and mental disorders [9]
Lobeline	*Lobelia inflata* (Lobelia)	Asthma and respiratory illness [10]

of the most potent effects on animals and humans; they demonstrate both therapeutic and toxic properties. Medicinal use of plants containing alkaloids are well established such as *Papaver somniferum* (opium), *Erythroxylum coca* (coca), *Aconitum napellus* (aconite), *Atropa belladonna* (belladonna), *Colchicum autumnale* (colchicum), as well as *Cinchona pubescens* (cinchona), and they have been used for centurics or millennia. Some of the best known phytomedicines are alkaloids, including atropine, emethine, capsaicin, cocaine, morphine, quinine, methylxanthines (i.e., caffeine and theophylline), strychnine, and nicotine.

Alkaloids have multiple therapeutic effects; some act directly on the central nervous system and some are too dangerous to be used in herbal practice [3]. Table 10.2 shows examples of therapeutic uses of alkaloids and their plant sources [4–10].

10.2.2 Flavonoids

Flavonoids occur both in the free state and as glycosides and lack nitrogen. Their structure is based on a C_{15} skeleton consisting of two benzene rings connected by a three-carbon chain, that is, C_6–C_3–C_6. The three-carbon chain is generally closed to form a heterocyclic ring, but 5-membered rings (aurone) and open-chain (chalcone) compounds are included (Fig. 10.3). Flavonoids are products of both the shikimic acid and acetate pathways, being formed by the condensation of a phenylpropanoid precursor with three malonyl coenzyme A units.

Flavonoids are plant phenolic compounds that cannot be synthesized by animals. They are widely distributed in higher plants and form part of the human diet. Some flavonoids are yellow compounds and contribute to the yellow colors of flowers and fruits where they are present as glycosides, dissolved in the cell sap. Highly colored flavonoids are the anthocyanidins and their glycosides (known as anthocyanins) are red, violet, or blue depending on the pH or the cell sap, as the color is associated with the distribution of positive charge throughout the arylsubsituted chroman ring system. Chelate formation with Fe^{3+} or Al^{3+} also influences the color of the anthocyanidines. Free

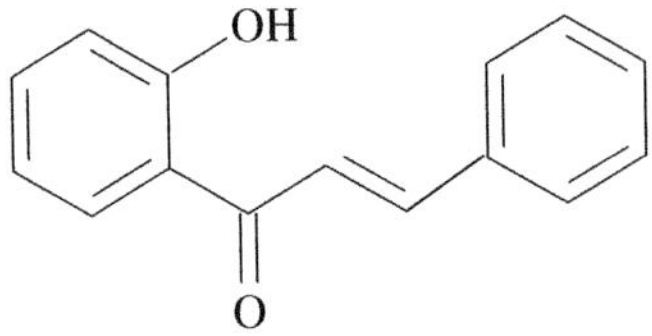

Name: 2-Hydroxyphenyl-styryl-ketone
Chalcones

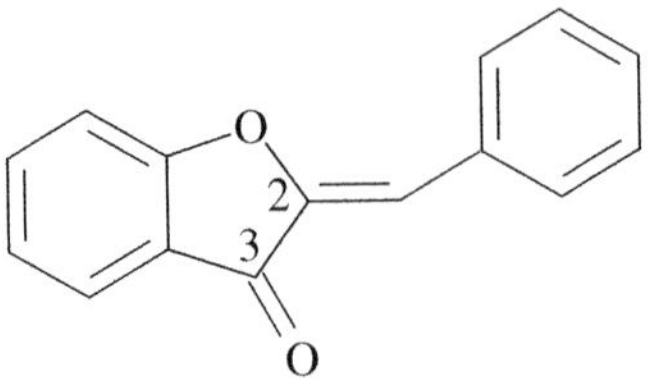

Name: 2-Benzylidene-2,3-dihydro-benzofuran-3-one
Aurones

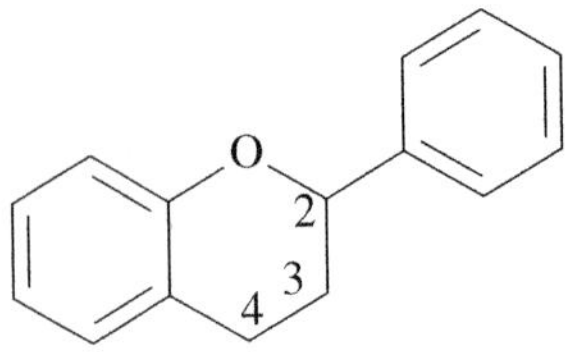

Name: 2-Phenylchroman (=Flavan)
Catechins (=flavan-3-ol)
Leucoanthocyanidines (=flavan-3,4-diol)
Anthocyanidines (=flavylium salts)
Flavanones (=flavan=4-on)
Flavanonols (=flavan-3-oil-4-on)
Flavones (=2,3-dehydroflavan-4-on)
Flavanols (=2,3-dehydro-3-hydroxy-flavan-4-on)

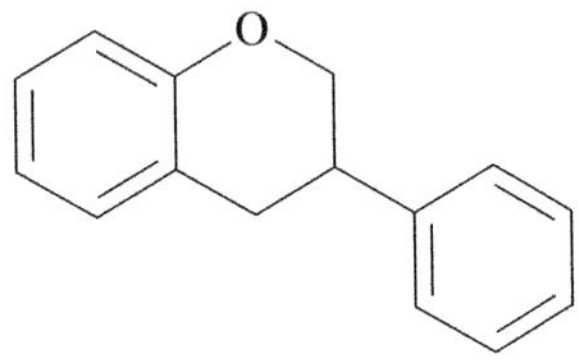

Name: 3-Phenylchroman
Isoflavonoids

FIGURE 10.3 The four basic structures of flavonoids.

flavonoids, the aglycones without glycosidic groups, are more lipophilic and may occur dissolved in essential oils.

Thousands of flavonoids have been isolated and characterized so far and they form one of the largest groups of naturally occurring phenols. They differ in their substituents, mostly hydroxyl or methoxy groups and in the nature and position the sugar residues are bonded to the aglycones. Flavanones, flavones, flavanols, isoflavonoids, anthocyanins, and flavans are the major classes of flavonoids that vary in their structural characteristics around the heterocyclic oxygen ring (Fig. 10.4). Table 10.3 shows the major classes and examples of flavonoids. The main flavonoids are the very common and structurally variable flavones (which are oxidation products of flavanones), flavanols (which are often glycosides), and the often bitter flavanones (including related dihydroflavonols). The other major constituents in this group are the ubiquitous colorful plant pigments, anthocyanins, and their glycosides, the anthocyanidins.

Flavanones occur predominately in citrus fruits, flavones in herbs, isoflavonoids in legumes, anthocyanins and catechins in fruits, and flavonols in all fruits and vegetables. They are found in high concentrations in many flowers and in foods such as citrus fruits, tomatoes, red wine, onions, and tea. Flavonoids commonly are anti-oxidants

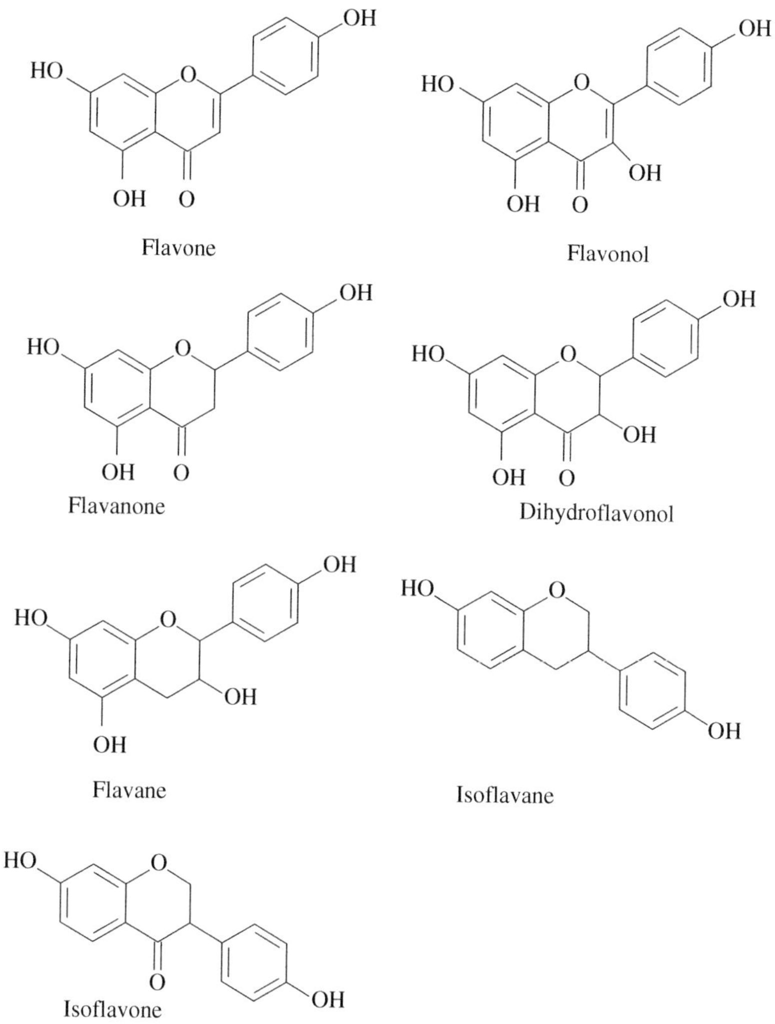

FIGURE 10.4 Structures of major classes of flavonoids.

with free-radical scavenging properties. They are believed to be anti-inflammatory and protective against various cancers, usually water soluble, and are present in many fruit and vegetable juices.

Table 10.4 shows some examples of flavonoids and their plant sources with therapeutic actions [11–18]. The best-known flavonoids are those in green tea (epigallo-catchin-3-gallate, catechin, etc.) and citrus fruits (kaempferol, quercetin and its derivatives, rutin, and hesperidin). Soy bioflavonoids (especially genistein and daidzein) are well-known food supplements that are being promoted as phytoestrogens.

TABLE 10.3 Classifications of Flavonoids

	Subclasses	Examples
Flavonoids	Flavone	Apigenin and luteolin
	Flavonols and dihydroflavonol	Quercetin, kaempterol, and rutin
	Flavanone	Purpurin and naringenin
	Flavans	Epigallocatechin-3-gallate
Isoflavonoids	Isoflavones	Daidzein and genistein
	Isoflavanones	Cyclokievitone
	Isoflavans	Licoricidin
	Isoflavanol	Ambanol
	Rotenoids	Rotenone and purarin
Biflavonoids		Amentoflavone
Other flavonoids	Anthocyanin	Malvidin
	Aurones	Hispidol
	Chalcone	Isoliquiritigenin

TABLE 10.4 Common Flavonoids, Plant Sources, and Major Uses

Flavonoids	Plants	Major Uses
Apigenin	*Matricaria recutita* (chamomile)	Anti-anxiety disorder [11] and cancer chemoprevention [12]
Quercetin	*Allium cepa* (onion)	Anti-allergy and anti-inflammation [13]
Rutin	*Punica granatum* (pomegranate)	Anti-oxidant [14]
Liquiritin	*Glycyrrhiza glabra* (licorice)	Anti-tussive [15]
Epigallocatechin-3-gallate	*Camellia sinensis* (tea)	Anti-oxidant [16]
Daidzein	*Glycine max* (soy)	Estrogenic [17]
Genistein	*Trifolium pratense* (red clover)	Estrogenic [18]

10.2.3 Glycosides and Saponins

Glycosides are a group of compounds characterized as consisting of a sugar portion (or moiety) attached by a special bond to one or more non-sugar portions (genin or aglycone). There are four classes of glycosides (*O*-, *C*-, *S*-, and *N*-glycosides). *O*-glycosides are formed when the bond between the two moieties may involve a phenolic hydroxyl group. *C*-glycosides are formed when the sugar molecule is attached to carbon atom of the genin or aglycone. Attachment of the sugar molecule to a thiol group forms *S*-glycosides and to an amino group forms *N*-glycosides (Fig. 10.5). Glycosylated compounds exhibit two main properties, which are different from the free aglycones, they are water soluble and decrease chemical reactivity. Glycosides are usually bitter-tasting, however some are sweet.

Most glycosides may be classed as "prodrugs" since they remain inactive until they are hydrolyzed in a large intestine with the help of specialized bacteria, leading

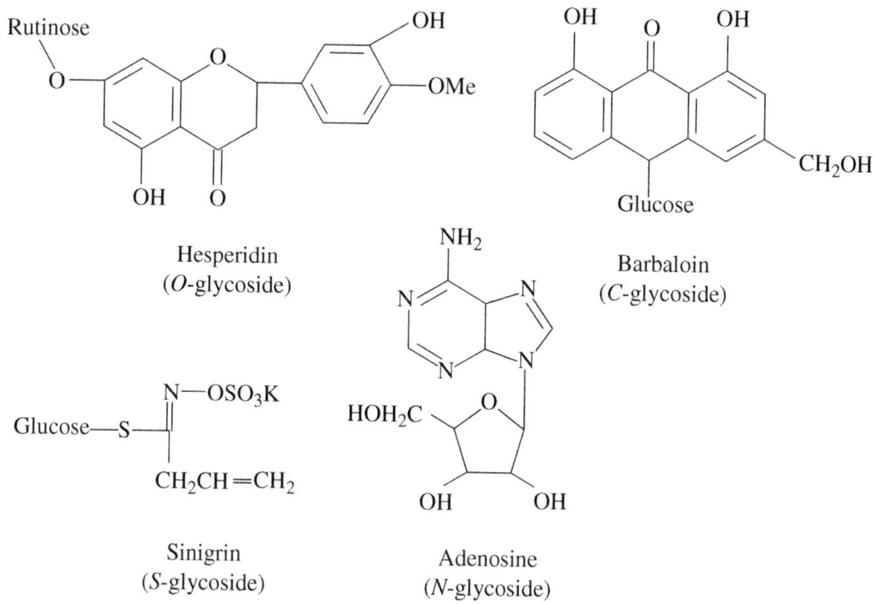

FIGURE 10.5 Structures of major classes of glycosides.

to the release of the aglycone, the truly active constituent. Classification of the plant glycosides is based on the structure of the aglycone, which range in molecular types, including phenols, quinones, terpenes, steroids, and others.

Glycosides are difficult to classify as their aglycone components are variable, but a common therapeutic classification is shown in Table 10.5, while Table 10.6 shows examples of glycosides with therapeutic uses and their plant sources [19–29]. Among the glycosides, cardiac and anthraquinone glycosides are those that have significant therapeutic use. The cardiac glycosides, such as digoxin, are major drugs in allopathic medicine. Anthraquinone glycosides are in common use as laxatives.

Saponins are a highly diverse group of glycosides with terpenoid aglycone components; they are high-molecular-weight glycosides, consisting of a sugar moiety (hexose, pentose, or saccharic acid) linked to a steroid or triterpene aglycone (amphiphilic). They possess two major characteristics and have soap-like surfactant effects and they cause hemolysis when directly introduced into the blood stream. They are usually poorly absorbed and do not cause hemolysis when taken orally. Three major classes of saponins exist: sapogenins (steroidal), triterpenes (terpenoid), and alkaloid glycosides (glycoalkaloids). Many possess potential medicinal effects, including anti-inflammatory, anti-venous permeability, and expectorant properties (Table 10.6).

10.2.4 Phytosterols

Plant sterols also called phytosterols are steroid alcohols; these have been reported to include over 250 different sterols and related compounds in various plant and marine species [30]. The most common representatives are β-sitosterol, stigmasterol, and

TABLE 10.5 Examples of Glycosides

Glycoside	Aglycone component	Examples
Cardiac	Steroids	Digitoxin and hellebrin
Anthraquinone	Hydroxyanthraquinone	Dianthrones, hypericin, and sennosides
Cyanogenic	Hydrocyanic acid	Amygdalin, prunasin, and sambunigrin
Glucosinolate	Isothiocyanate	Alliin and sinigrin
Aldehydic phenol	Aldehyde	Vanillin
Phenolics	Phenols	Hydroquinone
Alcoholic	Salicyl alcohol	Salicin
Phenolic	Z-2-Hydroxycinnamic acid	Coumarin and umbelliferone

TABLE 10.6 Examples of Glycosides and Saponins with Therapeutic Use and Plant Sources

Glycosides/Saponins	Plant Sources	Claimed Effects
Digitoxin	*Digitalis purpurea* (Foxglove)	Cardiotonic and anti-cancer [19]
Hellebrin	*Helleborus* (Hellebore)	Cardiotonic and anti-cancer [20]
Barbaloin	*Aloe barbadensis* (Aloes)	Laxative [21]
Sennosides	*Cassia Senna* (Senna)	Laxative [22]
Prunasin	*Prunus domestica* (Prunes)	Anti-oxidant [23]
Alliin	*Allium sativum* (Garlic)	Cholesterol reduction [24]
Sinigrin	*Brassica juncea* (Mustard)	Expectorant [25]
Cycloartane	*Cimicifuga racemosa* (Black cohosh)	Estrogenic [26]
Sarsasapogenin	*Yucca alofolia* (Yucca)	Anti-inflammatory and anti-oxidant [27]
Diosgenin	*Dioscorea villosa* (Yam)	Steroid precursor and anti-oxidant [28]
Astragalosides	*Astragalus membranaceus* (Astragalus)	Immunostimulant [29]

campesterol [31]. Chemical structures of these sterols (Fig. 10.6) are similar but differ in the side chain to cholesterol (Fig. 10.7), which is the predominant sterol found in animals. For instance, β-sitosterol and stigmasterol have an ethyl group at C-24 and campesterol, a methyl group at the same position. Dehydrogenation of the carbon 22–23 bond of β-sitosterol leads to stigmasterol, which is another common phytosterol. Chemical saturation of the delta 5 double bond of each of the aforementioned plant sterols leads to the formation of saturated phytosterols such as campestanol or sitostanol; Figure 10.8 shows the structures of sterols and stanols in comparison to chemical structures of cholesterol (Fig. 10.7).

In general, vegetable oils and products derived from vegetable oils are regarded as the richest natural sources of phytosterols [32]. Humans are not able to synthesize phytosterols, and dietary consumption is the only source of phytosterols in tissue and

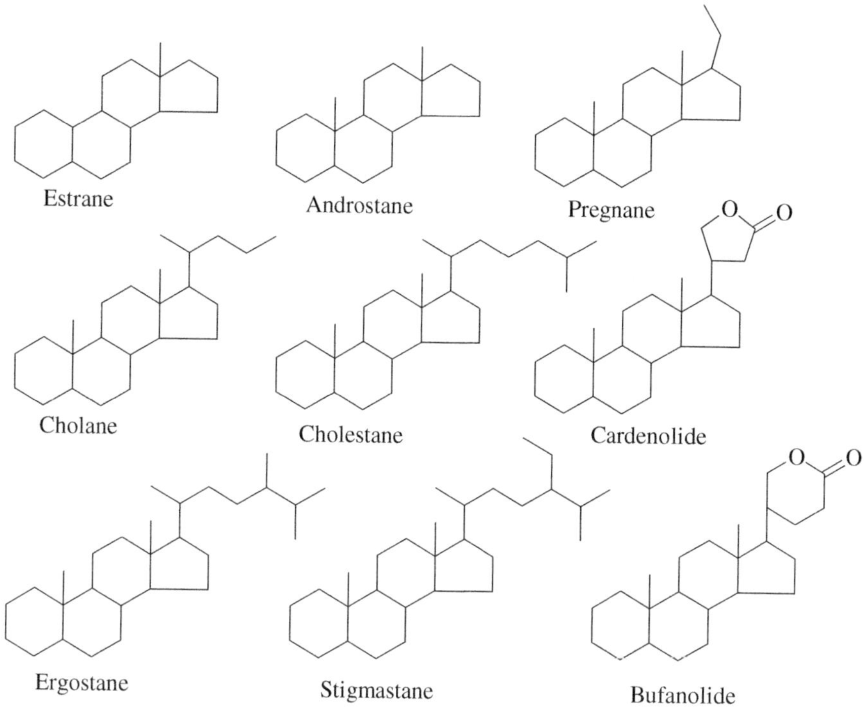

FIGURE 10.6 Structures of common sterols.

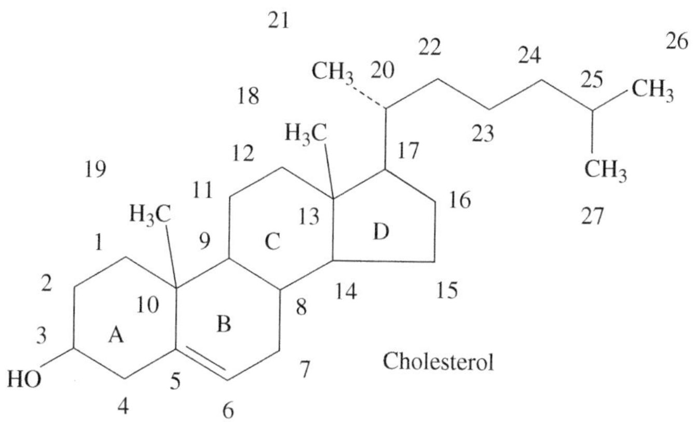

FIGURE 10.7 Structure of cholesterol.

plasma. Since the morbidity and mortality from cardiovascular disease have been dramatically reduced using cholesterol-lowering drugs (statins), interest in plant sterols lies in their potential to act as natural preventive dietary products. The important phytosterols are β-sitosterol, stigmasterol, and campesterol; they are found in popular

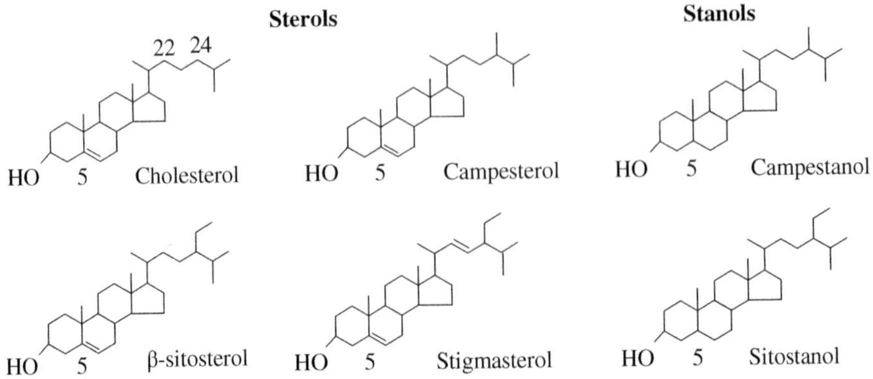

FIGURE 10.8 Chemical structures of sterols and stanols. Unsaturated phytosterols (campesterol, β-sitosterol, stigmasterol) and hydrogenated phytosterols (campestanol and sitostanol).

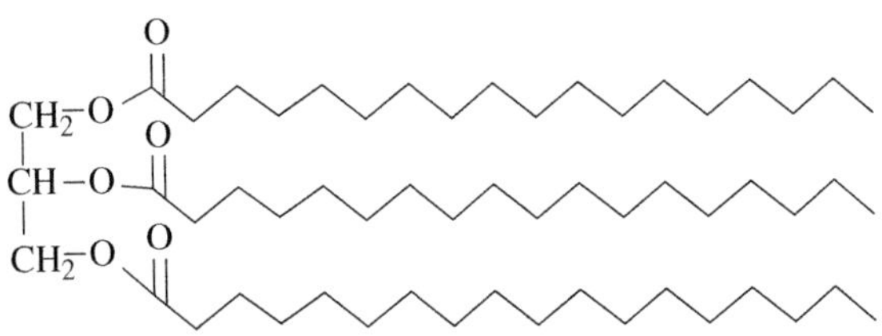

Glyceryl tristearate or tristearin

FIGURE 10.9 Structure of triacylglycerols.

sources such as olive oil and soy. They may reduce low-density lipoproteins and lower serum cholesterol [33, 34]. Other claimed phytosterol benefits include anti-inflammatory properties, immune enhancement, anti-cancer activity, anti-oxidant effects, and reduction of prostatism symptoms [35].

10.2.5 Fatty Acids

Lipids are broadly defined as substances found in living organisms that are insoluble in water but can be extracted from cells with a low polarity solvent like ether or chloroform. The lipids are non-volatile and referred to as "fixed" oils, as opposed to "essential" oils that are volatile. Fats and oils from common vegetables are triacylglycerols (Fig. 10.9), which are long chain carboxylic acids combined with glycerol through ester linkages. Solid or semi-solid triacylglycerols are called "fats" and liquid triacylglycerols are called "oils."

Saturated and unsaturated fatty acids are the two major types of fatty acids; examples of saturated fatty acids are given in Table 10.7. The fatty acids are often abbreviated as Cn:m, where n denotes the chain length and m the degree of unsaturation in the fatty acid chain. The most important unsaturated fatty acids are the C18 series,

TABLE 10.7 Summary of Common Saturated Fatty Acids

Fatty Acids	Chemical Formula	Abbreviation
Hexanoic (caproic)	$C_6H_{12}O_2$	C6:0
Octanoic (caprylic)	$C_8H_{16}O_2$	C8:0
Decanoic (capric)	$C_{10}H_{20}O_2$	C10:0
Dodecanoic (lauric)	$C_{12}H_{24}O_2$	C12:0
Tetradecanoic (myristic)	$C_{14}H_{28}O_2$	C14:0
Hexadecanoic (palmitic)	$C_{16}H_{32}O_2$	C16:0
Octadecanoic (stearic)	$C_{18}H_{36}O_2$	C18:0
Eicosanoic (arachidic)	$C_{20}H_{40}O_2$	C20:0
Docosanoic (behenic)	$C_{22}H_{44}O_2$	C22:0

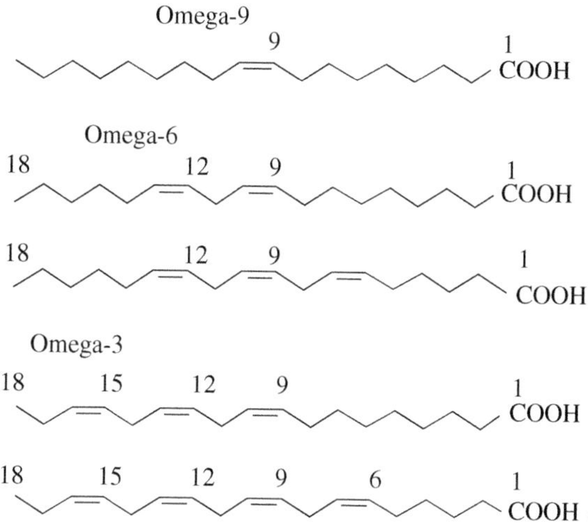

Oleic acid C18:1 (9)

Linolenic acid C18:2 (9,12)

γ-Linolenic acid C18:3 (6,9,12)

α-Linolenic acid C18:3 (9,12,15)

Stearidonic acid C18:4 (6,9,12,15)

FIGURE 10.10 Structures of omega-9, omega-6, and omega-3 carbon 18 polyunsaturated fatty acids.

the configuration of the unsaturation is Z as a general rule, and in polyunsaturated molecules, the double bonds occur in a 1,4-diene pattern (Fig. 10.10). Mammals do not have the enzymes capable of introducing double bonds after C-9 in the fatty acid chain, therefore humans cannot synthesize omega-6 ($n-6$) and omega-3 ($n-3$) fatty acids; however, these fatty acids are necessary precursors of other fatty acids, which are the starting compounds for the synthesis of prostaglandins and related compounds. Therefore, linoleic acid and α-linolenic acid are called essential fatty acids and must be supplied in the diet.

In humans, linoleic acid and α-linolenic acid are metabolized to polyunsaturated fatty acids (PUFAs) and other 20-carbon eicosanoids, which play vital roles in physiological functions [36]. Medicinal properties are attributed to oils with high omega-6 polyunsaturated fatty acid content. For example, herbal products containing evening primrose oil are used for the treatment of breast pain, premenstrual syndrome, eczema, etc. The important functions of omega-3 fatty acids have led to intervention studies and clinical trials in coronary heart disease, hypertension, diabetes, rheumatoid arthritis, psoriasis, and other diseases [37].

10.2.6 Essential Oils

Essential or volatile oils are oils with characteristic scent, odor, or smell obtained by steam distillation of plant materials. They are insoluble in water and soluble in organic solvents. These volatile compounds constitute the attractant odors of flowers and the attractant or defensive components of the other parts of the plant. They are sometimes referred to as volatile oils or essences, as they contain double bonds, alcohols, aldehydes, ketones, acids, and esters. They can undergo addition, oxidation, dehydrogenation, and esterification reactions.

They are commercially important as the basis of medicines, natural perfumes, and also as spices and flavorings in the food industry. Plant families particularly rich in essential oils include the Compositae *Artemisia*, *Chrysanthemum*; Lamiaceae, for example, *Mentha*, *Agastache*, *Lavandula*; Myrtaceae *Eucalyptus*, *Melaleuca*, *Eugenia*; Pinaceae *Pinus*; Rosaceae *Rosa*; Rutaceae Citrus, *Ruta*, *Evodia*, *Zanthoxylum*; Apiaceae (Umbelliferae) *Angelica*, *Foeniculum*, *Pimpinella anisum*; Lauraceae *Litsea*; Magnoliacea *Illicium*, Magnolia; Valerianaceae *Valeriana*; Verbenaceae *Verbena*.

Odiferous or pungent essential oils evoke responses varying from mucus secretion in the airways to rubefaction, manifested by reddening and tingling of the skin. Essential oils have a less important role in herbal therapeutics, although several specific agents, such as the essential oils of the mint family, are extensively used as flavors and drugs. Some essential oils are in popular use, for example, essential oils of thyme serve as mild anti-bacterials in products such as mouthwashes. The chemical composition of essential oils includes terpenes; the more pleasurable oils are mainly derived from terpenes and alcohols, while the pungent ones used to flavor food include sulfides and their derivatives.

10.2.7 Terpenes

Terpenoids or terpenes are made up of 5-carbon isoprene units containing two unsaturated bonds (Fig. 10.11), which are synthesized from acetyl-CoA via the mevalonic acid pathway. They are among the most common phytochemicals, comprising over 20,000 structures. They occur commonly in green vegetables, soy, and grains and many have anti-oxidant properties. Table 10.8 shows the classification and characteristics of terpenes and their plant sources. The low-molecular-weight monoterpenes and sesquiterpenes are volatile and many are found in the odorous essential oils of plants. The most important of the terpenes are 40-carbon atom tetraterpenes that encompass the large group of over 600 carotenoid pigments, which are anti-oxidants and vitamins.

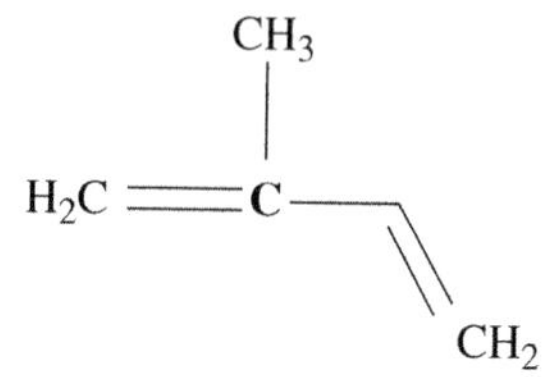

Isoprene — C_5H_8

FIGURE 10.11 Structure of isoprene.

TABLE 10.8 Terpenes: Classifications and Characteristics

Terpene	Chemical Formula	Characteristics/Examples
Isoprene	C_5H_8	
Terpene	$(C_5H_8)n$	
Monoterpene	$C_{10}H_{16}$	Essential oils, e.g., menthol and iridoids
Sesquiterpene	$C_{15}H_{24}$	Bitter principles, especially sesquiterpene lactones
Diterpene	$C_{20}H_{32}$	Resin acids and bitter principles
Triterpene	$C_{30}H_{48}$	Saponins and steroids
Tetraterpene	$C_{40}H_{64}$	Carotenoids
Polyterpene	$(C_5H_8)n$	Rubber

10.3 OTHER PHYTOCHEMICALS

Polysaccharides are defined as high-molecular-weight polymers resulting from the condensation of a large number of monosaccharide molecules. Each sugar is linked to its neighbor via a glycosidic linkage formed by the elimination of water molecule between the hemi-acetal hydroxyl group on C-1 of one sugar (or hemi-ketal hydroxyl group) and any of the hydroxyl groups on the other sugar molecule [1]. As natural molecules, polysaccharides are virtually universal and they are essential for a large number of vital functions in living organisms, such as the rigidity of cell walls in higher plants (cellulose), energy storage (starch, etc., in plants and glycogen in animals), and protection of tissues from dehydration (hydrophilic properties).

Starch is produced in plants as a storage form of glucose. Typically, it is a mixture of two polysaccharides, amylose, and amylopectin. Amylose is a straight chain molecule of 250–300 glucose molecules linked with α-1,4 glucoside bonds as shown in Figure 10.12. Amylopectin is the branched form and has one α-1,6 glucoside bond to about thirty α-1,4 linkages as shown in Figure 10.13 [38]. The structural difference between amylose and amylopectin allow them to be separated because amylose is more soluble in water. Most starches are a mixture of 25% amylose and 75% amylopectin.

Starch is obtained from the mature grain of corn (*Zea mays* L.), wheat (*Triticum aestivum* L.), or potato tubers (*Solanum tuberosum* L.). It is used in dusting powders because of its absorbent properties and in tablets as a disintegrant, filler, and binder.

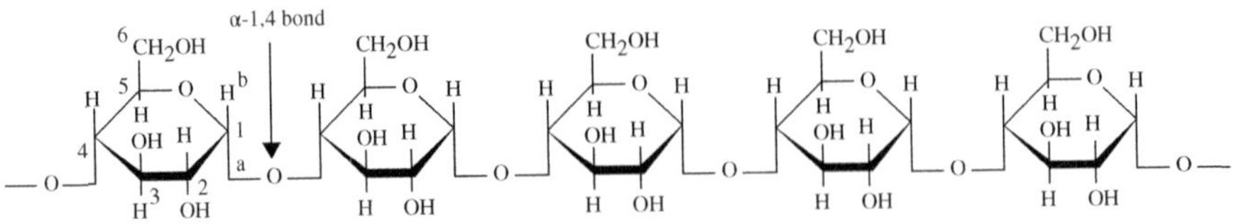

FIGURE 10.12 Structure of amylose linking glucose units together with α-1,4 glucosidic bonds.

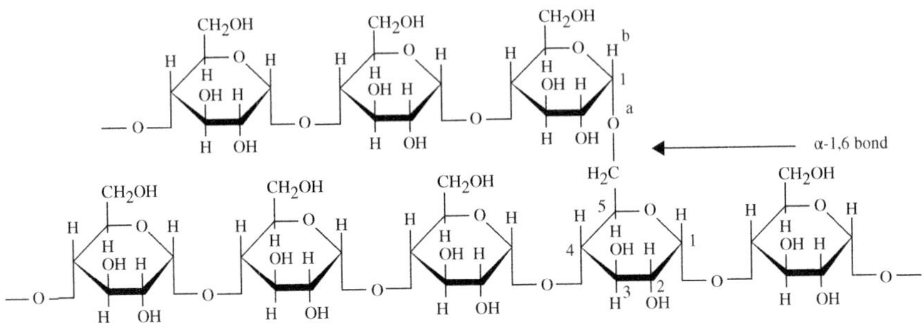

FIGURE 10.13 Structure of amylopectin linking glucose units together with a-1,4 and a-1,6 glucosidic bonds.

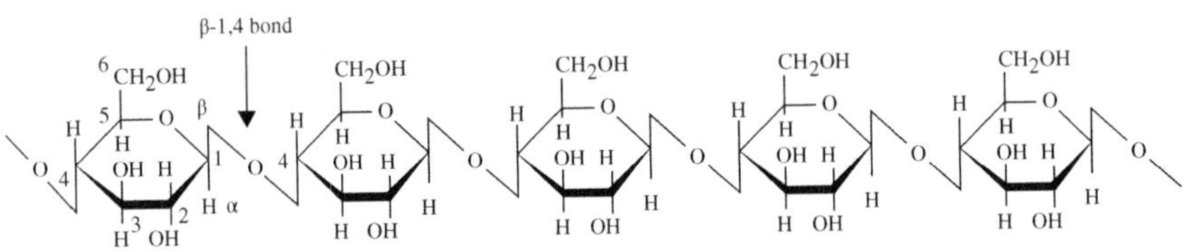

FIGURE 10.14 Structure of cellulose linking glucose units together with β-1,4-glucosidic bonds.

When a suspension of starch in cold water is heated while stirring, the opaque granules swell and burst to give a translucent hydrosol, a liquid colloidal dispersion [39].

Inulin is a linear molecule comprising about 30 fructose molecules linked by a β-1,2 bond. It is obtained from dahlia tubers, burdock root, chicory, and echinacea where it is biosynthesized as a carbohydrate reserve. It is excreted unchanged in the urine because mammalian systems do not have an enzyme to digest it [39].

Cellulose is an unbranched glucose polysaccharide and is composed of glucose units joined by β-1,4 linkages (Fig. 10.14). Cellulose is the structural material in plants giving them rigidity and form; cotton is a pure form of cellulose [39].

A gum is a hydrophilic plant polysaccharide or derivative that swells to produce a viscous dispersion or solution. It comprises a heterogeneous mixture of sugars (mainly glucose, mannose, xylose, arabinose, and galactose) and uronic acids (e.g., glucuronic acid). It is used as an emulsifier, suspending agent, binder, gelatinizing agent, and thickener in pharmaceutical preparations [39].

TABLE 10.9 Examples of Dietary Intake Components and Their Medicinal Effects

Components	Medicinal Effect	Examples
Flavonoids	Anti-oxidant	Quercetin
Statins	Cholesterol lowering	Mevastatin
Phytosterols	Blood lipid lowering	Sitostanol
Guanidine	Blood sugar lowering	Galegine
Peptide from a pit viper, *Bothrops jararaca*	Blood pressure lowering	Developed to the drug captopril
Purine alkaloids	Stimulant	Caffeine

A mucilage is a thick, viscous, adhesive liquid made by dispersing a gum in water or by extracting the mucilaginous compounds from a plant material with water. Mucilages are mainly used to suspend insoluble substances in liquids [39].

10.4 MEDICINAL EFFECTS RELATING TO DIETARY INTAKE

Dietary intake consists of a food or material that provides energy such as carbohydrates, proteins, and fats or essential material that maintains life such as minerals, vitamins, and essential fatty acids. Other components of dietary intake, while being non-essential for day-to-day living, are recognized to have health and medicinal effects (Table 10.9).

10.4.1 Anti-oxidants

Anti-oxidants have been shown or implicated for general health and well-being and also for the treatment of a wide range of diseases, for example, cancer, cardiovascular disease, diabetes, pain, and inflammation. A prominent group of natural anti-oxidants is the polyhydroxyphenols and related substances, for example, quinones. Polyhydroxyphenols include hydroxycinnamic acids, flavonoids, stilbenoids, lignins, and tannins (Fig. 10.15).

Changes to molecules as the result of digestion and metabolism by, for example, gut microflora may be critically important for medicinal effects. Many dietary substances are a part of a complex polymeric matrix and the active components are only released through digestive and metabolic processes. It is important to keep in mind that substances with anti-oxidant and other chemical properties may give rise to medicinal effects that are completely unrelated to and independent of their anti-oxidant properties. In some examples, the anti-oxidant property is clearly important for medicinal effect while in other cases the chemical structure, independent of anti-oxidant property, is important.

Q. *Why are anti-oxidants of biological and medicinal significance?*

Oxygen exposure in the body has both essential and destructive properties. Oxygen is essential in respiration to provide energy through metabolism. The metabolic process is tightly controlled in healthy individuals but can be very destructive

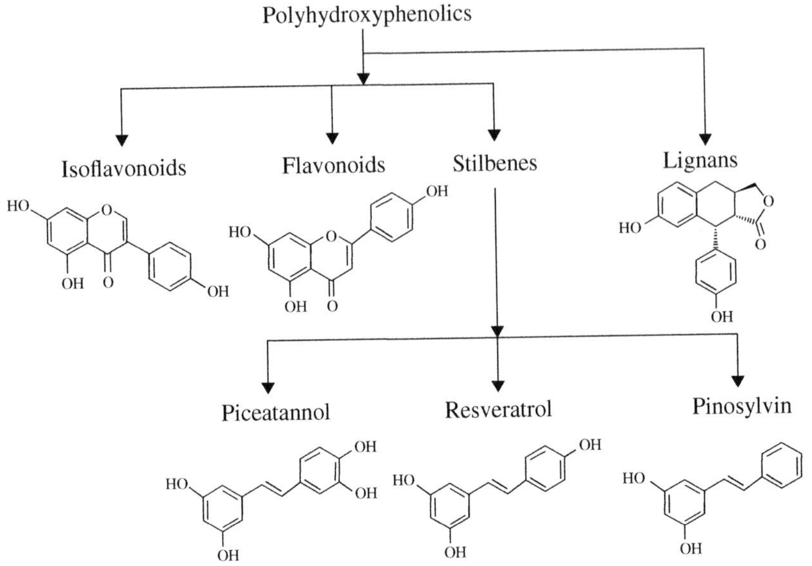

FIGURE 10.15 Structures of polyhydroxyphenols and related compounds.

with loss of control in injury or disease. Free radicals are essential in normal energy-related metabolism and also in other oxidative metabolic processes such as those working via the mixed functional oxidase systems. However, free radicals are very damaging when not controlled.

The actions of anti-oxidants are to neutralize and destroy free-radicals, to maintain cellular molecular components in the required reduced state, for example, to maintain cysteine amino acid residues as thiols (–SH) rather than as the oxidized disulfide (–SS–) linkages. Anti-oxidant effects may be measured using physical methods (electron spin resonance), chemical assays (colorimetric assays for unpaired electron), and through biological assays (enzyme activities).

10.4.1.1 Polyhydroxystilbenes (Resveratrol and Piceatannol) Polyhydroxyphen olic substances are biosynthetically related to flavonoids, lignans, and tannins. Stilbenes are compounds with the core structure made up of two benzene rings (rings A and B) linked via an E (*trans*) double bond (Fig. 10.16). The most well-known stilbene is *E*–3,4′,5-trihydroxystilbene, which is commonly referred to as resveratrol. Resveratrol is considered to be the constituent in red wine largely responsible for the "French paradox." Piceatannol is also found in red wine, but at a much lower concentration than resveratrol, it is the hydroxylated analogue, *E*–3,3′,4′,5-tetrahydroxystilbene of resveratrol. Resveratrol and piceatannol are stilbenoid substances, relatively rare in foods but widespread over the plant kingdom with the main dietary intake being from grapes, grape products, and wine.

Cardiovascular benefits observed for a "Mediterranean diet" from epidemiological studies have been partly attributed to resveratrol. Resveratrol is present in

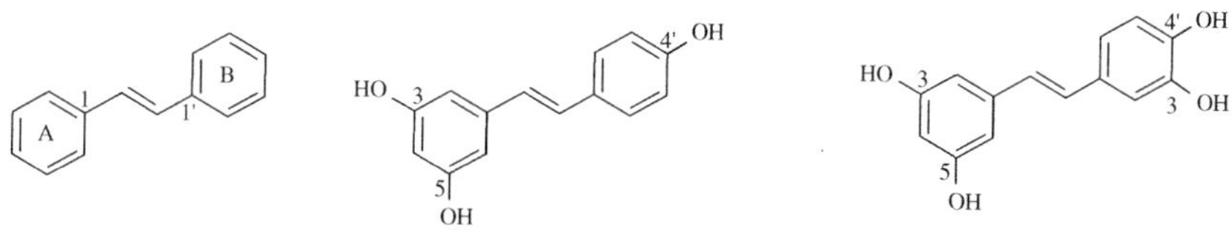

E-stilbene | *E*-3,4',5-Trihydoxystilbene (resveratrol) | *E*-3,3',4'5-Tetrahydroxystilbene (piceatannol)

FIGURE 10.16 Chemical structures of resveratrol and piceatannol.

concentrations up to 10 μM in red wine but is present at much lower concentrations in white wine [38]. Numerous studies have been carried out *in vitro* and *in vivo* to evaluate resveratrol as a dietary supplement or as a medicinal active ingredient.

Resveratrol displays a broad spectrum of biological activities including anti-oxidant, anti-aging, anti-atherosclerosis, anti-angiogenic, anti-diabetic, anti-viral, anti-proliferative, anti-inflammatory, and anti-cancer activities [40, 41]. Piceatannol is the hydroxylated metabolite of resveratrol formed through catalytic oxidation by cytochrome P450 1B1 (CYP 1B1). Cyp 1B1 is a member of the inducible phase I enzymes (CYP1) responsible for xenobiotic detoxification. Piceatannol exhibits a similar broad spectrum of biological activities to resveratrol and plays a role in the mediation of the anti-cancer activities of resveratrol [42]; however, piceatannol has not been investigated extensively [43].

Polyhydroxyphenols including resveratrol have attracted wide interest in cancer research; special interest has been given to polyhydroxyphenols because they appear to target cancer cells exclusively. Oxidative stress is well documented for causing damage to DNA and signaling proteins. Although there may not be damage to the primary DNA sequence initially, re-programming of DNA transcription eventually leads to carcinogenesis. These early changes occur without alteration in primary DNA sequence and are referred to as epigenetic changes. Epigenetic events (markers) are involved in the control of gene expression, the processes in the transcription and translation of genes. Epigenetic mechanisms in the anti-cancer actions of bioactive food components including resveratrol and their implications in cancer prevention have been reviewed recently [44, 45].

Since the discovery of its cardioprotective property in 1992, interest in resveratrol has grown and accelerated following the discovery of its cancer chemopreventive properties and there is a large volume of scientific data on resveratrol. It is capable of suppressing all stages of carcinogenesis including initiation, promotion, and progression [46, 47]. The catechol moiety in piceatannol is responsible for this compound's ability to chelate metal ions. This may be part of the reason for piceatannol being more potent at inhibition of zinc-dependent histone deacetylase (HDAC) enzymes than resveratrol. In intact cells, resveratrol can be metabolized to piceatannol. Blackwell et al. reported IC_{50} values of resveratrol and piceatannol in the inhibition of zinc-dependent HDAC enzymes; these data revealed that both resveratrol and piceatannol exhibited the highest potency against HDAC6. HDAC6 preferentially targeting non-histone substrates such as HSP60 and α-tubulin; HSP90, p53, and pRb are all influenced by acetylation and it is through these substrates that HDAC inhibitors may modulate their anti-proliferative effects [48].

10.4.1.2 Resveratrol Studies have been carried out to identify clinically useful properties and effects in patients given resveratrol. Clinical studies have shown that resveratrol at a daily dose of up to 5 g is generally well tolerated in patients; however, mild to moderate side effects may limit the dose to less than 1 g per day [49]. Resveratrol is efficiently absorbed orally; however, rapid Phase II metabolism severely limits is systemic availability. In a randomized double-blind placebo-controlled cross-over study, single oral doses of resveratrol (250 or 500 mg) were associated with a dose-dependent pattern of higher cerebral blood flow in the prefrontal cortex during task performance. This effect may prove to be useful for improved brain function.

Lack of *in vivo* activity may be due to poor bioavailability; rapid metabolism and excretion may result in an inability to achieve sufficiently high resveratrol concentration at the site of action. Resveratrol is relatively polar and readily metabolized and excreted, so clinically useful action may not be achievable. Similar limitations are expected for resveratrol's metabolite, piceatannol. Therefore, there is a need to view both molecules as potential drugs or as drug design lead molecules.

10.4.2 Omega-3 Long Chain Fatty Acids and Derivatives

The major cause of death from breast, prostate, and other cancers is metastatic disease in which cells migrate from the original tumor site to distant organs and establish secondary tumors [50]. Metastases also frequently render the tumor unresponsive to subsequent drug treatment, and because of their disseminated nature the surgical removal of metastases is not feasible [51].

$N-3$ and $n-6$ polyunsaturated fatty acids (PUFAs) are two major classes from essential dietary fatty acids—examples include eicosapentaenoic acid (EPA; $n-3$ C20:5) and arachidonic acid (AA; $n-6$, C20:4), respectively. EPA and AA are structural analogous, but EPA has an additional olefinic double bond at the $n-3$ position located between carbons 17 and 18. Numerous epidemiological studies in humans and experimental studies in cells and animals have suggested that $n-3$ poly-unsaturated fatty acids (PUFA) like EPA offer potential benefits in cancer by decreasing tumor growth and metastases, whereas n-6 PUFAs are tumor-promoting. In a large cohort of French patients, increased content of docosahexaenoic acid (DHA) or alpha-linolenic acid in breast adipose tissue was associated with improved survival compared with women in the lower intake group [52]. In xenograft mouse models, the spread of lung metastases produced by human breast cancer cells is decreased by EPA and DHA [53, 54]. Dietary PUFA are not only important risk factors for tumor progression but may also offer opportunities for the development of new cancer therapeutics.

$N-3$ and $n-6$ PUFAs are substrates for cyclooxygenase (COX), lipoxygenase, and cytochrome P450 enzymes and undergo biotransformation to parallel series of eicosanoid metabolites, including prostaglandins (PGs), leukotrienes, and epoxides.

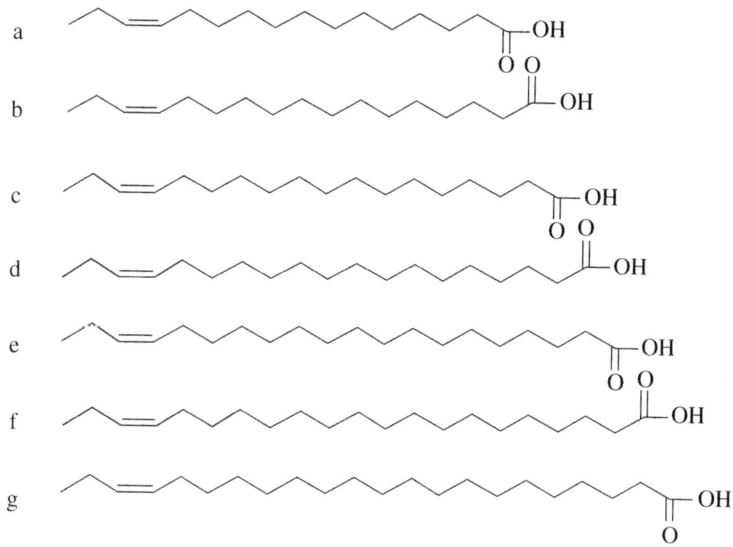

FIGURE 10.17 Structural analogues of $n-3$ mono-unsaturated fatty acids with different carbon chain lengths: 16 (a), 17 (b), 18 (c), 19 (d), 20 (e), 21 (f), and 22 (g) [63].

$n-6$-derived PGs, especially PGE_2, are strongly implicated in tumor growth and metastases [55, 56]. The constitutive expression of COX-2 is strongly induced by pro-inflammatory cytokines and extracellular stress stimuli [57, 58]. COX-2 is over expressed in many human breast and epithelial cell cancers and is associated with increased rates of PGE_2 production, which augments tumor progression and metastases, leading to a poor prognosis [59].

COX-2 inhibitors such as celecoxib have been shown to decrease mammary tumor burden in animal models by increasing tumor cell apoptosis, decreasing proliferation, and suppressing angiogenesis [60–62]. However, while nonsteroidal anti-inflammatory drugs effectively inhibit tumorigenesis, the use of these agents in longer term preventive treatment for cancer is impractical due to their toxicity. There is a need for alternative molecules that inhibit the action of PGE2 and minimize tumor spread; the identification of alternate agents that decrease PGE_2-dependent tumor cell migration may assist the development of novel anti-metastatic therapies. From studies in animals, the anti-metastatic actions of $n-3$ PUFA were found to be mediated in part by inhibiting COX-2 activity and PGE2 formation [54]. Thus, increasing the dietary intake of $n-3$ PUFAs could be a viable strategy to impair PGE_2 production by tumors and improve health outcomes. However, because long-term patient compliance with altered dietary regimens is rarely successful, there is a need to develop effective new chemical entities.

Accordingly, Cui et al. synthesized a series of novel $n-3$ monounsaturated fatty acid (MUFA) derivatives (Fig. 10.17) and evaluated their potential as inhibitors of breast cancer cell growth and migration (Table 10.10). The principal finding was that the longer chain $n-3$ MUFA analogues (C20–C22) effectively decreased the growth

TABLE 10.10 Anti-proliferative and Pro-apoptotic Activities of *n* – 3 MUFAs in COX-2 Overexpressed Breast Cancer Cells[a]

Analogue	Carbon Chain Length	Anti-proliferative Activity[b] (μM)	Pro-apoptotic Activity[c] (μM)
a	16	>100	>100
b	17	>100	69 ± 19
c	18	>100	54 ± 12
d	19	36 ± 4	43 ± 9
e	20	31 ± 8	43 ± 0.2
f	21	9.3 ± 2.9	12 ± 3
g	22	21 ± 1	26 ± 5
AA	20	>100	>100
EPA	20	>100	>100

[a] From Cui et al. [63].
[b] MTT activity.
[c] Caspase-3 activity.

TABLE 10.11 Effect of *n* – 3 MUFAs on PGE2 Secretion by COX-2 Overexpressed Breast Cancer Cells[a]

Analogue	PGE_2 Secretion	
	Basal (pmol/10^4 cells/48 h)	AA-Stimulated (nmol/10^4 cells/48 h)
Control	36 ± 13	1.09 ± 0.17
a	62 ± 12	ND
b	56 ± 19	ND
c	56 ± 22	ND
d	39 ± 6	0.94 ± 0.58
e	24 ± 6[b]	1.05 ± 0.11
f	20 ± 6[b]	0.46 ± 0.14[c]
g	18 ± 4[b]	0.58 ± 0.22[c]

[a] From Cui et al. [63].
ND, not detected.
Different from control:
[b] $P < 0.05$.
[c] $P < 0.01$.

and invasion capacity of breast cancer cells that over-expressed COX-2 and markedly decreased PGE_2 production by these cells (Table 10.11). Thus, the inhibition of COX-2-derived PGE_2 formation may be a strategy that may be exploited to minimize tumor growth and metastases [63].

The anti-tumor actions of *n* – 3 PUFA, such as EPA, have been attributed to the decreased formation of COX-2-derived PGE2 [64]. However, potencies of naturally occurring *n* – 3 PUFA may be limited by their intracellular availability, which would decline rapidly in cells because they also undergo COX-2-mediated biotransformation. Some reports indicate that the inhibitory effects of EPA and other PUFAs may be associated with lipid peroxidation [65]; by contrast, the saturated fatty acids did not increase lipid peroxidation or change cell viability [66]. The longer chain $n-3$ MUFA analogues effectively inhibited the proliferation and invasion potential of breast cancer cells that were engineered to overexpress COX-2; these agents also activated apoptotic cell death. The anti-proliferative potency of compound f (Fig. 10.17), $n-3$ mono unsaturated C21 fatty acid, was similar to that of clinically effective COX-2 inhibitors in cell-based assays. Thus, non-steroidal anti-inflammatory agents inhibited the formation of PGE2 in lipopolysaccharide-activated erythrocytes (IC_{50} values between 0.6 and 3.6 μM) [67] and ibuprofen and naproxen inhibited COX-2 directly with IC_{50} values of 10 and 18 μM, respectively; other agents of this type were more or less potent [68].

10.5 NATURAL PRODUCTS AS LEADS FOR DRUG DEVELOPMENT

Originally, all medicinal active ingredients were of natural origin and therefore "natural products," whose useful medicinal actions have arisen through changes in plant and animal metabolic processes over millions of years. Medicinal actions can vary from mild and nonspecific to extremely potent and specific with a very diverse range of chemical structures. Plants and microorganisms are the main source of medicinally useful natural products as they are relatively immobile and are very dependent on "chemical defense" for their survival. Plants and microorganisms are much more metabolically advanced and versatile than animals and form a great diversity of drug-like molecules to aid in their survival (Table 10.12). Natural products may prove to be useful medicines and a significant proportion of active ingredients in pharmaceutical products are original natural molecules. Some natural products, however, show *in vitro* biological properties that indicate potential medicinal benefit but are not medicinally effective when given to patients due to unforeseen toxicity and/or lack of efficacy.

Structure–activity relationship (SAR) studies may be carried out with natural and synthetic analogues to determine chemical structural changes required to reduce toxicity and increase efficacy. This may result in a synthetic analogue that is developed

TABLE 10.12 Survival Strategies are Wide and Include Plants and Microorganisms

Poison	Deadly toxins, vomiting, diarrhea
Deterrents	Bad taste, foul odor, nausea, and loss of appetite
Anti-fertility	Lose of reproductive capacity
Behavioral changes	Neurological actions, nerves, and brain
Immune system actions	Allergic reactions and anaphylactic shock
Anti-vitamins	Block actions of vitamins
Atypical amino acids	Mimic essential amino acids and give rise to dysfunctional peptides and proteins

to a useful pharmaceutical active ingredient. Examples include developing resveratrol and piceatannol analogues that required structural changes to improve bioavailability and potency toward biological targets and to minimize toxicity.

10.5.1 Catechol Moiety of Piceatannol: Implication and Significance

10.5.1.1 DPPH* as a Model for Lipid Peroxyl Radicals DMPO/ESR spin-trap experiment evaluates total H-atom-donating capacities of anti-oxidants: (Fig. 10.18). A DPPH• experiment assesses the ability of anti-oxidants to transfer labile H atoms to radicals and is widely used to assess the ability of polyphenols to transfer labile H atoms to radicals, a common mechanism for anti-oxidant protection [69].

10.5.1.2 Anti-oxidant Activity The anti-oxidant activities of hydroxystilbenes was determined in relation to the number of hydroxyl groups and their positions on the aromatic rings (Fig. 10.19) [38, 70, 71]. 3′-Hydroxylation of resveratrol **1** generates a catechol moiety (3′,4′-dihydroxy) in piceatannol **4**. Substitution of resveratrol **1** at position 3′, 4, and 5′, compounds **2**, **4**, **6**, increases the electron density of the hydroxyl groups and decreases the dissociation energy of the oxygen–hydrogen bond which correlates with anti-oxidant activity. Radical scavenging experiments with $O_2^{\bullet-}$ (5,5-dimethylpyrroline-*N*-oxide/electron spin resonance (DMPO/ESR) and 2,2-diphenyl-1-picrylhydrazide (DPPH•) (photometry) revealed that anti-radical activities of piceatannol **4**, 3,4,4′,5-tetrahydroxystilbene **2**, 3,3′,4,5,5′-tetrahydroxystilbene **5**, and 3,3′,4,4′,5,5′-hexahydroxystibene **6** are significantly higher than that of resveratrol **1** [38].

10.5.1.3 Anti-oxidant/Pro-oxidant Activities of Resveratrol and Hydroxylated Analogues Piceatannol is a better radical scavenger than resveratrol, ortho-semiquinones, quinones, and redox-cycling. In the study by Murias and coworkers [38], the presence of the signal with similar splitting in the ESR spectra indicating the presence of ortho-semiquinones was observed for compounds **2**, **4–6** and not for resveratrol or compound **3**. These ESR signals were not observed in the absence of microsomes or the nicotinamide adenine dinucleotide phosphate (reduced form) (NADPH) generating system. In addition, these ESR signals could also be observed with the addition of superoxide dismutase (20U) to the reaction mixture indicating a direct interaction of the compounds with the microsomal electron carriers and not an indirect interaction via microsomal $O_2^{\bullet-}$ production. Furthermore, these authors showed compounds **2**, **4–6** and not resveratrol or **3** consumed additional O_2 in the NADPH-generating system. This led the authors to conclude that polyhydroxystilbenes with catechol and/or pyrogallol groups induce oxidative stress by redox-cycling of *ortho*-semiquinones (Fig. 10.20).

10.5.1.4 Pro-oxidant Activities Correlate with Cytotoxicities In addition to their anti-oxidant property, polyhydroxystilbenes with catechol and/or pyrogallol groups have been shown in a number of studies to possess cytotoxic activity that may be attributed to pro-oxidant actions [38, 70, 72, 73]. The central role of the catechol/pyrogallol group was shown by variation of the hydroxyl substituents in the A ring

$$O_2^{\bullet -} + \text{DMPO} + H^+ \longrightarrow \text{DMPO/}^{\bullet}\text{OOH}$$

FIGURE 10.18 H-atom-donating reaction [38].

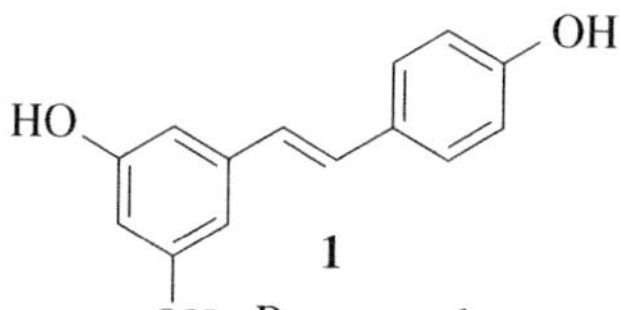

1 Resveratrol

IC_{50} 3.39 μM
$k_2 = 1.32 \times 10^2$ M^{-1}/S^{-1}

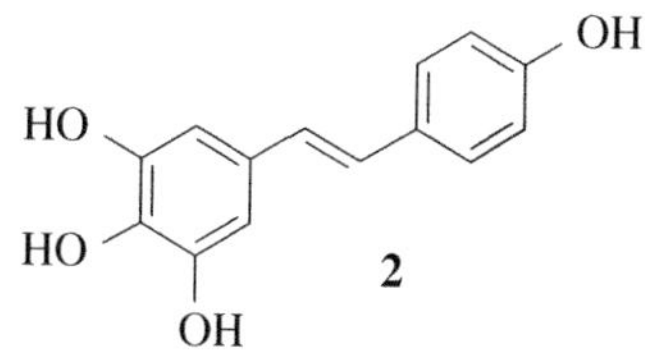

2

IC_{50} 41.5 μM
$k_2 = 2.22 \times 10^5$ M^{-1}/S^{-1}

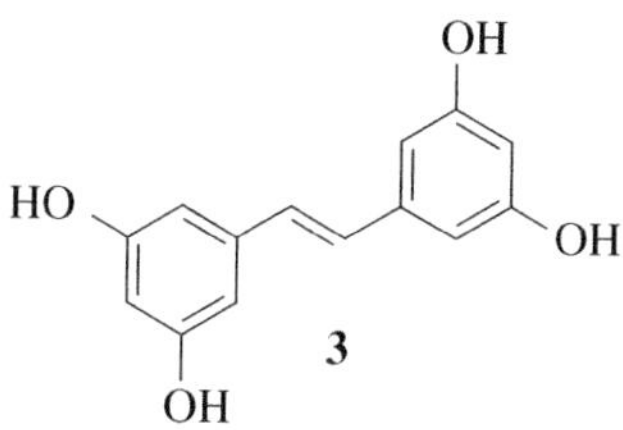

3

14% of antiradical activity of resveratrol

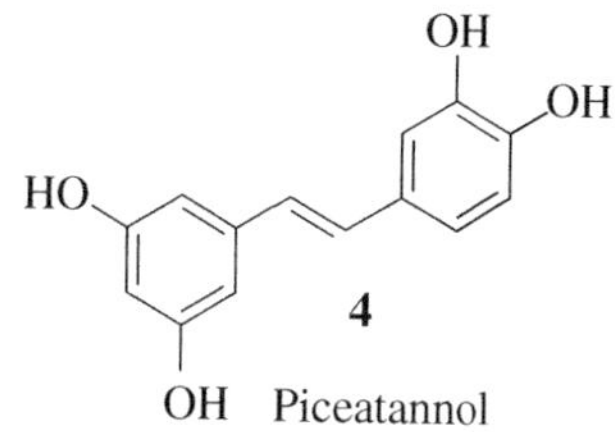

4 Piceatannol

IC_{50} 2.69×10^{-3} μM
$k_2 = 4.43 \times 10^5$ M^{-1}/S^{-1}

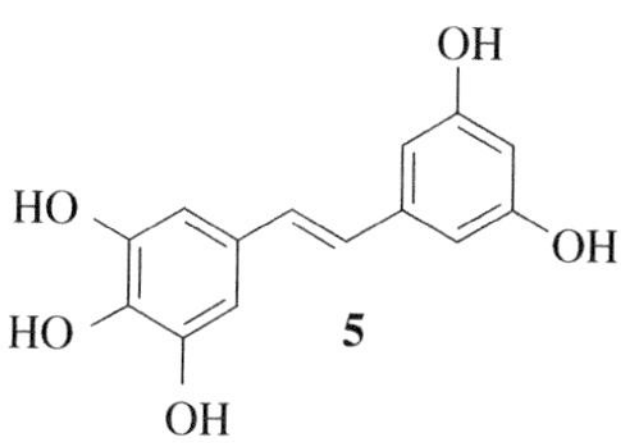

5

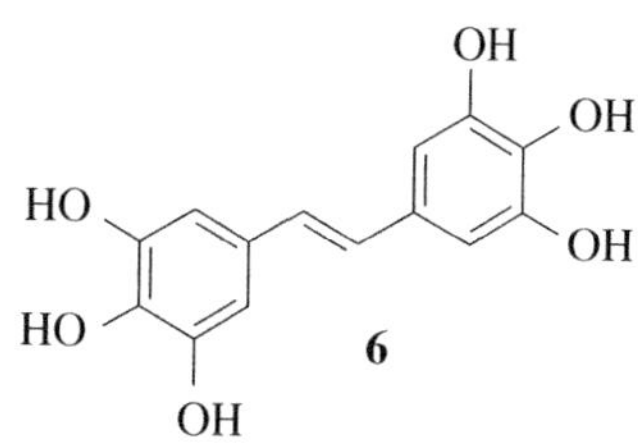

6

IC_{50} 5.02×10^{-3} μM

FIGURE 10.19 Structures and anti-oxidant activities of hydroxystilbenes. IC_{50}: concentrations causing 50% quenching of DMPO/.OOH signal in the ESR experiments. k_2: second-order rate constants for the abstraction of H-atoms from hydroxystilbenes by DPPH. [38].

$$\text{O=Stilb=O} + e^- \xrightarrow{\text{Microsomes}} {}^-\text{O-Stilb-O}^{\cdot}$$
$$^-\text{O-Stilb-O}^{\cdot} + O_2 \longrightarrow \text{O=Stilb=O} + O_2^{\cdot -}$$

FIGURE 10.20 Redox-cycling of *ortho*-semiquinones [38].

using resveratrol as the comparison. The results showed that the catechol/pyrogallol group not only increased the anti-oxidant activity but also may be central to cytotoxicity of the hydroxystilbenes [70].

Cytotoxicity correlates with oxidative stress, implying that oxidation is also involved. This was explained in terms of the pro-oxidant property of anti-oxidant-derived radicals formed during the reactions of anti-oxidants. The catechol and pyrogallol moieties are necessary not only for binding of metal ions but also for the formation of *ortho*-quinoid structures. On the other hand, in relation to anti-oxidant activities of compounds with catechol/pyrogallol moieties, piceatannol and 3,3′,4,4′,5,5′-hexahydroxystilbene were found to be more potent than resveratrol in inhibiting H_2O_2-induced DNA single strand break in leukemic L1210, K562, and HL-60 cells [72]. Increased protection of DNA against hydroxyl radicals (OH•) was interpreted as piceatannol and 3,3′,4,4′,5,5′-hexahydroxystilbene having higher anti-oxidant activity compared with resveratrol and this was also due to the presence of the catechol/pyrogallol moieties in these compounds.

10.5.2 SAR Studies for Drug Development

10.5.2.1 Resveratrol Analogues Naturally substituted derivatives of resveratrol are known and some show more potent biological activities and evidence of higher bioavailability. Common substituents for natural products (Fig. 10.21) are methyl and prenyl groups. Biological studies of tetrahydroxystilbenes (THOS) have revealed that they possess tyrosinase inhibitor, anti-herpetic, anti-HIV, anti-inflammatory, anti-oxidative, anti-apoptotic, and neuroprotective activities [38, 73–77].

Cytotoxicity activities of compounds **1–6** toward K562 leukemic cancer cell line are summarized in Figure 10.22 and Table 10.13 [78]. The conjugated double bond, between the two aromatic rings is an important structural feature in the inhibition of cancer cell proliferation. Reduction of this bond in **5** to give **6** resulted in about a 100-fold reduction in the potency. Maintenance of the co-planar orientation is attributed to the presence of this double bond. This co-planarity is significant as the potent anti-oxidant properties of stilbenes depend on their resonance effects [71, 79, 80], where the unpaired electrons are mainly distributed to the oxygen atom in the *para* position, double bond, and B-benzene ring. In relation to isoprenyl 2-C or 3-O prenyl substitution of the A-benzene ring, 2-C-linkage appears to enhance significantly the inhibitory potency of the compound in comparison with 3-*O*-linkage.

For the B-benzene ring, the vanilloid moiety (3-methoxy-4-hydroxyphenyl) appeared to contribute to the inhibitory activity of the stilbenes. Exchanging the methoxy and hydroxyl groups from vanilloid as in **1** to isovanilloid moiety, as in **2**, reduced the inhibition approximately three-fold. The hydroxyl group at the 4′-position contributes more toward cytotoxic activity than the other hydroxyl groups [78], indicating that the stronger H-donor anti-oxidant property of the 4′-OH compared with the 3′-OH and other hydroxyl groups may be important for the cytotoxic action observed.

10.5.2.2 n − 3 Mono-unsaturated Fatty Acid Analogues In the Cui et al. (2013) study [63], to understand the interactions of the $n-3$ MUFAs with COX-2 in greater detail, compounds **a–g** were docked into the active site of the enzyme (PDB code: 3HS5). When the $n-3$ MUFAs **a–g** are docked with the active site of COX-2 (Figs. 10.23 and 10.24), they adopted an L-shaped configuration similar to that found with the naturally occurring $n-6$ PUFA AA and $n-3$ PUFA EPA [68, 81, 82]. The carboxylate groups in $n-6$ and $n-3$ PUFAs are ion paired with the residues Arg120 and

Artocarpus chama, Moraceae
(E) – 2,6–bis (3–methyl–2–buten–1–yl) –3, 3′, 4′, 5–tetrahydroxystilbene: antifungal, free–radical scavenging

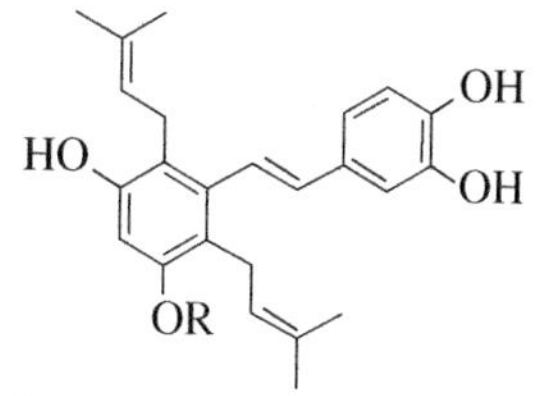

R = H, CH_3

Morus alba, Moraceae
(E) – 4′ – (3–methyl– E –but–1–enyloxy) –2′, 3, 5–trihydroxystilbene

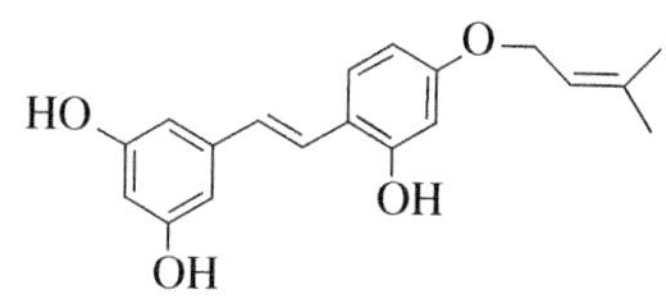

Artocarpus dadah, Moraceae
3–0–(3–methyl–2–buten–1–yl) oxyresveratrol: potassium channel modulator

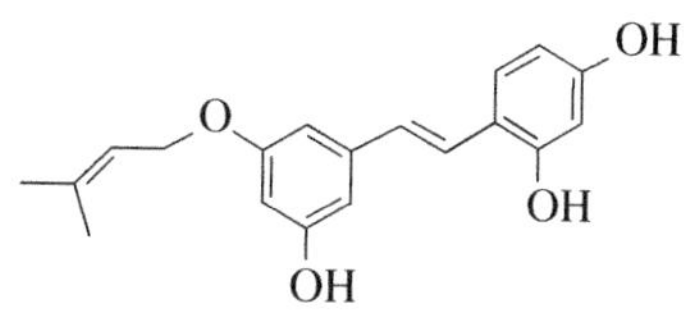

Artocarpus integer, A. incisus, A. nobilis, Moraceae
(E)–4–(3–methyl–(E)–but–1–enyl) –2′, 3,4′,5–tetrahydroxystilbene: antimalarial, tyrosinase inhibition

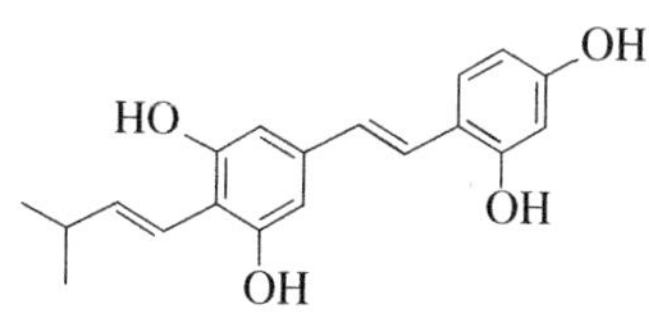

Arachis hypogaea L. peanut, Leguminosae
(E)–4–(3–methyl–1–buten–1–yl)–3, 3′, 4′, 5–tetrahydroxystilbene: anti-inflammatory antioxidant

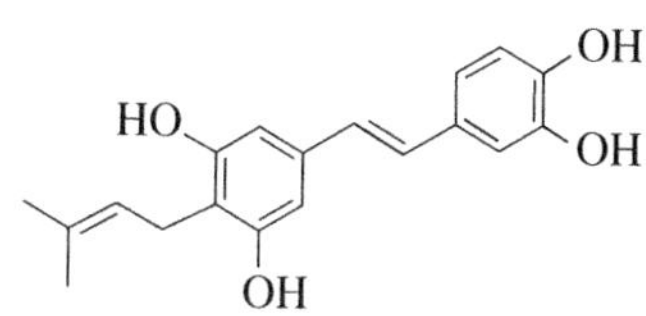

FIGURE 10.21 Studies of prenylated tetrahydroxystilbenes showing a range of biological activities [38, 73–77].

Tyr355 that are located at the entrance of the channel that controls the access of substrate to the active region of the enzyme. The omega-ends of the PUFA extend above Ser530 into an upper channel adjacent to Gly533 that is stabilized by Phe205, Phe209, Val228, and Leu534. Substrate binding is apparently stabilized by van der Waals interactions with methylene units in the aliphatic chain [81]. Thus, 54 and 56 contacts have been identified in modeling studies for interaction of AA and EPA, respectively, with residues that line with the COX-2 active site.

Similar to these findings, the aliphatic omega-ends of the $n-3$ MUFAs also extend into the hydrophobic groove that is adjacent to Ser530. The longer chain analogues (C20–C22) projected into the groove, whereas these contacts were minimal in the shorter chain analogues. In contrast, Dong et al. [83] found that MUFA analogues of

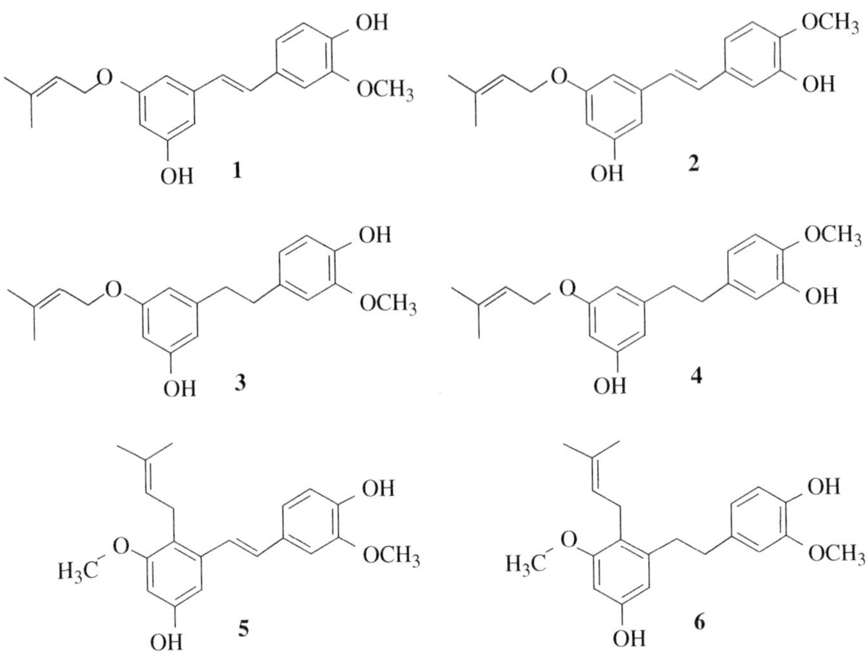

FIGURE 10.22 Structure of analogues used in cytotoxicity test toward K562 leukemic cancer cells [78].

TABLE 10.13 IC_{50} Values for Cell Growth Inhibition in K562 Cells by Compounds 1–6[a]

Compound	IC_{50} (μM)
1	21.0
2	46.8
3	63.0
4	31.6
5	0.10
6	10.0

[a] From Koolaji et al. [78].

short to intermediate chain length (C16–C20), which carried D^9 or D^{11} olefinic bonds, did not inhibit COX-2 activity, which suggests that they are unable to coordinate effectively into the substrate binding site of COX-2. Previous modeling studies with a range of clinically effective as well as experimental COX-2 inhibitors identified interactions with the active region of the enzyme that were similar to those of the present $n-3$ MUFA. Thus, from crystallography and molecular modeling evidence, carboxylate substituents in these agents also interacted with Arg120 and Tyr355 and projected inward into hydrophobic clefts in the enzyme, which enabled interactions with amino acid residues including Val349, Val523, Ala527, Ser530, and Leu534

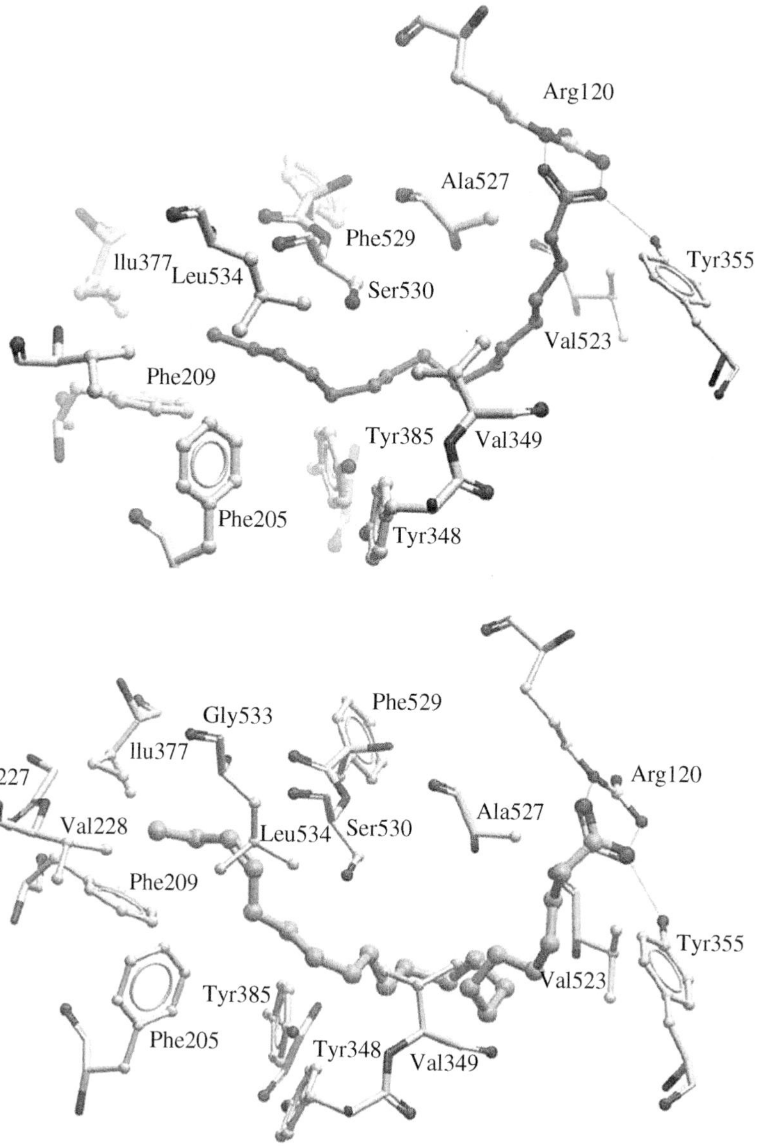

FIGURE 10.23 Docking of the $n-3$ MUFA analogues carbon 16 (upper) and carbon 22 (lower) into the COX-2 active site [63].

[84, 85]. Comparable DG_{bind} values were calculated for the interaction of COX-2 with the present $n-3$ MUFA and a range of non-steroidal anti-inflammatory agents (−10 to −14 kcal/mol).

Together, the findings from modeling studies reflect the experimentally observed inhibitory effects of the longer chain $n-3$ MUFAs on COX-2 activity as well as the lower potencies of the shorter chain analogues. Thus, the present molecules emerge as prototypes of a novel class of anti-cancer agents that modulate cell metastases; additional structural modifications may yield analogues with enhanced potency and efficacy *in vivo*.

10.6 SUMMARY

Current research has shown that the biological properties of phytochemicals are dramatically affected by chemical modifications. Some modifications greatly increase the activity toward biological targets such that clinically relevant actions may be developed by use of techniques to greatly improve potency, bioavailability, and metabolic pharmacokinetics [86].

REFERENCES

[1] Bruneton J (1999) *Pharmacognosy Phytochemistry Medicinal Plants*. Lavoisier Publishing: Secaucus.

[2] Kapoor LD (1995) *Opium Poppy: Botany, Chemistry and Pharmacology*. Food Products Press, New York, pp. 96.

[3] Chan TY (2009) Aconite poisoning. *Clinical Toxicology* 47: 279–285.

[4] Schlesinger N, Schumacher R, Catton M, Maxwell L (2006) Colchicine for acute gout. *Cochrane Database of Systematic Reviews*. doi: 10.1002/14651858.CD006190.

[5] Abourashed EA, El-Alfy A, Khan IA, Walker L (2003) Ephedra in perspective – a current review. *Phytotherapy Research* 17: 703–712.

[6] Kulkarni SK, Dhir A (2010) Berberine: a plant alkaloid with therapeutic potential for central nervous system disorders. *Phytotherapy Research* 24: 317–324.

[7] Singh IP, Mahajan S (2013) Berberine and its derivatives: a patent review (2009-2012). *Expert Opinion on Therapeutic Patents* 23: 215–231.

[8] Tfelt-Hansen P, Saxena PR, Dahlof C (2000) Ergotamine in the acute treatment of migraine – European Consensus. *Brain* 123: 9–18.

[9] Shamon SD, Perez MI (2009) Blood pressure lowering efficacy of reserpine for primary hypertension (review). *Cochrane Database of Systematic Reviews*. doi: 10.1002/14651858.CD007655.pub2.

[10] Stansbury JS, Richard P, Eugene ZR (2013) The use of lobelia in the treatment of asthma and respiratory illness. *Journal of Restorative Medicine* 2: 94–100.

[11] Jay D, Amsterdam YL Soeller I, Rockwell K, Mao JJ, Shults J (2009) A randomized, double-blind, placebo-controlled trial of oral Matricaria recutita (chamomile) extract therapy of generalized anxiety disorder. *Journal of Clinical Psychopharmacology* 29: 378–382.

[12] Patel D, Gupta S (2007) Apigenin and cancer chemoprevention: progress, potential and promise (review). *International Journal of Oncology* 30: 233–245.

[13] Bischoff SC (2008) Quercetin: potentials in the prevention and therapy of disease. *Current Opinion in Clinical Nutrition and Metabolic Care* 11: 733–740.

[14] Jurenka J (2008) Therapeutic applications of pomegranate (Punica granatum L.): a review. *Alternative Medicine Review* 13: 128–144.

[15] Kao TC, Wu CH, Yen GC (2014) Bioactivity and potential health benefits of licorice. *Journal of Agricultural and Food Chemistry* 62: 542–553.

[16] Sharangi AB (2009) Medicinal and therapeutic potentialities of tea (Camellia sinensis L.)-a review. *Food Research International* 42: 529–536.

[17] Huntley A, Edzard H (2004) Soy for the treatment of perimenopausal symptoms – a systematic review. *Maturitas* 47: 1–9.

[18] Beck V, Rohr U, Jungbauer A (2005) Phytoestrogens derived from red clover: an alternative to estrogen replacement therapy? *The Journal of Steroid Biochemistry and Molecular Biology* 94: 499–518.

[19] Gheorghiade M, Zarowitz B (1992) Review of randomized trials of digoxin therapy in patients with chronic heart failure. *The American Journal of Cardiology* 69: 48–63.

[20] Mijatovica T, Quaquebeke EV, Delest B, Debeir O, Darro F, Kiss R (2007) Cardiotonic steroids on the road to anti-cancer therapy. *Biochimica et Biophysica Acta* 1776: 32–57.

[21] Ulbricht C, Basch E, Bash S, Bent S, Dacey C, Dalton S, Foppa I, Giese N, Hammerness P, Kirkwood C, Sollars D, Tanguay-Colucci S, Weissner W (2008) An evidence-based systematic review of aloe vera by the natural standard research collaboration. *Journal of Herbal Pharmacotherapy* 7: 279–323.

[22] Ulbricht C, Conquer J, Costa D, Hamilton W, Higdon ER, Isaac R, Rusie E, Rychlik I, Serrano JM, Tanguay-Colucci S, Theeman M, Varghese M (2011) An evidence-based systematic review of aloe vera by the natural standard research collaboration. *Journal of Dietary Supplements* 8: 189–238.

[23] Jabee Q, Aslam N (2011) The pharmacological activities of prunes: the dried plums. *Journal of Medicinal Plants Research* 5: 1508–1511.

[24] Stevinson C, Pittler MH, Ernst E (2000) Garlic for treating hypercholesterolemia: a meta-analysis of randomized clinical trials. *Annals of Internal Medicine* 133: 420–429.

[25] Gupta M (2010) Pharmacological properties and traditional therapeutic uses of important Indian spices: a review. *International Journal of Food Properties* 13: 1092–1116.

[26] Lieberman S (1998) A review of the effectiveness of Cimicifuga racemosa (black cohosh) for the symptoms of menopause. *Journal of Women's Health* 7: 525–529.

[27] Patel S (2012) Yucca: a medicinally significant genus with manifold therapeutic attributes. *Natural Products and Bioprospecting* 2: 231–234.

[28] Raju J, Mehta R (2008) Cancer chemopreventive and therapeutic effects of diosgenin, a food saponin. *Nutrition and Cancer* 61: 27–35.

[29] Block KI (2003) Immune system effects of Echinacea, Ginseng and Astragalus: a review. *Integrative Cancer Therapies* 2: 247–267.

[30] Akihisa T, Kokke W., Tamura T (1991) Naturally occurring sterols and related compounds from plants. In: Patterson GW, Nes WD (Eds.) *Physiology and Biochemistry of Sterols*. American Oil Chemists' Society, Champaign, pp. 172–228.

[31] Johansson A (1979) The content and composition of sterols and sterol esters in sunflower and poppy seed oils. *Lipid* 14: 285–291.

[32] Piironen V, Miettinen T, Toivo J, Anna-Maija L (2000) Plant sterols: biosynthesis, biological function and their importance to human nutrition. *Journal of the Science of Food and Agriculture* 80, 939–966.

[33] Patel MD, Thompson PD (2006) Phytosterols and vascular disease. *Atherosclerosis* 186: 12–19.

[34] Moghadasian M, Jiri F (1999) Effects of dietary phytosterols on cholesterol metabolism and atherosclerosis: clinical and experimental evidence. *The American Journal of Medicine* 107: 588–594.

[35] Othman R, Mohammed H (2011) Beyond cholesterol-lowering effects of plant sterols: clinical and experimental evidence of anti-inflammatory properties. *Nutrition Reviews* 69: 371–382.

[36] Fenton WS, Hibbeln J, Knable, M (2000) Essential fatty acids, lipid membrane abnormalities and the diagnosis and treatment of schizophrenia. *Biological Psychiatry* 47: 8–21.

[37] Fetterman J, James W, Zdanowicz MM (2009). Therapeutic potential of n-3 polyunsaturated fatty acids in disease. *American Journal of Health-System Pharmacy* 66: 1169–1179.

[38] Murias M, Jaeger W, Handler N, Erker T, Horvath Z, Szekeres T, Nohl H, Gille L (2005) Antioxidant, prooxidant and cytotoxic activity of hydroxylated resveratrol analogues: structure-activity relationship. *Biochemical Pharmacology* 69: 903–912.

[39] Thadani MB (1996) *Medicinal and Pharmaceutical Uses of Natural Products*. First Edition. Context Publications, Winnipeg.

[40] Smoliga JM, Baur JA, Hausenblas HA (2011) Resveratrol and health – a comprehensive review of human clinical trials. *Molecular Nutrition & Food Research* 55: 1129–1141.

[41] Catalgol B, Batirel S, Taga Y, Ozer NK (2012) Resveratrol: French paradox revisited. *Frontiers in Cardiovascular and Smooth Muscle Pharmacology* 3(July): 141.

[42] Potter GA, Patterson LH, Wanogho E, Perry PJ, Butler PC, Ijaz T, Ruparelia KC, Lamb JH, Farmer PB, Stanley LA, Burk MD (2002) The cancer preventative agent resveratrol is converted to the anticancer agent piceatannol by the cytochrome P450 enzyme CYP1B1. *British Journal of Cancer* 86: 774–778.

[43] Piotrowska H, Kucinska M, Murias M (2012) Biological activity of piceatannol: leaving the shadow of resveratrol. *Mutation Research, Reviews in Mutation Research* 750: 60–82.

[44] Stefanska B, Karlic H, Varga F, Fabianowska-Majewska K, Haslberger AG (2012) Epigenetic mechanisms in anti-cancer actions of bioactive food components – the implications in cancer prevention. *British Journal of Pharmacology* 167: 279–297.

[45] Vanden Berghe W (2012) Epigenetic impact of dietary polyphenols in cancer chemoprevention: lifelong remodeling of our epigenomes. *Pharmacological Research* 65: 565–576.

[46] Kundu JK, Surh YJ (2009) Molecular basis of chemoprevention with dietary phytochemicals: redox-regulated transcription factors as relevant targets. *Phytochemistry Reviews* 8: 333–347.

[47] Hsieh TC, Wu JM (2010) Resveratrol: biological and pharmaceutical properties as anticancer molecule. *BioFactors* 36: 360–369.

[48] Blackwell L, Norris J, Suto CM, Janzen WP (2008) The use of diversity profiling to characterize chemical modulators of the histone deacetylases. *Life Sciences* 82: 1050–1058.

[49] Patel KR, Scott E, Brown VA, Gescher AJ, Steward WP, Brown K (2011) Clinical trials of resveratrol. *Annals of the New York Academy of Sciences* 1215: 161–169.

[50] Mehlen P, Puisieux A (2006) Metastasis: a question of life or death Nat. *Nature Reviews. Cancer* 6: 449–458.

[51] Borst P, Jonkers J, Rottenberg S (2007) What makes tumors multidrug resistant? *Cell Cycle* 6: 2782–2787.

[52] Maillard V, Bougnoux P, Ferrari P, Jourdan ML, Pinault M, Lavillonniere F, Body G, Le Floch O, Chajes V (2002) N-3 and N-6 fatty acids in breast adipose tissue and relative risk of breast cancer in a case-control study in Tours, France *International Journal of Cancer* 98: 78–83.

[53] Rose DP, Connolly JM, Rayburn J, Coleman M (1995) Influence of diets containing eicosapentaenoic or docosahexaenoic acid on growth and metastasis of breast cancer cells in nude mice *Journal of the National Cancer Institute* 87: 587–592.

[54] Rose DP, Connolly JM, Coleman M (1996) Effect of omega-3 fatty acids on the progression of metastases after the surgical excision of human breast cancer cell solid tumors growing in nude mice. *Clinical Cancer Research* 2: 1751–1756.

[55] Schrey MP, Patel KV (1995) Prostaglandin E2 production and metabolism in human breast cancer cells and breast fibroblasts. Regulation by inflammatory mediators. *British Journal of Cancer* 72: 1412–1419.

[56] Chang SH, Liu CH, Conway R, Han DK, Nithipatikom K, Trifan OC, Lane TF, Hla, T (2004) Role of prostaglandin E2-dependent angiogenic switch in cyclooxygenase 2-induced breast cancer progression. *Proceedings of the National Academy of Sciences of the United States of America* 101: 591–596.

[57] Hwang D, Scollard D, Byrne J, Levine E (1998) Expression of cyclooxygenase-1 and cyclooxygenase-2 in human breast cancer *Journal of the National Cancer Institute* 90: 455–460.

[58] Hla T, Bishop-Bailey D, Liu CH, Schaefers H, Trifan OC (1999) Cyclooxygenase-1 and -2 isoenzymes. *The International Journal of Biochemistry & Cell Biology* 31: 551–557.

[59] Half E, Tang XM, Gwyn K, Sahin A, Wathen K, Sinicrope FA (2002) Cyclooxygenase-2 expression in human breast cancers and adjacent ductal carcinoma in situ *Cancer Research* 62: 1676–1681.

[60] Rozic JG, Chakraborty C, Lala PK (2001) Cyclooxygenase inhibitors retard murine mammary tumor progression by reducing tumor cell migration, invasiveness and angiogenesis. *International Journal of Cancer* 93: 497–506.

[61] Basu GD, Pathangey LB, Tinder TL, Lagioia M, Gendler SJ, Mukherjee P (2004) Cyclooxygenase-2 inhibitor induces apoptosis in breast cancer cells in an *in vivo* model of spontaneous metastatic breast cancer. *Molecular Cancer Research* 2: 632–642.

[62] Connolly EM, Harmey JH, O'Grady T, Foley D, Roche-Nagle G, Kay E, Bouchier-Hayes DJ (2002) Cyclo-oxygenase inhibition reduces tumor growth and metastasis in an orthotopic model of breast cancer. *British Journal of Cancer* 87: 231–237.

[63] Cui PH, Rawling T, Bourget K, Kim T, Duke CC, Doddareddy MR, Hibbs D, Zhou F, Tattam BN, Petrovic N, Murray M (2012) Antiproliferative and antimigratory actions of synthetic long chain n-3 monounsaturated fatty acids in breast cancer cells that overexpress cyclooxygenase-2. *Journal of Medicinal Chemistry* 55: 7163–7172.

[64] Terano T, Salmon JA, Higg GA, Moncada S (1986). Eicosapentaenoic acid as a modulator of inflammation effect on prostaglandin and leukotriene synthesis. *Biochemical Pharmacology* 35: 779–785.

[65] Bégin ME, Ells G, Horrobin DF (1988) Polyunsaturated fatty acid-induced cytotoxicity against tumor cells and its relationship to lipid peroxidation *Journal of the National Cancer Institute* 80: 188–194.

[66] Falconer JS, Ross JA, Fearon KC, Hawkins RA, O'Riordain MG, Carter DC (1994) Effect of eicosapentaenoic acid and other fatty acids on the growth *in vitro* of human pancreatic cancer cell lines. *British Journal of Cancer* 69: 826–832.

[67] Palomer A, Cabré F, Pascual J, Campos J, Trujillo MA, Entrena A, Gallo MA, García L, Mauleón D, Espinosa A (2002) Identification of novel cyclooxygenase-2 selective inhibitors using pharmacophore models *Journal of Medicinal Chemistry* 45: 1402–1411.

[68] Rowlinson SW, Crews BC, Lanzo CA, Marnett LJ (1999) The binding of arachidonic acid in the cyclooxygenase active site of mouse prostaglandin endoperoxide synthase-2 (COX-2). A putative L-shaped binding conformation utilizing the top channel region. *The Journal of Biological Chemistry* 274: 23305–23310.

[69] Brand-Williams W, Cuvelier ME, Berset C (1995). Use of a free radical method to evaluate antioxidant activity. *Food Science & Technology (London)* 28: 25–30.

[70] Cai YJ, Wei QY, Fang JG, Yang L, Liu ZL, Wyche JH Han ZY (2004) The 3,4-dihydroxyl groups are important for trans-resveratrol analogs to exhibit enhanced antioxidant and apoptotic activities. *Anticancer Research* 24: 999–1002.

[71] Queiroz AN, Gomes BAQ, Moraes WM, Borges RS (2009) A theoretical antioxidant pharmacophore for resveratrol. *European Journal of Medicinal Chemistry* 44: 1644–1649.

[72] Ovesna Z, Kozics K, Bader Y, Saiko P, Handler N, Erker T, Szekeres T (2006) Antioxidant activity of resveratrol, piceatannol and 3, 3′, 4,4′,5,5′-hexahydroxy-trans-stilbene in three leukemia cell lines. *Oncology Reports* 16: 617–624.

[73] Kim YM, Yun J, Lee CK, Lee H, Min KR, Kim Y (2002) Oxyresveratrol and hydroxystilbene compounds: Inhibitory effect on tyrosinase and mechanism of action. *Journal of Biological Chemistry* 277: 16340–16344.

[74] Mouihate A, Horn TF, Pittman QJ (2006) Oxyresveratrol dampens neuroimmune responses in vivo: a selective effect on TNF-α. *American Journal of Physiology* 291: R1215–R1221.

[75] Jayasinghe ULB, Puvanendran S, Hara N, Fujimoto Y (2004) Stilbene derivatives with antifungal and radical scavenging properties from the stem bark of Artocarpus nobilis. *Natural Product Research* 18: 571–574.

[76] Patel B, Patel S, Hoffman R (2005) Inhibition of cyclo-oxygenase-2 expression in mouse macrophages by 4-(3-methyl-but-1-enyl)-3,5,3′,4′-tetrahydroxystilbene, a resveratrol derivative from peanuts. *Phytotherapy Research* 19: 552–555.

[77] Su BN, Cuendet M, Hawthorne ME, Kardono LBS, Riswan S, Fong HHS, Mehta R, Pezzuto JM, Kinghorn AD (2002) Constituents of the bark and twigs of Artocarpus dadah with cyclooxygenase inhibitory activity. *Journal of Natural Products* 65: 163–169.

[78] Koolaji N, Abu-Mellal A, Tran VH, Duke RK, Duke CC (2013) Synthesis of C-and O-prenylated tetrahydroxystilbenes and O-prenylated cinnamates and their action towards cancer cells. *European Journal of Medicinal Chemistry* 63: 415–422.

[79] Wang M, Jin Y, Ho CT (1999) Evaluation of resveratrol derivatives as potential antioxidants and identification of a reaction product of resveratrol and 2,2-diphenyl-1-picryhydrazyl radical. *Journal of Agricultural and Food Chemistry* 47: 3974–3977.

[80] Fukuhara K, Nagakawa M, Nakanishi I, Ohkubo K, Imai K, Urano S, Fukuzumi S, Ozawa T, Ikota N, Mochizuki M, Miyata N, Okuda H (2006) Structural basis for DNA-cleaving activity of resveratrol in the presence of Cu(II). *Bioorganic & Medicinal Chemistry* 14: 1437–1443.

[81] Vecchio AJ, Simmons DM, Malkowski MG (2010) Structural basis of fatty acid substrate binding to cyclooxygenase-2J. *Biological Chemistry* 285: 22152–22163.

[82] Kiefer JR, Pawlitz JL, Moreland KT, Stegeman RA, Hood WF, Gierse JK, Stevens AM, Goodwin DC, Rowlinson SW, Marnett LJ, Stallings WC, Kurumbail RG (2000) Structural insights into the stereochemistry of the cyclooxygenase reaction. *Nature* 405: 97–101.

[83] Dong L, Vecchio AJ, Sharma NP, Jurban BJ, Malkowski MG, Smith WL (2011) Human cyclooxygenase-2 is a sequence homodimer that functions as a conformational heterodimer *The Journal of Biological Chemistry* 286: 19035–19046.

[84] Picot D, Loll PJ, Garavito RM (1994) The X-ray crystal structure of the membrane protein prostaglandin H2 synthase-1. *Nature* 367: 243–249.

[85] Kurumbail RG, Stevens AM, Gierse JK, McDonald JJ, Stegeman RA, Pak JY, Gildehaus D, Miyashiro JM, Penning TD, Seibert K, Isakson PC, Stallings WC (1996) Structural basis for selective inhibition of cyclooxygenase-2 by anti-inflammatory agents. *Nature* 384: 644–648.

[86] Epifano F, Genovese S, Menghini L, Curini M (2007) Chemistry and pharmacology of oxyprenylated secondary plant metabolites. *Phytochemistry (Elsevier)* 68: 939–953.

11

THERAPEUTIC POTENTIAL OF GINSENOSIDES IN MANAGEMENT OF ATHEROSCLEROSIS

Xiao-Jing Zhang, Huanxing Su, Yi-Tao Wang, and Jian-Bo Wan

State Key Laboratory of Quality Research in Chinese Medicine, Institute of Chinese Medical Sciences, University of Macau, Taipa, Macau, China

11.1 INTRODUCTION

Atherosclerosis (AS), a progressive disease characterized by the accumulation of lipids and fibrous elements in the large arteries, underlies the vast majority of cardiovascular diseases (CVDs) that accounts for 16.7 million deaths per year worldwide [1–3]. Atherosclerosis has now been widely recognized as a complex and multifactorial disease that might be caused by a combination of various environmental, behavioral, physiologic, and genetic factors [4]. Increasing evidence indicates that multiple mechanisms, including elevated and modified low-density lipoprotein (LDL), endothelial dysfunction, chronic vascular inflammation, and oxidative stress, have been implicated in the initiation and progression of AS [5]. Despite statins, a class of cholesterol-lowering drugs, inhibiting the 3-hydroxy-3-methylglutaryl coenzyme A (HMG-CoA) reductase, serving as the first-line therapy for the prevention and treatment of AS, the long-term use or overdose by statins may increase the risk of severe adverse effects, particularly muscle damage and mononeuropathy [6]. Multitarget drugs with less toxicity or adverse effects are the current trend of drug research and development [7]. The use of herbal medicines to prevent or treat various chronic diseases, including CVD, has been a common clinical practice for years in Asian countries. The therapeutic effect of herbal medicine is the comprehensive and

Phytotherapies: Efficacy, Safety, and Regulation, First Edition. Edited by Iqbal Ramzan.

integrated outcomes of their active components. Thus, the multitarget herbal medicines, acting on diverse mechanisms involved in AS, might provide an alternative approach to preventing or treating this disease. Numerous herbal medicines, when used alone or in combination, have been shown to be quite effective and are being widely used currently for cardiovascular care.

Foremost among these herbs is ginseng (*Panax* species, Araliaceae family), a set of highly valued medicinal herbs that have been traditionally used clinically for preventive and therapeutic purposes, especially for CVD, since ancient times [8]. The botanic name *Panax* is derived from the same ancient Greek root as the term "Panacea," meaning a remedy for all diseases or "all-healing" reagent. According to incomplete statistics, more than 10 species are used as traditional Chinese medicine or folk medicine. Among them, the roots of *Panax ginseng* C.A. Mayer (commonly referred to as Asian, Chinese or Korean ginseng), *Panax quinquefolius* L. (American ginseng), and *Panax notoginseng* (Burk.) F.H. Chen (Notoginseng or Sanqi) are very well recognized worldwide. Ginsengs exhibit potent and extensive beneficial effects on the immune, central nervous, and cardiovascular systems, cancer and diabetes, which are primarily attributed to the presence of triterpenoid saponins (also referred to as ginsenosides) [9, 10]. Over the past two decades, ginsenosides have attracted increasing attention due to their chemical diversities and various pharmacological properties. The number of publications on ginsenosides has been growing steadily since 2005 (Fig. 11.1).

Ginsengs and ginsenosides have been widely prescribed for the prevention and treatment of atherosclerosis and other cardiovascular diseases in Asian countries, either alone or in combination [9]. Xue-Se-Tong injection and Xue-Shuan-Tong

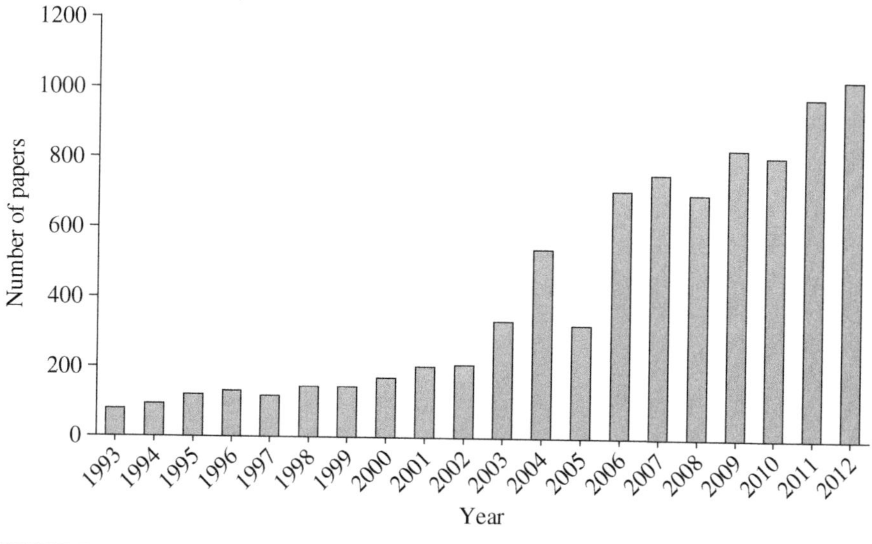

FIGURE 11.1 Yearly trend of publications on ginsenoside-related studies in the recent two decades. Data were acquired from SciFinder Scholar using the keyword "ginsenoside".

injection, derived from total saponins of rhizomes and roots of *P. notoginseng* saponins, respectively, have been approved by China Food and Drug Administration for the treatment of patients with coronary heart disease, angina pectoris, arrhythmia cordis, hypertension, and arteriosclerosis. Fufang Danshen dripping pill, a well-known composite formula consisting of *Radix Notoginseng*, *Radix Salviae Miltiorrhizae*, and *Borneolum*, has completed Phase II clinical trial in patients with chronic stable angina pectoris in the United States (see http://www.clinicaltrials.gov/ct2/show/NCT00797953). However, such rigorously designed and randomized controlled clinical trial of ginsengs and ginsenosides is rare. Nonetheless, the accumulated experimental evidence has demonstrated the potential benefits of ginsenosides on CVD, including atherosclerosis [10–14]. Therefore, the aim of the present chapter is to comprehensively summarize the experimental data that outline the evidence for ginsenosides as anti-atherosclerotic agents and, if appropriate, to identify the most effective components and the underlying mechanisms for regulating atherosclerosis.

11.2 CHEMICAL DIVERSITY OF GINSENOSIDES AND DISTRIBUTION

Ginsenosides are found exclusively in *Panax* species and have been generally considered as the major bioactive ingredients behind the claims of ginsengs efficacy [9, 15]. Most ginsenosides belong to a family of four or five trans-ring triterpene with sugar moieties attached at C-3, C-6, and/or C-20 positions. To date, nearly 300 naturally occurring ginsenosides and their derivatives have been isolated and elucidated from the roots, stems, leaves, flower buds, berries, and seeds of *Panax* species, as well as the steam or heat-processed ginseng samples [15]. Based on the skeleton of their aglycones or sapogenin, ginsenosides are generally classified into groups as dammarane-type (4-ring) and oleanane-type (5-ring) saponins [15–18]. Twenty (S)-protopanaxatriol saponins (PTS) and 20 (S)-protopanaxadiol saponins (PDS) are two major subcategories of dammarane saponins (Fig. 11.2). The PTS group or its analogs have the sugar moieties at C-3 and/or C-20 positions, for example, ginsenosides Re, Rf, Rg1, Rg2, Rh1, and notoginsenoside R1, whereas the PDS group has sugar moieties attached to C-6 and/or C-20 positions, for example, ginsenosides Rb1, Rb2, Rb3, Rc, and Rg3. Pseudoginsenoside F11 unique to *P. quinquefolius* and referred to as ocotillol type saponin in some publications is viewed as a derivative of the PTS group in that the carbon chain at the 20-position is substituted by a five-membered epoxy ring. Ginsenoside Ro is the only oleanane-type saponin identified in *Panax* species. Several ginsenosides are also present as stereoisomers depending on the carbon configuration at 20-position, such as 20 (R)-ginsenoside Rg3.

Chemical differences in ginsenoside composition of notoginseng, Asian and American ginseng are shown in Table 11.1 [19]. Total and individual saponin content varied greatly in three species, suggesting that each species has its own chemical pattern of ginsenosides. The total content of saponins is highest in notoginseng (80–140 mg/g) and lowest in Asian ginseng (10–25 mg/g). Ginsenosides Rg1, Re, Rb1, Rc, and Rd are found in all of the three herbs. However, notoginsenoside R1

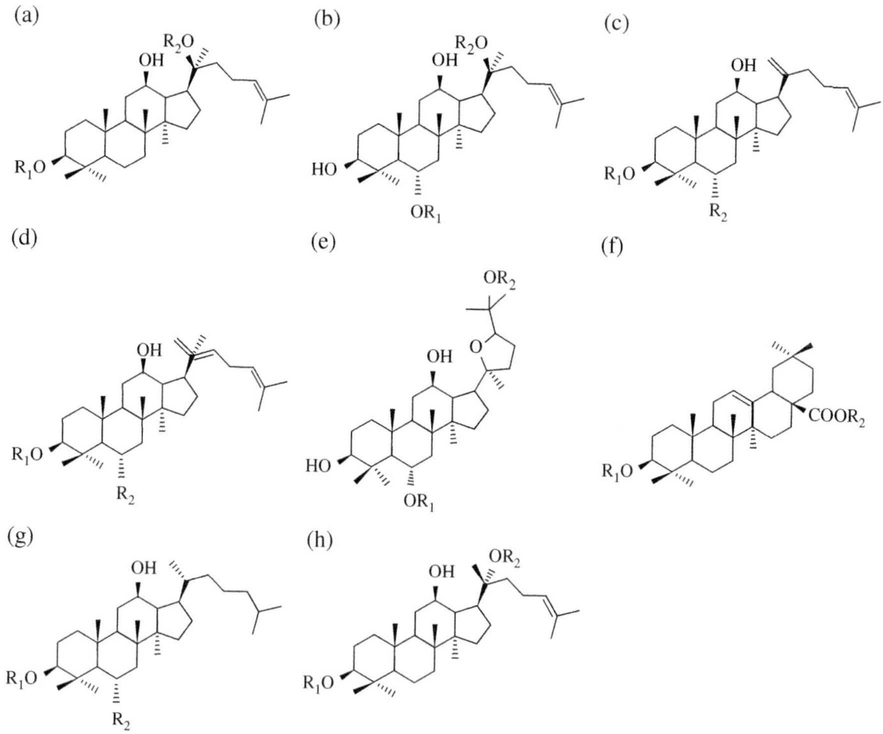

Parent nucleus	Saponins	R1	R2	Parent nucleus	Saponins	R1	R2
A	Ginsenoside Rb1	Glc2Glc	Glc6Glc		Ginsenoside Rf	Glc2Glc	H
A	Ginsenoside Rb2	Glc2Glc	Glc6Arap	B	Ginsenoside Rg1	Glc	Glc
A	Ginsenoside Rb3	Glc2Glc	Glc6Xyl	B	Ginsenoside Rh1	Glc	H
A	Ginsenoside Rc	Glc2Glc	Glc6Araf	B	Protopanaxatriol	H	H
A	Ginsenoside Rd	Glc2Glc	Glc	C	Ginsenoside Rk1	Glc2Glc	H
A	Ginsenoside Rg3	Glc2Glc	H	C	Ginsenoside Rk3	H	O-Glc
A	Ginsenoside Rh2	Glc	H	D	Ginsenoside Rh4	H	O-Glc
A	Compound K	H	Glc	D	Ginsenoside Rg5	Glc2Glc	H
A	Protopanaxadiol	H	H	E	Pseudoginsenoside F11	Glc2Rha	H
	Notoginsenoside R1	Glc2Xyl	Glc	F	Ginsenoside Ro	GlcUA2Glc	Glc
	Ginsenoside F1	H	Glc	G	Ginsenoside Rp1	Glc2Glc	H
	Ginsenoside Re	Glc2Rha	Glc	H	20R-ginsenoside Rg3	Glc2Glc	H

FIGURE 11.2 Chemical structures of selected ginsenosides. Araf, α-L-arabinofuranosyl; Arap, α-L-arabinopyranosyl; Glc, β-D-glucopyranosyl; GlcUA, β-D-glucuronic acid; Rha, α-L-rhamnopyranosyl; Xyl, β-D-xylopyranosyl.

TABLE 11.1 Chemical Differences in Ginsenoside Composition of Notoginseng (*P. notoginseng*), Asian Ginseng (*P. ginseng*), and American Ginseng (*P. quinquefolius*)

Ginsenosides	Notoginseng	Asian Ginseng	American Ginseng
Major ginsenosides	Rg1>Rb1>Rd> R1>Re	Rb1>Rg1>Rc> Re>Rd	Rb1>Re>Rd>Rg1> Rc>F11
Total ginsenosides (mg/g)[a]	80–140	10–25	35–65
Pseudo-ginsenoside F11 (mg/g)	—[b]	—	1–2
Notoginsenoside R1 (mg/g)	5–15	—	—
PTS group	45–70	4–11	15–22
PDS group	35–85	6–12	20–40
PDS/PTS	0.6–1.6	0.9–1.5	1.1–2.1
Rb1/Rg1	0.6–2.1	0.8–1.5	3.0–6.5
Rg1/Re	3.5–7.5	1.5–3.0	0.3–0.7

Data are adapted from Ref. [19] and reproduced with permission of Wiley-VCH Verlag GmbH & Co. KGaA.
[a]Total amount of 12 major saponins investigated, including notoginsenoside R1, pseudo-ginsenoside F11, ginsenosides Rg1, Re, Rf, Rb1, Rg2, Rc, Rb2, Rb3, Rd, and Rg3. PDS: protopanaxadiol saponins; PTS: protopanaxatriol saponins.
[b]Below the limit of quantification.

and pseudo-ginsenoside F11 are unique to notoginseng and American ginseng, respectively. Additionally, total amounts of PTS, PDS, and their ratios, as well as the ratio of other ginsenosides are also obviously diverse in the three species.

Numerous recent studies have demonstrated that ginsenoside metabolites, produced by heating or by intestinal bacteria, had greater biological effects than naturally occurring ginsenosides [20–23]. Post-processing of ginseng by steam has shown to dramatically alter the chemical profile, accompanied by reduced content of ginsenosides Rg1, Re, Rb1, Rd and increased new marker components, such as ginsenosides Rh1, Rg3, Rk3, Rh4, Rk1, and Rg5 [22, 24]. Oral ingestion of ginseng exposes ginsenosides to intestinal bacteria that cleave the oligosaccharide connected to the aglycone successively from the terminal sugar [21, 25]. After biodegradation, protopanaxatriol (PPT) and ginsenoside F1 are the major metabolites of PTS, whereas PDS are transformed to protopanaxadiol (PPD) and compound K. Differences in the position, type, and amount of sugar moieties, as well as the type of sapogenin, may lead to various pharmacological properties, including those on atherosclerosis.

11.3 ANTI-ATHEROSCLEROTIC EFFECTS OF GINSENOSIDES

In the recent two decades, ginsengs and ginsenosides have attracted greater attention in the prevention of atherosclerosis. A total of 20 *in vivo* studies involving extracts of ginsengs and individual ginsenosides exhibit their potentially protective effects against atherosclerosis (Table 11.2). Most of the samples have been shown to reduce

TABLE 11.2 ***In Vivo* Anti-atherosclerotic Studies of Ginsengs and Ginsenosides**

Dosage (mg/kg/day)	Animal/Model	Plaque Evaluation Method	Parameters	Proposed Mechanisms	Reference
Notoginseng (*Panax notoginseng*)					
Total saponins (PNS)					
60 (i.g.)	ApoE$^{-/-}$ mice/HCD	Cross-section	↓Plaque area, ↓TC, ↓TG, ↓LDL, ↓Ox-LDL, ↓CD40, ↓MMP-9	Decrease in serum lipid level Decrease in vascular CD40 and MMP-9	[26]
4/12 mg/mouse (i.g.)	ApoE$^{-/-}$ mice/HCD	*En face*; cross-section	↓Plaque area,↓ TC, ↓ TG, ↓ LDL ↓IL-6, ↓TNF-α,↓ICAM-1, ↓VCAM-1	Decrease in plasma lipid level Anti-vascular inflammation	[27]
60 (i.g.)	ApoE$^{-/-}$ mice/HCD	*En face*; cross-section	↓Plaque area, ↓MDA, ↑SOD, ↑GSH, ↓ROS, ↓VCAM-1, ↓ICAM-1, ↓MCP-1, ↓RAGE, ↓NF-κB, ↓JNK, ↓p38MAPK, ↓ERK1/2.	Anti-oxidation and anti-inflammation via suppressing RAGE/MAPK and NF-κB signaling pathways	[28]
100 (i.p.)	Wistar rats/zymosan A	*En face*	↓Plaque area, ↑LXRα	Anti-inflammation	[29]
100 (i.p.)	Wistar rats/zymosan A	*En face*	↓Plaque area, ↓TC, ↓TG, ↓blood viscosity, ↓IL-18, ↓IL-1β, ↓MMP-2, ↓MMP-9, ↓p65	Improvement in blood lipid profiles Anti-inflammation	[30]
120 (i.g.)	Rabbit/zymosan A	*En face*	↓Plaque area, ↓TC, ↓TG, ↓LDL-C, ↑HDL-C, ↓IL-1β, ↓MCP-1, ↓CRP	Improvement in blood lipid profiles Anti-inflammation	[31]
100 (i.p.)	Wistar rats/zymosan A	*En-face*	↓Plaque area, ↓foam cell formation, ↓integrin families, ↓pFAK, ↓NF-κB translocation	Anti-inflammation	[32]

(*Continued*)

TABLE 11.2 *(Continued)*

Dosage (mg/kg/day)	Animal/Model	Plaque Evaluation Method	Parameters	Proposed Mechanisms	Reference
Root powder 0.25%/0.5%/1% (w/w)	SD rats/HCD	—	↓TC, ↓TG, ↓LDL-C, ↑HDL-C, ↓HMG-CoA reductase, ↓MDA, ↑SOD, ↑GPx	Improvement in blood lipid profiles Inhibition of lipid peroxidation Increase in activity of anti-oxidant enzymes	[33]
Aqueous root extract 40 (i.g.)	Poloxamer-407 induced-hyperlipidemic Wistar rats	—	↓TC, ↓TG, –LDL,–HDL, –IL-1, –TNF-α, –COX-2, –ICAM	Decrease in serum lipid level	[34]
Asian ginseng (*Panax ginseng*)					
Aqueous root extract 200 (i.g.)	Rabbit/HCD	*En face*	↓Plaque area, ↓platelet aggregation, ↓diacylglycerol liberation, –AA, –TC	Anti-platelet aggregation	[35]
10 mg/mouse (i.g.)	$LDLR^{-/-}$/HCD	—	↓Arginase activity, ↑NO, ↑vascular tension	Improvement in endothelial function	[36]
Total saponins 0.8 (i.g.)	SD rats/L-methionine	—	↓COX-2, ↓iNOS, ↓NT	Anti-oxidation and anti-inflammation	[37]
0.05%, 0.075% (w/w)	$ApoE^{-/-}$ mice/HCD	*En face*; cross-section	↓Plaque area, ↓TNF-α, ↓IL-1β, ↓iNOS, ↓COX-2, ↓ICAM-1, ↓VCAM-1, ↓Phase II Enzymes, ↑HO-1, ↑GCLC – serum cholesterol	Anti-inflammation	[38]

PDS/PTS (1:1)					
100 (i.p.)	C57BL/6 mice/HCD	—	↓TG, ↓TC, ↓HDL, ↓atherogenic indices, ↓LPO, ↓MDA	Anti-oxidation	[17]
PDS/PTS (2:1)					
100 (i.p.)	C57BL/6 mice/HCD	—	↓TG, ↓HDL, ↓atherogenic indices, ↓MDA	Anti-oxidation	[17]
Ginsenoside Rb1					
0.35 mM (i.p.)	C57BL/6J mice/guidewire, Hcy	Cross-section	↓Intimal-medium thickness, ↓macrophage content	Inhibition of intimal hyperplasia Anti-inflammation	[39]
1.2 g/kg/day (i.g.)	SD rats/Hcy	—	↑NO, ↓vWF, ↓endothelial dysfunction and relaxation	Prevention of endothelial dysfunction	[40]
10, 30 (i.p.)	SD rat/carotid balloon	Cross-section	↓Neointimal hyperplasia, ↓PCNA, ↓pERK1/2, ↓c-myc, ↓p38, ↓embryonic smooth muscle myosin heavy chain, ↑smooth muscle α-actin	Inhibition of the vascular neointimal hyperplasia	[41]
Ginsenoside Rd					
20 (i.p.)	ApoE$^{-/-}$ mice/HCD	*En face*	↓Plaque area	Suppression of cholesterol accumulation	[42]
Ginsenoside Rg1					
4, 8, 16 (i.p.)	SD rat/balloon	Cross-section	↓Neointimal hyperplasia, ↓PCNA, ↓pERK1/2, ↓c-myc, ↓p38, ↓embryonic smooth muscle myosin heavy chain, ↑smooth muscle α-actin	Inhibits the vascular neointimal hyperplasia	[43]

Abbreviations: i.g.= intra gastric; i.p. = intra peritoneal

the plaque area using various animal models, including ApoE-deficient mice fed high cholesterol diet (HCD) [26–28, 38, 42], zymosan A-induced atherosclerosis [29–32], high L-methionine-induced vascular endothelial lesion [37], and guidewire or balloon catheter-induced endothelial injury [39, 41]. In addition, poloxamer-407-induced hyperlipidemia [34], homocysteine-induced endothelial dysfunction [40], and high cholesterol-fed normal animals [17, 33, 35] were also used to clarify the efficacy and underlying mechanisms of ginsengs on the critical components of atherogenesis. As one of the most extensively used models in atherosclerosis studies, the ApoE-deficient mice have been shown to spontaneously develop severe atherosclerotic lesions with higher consistency and more characteristic in appearance and distribution to human atheromas, including fatty streaks, necrotic cores, and fibrous caps [44]. Cross-sectional and *en face* methods have been commonly used to assess the extent of atherosclerosis and to evaluate the anti-atherosclerotic efficacy of ginsengs and/or ginsenosides. The *en face* method consists of dissecting the entire aorta, opening it longitudinally to expose the intraluminal side and staining with Sudan IV or oil red O to reveal lipid-rich lesions, which allows one to observe the entire surface of the endothelium with a clear focused image and thus to quantify the extent of atherosclerotic lesion. Therefore, this method has been one of the most appropriate methods for evaluating the atherosclerosis in the artery [45]. We demonstrated previously that PNS, administered orally at dosages of 4 and 12 mg/day/mouse, notably decreased the formation of atherosclerotic plaques in the entire aorta by 61.4 and 66.2%, respectively, in ApoE-deficient mice using *en face* evaluation method [27]. Li et al. [42] demonstrated that the preventive and therapeutic treatment of ginsenoside Rd, administered i.p. at a dose of 20 mg/kg, remarkably reduced the atherosclerotic plaque in the entire aorta of ApoE-deficient mice to $7.3 \pm 2.1\%$ and $13.7 \pm 3.5\%$, respectively, from $35.8 \pm 6.1\%$ in the control group. These *in vivo* data indicate undeniable anti-atherogenesis activity of ginsenosides.

11.4 UNDERLYING MECHANISMS OF GINSENOSIDES AGAINST ATHEROSCLEROSIS

The representative molecular targets and proposed mechanisms involved in the atherogenesis and targeted by ginsengs/ginsenosides are summarized in Table 11.2 (*in vivo* studies) and Table 11.3 (*in vitro* studies). The mechanisms underlying the protective effects of ginseng extracts and ginsenosides on atherosclerosis have shown to be multifaceted and associated with regulation of lipid profile, anti-oxidation, anti-inflammation, amelioration of endothelial function, inhibition of vascular smooth muscle cell proliferation, and suppression of platelet aggregation.

11.4.1 Regulation of Blood Lipid Profile

Numerous clinical and epidemiological studies have indicated that hyperlipidemia is a well-established atherogenic risk factor [92–97]. High serum/plasma levels of LDL act as a key contributor to the initiation and progression of atherosclerosis [98]. It has

TABLE 11.3 ***In Vitro*** **Evidence for Anti-atherosclerotic Effects of Ginsengs and Ginsenosides**

Extract or Ginsenoside Effective Dose	Cells/Model	Parameter	Proposed Mechanisms	Reference
Notoginseng (*Panax notoginseng*)				
Total saponins (PNS)				
25, 50, 100 mg/L	THP-1/LPS	↑LXRα, ↑ABCA1, ↑ABCG1, ↓NF-κB, ↓IL-6, ↓MCP-1	Anti-inflammation	[29]
100, 300 μg/mL	HUVEC/ TNF-α	↓ICAM-1, ↓VCAM-1, ↓NF-κB	Inhibition of adhesion Anti-vascular inflammation	[46]
100, 300 μg/mL	HCAECs/TNF-α	↓ICAM-1, ↓VCAM-1	Anti-vascular inflammation	[27]
100 μg/mL	PBMC/ox-LDL, LPS	↓CD40, ↓CD86, ↓HLA-DR, ↓CD1a, ↑endocytosic function, ↓IL-12, ↓TNF-α	Inhibition of the maturation of DCs	[47]
10, 30, 100 μg/mL	Peritoneal macrophage/LPS	↓Phosphorylation of FAK, ↓NF-κB	Anti-inflammation	[32]
200, 400, 800 μg/mL	VSMCs/ PDGF	↑P53, ↑Bax, ↑caspase-3, ↓Bcl-2, ↑Bax/Bcl-2	Inhibition of VSMCs proliferation Induction of VSMCs apoptosis	[48]
Aqueous root extract				
62.5, 125, 250, 500, 1000 μg/mL	Platelet aggregation/ADP	↓Platelet aggregation	Anti-platelet aggregation	[49]
Ethanol root extract				
5, 25, 50 μg/mL	DC2.4/ LPS, CpG, poly(I:C)	↓TNF-α, ↓IL-6, ↓COX-2, ↓CD40, ↓CD86	Anti-inflammation Inhibition of immune responsiveness	[50]
5, 25, 50 μg/mL	RAW264.7/LPS	↓TNF-α, ↓IL-6, ↓IL-1, ↓COX-2, ↓CD40, ↓CD86	Anti-inflammation Immunosuppression	[51]
Ethanol flower extract				
25, 50, 100 μg/mL	RAW264.7/LPS	↓NO, ↓PGE2, ↓TNF-α, ↓IL-1β, ↓iNOS, ↓COX-2	Anti-inflammation via blocking NF-κB signaling pathway	[52]

(*Continued*)

TABLE 11.3 *(Continued)*

Extract or Ginsenoside Effective Dose	Cells/Model	Parameter	Proposed Mechanisms	Reference
PDS 50 μg/mL	HUVEC/TNF-α	↓ICAM-1, ↓VCAM-1, ↓NF-κB	Anti-inflammation	[46]
PTS 100 μg/mL	HUVEC/ TNF-α	↓ICAM-1, ↓VCAM-1, ↓NF-κB	Anti-inflammation	[46]
American ginseng (*Panax quinquefolium*)				
Root 50 mg/mL	RASMCs/FBS, PDGF, insulin, AngII	↓VSMC proliferation, ↓Jak, ↓Stat3	Inhibition of VSMC proliferation	[53]
Stems and leaves 0.25–1.0 mg/mL	Aortic ring/Ox-LDL	↓LDL oxidation, ↓ phosphatidylcholine of LDL	Inhibition of lipid peroxidation	[54]
Asian ginseng (*Panax ginseng*)				
Aqueous root extract 500 μg/mL	HUVEC/VEGF	↑Proliferation, ↑migration, ↑tube formation, ↑angiogenesis, ↑phosphorylation of ERK1/2, ↑Akt, ↑eNOS, ↑NO	Stimulation of angiogenesis	[55]
15, 25 μg/mL	HUVEC	↓Arginase activity, ↑NO, ↑dimer/monomer eNOS	Improvement in endothelium function	[36]
15 μg/mL	Aortic rings	↑NO, ↓ROS, ↓O2•−,↓arginase activity, ↑vessel relaxation	Vasorelaxation Anti-oxidation	[36]
Berry extract 50, 100 μg/mL	Primary macrophages/LPS	↑Phase II gene, ↓NF-κB, ↓Phase II Enzymes, ↓COX-2, ↓PGE2, ↓TNF-α, ↓IL-1β,	Anti-inflammation	[38]
50, 100 μg/mL	RAW 264.7/LPS	↓PGE2, ↓nitrite, ↓TNF-α, ↓IL-1β,	Anti-inflammation	[38]
100 μg/mL	HUVEC/TNF-α	↓ICAM-1, ↓VCAM-1	Anti-inflammation	[38]

Ginsenoside Rb1				
10 µg/mL	RAW264.7/LPS	↓IL-6, ↓TNF-α	Anti-inflammation	[56]
10 mg/mL	HUVECs/ OxLDL	↑LDH activity, ↑eNOS, ↑t-PA, ↑PAI-1	Protection of endothelial cells	[57]
20, 40 µM	HUVECs/H_2O_2	↑SOD, ↓MDA, ↓ ROS	Anti-oxidation Protection of endothelial cells	[58]
10 µM	SVEC4-10, HUVECs/Hcy	↑Endothelial cell proliferation, ↓superoxide anion	Prevention of endothelial dysfunction	[59]
10 µM	HUVECs/Hcy	↑NO, ↑phosphorylation of eNOS and Akt	Prevention of endothelial dysfunction	[40]
50 µM	HUVEC/TNF-α	↓ICAM-1, ↓VCAM-1, ↓NF-κB	Anti-vascular inflammation	[46]
1, 10 µM	Coronary arteries/Hcy	↑Endothelial function, ↑eNOS	Improvement in endothelial function	[60]
0.25, 0.5, 1.0, 2.5 µM	VSMCs/FBS, TNF-α	↓Proliferation, ↓G1 cell cycle arrest, ↓cell cycle proteins, ↓ERK1/2 phosphorylation, ↓inflammatory responses	Inhibition of VSMCs proliferation Anti-inflammation	[61]
5, 10 µM	Peritoneal macrophages/LPS	↓IL-1β, ↓TNF-α, ↓IL-6, ↓IRAK-1, ↓IKK-β, ↓NF-κB, ↓MAP kinases	Anti-inflammation	[62]
Ginsenoside Rg1				
50, 150, 300 nM	HUVECs	↑Phosphorylation of GR, ↑PI3K, ↑Akt/PKB, ↑eNOS, NO, β-catenin	Increase in NO production Pro-angiogenesis	[63, 64]
30 µM	HUVEC/TNF-α	↓ICAM-1, ↓VCAM-1, ↓NF-κB	Anti-inflammation	[46]
0.1, 1.0, 10 µM	HASMCs/TNF-α	↓HASMCs proliferation, ↓G1 phase, ↓p53, ↓$p21^{WAF/CIP1}$, ↓$p27^{KIP1}$, ↓ERK1/2, ↓PKB	Inhibition of SMC proliferation	[65]
20, 40, 80, 160 mg/L	VSMCs/TNF-α	↓Proliferation, ↓PKC-ζ, ↓N-ras protein, ↑p21	Inhibition of VSMC proliferation	[66]

(Continued)

TABLE 11.3 *(Continued)*

Extract or Ginsenoside Effective Dose	Cells/Model	Parameter	Proposed Mechanisms	Reference
30 μg/mL	HUVECs	↑Proliferation, ↑migration, ↑tube formation	Pro-angiogenesis	[67]
150, 300, 600 nM	HUVECs/VEGF	↑Proliferation, ↑tube formation	Pro-angiogenesis	[68]
0.5, 1.0, 5.0 μM	EPC	↑Migration, ↑adhesion, ↑proliferation, ↑VEGF	Protection of endothelial progenitor cells	[69]
150 nM	HUVECs	↑HIF-1α, ↑PI3K/Akt, ↑$p70^{S6K}$, ↑VEGF	Pro-angiogenesis	[70]
10^{-5} to 10^{-3} g/mL	Thoracic aorta rings/ phenylephrine	↑Ring relaxation, ↑cGMP	Increase in aorta ring relaxation	[71]
Ginsenoside Re				
0.05, 0.1, 0.5 mg/mL	Chick embryonic ventricular myocytes/ H_2O_2, antimycin A	↓Cell death, ↓intracellular oxidants	Anti-oxidation	[72]
5, 10 μM	Peritoneal macrophages/ PGN, LPS, TNF-α	↓IKK-β phosphorylation, ↓NF-κB, ↓TNF-α, ↓IL-1β, ↓IRAK-1 phosphorylation, ↓binding of LPS to TLR4	Anti-inflammation	[73]
10^{-5} to 10^{-3} g/mL	Thoracic aorta rings/ phenylephrine	↑Ring relaxation	Increase in relaxation of aorta ring	[71]
30 μg/mL	HUVECs	↑Proliferation, ↑migration, ↑tube formation	Pro-angiogenesis	[67]
Ginsenoside Rg3				
1, 3, 10 μg/mL	ECV 304/TNF-α	↓ICAM-1, ↓VCAM-1, ↓NF-κB, ↓IL-1β, ↓AP-1	Anti-vascular inflammation	[74]
10 μM	HepG2	↓Intracellular cholesterol, ↓TG, ↓SREBP-2, ↓HMGCR, ↑AMPK	Regulation of dyslipidemia	[75]
25 μM	RAW264.7/LPS	↓PGE2, ↓ROS, ↓MMP-9, ↓COX-2, ↓TNF-α, ↓IL-1, ↓IL-6	Anti-inflammatory Anti-oxidation	[76]

10 μg/mL	HUVECs/serum deprivation	↓Caspase-9, ↓caspase-3, ↑Akt, ↓mitochondrial cytochrome c release	Prevention of endothelial cells	[77]
1, 3, 10 μg/mL	ECV 304	↑NO, ↑PI3-kinase, ↑JNK, ↑p38, ↑ER and GR-dependent reporter gene transcriptions, ↑CaM kinase II	Improvement of endothelial function	[78]
60, 300 ng/mL	$CD34^+$/VEGF	↓Proliferation, ↓differentiation, ↓migration, ↓tube formation, ↓Akt/eNOS	Anti-angiogenesis	[79]
10 μg/mL	Endothelium-denuded aortic ring/phenylephrine	↓Vasocontraction, ↑NO, ↑iNOS	Vessel relaxation	[80]
Compound K				
20 μM	HepG2	↓SREBP1c, ↓fatty acid synthesis, ↓stearoyl-CoA desaturase 1, ↑PPAR-α, ↑CD36	Regulation of dyslipidemia	[81]
25 μM	HUVECs, THP-1/TNF-α	↓VCAM-1, ↓ICAM-1, ↓E-selectin, ↓IL-8, ↓MMP-9, ↓VLA-4, ↓αL/β2, ↓integrin ↓LFA-1, ↓CXCR1	Anti-vascular inflammation	[82]
5, 10 μM	Peritoneal macrophages/LPS	↓IL-1β, ↓TNF-α, ↓IL-6, ↓IRAK-1, ↓IKK-β, ↓NF-κB, ↓MAP kinases	Anti-inflammation	[62]
Notoginsenoside R1				
0.1, 1, 10 μM	HASMCs/TNF-α	↓Fibronectin, ↓TNF-α, ↓ERK, ↓H_2O_2-induced migration, ↓ROS	Anti-oxidation Anti-inflammation	[83]
0.1, 1, 10 μM	HASMCs/TNF-α	↓PAI-1, ↓ERK, ↓PKB	Protection of vascular function	[84]
100 μg/mL	HUVEC, THP-1, human whole-blood cells/LPS	↓PAI-1, ↓TF, ↑IκB-α, ↓TNF-α	Protection of endothelial function Anti-inflammation	[85]
1, 10, 100 μg/mL	HUVEC	↑tPA, ↓PAI-1	Protection of endothelial function	[86]

(Continued)

TABLE 11.3 (*Continued*)

Extract or Ginsenoside Effective Dose	Cells/Model	Parameter	Proposed Mechanisms	Reference
20(S)-Protopanaxatriol				
10, 15, 20 μM	RAW264.7/LPS	↓iNOS, ↓COX-2, ↓phosphorylation of IKK/IκBα	Anti-inflammation	[87]
10^{-5} to 10^{-3} g/mL	Thoracic aorta rings/ phenylephrine, L-NMMA, methylene blue	↑Endothelium dependent relaxation, ↑cGMP	Increase in endothelium-dependent relaxation	[71]
10 μM	HUVECs/H_2O_2	↓DNA damage, ↑PARP-1, ↓NAD+, ↑ATP/ADP, ↑GSH/GSSG	Anti-oxidation	[88]
Other ginsenosides				
Ginsenoside Rb3 (0.1, 1 μM)	VSMCs/Ang II	↑proliferation, ↓G0/G1, ↓c-myc, ↓c-fos, ↓c-JUN	Inhibition of VSMCs proliferation	[89]
Ginsenoside Rd (20 μM)	Peritoneal macrophage RAW264.7/OxLDL	↓Ca2+ influx, ↓OxLDL uptake, ↓cholesterol accumulation, ↓SR-A	Inhibition of foam cell formation	[42]
Ginsenoside Rh1 (25, 50 μM)	THP-1 monocytes	↓MCP-1, ↓CCR-2, ↓VLA5, ↓activated β1 integrin, ↓phosphorylation of MAPKs	Anti-inflammation	[90]
Ginsenoside Rg5 (5, 10 μM)	C57BL/6 mice alveolar microphage cells/LPS	↓IL-1β, ↓TNF-α, ↓COX-2, ↓iNOS, ↓IRAK-1, ↓IRAK-4, ↓IKK-β, ↓NF-κB, ↓p65	Anti-inflammation	[91]

been proposed that the entry and retention of LDL in the subendothelial layer mainly depends on its elevated serum/plasma levels and the accumulated LDL is susceptible to oxidation during passage through resident vascular cells and recruitment of monocytes and macrophages. The oxidized LDL is recognized by scavenger receptors on macrophages and internalized to form so-called foam cells, which is the early histologic feature of atherosclerotic plaques [98]. However, high-density lipoprotein (HDL) is atheroprotective and interrupts the process of atherogenesis by transferring cholesterol from vasculature to liver for disposal and inhibiting the oxidation of LDLs and by limiting the inflammatory processes that underlie atherosclerosis [98]. During recent years, other properties of HDL have been identified to contribute to its overall anti-atherosclerotic effects, including anti-inflammatory, immunomodulatory, anti-oxidant, anti-thrombotic, and endothelial cell repair effects [99]. In addition, more evidence has indicated that elevated triglyceride led to increased risk of plaque formation and cardiovascular diseases, independent of HDL level [100, 101]. Thus, reducing the serum/plasma levels of LDL and triglyceride or elevating the HDL level during early atherosclerosis may block or delay the progress of atherosclerosis.

Notoginseng and notoginseng saponins (PNS) have been shown to exert anti-atherogenic effects through, at least in part, improving the blood lipid profiles, including decrease of serum/plasma levels of total cholesterol, triglycerides, and LDL, in $ApoE^{-/-}$ mice, rats, and rabbits fed a high-cholesterol diet [26, 27, 30, 31, 33, 34]. Moreover, the treatment with PNS and root powder of notoginseng may also elevate serum HDL levels in hyperlipidemic rabbits [31] and rats [33], respectively. Sterol regulatory element binding protein-2 (SREBP-2) is a critical marker of the regulation of cholesterol homeostasis. It can bind and activate SREBP-2-regulated gene promoters, such as LDL receptor, 3′-hydroxylmethyl glutaryl CoA synthase (HMGCS), and 3′-hydroxylmethyl glutaryl coenzyme A reductase (HMGCR) [102]. Treatment with PNS and ginsenoside Rg3 elicited significantly reduced levels of hepatic SREBP-2 and HMGCR expression [33, 75]. Another SREBP isoform, SREBP1c, was also identified as an important transcriptional factor for lipid and cholesterol homeostasis. Compound K, a major intestinal metabolite of ginsenosides, may attenuate the gene expression of SREBP1c in time- and dose-dependent manner and its target molecules include stearoyl-CoA desaturase 1 (SCD1) and fatty acid synthase (FAS). This lipid-regulating effect of SREBP transcriptional factors appears to be mediated via AMP-activated protein kinase (AMPK) signaling pathway [81]. These findings suggest that the lipid-lowering property of ginsenosides may be associated with reduced lipid synthesis.

11.4.2 Anti-oxidant Activity

Several lines of evidence have demonstrated that oxidative stress is involved in the pathogenesis of atherosclerosis. Oxidation of polyunsaturated fatty acids in LDL may be one mechanism by which LDL is modified, leading to uptake by macrophages more rapidly than native LDL and the formation of cholesterol-laden foam cells [103]. Furthermore, oxidation of other lipids and proteins within the vessel wall could promote inflammation, endothelial dysfunction, and atherogenesis [104].

Evidence from numerous epidemiologic studies also indicated that the intake of anti-oxidants, such as vitamin E and vitamin C, is associated with a reduced risk of atherosclerotic vascular disease [5]. Thus, the suppression of oxidative stress may be beneficial for slowing the progress of atherosclerosis.

In vivo and *in vitro* studies have demonstrated that ginsenosides exhibit anti-oxidative effects in atherogenesis, which is achieved through decreasing oxidized low-density lipoprotein (ox-LDL) level, reducing the vascular production of reactive oxygen species (ROS) and enhancing endogenous anti-oxidant systems. Oral administration of PNS significantly decreased serum ox-LDL level in ApoE-KO mice fed high cholesterol diet [26], and the saponins extracted from the stems and leaves of American ginseng protected native LDL from oxidation and reduced the extent of ox-LDL impairment in endothelium-dependent relaxation of rat aortic rings [54]. The pathogenesis of atherosclerosis has been shown to involve a gross imbalance between oxidant stress and anti-oxidant defense mechanisms leading to increased ROS production [105]. This increase was restored by ginsenosides in the various models. After PNS administration for 4 weeks, $ApoE^{-/-}$ mice displayed impaired ROS generation in the aortic root [28]. Aqueous root extract of Asian ginseng decreased ROS production in isolated mice aorta [36]. Several *in vitro* studies also indicated that the increased cellular ROS production induced by LPS or tumor necrosis factor -α (TNF-α), could be normalized by notoginsenoside R1 [83], ginsenosides Rb1 [58], and Rg3 [76]. Lipid peroxidation (LPO), the product resulting from interaction of ROS with lipids, particularly with polyunsaturated fatty acids, leads to cell membrane damage. Malonyldialdehyde (MDA), an end-product of LPO, has been widely used as an indicator of LPO and a marker for oxidative stress status. Notoginseng [33], PNS [28], PDS/PTS (1:1 and 2:1) [17], and ginsenoside Rb1 [58] have been shown to reverse the elevation of serum or cellular MDA level in atherosclerosis. Endogenous enzymatic/non-enzymatic anti-oxidant systems, for example, superoxide dismutase (SOD), glutathione peroxidase (GPx), and glutathione (GSH), are involved in ROS elimination, which serve as the first line of defense against oxidative damage [106]. Notoginseng [33] and its PNS [28], ginsenoside Rb1 [48] and 20 (S)-protopanaxatriol [88], have been shown to enhance the activities of serum endogenous enzymatic anti-oxidants, such as SOD or GPx, and to increase the concentration of non-enzymatic anti-oxidants, such as GSH. These data demonstrate that ginsengs and ginsenosides suppress oxidative stress, at least in part, contributing to their anti-atherosclerotic activities.

11.4.3 Anti-vascular Inflammation

In the past decade, many lines of compelling evidence have indicated that inflammation is the ultimate cause of atherosclerosis and not merely the passive accumulation of lipids within artery walls [107–109]. Inflammatory mechanisms are involved in all stages of atherosclerosis [110, 111]. Thus, reducing inflammation may postpone the initiation and progression of atherosclerosis. In a variety of animal and cell models of atherosclerosis, ginsengs and their ginsenosides exhibit extensive inhibitory effects on the vascular inflammatory components during atherosclerosis, including adhesive molecules, cytokines, chemokines, and nuclear factor-κB (NF-κB) activity.

11.4.3.1 Adhesive Molecules Atherogenic lipoproteins, such as ox-LDL, act on endothelial cells to up-regulate leukocyte-specific cell-adhesion molecules (CAMs), such as selectins (e.g., E-selectin and P-selectin), intercellular cell adhesion molecule-1 (ICAM-1) and vascular cell adhesion molecule-1 (VCAM-1), leading to circulating monocyte adhesion and transmigration into inflammatory sites and delayed maturation into macrophages, which initiates vascular inflammation [112, 113]. Up-regulation of these adhesion molecules in endothelium is important in the initial stages of inflammatory response in atherosclerosis [110, 111].

Effects of saponin fractions from notoginseng (PNS, PDS, and PTS) and major individual ginsenosides (Rg1 and Rb1) on the expression of monocyte adhesion have been systematically described by Wang and colleagues [46]. All tested saponin samples exhibited inhibitory effects on monocyte adhesion on TNF-α-activated human coronary artery endothelial cells (HCAECs) and the expressions of ICAM-1 and VCAM-1 at both mRNA and protein levels *in vitro*. Furthermore, three saponin fractions exhibited a similar trend of inhibitory effects on the mRNA expression of CAMs in the aorta of HCD-fed rat *in vivo*. Among the tested fractions, PDS was the most potent fraction against TNF-α-induced expression of adhesion molecules [46]. Furthermore, these *in vivo* studies also indicated that PNS and total saponins of Asian ginseng decrease the expressions of ICAM-1 and VCAM-1 [28, 38]. The receptor of advanced glycation end-products (RAGE), a key regulator of adhesion molecules and chemokines [114, 115], could be suppressed by PNS (60 mg/kg) and thus inactivate NF-κB and reduce expression of adhesion molecules, such as VCAM-1 and ICAM-1, in atherosclerotic lesions of $ApoE^{-/-}$ mice [28]. The major ginsenoside Rb1 effectively blocked the TNF-α-induced over-expression of VCAM-1 and the activated monocytic THP-1 cells adhesion to human umbilical vein endothelial cells (HUVECs), suggesting its potential therapeutic effects in controlling inflammation in vascular disease [116]. Hien et al. [74] identified that the anti-atherosclerotic activities of ginsenoside Rg3 in vasculature is mediated partly through down-regulation of cell adhesion molecule expression and pro-inflammatory cytokines in endothelial cells as well. Treatment with ginsenoside Rh1, a 20(S)-protopanaxatriol saponin with a single sugar moiety, decreases levels of MCP-1, CCR2, α5/β1 integrins (VLA-5) and activates β1 integrin on the cell surface of THP-1 [90]. Choi et al. [117] demonstrated that ginsenoside metabolites, compound K and ginsenoside Rh2, showed a significant inhibitory effect on TNF-α-induced expression of ICAM-1 in human astroglial cells.

11.4.3.2 Inflammatory Cytokines, Chemokines, and Lipid Mediators Pro-inflammatory cytokines (e.g., IL-1β, IL-6, and TNF-α) and lipid mediators (e.g., prostaglandins, thromboxanes, and platelet-activating factor), expressed by leukocytes, endothelial cells, and smooth muscle cells in atherosclerotic lesions, exhibit redundant and pleiotropic effects that together contribute to the inflammatory response during atherosclerosis, such as increasing vascular permeability, promoting adhesion molecules on vascular endothelium, chemokine induction, T-cell and B-cell activation, chemo-attraction of leukocytes, and the induction of cell death [110, 111]. Cytokines are also thought to contribute to destabilization of the atherosclerotic plaques [118]. As shown in Tables 11.1, 11.2, and 11.3, studies have demonstrated that many ginsengs or individual ginsenosides inhibit the expression/production of inflammatory cytokines or lipid mediators both *in vivo* and *in vitro*. PNS treatment significantly decreased the gene expression of IL-1β in zymosan A-treated rats [30] and rabbits [31]. Treatment of Asian ginseng berry extract decreased the production of TNF-α, IL-1β, and PGE2, as well as inhibited the expression of its rate-limiting enzyme, cyclooxygenase-2 (COX-2), in LPS-induced RAW 264.7 cells and primary macrophages [38]. Single compounds, such as notoginsenoside R1 [83, 85], compound K [62], ginsenosides Rb1 [56, 62], Re [73], Rg3 [76], and Rg5 [91] have been shown to decrease the production/expression of cytokines/lipid mediators in various *in vitro* models. For example, ginsenoside Rg5, a major constituent of steamed ginseng, ameliorated inflammation possibly via TLR-4 on macrophages. It inhibited the expression of interleukin-1β (IL-1β), and TNF-α, as well as inflammatory enzymes, COX-2 and iNOS (inducible nitric oxide synthase) in LPS-stimulated alveolar macrophages [91]. Bi et al. [119] reported that sulfated derivative B2 of ginsenoside Rh2 (Rh2-B2) blocked inflammatory cytokine production induced by LPS through mitogen-activated protein kinase (MAPK) and NF-κB signaling pathways. This compound (1–5 mg/L) significantly inhibited the protein and mRNA levels of TNF-α, IL-6, and IL-1β and increased IL-10 production.

Chemokines, such as monocyte chemoattractant protein-1 (MCP-1), have been shown to overexpress in endothelial cells during inflammatory response [120]. These molecules are potent chemo-attractants and activators of mononuclear phagocytes and play crucial roles in the formation of atheromatous lesions [110]. PNS downregulates the expression of MCP-1 in LPS-activated THP-1 monocytes and in the arteries of $ApoE^{-/-}$ mice [28] and zymosan A-treated rabbits [31]. Treatment with ginsenoside Rh1 reduces chemotactic activity of THP-1 monocytes via suppressing the expressions of MCP-1 and its chemokine (C–C motif) receptor 2 (CCR2) in nonactivated THP-1 cells [90].

11.4.3.3 NF-κB Activity NF-κB controls the transcription of many inflammatory genes with an established role in atherosclerosis, such as cytokines, chemokines, adhesion molecules, and macrophage infiltration [121]. The NF-κB family contains five members: p65, RelB, c-Rel, p105/p50, and p100/p52, responsible for homo- and hetero-dimerization, nuclear translocation, and association with inhibitory proteins. Among these members, only p65, RelB, and c-Rel contain a C-terminal transcription activation domain, which is essential for transcriptional activation of target genes. In normal cells, NF-κB dimers bind to IκBs and remain inactivated. Once stimulated by LPS, OxLDL, TNF-α, and/or other factors, NF-κB is released from IκB usually by ubiquitination and degradation of IκB [122, 123]. This active process is regulated by members of the MAPK, including ERK1/2, p38, and c-JNK, and IκB kinases (IKKs), which promote the phosphorylation of IκBα and regulate NF-κB translocation to the nucleus, respectively [124, 125]. Accumulating *in vivo* and *in vitro* evidence indicates that the most plausible underlying mechanism for the anti-inflammatory effect of various ginsengs or ginsenosides might be the inhibition of NF-κB signaling pathway.

It has been reported that PNS and flower extract of notoginseng attenuate atherogenesis mainly via an anti-inflammatory action by blocking NF-κB signaling pathway in macrophages. PNS treatment significantly decreased the gene expression of inflammatory factors via increasing the expression of IκBα and inhibiting the expression of NF-κB/p65 [30, 31, 52], phosphorylation of FAK, and translocation of NF-κB [32]. Joh et al. [62, 116] reported that ginsenoside Rb1 significantly inhibited the activation of interleukin-1 receptor-associated kinase (IRAK-1), IKK-β, NF-κB, and MAP kinases. The relaxing effect of ginsenoside Rg3 on vessels, as a consequence of iNOS and NO production, accompanied by the inhibition of NF-κB activation, is derived from phosphorylation and degradation of IκBα and nuclear translocation of p65 [80]. Rg5 also reduced LPS-induced phosphorylation of IRAK-1 and IKK-β, as well as the degradation of IRAK-1 and IRAK-4 and thus inhibited the translocation of p65 into the nucleus [91]. Compound K and ginsenoside Rh2 exerted anti-inflammatory effects by inhibiting TNF-α-induced activation of both NF-κB and JNK pathways in human astroglial cells [117].

11.4.4 Effect on Vascular Cells

Atherosclerosis process is a multifactorial sequence of events involving multicellular dysfunctional responses, which is initiated with injured vascular endothelial cells, the primary defense to keep vascular integrity and endothelial function. Once endothelial cells are impaired, the vascular homeostasis is disturbed and subsequently induces smooth muscle cell abnormal migration, proliferation, and apoptosis, leading to the formation of fibrous cap and the ultimate instability, even rupture, of plaque. Therefore, protective effects of ginsengs and/or corresponding ginsenosides on vascular endothelial cells and smooth muscle cells might be partly responsible for their therapeutic potential against atherosclerosis.

11.4.4.1 Vascular Endothelial Cells Vascular endothelium plays a vital role in vascular homeostasis through regulating vascular tone and structure, as well as exerting anti-platelet, anti-coagulant, and fibrinolytic effects [126]. Endothelial dysfunction has been shown to precede the development of clinically detectable atherosclerotic plaques and is now considered as an early marker for atherosclerosis [126, 127]. Endothelial nitric oxide (NO), an important vasoprotective molecule, mediates endothelium-dependent vasodilation, which is a hallmark of endothelial function. A decrease in NO production has been proposed as a major mechanism of endothelial dysfunction and a contributor to atherosclerosis [126]. Impaired vascular function is observed in atherogenic model mice fed high-cholesterol diet and could be improved by treatment with aqueous root extract of Asian ginseng, the vasoprotective effects of which are associated with augmentation of NO production by inhibiting arginase and increasing dimerization of endothelial NO synthase (eNOS) [36]. Hien et al. [78] revealed that several ginsenosides possess the capability of increasing NO production in endothelium, with ginsenoside Rg3 being the most effective. Rg3-stimulated NO production in ECV 304 human endothelial cells is associated with increased phosphorylation and the expression of eNOS via the activation of phosphatidylinositol 3-kinase (PI3K), c-Jun N-terminal kinase (JNK), p38 kinase, AMPK, and CaM kinase II [78]. Ginsenoside Rg1 also served as an agonist ligand for glucocorticoid receptor (GR) and the activated GR-induced rapid production of

NO from eNOS via PI3K pathway [63, 64]. Hyper-homocysteinemia (Hcy) has been recognized as an independent risk for inducing atherosclerosis with the inhibitory effects of endothelial proliferation [59, 128]. Increasing evidence demonstrates that the impairment of endothelial function plays an important role in Hcy-induced pathophysiological process of angioplasty and atherosclerosis [129–132]. Nitrate reductase detection showed that ginsenoside Rb1 could reverse Hcy-induced reduction of NO production in a dose-dependent manner. Rb1 activated serine-1177 and threonine-495 phosphorylation of eNOS and serine-473 phosphorylation of Akt. These results suggest that the anti-oxidative capacity of Rb1 is via its beneficial effects on Hcy-induced endothelial damage via PI3K/Akt activation and PKC inhibition [39, 40, 59, 60].

Plasminogen activator inhibitor-1 (PAI-1) plays a critical role in the regulation of vascular function and tissue remodeling by modulating thrombosis, inflammation, and the extracellular matrix [84]. Notoginsenoside R1 counteracts PAI-1 production as induced by TNF-α in human aortic smooth muscle cells [84] and by endotoxin in endothelial cells [85]. In addition, the apoptosis [133] and senescence [134] of vascular endothelial cells might induce alteration in vascular integrity and endothelial function, which contribute to pathogenesis of atherosclerosis. Ginsenoside Rg3 prevents serum deprivation-induced apoptosis of endothelial cells via Akt-dependent inhibition of the mitochondrial apoptotic signaling pathway [77]. The anti-atherosclerotic effects of ginsenoside Rb1 may partly contribute to antagonism of hydrogen peroxide-induced HUVECs senescence by modulating redox status [58].

11.4.4.2 Vascular Smooth Muscle Cells Abnormal proliferation and migration of vascular smooth muscle cells (VSMCs) are the key features of diverse vascular complication and dysfunction and are key elements in atherosclerosis [135]. Furthermore, VSMC apoptosis might promote plaque rupture in advanced atherosclerotic plaques [136]. Thus, the regulatory effects on the balance of VSMCs proliferation and apoptosis may postpone the progression of atherosclerosis.

American ginseng crude extract inhibits fetal bovine serum (FBS)-, PDGF-, or Ang II-induced VSMC proliferation via selectively suppressing JAK/STAT (janus kinase/signal transducer and activator of transcription pathway [53]. PNS may exert anti-atherosclerotic action by inhibition of VSMCs proliferation and by initiation of apoptosis via up-regulating expressions of p53, Bax, and caspase-3, as well as down-regulating Bcl-2 expression [48]. TNF-α and FBS-enhanced proliferation of VSMC could also be reversed by ginsenosides Rg1 and Rb1. Subsequent proteomic analysis suggested that PKC-ζ and p21 pathway might be involved in the underlying inhibitory effects of ginsenoside Rg1 [65, 66]. The inhibitory effects of ginsenoside Rb1 were associated with G1 cell cycle arrest and downregulation of ERK1/2 phosphorylation [61]. Additional studies demonstrated that ginsenosides Rg1 and Rb1 inhibited the vascular neointimal hyperplasia induced by balloon injury in rats via suppressing VSMC proliferation *in vivo* and the mechanism may involve the decreased expressions of p-ERK2 protein, ERK2, and c-Fos mRNA in vessel wall and upregulation of mitogen-ativated protein kinase phosphatase-1 (MKP-1) expression [41, 43]. In addition, ginsenoside Rb3 decreased cell cycle progression from G0/G1 to S phase and the expression of mRNA of proto-oncogene c-Myc, c-Fos, and c-Jun to inhibit Ang II-induced VSMCs proliferation [89].

11.4.5 Anti-platelet Effects

Atherosclerotic plaques consist of concentrated platelets that weaken the arterial wall and may intrude into the arteries to limit blood flow. These abnormalities of mushy material are often encrusted or hardened by calcium deposition [137]. Ginsengs and ginsenosides have been shown to have anti-platelet and anticoagulant effects *in vitro*, which have been translated into prolongation of bleeding times in rats. Steamed notoginseng had the most potent activity on the inhibition of platelet aggregation and plasma coagulation compared with steamed Asian ginseng and American ginseng. At a dose of 500 mg/kg, steamed notoginseng significantly suppressed platelet accumulation and this effect was positively associated with steaming duration [24, 138]. Red ginseng exhibited a potent anti-thrombotic effect on rat carotid artery, which may be due to anti-platelet rather than anti-coagulant activity *in vivo* [139]. Total saponins of Asian ginseng and notoginseng have been identified as the main active constituents that inhibit platelet aggregation and adhesion [140–143] and increase the fluidity of platelet membranes [144]. Ginsenoside Rp1 controlled collagen-induced platelet accumulation and thrombus formation via regulation on early GPVI signaling involving VASP stimulation and inhibitory effects on ERK2 and p38-MAPK [145]. In addition, four components of notoginseng, including adenosine, guanosine, ginsenosides Rh1, and F1, were identified as potential anti-platelet aggregation agents using human platelet extraction and HPLC-DAD-ESI-MS/MS analysis [146].

11.4.6 Anti-angiogenesis Effects

The role of angiogenesis has become one of the outstanding pieces of the puzzle of atherogenesis. Pro-angiogenesis therapy has been widely considered as an attractive treatment modality for ischemic heart disease [147, 148]. However, a growing body of evidence indicates that angiogenesis is a feature of advanced human atherosclerotic plaques and contributes to plaque rupture [149, 150]. Different ginsenosides exhibit diverse effects on angiogenesis. For example, ginsenoside Re and Rg1 enhanced angiogenesis, while Rb1, Rg3, and Rh2 showed opposing effects [67, 151, 152]. The pro-angiogenesis effects of ginsenoside Re on HUVECs proliferation, migration, and tube formation were dose-dependent and reached a maximal at a concentration of about 30 μg/mL [153]. Ginsenoside Rg1 has the ability to promote angiogenesis based on mechanisms involving the promotion of endothelial progenitor cell (EPC) proliferation, increased expression of VEGF and eNOS activation, inhibition of apoptosis, and hypoxia-independent activation of HIF-1α [68–70, 154–156]. In additional, Korean red ginseng water extract increased the proliferation, migration, and tube formation of HUVECs, as well as stimulated angiogenesis *in vivo* via the activation of the PI3K/Akt-dependent ERK1/2 and eNOS signaling pathways [55]. However, 20 (R)-ginsenoside Rg3 interfered with various steps of angiogenesis. Rg3 (1–103 nM) dose-dependently suppressed proliferation, chem invasion, and capillary tube formation of HUVEC, as well as the microvascular sprouting through inhibition of MMP-2 and MMP-9 in rat aortic ring assay. Meanwhile, Rg3 (150 and 600 nM) remarkably abolished bFGF-induced angiogenesis in the *in vivo* Matrigel plug assay [157].

11.5 CONCLUSIONS AND FUTURE PERSPECTIVES

In conclusion, we have summarized and integrated the efficacy and underlying mechanisms of action(s) of ginsengs and ginsenosides for the prevention and treatment of atherosclerosis. Atherosclerosis is increasingly recognized as a complex and multifaceted disease; drugs that act on multiple targets should be more effective than those having a narrow spectrum of action. As summarized in Tables 11.2 and 11.3, ginsengs and ginsenosides could alleviate atherosclerosis via multiple mechanisms involving in regulating lipid profiles, anti-oxidation, anti-vascular inflammation, anti-platelet effects, and protect vascular cells, which collectively suggests that ginsengs and ginsenosides are promising candidate molecules for the prevention and treatment of atherosclerosis. However, the findings from animal and cell studies are insufficient to completely support their clinical application currently. The paucity of rigorously designed, randomized, and/or controlled clinical trials limits the development of ginseng and/or ginsenosides for their full potential clinical therapeutic effects against atherosclerosis. In addition, the specific molecular mechanisms underlying their preventative effects against atherosclerosis need to be elucidated further.

ACKNOWLEDGMENTS

Research reported in this chapter was supported by the grants from the Research Committee of the University of Macau (SRG009-ICMS12, MYRG123-ICMS12, and MYRG111-ICMS13 to JB Wan) and from Macao Science and Technology Development Fund (010/2013/A1 to JB Wan).

REFERENCES

[1] Gui T, Shimokado A, Sun Y, Akasaka T, Muragaki Y (2012) Diverse roles of macrophages in atherosclerosis: from inflammatory biology to biomarker discovery. *Mediators Inflamm* 2012: 693083.

[2] Lloyd-Jones DM (2010) Cardiovascular risk prediction: basic concepts, current statusand future directions. *Circulation* 121: 1768–1777.

[3] Dahlof B (2010) Cardiovascular disease risk factors: epidemiology and risk assessment. *Am J Cardiol* 105: 3A–9A.

[4] Libby P (2002) Inflammation in atherosclerosis. *Nature* 420: 868–874.

[5] Lonn ME, Dennis JM, Stocker R (2012) Actions of "anti-oxidants" in the protection against atherosclerosis. *Free Radic Biol Med* 53: 863–884.

[6] Fadini GP, Manzato E, Crepaldi C, de Kreutzenberg S, Tiengo A, et al. (2010) Two cases of statin-induced rhabdomyolysis associated with mononeuropathy. *Clin Drug Investig* 30: 347–350.

[7] Lu JJ, Pan W, Hu YJ, Wang YT (2012) Multi-target drugs: the trend of drug research and development. *PLoS One* 7: e40262.

[8] Takagi K, Sato H, Nabata H (1972) Pharmacological studies of *Panax* ginseng roots: estimation of pharmacological actions of *Panax* ginseng roots. *Jpn J Pharmacol* 22: 245–249.

[9] Wan JB, Wang YT, Li SP (2008) *Panax notoginseng* (Sanqi). In: Li SP, Wang YT, editors. *Pharmacological Activity-Based Quality Control of Chinese Herbs*. New York: Nova Science Publishers, Inc. pp. 179–203.

[10] Xiang YZ, Shang HC, Gao XM, Zhang BL (2008) A comparison of the ancient use of ginseng in traditional Chinese medicine with modern pharmacological experiments and clinical trials. *Phytother Res* 22: 851–858.

[11] Karmazyn M, Moey M, Gan XT (2011) Therapeutic potential of ginseng in the management of cardiovascular disorders. *Drugs* 71: 1989–2008.

[12] Peng L, Sun S, Xie LH, Wicks SM, Xie JT (2012) Ginsenoside Re: pharmacological effects on cardiovascular system. *Cardiovasc Ther* 30: e183–e188.

[13] Buettner C, Yeh GY, Phillips RS, Mittleman MA, Kaptchuk TJ (2006) Systematic review of the effects of ginseng on cardiovascular risk factors. *Ann Pharmacother* 40: 83–95.

[14] Zhou W, Chai H, Lin PH, Lumsden AB, Yao Q, et al. (2004) Molecular mechanisms and clinical applications of ginseng root for cardiovascular disease. *Med Sci Monit* 10: RA187–RA192.

[15] Wan JB, Chen Y, Li P, Huang WH, Zhang QW (2012) Saponins from *Panax* species: chemistry, isolation and analysis. In: Koh R, Tay I, editors. *Saponins: Properties, Applications and Health Benefits*. New York: Nova Science Publishers, Inc. pp. 51–98.

[16] Liu ZQ (2012) Chemical insights into ginseng as a resource for natural antioxidants. *Chem Rev* 112: 3329–3355.

[17] Qi LW, Wang CZ, Yuan CS (2011) Isolation and analysis of ginseng: advances and challenges. *Nat Prod Rep* 28: 467–495.

[18] Christensen LP (2009) Ginsenosides chemistry, biosynthesis, analysisand potential health effects. *Adv Food Nutr Res* 55: 1–99.

[19] Wan JB, Li SP, Chen JM, Wang YT (2007) Chemical characteristics of three medicinal plants of the *Panax* genus determined by HPLC-ELSD. *J Sep Sci* 30: 825–832.

[20] Liu Y, Zhang JW, Li W, Ma H, Sun J, et al. (2006) Ginsenoside metabolites, rather than naturally occurring ginsenosides, lead to inhibition of human cytochrome P450 enzymes. *Toxicol Sci* 91: 356–364.

[21] Bae EA, Han MJ, Kim EJ, Kim DH (2004) Transformation of ginseng saponins to ginsenoside Rh2 by acids and human intestinal bacteria and biological activities of their transformants. *Arch Pharm Res* 27: 61–67.

[22] Kim WY, Kim JM, Han SB, Lee SK, Kim ND, et al. (2000) Steaming of ginseng at high temperature enhances biological activity. *J Nat Prod* 63: 1702–1704.

[23] Toh DF, New LS, Koh HL, Chan EC. (2010) Ultra-high performance liquid chromatography/time-of-flight mass spectrometry (UHPLC/TOFMS) for time-dependent profiling of raw and steamed Panax notoginseng. *J Pharm Biomed Anal* 52: 43–50.

[24] Lau AJ, Toh DF, Chua TK, Pang YK, Woo SO, et al. (2009) Antiplatelet and anti-coagulant effects of *Panax notoginseng*: comparison of raw and steamed *Panax notoginseng* with Panax ginseng and Panax quinquefolium. *J Ethnopharmacol* 125: 380–386.

[25] Hasegawa H (2004) Proof of the mysterious efficacy of ginseng: basic and clinical trials: metabolic activation of ginsenoside: deglycosylation by intestinal bacteria and esterification with fatty acid. *J Pharmacol Sci* 95: 153–157.

[26] Liu G, Wang B, Zhang J, Jiang H, Liu F (2009) Total panax notoginsenosides prevent atherosclerosis in apolipoprotein E-knockout mice: role of downregulation of CD40 and MMP-9 expression. *J Ethnopharmacol* 126: 350–354.

[27] Wan JB, Lee SM, Wang JD, Wang N, He CW, et al. (2009) Panax notoginseng reduces atherosclerotic lesions in ApoE-deficient mice and inhibits TNF-alpha-induced endothelial adhesion molecule expression and monocyte adhesion. *JAgri Food Chem* 57: 6692–6697.

[28] Dou L, Lu Y, Shen T, Huang X, Man Y, et al. (2012) Panax notogingseng saponins suppress RAGE/MAPK signaling and NF-kappaB activation in apolipoprotein-E-deficient atherosclerosis-prone mice. *Cell Physiol Biochem* 29: 875–882.

[29] Fan JS, Liu DN, Huang G, Xu ZZ, Jia Y, et al. (2012) *Panax notoginseng* saponins attenuate atherosclerosis via reciprocal regulation of lipid metabolism and inflammation by inducing liver X receptor alpha expression. *J Ethnopharmacol* 142: 732–738.

[30] Zhang YG, Zhang HG, Zhang GY, Fan JS, Li XH, et al. (2008) *Panax notoginseng* saponins attenuate atherosclerosis in rats by regulating the blood lipid profile and an anti-inflammatory action. *Clin Exp Pharmacol Physiol* 35: 1238–1244.

[31] Liu Y, Zhang HG, Jia Y, Li XH (2010) *Panax notoginseng* saponins attenuate atherogenesis accelerated by zymosan in rabbits. *Biol Pharm Bull* 33: 1324–1330.

[32] Yuan ZM, Liao Y, Tian G, Li H, Jia Y, et al. (2011) *Panax notoginseng* saponins inhibit Zymosan A induced atherosclerosis by suppressing integrin expression, FAK activation and NF-κB translocation. *J Ethnopharmacol* 138: 150–155.

[33] Xia W, Sun CH, Zhao Y, Wu LJ (2011) Hypolipidemic and anti-oxidant activities of Sanchi (Radix Notoginseng) in rats fed with a high fat diet. *Phytomedicine* 18: 516–520.

[34] Joo IW, Ryu JH, Oh HJ (2010) The influence of Sam-Chil-Geun (*Panax notoginseng*) on the serum lipid levels and inflammations of rats with hyperlipidemia induced by poloxamer-407. *Yonsei Med J* 51: 504–510.

[35] Hwang SY, Son DJ, Kim IW, Kim DM, Sohn SH, et al. (2008) Korean red ginseng attenuates hypercholesterolemia-enhanced platelet aggregation through suppression of diacylglycerol liberation in high-cholesterol-diet-fed rabbits. *Phytother Res* 22: 778–783.

[36] Shin W, Yoon J, Oh GT, Ryoo S (2013) Korean red ginseng inhibits arginase and contributes to endotheliumdependent vasorelaxation through endothelial nitric oxide synthase coupling. *J Ginseng Res* 37: 64–73.

[37] Li YN, Wu YL, Jia ZH, Qi JS (2008) Interaction between COX-2 and iNOS aggravates vascular lesion and antagonistic effect of ginsenoside. *J Ethnopharmacol* 119: 305–311.

[38] Kim CK, Cho DH, Lee KS, Lee DK, Park CW, et al. (2012) Ginseng berry extract prevents atherogenesis via anti-inflammatory action by upregulating phase II gene expression. *Evid Based Complement Altern Med* 2012: 490301.

[39] Chai H, Dong YL, Wang XW, Zhou W (2009) Ginsenoside Rb1 attenuates homocysteine-augmented guidewire injury-induced intimal hyperplasia in mice. *J Surg Res* 157: 193–198.

[40] Lan TH, Xu ZW, Wang Z, Wu YL, Wu WK, et al. (2011) Ginsenoside Rb1 prevents homocysteine-induced endothelial dysfunction via PI3K/Akt activation and PKC inhibition. *Biochem Pharmacol* 82: 148–155.

[41] Zhang S, Deng J, Gao Y, Yang DL, Gong QH, et al. (2012) Ginsenoside Rb(1) inhibits the carotid neointimal hyperplasia induced by balloon injury in rats via suppressing the phenotype modulation of vascular smooth muscle cells. *Eur J Pharmacol* 685: 126–132.

[42] Li J, Xie ZZ, Tang YB, Zhou JG, Guan YY (2011) Ginsenoside-Rd, a purified component from *Panax notoginseng* saponins, prevents atherosclerosis in apoE knockout mice. *Eur J Pharmacol* 652: 104–110.

[43] Gao Y, Deng J, Yu XF, Yang DL, Gong QH, et al. (2011) Ginsenoside Rg1 inhibits vascular intimal hyperplasia in balloon-injured rat carotid artery by down-regulation of extracellular signal-regulated kinase 2. *J Ethnopharmacol* 138: 472–478.

[44] Zhang SH, Reddick RL, Piedrahita JA, Maeda N (1992) Spontaneous hypercholesterolemia and arterial lesions in mice lacking apolipoprotein E. *Science* 258: 468–471.

[45] Azuma K, Kawamori R, Toyofuku Y, Kitahara Y, Sato F, et al. (2006) Repetitive fluctuations in blood glucose enhance monocyte adhesion to the endothelium of rat thoracic aorta. *Arterioscler Thromb Vasc Biol* 26: 2275–2280.

[46] Wang N, Wan JB, Chan SW, Deng YH, Yu N, et al. (2011) Comparative study on saponin fractions from *Panax notoginseng* inhibiting inflammation-induced endothelial adhesion molecule expression and monocyte adhesion. *Chin Med* 6: 37.

[47] Stushnoff C, Ducreux LJ, Hancock RD, Hedley PE, Holm DG, et al. (2010) Flavonoid profiling and transcriptome analysis reveals new gene-metabolite correlations in tubers of *Solanum tuberosum* L. *J Exp Bot* 61: 1225–1238.

[48] Xu L, Liu JT, Liu N, Lu PP, Pang XM (2011) Effects of *Panax notoginseng* saponins on proliferation and apoptosis of vascular smooth muscle cells. *J Ethnopharmacol* 137: 226–230.

[49] Yao Y, Wu WY, Liu AH, Deng SS, Bi KS, et al. (2008) Interaction of salvianolic acids and notoginsengnosides in inhibition of ADP-induced platelet aggregation. *Am J Chin Med* 36: 313–328.

[50] Rhule A, Rase B, Smith JR, Shepherd DM (2008) Toll-like receptor ligand-induced activation of murine DC2.4 cells is attenuated by *Panax notoginseng*. *J Ethnopharmacol* 116: 179–186.

[51] Rhule A, Navarro S, Smith JR, Shepherd DM (2006) *Panax notoginseng* attenuates LPS-induced pro-inflammatory mediators in RAW264.7 cells. *J Ethnopharmacol* 106: 121–128.

[52] Jung HW, Seo UK, Kim JH, Leem KH, Park YK (2009) Flower extract of *Panax notoginseng* attenuates lipopolysaccharide-induced inflammatory response via blocking of NF-kappaB signaling pathway in murine macrophages. *J Ethnopharmacol* 122: 313–319.

[53] Wu Q, Wang W, Li S, Nagarkatti P, Nagarkatti M, et al. (2012) American ginseng inhibits vascular smooth muscle cell proliferation via suppressing Jak/Stat pathway. *J Ethnopharmacol* 144: 782–785.

[54] Li J, Huang M, Teoh H, Man RY (1999) Panax quinquefolium saponins protects low density lipoproteins from oxidation. *Life Sci* 64: 53–62.

[55] Kim YM, Namkoong S, Yun YG, Hong HD, Lee YC, et al. (2007) Water extract of Korean red ginseng stimulates angiogenesis by activating the PI3K/Akt-dependent ERK1/2 and eNOS pathways in human umbilical vein endothelial cells. *Biol Pharm Bull* 30: 1674–1679.

[56] Smolinski AT, Pestka JJ (2003) Modulation of lipopolysaccharide-induced proinflammatory cytokine production *in vitro* and *in vivo* by the herbal constituents apigenin (chamomile), ginsenoside Rb1(ginseng) and parthenolide (feverfew). *Food Chem Toxicol* 41: 1381–1390.

[57] He F, Guo R, Wu SL, Sun M, Li M (2007) Protective effects of ginsenoside Rb1 on human umbilical vein endothelial cells *In vitro*. *J Cardiovasc Pharmacol* 50: 314–320.

[58] Liu DH, Chen YM, Liu Y, Hao BS, Zhou B, et al. (2011) Rb1 protects endothelial cells from hydrogen peroxide-induced cell senescence by modulating redox status. *Biol Pharm Bull* 34: 1072–1077.

[59] Ohashi R, Yan S, Mu H, Chai H, Yao Q, et al. (2006) Effects of homocysteine and ginsenoside Rb1 on endothelial proliferation and superoxide anion production. *J Surg Res* 133: 89–94.

[60] Zhou W, Chai H, Lin PH, Lumsden AB, Yao Q, et al. (2005) Ginsenoside Rb1 blocks homocysteine-induced endothelial dysfunction in porcine coronary arteries. *J Vasc Surg* 41: 861–868.

[61] Li QY, Chen L, Fu WH, Li ZD, Wang B, et al. (2011) Ginsenoside Rb1 inhibits proliferation and inflammatory responses in rat aortic smooth muscle cells. *J Agric Food Chem* 59: 6312–6318.

[62] Joh EH, Lee IA, Jung IH, Kim DH (2011) Ginsenoside Rb1 and its metabolite compound K inhibit IRAK-1 activation-the key step of inflammation. *Biochem Pharmacol* 82: 278–286.

[63] Leung KW, Cheng YK, Mak NK, Chan KK, Fan TP, et al. (2006) Signaling pathway of ginsenoside-Rg1 leading to nitric oxide production in endothelial cells. *FEBS Lett* 580: 3211–3216.

[64] Leung KW, Pon YL, Wong RN, Wong AS (2006) Ginsenoside-Rg1 induces vascular endothelial growth factor expression through the glucocorticoid receptor-related phosphatidylinositol 3-kinase/Akt and beta-catenin/T-cell factor-dependent pathway in human endothelial cells. *J Biol Chem* 281: 36280–36288.

[65] Zhang HS, Wang SQ (2006) Ginsenoside Rg1 inhibits tumor necrosis factor-alpha (TNF-alpha)-induced human arterial smooth muscle cells (HASMCs) proliferation. *J Cell Biochem* 98: 1471–1481.

[66] Ma ZC, Gao Y, Wang YG, Tan HL, Xiao CR, et al. (2006) Ginsenoside Rg1 inhibits proliferation of vascular smooth muscle cells stimulated by tumor necrosis factor-alpha. *Acta Pharmacol Sin* 27: 1000–1006.

[67] Yu LC, Chen SC, Chang WC, Huang YC, Lin KM, et al. (2007) Stability of angiogenic agents, ginsenoside Rg1 and Re, isolated from Panax ginseng: *in vitro* and *in vivo* studies. *Int J Pharm* 328: 168–176.

[68] Yue PY, Wong DY, Ha WY, Fung MC, Mak NK, et al. (2005) Elucidation of the mechanisms underlying the angiogenic effects of ginsenoside Rg(1) *in vivo* and *in vitro*. *Angiogenesis* 8: 205–216.

[69] Shi AW, Wang XB, Lu FX, Zhu MM, Kong XQ, et al. (2009) Ginsenoside Rg1 promotes endothelial progenitor cell migration and proliferation. *Acta Pharmacol Sin* 30: 229–306.

[70] Leung KW, Ng HM, Tang MK, Wong CC, Wong RN, et al. (2011) Ginsenoside-Rg1 mediates a hypoxia-independent upregulation of hypoxia-inducible factor-1 alpha to promote angiogenesis. *Angiogenesis* 14: 515–522.

[71] Kang SY, Schini-Kerth VB, Kim ND (1995) Ginsenosides of the protopanaxatriol group cause endothelium-dependent relaxation in the rat aorta. *Life Sci* 56: 1577–1586.

[72] Xie JT, Shao ZH, Vanden Hoek TL, Chang WT, Li J, et al. (2006) Antioxidant effects of ginsenoside Re in cardiomyocytes. *Eur J Pharmacol* 532: 201–207.

[73] Lee IA, Hyam SR, Jang SE, Han MJ, Kim DH (2012) Ginsenoside Re ameliorates inflammation by inhibiting the binding of lipopolysaccharide to TLR4 on macrophages. *J Agric Food Chem* 60: 9595–9602.

[74] Hien TT, Kim ND, Kim HS, Kang KW (2010) Ginsenoside Rg3 inhibits tumor necrosis factor-alpha-induced expression of cell adhesion molecules in human endothelial cells. *Pharmazie* 65: 699–701.

[75] Lee S, Lee MS, Kim CT, Kim IH, Kim Y (2012) Ginsenoside Rg3 reduces lipid accumulation with AMP-activated protein kinase (AMPK) activation in HepG2 cells. *Int J Mol Sci* 13: 5729–5739.

[76] Shin YM, Jung HJ, Choi WY, Lim CJ (2013) Antioxidative, anti-inflammatory and matrix metalloproteinase inhibitory activities of 20(S)-ginsenoside Rg3 in cultured mammalian cell lines. *Mol Biol Rep* 40: 269–279.

[77] Min JK, Kim JH, Cho YL, Maeng YS, Lee SJ, et al. (2006) 20(S)-Ginsenoside Rg3 prevents endothelial cell apoptosis via inhibition of a mitochondrial caspase pathway. *Biochem Biophys Res Commun* 349: 987–994.

[78] Hien TT, Kim ND, Pokharel YR, Oh SJ, Lee MY, et al. (2010) Ginsenoside Rg3 increases nitric oxide production via increases in phosphorylation and expression of endothelial nitric oxide synthase: essential roles of estrogen receptor-dependent PI3-kinase and AMP-activated protein kinase. *Toxicol Appl Pharmacol* 246: 171–183.

[79] Kim JW, Jung SY, Kwon YH, Lee SH, Lee JH, et al. (2012) Ginsenoside Rg3 inhibits endothelial progenitor cell differentiation through attenuation of VEGF-dependent Akt/eNOS signaling. *Phytother Res* 26: 1286–1293.

[80] Kim ND, Kim EM, Kang KW, Cho MK, Choi SY, et al. (2003) Ginsenoside Rg3 inhibits phenylephrine-induced vascular contraction through induction of nitric oxide synthase. *Br J Pharmacol* 140: 661–670.

[81] Kim DY, Yuan HD, Chung IK, Chung SG (2009) Compound K, intestinal metabolite of ginsenoside, attenuates hepatic lipid accumulation via AMPK activation in human hepatoma cells. *J Agric Food Chem* 57: 1532–1537.

[82] Lee ES, Choi JS, Kim MS, You HJ, Ji GE, et al. (2011) Ginsenoside metabolite compound K differentially antagonizing tumor necrosis factor-α-induced monocyte–endothelial trafficking. *Chem Biol Interact* 194: 13–22.

[83] Zhang HS, Wang SQ (2006) Notoginsenoside R1 inhibits TNF-α-induced fibronectin production in smooth muscle cells via the ROS/ERK pathway. *Free Radic Biol Med* 40: 1664–1674.

[84] Zhang HS, Wang SQ (2006) Notoginsenoside R1 from *Panax notoginseng* inhibits TNF-alpha-induced PAI-1 production in human aortic smooth muscle cells. *Vascul Pharmacol* 44: 224–230.

[85] Zhang WJ, Wojta J, Binder BR (1997) Notoginsenoside R1 counteracts endotoxin-induced activation of endothelial cells *in vitro* and endotoxin-induced lethality in mice in vivo. *Arterioscler Thromb Vasc Biol* 17: 465–474.

[86] Zhang W, Wojta J, Binder BR (1994) Effect of notoginsenoside R1 on the synthesis of tissue-type plasminogen activator and plasminogen activator inhibitor-1 in cultured human umbilical vein endothelial cells. *Arterioscler Thromb* 14: 1040–1046.

[87] Oh GS, Pae HO, Choi BM, Seo EA, Kim DH, et al. (2004) 20(S)-Protopanaxatriol, one of ginsenoside metabolites, inhibits inducible nitric oxide synthase and cyclooxygenase-2 expressions through inactivation of nuclear factor-kappaB in RAW 264.7 macrophages stimulated with lipopolysaccharide. *Cancer Lett* 205: 23–29.

[88] Kwok HH, Ng WY, Yang MSM, KiMak N, Wong RNS, et al. (2010) The ginsenoside protopanaxatriol protects endothelial cells from hydrogen peroxide-induced cell injury

and cell death by modulating intracellular redox status. *Free Radic Biol Med* 48: 437–445.

[89] Wang T, Yu XF, Qu SC, Xu HL, Sui DY (2010) Ginsenoside Rb3 inhibits angiotensin II-induced vascular smooth muscle cells proliferation. *Basic Clin Pharmacol Toxicol* 107: 685–689.

[90] Choi YJ, Yoon JH, Cha SW, Lee SG (2011) Ginsenoside Rh1 inhibits the invasion and migration of THP-1 acute monocytic leukemia cells via inactivation of the MAPK signaling pathway. *Fitoterapia* 82: 911–919.

[91] Kim TW, Joh EH, Kim B, Kim DH (2012) Ginsenoside Rg5 ameliorates lung inflammation in mice by inhibiting the binding of LPS to toll-like receptor-4 on macrophages. *Int Immunopharmacol* 12: 110–116.

[92] Smith SC, Jr., Jackson R, Pearson TA, Fuster V, Yusuf S, et al. (2004) Principles for national and regional guidelines on cardiovascular disease prevention: a scientific statement from the World Heart and Stroke Forum. *Circulation* 109: 3112–3121.

[93] Kannel WB (1987) New perspectives on cardiovascular risk factors. *Am Heart J* 114: 213–219.

[94] Baker RG, Hayden MS, Ghosh S (2011) NF-kappaB, inflammation and metabolic disease. *Cell Metab* 13: 11–22.

[95] Saku K, Zhang B, Ohta T, Arakawa K (1999) Quantity and function of high density lipoprotein as an indicator of coronary atherosclerosis. *J Am Coll Cardiol* 33: 436–443.

[96] Korhonen T, Savolainen MJ, Koistinen MJ, Ikaheimo M, Linnaluoto MK, et al. (1996) Association of lipoprotein cholesterol and triglycerides with the severity of coronary artery disease in men and women. *Atherosclerosis* 127: 213–220.

[97] Plutzky J (2000) Peroxisome proliferator-activated receptors in vascular biology and atherosclerosis: emerging insights for evolving paradigms. *Curr Atheroscler Rep* 2: 327–335.

[98] Badimon L, Vilahur G (2012) LDL-cholesterol versus HDL-cholesterol in the atherosclerotic plaque: inflammatory resolution versus thrombotic chaos. *Ann N Y Acad Sci* 1254: 18–32.

[99] Barter P (2005) The role of HDL-cholesterol in preventing atherosclerotic disease. *Eur Heart J Supp* 7: F4–F8.

[100] Austin MA (1998) Plasma triglyceride as a risk factor for cardiovascular disease. *Can J Cardiol* 14 Suppl B: 14B–17B.

[101] Hokanson JE, Austin MA (1996) Plasma triglyceride level is a risk factor for cardiovascular disease independent of high-density lipoprotein cholesterol level: a meta-analysis of population-based prospective studies. *J Cardiovasc Risk* 3: 213–219.

[102] Brown MS, Goldstein JL (1997) The SREBP pathway: regulation of cholesterol metabolism by proteolysis of a membrane-bound transcription factor. *Cell* 89: 331–340.

[103] Mitra S, Goyal T, Mehta JL (2011) Oxidized LDL, LOX-1 and atherosclerosis. *Cardiovasc Drugs Ther* 25: 419–429.

[104] Meagher E, Rader DJ (2001) Antioxidant therapy and atherosclerosis: animal and human studies. *Trends Cardiovasc Med* 11: 162–165.

[105] Thomson MJ, Puntmann V, Kaski JC (2007) Atherosclerosis and oxidant stress: the end of the road for anti-oxidant vitamin treatment? *Cardiovasc Drugs Ther* 21: 195–210.

[106] Ding RB, Tian K, Huang LL, He CW, Jiang Y, et al. (2012) Herbal medicines for the prevention of alcoholic liver disease: a review. *J Ethnopharmacol* 144: 457–465.

[107] Wong BW, Meredith A, Lin D, McManus BM (2012) The biological role of inflammation in atherosclerosis. *Can J Cardiol* 28: 631–641.

[108] Johnson LN, Koval M (2009) Cross-talk between pulmonary injury, oxidant stress and gap junctional communication. *Antioxid Redox Signal* 11: 355–367.

[109] Rocha VZ, Libby P (2009) Obesity, inflammation and atherosclerosis. *Nat Rev Cardiol* 6: 399–409.

[110] Ross R (1999) Atherosclerosis—an inflammatory disease. *N Engl J Med* 340: 115–126.

[111] Ross R (1986) The pathogenesis of atherosclerosis—an update. *N Engl J Med* 314: 488–500.

[112] Weisberg SP, Hunter D, Huber R, Lemieux J, Slaymaker S, et al. (2006) CCR2 modulates inflammatory and metabolic effects of high-fat feeding. *J Clin Invest* 116: 115–124.

[113] Desai A, Darland G, Bland JS, Tripp ML, Konda VR (2012) META060 attenuates TNF-a-activated inflammation, endothelialemonocyte interactionsand matrix metalloproteinase-9 expression and inhibits NF-kB and AP-1 in THP-1 monocytes. *Atherosclerosis* 223: 230–236.

[114] Sun L, Ishida T, Yasuda T, Kojima Y, Honjo T, et al. (2009) RAGE mediates oxidized LDL-induced pro-inflammatory effects and atherosclerosis in non-diabetic LDL receptor-deficient mice. *Cardiovasc Res* 82: 371–381.

[115] Harja E, Bu DX, Hudson BI, Chang JS, Shen X, et al. (2008) Vascular and inflammatory stresses mediate atherosclerosis via RAGE and its ligands in apo$E^{-/-}$ mice. *J Clin Invest* 118: 183–194.

[116] Chai H, Wang Q, Huang L, Xie T, Fu Y (2008) Ginsenoside Rb1 inhibits tumor necrosis factor-alpha-induced vascular cell adhesion molecule-1 expression in human endothelial cells. *Biol Pharm Bull* 31: 2050–2056.

[117] Choi K, Kim M, Ryu J, Choi C (2007) Ginsenosides compound K and Rh(2) inhibit tumor necrosis factor-alpha-induced activation of the NF-kappaB and JNK pathways in human astroglial cells. *Neurosci Lett* 421: 37–41.

[118] Shin J, Edelberg JE, Hong MK (2003) Vulnerable atherosclerotic plaque: clinical implications. *Curr Vasc Pharmacol* 1: 183–204.

[119] Bi WY, Fu BD, Shen HQ, Wei Q, Zhang C, et al. (2012) Sulfated derivative of 20(S)-ginsenoside Rh2 inhibits inflammatory cytokines through MAPKs and NF-kappa B pathways in LPS-induced RAW264.7 macrophages. *Inflammation* 35: 1659–1668.

[120] Gu L, Okada Y, Clinton SK, Gerard C, Sukhova GK, et al. (1998) Absence of monocyte chemoattractant protein-1 reduces atherosclerosis in low density lipoprotein receptor-deficient mice. *Mol Cell* 2: 275–281.

[121] Pamukcu B, Lip GY, Shantsila E (2011) The nuclear factor—kappa B pathway in atherosclerosis: a potential therapeutic target for atherothrombotic vascular disease. *Thromb Res* 128: 117–123.

[122] Heiden KVD, Cuhlmann S, Luong LA, Zakkar M, Evans PC (2010) Role of nuclear factorκB in cardiovascular health and disease. *Clin Sci* 118: 593–605.

[123] Lou XW, Sun SG, Wang C (2009) Regulation of NF-κB activity in nucleus. *Chin J Cell Biol* 31: 741–748.

[124] Dudhgaonkar S, Thyagarajan A, Sliva D (2009) Suppression of the inflammatory response by triterpenes isolated from the mushroom Ganoderma lucidum. *Int Immunopharmacol* 9: 1272–1280.

[125] Wang TM, Chen CJ, Lee TS, Chao H-Y, Wu WH, et al. (2011) Docosahexaenoic acid attenuates VCAM-1 expression and NF-κB activation in TNF-α-treated human aortic endothelial cells. *J Nutr Biochem* 22: 187–194.

[126] Stocker R, Keaney JF, Jr. (2004) Role of oxidative modifications in atherosclerosis. *Physiol Rev* 84: 1381–1478.

[127] Mudau M, Genis A, Lochner A, Strijdom H (2012) Endothelial dysfunction: the early predictor of atherosclerosis. *Cardiovasc J Afr* 23: 222–231.

[128] McCully KS (1969) Vascular pathology of homocysteinemia: implications for the pathogenesis of arteriosclerosis. *Am J Pathol* 56: 111–128.

[129] Ross R (1993) The pathogenesis of atherosclerosis: a perspective for the 1990s. *Nature* 362: 801–809.

[130] Lentz SR, Sobey CG, Piegors DJ, Bhopatkar MY, Faraci FM, et al. (1996) Vascular dysfunction in monkeys with diet-induced hyperhomocyst(e)inemia. *J Clin Invest* 98: 24–29.

[131] Liu X, Shen J, Zhan R, Wang X, Zhang Z, et al. (2009) Proteomic analysis of homocysteine induced proliferation of cultured neonatal rat vascular smooth muscle cells. *Biochim Biophys Acta* 1794: 177–184.

[132] Ventura E, Durant R, Jaussent A, Picot MC, Morena M, et al. (2009) Homocysteine and inflammation as main determinants of oxidative stress in the elderly. *Free Radic Biol Med* 46: 737–746.

[133] Kam PC, Ferch NI (2000) Apoptosis: mechanisms and clinical implications. *Anaesthesia* 55: 1081–1093.

[134] Minamino T, Miyauchi H, Yoshida T, Ishida Y, Yoshida H, et al. (2002) Endothelial cell senescence in human atherosclerosis: role of telomere in endothelial dysfunction. *Circulation* 105: 1541–1544.

[135] Rosamond W, Flegal K, Furie K, Go A, Greenlund K, et al. (2008) Heart disease and stroke statistics—2008 update: a report from the American Heart Association Statistics Committee and Stroke Statistics Subcommittee. *Circulation* 117: e25–e146.

[136] Rudijanto A (2007) The role of vascular smooth muscle cells on the pathogenesis of atherosclerosis. *Acta Med Indones* 39: 86–93.

[137] Pugliese G, Iacobini C, Fantauzzi CB, Menini S (2015) The dark and bright side of atherosclerotic calcification. Atherosclerosis 238: 220–230.

[138] Darwin L, editor. (2010) *Epidemiology and Prevention of Cardiovascular Diseases: A Global Challenge*. Sudbury: Jones and Baetkett Publishers. pp. 41.

[139] Jin YR, Yu JY, Lee JJ, You SH, Chung JH, et al. (2007) Antithrombotic and anti-platelet activities of Korean red ginseng extract. *Basic Clin Pharmacol Toxicol* 100: 170–175.

[140] Mo ZX, Huang YH, Li XF (1989) Effects of ginsenosides on the function, morphology and cAMP level of rabbit platelets. *Asia Pac J Pharmacol* 4: 213–218.

[141] Liao FL, Bin L (1997) Inhibition of shear-induced platelet aggregation by Chinese herbal medicines. *Clin Hemorheol Microcirc* 17: 315–318.

[142] Ma LY, Xiao PG (1998) Effects of *Panax notoginseng* saponins on platelet aggregation in rats with middle cerebral artery occlusion or *in vitro* and on lipid fluidity of platelet membrane. *Phytother Res* 12: 138–140.

[143] Wang J, Xu J, Zhong JB, Liu JG (2004) Effect of Radix notoginseng saponins on platelet activating molecule expression and aggregation in patients with blood hyperviscosity syndrome. *Chin J Integr Tradit West Med* 24: 312–316.

[144] Su Y, Zhao YG, Zhang ZP, Yue N (1996) Effects of panaxatriol saponins isolated from sanchi (Panax pseudo-ginseng var.notoginseng) on animal platelet function and thrombosis. *Chin Tradit Herb Drugs* 27: 666–669.

[145] Endale M, Lee WM, Kamruzzaman SM, Kim SD, Park JY, et al. (2012) Ginsenoside-Rp1 inhibits platelet activation and thrombus formation via impaired glycoprotein VI signalling pathway, tyrosine phosphorylation and MAPK activation. *Br J Pharmacol* 167: 109–127.

[146] Wang J, Huang ZG, Cao H, Wang YT, Hui P, et al. (2008) Screening of anti-platelet aggregation agents from *Panax notoginseng* using human platelet extraction and HPLC-DAD-ESI-MS/MS. *J Sep Sci* 31: 1173–1180.

[147] Ahn A, Frishman WH, Gutwein A, Passeri J, Nelson M (2008) Therapeutic angiogenesis: a new treatment approach for ischemic heart disease—Part II. *Cardiol Rev* 16: 219–229.

[148] Nelson MA, Passeri J, Frishman WH (2000) Therapeutic angiogenesis: a new treatment modality for ischemic heart disease. *Heart Dis* 2: 314–325.

[149] Michel JB, Virmani R, Arbustini E, Pasterkamp G (2011) Intraplaque haemorrhages as the trigger of plaque vulnerability. *Eur Heart J* 32: 1977–1985, 1985a, 1985b, 1985c.

[150] Virmani R, Kolodgie FD, Burke AP, Finn AV, Gold HK, et al. (2005) Atherosclerotic plaque progression and vulnerability to rupture: angiogenesis as a source of intraplaque hemorrhage. *Arterioscler Thromb Vasc Biol* 25: 2054–2061.

[151] Yuan CS, Wang CZ, Wicks SM, Qi LW (2010) Chemical and pharmacological studies of saponins with a focus on American ginseng. *J Ginseng Res* 34: 160–167.

[152] Lin KM, Hsu CH, Rajasekaran S (2008) Angiogenic evaluation of ginsenoside Rg 1 from Panax ginseng in fluorescent transgenic mice. *Vascul Pharmacol* 49: 37–43.

[153] Heeney MM, Whorton MR, Howard TA, Johnson CA, Ware RE (2004) Chemical and functional analysis of hydroxyurea oral solutions. *J Pediatr Hematol Oncol* 26: 179–184.

[154] Liang HC, Chen CT, Chang Y, Huang YC, Chen SC, et al. (2005) Loading of a novel angiogenic agent, ginsenoside Rg1 in an acellular biological tissue for tissue regeneration. *Tissue Eng* 11: 835–846.

[155] Pajusola K, Kunnapuu J, Vuorikoski S, Soronen J, Andre H, et al. (2005) Stabilized HIF-1alpha is superior to VEGF for angiogenesis in skeletal muscle via adeno-associated virus gene transfer. *FASEB J* 19: 1365–1367.

[156] Yang N, Chen P, Tao Z, Zhou N, Gong X, et al. (2012) Beneficial effects of ginsenoside-Rg1 on ischemia-induced angiogenesis in diabetic mice. *Acta Biochim Biophys Sin (Shanghai)* 44: 999–1005.

[157] Yue PY, Wong DY, Wu PK, Leung PY, Mak NK, et al. (2006) The angiosuppressive effects of 20(R)-ginsenoside Rg3. *Biochem Pharmacol* 72: 437–445.

12

PHYTOTHERAPY PHARMACOPHORES FOR MAJOR CELLULAR DRUG TARGETS

Jennifer A. Ong, Paul W. Groundwater, and David E. Hibbs
Faculty of Pharmacy, The University of Sydney, Sydney, New South Wales, Australia

12.1 INTRODUCTION

In this chapter, the use of computational determination of pharmacophores to illustrate and predict interactions of active constituents of phytotherapies with novel drug targets will be examined. Pharmacophore modeling, which is an *in silico* method of molecular modeling, allows the identification of chemical features and its three-dimensional orientation necessary for an active ligand. It has been shown to be particularly useful for virtual screening of natural compound libraries in the absence of a target crystal structure, such as that for the downstream effector in Hedgehog signaling. In a similar fashion, natural compounds from the *Veratrum* species have been studied and used to generate a ligand-based pharmacophore model. For many targets including PPAR, ACC1 and -2 dual inhibitors, FXR, 11β-HSD1, androgen receptor, mPGES-1, GABA-A, NNRT, KAS III, PDE-5, and CYP1A2, for which crystal structures are available for structure-based docking, pharmacophore modeling is typically used in conjunction with X-ray structural data, to elucidate key protein interactions and support the findings from docking studies, and vice versa. Alternatively, pharmacophore models have been generated and manipulated to restrict findings to target compounds with specific properties, such as chemical structure in the case of some morphinan-based AChE inhibitors,

Phytotherapies: Efficacy, Safety, and Regulation, First Edition. Edited by Iqbal Ramzan.

or activity for new anticonvulsant medications. The use of multiple pharmacophore models in parallel screening has been shown to be advantageous for *in silico* high-throughput screening for targets with numerous binding conformations (e.g., COX-1 and -2), or for compounds that potentially target multiple targets. As a simple yet valuable tool, there are many opportunities for which pharmacophore modeling can be utilized in the search for lead active compounds in the drug discovery process. For this reason, a variety of case studies are presented in this chapter as potential phytotherapies for the treatment of disease.

12.2 WHAT IS A PHARMACOPHORE?

In its simplest form, a pharmacophore is the three-dimensional arrangement in space of chemical functional groups, in a molecule, deemed as essential for biological activity. The International Union of Pure and Applied Chemistry (IUPAC) defines a pharmacophore a little more strictly: A pharmacophore is the ensemble of steric and electronic features that are necessary to ensure the optimal supramolecular interactions with a specific biological target structure and to trigger (or to block) its biological response. Each discrete functional group within a drug molecule can be assigned as one of the following pharmacophoric features: hydrophobic groups, aromatic and ring systems, hydrogen bond acceptors, hydrogen bond donors, cations, anions, and exclusion volumes (which represent areas occupied by the target). These features represent potential interaction sites with the drug's receptor. The pharmacophore concept is based on the premise that different chemical groups can have the same types of interactions with a target. For example, carboxylic acids, certain sulfonamides, and tetrazoles are acidic and so can be considered to be equivalent in a bioisosteric context. In order to achieve this equivalence, however, the user must have a means for defining and identifying pharmacophorically similar groups, as well as the means to align them on each molecule. Importantly, a pharmacophore represents not only the types of functional groups present in a molecule but also the distances and angles between these groups. Figure 12.1 shows a basic pharmacophore for (+)-6-aminopenicillanic acid.

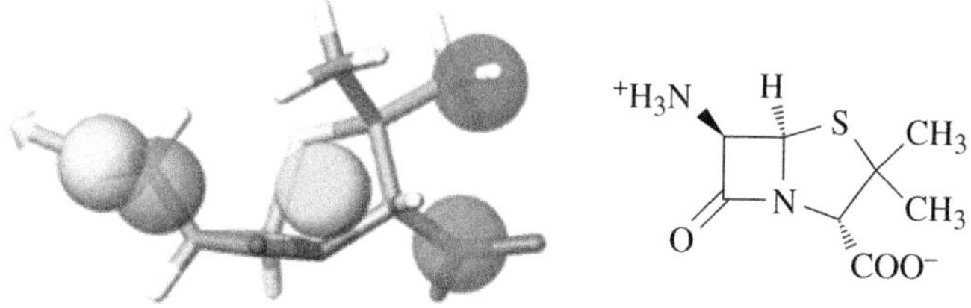

FIGURE 12.1 Pharmacophoric features in (+)-6-aminopenicillanic acid. Key functional groups contain negatively ionizable (anion, carboxylate) features, the positively ionizable (cation, amine) H-bond donor (amine) features, and the ring and the methyl (hydrophobe) features.

12.3 PHARMACOPHORE MODELS OF CARDIOVASCULAR DRUGS

In recent years, great attention has been paid to the agonism of peroxisome proliferator-activated receptor (PPAR) subtype gamma (γ), in the treatment of metabolic diseases such as Type 2 diabetes and other metabolic diseases. Using several agonist-bound X-ray crystal structures of PPARγ (PDB 1NYX, 1KNU, 1I7I, 1ZGY), Tanrikulu et al. [1] developed a "fuzzy" structure-based pharmacophore model and screened for naturally derived compounds in the AnalytiCon Discovery collection of natural products. The majority of the top-scoring hits shared common chemical scaffolds (**S1** and **S2**) that are derived from sesquiterpene lactone, α-santonin (Fig. 12.2). The authors manually selected eight compounds containing scaffold **S1**, which is believed to be more efficacious than **S2**, with the intention of retaining carboxyl groups at positions R^1 and R^2, as an acidic ligand moiety has been shown to be important for PPARγ activation (Table 12.1). At a concentration of 30 μM, compounds **1** and **2** were the only compounds shown to agonize PPARγ (110 ± 31% and 33 ± 8% relative to pioglitazone, respectively). Compound **1** was also noted to have some activity at PPARα, whereas **2** had none.

Fakhrudin et al. [2] utilized a single X-ray crystal structure of PPARγ (PDB 2G0G) to generate a different structure-based pharmacophore model (Fig. 12.3), in an attempt to identify partial agonists from the *DIOS* (described in Dioscorides' *De material medica*) and Chinese Herbal Medicine (CHM) databases. The shift in focus from full PPARγ agonism has arisen in an attempt to alleviate serious side effects associated with the use of full PPARγ agonists such as thiazolidinediones, which include weight gain, fluid retention, heart failure, increased bone fractures, and hepatotoxicity. Using PPARγ luciferase reporter gene assays, the maximal activity of neolignans dieugenol, tetrahydrodieugenol, and magnolol were observed to be several fold lower than pioglitazone (Fig. 12.4). Furthermore, dieugenol and tetrahydrodieugenol

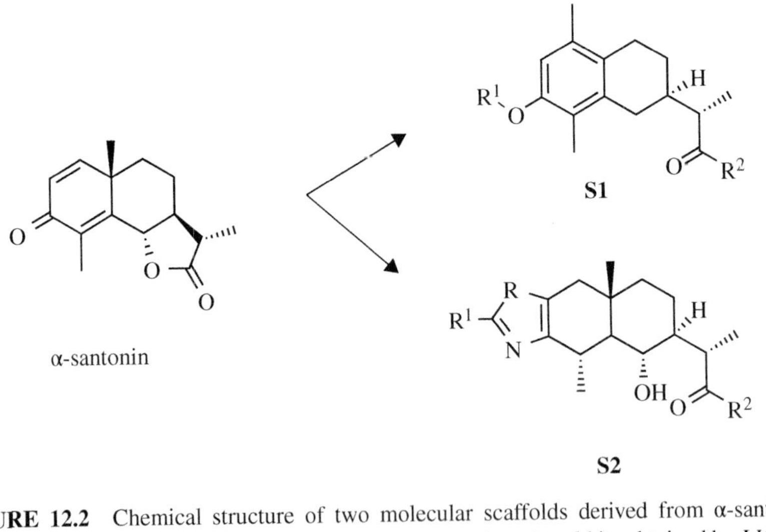

FIGURE 12.2 Chemical structure of two molecular scaffolds derived from α-santonin, **S1** and **S2**. Both scaffolds were found among the top-scoring virtual hits obtained by *LIQUID* pharmacophore searching the *AnalytiCon Discovery* compound collection.

TABLE 12.1 Chemical structures and activity of eight compounds chosen for cellular reporter gene assay at a concentration of 30 μM

No.	Chemical Structure	PPARα (%)	PPARγ (%)
1.	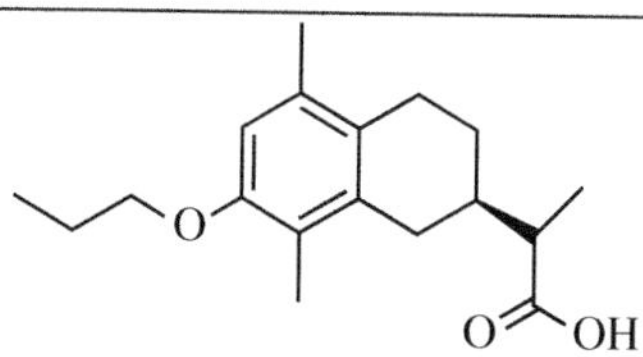	16±3	110±31
2.	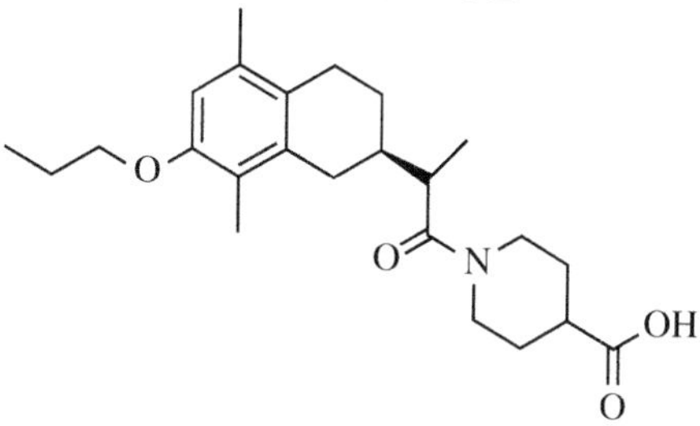	0	33±8
3.	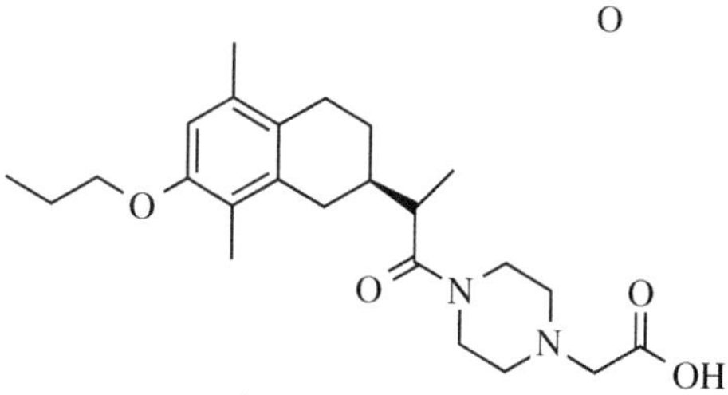	Inactive	
4.	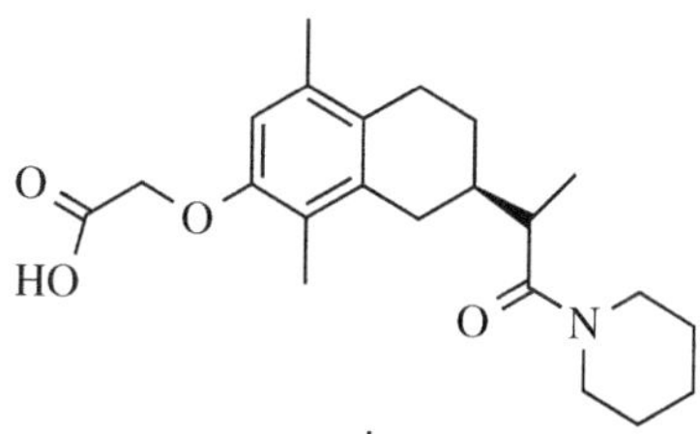	Inactive	
5.	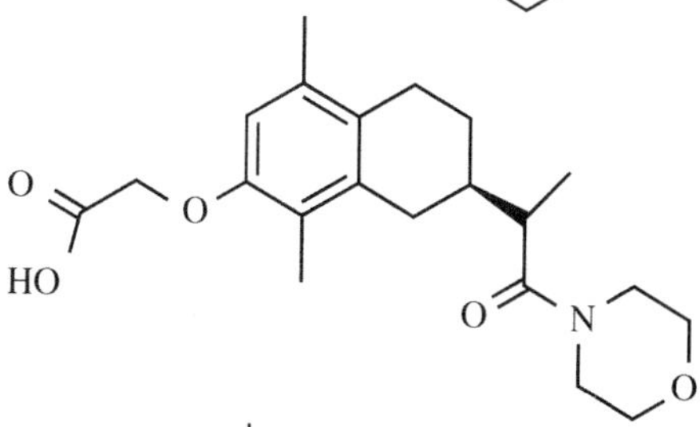	Inactive	
6.	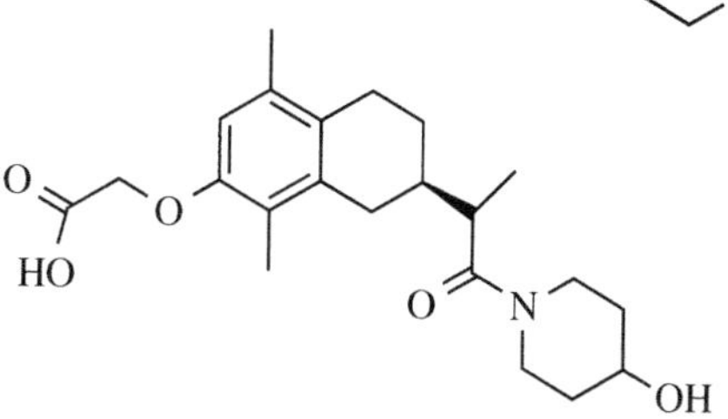	Inactive	

(*Continued*)

TABLE 12.1 (*Continued*)

No.	Chemical Structure	PPARα (%)	PPARγ (%)
7.	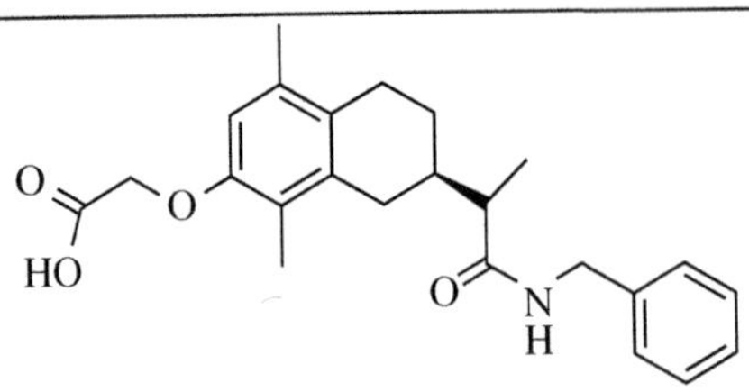	Inactive	
8.	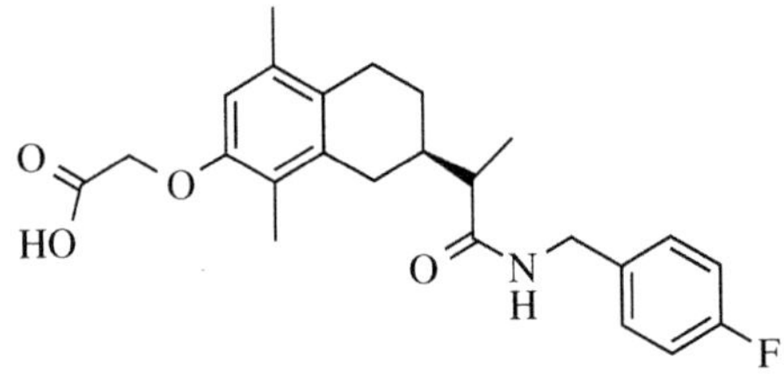	Inactive	

Values give PPARα activation relative to the selective agonist GW7647 (EC_{50} = 6 nM) and PPARγ activation relative to the selective agonist pioglitazone (EC_{50} = 0.27 μM).

displayed similar potency as pioglitazone as maximal activations were achieved at similar concentrations (~1 μM). Interestingly, the substructure eugenol itself does not bind to PPARγ and showed no activity *in vitro* (Fig. 12.5).

From ligand pharmacophore screening generated using known PPARγ partial agonists (n = 13) (Fig. 12.6), Petersen et al. [3] screened natural compounds from the Chinese Natural Product Database (CNPD) and selected methyl oleanonate (me-ester-3-oxo-olean-12-en-28-oic acid) for further investigation (Fig. 12.7). Bioassay-guided chromatographic fractionation of *Pistacia lentiscus* var. Chia oleoresin (Chios mastic gum), a natural source of methyl oleanonate, led to the isolation of oleanonic acid (from acidic fraction *6-II*), which was shown to have 20% of the activity of full PPARγ agonist, rosiglitazone at saturating concentration.

Much attention has also been given to pan-PPAR agonism, which targets subtypes alpha (α), delta (δ), as well as γ, as a means of countering the aforementioned side effects. From the TCM Database@Taiwan, natural compounds (S)-tryptophan-betaxanthin and berberrubine (Fig. 12.8) rendered high dock scores across all three homology models of PPAR subtypes. Chen et al. [4] generated the models from the alignment of protein sequence UnitProtKB Q07869 (PPARα), Q03181 (PPARδ), and P37231 (PPARγ) with crystal structures of human PPAR PDBs 1K7L and 3KDT (PPARα); 3ET2 and 2GWX (PPARδ); and 3LMP and 3K8S (PPARγ). Unspecified protein–ligand complexes from the virtual screening were subsequently utilized to generate structure-based pharmacophore models as a tool to identify key interactions for each PPAR subtype. PPARα was found to consist of four features: two HA and two HY, of which the former were complementary to Tyr314 and Ala333. PPARδ contained eight features: four HA, one HD, and three HY. The four HA and one HD

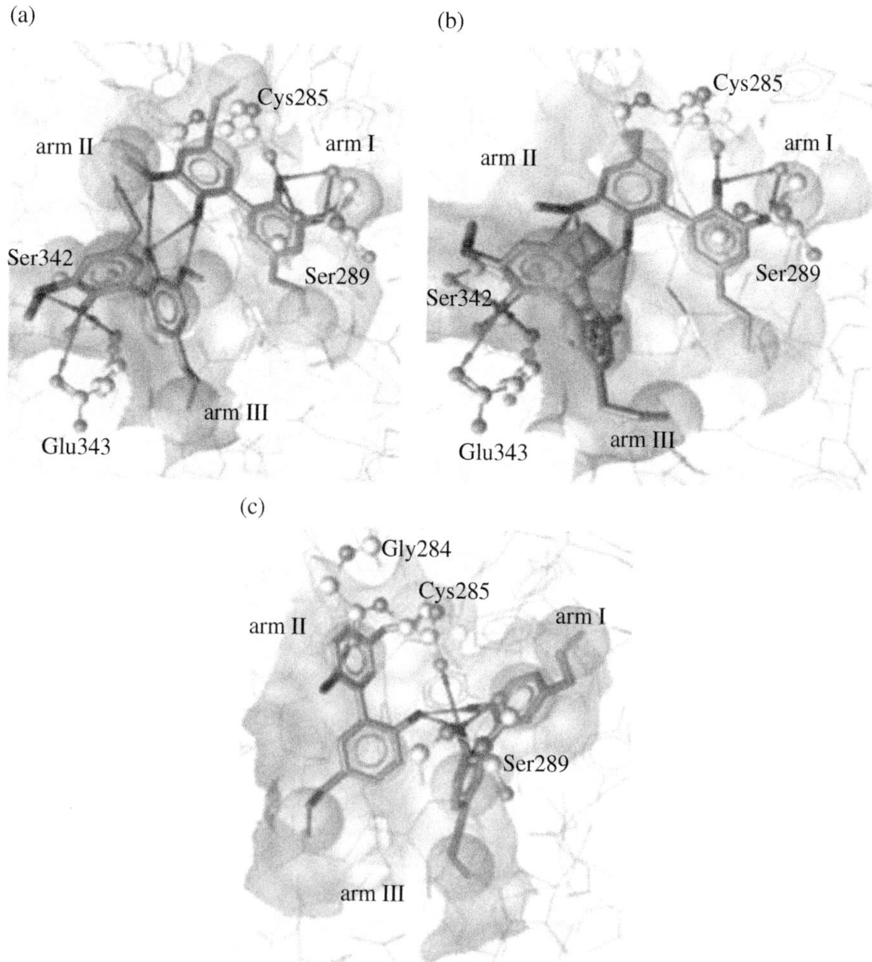

FIGURE 12.3 Neolignans: dieugenol (a); tetrahydrodieugenol (b); magnolol (c), aligned with a structure-based pharmacophore model consisting of hydrogen bond acceptor (red arrow), hydrogen bond donor (green arrow), hydrophobic interaction (yellow sphere), and aromatic interaction (blue rings). *For color details, please see color plate section.*

features were generated as complementary features to Thr253, Lys22, Tyr437, His413, and Leu304, respectively. PPARγ contained three HA, one HD, and two HY. The three HA and one HD features were generated as complementary features to Arg316, Glu371, Leu256, and Glu323 (Fig. 12.9). Further molecular dynamics studies supported stable H-bonds between (S)-tryptophan-betaxanthin and PPARα Ala333, Lys275, Thr279, and Met355; PPARδ Lys331, Leu304, Thr252, Thr253; and PPARγ Lys395. A pi-cation interaction with Phe359 was maintained after MD between berberrubine and PPARα and H-bonds were retained with PPARδ Thr252 and Thr253 and PPARγ Ser370. Generally, carboxylic acid and tetrahydropyridine

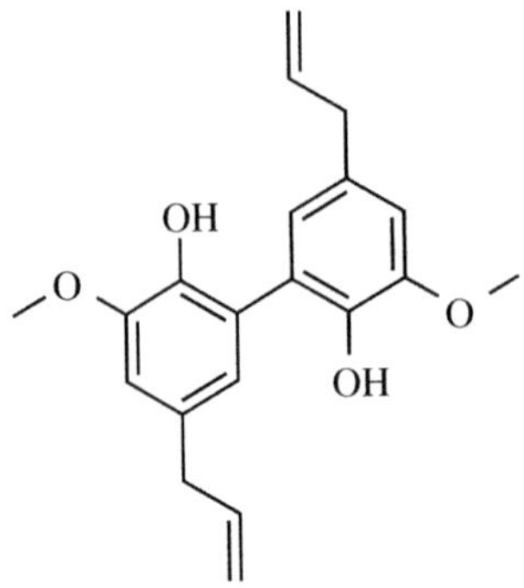

Dieugenol (1)

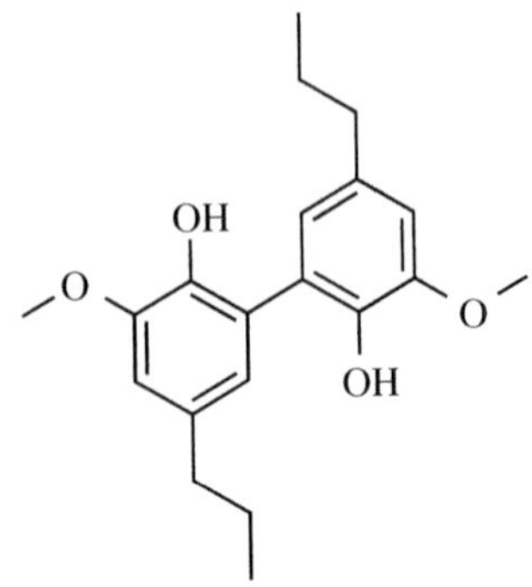

Tetrahydrodieugenol (2)

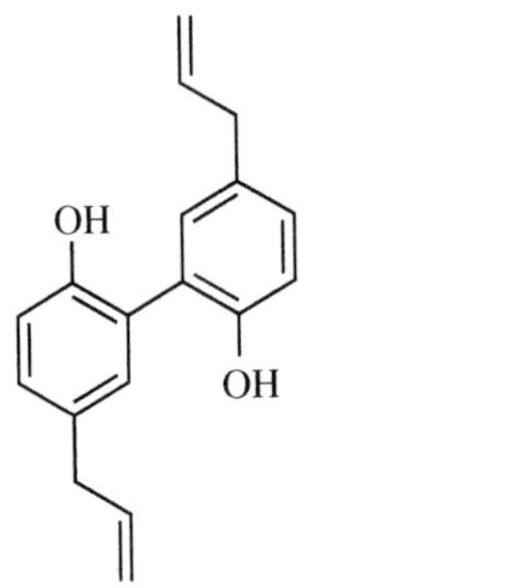

Magnolol (3)

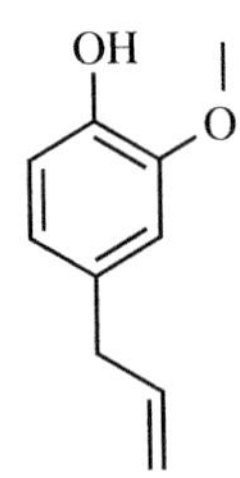

Eugenol (4)

FIGURE 12.4 Chemical structures of neolignans dieugenol (1), tetrahydrodieugenol (2), magnolol (3), and eugenol (4).

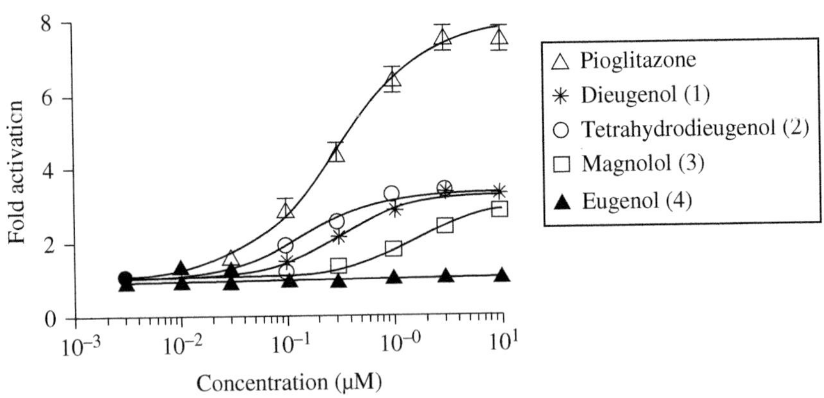

FIGURE 12.5 Influence of the neolignans on human PPAR**γ**-mediated reporter gene transactivation. Results are expressed as fold activation compared with negative control (dimethyl sulfoxide (DMSO) vehicle treatment). Data are shown as means ± S.D. of three independent experiments each performed in four replicates.

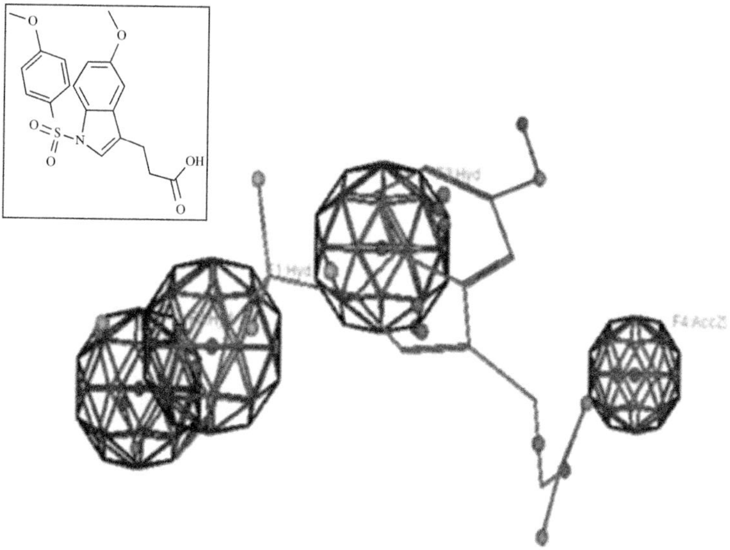

FIGURE 12.6 Four-point pharmacophore model for known PPARγ partial agonists superimposed on indeglitazar (RMSD = 0.50 Å). Pharmacophoric features are represented by a point encased in a sphere: hydrogen bond donor (blue), hydrophobic region (green), aromatic center (orange), and CO_2 centroid (red). Points not encased in spheres are other potential pharmacophore features on the indeglitazar structure (2D structure shown on top left). *For color details, please see color plate section.*

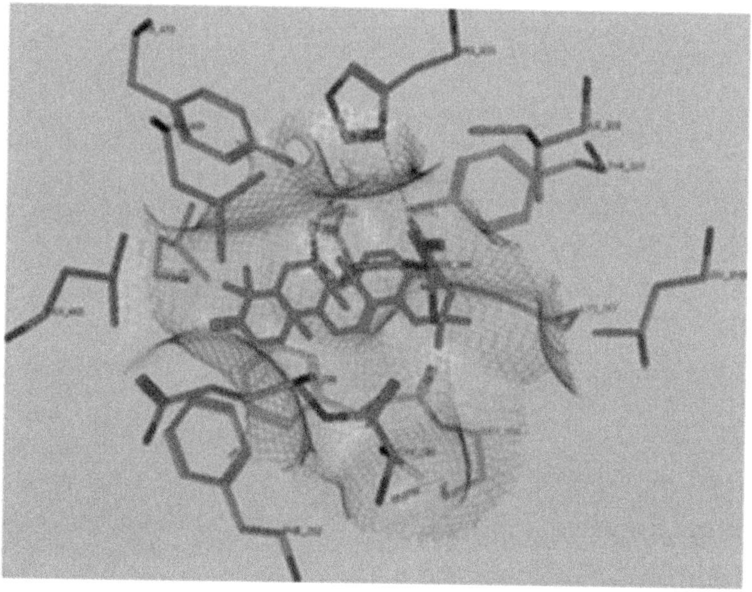

FIGURE 12.7 Oleanonic acid docked in the ligand-binding domain of PPARγ, which shows a hydrogen bond between the carboxylic moiety with His323 and Thr327 on helix 4/5 on arm I. The remainder of the ligand is stabilized within a hydrophobic pocket formed by residues Gln286, Met364, Leu453, and Leu469. *For color details, please see color plate section.*

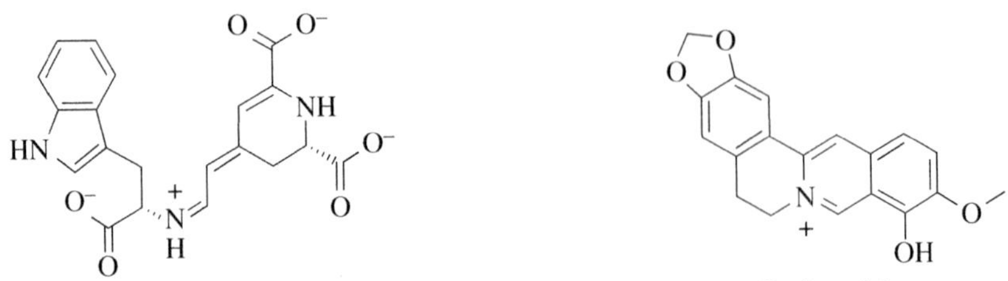

FIGURE 12.8 Chemical structures of (S)-tryptophan-betaxanthin and berberrubine.

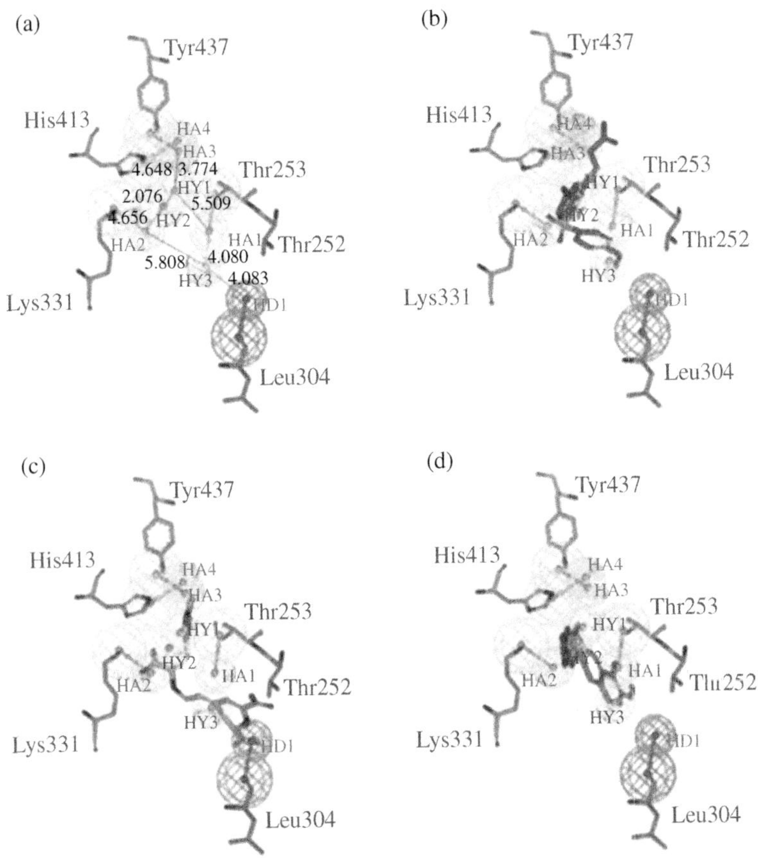

FIGURE 12.9 Generated structure-based pharmacophore models for PPARδ shown with inter-feature distance constraints only (a), control, ET1 (b), (S)-tryptophan-betaxanthin (c), and berberrubine (d). Pharmacophoric features are shown for hydrogen bond acceptors (green), hydrogen bond donors (magenta), and hydrophobic feature (blue). *For color details, please see color plate section.*

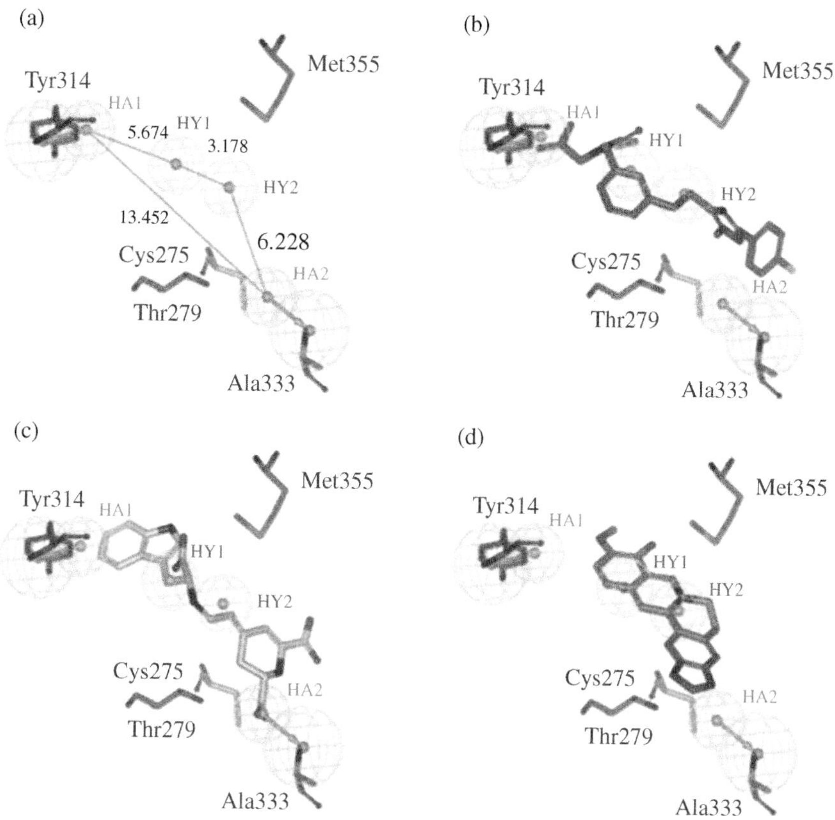

FIGURE 12.10 Generated structure-based pharmacophore models for PPARα shown with inter-feature distance constraints only (a), control, 7HA (b), (S)-tryptophan-betaxanthin (c), and berberrubine (d). Pharmacophoric features are shown for hydrogen bond acceptors (green), hydrogen bond donors (magenta), and hydrophobic feature (blue). *For color details, please see color plate section.*

moiety of (S)-tryptophan-betaxanthin are necessary for H-bond formation with PPARs. Its secondary amine moiety and pi-cation interaction leads to PPARα and PPARδ selectivity, respectively. For berberrubine, the phenol moiety and methoxy moiety are critical for H-bond formation with PPARs (Figs. 12.9, 12.10, and 12.11).

Farnesoid X receptor (FXR) is another nuclear hormone receptor that has also been implicated in glucose and lipid metabolism through its maintenance of bile acid homeostasis. Initial screening of an in-house Chinese herbal medicines database using structure-based pharmacophore model based on crystal structure PDB 1OSH (*1osh-1*, Fig. 12.12) by Grienke et al. [5] resulted in 572 virtual hits that were generally found to be constituents of *Ganoderma lucidum*, *Ginkgo biloba* leaves, *Vitex agnus-castus* fruits, *Ruta graveolens* roots and leaves, and *Capsicum annum* fruits. Pharmacological evaluation of each extract revealed that fruiting bodies of *G. lucidum* induced FXR by about 150% at test concentration of 100 μg/ml and was a promising starting point for further investigation. These known constituents and

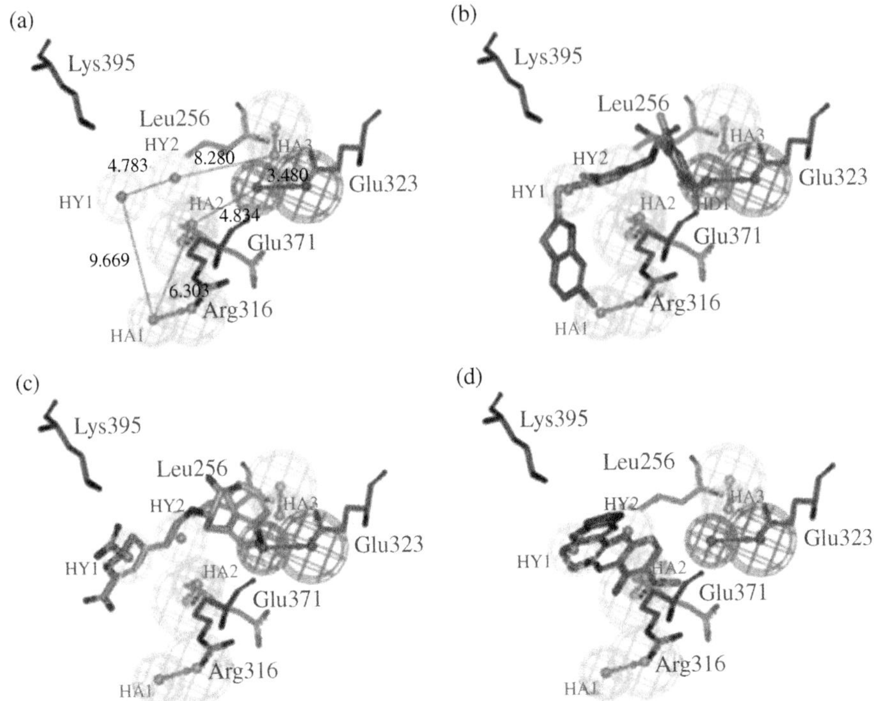

FIGURE 12.11 Generated structure-based pharmacophore models for PPARγ shown with inter-feature distance constraints only (a), control, T2384 (b), (S)-tryptophan-betaxanthin (c), and berberrubine (d). Pharmacophoric features are shown for hydrogen bond acceptors (green), hydrogen bond donors (magenta), and hydrophobic feature (blue). *For color details, please see color plate section.*

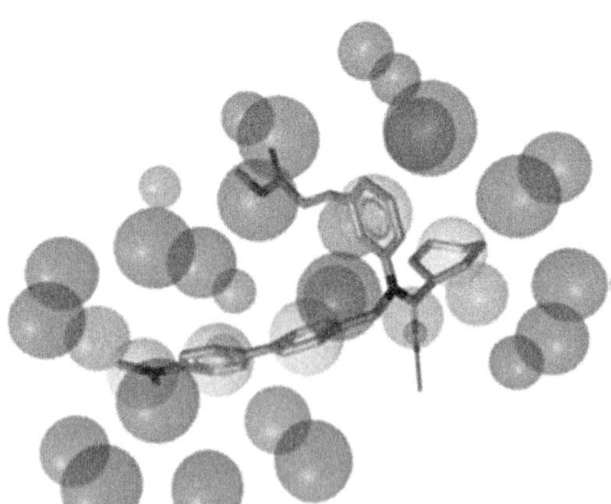

FIGURE 12.12 Pharmacophore model *1osh-1* comprising five hydrophobic features, 1 hydrogen bond acceptor with His294, and 27 exclusion volume spheres aligned with native ligand fexaramine.

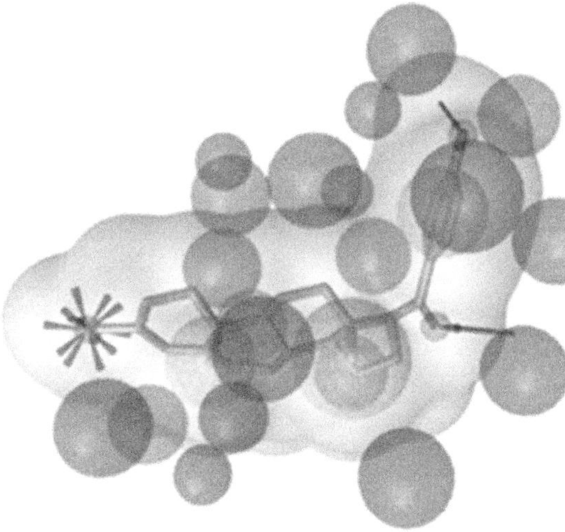

FIGURE 12.13 Shape-constrained pharmacophore model *3bej-1-s* comprising three hydrophobic features, 2 hydrogen bond acceptors anchoring the ligand with His294 and Thr288, a negatively ionizable feature representing the interaction with Arg331, and 25 exclusion volume sphere, aligned with native ligand MFA-1.

associated virtual hits derived from genus *Ganoderma* consist of a certain structural class of triterpenoids, namely, steroids. With the intention of further screening for more triterpenoids, the rational use of crystal structure of FXR co-crystallized with steroidal ligand, MFA-1 (PDB 3BEJ) was incorporated for another structure-based pharmacophore model (*3bej-1-s*) that comprised three hydrophobic features, two HBA anchoring the ligand with His294 and Thr288, negative ionizable feature representing the interaction with Arg331, 25 exclusion spheres, and a shape constraint (-s) to restrict hits to steric constraints (Fig. 12.13), which had rendered 10 more different hits. Five of the hits were identified to be secondary metabolites of *G. lucidum* and chemically described by an unsaturated lanostane scaffold with a double bond in position 8, which is in conjugation with one to two oxo groups in positions 7 and 11.

The authors of this study continued to evaluate all 25 previously isolated compounds from the fruit body of *G. lucidum* that were accessible at that time, which comprised of fatty acid derivatives and lanostane-type triterpines. Ultimately, the group found ergosterol peroxide (compound **2**), ganodermanontriol (compound **13**), and ganoderiol F (compound **14**) to have significant FXR-inducing activity (EC_{50} 0.855, 2.5, and 5.0 μM, respectively) (Table 12.2). Finally, with retrospective screening of the original pharmacophore models generated from crystal structure PDB 3BEJ for the 25 *Ganoderma* constituents, the pharmacophore model *3bej-2* rendered the highest number of hits and was therefore determined to be the best predictive model for triterpine compounds. It is similar to *3bej-1-s* except the ionic interaction with Arg331 is represented by a HBA instead of a negatively ionizable feature (Fig. 12.14).

Triterpines derived from loquat (*Eriobotrya japonica*) have also been identified for their ability to inhibit 11β-hydroxysteroid dehydrogenase 1 (11β-HSD1), which

TABLE 12.2 Chemical structures of tested *G. Lucidum* constituents

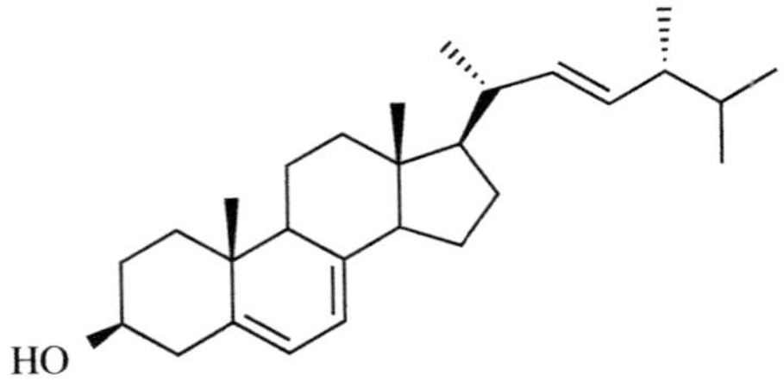

Ergosterol (**1**)

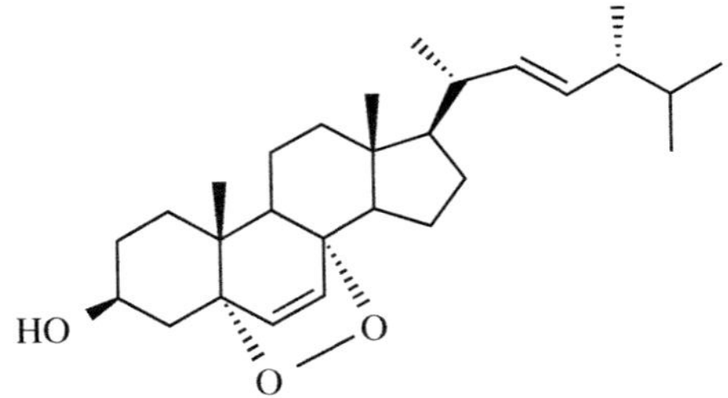

Ergosterol peroxide (**2**)

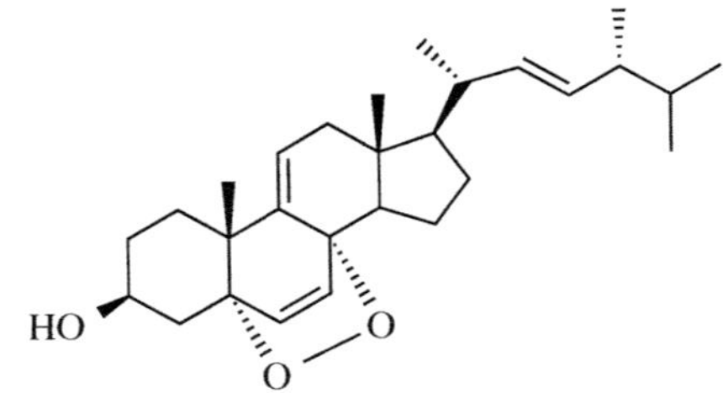

(22E,24R)-5α,8α-epidioxyergosta-6,9,22-trien-3β-ol (**4**)

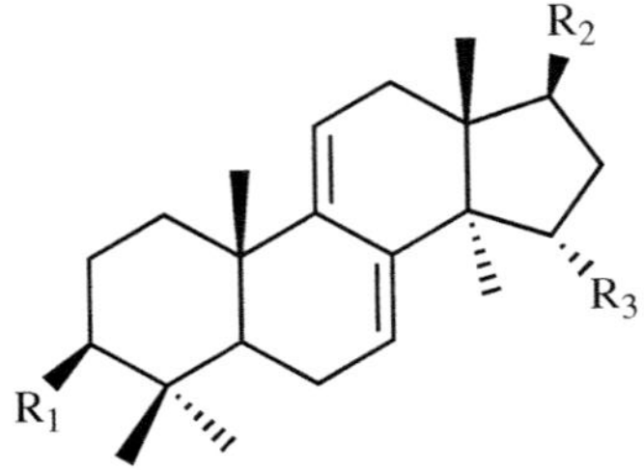

Ganoderol B (**3**)

	R1	R2	R3
Ganoderol B (**3**)	OH	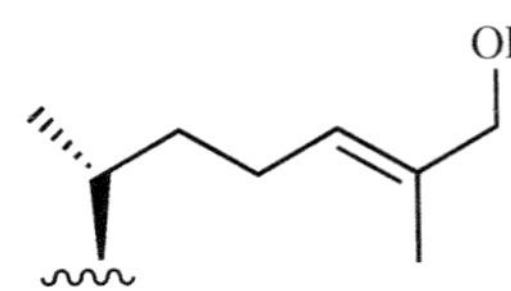	H

Ganodermanondiol (**9**)	O	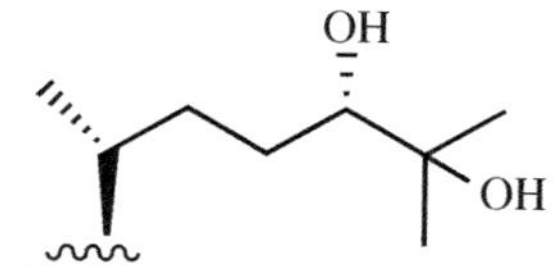	H
Ganoderic acid (**12**)	O	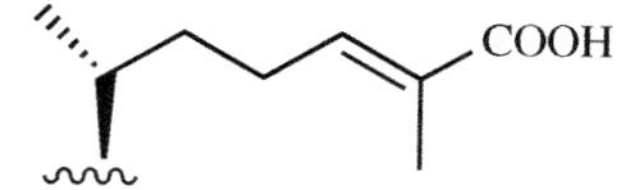	H
Ganodermanontriol (**13**)	O	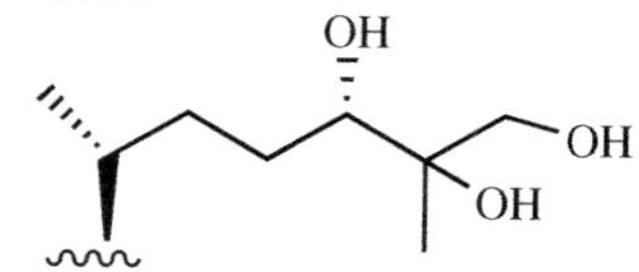	H
Ganoderiol F (**14**)	O	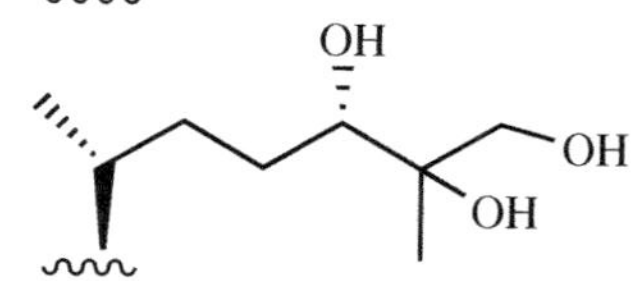	H

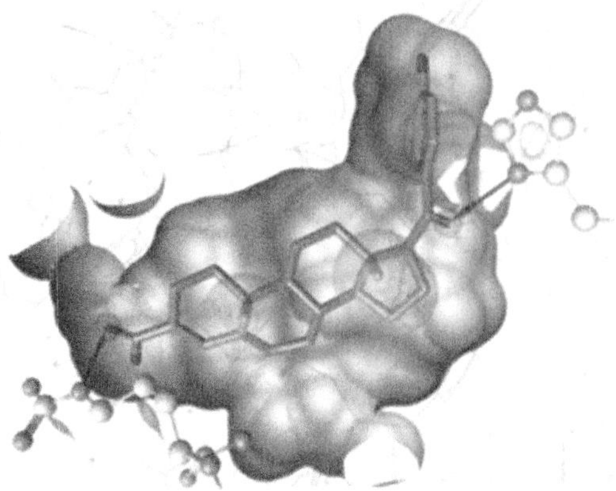

FIGURE 12.14 Best-fitting FXR pharmacophore model *3bej-2* for *Ganoderma* constituents investigated. Crucial interactions of native ligand MFA-1 with Arg331 and His447 are highlighted in ball-and-stick style.

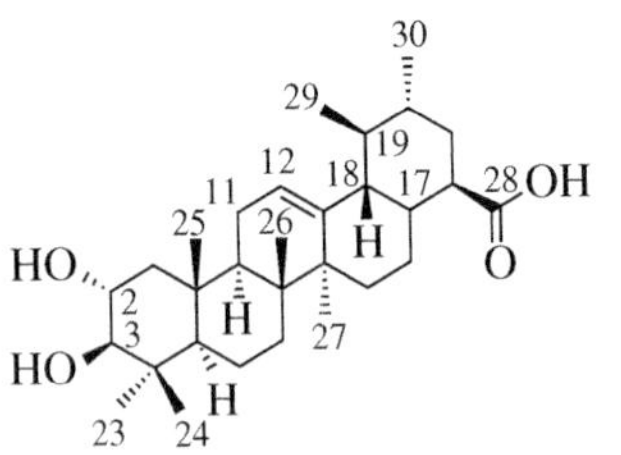

1

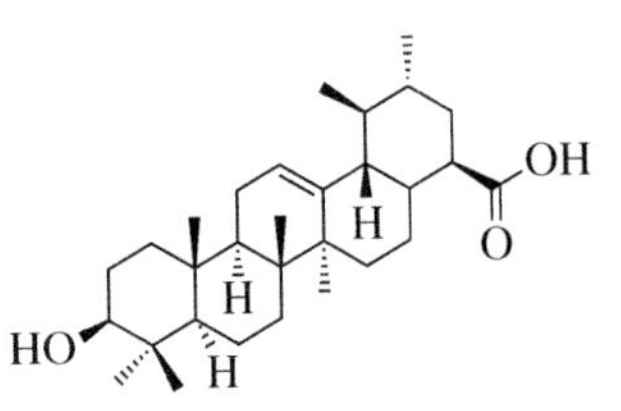

9

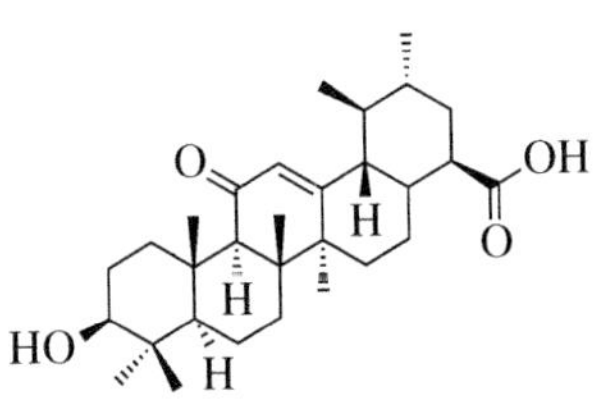

17

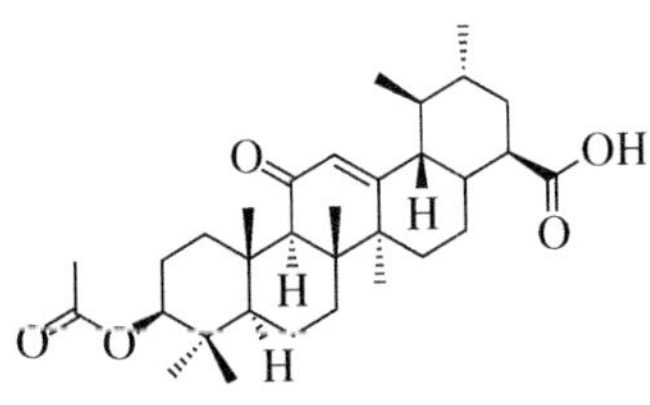

18

FIGURE 12.15 Chemical structures of the four most active triterpenoids, compound **1**, **9**, **17**, and **18**.

is implicated in the development of metabolic syndromes including obesity and Type 2 diabetes. Rollinger et al. [6] utilized a ligand-based pharmacophore model generated previously to virtually screen the *DIOS* database for 11β-HSD1 inhibitors. Corosolic acid (compound **1**, $IC_{50} = 0.81 \pm 0.06\,\mu M$) (Fig. 12.15) was among one of the highest scored triterpines and was shown to inhibit 11β-HSD in a dose-dependent

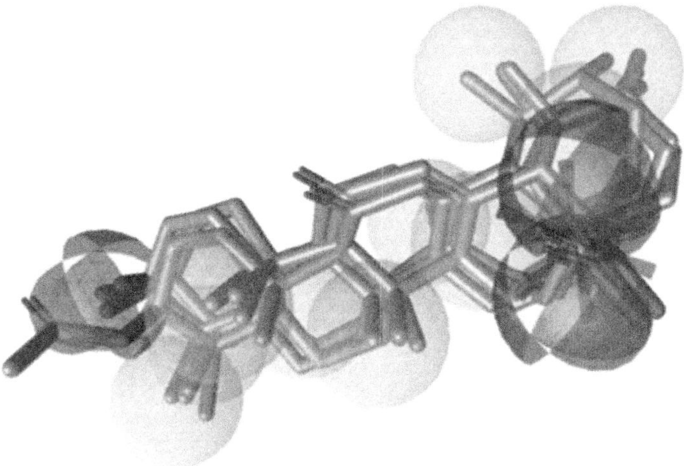

FIGURE 12.16 Merged features pharmacophore model derived from compounds **1**, **9**, **17**, and **18**. Chemical features include hydrophobic features (yellow) and hydrogen bond acceptors (red).

manner *in vivo*. It accounts for the bioactivity in the methanol leaf extract of *E. japonica*. The authors also noted that the dichloromethane (DCM) component of the extract, which contained only traces of corosolic acid, also showed activity. Bioassay-guided phytochemical investigations of the DCM component and biological evaluation of known triterpenic acids led to the identification of three additional active compounds from the resin of *Bursera delpechiana Poiss* (Indian linaloe): ursolic acid (**9**), 11-keto-ursolic acid (**17**), and 3-acetyl-11-keto-ursolic acid (**18**) (IC_{50} = 1.90 ± 0.25, 2.06 ± 0.44, and 1.35 ± 0.52 μM, respectively). A merged pharmacophore generated by the alignment of pharmacophoric points identified in the four active compounds contained 11 hydrophobic features, representing the triterpine core structure, a negatively ionizable feature, and two HBAs placed on the carboxylic acid structure on C28 and on the 3R position (Fig. 12.16), which supported observations from the docking experiments involving a crystal structure of 11β-HSD1 (PDB 2BEL).

Bhadauriya et al. [7] used ligand-based pharmacophore models for virtual screening of the Natural Product Database, among other compound databases to design dual inhibitors of acetyl-CoA carboxylase isoform 1 and 2 (ACC -1 and -2, respectively), which is implicated in obesity-induced diabetes. The pharmacophore model for each ACC isoform was generated using a large number of known ACC inhibitors (n = 111 for ACC1, n = 109 for ACC2) and extensively validated using methods including cost analysis, test set prediction, Fischer randomization, and decoy testing. Hit compounds used from initial screening using pharmacophore model for ACC1 (*Hypo1_ACC1*) were filtered by satisfying criteria as determined by further screening by ACC2 pharmacophore model (*Hypo1_ACC2*), Lipinki's

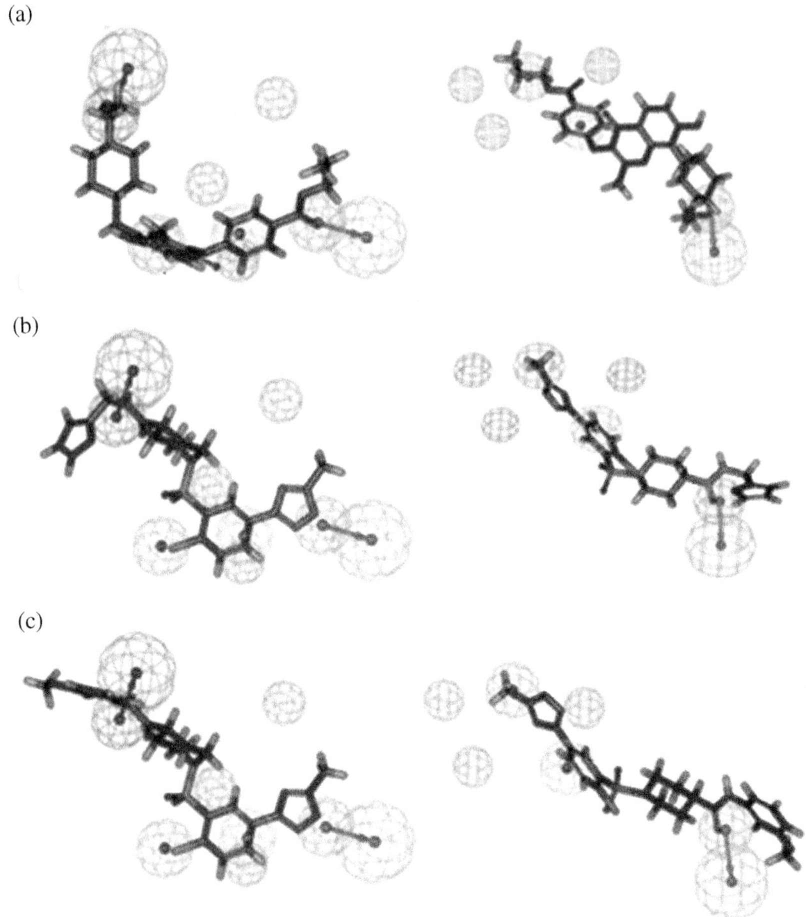

FIGURE 12.17 Alignment of top three virtual hits ZINC19924472 (a), ZINC33086616 (b), and ZINC33086598 (c) with best pharmacophore hypotheses for ACC1 (*Hypo1_ACC1*, left) and ACC2 (*Hypo1_ACC2*, right). Chemical features include hydrogen bond acceptors (green), hydrogen bond donors (magenta), hydrophobic features (blue), and exclusion volumes (gray). *For color details, please see color plate section.*

rule of five, drug-likeness (by QED values), and docking using X-ray crystal structure of human ACC2 (PDB 3FF6). *Hypo1_ACC1* consists of one hydrogen bond acceptor, one hydrogen bond donor, one hydrophobic aliphatic feature, one aromatic feature, and three excluded volumes, while *Hypo1_ACC2* is similar to *Hypo1_ACC1*, it lacks one hydrogen bond donor. Top three virtual hits (ZINC33086616, ZINC19924472, and ZINC33086598) were found with fit values of 6.67, 6.99, and 6.51, respectively, for *Hypo1_ACC1* and 7.25, 7.13, and 6.65, respectively, for *Hypo1_ACC2*. Mapping of these compounds aligned to both pharmacophore models are shown in Figure 12.17.

12.4 PHARMACOPHORE MODELS FOR ANTICANCER DRUGS

In the treatment of prostate cancer, Khanfar and El Sayed [8] have investigated the activity of steroidal alkaloids from the *Veratrum* species against Hedgehog (Hh) signaling. As no crystal structure existed for downstream effector Smoothened (Smo), which in turn transmits the signal to downstream effectors in the cytoplasm at the time; this work details results from a ligand based approach, using compounds that included cyclopamine (compound 1), jervine derivatives (**2–5**, **10**), and veratramine derivatives (**6–9**) formed the training set for the generation of ligand-based pharmacophore model (Fig. 12.18). The best pharmacophore model generated, *Hypo2*, consisted of 3 hydrophobic areas, 2 HBAs, 1 ionizable feature, and 23 excluded volumes and was validated by test set testing and receiver-operating characteristic (ROC) analysis (Fig. 12.19). The test set comprised known cyclopamine-derived Smo inhibitors and decoys from the *ZINC* database ($n=544$). *Hypo2* was shown to incorporate important interactions known to be required for this series of Hh signaling inhibitors and performed quite adequately in the ROC analysis (area under the curve, AUC = 0.848; overall false negative rate, FNR = 0.0021; overall true positive rate, TPR = 0.0625; overall specificity, SPC = 0.998; overall accuracy, ACC = 0.968).

Compounds	R_1	R_2	Other	Compounds	R	Other
Cyclopamine (1)	β-OH	H_2	$\Delta^{5,6}$	Veratramine (6)	β-OH	$\Delta^{5,6}$
Jervine (2)	β-OH	O	$\Delta^{5,6}$	Dihydroveratram-3-dione (7)	=O	5α-H
Pseudojervine (3)	β-O-glucoside	O	$\Delta^{5,6}$	8	=O	$\Delta^{4,5}$
Jervinone (4)	=O	O	$\Delta^{4,5}$	9	=O	$\Delta^{1,2}$, $\Delta^{4,5}$
Veratvirine (5)	β-OH	O	1α-OH, 5α-H			
Dihydrojervine (10)	β-OH	O	5α-H			

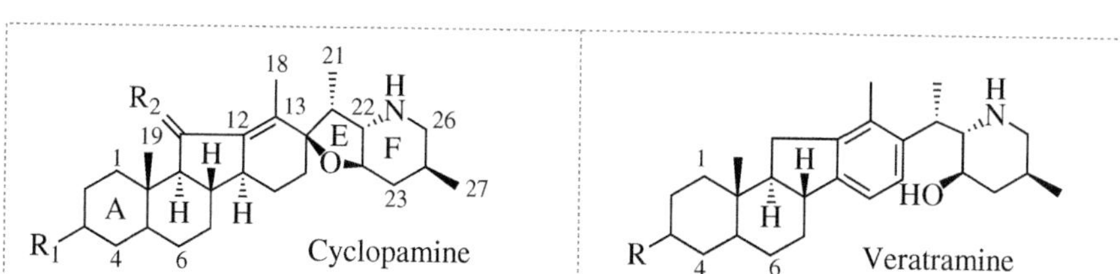

FIGURE 12.18 Chemical structures of natural, biocatalytic, and semis-synthetic *Veratrum alkaloids*.

(a)

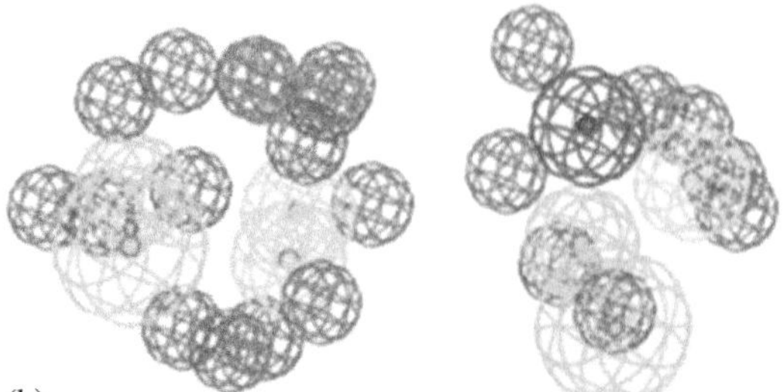

(b)

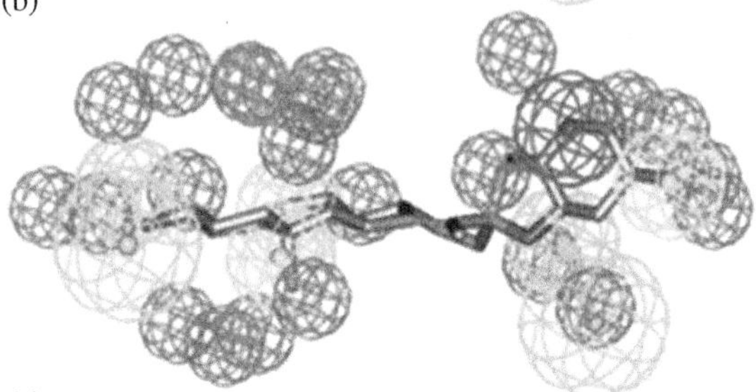

(c)

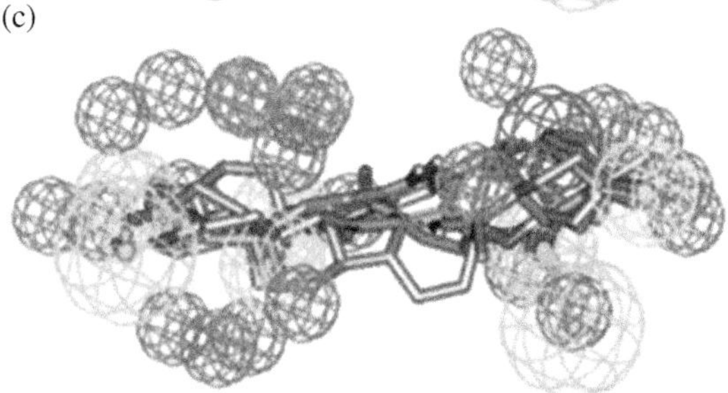

FIGURE 12.19 Pharmacophore model *Hypo2* (a), alignment with cyclopamine (b), alignment with active compounds 4, 6, 7, and 10. Chemical features include hydrogen bond acceptors (green), hydrophobic features (blue), positive ionizable features (red), and exclusion volumes (gray). *For color details, please see color plate section.*

Purushottamachar et al. [9] have also developed a ligand-based pharmacophore model using natural compounds, (−)-epicatechin, quercetin, curcumin, vitamin E succinate, and flufenamic acid (Fig. 12.20), which have been found to decrease androgen receptor (AR) protein expression (EC_{50} = 13.0, 25.0, 35.0, 38.0, and ~200 µM, respectively). *Hypo1* that incorporated two hydrogen bond acceptors, one hydrophobic feature and one ring aromatic, was found to be the best statistically, mapped all important features of active compounds, and showed some correlation between best fit values, conformational energies, and actual activities of training set compounds (Fig. 12.21). Using this model, the authors identified six AR down-regulating agents (ARDAs) from the Maybridge and National Cancer Institute databases: KM 06622, NCI-0001009, NCI-0002091, NCI-0002815, NCI-0004355, and BTB 01434 (Fig. 12.22). Interestingly, the most potent compound KM 06622

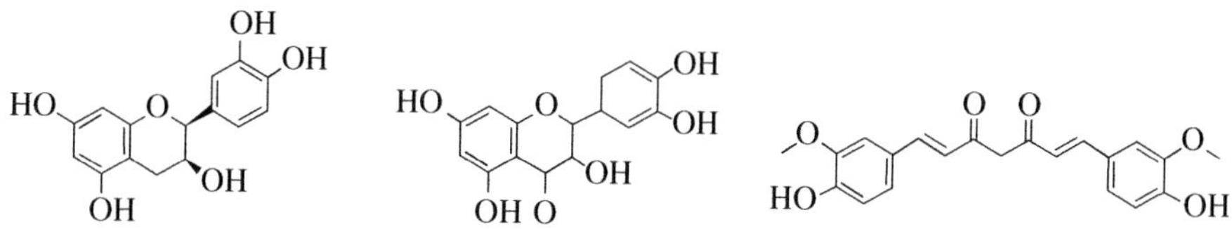

(−) Epicatechine
$EC_{50} = 13\,\mu M$

Quercetin
$EC_{50} = 25\,\mu M$

Curcumin
$EC_{50} = 35\,\mu M$

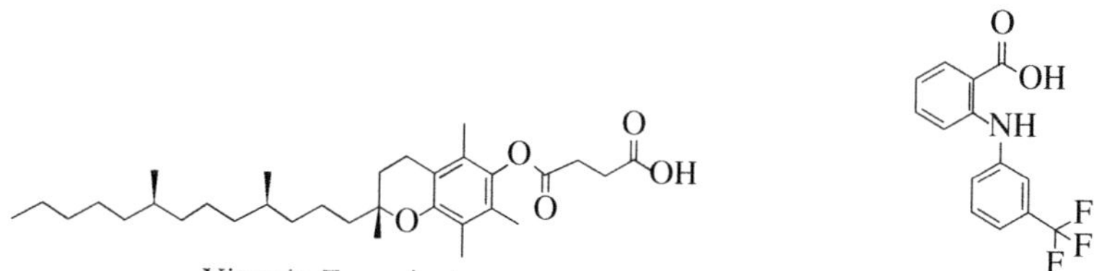

Vitamin E succinate
$EC_{50} = 38\,\mu M$

Flufenamic acid
$EC_{50} = 100–200\,\mu M$

FIGURE 12.20 Chemical structures of known androgen receptor down-regulating agents (ARDAs) used to generate the pharmacophore model *Hypo1*.

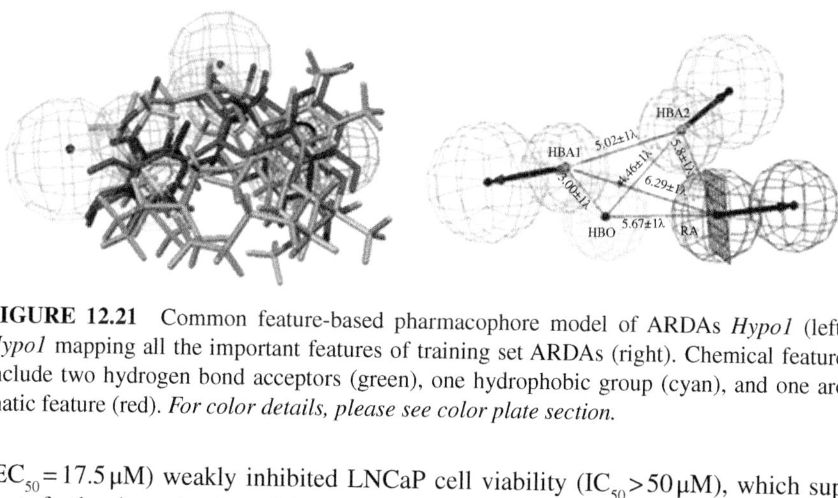

FIGURE 12.21 Common feature-based pharmacophore model of ARDAs *Hypo1* (left). *Hypo1* mapping all the important features of training set ARDAs (right). Chemical features include two hydrogen bond acceptors (green), one hydrophobic group (cyan), and one aromatic feature (red). *For color details, please see color plate section.*

($EC_{50} = 17.5\,\mu M$) weakly inhibited LNCaP cell viability ($IC_{50} > 50\,\mu M$), which supports further investigation of the mechanism by which ARDAs inhibit the growth of human prostate cancer cells and tumors (Table 12.3).

Extensive work has gone into gathering structure–activity relationships (SARs) of "caged" xanthones derived from the lowland rainforest tree of the *Garcinia* genus, which has been shown to selectively inhibit tumor cell proliferation and exhibit potent cytotoxicity for the purpose of elucidating a pharmacophoric motif [10, 11]. This family of natural products is characterized by a xanthone backbone in which the C ring has been converted into a tricyclic structure, whereby plant-specific substitutions and oxidations of this motif give rise to several subfamilies, such as morellins, gaudichaudiones, and gambogins. Chantarasriwong et al. [10, 11] established that

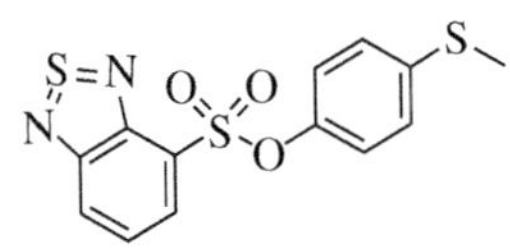

KM06622
3.94

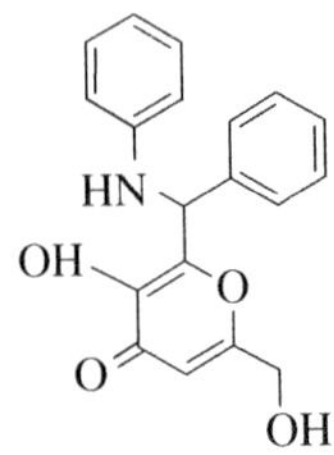

NCI0001009
3.70

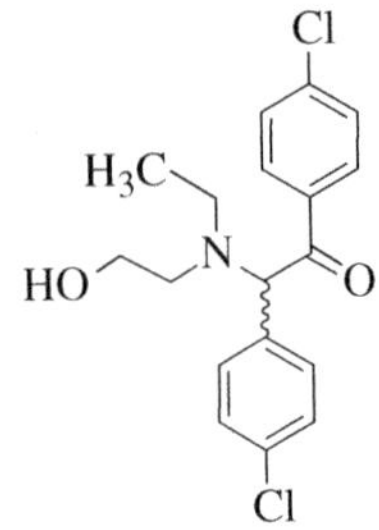

NCI0002091
3.89

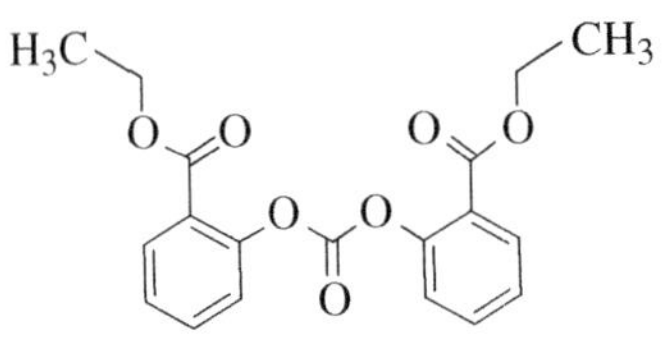

NCI0002815
3.95

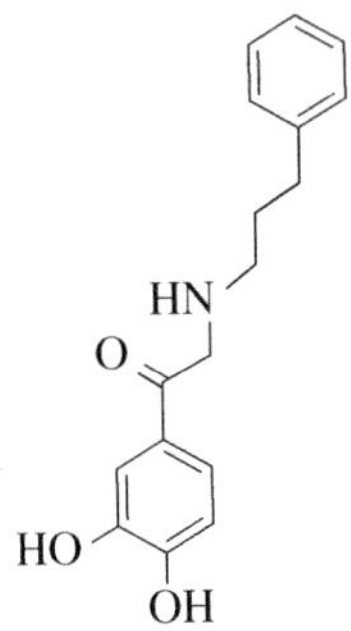

NCI0004355
3.94

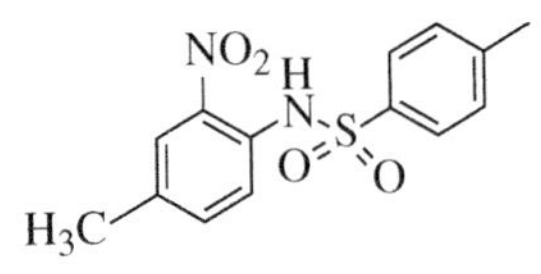

BTB01434
3.37

FIGURE 12.22 Chemical structures of six compounds, KM 06622, NCI-0001009, NCI-0002091, NCI-0002815, NCI-0004355, and BTB 01434, which were found to possess significant androgen down-regulating activities.

TABLE 12.3 EC_{50} and IC_{50} values for six compounds that were found to have significant androgen receptor down-regulating activities

Compound	EC_{50}^{a} (μM)	IC_{50}^{b} nμM)
NCI-0001009	212	20.9
NCI-0002091	39.5	8.31
NCI-0002815	65.5	4.5
NCI-0004355	43.5	26.9
BTB 01434	76	39.8
KM 06622	17.5	>50

[a] EC_{50} values represent the ability of the compounds to down-regulate AR protein expression.
[b] IC_{50} values are indicative of the effect of the molecules on LNCaP cell viability. All values are indicated as percent of control.

an intact ring system containing the C-ring structure is essential for bioactivity. The C9–C10 enone functionality was also shown to be important, whereas C5 prenyl group oxidization did not lead to a loss of activity and therefore be modified for optimal pharmacokinetics. It is worth mentioning that cluvenone (Fig. 12.23), a

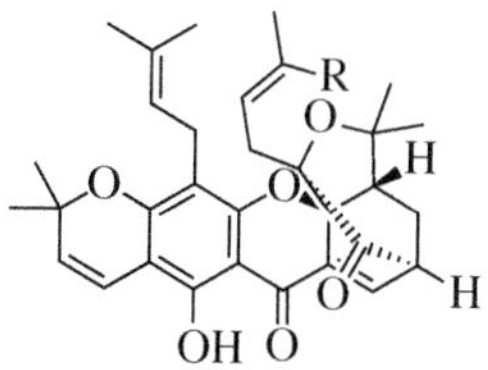

1: (R=CHO) morellin
2: (R=H) desoxymorellin

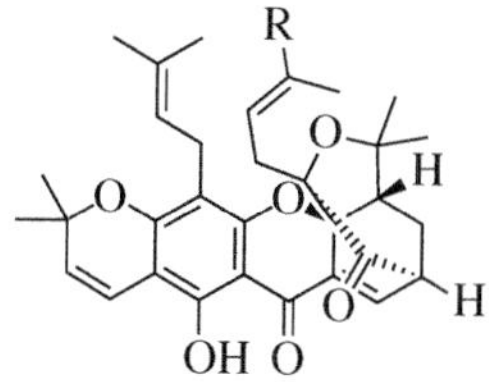

3: (R=CHO) gaudichaudione A
4: (R=H) desoxygaudichaudione A

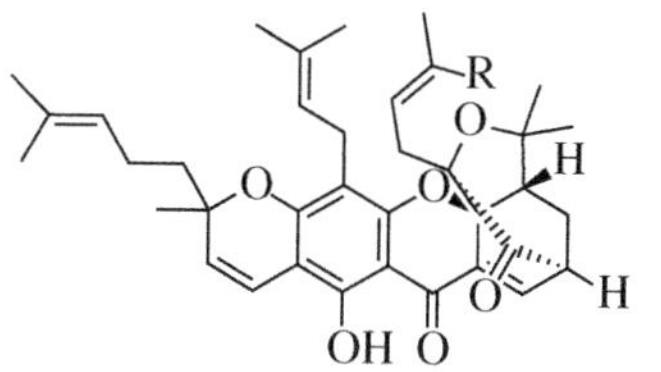

5: (R=CHO) gambogic acid
6: (R=H) gambogin

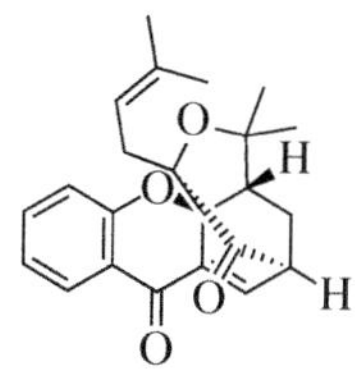

7: Cluvelone

FIGURE 12.23 Chemical structures of selected caged *Garcinia* xanthones.

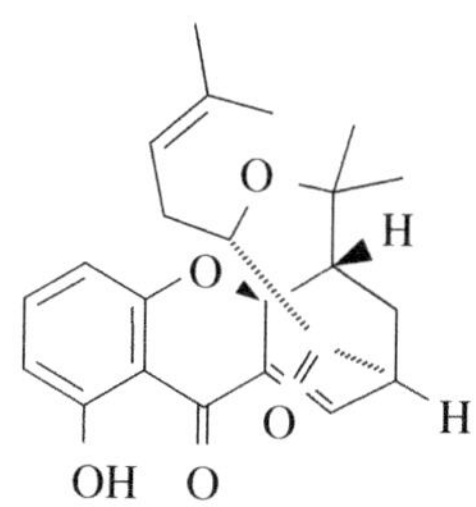

11a

FIGURE 12.24 Chemical structure of compound **11a**.

synthetic analogue of gambogic acid (GA) that has entered clinical trials in China, also has significant cytotoxicity at nanomolar concentrations against a broad range of tumor types including multi-drug-resistant leukemia, colon, breast, prostate, and renal cancers. In further SAR analyses, Wang et al. found that additional hydroxyl group at C1 of compound **11a** (Fig. 12.24) significantly increased antiproliferative activity in human breast carcinoma MCF-7, human gastric carcinoma BGC-823, human hepatocellular carcinoma SMMC-7721, and HepG2 cell lines [12].

Similarly, SAR studies on epoxomicin-derived small peptides have highlighted the importance of the α,β-epoxyketone pharmacophore for the nucleophilic attack of the Thr1 hydroxy groups (Thr1Oγ) that is necessary for 20S proteasome inhibition, from which YU101 and YU102 (Fig. 12.25) exhibit higher potency and caspase-like specific activity, respectively [13]. From secondary metabolites of *Salinispora*

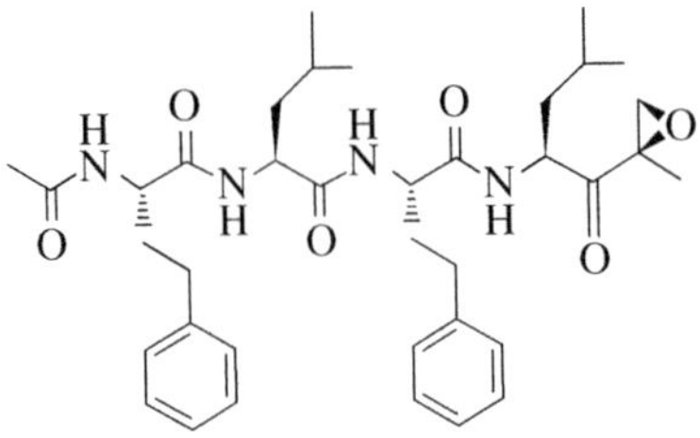

YU101 (14)

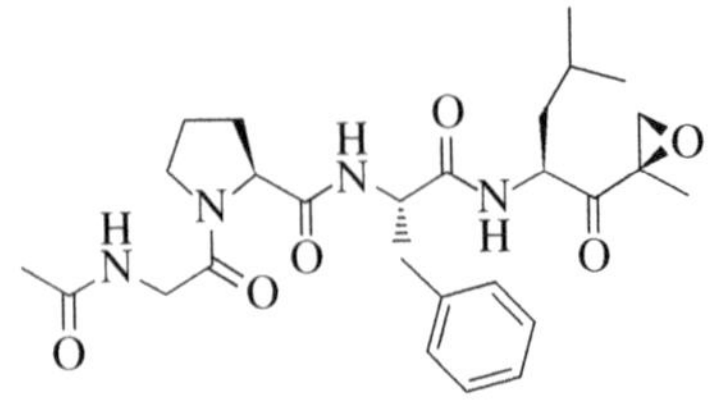

YU102 (15)

FIGURE 12.25 Chemical structures of synthetic **α,β**-epoxyketone proteasome inhibitors YU101 and YU102 that are selective to chymotrypsin-like and caspase-like activity, respectively.

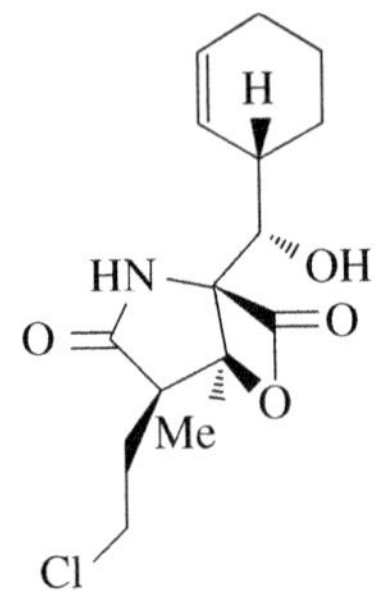

4

FIGURE 12.26 Chemical structure of salinosporamide A.

tropica such as cyanosporaside, saliniketal A, and sporolide A, a γ-lactam-β-lactone structural motif was revealed to be well suited to nonpeptidic natural proteasome inhibitors. In particular, salinosporamide A (Fig. 12.26) delivers IC_{50} values in the low nanomolar range in caspase-like, trypsin-like, and chymotrypsin-like inhibition assays (2.6 ± 0.2, 21 ± 3, and 430 ± 60 nM, respectively) [14].

12.5 PHARMACOPHORE MODELS FOR ANTI-INFLAMMATORY DRUGS

A novel method termed "pharmacophore-based parallel screening" (PS), involving a collection of structure-based pharmacophore models, and therefore accounting for various possible binding modes, has been developed and utilized by Schuster et al. for the investigation of cyclooxygenase (COX) inhibitors present in the "Prasaplai" formulation used in traditional medicine derived from Thai cultures [15, 16]. The development of the collection of pharmacophore models for PS initially involved the generation of 39 structure-based pharmacophores from 11 X-ray crystal structures of COX-1 and -2 isoforms bound to inhibitors, which was extensively validated by a number of test sets (WDI_{2005}, $n = 67{,}050$; WDI_{MA}, $n = 20, 994$; most commonly used COX-inhibitor test set, *47-COX-DB*, $n = 47$) and decoy testing (enrichment factor,

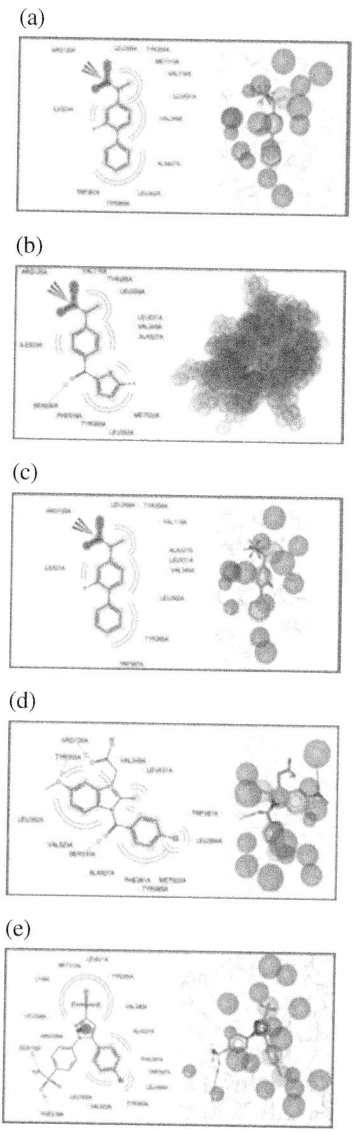

FIGURE 12.27 Observed interactions between 3D X-ray crystal structures with its native ligand (left) and its corresponding structure-based pharmacophore model (right), which form the final model collection for virtual screening: *1cqe-1* (a), *1pge-2-s* (b), *2ayl-1* (c), *4cox-2* (d), and *6cox-1-s* (e). *For color details, please see color plate section.*

EF ≥ 3.75). The elimination of multiple models originating from same X-ray crystal structure show the same % hit retrieval from *47-COX-DB* or comparable EFs in the DUD dataset prediction, ultimately revealed the final model collection (EF = 10.55), which consisted of five models: *1cqe-1*, *1pge-2-s*, *2ayl-1*, *4cox-2*, *6cox-1-s* (Fig. 12.27).

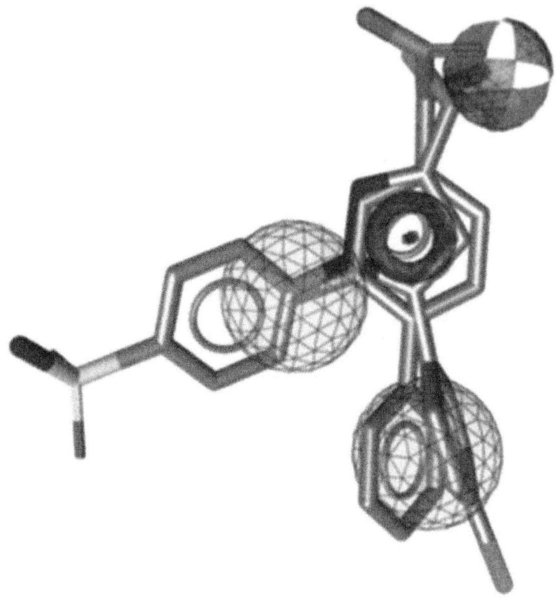

FIGURE 12.28 Ligand-based pharmacophore model for COX inhibitors based on (S)-flurbiprofen and SC-558. Chemical features include hydrogen bond acceptor (tiled sphere), hydrophobic features (full sphere), and an aromatic ring (ring).

An even less restrictive, ligand-based pharmacophore model was generated by the alignment of (S)-flurbiprofen and SC-558 (79% *47-COX-DB* hit retrieval rate vs. 53% achieved by all 39 structure-based pharmacophore models) and used in conjunction with the final model collection to investigate the so-called Prasaplai database (*Prasaplai DB*) (Fig. 12.28). Prasaplai is known to be used for primary dysmenorrhea, which is associated with a significant increase in prostaglandin (PG) release and is a formulation of 12 ingredients: roots of *Acorus calamus* L., bulbs of *Allium sativum* L., pericarps of *Citrus hystrix* DC., rhizomes of *Curcuma zedoaria* Roscoe, bulbs of *Eleutherine americana* (L.) Merr., seeds of *Nigella sativa* L., fruits of Piper *chaba* Hunt, fruits of *Piper nigrum* L., rhizomes of *Zingiber cassumunar* Roxb., rhizomes of *Zingiber officinale* Roscoe), sodium chloride, and camphor. Waltenberge et al. [16] found 25 (out of 166) hits from constituents that are derived from five Prasaplai ingredients, that is, *Acorus calamus*, *Nigella sativa*, *Piper nigrum*, and *Zingiber officinale*, that warranted *in vitro* testing. Testing of all known constituents of the five ingredients mentioned revealed an average rate of correct prediction of 56% using all six pharmacophore models. The number and percentage of virtual hit predictions of compounds from each ingredient is shown in Figure 12.29.

Ehrman et al. [17] have also utilized PS incorporating multiple pharmacophore models of different protein targets, namely, COX-1, COX-2, p38α, c-Jun terminal-NH2 kinase 1 (JNK-1), and type 4 cAMP-specific phosphodiesterase B and D

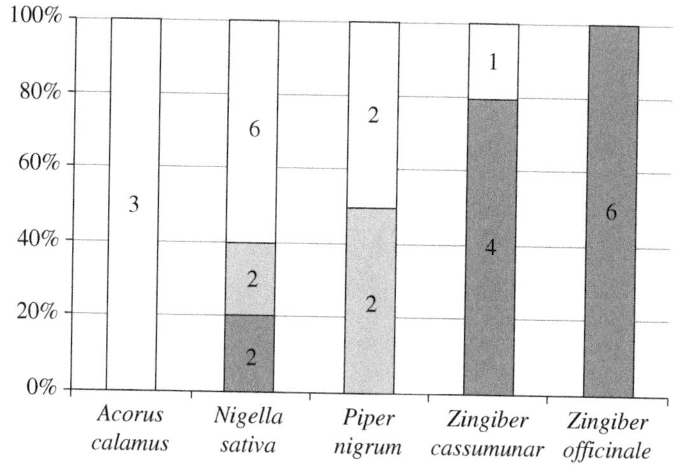

FIGURE 12.29 Number of percentage of virtual hits that are classified as highly active (IC_{50}<25.0 μM), moderately active (IC_{50} = 25.0–150.0 μM), and inactive (IC_{50}>150.0 μM).

(PDE4-B and -D, respectively), to elucidate herbal medicines with multitarget anti-inflammatory activity. A total of 58 structure-based pharmacophore models were used to identify Lipinski's drug-like compounds from the Chinese Herbal Constituents Database (CHCD). Hit compounds for each model were ranked by docking scores rendered from virtual docking with respective crystal structures and categorized by phytochemical class within each herb. For each target, the authors analyzed the top 50 herb/class groups containing the highest number of hits that render better docking scores than the median value rendered by co-crystallized ligands. Generally, smaller terpenoids (sesqui- and monoterpenes), simple phenolics, coumarins, flavonoids, and lignans are represented across all targets (Fig. 12.30). Compounds from various phytochemical classes that show activity against three or more targets include lithospermidin A, andrographolide, capillartemisin B, loganin, precatorin I, xanthumanol, magnolignan D, 3-hexenyl glucoside, gancaonin R, cnidioside C, regaloside A, and platydesmine (Fig. 12.31). The authors noted hydrophobic prenyl groups such as those present in capillartemisin B, xanthumanol, and cnidioside C were recurring chemical features in higher ranking hits. Similarly, small glycosides, such as those in loganin, precatorin I, 3-hexenyl glucoside, cnidioside C, and regaloside A, were noted to confer greater capacity for hydrogen binding in many monoterpenes and simple ring phenolics.

In the interest of sparing the suppression of prostanoids with homeostatic functions, such as gastro-protection, unlike conventional nonsteroidal anti-inflammatory drugs (NSAIDs) that target COX enzymes, various research groups have investigated the selective interference of PG E2 production by microsomal PG E2 synthase-1 (mPGES-1), an inducible enzyme that is upregulated at various pathophysiological phases. Through the use of previously developed pharmacophore models (*M1* and *M2*), Bauer et al. [18] have identified active phenolic compounds from the Chinese

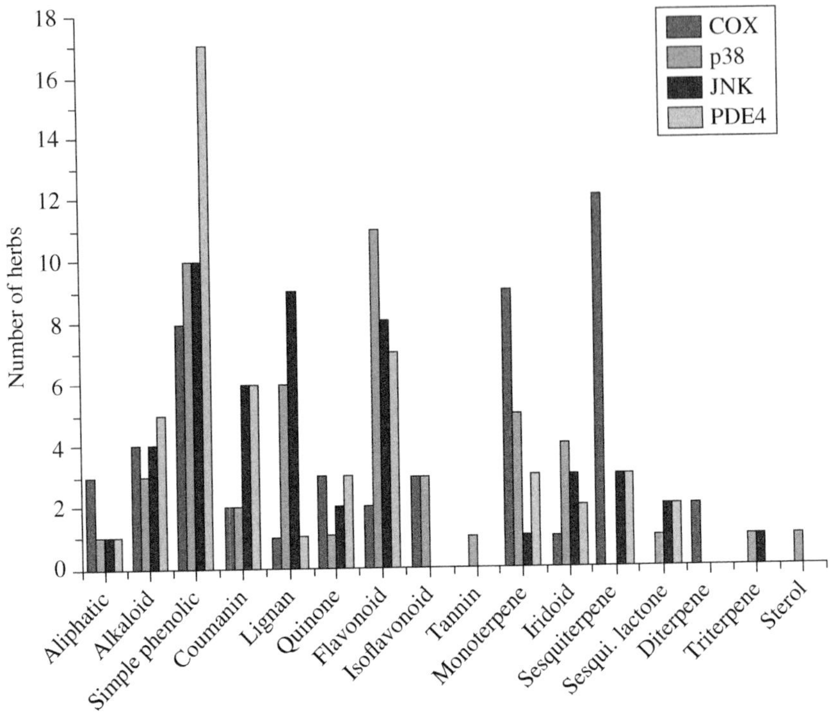

FIGURE 12.30 Distribution of phytochemical classes from herbs containing predicted inhibitors against COX, p38, JNK, and PDE4.

Herbal Medicine database, namely, depsides from lichen: physodic acid (**2**), perlatolic acid (**8**), and olivetoric acid (**9**) (Fig. 12.32). *In vivo*, these compounds were shown to have IC_{50} values of 0.43, 0.4, and 1.15 µM, respectively; moderate inhibition of COX-1 (IC_{50}>30 µM); and no significant inhibition of COX-2. *M1* consisted of a negatively ionizable feature, one aromatic ring feature, four hydrophobic features, and a shape restriction. *M2* was a partial query of *M1* where the aromatic ring feature or one hydrophobic feature is allowed to be omitted (Fig. 12.33). The authors noted that compounds require a free acidic group and hydrophobic constituents at position 6 or 6′ for potent activity.

From the TCM *Database@Taiwan*, Chen et al. [19] have identified other potential inhibitors of mPGES-1 shanciol A and B from the pseudobulb of *Appendiculate cremastra*, castilliferol from *Centella asiatica*, and *Aurantiamide acetate* from *Daphne genkwa* (Fig. 12.34). The pharmacophore model utilized was generated using a training set of known mPGES-1 inhibitors ($n = 34$) and validated by cost analysis and Fischer randomization. Pharmacophore features identified one HBD, one hydrophobic feature, and two aromatic rings for pi-interaction (Fig. 12.35).

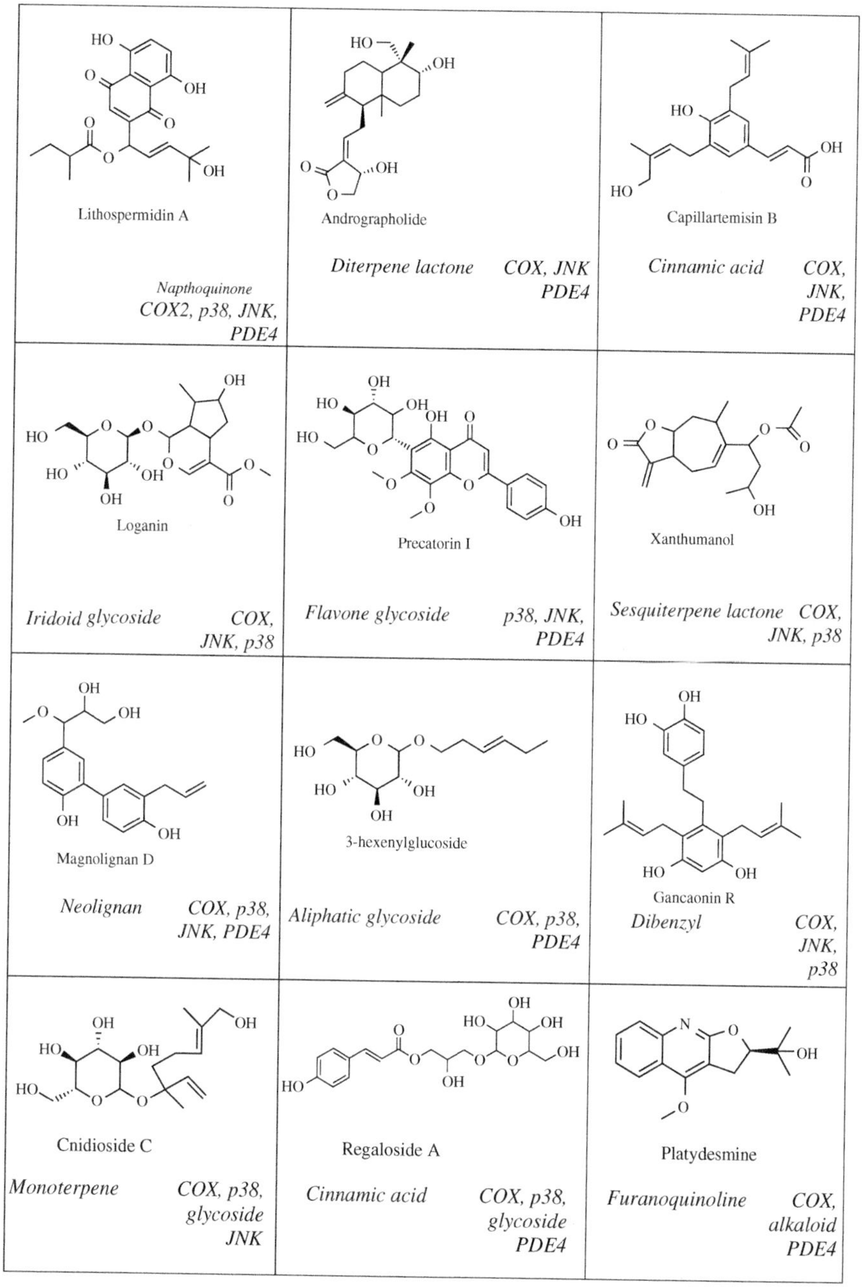

FIGURE 12.31 Some predicted multi-target inhibitors from Chinese herbs. Phytochemical class and targets are shown in the bottom left and right corners of each structure, respectively.

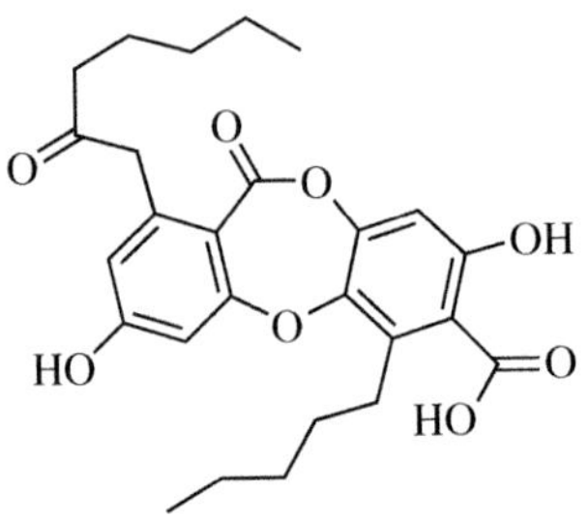

Physodic acid (2)

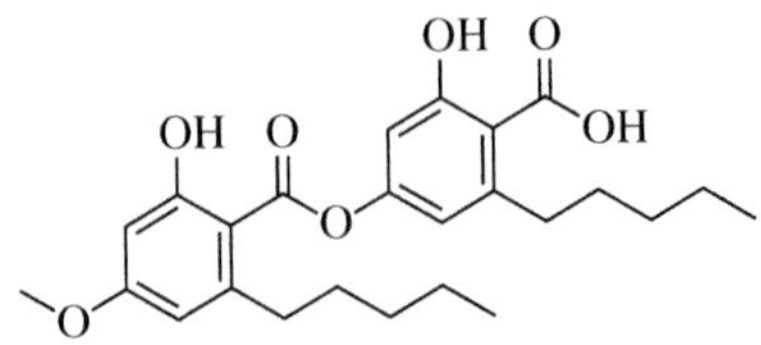

Perlatolic acid (8)

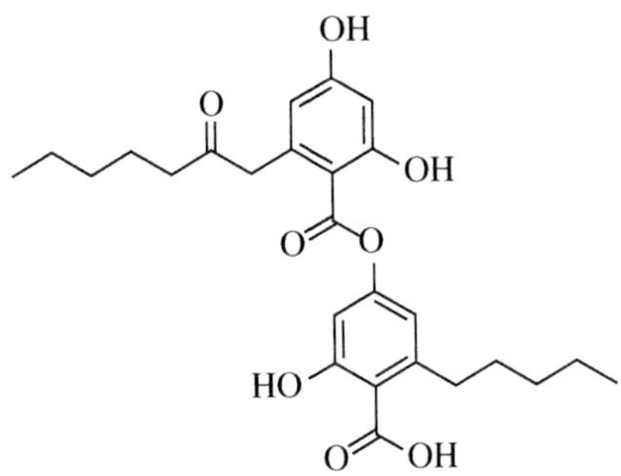

Olivetoric acid (9)

FIGURE 12.32 Chemical structures of active depsidones: physodic acid, perlatolic acid, and olivertoric acid.

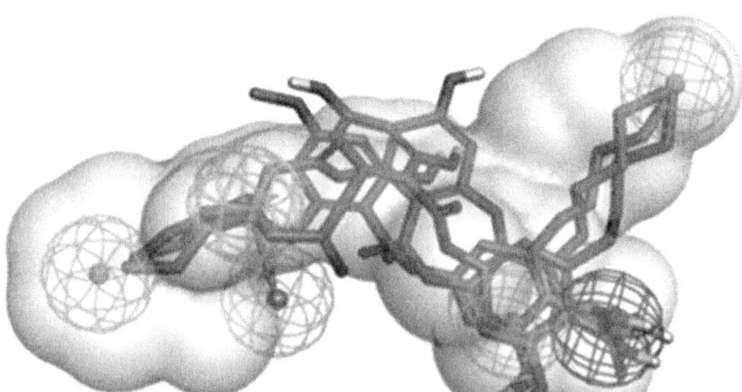

FIGURE 12.33 Pharmacophore model for acidic mPGES-1 inhibitors. For *M1*, it is necessary to satisfy all chemical features consisting of four hydrophobic features (cyan), one aromatic ring (gold), one negatively ionizable feature (blue), and a spatial shape restriction (gray). Whereas screening with *M2* allows the omission of one hydrophobic group or aromatic ring features, inhibitors of mPGES-1, namely, 2 (green), 8 (blue), and 9 (gray) map two of the hydrophobic features with their alkyl chains. *For color details, please see color plate section.*

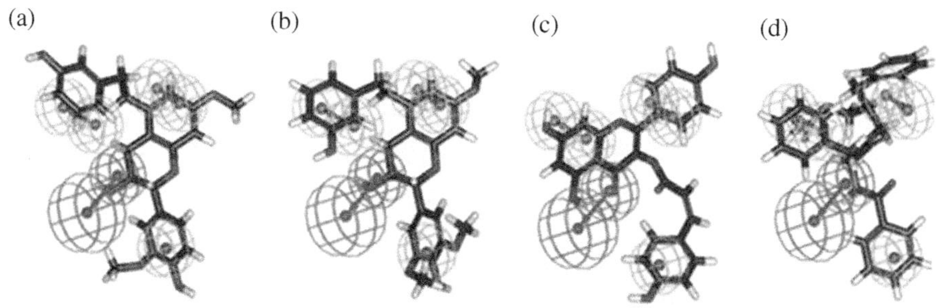

FIGURE 12.34 Pharmacophore mapping for (a) shanciol A and (b) shanciol B, (c) castilliferol, and (d) aurantiamide acetate. Chemical features include one hydrogen bond donor (purple), one hydrophobic group (light blue), and two aromatic rings (orange). *For color details, please see color plate section.*

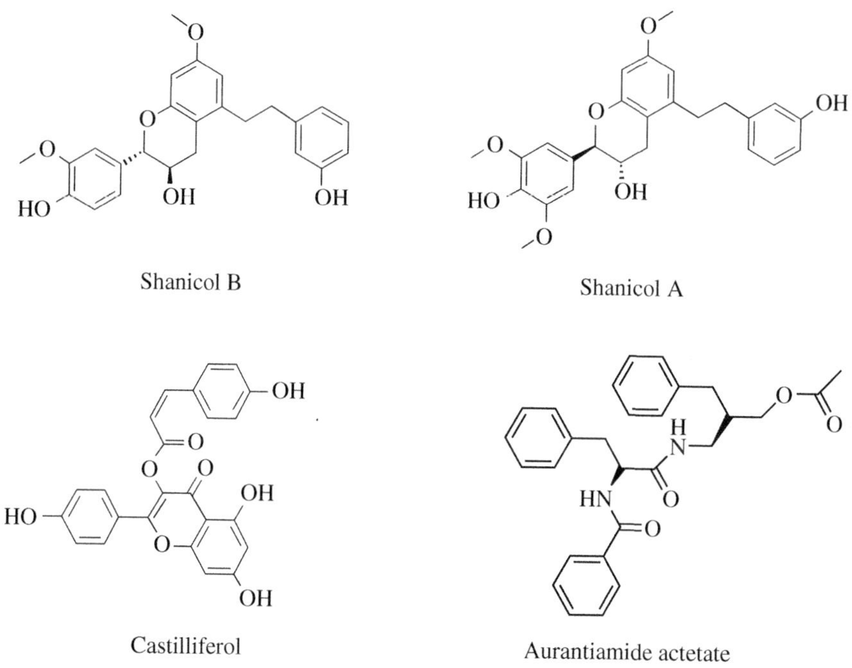

FIGURE 12.35 Chemical structures of shanciol A and B, castilliferol, and aurantiamide acetate.

12.6 PHARMACOPHORE MODELS FOR ANTI-INFECTIVE DRUGS

In the search of non-nucleoside reverse transcriptase inhibitors (NNRTIs) of HIV-1, Liu et al. [20] have identified a flavone (**M4753**) (Fig. 12.36) from the Traditional Chinese Medicine Database using a ligand-based pharmacophore model characterized by second-generation NNRTIs. Virtual docking revealed the importance of a butterfly-like shape constraint and hydrophobicity, which are both features observed

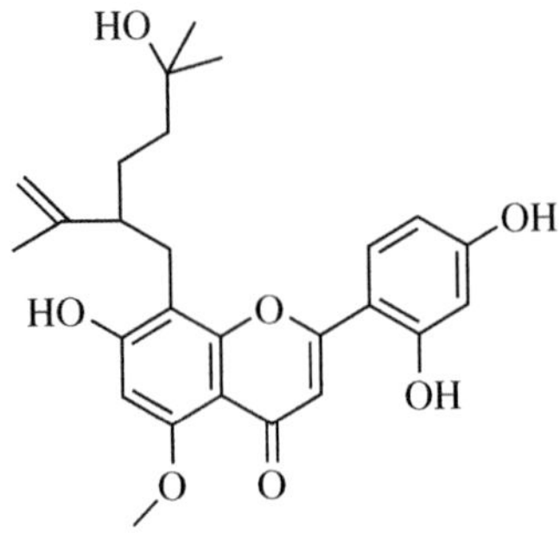

M4753

FIGURE 12.36 Chemical structure of flavone M4753.

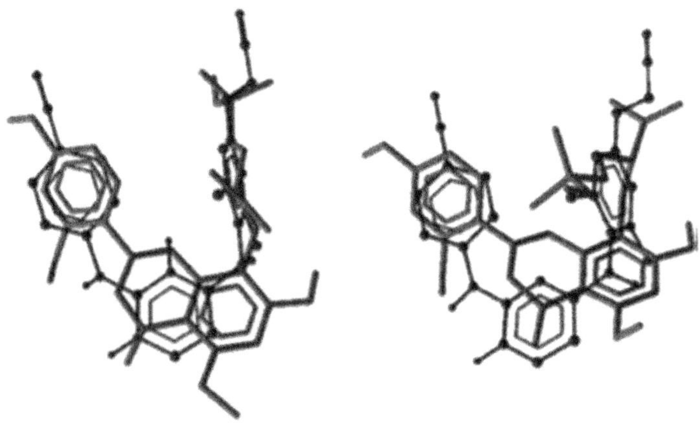

FIGURE 12.37 Superposition of M4753 (red) and native ligands (blue) co-crystallized with WT (PDB 2ZD1, left) and mutant HIV-1 RT (PDB 3BGR, right).

with the native ligand co-crystallized in the wild-type and mutant crystal structures of HIV-1 RT (PDB 2ZD1 and 3BGR, respectively) (Fig. 12.37). Further MD showed stability of **M4753** binding to the NNI-binding pocket of HIV-1 RT.

Pharmacophore-based screening has also been utilized by Lee et al. [21] to identify four flavonoids (YKAF01, YKAF02, YKAF03, and YKAF04) (Fig. 12.38) from the *BIOMOL International LP* and *Indofine Chemical Company*, with the aim of inhibiting β-ketoacyl-acyl carrier protein synthase III (KAS III), which in turn inhibits fatty acid biosynthesis in bacteria, namely, *Staphylococcus aureus* and *methicillin*-resistant *Staphylococcus aureus* (MRSA). Screening involved two pharmacophore models (*Map I* and *III*) generated previously. *Map I* represented the binding model of CoA-KAS III, which featured two HBDs for the binding to the backbone oxygen of Phe304 and Gly209 and one lipophilic feature for interaction with Ile156, Phe157, and Met207. Map *III* highlighted interactions between KAS III and a known inhibitor that is characterized by one HBA to form a H-bond with Arg36; two lipophilic

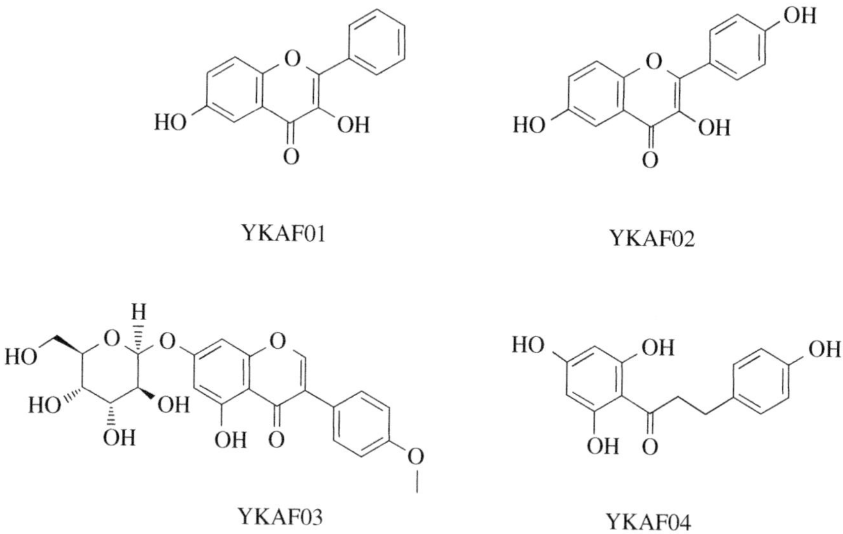

FIGURE 12.38 Chemical structures of most active flavonoids YKAF01, YKAF02, YKAF03, and YKAF04.

features for hydrophobic interactions with Phe157, Ile156, Leu189, and Met207; and a shape constraint (Fig. 12.39). Binding affinity, antibacterial and tolerable cytotoxicity was demonstrated for YKAF01 and YKAF04 *in vivo*. YKAF02 and YKAF03 did not display any antibacterial activity against tested strains. MIC for YKAF01 showed against gram-positive bacteria *S. aureus* and MRSA was 16 μg/ml. YKAF04 also showed activity against *S. aureus* and MRSA (MIC = 32, 16 μg/ml, respectively), as well as activity against gram-negative bacteria *Escherichia coli*.

12.7 PHARMACOPHORE MODELS FOR NEUROLOGICAL DRUGS

Previous findings from a study conducted by Sim and Chua [22] and preliminary studies support the use of morphinan-based compounds as potential inhibitors of acetylcholinesterase (AChE) in the treatment of Alzheimer's disease [22]. As a result, Schuster et al. [23] have developed a ligand-based pharmacophore model generated exclusively from biologically tested morphinan-based compounds found from an in-house database ($n = 416$). This model was shown to be more effective for the identification of potential morphinan-based AChE inhibitors in comparison to a structure-based pharmacophore model generated from the crystal structure of *Torpedo californica* AChE (PDB 1QTI). Based on the retrieval of highly active compounds from a library of 481 tested morphinan and isoquinolines (>80% AChE inhibition), enrichment factors for rigid search by ligand- and structure-based pharmacophore models are 1.56 and 5.37, respectively. Twelve morphinans and isoquinolines were identified to be highly active compounds, that is, compounds **5–9**, **25–27**, **33**, **34**, **37**,

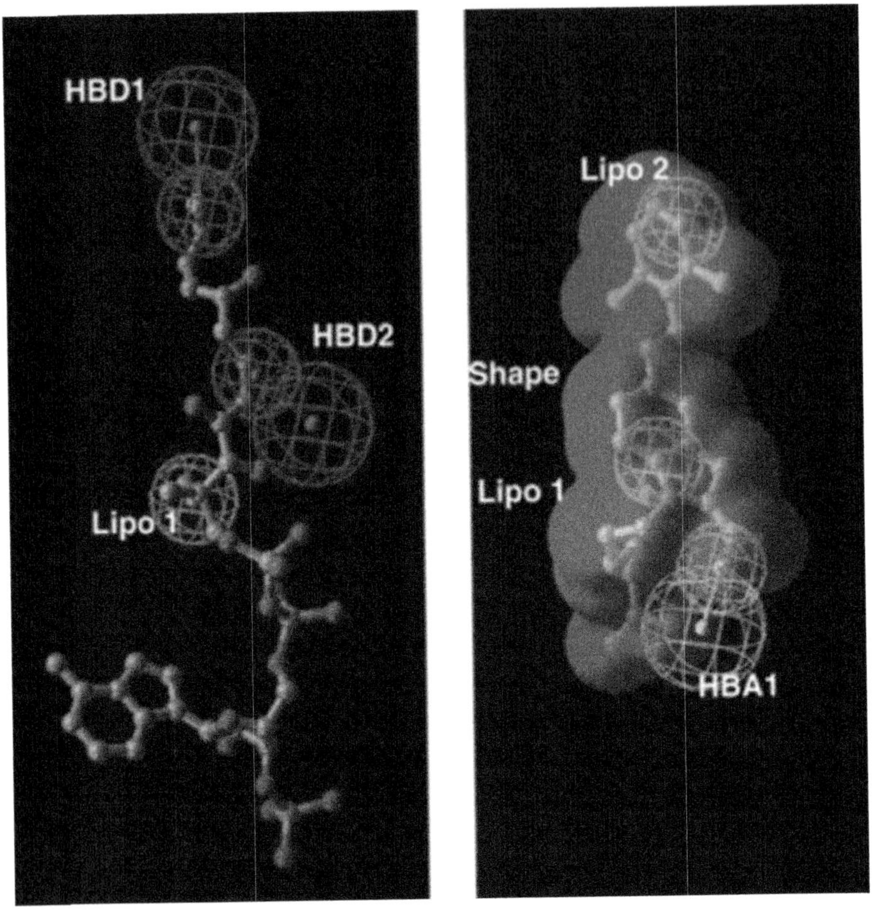

FIGURE 12.39 Pharmacophore maps, previously determined using receptor-oriented pharmacophore-based *in silico* screening.

and **43** (Fig. 12.40). Further opioid receptor binding assays of these compounds reveal compounds **9**, **33**, **34**, **37**, and **43** to show low or no binding to μ, δ, nor κ opioid receptors (K_i > 10,000 nM).

In search of new anticonvulsants from the *InterBioScreen Natural Compounds Collection*, Gavernet et al. [24] employed a conservative approach by ensuring satisfactory pharmacophore mapping of hit compounds, that is, RMS < 0.200 upon superposition with pharmacophore template, following 2D virtual screening based on discriminant function (df) and suitable ADME filters (i.e., Lipinski's rule of five, Veber rules, and Moriguchi log *P* value between 1.0 and 3.0). Finally, energy difference between the global minimum and the active conformation was calculated for the seven structures identified by pharmacophore matching (Fig. 12.41), and four compounds showing small energy difference (<7 kcal/mol) were considered for *maximal electroshock*

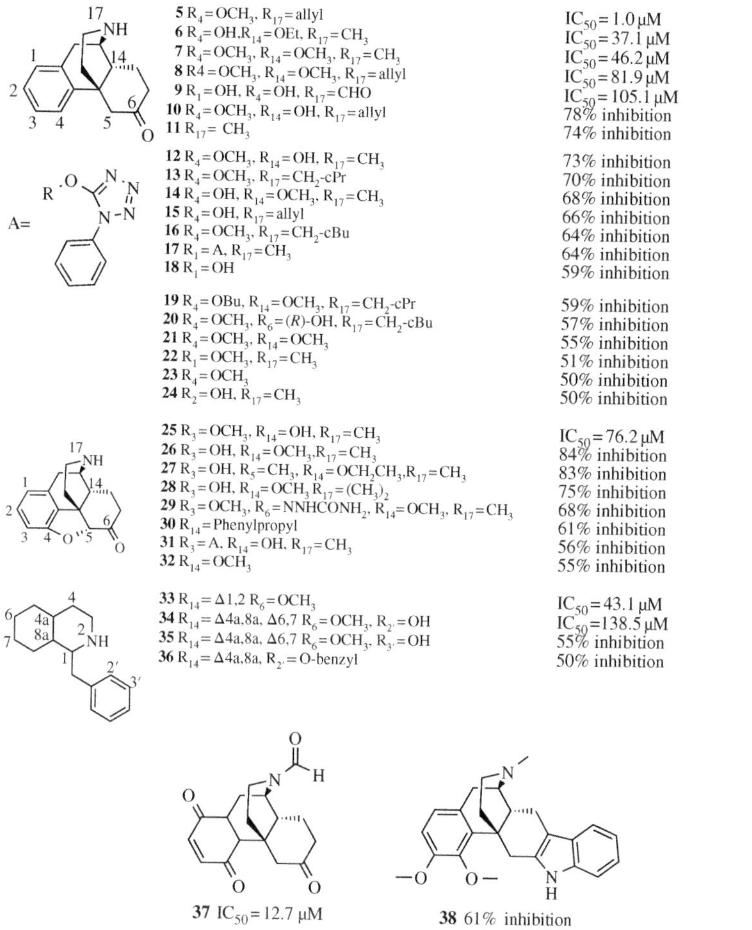

FIGURE 12.40 Chemical structures of highly active morphinans and isoquinolines.

seizures (MES) and RotoRod tests. The compound 2-(4,6-dimethyl-1-benzofuran-3-yl)-acetic acid shows anticonvulsant activity in mice at 30 and 100 mg/kg (after 0.5 and 4 h) without signs of neurotoxicity. The pharmacophore identified for anticonvulsant activity in the MES test is shown in Figure 12.42.

Chen [25] employed pharmacophore analysis to investigate the interaction between compounds derived from Chinese herb Semen *ziziphi spinosae* (suanzaoren), including betulin, betulin acid, jujuboside A, jujuboside B, jujubogenin, *cis*-ebelin lactone, and *trans*-ebelin lactone (Fig. 12.43), with GABA-A at the α1/γ2 interface. The pharmacophore comprised of HBAs that corresponded to NH1 and NH2 of γ2-Arg132 and γ2-Arg138 and NZ of γ2-Lys184, and HBDs from the O and OH on α1-Tyr 160 and O on α1-Val203. Electronegative atoms on the C=O group, tetrahydro-2H-pyran group, and dihydrofuran-2-(3H)-one group contained strong

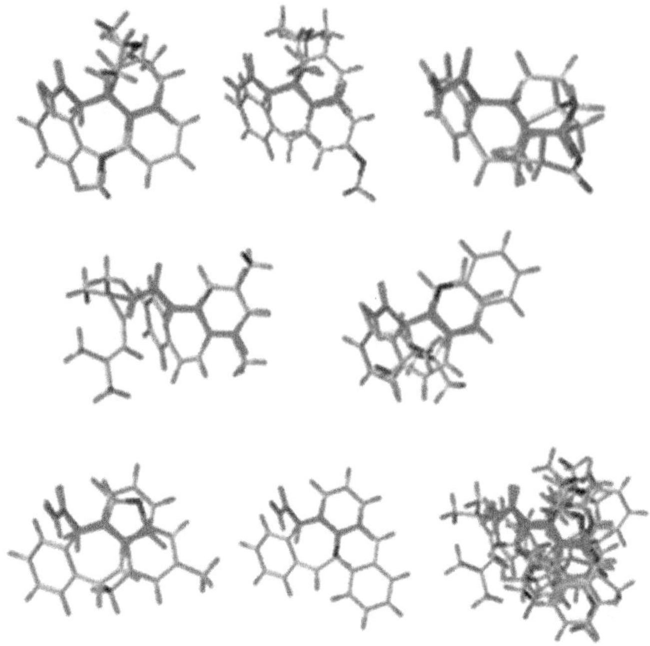

FIGURE 12.41 Superposition of seven structures matching the pharmacophore template with RMS < 0.200.

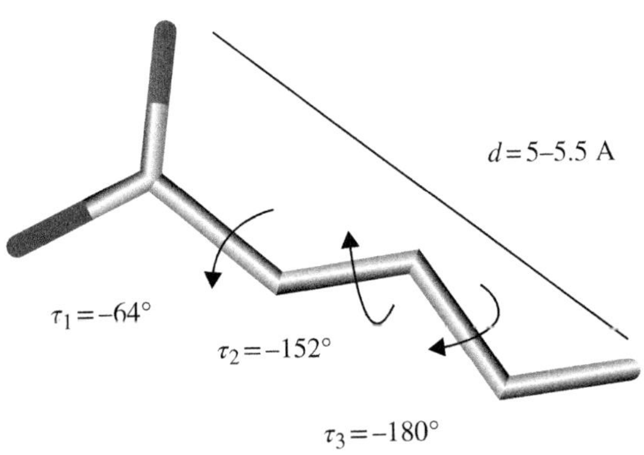

FIGURE 12.42 Identified pharmacophore template for anti-convulsant activity in the MES test.

electronic attraction and were observed to contribute to HBD properties of the compounds. HBA, –OH group at position R1 favored areas associated with HBD features. Ebelin lactone isomers experience steric hindrance due to the chain of the 2,6-dimethyl heptyl group (Fig. 12.44).

Compounds	R_1	A	B	A–B
Betulin	OH	-	-	(a)
Betulin acid	OH	-	-	(b)
Jujuboside A	(d)	-	-	(c)
Jujuboside B	(e)	-	-	(c)
Jujubogenin	OH	-	-	(c)
cis-Ebelin lactone	OH	(f)	(g)	-
trans-Ebelin lactone	OH	(h)	(g)	-

Scaffold A

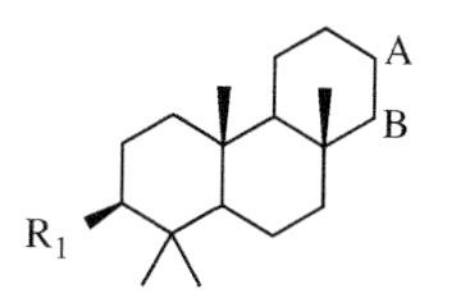

R groups

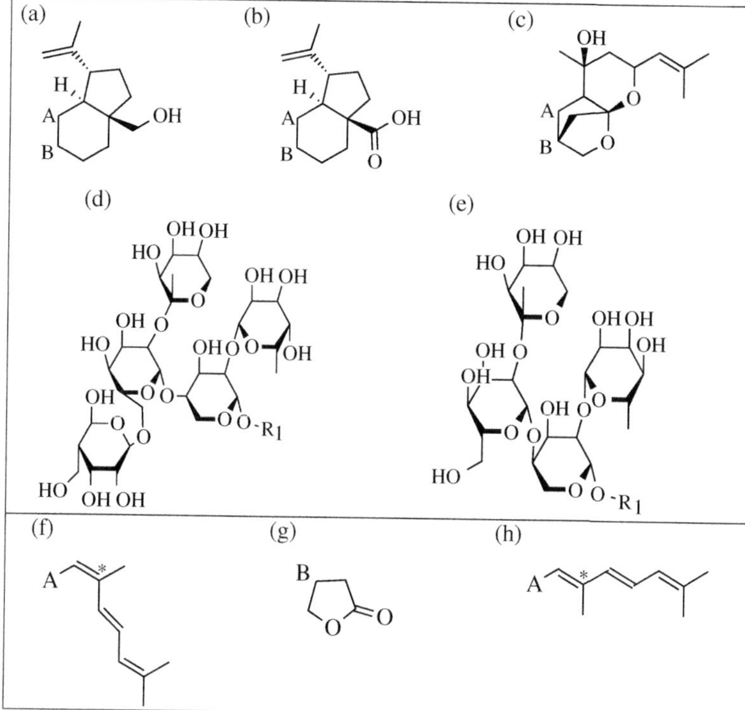

FIGURE 12.43 Chemical structures of compounds derived from Chinese herb, Semen *ziziphi spinosae* (suanzaoren), including betulin, betulin acid, jujuboside A, jujuboside B, jujubogenin, *cis*-ebelin lactone, and *trans*-ebelin lactone.

Major pharmaceutical companies have also investigated histamine H_3 receptor antagonists that have demonstrated efficacy in a number of CNS pathologies including schizophrenia, epilepsy, depression, pain, decreasing food intake, drug abuse and addiction, sleep disorders/narcolepsy, and cognitive enhancement. Such efforts have resulted in a refined H_3 antagonist pharmacophore model (Fig. 12.45) that has been utilized by Phillip et al. [26] to support the iterative parallel synthesis of compounds based on lead molecule dispyrin, which was identified previously from bromotyramine, a bromopyrrole alkaloid extracted from marine sponge *Agelas dispar*. Analogues have modest activity with K_i and IC_{50} values in the micromolar

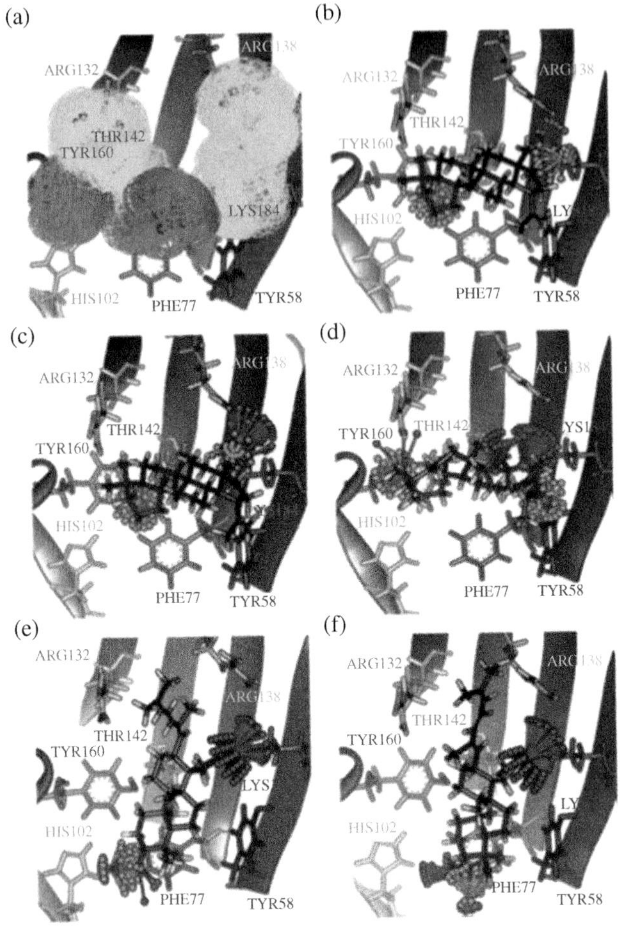

FIGURE 12.44 Overall pharmacophore model (a), and pharmacophore mapping of betulin (b), betulic acid (c), jujubogenin (d), *cis*-ebelin lactone (e), and *trans*-ebelin lactone (f). Pharmacophore features include hydrogen bond acceptors (green) and hydrogen bond donors (purple). *For color details, please see color plate section.*

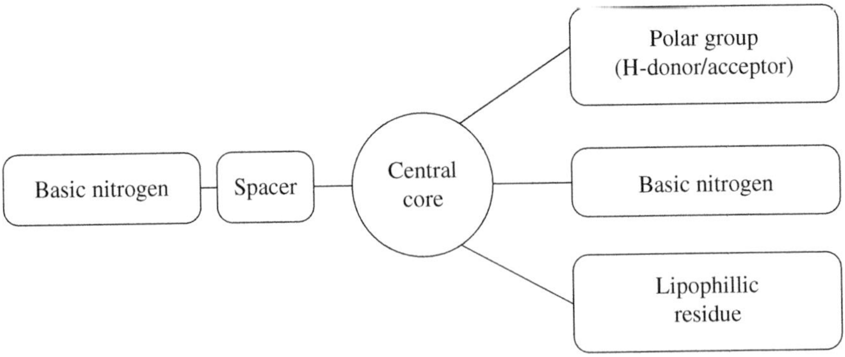

FIGURE 12.45 Schematic illustration of the refined H_3 pharmacophore model derived from imidazole and non-imidazole H_3 antagonists.

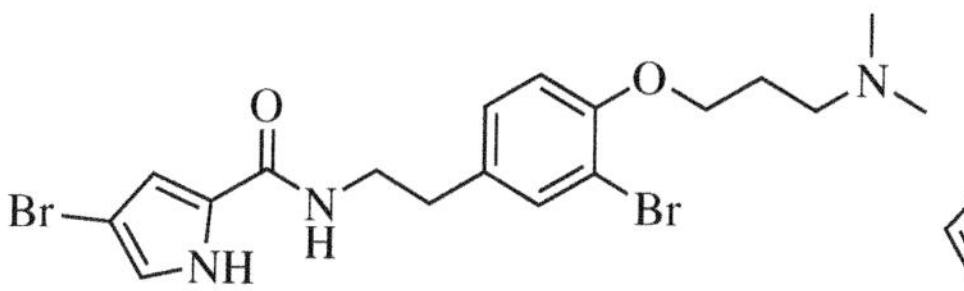

10

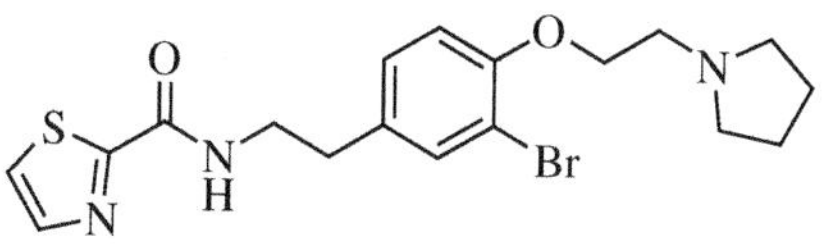

24d

FIGURE 12.46 Chemical structures of dispyrin (left) and compound **24d** (right).

range with compound **24d** (Fig. 12.46) showing the greatest potency in the class (K_i=0.03 μM, IC_{50}=0.07 μM).

12.8 PHARMACOPHORE MODELS FOR MISCELLANEOUS DRUGS

Against the backdrop of alternative phosphodiesterase-5 (PDE-5) inhibitors for erectile dysfunction, Chen [27] has examined traditional treatments used in ancient China: *Epimedium sagittatum* (ES series), *Cnidium monnieri* (CM series), and *Semen cuscutae* (SC series) using *in silico* methods. Active constituents found in each herbal medicine mentioned were docked into a crystal structure of PDE-5 (PDB: 1UDT).The binding site for PDE-5 is surrounded by hydrophobic regions, favors HBAs and HBDs. Compounds *CS01*, *CS03*, *ES03a*, and *ES03b* all contained an optimal distribution of HBA (Fig. 12.47), which appears to have an important role in complex formation. Chen noted that *ES03b* that had the highest docking score also perfectly satisfied the pharmacophore model. The highest-scoring compounds resulting from docking (*SC01*, *SC03*, and *ES03b*) were further modified to generate analogues (*Evo* series), that are optimized to improve PDE-5 affinity and support findings from docking, pharmacophore, and multiple linear regression analyses. *SC01-Evo4*, *SC03-Evo1*, and *ES03b-Evo48* (Fig. 12.48) rendered higher docking scores than their parent compounds (*SC01*, *SC03*, and *ES03b*), respectively. In conclusion, the author noted that compounds that are hydrophobic contain HBAs and no more than six ring groups (or four aromatic rings) are expected to be more potent inhibitors. Larger molecules (>MW 500) can also be expected to retain activity due to the large binding site of PDE-5.

To study the inhibitory effects of 56 herbal compounds from the National Institute for the Control of Pharmaceutical and Biological Products (Beijing, China) on human cytochrome P450 1A2 (CYP1A2), Yang et al. [28] utilized a ligand-based pharmacophore model to investigate the competence of computational methods for predicting such interactions against *in vitro* methods. Both pharmacophore models (*Hopyo-1* and its modified model *Hopyo-1m*, Fig. 12.49) rendered low true positive results for the identification of active herbal compounds (23 and 26% of hits were found to be active *in vitro*, respectively). Fit values for these compounds in both models ranged from 0.126 to 3.719. A small training set (n=5) and test set (n=9) of CYP1A2 inhibitors and unconventional assignment of ligand activity (by molecular weight rather than observed biological activity) utilized to generate the models may have contributed to reduced accuracy. However, results appeared to improve when considered in conjunction with docking studies in X-ray crystal structures of human CYP1A2

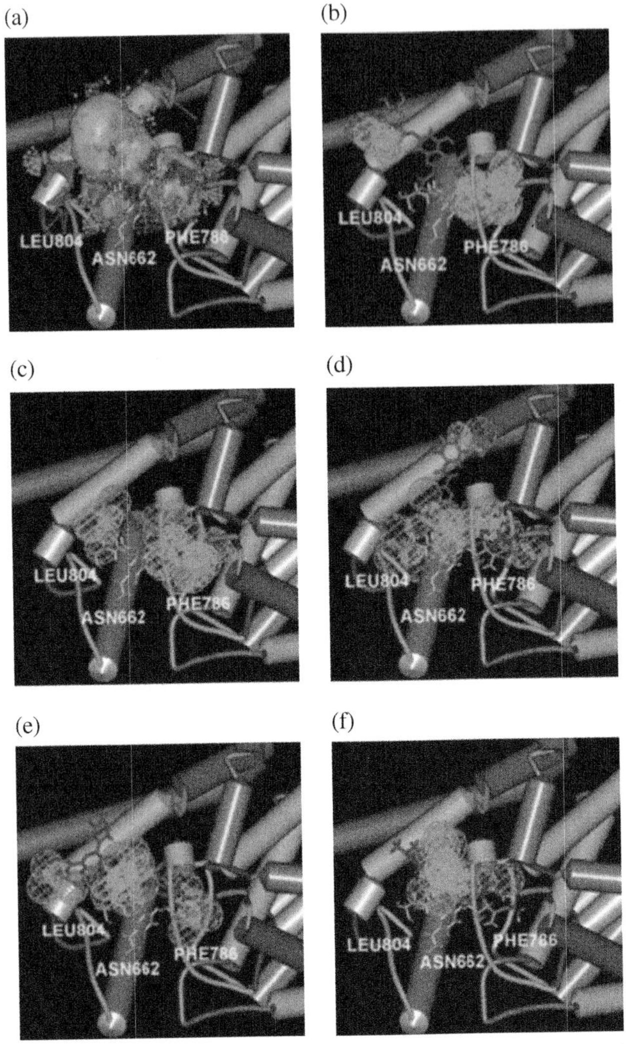

FIGURE 12.47 Overall pharmacophore of PDE5 binding site (a), CS01 (b), CS03 (c), ES03a (d), ES03b (e), and natural substrate cGMP (f). Chemical features include hydrogen bond acceptors (green), hydrogen bond donors (purple), and hydrophobic regions (blue). *For color details, please see color plate section.*

(PDB: 2HI4), which increased to 58–62%. Nonetheless, the authors noted the importance of planarity of aromatic polycyclic compounds (i.e., rutaecarpine, IC_{50}: 22 nM; tanshinone I, IC_{50}: 27 nM, figure Y) for potent CYP1A2 inhibitory activity, whereby a departure from the planar conformation, such as that seen with analogues of tanshinone I, tanshinone IIA, and cryptotanshinone, indeed impairs CYP1A2 activity (IC_{50} 187 and 950 nM, respectively). This was further highlighted by examining the less planar piperidines, namely, matrine, sophoridine, and oxymatrine (Fig. 12.50). Herbal

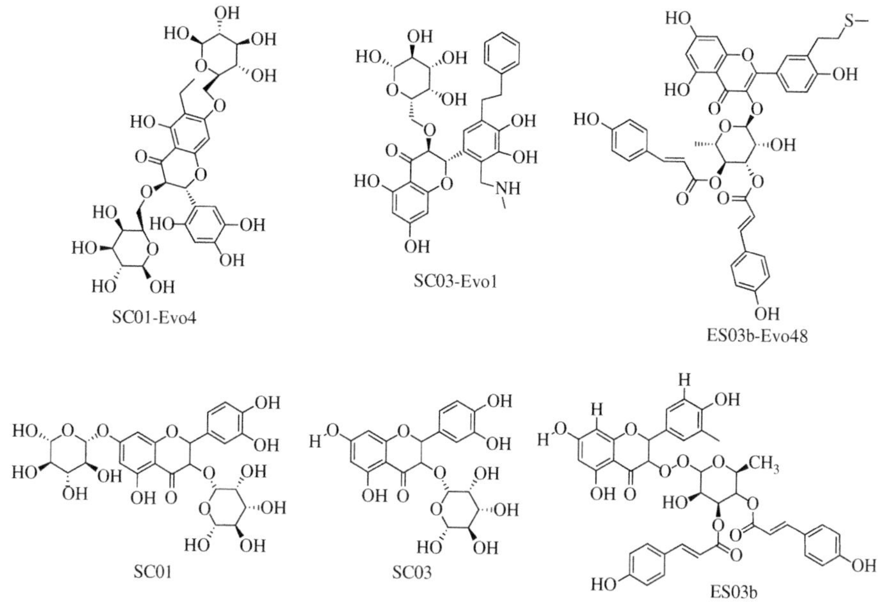

FIGURE 12.48 Chemical structures of the most active Evo compounds *SC01-Evo4*, *SC03-Evo1*, and *ES03b-Evo48* and respective parent compounds *SC01*, *SC03*, and *ES03b*.

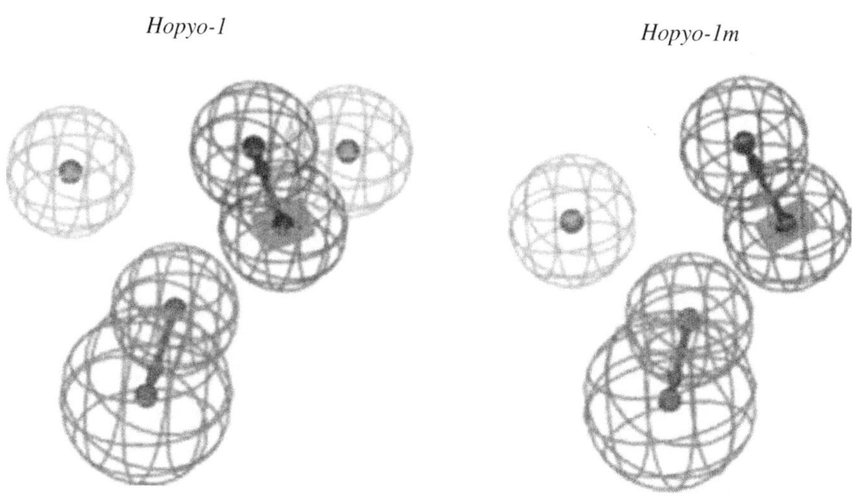

FIGURE 12.49 Pharmacophore models *Hopyo-1* (left) and *Hopyo-1m* (right) that consist of chemical features, which include hydrogen bond acceptors (green), hydrophobic features (blue), and aromatic rings (yellow). *For color details, please see color plate section.*

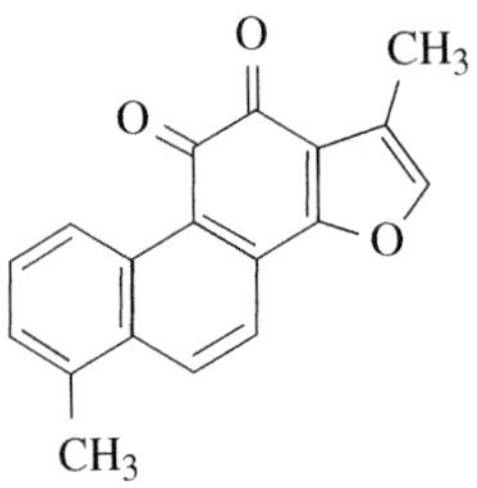

Tanshinone I

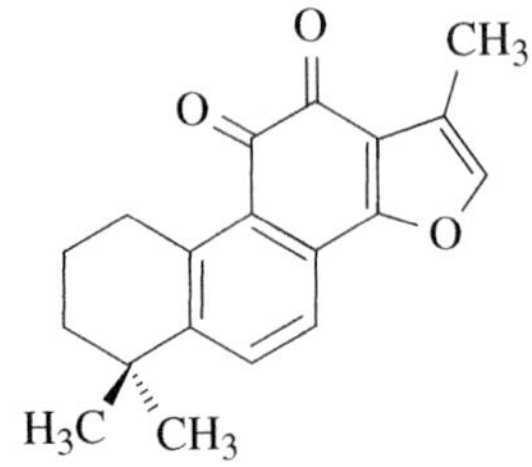

Tanshinone IIA

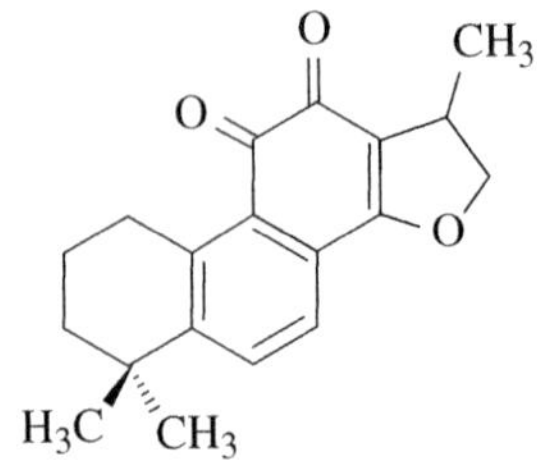

Cryptotanshinone

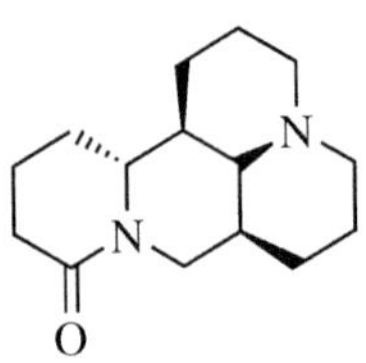

Matrine

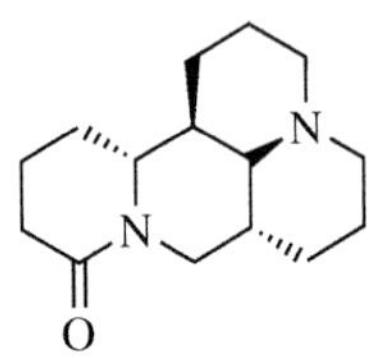

Sophoridine

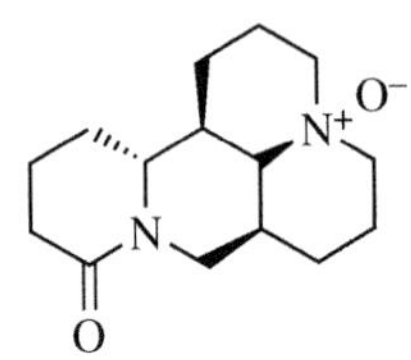

Oxymatrine

FIGURE 12.50 Chemical structures of tanshinone I, tanshinone IIA, cryptotanshinone, matrine, sophoridine, and oxymatrine.

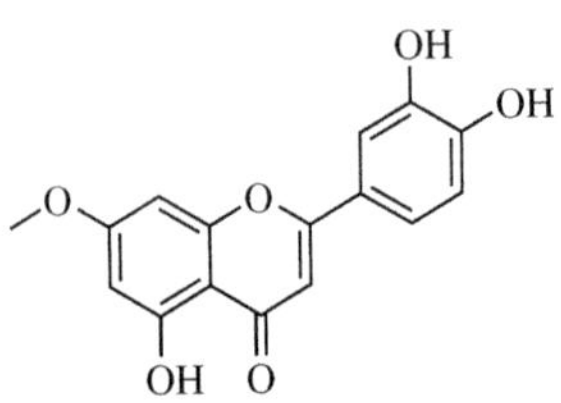

zhang272

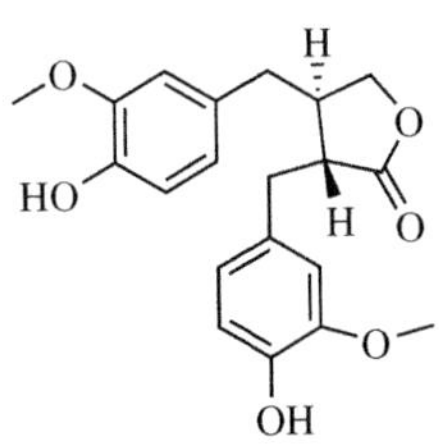

zhang284

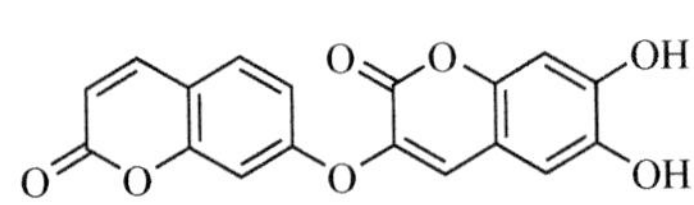

zhang300

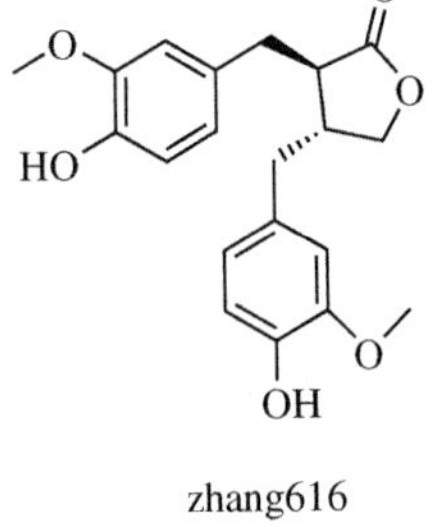

zhang616

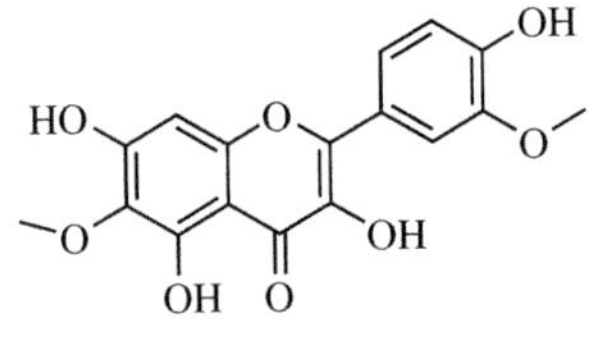

zhang817

FIGURE 12.51 Chemical structures of five herbal compounds 272, 284, 300, 616, and 817 with CYP1A2 inhibitory activity.

compounds examined in this study spanned across many different classes including triterpenoids, flavonoids, saponins, lactones, and alkaloids.

In a similar study, Zhu et al. [29] developed a pharmacophore model using a similar sized training set that consisted of reported CYP1A2 inhibitors and co-crystallized ligands extracted from X-ray crystal structure of CYP1A2 (PDB 2HI4) and its homologue, CYP2B4 (PDB 2BDM) ($n=6$), and a substantially large test set of herbal compounds with known biological activity ($n=202$). Negative data were also incorporated in the validation and refinement of the model to highlight excluded volumes. Following further screening by molecular docking and *in vitro* testing, five active compounds were identified from the in-house database of over a thousand herbal compounds: two flavonoids (compound 272 and 817), one coumarin (compound 300), and two lignanoids (compound 284 and 616) (Fig. 12.51).

12.9 CONCLUSIONS

It is apparent that pharmacophore modeling has an invaluable role in virtual screening alone when a crystal structure is not available or for the design of compounds with specific properties such as a particular chemical structure or pharmacologic activity. As an adjunct to other methods such as structure-based virtual screening, pharmacophore modeling may also provide information that confirms key ligand–protein interactions by highlighting the necessary 3D chemical features for optimal binding and therefore be used to further refine hit compounds. Furthermore, multiple binding conformations and multiple targets for any given ligand can also be accounted for through the use of PS that employs multiple pharmacophore models. This is particularly useful for identifying active constituents of phytotherapies with unknown targets and mechanisms of action. Furthermore, the versatile nature of pharmacophore modeling allows it to continue to be applied to any target with a crystal structure (structure-based pharmacophore modeling) or known ligands (ligand-based pharmacophore modeling).

REFERENCES

[1] Tanrikulu Y, Rau O, Schwarz O, Proschak E, Siems K, et al. (2009) Structure-based pharmacophore screening for natural product-derived PPARγ agonists. *ChemBioChem* 10: 75–78.

[2] Fakhrudin N, Ladurner A, Atanasov AG, Heiss EH, Baumgartner L, et al. (2010) Computer-aided discovery, validation, and mechanistic characterization of novel neolignan activators of peroxisome proliferator-activated receptor. *Mol Pharmacol* 77: 559–566.

[3] Petersen RK, Christensen KB, Assimopoulou AN, Frette X, Papageorgiou VP, et al. (2011) Pharmacophore-driven identification of PPARγ agonists from natural sources. *J Comput Aided Mol Des* 25: 107–116.

[4] Chen K-C, Chang S-S, Huang H-J, Lin T-L, Wu Y-J, et al. (2012) Three-in-one agonists for PPAR-α, PPAR-γ, and PPAR-δ from traditional Chinese medicine. *J Biomol Struct Dyn* 30: 662–683.

[5] Grienke U, Mihaly-Bison J, Schuster D, Afonyushkin T, Binder M, et al. (2011) Pharmacophore-based discovery of FXR-agonists. Part II: identification of bioactive triterpenes from *Ganoderma lucidum*. *Bioorg Med Chem* 19: 6779–6791.

[6] Rollinger JM, Kratschmar DV, Schuster D, Pfisterer PH, Gumy C, et al. (2010) 11β-Hydroxysteroid dehydrogenase 1 inhibiting constituents from *Eriobotrya japonica* revealed by bioactivity-guided isolation and computational approaches. *Bioorg Med Chem* 18: 1507–1515.

[7] Bhadauriya A, Dhoke GV, Gangwal RP, Damre MV, Sangamwar AT (2013) Identification of dual Acetyl-CoA carboxylases 1 and 2 inhibitors by pharmacophore based virtual screening and molecular docking approach. *Mol Divers* 17: 139–149.

[8] Khanfar MA, El Sayed KA (2013) The *Veratrum* alkaloids jervine, veratramine, and their analogues as prostate cancer migration and proliferation inhibitors: biological evaluation and pharmacophore modeling. *Med Chem Res* 22: 4775–4786.

[9] Purushottamachar P, Khandelwal A, Chopra P, Maheshwari N, Gediya LK, et al. (2007) First pharmacophore-based identification of androgen receptor down-regulating agents: discovery of potent anti-prostate cancer agents. *Bioorg Med Chem* 15: 3413–3421.

[10] Chantarasriwong O, Cho WC, Batova A, Chavasiri W, Moore C, et al. (2009) Evaluation of the pharmacophoric motif of the caged *Garcinia* xanthones. *Org Biomol Chem* 7: 4886–4894.

[11] Batova A, Altomare D, Chantarasriwong O, Ohlsen KL, Creek KE, et al. (2010) The synthetic caged *Garcinia* xanthone cluvenone induces cell stress and apoptosis and has immune modulatory activity. *Mol Cancer Ther* 9: 2869–2878.

[12] Wang X, Lu N, Yang Q, Gong D, Lin C, et al. (2011) Studies on chemical modification and biology of a natural product, gambogic acid (III): determination of the essential pharmacophore for biological activity. *Eur J Med Chem* 46: 1280–1290.

[13] Kim KB, Fonseca FN, Crews CM (2005) Development and characterization of proteasome inhibitors. *Methods Enzymol* 399: 585–609.

[14] Gulder TAM, Moore BS (2010) Salinosporamide natural products: potent 20S proteasome inhibitors as promising cancer chemotherapeutics. *Angew Chem Int Ed Engl* 49: 9346–9367.

[15] Schuster D, Waltenberger B, Kirchmair J, Distinto S, Markt P, et al. (2010a) Predicting cyclooxygenase inhibition by three-dimensional pharmacophoric profiling. Part I: model generation, validation and applicability in ethnopharmacology. *Mol Inf* 29: 75–86.

[16] Waltenberger B, Schuster D, Paramapojn S, Gritsanapan W, Wolber G, et al. (2011) Predicting cyclooxygenase inhibition by three-dimensional pharmacophoric profiling. Part II: identification of enzyme inhibitors from Prasaplai, a Thai traditional medicine. *Phytomedicine* 18: 119–133.

[17] Ehrman TM, Barlow DJ, Hylands PJ (2010) In silico search for multi-target anti-inflammatories in Chinese herbs and formulas. *Bioorg Med Chem* 18: 2204–2218.

[18] Bauer J, Waltenberger B, Noha SM, Schuster D, Rollinger JM, et al. (2012) Discovery of depsides and depsidones from lichen as potent inhibitors of microsomal prostaglandin E2 synthase-1 using pharmacophore models. *ChemMedChem* 7: 2077–2081.

[19] Chen K-C, Sun M-F, Yang S-C, Chang S-S, Chen H-Y, et al. (2011) Investigation into potent inflammation inhibitors from traditional Chinese medicine. *Chem Biol Drug Des* 78: 679–688.

[20] Liu T, Li AX, Miao YP, Wu KZ, Ma Y (2009) Screening of new non-nucleoside reverse transcriptase inhibitors of HIV-1 based on traditional Chinese medicines database. *Chin Chem Lett* 20: 1386–1388.

[21] Lee J-Y, Jeong K-W, Shin S, Lee J-U, Kim Y (2009) Antimicrobial natural products as β-ketoacyl-acyl carrier protein synthase III inhibitors. *Bioorg Med Chem* 17: 5408–5413.

[22] Sim MK, Chua ME (1986) Inhibition of acetylcholinesterase by various opioids. *Clin Exp Pharmacol Physiol* 13: 159–162.

[23] Schuster D, Spetea M, Music M, Rief S, Fink M, et al. (2010b) Morphinans and isoquinolines: acetylcholinesterase inhibition, pharmacophore modeling, and interaction with opioid receptors. *Bioorg Med Chem* 18: 5071–5080.

[24] Gavernet L, Talevi A, Castro EA, Bruno-Blanch LE (2008) A combined virtual screening 2D and 3D QSAR methodology for the selection of new anticonvulsant candidates from a natural product library. *QSAR Comb Sci* 27: 1120–1129.

[25] Chen CYC (2009a) Chemoinformatics and pharmacoinformatics approach for exploring the GABA-A agonist from Chinese herb suanzaoren. *J Taiwan Inst Chem Eng* 40: 36–47.

[26] Phillip KJ, Jeffrey CP, Lindsley CW (2009) A novel class of H3 antagonists derived from the natural product guided synthesis of unnatural analogs of the marine bromopyrrole alkaloid dispyrin. *Bioorg Med Chem Lett* 19: 3204–3208.

[27] Chen CYC (2009b) Computational screening and design of traditional Chinese medicine (TCM) to block phosphodiesterase-5. *J Mol Graph Model* 28: 261–269.

[28] Yang L-P, Zhou Z-W, Chen X-W, Li CG, Sneed KB, et al. (2012) Computational and in vitro studies on the inhibitory effects of herbal compounds on human cytochrome P450 1A2. *Xenobiotica* 42: 238–255.

[29] Zhu R, Hu L, Li H, Su J, Cao Z, et al. (2011) Novel natural inhibitors of CYP1A2 identified by in silico and in vitro screening. *Int J Mol Sci* 12: 3250–3262.

13

USE OF KAVA AS A PHYTOTHERAPEUTIC AGENT AND KAVA-RELATED HEPATOTOXICITY

DONG FU AND IQBAL RAMZAN
Faculty of Pharmacy, The University of Sydney, Sydney, New South Wales, Australia

13.1 INTRODUCTION

Kava (*Piper methysticum*) is a commonly consumed crop in western Pacific regions including Melanesia, Micronesia, Hawaii, Vanuatu, and Polynesia. The traditional kava root extract is prepared using water or coconut milk, which is used to produce sedative and anesthetic beverages with a bitter, acrid taste described as "awa" in the Polynesian language. Kava consumption has been an important part of local life in Polynesia. Traditionally, kava is consumed as a drink, during religious or cultural ceremonies, as well as at social gatherings; it promotes physiological and psychological relaxation without disrupting mental clarity [1, 2]. Beginning in the 1990s, kava became more popular worldwide. Kava has been widely used as a herbal medicine for alternative treatment of anxiety and insomnia. In Europe, kava has been used for the treatment of anxiety and nervous disorders such as stress and restlessness, and in the United States kava is used as a natural alternative to anti-anxiety drugs and sleeping remedies [3, 4].

Although the kava root and rhizome (underground stem) is traditionally prepared as an extract using water or coconut milk in many south Pacific countries, extraction by organic solvents, such as supercritical carbon dioxide, ethanol, and acetone, is also used for kava products produced by pharmaceutical companies [5]. In addition, plant parts other than the root and rhizome, such as stems or peelings, are exported

Phytotherapies: Efficacy, Safety, and Regulation, First Edition. Edited by Iqbal Ramzan.

to manufacturers for kava extraction, even though Pacific Islanders avoid stem peelings and aerial parts in preparing aqueous extracts [6]. Along with drinks, other forms of kava extracts are available, including tinctures or standardized extracts, topical lotion/cream, topical solution, and powder in capsules or tablets. These kava formulations are used for many conditions such as anxiety, insomnia, menopausal symptoms, fatigue, asthma, and as a topical numbing agent.

13.2 ACTIVE COMPONENTS IN KAVA

More than 40 compounds have been identified from kava. The main pharmacological active ingredients in kava are called kavalactones [3, 7]. There are 18 kavalactones identified from organic extracts of kava root, which include kavain, methysticin, 10-methoxyyangonin, 7,8-dihydrokavain, 7,8-dihydromethysticin, 11-methoxyyangonin, 5,6-dehydrokavain, 5,6-dehydromethysticin, 11-hydroxyyangonin, yangonin, 5,6-dihydroyangonin, 5-hydroxykavain, 5,6,7,8-tetrahydroyangonin, 7,8-dihydroyangonin, 11-methoxy-12-hydroxydehydrokavain, and 11-methoxy-5,6-dihydroyangonin. In addition, the rootstock also contains flavokawains A, B, and C, piperidine alkaloids, pipermethystine, cepharadione A, and awaine.

Kavain, yangonin, methysticin, 5,6-dehydrokavain, 7,8-dihydrokavain, and 7,8-dihydromethysticin are the six main kavalactones in kava root, which account for approximately 95% of the organic extract (Fig. 13.1). The amounts of these kavalactones vary depending on the preparation technique, as well as the age and cultivar of the kava plant [8, 9]. There are 3–20% of kavalactones in dry weight of the plant. Given relatively low solubility in water, kavalactones account for nearly 5% in the aqueous extract. Commercial products are primarily prepared from organic extracts, which contain 30–70% of kavalactones [1, 7].

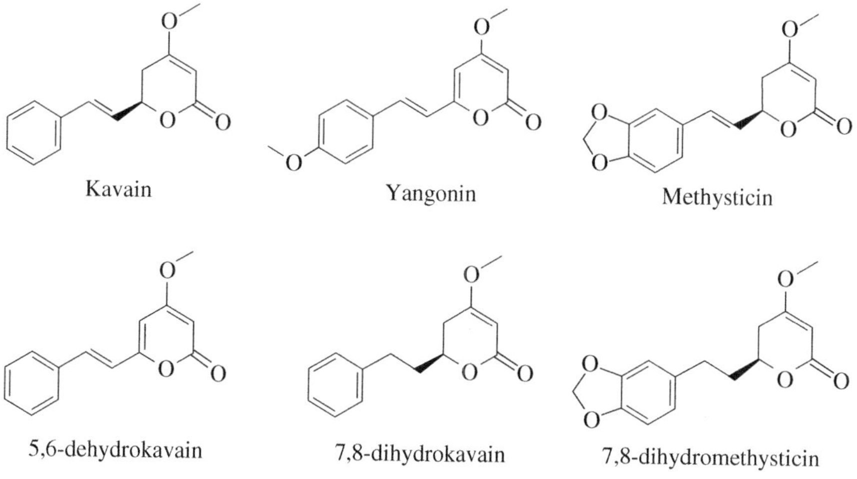

FIGURE 13.1 Chemical structures of six main kavalactones in kava root extract. These are kavain, yangonin, methysticin, 5,6-dehydrokavain, 7,8-dihydrokavain, and 7,8-dihydromethysticin.

13.3 THERAPEUTIC APPLICATIONS OF KAVA

Kava has been used for various conditions. The most important medical application of kava is associated with its anti-psychotic effects. Kava is best known for its powerful anti-anxiety profile [3]. Studies suggest that kava extracts decrease anxiety and might work as well as prescription anti-anxiety drugs, benzodiazepines [4, 10–17]. Kava consumption is also shown to induce and improve sleep [18–23], and reduce stress [24–26], restlessness [27–30], nervousness [3], and headaches [31, 32]. In addition, other pharmacological effects of kava were reported including anti-convulsant [32–36], anti-depressive [37–40], muscle relaxant [41, 42], anti-thrombotic [43], and mild anesthetic [32, 44]. Moreover, kava extracts were shown to have anti-cancer effects in various types of cancers [45–49] (Table 13.1).

TABLE 13.1 Pharmacological Effects of Kava

Anti-anxiety
Anti-sleeplessness
Anti-stress
Anti-restlessness
Anti-attention deficit-hyperactivity disorder (ADHD)
Anti-epilepsy
Anti-depression
Headaches relief
Anti-cancer
Muscle relaxant
Anti-thrombotic
Anesthetic

13.4 PHARMACOLOGY OF KAVA

Despite the various medical applications of kava, the anti-psychotic effects, especially the anti-anxiety profile, is shown to be the most significant effect and has been the focus of most investigations. In addition, recent studies reveal the exciting anti-cancer activity of kava. Thus, the pharmacology of anti-psychotic and anti-cancer effects will be discussed further.

13.4.1 Anti-psychotic Effects of Kava

Among all the anti-psychotic effects of kava, it is well documented that kava extract has significant anti-anxiety effect. Studies suggest that the anti-anxiety effects of kavalactones may be linked to the *gamma*-aminobutyric acid (GABA) pathway, and it may be involved in enhancing ligand binding to the GABA type A receptor [4, 50]. $GABA_A$ receptors are channel receptors and are the major inhibitory neurotransmitter-activated receptors in the mammalian brain. $GABA_A$ receptors are localized in central nociceptive circuits [51]. $GABA_A$ receptors can change shape slightly when GABA binds to them, allowing negatively charged chloride ions to pass through the central channel of the receptor and enter the neuron, which results in a reduction of neuron excitability [52]. It is still unclear how kava exactly affects GABA receptor binding. A study showed kavalactones could activate GABAergic effects via modulation of the GABA channels, enhancing ligand binding to GABA-binding sites [53]. Moreover, in a mouse behavioral test, kava extract exhibited similar anxiolytic and sedative effect as the $GABA_A$ receptor modulator diazepam [54]. A study in rats showed that the root extract rich in kavain (41%) partially substituted for chlordiazepoxide, the prototypic anxiolytic drug [55], confirming the GABAergic contribution to the anxiolytic properties of kava. Furthermore another, study showed that the root extract of kava inhibited neurotransmitter release due to blocked voltage-dependent sodium and calcium ion channels, and suppressed the release of endogenous glutamate [56], suggesting the mechanism for anti-convulsive and possible local anesthetic effects of kava.

A structure activity relationship (SAR) analysis suggested that yangonin, one of the main kavalactones, exhibited selectivity and strong binding affinity for the human recombinant cannabinoid receptor type-1 (CB1) with a $Ki = 0.72\,\mu M$. CB1, the major presynaptic cannabinoid receptor in subsets of inhibitory synapses, is expressed predominantly in the nervous system. However, yangonin did not exhibit strong inhibitory effects on two major metabolic enzymes of the endocannabinoid system, fatty acid amide hydrolase (FAAH) and monoacylglycerol lipase (MAGL) [57]. Although further studies are needed to establish the function and the nature of the interaction of yangonin with the CB1 receptor, the CB1 receptor affinity of yangonin reveals that the endocannabinoid system might contribute to psychopharmacology and the anxiolytic property of the kava plant [58].

Studies conducted in rats showed that the administration of kava extract resulted in behavioral changes and increased concentrations of dopamine in the nucleus accumbens [23]. These results suggest that the relaxing and slightly euphoric actions may be caused by the activation of the mesolimbic dopaminergic neurons [23]. Furthermore, some of the kavalactones, such as kavain, also caused a decrease in 5-hydroxytryptamine (5-HT, or serotonin) concentrations, which could explain the sleep-inducing action of kava [23]. However, the same study showed that other kavalactones (or kava pyrones), such as yangonin, dihydrokavain, and methysticin, did not have any effects on 5-HT concentration. The detailed mechanisms of the effect of kava on insomnia need to be further investigated.

Other than its anti-anxiety property, kava is well known for its anti-depressant effect. An *in vitro* study showed that, in comparison to anti-depressant drugs, amitriptyline, imipramine, and brofaromine, kava extract and pure synthetic kavalactones reversibly inhibited monoamine oxidase B (MAO-B) [59], an enzyme on the outer mitochondrial membrane that catalyzes the oxidation of arylalkylamine neurotransmitters such as dopamine [60, 61]. Consequently, kavalactones could prevent the breakdown of monoamine neurotransmitters and increase concentrations of the neurotransmitter in the brain; therefore, kava extracts exhibit their anti-depressant effects by inhibition of MAO-B. Different kavalactones displayed different inhibitory potency; and the two most potent kavalactones, desmethoxyyangonin and (+/−)-methysticin, showed a competitive inhibition pattern for MAO-B [59]. Since elevated MAO-B level in the brain is also associated with Alzheimer's and Parkinson's diseases as well as depression [61–63], the inhibition of MAO-B by kava extracts might be an important mechanism for the effects of kavalactones on depression and neuronal protection [59].

13.4.2 Anti-cancer Effects of Kava

Kava consumption among the natives of Fiji Islands has been suggested to be responsible for the low incidence of cancer on this island [3]. While most of the anti-psychotic effect of kava links to kavalactones, the anti-cancer effects are associated with another component in kava root extract: flavokawains, which are a class of chalconoids. Currently, three types of flavokawains, flavokawain A, B, and C, are identified (Fig. 13.2) [49]. Studies conducted both *in vivo* and *in vitro* demonstrated that flavokawains (e.g., flavokawain A, B) have anti-cancer effects in various cancer cells, including human urothelial cell cancer [64], human osteosarcoma [65], lung cancer [47, 66], prostate cancer [46, 48], sarcoma [67], colon cancer [45], oral adenoid cystic carcinoma [68], and human bladder cancer [69]. The anti-cancer effect of flavokawains is involved in many cellular pathways.

Using a transgenic mouse model that resembles human urothelial cell carcinoma with defects in the p53 and the retinoblastoma (Rb) protein pathways, a study demonstrated that flavokawain A induced apoptosis via activation of the pro-apoptotic pathway (p27 and DR5) and inhibited the expression of anti-apoptotic proteins, x-linked inhibitor of apoptotic proteins (XIAP), Bcl2, and survivin [64]. Consequently,

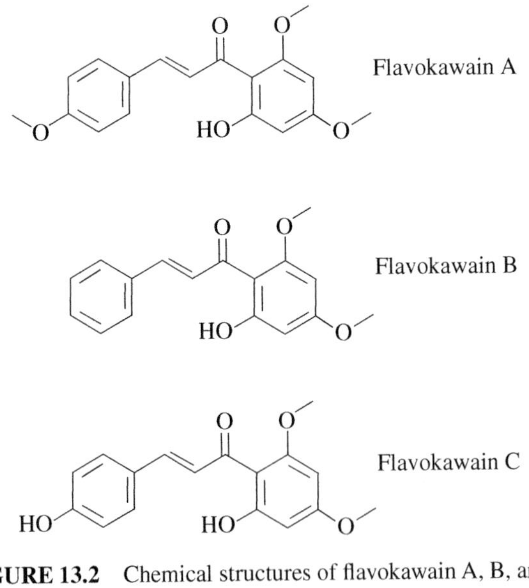

FIGURE 13.2 Chemical structures of flavokawain A, B, and C.

flavokawain A inhibited the occurrence of high-grade papillary urothelial cell cancer in mice [64]. Moreover, flavokawain A significantly decreased mitochondrial membrane potential resulting in the release of cytochrome c into the cytosol and activated apoptosis in an invasive bladder cancer cell line T24. In addition, the pro-apoptotic effects of flavokawain A were also mediated by a decrease in anti-apoptotic protein Bcl-x(L), and an increase in the active form of pro-apoptotic protein Bax [69].

Flavokawain B was also shown to induce apoptosis in human osteosarcoma cells by activating caspase-3/7, 8, and 9, down-regulating anti-apoptotic proteins, Bcl-2, and survivin [65]. Moreover, flavokawain B induced apoptosis in prostate cancer cells by activating pro-apoptotic proteins (e.g., death receptor-5, Bim, and Puma) and inhibiting apoptotic protein XIAP and survivin [48]. Similarly, in oral adenoid cystic carcinoma cells, treatment with flavokawain B caused upregulation of the pro-apoptotic proteins, Bim, Bak, and Bax, and downregulation of the anti-apoptotic Bcl-2 proteins. Furthermore, flavokawain B also promoted cytochrome c release from mitochondria into the cytosol, and activated the cleavage of poly (ADP-ribose) polymerase (PARP), which resulted in significant inhibition of proliferation of the cancer cells [68].

In addition to its anti-apoptotic effect, flavokawain B also caused cell cycle arrest in osteosarcoma cells by decreasing levels of cyclin B1, cdc2, and cdc25c [65]. Using lung cancer cells, studies revealed that flavokawain exhibits inhibition of nuclear factor kappa-light-chain-enhancer of activated B cells (NF-κB), activation of caspases, activation of extracellular signal regulated kinase 1/2 (ERK1/2), and c-Jun N-terminal kinase (JNK), which are important for cell fate and viability [66]. Furthermore, in the mice model, flavokawains showed significant reduction of lung tumor multiplicity [47]. Kava root extract and flavokawain B lowered the incidence of prostate cancer via inhibition of androgen receptor (AR) expression by decreasing transcription and accelerating its degradation [46]. Taken together, the potential anti-cancer effects of flavokawains in kava extract are involved in multiple cellular pathways including apoptosis, cell cycle and cell proliferation, and viability pathways.

13.5 SIDE EFFECTS OF KAVA

Kava consumption has been well tolerated and is non-addictive at therapeutic dosages. Allergic reactions are usually mild and include itchy skin or itchy throat, and hives on the skin usually prevalent on the abdominal region; however, potential side effects can occur when very high doses are taken for extended periods, including headache, dizziness, drowsiness, diarrhea, and occasionally dermatologic manifestations [70]. Heavy use of kava could cause malnutrition, weight loss, renal dysfunction, rashes, pulmonary hypertension, macrocytosis of red cells, lymphocytopenia, and decreasing platelet volumes [70–72]. More importantly, unexpected high liver toxicity/damage (e.g., increased serum γ-glutamyltransferase and high-density lipoprotein cholesterol levels) has been reported during kava consumption [28, 73]. Therefore, many countries took actions and regulated or restricted kava consumption [70].

13.6 HEPATOTOXICITY OF KAVA

Some reported cases suggest that ingestion of kava was associated with liver toxicity, which involved elevated liver enzymes (AST, ALT, and γ-GT), elevated bilirubin levels, hepatitis, and jaundice, which could lead to severe liver damage and liver failure placing a patient in need of a liver transplantation or in jeopardy of death [9, 74].

A study suggested components such as pipermethystine and flavokawain B in kava is associated with its hepatotoxicity, while the alkaloid pipermethystine was an unlikely cause for the observed hepatotoxicity because it was found in extremely low quantities in Western extracts [75]. However, another study suggested that the amount of flavokawain B, which is found in various preparations, was not large enough to cause liver damage [76], and kava hepatotoxicity may be due to contamination with aflatoxins or other mold hepatotoxins [76]. It is still uncertain which components of kava are directly linked to hepatotoxicity, and more studies are required to investigate the details. Studies indicated that the hepatotoxicity of kava appears to be an intrinsic toxicity, and possibly also an idiosyncratic toxicity [77–79]. A case study showed that hepatic necrosis or cholestatic hepatitis occurred with both alcoholic and acetonic kava extracts. Some patients developed fulminant liver failure and underwent liver transplantation [78, 80]. In addition, adverse effects were observed in patients using either ethanolic or acetonic kava extracts or traditional aqueous extracts, suggesting that toxicity is dependent on the kava plant itself rather than the extraction solvent(s) used [5]. Multiple cellular events, such as change in drug metabolism and detoxification, induction of inflammation, inhibition of transporter function, and damage to mitochondria, may cause hepatotoxicity. Studies suggested that kava may affect these pathways resulting in hepatotoxicity. The following sections will discuss the detailed mechanisms of kava-mediated hepatotoxicity, including (i) inhibition of cytochrome P450 enzyme activities that are important for drug metabolism; (ii) decreasing liver glutathione content that is linked to drug detoxification; (iii) inhibition of cyclooxygenase enzyme activity; (iv) induction of inflammatory responses; (v) inhibition of hepatic transporter function, which results in accumulation of drugs, toxins, and bile acids; and (vi) direct mitochondrial damage that plays a central role in drug-induced liver injury (Fig. 13.3).

13.6.1 Inhibition of Cytochrome P450 Enzymes Activities

Cytochrome P450 enzymes are the major enzymes participating in drug metabolism and bioactivation. Studies suggest that kava extracts have an inhibitory effect on the activities of many of cytochrome P450 enzymes. Using human liver microsomes, a study showed that kava extract decreases cytochrome P450 activities in a concentration-dependent manner, which significantly inhibited the activities of CYP1A2 (56% inhibition), 2C9 (92%), 2C19 (86%), 2D6 (73%), 3A4 (78%), and 4A9/11 (65%). However, CYP2A6, 2C8, and 2E1 activities were unaffected. Major kavalactones desmethoxyyangonin, methysticin, and dihydromethysticin significantly inhibited activation of CYP2C9, 2C19, 2D6, and 3A4, suggesting that kava

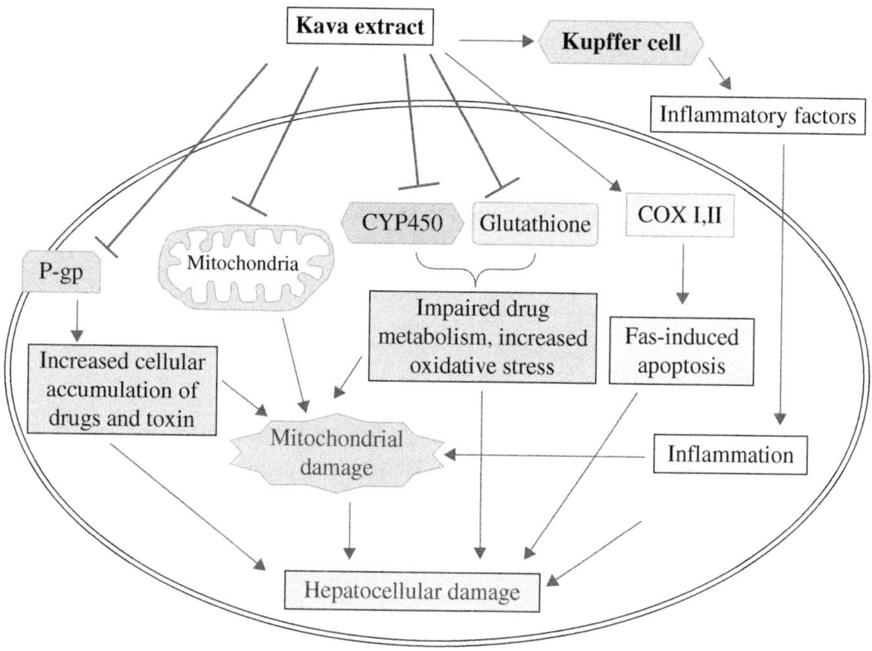

FIGURE 13.3 Diagram of cellular mechanisms of kava-induced hepatotoxicity.

has a high potential for causing drug interactions and increasing drug toxicity through inhibition of P450 enzymes [81, 82].

Another study showed that purified kavalactones, methysticin, desmethoxyyangonin, and yangonin inhibited the activity of CYPs 1A2, 2C9, 2C19, 2E1, and 3A4 and caused moderate cytotoxicity to human hepatocytes [83]. Given the significant inhibitory effect of kava on CYP450 enzyme activity, consumption of kava can markedly alter the metabolism of other drugs/alcohol that result in potential liver toxicity (Fig. 13.3) [82]. Several adverse interactions between kava and either prescription or non-prescription drugs have been reported, including anti-convulsants, alcohol, anti-anxiety medications (CNS depressants such as benzodiazepines), anti-psychotic medications, levodopa, and diuretics [6, 82].

13.6.2 Reduction of Liver Glutathione

Glutathione is the major cellular anti-oxidant and can directly neutralize free radicals and reactive oxygen compounds, thus preventing cellular damage. Glutathione depletion leads to oxidative stress and cytotoxicity in hepatocytes [84]. Using HepG2 cells to investigate the hepatotoxicity of kava, a study showed that flavokawain B, a main component in kava extract, induced cell death in HepG2 cells, and the hepatocellular toxicity of flavokawain B is mediated by the induction of oxidative stress and depletion of glutathione (GSH). Replenishment with exogenous GSH rescues hepatocytes from flavokawain-induced death, confirming that flavokawain B caused hepatotoxicity through depletion of glutathione, and flavokawain B is a potent GSH-sensitive hepatotoxin [85]. It is likely that kava extracts can conjugate to glutathione to produce water-soluble metabolites for excretion and result in the depletion of glutathione, which causes oxidative stress and hepatocellular damage (Fig. 13.3) [86]. However, a study suggested that there were no significant changes in reduced glutathione levels when rat liver was perfused with kavain or methysticin [36]. Similarly, depletion of glutathione was not found in HepG2 cells treated with kavain, methysticin, or yangonin [87]. It is likely that different kava constituents have differential effects on glutathione depletion, indicating that hepatotoxic constituents from kava may induce liver toxicity via different mechanisms.

13.6.3 Induction of Hepatic Inflammatory Responses

Studies showed that kava consumption caused liver inflammation and resulted in liver damage [88, 89]. Histopathology study revealed extensive, severe hepatocellular necrosis and infiltration with lymphocytes, eosinophils, and activated macrophages after kava consumption [88, 89]. Furthermore, electron microscope studies demonstrated that there were disruptions of hepatic vasculature with narrowing of blood vessels, constriction of sinusoidal blood vessels, and retraction of the endothelium in kavain-perfused rat livers [90]. Kupffer cells (liver macrophages) within the sinusoids became swollen with large cytoplasmic vacuoles and phagocytosed material after rat livers were perfused with kavain [36, 91], suggesting inflammatory responses participate in kava-mediated hepatotoxicity (Fig. 13.3). A recent study showed that water extractable active ingredients, but not isolated kavalactones, promoted calcium release, influx, and the secretion of pro-inflammatory mediators in mast cells, suggesting mast cell activation may be a mechanistic component of kava-related inflammation [92]. However, a study showed that kava and related kavalactones suppressed TNF-α secretion in THP-1 human monocytic cells and provided protection *in vivo* in a TNF-α driven model of inflammation [93]. Whether this observation could be extended to liver or hepatocytes is worth further investigation. Nevertheless, more studies are certainly needed to examine the effects and the mechanisms of kava-mediated inflammatory responses in either intact liver or co-culture of hepatocytes and Kupffer cells.

13.6.4 Inhibition of Cyclooxygenase Enzyme Activity

Another possible mechanism for kava-mediated liver injury is the inhibitory effect on cyclooxygenase (COX). Although COX inhibition is one of the main strategies for the treatment of inflammation, studies suggest that COX inhibition may also increase adverse drug effect. The significantly increased toxicity of acetaminophen was associated with COX-2 deficiency in COX(−/−) mice or in the presence of selective COX-2 inhibitor, suggesting that COX-2 serves an important hepato-protective function and that COX inhibition may contribute to the risk of drug-induced liver injury [94]. In another study, deletion of COX-2 aggravated Fas-induced liver injury, and pre-treatment with the COX-2 inhibitor NS-398 enhanced anti-Fas antibody

Jo2-induced liver injury, demonstrating that COX-2 in hepatocytes is important for cell survival and prevention of Fas-induced hepatocyte apoptosis and liver injury (Fig. 13.3) [95]. A number of studies showed that kava inhibited the COX-1, 2 enzyme activities. A study showed that extracts of kava selectively inhibited COX-2 enzyme as well as lipid peroxidation *in vitro* (5–99%) [96]. Dihydrokawain and yangonin from kava extract also showed the highly inhibitory effects on COX-I and COX-II [97]. Moreover, flavokawain B and 5,7-dimethoxyflavanone was able to inhibit COX-I and II activities [98]. Taken together, these studies reveal that many components from kava extract exhibit inhibitory effects on COX-I and II, which is potentially associated with the hepatotoxicity of kava.

13.6.5 Inhibition of Hepatic Transporters

Many transporters are expressed on the surface of hepatocytes. Through the efflux and influx activities, these transporters are critical for maintaining normal hepatic function. A number of ABC (ATP-binding cassette) transporters, such as P-glycoprotein (P-gp), multidrug resistance-associated protein 2 (MRP2), and bile salt export pump (BSEP), are localized on the apical membrane of hepatocytes and are important for the efflux of many endogenous metabolites, exogenous substrates (e.g., drugs), and bile acids [99–101]. Inhibition of the expressions and/or functions of these transporters results in intracellular accumulation of metabolites, exogenous substrates and bile, and consequently hepatocellular damage. A study suggested that kava extract kavalactones exhibited a moderate to potent inhibitory effect on the activity of P-glycoprotein (P-gp) [102]. P-gp is able to transport many chemically and structurally unrelated substrates out of the cells and acts as an efflux pump [103–105]. P-gp is involved in the absorption and excretion of many drugs, endogenous metabolites, and exogenous toxins in hepatocytes [104, 106]. The crude kava extract and the main kavalactones (kavain, dihydrokavain, methysticin, dihydromethysticin, yangonin, and desmethoxyyangonin) inhibited the P-gp-mediated efflux of calcein-acetoxymethyl ester [102]. Thus, the inhibition of P-gp function by kava can cause the cellular accumulation of drugs, endogenous metabolites, and toxins, which may be indirectly responsible for the hepatotoxicity of kava (Fig. 13.3). It is possible that kava may also inhibit functions of other ABC transporters, such as multidrug resistance-associated protein 2 (MRP2) and bile salt export pump (BSEP), which are important for bile acid secretion in hepatocytes [107], and consequently impairing bile secretion in hepatocytes, leading to bile acid-mediated hepatocellular damage. However, investigations are needed to examine the effects of kava on other transporters.

13.6.6 Damage of Hepatic Mitochondria

Despite various mechanisms (as stated previously) involved in hepatotoxicity, such as change of drug metabolism and detoxification, which results in increased hepatocellular drug accumulation, and induction of hepatic inflammation, drug-induced mitochondrial dysfunction and damage is postulated as the key mechanism of drug-induced liver damage [108–111]. Hepatotoxic drugs can directly or indirectly lead to mitochondrial dysfunction and damage, which initiates the apoptotic or necrotic signaling. As such, mitochondrial damage is responsible for cytolytic hepatitis, steatosis, steatohepatitis, liver failure, and even cirrhosis [109, 111]. Drug-induced mitochondrial dysfunction alters cellular metabolic pathways and damages mitochondrial components as well as altering mitochondrial morphology [112]. Many hepatotoxic drugs (e.g., paracetamol/acetaminophen) impair mitochondrial energy production by inhibiting mitochondrial oxidative phosphorylation, fatty acid oxidation [113, 114], damaging mitochondrial potential and inducing the release of mitochondrial pro-apoptotic proteins [114, 115]. The damaged mitochondria release apoptosis-inducing factors and excessive amounts of reactive oxygen species (ROS) that cause apoptosis or necrosis and result in liver damage [116]. A study used three different kava extracts (a methanolic root extract, an acetonic root extract, and a methanolic leaf extract) to analyze their toxicity in HepG2 liver cancer cells and isolated rat liver mitochondria. All three extracts decreased the mitochondrial membrane potential, and inhibited and/or uncoupled the mitochondrial respiratory chain. In addition, the mitochondrial beta-oxidation was inhibited by all extracts. These results demonstrate that kava extracts cause mitochondrial dysfunction resulting in decreased respiratory chain and mitochondrial potential, elevated ROS production, and eventually apoptosis and/or necrosis, revealing that kava targets hepatic mitochondria and leads to hepatocellular damage (Fig. 13.3) [117].

13.7 SUMMARY AND FUTURE CHALLENGES

Kava is widely used as a supplement or herbal medicine in many parts of world. It is well documented for its anxiolytic, analgesic, muscle relaxing, and anti-convulsant effects. Most recently, kava has also shown to have anti-cancer effect. While the anti-psychotic effects of kava (treatment of anxiety, insomnia, and stress) are associated with kavalactones (kavain, yangonin, methysticin, 7,8-dihydrokavain), the anti-cancer properties link to flavokawains. The detailed mechanisms of the pharmacological effects of the kava components require further elucidation to offer insights for the development of more potent therapeutic analogues. In addition, several factors, such as concentration, type of preparation, kava pyrone content, and the kava variety used may affect pharmacologic activity. Thus, additional studies are needed to analyze these variations, which is essential for understanding the pharmacological effects as well as the cellular and molecular mechanisms of actions of kava.

Although the therapeutic properties are well known, many reports/studies indicate problematic side effects of kava products, especially serious hepatotoxicity. Currently, regulatory organizations in many countries, such as the United States Centers for Disease Control and Prevention, the Food and Drug Administration (FDA), Australian Therapeutic Goods Administration (TGA), and Medicines Control Agency in the United Kingdom, have expressed reservations/concerns about the use of kava and its possible adverse effects (specifically severe liver toxicity) [70]. The possible hepatotoxicity of kava products has led some countries, including Germany, Switzerland,

France, Canada, and the United Kingdom, to take regulatory actions, ranging from consumer warnings to removing kava-containing products from the market [71].

Because of various pharmacologically active chemicals in kava products/extracts, it is not surprising that kava extracts may affect multiple cellular pathways that may be associated with kava-mediated hepatotoxicity (Fig. 13.3). This raises a great challenge to understand the detailed insights of kava-mediated hepatotoxicity. More investigations are needed to, firstly, identify the main pharmacological and/or toxicological chemicals in various kava extracts; secondly, to examine the effects of the main therapeutic components in kava on the cellular pathways that link to hepatotoxicity; and to determine which of these cellular pathways/mechanisms play a major role in its hepatotoxicity and how these pathways link to or interact with each other. All this information will be essential for understanding the pharmacology and toxicology of kava and thereby develop potential new therapeutic analogues with minimum or lower toxicity.

REFERENCES

[1] Denham A, McIntyre M, Whitehouse J (2002) Kava – the unfolding story: report on a work-in-progress. *J Altern Complement Med* 8: 237–263.

[2] Norton SA, Ruzc P (1994) Kava dermopathy. *J Am Acad Dermatol* 31: 89–97.

[3] Singh YN (1992) Kava: an overview. *J Ethnopharmacol* 37: 13–45.

[4] Singh YN, Singh NN (2002) Therapeutic potential of kava in the treatment of anxiety disorders. *CNS Drugs* 16: 731–743.

[5] Teschke R, Genthner A, Wolff A (2009) Kava hepatotoxicity: comparison of aqueous, ethanolic, acetonic kava extracts and kava-herbs mixtures. *J Ethnopharmacol* 123: 378–384.

[6] Anke J, Ramzan I (2004) Pharmacokinetic and pharmacodynamic drug interactions with Kava (*Piper methysticum* Forst. f.). *J Ethnopharmacol* 93: 153–160.

[7] Bilia AR, Scalise L, Bergonzi MC, Vincieri FF (2004) Analysis of kavalactones from *Piper methysticum* (kava-kava). *J Chromatogr B Analyt Technol Biomed Life Sci* 812: 203–214.

[8] Teschke R, Lebot V (2011) Proposal for a kava quality standardization code. *Food Chem Toxicol* 49: 2503–2516.

[9] Olsen LR, Grillo MP, Skonberg C (2011) Constituents in kava extracts potentially involved in hepatotoxicity: a review. *Chem Res Toxicol* 24: 992–1002.

[10] Lehmann E, Kinzler E, Friedemann J (1996) Efficacy of a special Kava extract (*Piper methysticum*) in patients with states of anxiety, tension and excitedness of non-mental origin – a double-blind placebo-controlled study of four weeks treatment. *Phytomedicine* 3: 113–119.

[11] Scherer J (1998) Kava-kava extract in anxiety disorders: an outpatient observational study. *Adv Ther* 15: 261–269.

[12] Pittler MH, Ernst E (2000) Efficacy of kava extract for treating anxiety: systematic review and meta-analysis. *J Clin Psychopharmacol* 20: 84–89.

[13] Boerner RJ (2001) Kava kava in the treatment of generalized anxiety disorder, simple phobia and specific social phobia. *Phytother Res* 15: 646–647.

[14] Pittler MH, Ernst E (2003) Kava extract for treating anxiety. *Cochrane Database Syst Rev* CD003383.

[15] Cagnacci A, Arangino S, Renzi A, Zanni AL, Malmusi S, et al. (2003) Kava-kava administration reduces anxiety in perimenopausal women. *Maturitas* 44: 103–109.

[16] Geier FP, Konstantinowicz T (2004) Kava treatment in patients with anxiety. *Phytother Res* 18: 297–300.

[17] Kinrys G, Coleman E, Rothstein E (2009) Natural remedies for anxiety disorders: potential use and clinical applications. *Depress Anxiety* 26: 259–265.

[18] Meolie AL, Rosen C, Kristo D, Kohrman M, Gooneratne N, et al. (2005) Oral nonprescription treatment for insomnia: an evaluation of products with limited evidence. *J Clin Sleep Med* 1: 173–187.

[19] Jacobs BP, Bent S, Tice JA, Blackwell T, Cummings SR (2005) An internet-based randomized, placebo-controlled trial of kava and valerian for anxiety and insomnia. *Medicine (Baltimore)* 84: 197–207.

[20] Shinomiya K, Inoue T, Utsu Y, Tokunaga S, Masuoka T, et al. (2005) Effects of kava-kava extract on the sleep-wake cycle in sleep-disturbed rats. *Psychopharmacology (Berl)* 180: 564–569.

[21] Wheatley D (2001) Stress-induced insomnia treated with kava and valerian: singly and in combination. *Hum Psychopharmacol* 16: 353–356.

[22] Wheatley D (2001) Kava and valerian in the treatment of stress-induced insomnia. *Phytother Res* 15: 549–551.

[23] Baum SS, Hill R, Rommelspacher H (1998) Effect of kava extract and individual kavapyrones on neurotransmitter levels in the nucleus accumbens of rats. *Prog Neuropsychopharmacol Biol Psychiatry* 22: 1105–1120.

[24] Stacy S (2011) Relaxation drinks and their use in adolescents. *J Child Adolesc Psychopharmacol* 21: 605–610.

[25] Feltenstein MW, Lambdin LC, Ganzera M, Ranjith H, Dharmaratne W, et al. (2003) Anxiolytic properties of *Piper methysticum* extract samples and fractions in the chick social-separation-stress procedure. *Phytother Res* 17: 210–216.

[26] Cropley M, Cave Z, Ellis J, Middleton RW (2002) Effect of kava and valerian on human physiological and psychological responses to mental stress assessed under laboratory conditions. *Phytother Res* 16: 23–27.

[27] Gastpar M, Klimm HD (2003) Treatment of anxiety, tension and restlessness states with Kava special extract WS 1490 in general practice: a randomized placebo-controlled double-blind multicenter trial. *Phytomedicine* 10: 631–639.

[28] Bilia AR, Gallon S, Vincieri FF (2002) Kava-kava and anxiety: growing knowledge about the efficacy and safety. *Life Sci* 70: 2581–2597.

[29] Sarris J, Kean J, Schweitzer I, Lake J (2011) Complementary medicines (herbal and nutritional products) in the treatment of Attention Deficit Hyperactivity Disorder (ADHD): a systematic review of the evidence. *Complement Ther Med* 19: 216–227.

[30] Larzelere MM, Campbell JS, Robertson M (2010) Complementary and alternative medicine usage for behavioral health indications. *Prim Care* 37: 213–236.

[31] Brown RP, Gerbarg PL (2001) Herbs and nutrients in the treatment of depression, anxiety, insomnia, migraine, and obesity. *J Psychiatr Pract* 7: 75–91.

[32] Jamieson DD, Duffield PH (1990) The antinociceptive actions of kava components in mice. *Clin Exp Pharmacol Physiol* 17: 495–507.

[33] Pearl PL, Drillings IM, Conry JA (2011) Herbs in epilepsy: evidence for efficacy, toxicity, and interactions. *Semin Pediatr Neurol* 18: 203–208.

[34] Dinh LD, Simmen U, Bueter KB, Bueter B, Lundstrom K, et al. (2001) Interaction of various *Piper methysticum* cultivars with CNS receptors in vitro. *Planta Med* 67: 306–311.

[35] Backhauss C, Krieglstein J (1992) Extract of kava (*Piper methysticum*) and its methysticin constituents protect brain tissue against ischemic damage in rodents. *Eur J Pharmacol* 215: 265–269.

[36] Zhang LY, Rowe A, Ramzan I (2011) Does inflammation play a role in kava hepatotoxicity? *Phytother Res* 25: 629–630.

[37] Sarris J, Kavanagh DJ, Adams J, Bone K, Byrne G (2009) Kava Anxiety Depression Spectrum Study (KADSS): a mixed methods RCT using an aqueous extract of *Piper methysticum*. *Complement Ther Med* 17: 176–178.

[38] Sarris J, Kavanagh DJ, Deed G, Bone KM (2009) St. John's wort and Kava in treating major depressive disorder with comorbid anxiety: a randomised double-blind placebo-controlled pilot trial. *Hum Psychopharmacol* 24: 41–48.

[39] van der Watt G, Laugharne J, Janca A (2008) Complementary and alternative medicine in the treatment of anxiety and depression. *Curr Opin Psychiatry* 21: 37–42.

[40] Cauffield JS, Forbes HJ (1999) Dietary supplements used in the treatment of depression, anxiety, and sleep disorders. *Lippincotts Prim Care Pract* 3: 290–304.

[41] Martin HB, Stofer WD, Eichinger MR (2000) Kavain inhibits murine airway smooth muscle contraction. *Planta Med* 66: 601–606.

[42] Gleitz J, Beile A, Peters T (1995) (+/-)-Kavain inhibits veratridine-activated voltage-dependent Na(+)-channels in synaptosomes prepared from rat cerebral cortex. *Neuropharmacology* 34: 1133–1138.

[43] Gleitz J, Beile A, Wilkens P, Ameri A, Peters T (1997) Antithrombotic action of the kava pyrone (+)-kavain prepared from *Piper methysticum* on human platelets. *Planta Med* 63: 27–30.

[44] Sullivan J, Romm J, Reilly M (2009) A brief report of student research: mechanism of analgesic effect and efficacy and anesthesia interactions of kava in the male Sprague-Dawley rat. *Dimens Crit Care Nurs* 28: 138–140.

[45] Triolet J, Shaik AA, Gallaher DD, O'Sullivan MG, Xing C (2012) Reduction in colon cancer risk by consumption of kava or kava fractions in carcinogen-treated rats. *Nutr Cancer* 64: 838–846.

[46] Li X, Liu Z, Xu X, Blair CA, Sun Z, et al. (2012) Kava components down-regulate expression of AR and AR splice variants and reduce growth in patient-derived prostate cancer xenografts in mice. *PLoS One* 7: e31213.

[47] Johnson TE, Hermanson D, Wang L, Kassie F, Upadhyaya P, et al. (2011) Lung tumorigenesis suppressing effects of a commercial kava extract and its selected compounds in A/J mice. *Am J Chin Med* 39: 727–742.

[48] Tang Y, Li X, Liu Z, Simoneau AR, Xie J, et al. (2010) Flavokawain B, a kava chalcone, induces apoptosis via up-regulation of death-receptor 5 and Bim expression in androgen receptor negative, hormonal refractory prostate cancer cell lines and reduces tumor growth. *Int J Cancer* 127: 1758–1768.

[49] Abu N, Ho WY, Yeap SK, Akhtar MN, Puad M, et al. (2013) The flavokawains: uprising medicinal chalcones. *Cancer Cell Int* 13: 102.

[50] Jussofie A, Schmiz A, Hiemke C (1994) Kavapyrone enriched extract from Piper methysticum as modulator of the GABA binding site in different regions of rat brain. *Psychopharmacology (Berl)* 116: 469–474.

[51] Munro G, Hansen RR, Mirza NR (2013) GABA(A) receptor modulation: potential to deliver novel pain medicines? *Eur J Pharmacol* 716: 17–23.

[52] Rudolph U, Knoflach F (2011) Beyond classical benzodiazepines: novel therapeutic potential of GABAA receptor subtypes. *Nat Rev Drug Discov* 10: 685–697.

[53] Yuan CS, Dey L, Wang A, Mehendale S, Xie JT, et al. (2002) Kavalactones and dihydrokavain modulate GABAergic activity in a rat gastric-brainstem preparation. *Planta Med* 68: 1092–1096.

[54] Garrett KM, Basmadjian G, Khan IA, Schaneberg BT, Seale TW (2003) Extracts of kava (*Piper methysticum*) induce acute anxiolytic-like behavioral changes in mice. *Psychopharmacology (Berl)* 170: 33–41.

[55] Bruner NR, Anderson KG (2009) Discriminative-stimulus and time-course effects of kava-kava (*Piper methysticum*) in rats. *Pharmacol Biochem Behav* 92: 297–303.

[56] Gleitz J, Friese J, Beile A, Ameri A, Peters T (1996) Anticonvulsive action of (+/-)-kavain estimated from its properties on stimulated synaptosomes and Na^+ channel receptor sites. *Eur J Pharmacol* 315: 89–97.

[57] Lichtman AH, Blankman JL, Cravatt BF (2010) Endocannabinoid overload. *Mol Pharmacol* 78: 993–995.

[58] Ligresti A, Villano R, Allara M, Ujvary I, Di Marzo V (2012) Kavalactones and the endocannabinoid system: the plant-derived yangonin is a novel CB(1) receptor ligand. *Pharmacol Res* 66: 163–169.

[59] Uebelhack R, Franke L, Schewe HJ (1998) Inhibition of platelet MAO-B by kava pyrone-enriched extract from *Piper methysticum* Forster (kava-kava). *Pharmacopsychiatry* 31: 187–192.

[60] Binda C, Hubalek F, Li M, Herzig Y, Sterling J, et al. (2004) Crystal structures of monoamine oxidase B in complex with four inhibitors of the N-propargylaminoindan class. *J Med Chem* 47: 1767–1774.

[61] Mallajosyula JK, Chinta SJ, Rajagopalan S, Nicholls DG, Andersen JK (2009) Metabolic control analysis in a cellular model of elevated MAO-B: relevance to Parkinson's disease. *Neurotox Res* 16: 186–193.

[62] Saura J, Luque JM, Cesura AM, Da Prada M, Chan-Palay V, et al. (1994) Increased monoamine oxidase B activity in plaque-associated astrocytes of Alzheimer brains revealed by quantitative enzyme radioautography. *Neuroscience* 62: 15–30.

[63] Dlugos AM, Palmer AA, de Wit H (2009) Negative emotionality: monoamine oxidase B gene variants modulate personality traits in healthy humans. *J Neural Transm* 116: 1323–1334.

[64] Liu Z, Xu X, Li X, Liu S, Simoneau AR, et al. (2013) KAVA chalcone, flavokawain A, inhibits urothelial tumorigenesis in the UPII-SV40T transgenic mouse model. *Cancer Prev Res (Phila)* 6: 1365–1375.

[65] Ji T, Lin C, Krill LS, Eskander R, Guo Y, et al. (2013) Flavokawain B, a kava chalcone, inhibits growth of human osteosarcoma cells through G2/M cell cycle arrest and apoptosis. *Mol Cancer* 12: 55.

[66] Warmka JK, Solberg EL, Zeliadt NA, Srinivasan B, Charlson AT, et al. (2012) Inhibition of mitogen activated protein kinases increases the sensitivity of A549 lung cancer cells to

the cytotoxicity induced by a kava chalcone analog. *Biochem Biophys Res Commun* 424: 488–492.

[67] Sakai T, Eskander RN, Guo Y, Kim KJ, Mefford J, et al. (2012) Flavokawain B, a kava chalcone, induces apoptosis in synovial sarcoma cell lines. *J Orthop Res* 30: 1045–1050.

[68] Zhao X, Chao YL, Wan QB, Chen XM, Su P, et al. (2011) Flavokawain B induces apoptosis of human oral adenoid cystic cancer ACC-2 cells via up-regulation of Bim and down-regulation of Bcl-2 expression. *Can J Physiol Pharmacol* 89: 875–883.

[69] Zi X, Simoneau AR (2005) Flavokawain A, a novel chalcone from kava extract, induces apoptosis in bladder cancer cells by involvement of Bax protein-dependent and mitochondria-dependent apoptotic pathway and suppresses tumor growth in mice. *Cancer Res* 65: 3479–3486.

[70] Ulbricht C, Basch E, Boon H, Ernst E, Hammerness P, et al. (2005) Safety review of kava (*Piper methysticum*) by the Natural Standard Research Collaboration. *Expert Opin Drug Saf* 4: 779–794.

[71] Fu PP, Xia Q, Guo L, Yu H, Chan PC (2008) Toxicity of kava kava. *J Environ Sci Health C Environ Carcinog Ecotoxicol Rev* 26: 89–112.

[72] Mathews JD, Riley MD, Fejo L, Munoz E, Milns NR, et al. (1988) Effects of the heavy usage of kava on physical health: summary of a pilot survey in an aboriginal community. *Med J Aust* 148: 548–555.

[73] Wooltorton E (2002) Herbal kava: reports of liver toxicity. *CMAJ* 166: 777.

[74] Anke J, Ramzan I (2004) Kava hepatotoxicity: are we any closer to the truth? *Planta Med* 70: 193–196.

[75] Lechtenberg M, Quandt B, Schmidt M, Nahrstedt A (2008) Is the alkaloid pipermethystine connected with the claimed liver toxicity of Kava products? *Pharmazie* 63: 71–74.

[76] Teschke R, Qiu SX, Lebot V (2011) Herbal hepatotoxicity by kava: update on pipermethystine, flavokavain B, and mould hepatotoxins as primarily assumed culprits. *Dig Liver Dis* 43: 676–681.

[77] Clouatre DL (2004) Kava kava: examining new reports of toxicity. *Toxicol Lett* 150: 85–96.

[78] Stickel F, Baumuller HM, Seitz K, Vasilakis D, Seitz G, et al. (2003) Hepatitis induced by Kava (*Piper methysticum* rhizoma). *J Hepatol* 39: 62–67.

[79] Schulze J, Raasch W, Siegers CP (2003) Toxicity of kava pyrones, drug safety and precautions – a case study. *Phytomedicine* 10 Suppl 4: 68–73.

[80] Humberston CL, Akhtar J, Krenzelok EP (2003) Acute hepatitis induced by kava kava. *J Toxicol Clin Toxicol* 41: 109–113.

[81] Mathews JM, Etheridge AS, Black SR (2002) Inhibition of human cytochrome P450 activities by kava extract and kavalactones. *Drug Metab Dispos* 30: 1153–1157.

[82] Li XZ, Ramzan I (2010) Role of ethanol in kava hepatotoxicity. *Phytother Res* 24: 475–480.

[83] Zou L, Henderson GL, Harkey MR, Sakai Y, Li A (2004) Effects of kava (Kava-kava, 'Awa, Yaqona, Piper methysticum) on c-DNA-expressed cytochrome P450 enzymes and human cryopreserved hepatocytes. *Phytomedicine* 11: 285–294.

[84] Zhang H, Forman HJ (2012) Glutathione synthesis and its role in redox signaling. *Semin Cell Dev Biol* 23: 722–728.

[85] Zhou P, Gross S, Liu JH, Yu BY, Feng LL, et al. (2010) Flavokawain B, the hepatotoxic constituent from kava root, induces GSH-sensitive oxidative stress through modulation of IKK/NF-kappaB and MAPK signaling pathways. *FASEB J* 24: 4722–4732.

[86] Whitton PA, Lau A, Salisbury A, Whitehouse J, Evans CS (2003) Kava lactones and the kava-kava controversy. *Phytochemistry* 64: 673–679.

[87] Tang J, Dunlop RA, Rowe A, Rodgers KJ, Ramzan I (2011) Kavalactones Yangonin and Methysticin induce apoptosis in human hepatocytes (HepG2) in vitro. *Phytother Res* 25: 417–423.

[88] Escher M, Desmeules J, Giostra E, Mentha G (2001) Hepatitis associated with Kava, a herbal remedy for anxiety. *BMJ* 322: 139.

[89] Russmann S, Lauterburg BH, Helbling A (2001) Kava hepatotoxicity. *Ann Intern Med* 135: 68–69.

[90] Fu S, Korkmaz E, Braet F, Ngo Q, Ramzan I (2008) Influence of kavain on hepatic ultrastructure. *World J Gastroenterol* 14: 541–546.

[91] Rowe A, Zhang LY, Ramzan I (2011) Toxicokinetics of kava. *Adv Pharmacol Sci* 2011: 326–724.

[92] Shimoda LM, Park C, Stokes AJ, Gomes HH, Turner H (2012) Pacific island 'Awa (Kava) extracts, but not isolated kavalactones, promote proinflammatory responses in model mast cells. *Phytother Res* 26: 1934–1941.

[93] Pollastri MP, Whitty A, Merrill JC, Tang X, Ashton TD, et al. (2009) Identification and characterization of kava-derived compounds mediating TNF-alpha suppression. *Chem Biol Drug Des* 74: 121–128.

[94] Reilly TP, Brady JN, Marchick MR, Bourdi M, George JW, et al. (2001) A protective role for cyclooxygenase-2 in drug-induced liver injury in mice. *Chem Res Toxicol* 14: 1620–1628.

[95] Li G, Han C, Xu L, Lim K, Isse K, et al. (2009) Cyclooxygenase-2 prevents fas-induced liver injury through up-regulation of epidermal growth factor receptor. *Hepatology* 50: 834–843.

[96] Raman P, Dewitt DL, Nair MG (2008) Lipid peroxidation and cyclooxygenase enzyme inhibitory activities of acidic aqueous extracts of some dietary supplements. *Phytother Res* 22: 204–212.

[97] Wu D, Yu L, Nair MG, DeWitt DL, Ramsewak RS (2002) Cyclooxygenase enzyme inhibitory compounds with antioxidant activities from *Piper methysticum* (kava kava) roots. *Phytomedicine* 9: 41–47.

[98] Wu D, Nair MG, DeWitt DL (2002) Novel compounds from *Piper methysticum* Forst (Kava Kava) roots and their effect on cyclooxygenase enzyme. *J Agric Food Chem* 50: 701–705.

[99] Nicolaou M, Andress EJ, Zolnerciks JK, Dixon PH, Williamson C, et al. (2012) Canalicular ABC transporters and liver disease. *J Pathol* 226: 300–315.

[100] Wakabayashi Y, Kipp H, Arias IM (2006) Transporters on demand: intracellular reservoirs and cycling of bile canalicular ABC transporters. *J Biol Chem* 281: 27669–27673.

[101] Kipp H, Arias IM (2002) Trafficking of canalicular ABC transporters in hepatocytes. *Annu Rev Physiol* 64: 595–608.

[102] Weiss J, Sauer A, Frank A, Unger M (2005) Extracts and kavalactones of *Piper methysticum* G. Forst (kava-kava) inhibit P-glycoprotein in vitro. *Drug Metab Dispos* 33: 1580–1583.

[103] Endicott JA, Ling V (1989) The biochemistry of P-glycoprotein-mediated multidrug resistance. *Annu Rev Biochem* 58: 137–171.

[104] Gottesman MM, Pastan I, Ambudkar SV (1996) P-glycoprotein and multidrug resistance. *Curr Opin Genet Dev* 6: 610–617.

[105] Fu D (2013) Where is it and how does it get there – intracellular localization and traffic of P-glycoprotein. *Front Oncol* 3: 321.

[106] Fu D, Arias IM (2012) Intracellular trafficking of P-glycoprotein. *Int J Biochem Cell Biol* 44: 461–464.

[107] Halilbasic E, Claudel T, Trauner M (2013) Bile acid transporters and regulatory nuclear receptors in the liver and beyond. *J Hepatol* 58: 155–168.

[108] Porceddu M, Buron N, Roussel C, Labbe G, Fromenty B, et al. (2012) Prediction of liver injury induced by chemicals in human with a multiparametric assay on isolated mouse liver mitochondria. *Toxicol Sci* 129: 332–345.

[109] Russmann S, Kullak-Ublick GA, Grattagliano I (2009) Current concepts of mechanisms in drug-induced hepatotoxicity. *Curr Med Chem* 16: 3041–3053.

[110] Begriche K, Massart J, Robin MA, Borgne-Sanchez A, Fromenty B (2011) Drug-induced toxicity on mitochondria and lipid metabolism: mechanistic diversity and deleterious consequences for the liver. *J Hepatol* 54: 773–794.

[111] Pessayre D, Mansouri A, Berson A, Fromenty B (2010) Mitochondrial involvement in drug-induced liver injury. *Handb Exp Pharmacol* 196: 311–365.

[112] Ruepp SU, Tonge RP, Shaw J, Wallis N, Pognan F (2002) Genomics and proteomics analysis of acetaminophen toxicity in mouse liver. *Toxicol Sci* 65: 135–150.

[113] Vickers AE (2009) Characterization of hepatic mitochondrial injury induced by fatty acid oxidation inhibitors. *Toxicol Pathol* 37: 78–88.

[114] Fromenty B, Pessayre D (1995) Inhibition of mitochondrial beta-oxidation as a mechanism of hepatotoxicity. *Pharmacol Ther* 67: 101–154.

[115] Jones DP, Lemasters JJ, Han D, Boelsterli UA, Kaplowitz N (2010) Mechanisms of pathogenesis in drug hepatotoxicity putting the stress on mitochondria. *Mol Interv* 10: 98–111.

[116] McGill MR, Yan HM, Ramachandran A, Murray GJ, Rollins DE, et al. (2011) HepaRG cells: a human model to study mechanisms of acetaminophen hepatotoxicity. *Hepatology* 53: 974–982.

[117] Lude S, Torok M, Dieterle S, Jaggi R, Buter KB, et al. (2008) Hepatocellular toxicity of kava leaf and root extracts. *Phytomedicine* 15: 120–131.

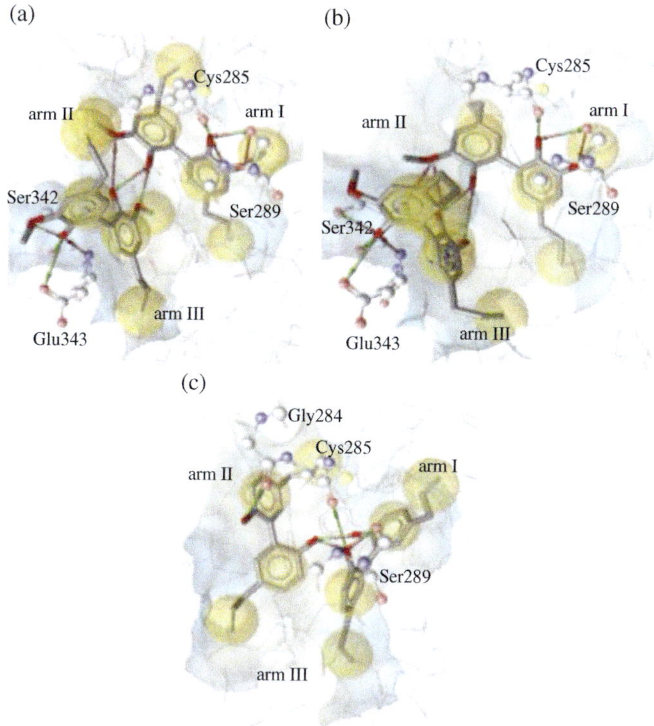

FIGURE 12.3 Neolignans: dieugenol (a); tetrahydrodieugenol (b); magnolol (c), aligned with a structure-based pharmacophore model consisting of hydrogen bond acceptor (red arrow), hydrogen bond donor (green arrow), hydrophobic interaction (yellow sphere), and aromatic interaction (blue rings).

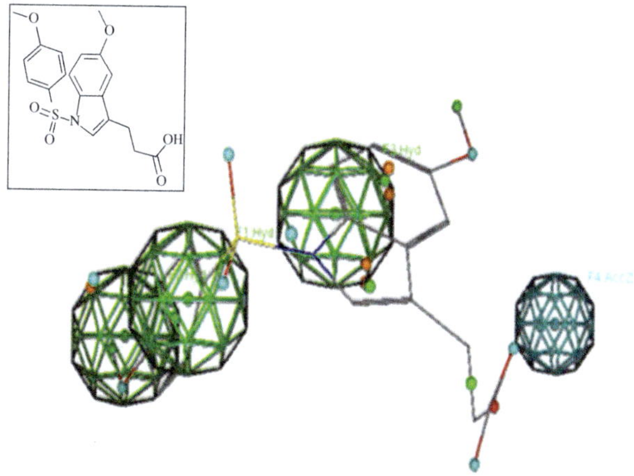

FIGURE 12.6 Four-point pharmacophore model for known PPARγ partial agonists superimposed on indeglitazar (RMSD = 0.50 Å). Pharmacophoric features are represented by a point encased in a sphere: hydrogen bond donor (blue), hydrophobic region (green), aromatic center (orange), and CO_2 centroid (red). Points not encased in spheres are other potential pharmacophore features on the indeglitazar structure (2D structure shown on top left).

Phytotherapies: Efficacy, Safety, and Regulation, First Edition. Edited by Iqbal Ramzan.

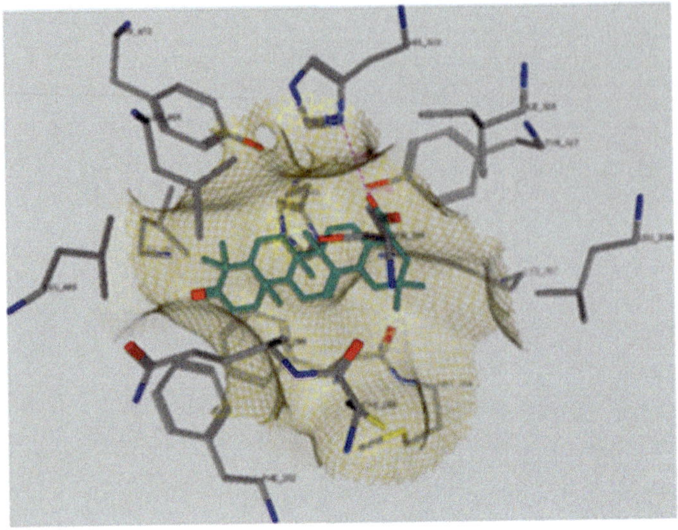

FIGURE 12.7 Oleanonic acid docked in the ligand-binding domain of PPARγ, which shows a hydrogen bond between the carboxylic moiety with His323 and Thr327 on helix 4/5 on arm I. The remainder of the ligand is stabilized within a hydrophobic pocket formed by residues Gln286, Met364, Leu453, and Leu469.

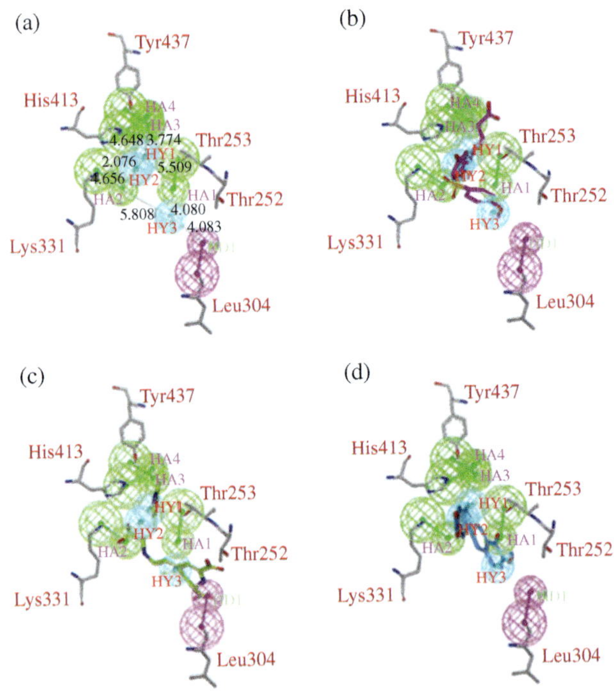

FIGURE 12.9 Generated structure-based pharmacophore models for PPARδ shown with inter-feature distance constraints only (a), control, ET1 (b), (S)-tryptophan-betaxanthin (c), and berberrubine (d). Pharmacophoric features are shown for hydrogen bond acceptors (green), hydrogen bond donors (magenta), and hydrophobic feature (blue).

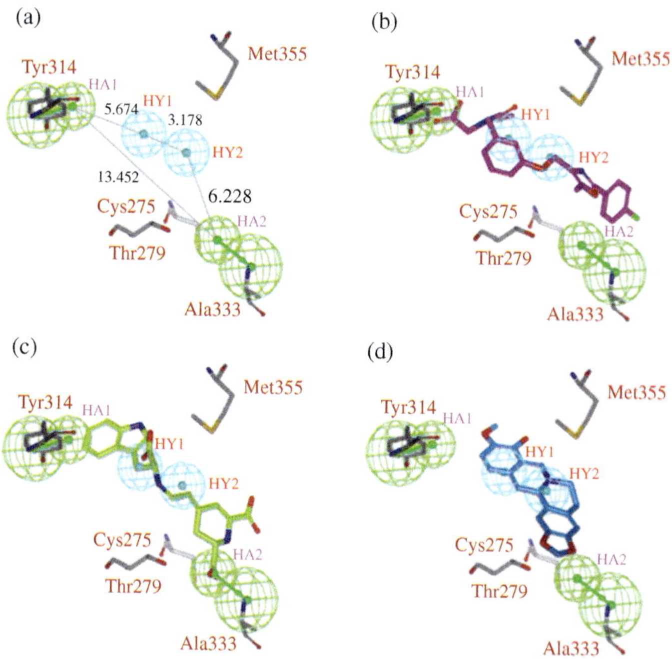

FIGURE 12.10 Generated structure-based pharmacophore models for PPARα shown with inter-feature distance constraints only (a), control, 7HA (b), (S)-tryptophan-betaxanthin (c), and berberrubine (d). Pharmacophoric features are shown for hydrogen bond acceptors (green), hydrogen bond donors (magenta), and hydrophobic feature (blue).

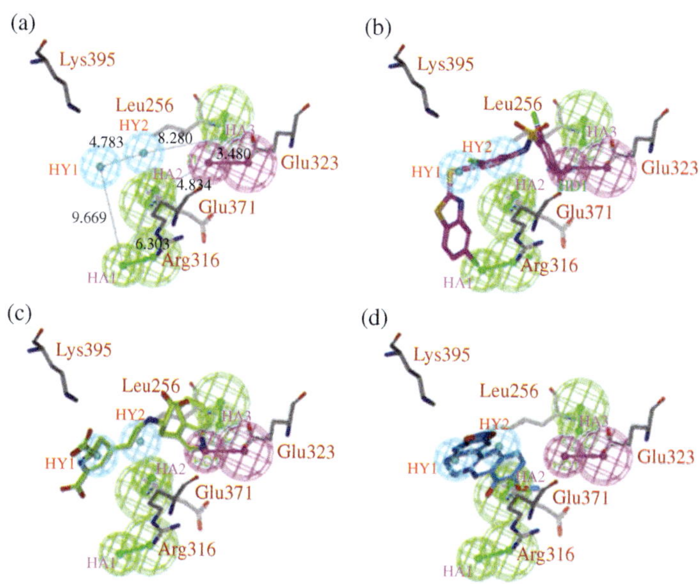

FIGURE 12.11 Generated structure-based pharmacophore models for PPARγ shown with inter-feature distance constraints only (a), control, T2384 (b), (S)-tryptophan-betaxanthin (c), and berberrubine (d). Pharmacophoric features are shown for hydrogen bond acceptors (green), hydrogen bond donors (magenta), and hydrophobic feature (blue).

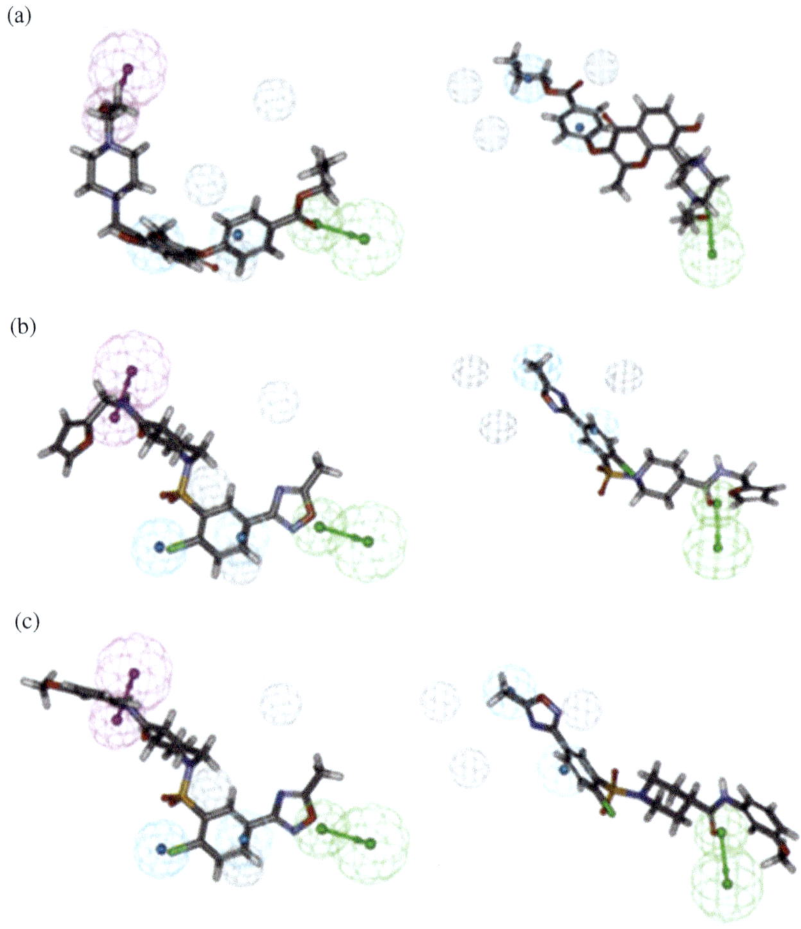

FIGURE 12.17 Alignment of top three virtual hits ZINC19924472 (a), ZINC33086616 (b), and ZINC33086598 (c) with best pharmacophore hypotheses for ACC1 (*Hypo1_ACC1*, left) and ACC2 (*Hypo1_ACC2*, right). Chemical features include hydrogen bond acceptors (green), hydrogen bond donors (magenta), hydrophobic features (blue), and exclusion volumes (gray).

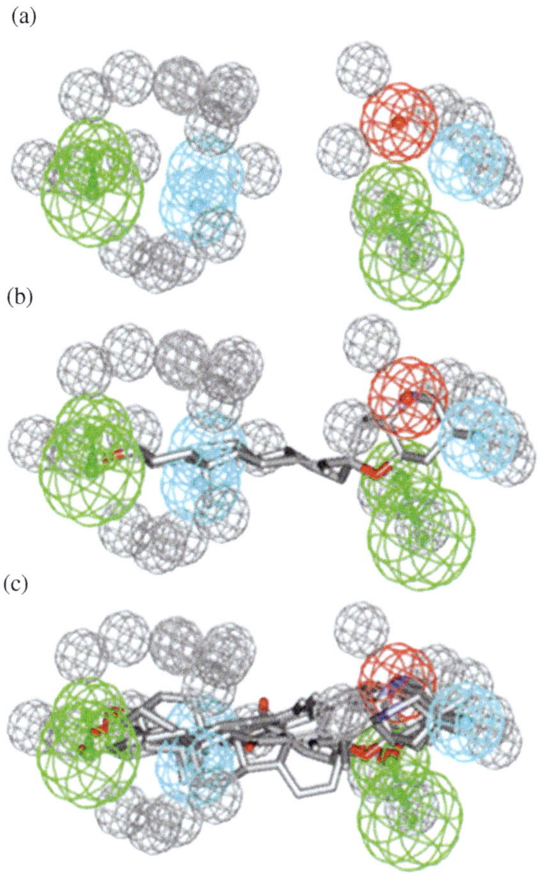

FIGURE 12.19 Pharmacophore model *Hypo2* (a), alignment with cyclopamine (b), alignment with active compounds 4, 6, 7, and 10. Chemical features include hydrogen bond acceptors (green), hydrophobic features (blue), positive ionizable features (red), and exclusion volumes (gray).

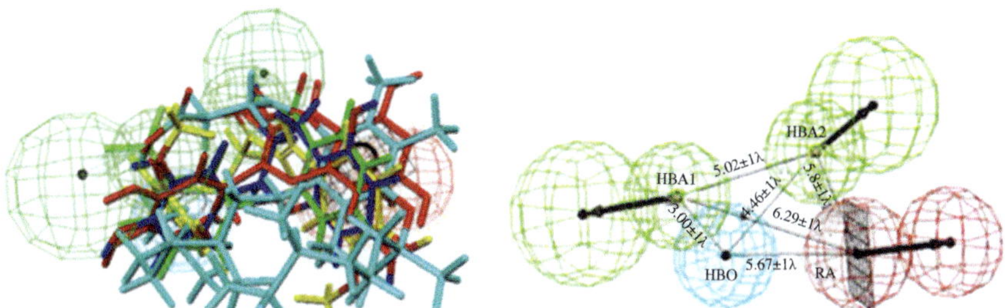

FIGURE 12.21 Common feature-based pharmacophore model of ARDAs *Hypo1* (left). *Hypo1* mapping all the important features of training set ARDAs (right). Chemical features include two hydrogen bond acceptors (green), one hydrophobic group (cyan), and one aromatic feature (red).

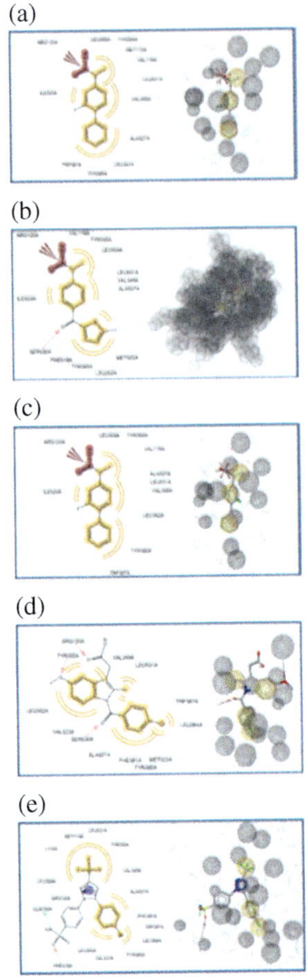

FIGURE 12.27 Observed interactions between 3D X-ray crystal structures with its native ligand (left) and its corresponding structure-based pharmacophore model (right), which form the final model collection for virtual screening: *1cqe-1* (a), *1pge-2-s* (b), *2ayl-1* (c), *4cox-2* (d), and *6cox-1-s* (e).

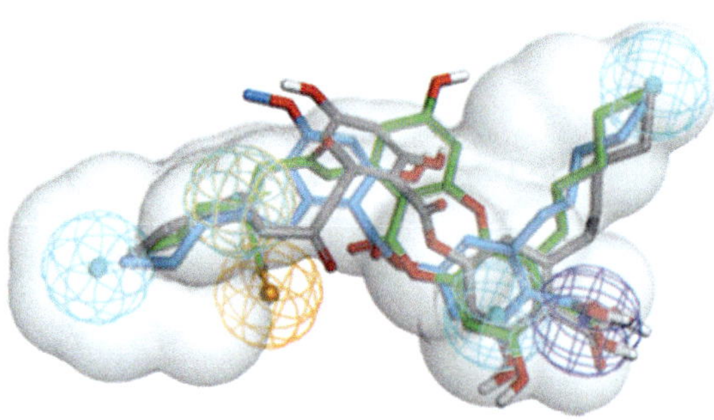

FIGURE 12.33 Pharmacophore model for acidic mPGES-1 inhibitors. For *M1*, it is necessary to satisfy all chemical features consisting of four hydrophobic features (cyan), one aromatic ring (gold), one negatively ionizable feature (blue), and a spatial shape restriction (gray). Whereas screening with *M2* allows the omission of one hydrophobic group or aromatic ring features, inhibitors of mPGES-1, namely, 2 (green), 8 (blue), and 9 (gray) map two of the hydrophobic features with their alkyl chains.

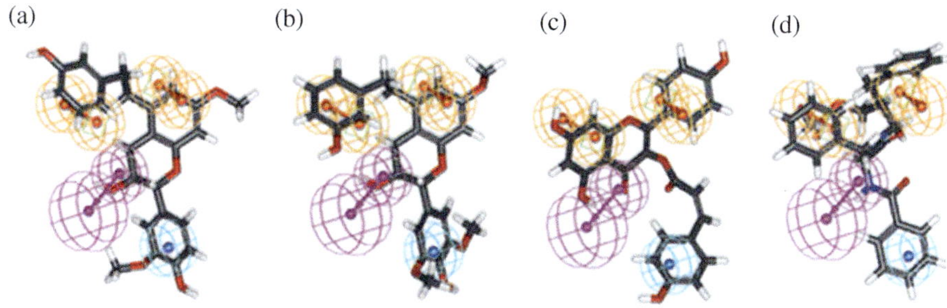

FIGURE 12.34 Pharmacophore mapping for (a) shanciol A and (b) shanciol B, (c) castilliferol, and (d) aurantiamide acetate. Chemical features include one hydrogen bond donor (purple), one hydrophobic group (light blue), and two aromatic rings (orange).

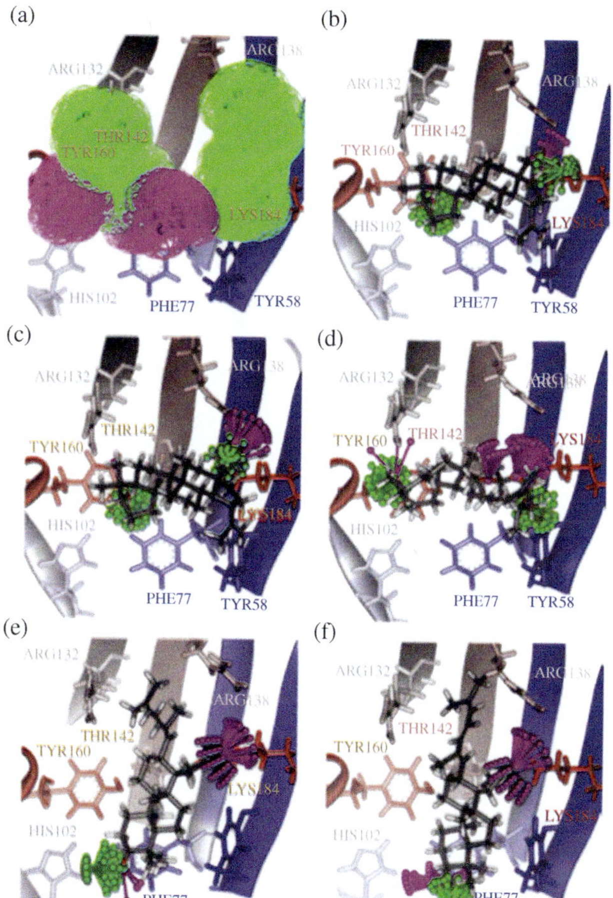

FIGURE 12.44 Overall pharmacophore model (a), and pharmacophore mapping of betulin (b), betulic acid (c), jujubogenin (d), *cis*-ebelin lactone (e), and *trans*-ebelin lactone (f). Pharmacophore features include hydrogen bond acceptors (green) and hydrogen bond donors (purple).

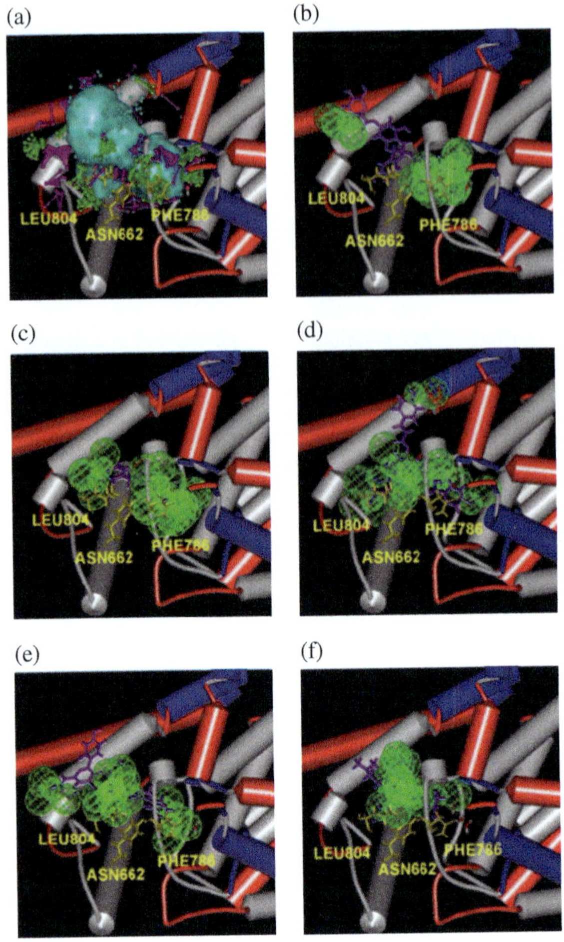

FIGURE 12.47 Overall pharmacophore of PDE5 binding site (a), CS01 (b), CS03 (c), ES03a (d), ES03b (e), and natural substrate cGMP (f). Chemical features include hydrogen bond acceptors (green), hydrogen bond donors (purple), and hydrophobic regions (blue).

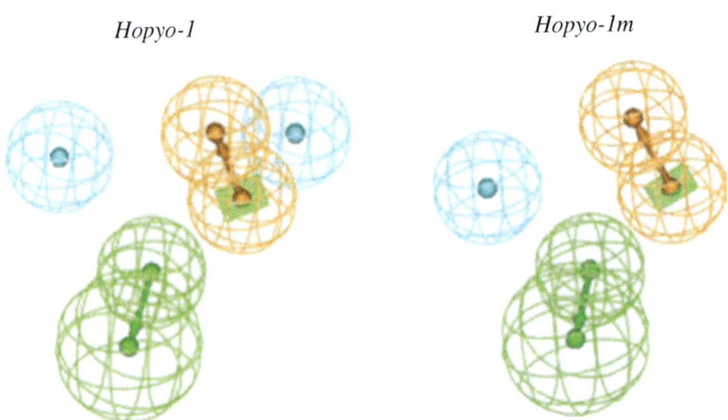

FIGURE 12.49 Pharmacophore models *Hopyo-1* (left) and *Hopyo-1m* (right) that consist of chemical features, which include hydrogen bond acceptors (green), hydrophobic features (blue), and aromatic rings (yellow).

14

PHYTOTHERAPIES AS NEW DRUG SOURCES: GOSSYPOL AND CURCUMIN

Vivian Wan Yu Liao[1], Rajeshwar Narlawar[2], David E. Hibbs[1], and Paul W. Groundwater[1]

[1]*Faculty of Pharmacy, The University of Sydney, Sydney, New South Wales, Australia*
[2]*School of Chemistry, The University of Sydney, Sydney, New South Wales, Australia*

14.1 BOTANICAL SOURCES OF GOSSYPOL AND CURCUMIN

Gossypol (1,1′,6,6′,7,7′-hexahydroxy-5,5′-diisopropyl-3,3′-dimethyl-2,2′-binaphthalene-8,8′-dicarboxaldehyde) **1** is a yellow polyphenolic aldehyde that is obtained from the glands of cottonseeds (*Gossypium*), for example, *Gossypium hirsutum* and the tropical tree *Thespesia populnea*, both members of the Malvaceae family [1]. As a result of axial chirality about the central C2–C2′ bond, gossypol exists in two stereoisomeric forms; an excess of (+)-gossypol has been found in *Gossypium arboreum*, *G. herbaceum*, *G. hirsutum*, and *T. populnea*, whereas (−)-gossypol is found in excess in *G. barbadense* (Table 14.1 [2]). Meyers and Willemsen have completed the total synthesis of (−)-gossypol **1** in 26 steps from commercially available starting materials [3].

Phytotherapies: Efficacy, Safety, and Regulation, First Edition. Edited by Iqbal Ramzan.

TABLE 14.1 Examples of the Enantiomeric (Atropisomeric) Excesses of Gossypol in Cottonseeds (*Gossypium* Species) and *T. Populnea* [2]

Species/Variety	Gossypol % (−)	Gossypol % (+)	Enantiomeric Excess (ee) %
G. hirsutum			
latifolium Hutch	46	54	8 (+)
marie-galante	5	95	90 (+)
G. barbadense			
brasiliense Tussac	65	35	30 (−)
brasiliense Tussac	59	41	18 (−)
G. herbaceum			
africanum Hutch	37	63	26 (+)
T. populnea	28	71	43 (+)

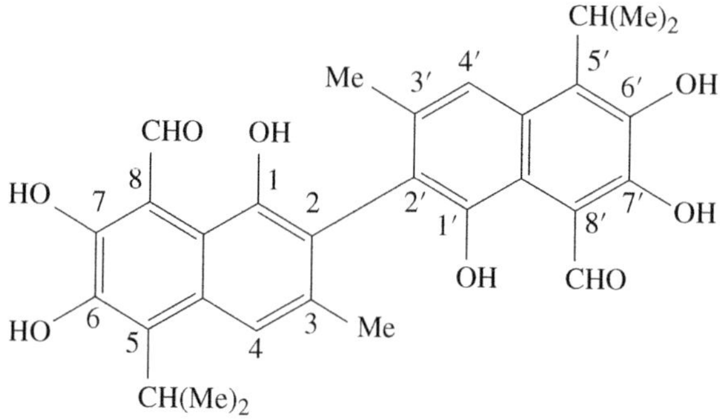

1

Curcumin ((1*E*,6*E*)-1,7-bis(4-hydroxy-3-methoxyphenyl)-1,6-heptadiene-3,5-dione or diferuloylmethane) **2** is a bright yellow/orange solid that is responsible for the color of the Indian spice turmeric (*Curcuma longa*), a member of the ginger family Zingiberaceae [4].

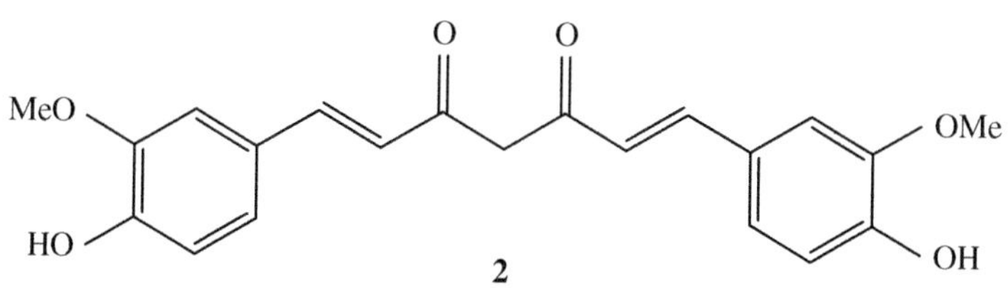

2

Curcumin is the major bioactive constituent of turmeric, constituting approximately 3–5% of the mass of the dried rhizome [5, 6]. Turmeric has been part of the daily diet in many Asian countries and has been used in traditional remedies in Ayurvedic and traditional Chinese medicine for centuries [7].

Curcumin holds great therapeutic potential and has been a molecule of interest for many years, as it exhibits activity against a wide range of diseases, with no known toxicity. To date, no maximal tolerated dose has been defined, and up to 12 g of curcumin a day is well tolerated [8]. Curcumin has anticancer, anti-inflammatory, antioxidant, antiviral, and antimalarial activity, and these will be reviewed in Section 14.4.

14.2 STEREOISOMERISM, TAUTOMERISM, AND REACTIVITY

14.2.1 Stereoisomerism

As mentioned in Section 14.1, gossypol **1** exhibits atropisomerism, a form of stereoisomerism known as axial chirality, in which restricted rotation about a single bond (the C2–C2′ bond of gossypol is the chiral axis) gives rise to different conformations of a molecule that can be isolated (resolved). Atropisomerism in biphenyls (or binaphthyls, such as gossypol) is mostly the result of the steric clash of the *ortho*-substituents, with the height of the energy barrier between atropisomers being largely dependent on the steric bulk of these groups. If the sum of the van der Waals radii (r_{vdW}) of the *ortho*-substituents in a biphenyl/binaphthyl is greater than 290 pm, then the barrier to rotation is sufficiently large for the atropisomers to be resolved. In the case of gossypol, the substituents are methyl (r_{vdW} = 173 pm) and hydroxyl (r_{vdW} = 145 pm) so the two atropisomeric forms are effectively non-interconvertible, with the plane of one naphthyl unit being almost perpendicular to that of the other (Figure 14.1).

The stereochemical descriptors for atropisomers are the same as those used for helices, and the absolute configuration of the levorotatory (−)-isomer of gossypol has been assigned as *M* (for minus, i.e., counterclockwise), with the other atropisomer of the pair being (*P*)-(+)-gossypol [9].

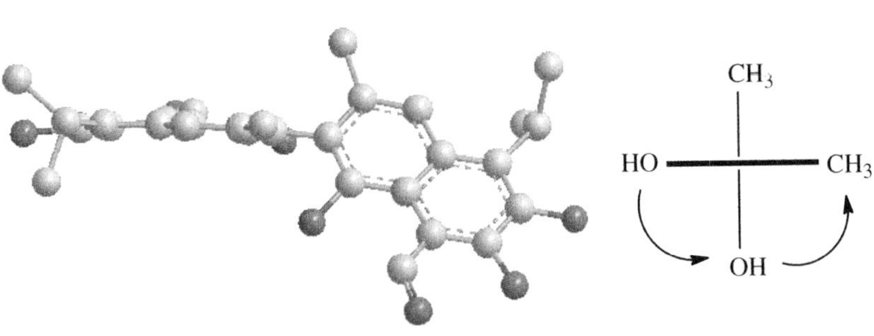

FIGURE 14.1 (*M*)-(−)-Gossypol **1**. In the right-hand figure, the lines represent the planes of the naphthyl groups, and the groups shown are those *ortho* to the C2–C2 bond. When viewed along the C2–C2 bond from the RHS, the horizontal OH group is on the LHS, and the vertical OH is below the plane of the horizontal naphthyl system.

14.2.2 Tautomerism

As a result of the functional groups present, and the fact that it is a dimer, gossypol can exist in a number of symmetrical and asymmetrical tautomeric forms, with the symmetrical forms (ald–ald **1**, ketol–ketol **1a**, lactol–lactol **1b**) shown in Scheme 14.1. The aldehyde (ald) tautomer **1** is the predominant form in nonpolar solvents such as chloroform ($CHCl_3$) and dichloromethane (CH_2Cl_2) solutions [10, 11]. The ketol tautomer **1a** is present in aqueous alkaline solutions [12], while in polar solvents an equilibrium exists between the aldehyde **1a** and lactol **1b** tautomers, which favors the lactol form **1b** when the nucleophilicity of the solvent increases, for example, DMSO solution [10]. The lactol–lactol tautomer **1b** contains additional stereogenic centers (*) as a result of the formation of the 5-membered ring. The biological activity of gossypol and its analogues, which will be discussed in future sections, will, of course, be influenced by the binding of the predominant tautomer to its cellular target.

Tautomerism is also possible for gossypol analogues, such as the Schiff's bases (imines) **3** formed on reaction with ammonia derivatives (Scheme 14.2) [13]. The position of the tautomeric equilibrium is dependent upon both the solvent and the substituents present, with both the diastereoisomers [i.e., the Schiff's bases with (+)- and (−)-gossypol] of the phenylalanine **3a** and indolyl **3c** derivatives existing as the enamines in $CDCl_3$, while the tyrosine ethyl ester **3b** exists as the imine form in $CDCl_3$ but as the enamine in DMSO-d_6.

Curcumin, as a β-diketone, also exhibits tautomerism between the diketone **2** and keto–enol **2a** forms (Scheme 14.3), with the keto–enol form predominating in all solvents, irrespective of their polarities, from chloroform to buffered aqueous DMSO (pH 3–9) [14]. The keto–enol form in this case is stabilized with respect to the diketo form as a result of the extended π-conjugation within the molecule and the intramolecular hydrogen bond reducing the dipole–dipole repulsion between the carbonyl groups. The predominance of the keto–enol tautomer may have implications in the binding of curcumin and its analogues to their cellular targets, and we will discuss this in greater detail in the following sections.

14.2.3 Reactivity

In the following sections, we will discuss the reactivity of both gossypol and curcumin in terms of the derivatives that have been synthesized and had their biological activity assessed. In this section, however, we will simply outline the characteristic reaction of each compound. Gossypol has a number of phenolic groups and so can form a range of aryl ethers, but its key reaction could be considered to be imine (Schiff's base) formation, through reaction of the aldehyde group(s) with amines [13]. A range of Schiff's bases have been prepared [15], and this reversible reaction has been widely employed in the resolution of the gossypol atropisomers via the chromatographic separation of their diastereoisomeric Schiff's bases, for example, **3a–c**, and their hydrolysis back to the individual atropisomers [2, 13]. Partial hydrolysis of

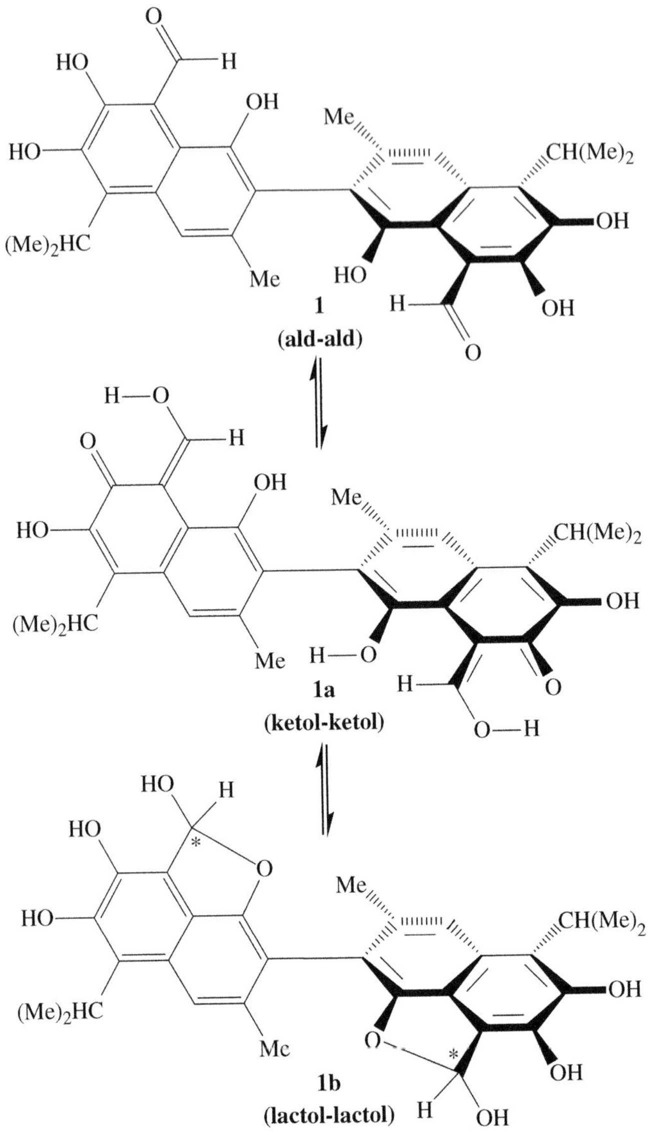

SCHEME 14.1 Symmetrical gossypol tautomers.

the bis-Schiff's bases can also be accomplished to give half-Schiff's bases **4** (Scheme 14.4) [13].

As mentioned earlier, the barrier to rotation about C2–C2′ means that the atropisomers of gossypol are noninterconvertible, and the only example of the apparent interconversion of gossypol is the photoepimerization of the bis-Schiff's base with L-phenylalanine methyl ester **3a** (Scheme 14.5) [16].

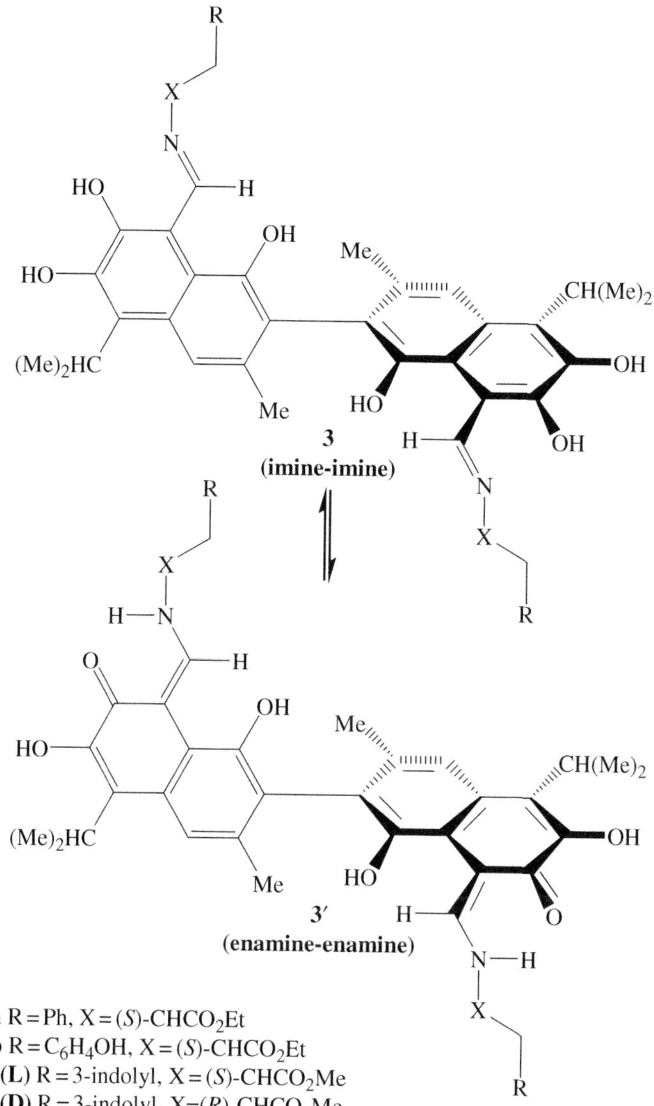

3a R = Ph, X = (*S*)-$CHCO_2Et$
3b R = C_6H_4OH, X = (*S*)-$CHCO_2Et$
3c(L) R = 3-indolyl, X = (*S*)-$CHCO_2Me$
3c(D) R = 3-indolyl, X=(*R*)-$CHCO_2Me$

SCHEME 14.2 Tautomerism of gossypol Schiff's bases; imine **3** and enamine **3′** tautomers.

The characteristic reaction of an α,β-unsubstituted enone is a nucleophilic addition, either at the carbonyl or at the β-position of the double bond (Michael addition), with the latter attack predominating for soft nucleophiles, such as the thiol of glutathione that has been proposed to form a bis-Michael adduct **5** with curcumin **2** (Scheme 14.6) [17].

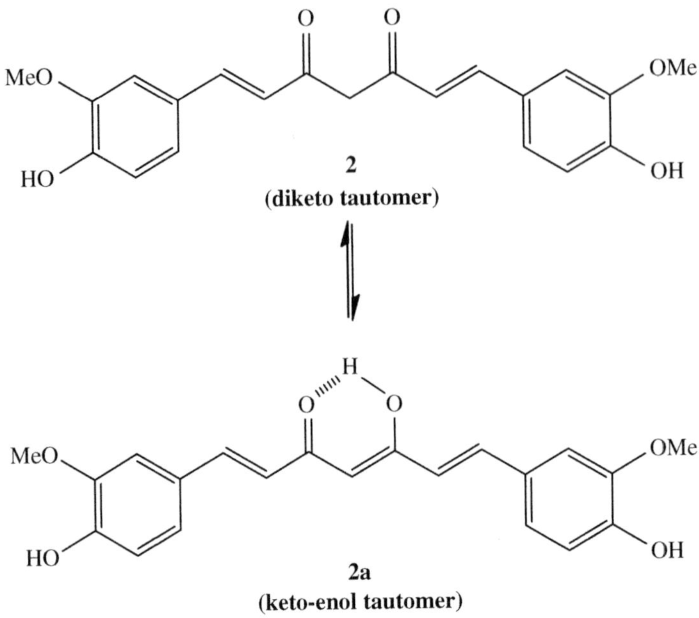

SCHEME 14.3 Tautomers of curcumin.

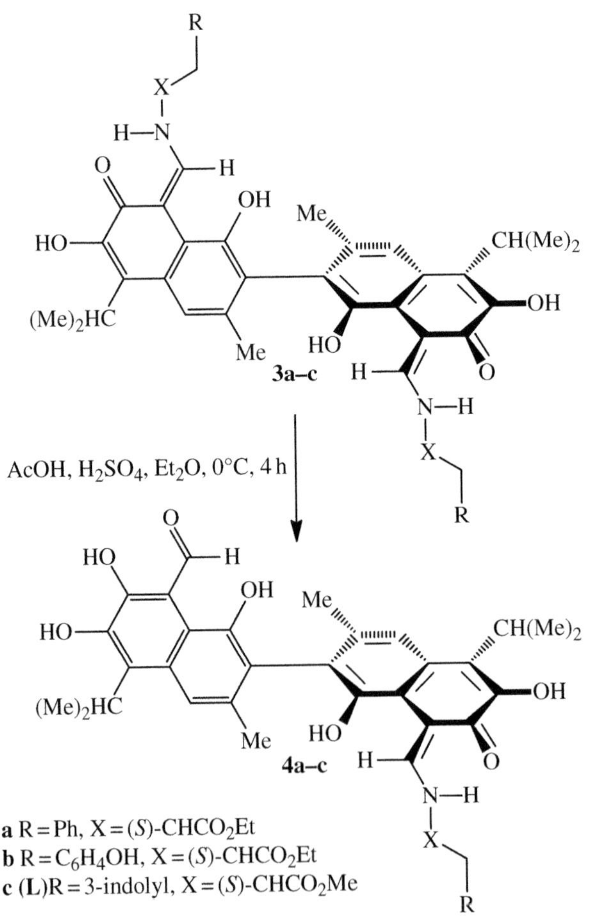

a R = Ph, X = (*S*)-$CHCO_2Et$
b R = C_6H_4OH, X = (*S*)-$CHCO_2Et$
c (**L**)R = 3-indolyl, X = (*S*)-$CHCO_2Me$

SCHEME 14.4 Partial hydrolysis of gossypol bis-Schiff's bases **3**.

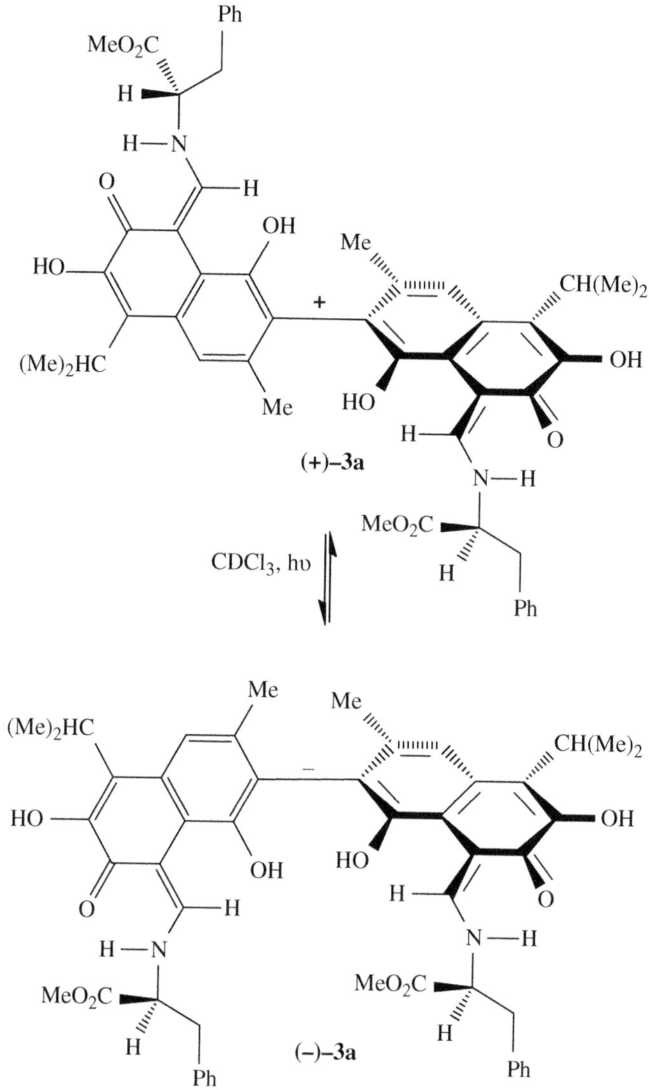

SCHEME 14.5 Photoepimerization of gossypol bis-Schiff's base **3a**.

14.3 BIOLOGICAL ACTIVITY OF GOSSYPOL AND ITS ANALOGUES

14.3.1 Antifertility

Perhaps, the most studied biological effect of gossypol is its antifertility activity, which was initially discovered as a result of an observation by Liu that there had been no children born in a decade in a rural community in Jiangsu, P.R. China [18]. This observation was attributed to the fact that the community had switched to crude

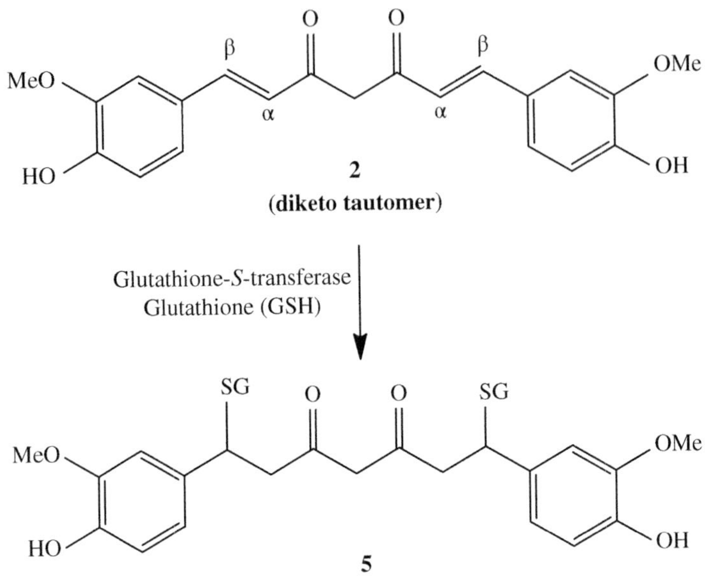

5

SCHEME 14.6 Proposed formation of a bis-Michael adduct **5** from curcumin and glutathione.

cottonseed oil for cooking and the study of the antispermatogenic effects of gossypol was initiated. Clinical trials followed in China, in which approximately 8800 men were treated with gossypol or gossypol acetic acid (20 mg/day for 60–70 days, usually followed by a maintenance dose of 60 mg/week), resulting in azoospermia or sperm counts of below four million per milliliter in 99% of cases. Despite gossypol's obvious efficacy, in 1986, the WHO announced a cessation in funding for research on its antifertility effects as a result of the occurrence of irreversible testicular damage and hypokalemia. Later studies showed that the hypokalemia was related to dietary intake of potassium in China, as no changes were observed in serum potassium levels in males who were administered gossypol in a Brazilian study [19, 20]. This finding prompted some research groups to propose gossypol as an alternative to vasectomy in men who have completed their families, due to the high levels of irreversibility (20%); indeed, a male contraceptive pill ("Nofertil") was developed by Brazilian pharmaceutical company, Hebron S.A., but never marketed [21].

This area has been extensively reviewed [20, 22, 23] and so will not be considered further here.

14.3.2 Anticancer

Gossypol and its derivatives act on a number of cellular targets, and their growth inhibitory activity has thus been investigated in a number of human tumor cell lines [20, 24]. As can be seen from the examples given in Table 14.2, the growth

TABLE 14.2 Examples of the Growth Inhibitory Activity of Gossypol Atropisomers in Human Cancer Cell Lines

	Gossypol GI_{50} (μM)		Relative Potency
Cell Line	(−)-1	(+)-1	(+)-1/(−)-1
OVCA-429 (**ovarian**)	0.91	11.28	12.40 [25]
OVCA-432 (**ovarian**)	1.10	10.16	9.24 [25]
OVCAR-3 (**ovarian**)	0.62	5.80	9.35 [25]
SK-MEL-19 (**melanoma**)	21.0	>30	>1.43 [26]
SK-MEL-28 (**melanoma**)	16.5	>30	1.82 [26]
HL-60 (**leukemia**)	0.64	3.20	5.00 [25]
HeLa (**cervical**)	17.8	31.3	1.76 [27]
U87 (**glioblastoma**)	30.2	59.6	1.99 [27]
M85 (**gastric**)	18.4	39.7	2.16 [27]

inhibitory activity of (−)-gossypol against human cancer cell lines is usually greater than that of its atropisomer, but despite its low micromolar activity against a range of cell lines, gossypol alone initially seemed to have limited potential for development as an anticancer agent as data from *in vivo* studies did not support its use. For example, a study by Stein et al., in which a total of 34 patients with advanced cancers were treated with weekly or daily escalating doses of oral gossypol, concluded that gossypol was safe (the dose-related limiting toxicity being emesis) but was unlikely to be clinically useful in the treatment of advanced cancers as there was no evidence of tumor regression in any of the patients [28]. In a phase I/II clinical trial, in which racemic gossypol was administered to women with refractory metastatic breast cancer, despite gossypol being well tolerated (two patients receiving 50 mg/day gossypol experienced dose-limiting dermatological toxicity, while 30% of patients experienced nausea, 15% emesis, 15% fatigue, 15% altered taste sensations, and 10% diarrhea), no therapeutic responses were observed—only 1 of the 20 patients showed a partial response (and 2 of the 20 had stable disease) [29]. In this limited study, gossypol administration appeared to lead to a decrease in cyclin D1 expression and an increase in nuclear retinoblastoma (Rb) protein expression, suggesting a role for gossypol as a modulator of cell cycle regulation.

Schiff's bases are often more potent than the individual gossypol atropisomers in cell proliferation studies but none have progressed into clinical trials [24, 27, 30–32].

More recent studies showing that it is a potent inhibitor of the B-cell lymphoma/leukemia-2 (Bcl-2) protein family seem to hold more promise for the anticancer activity of gossypol and its derivatives. (−)-Gossypol **1** (AT-101, Ascenta Therapeutics) is a pan-bcl-2 inhibitor, promoting apoptosis by acting as a BH3 mimic, and is currently in a number of phase II clinical trials, particularly in combination with established anticancer agents [33]. In a randomized phase II trial of docetaxel plus prednisone, in combination with AT-101 or placebo, for the treatment of metastatic castration-resistant prostate cancer (mCRPC), gossypol was well tolerated but did not lead to greater overall survival [34]. In a phase I/II clinical trial in chemotherapy-naïve CRPC patients, AT-101 administered at 20 mg/day was well tolerated and resulted in a greater than 50% decline in prostate-specific antigen (PSA) levels in 2 (of 23) patients, with no objective responses seen, suggesting that AT-101 has modest clinical activity [35]. Other clinical trials, as a single agent or in combination, are currently underway.

A mitochondrial-mediated mechanism of (−)-gossypol **1**-induced apoptosis has been proposed, with the treatment of isolated mitochondria resulting in cytochrome *c* release, suggesting its direct action on Bcl-2 in the mitochondrial outer membrane [36].

As mentioned previously, gossypol contains reactive aldehyde groups that may be responsible for its toxicity and targeting of other proteins via the formation of Schiff's bases. For this reason, Pellecchia and coworkers have investigated the Bcl-2 inhibitory activity of apogossypol analogues [37, 38]. Apogossypol **6**, which is simply prepared by refluxing gossypol in 40% aqueous sodium hydroxide, was shown to be capable of inhibiting the antiapoptotic proteins Bcl-2 and Bcl-X_L (Bcl-X_L gene expression shows strong correlation with resistance to cytotoxic anticancer agents) [39]. Apogossypol **6** induced apoptosis in primary chronic lymphocytic leukemia (CLL) cells either as a single agent or in combination with 9-β-D-arabinosyl-2-fluoroadenine (F-ara-A), the free nucleoside active metabolite of fludarabine [37].

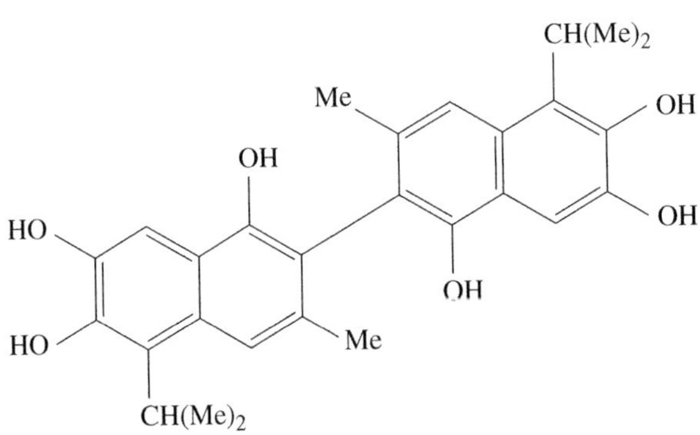

6

In further extensions of these studies, this group has studied apogossypol **6** and apogossypolone analogues [38, 40], which resulted in the apogossypol analogue BI-97C1 (Sabutoclax) **7**, which can be synthesized in 10 steps from gossypol via apogossypol using a chiral separation as the final step [40].

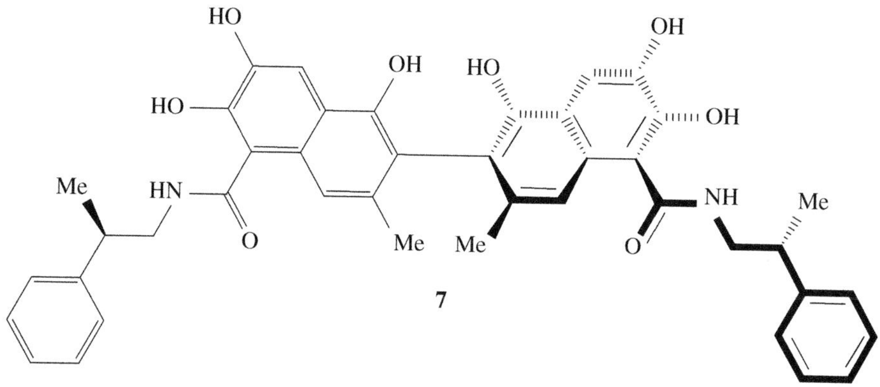

7

BI-97C1 **7** is a potent inhibitor of the binding of BH3 to the Bcl-2 protein family (Bcl-2, Bcl-X_L, Mcl-1, and Bfl-1), with IC_{50} values in the 0.2–0.6 μM range; inhibits the growth of prostate cancer (PC3; EC_{50} 0.13 μM), lung cancer (H460; EC_{50} 0.42 μM), and lymphoma cells (BP3; EC_{50} 0.049 μM); and induces apoptosis in the BP3 cell line in a dose-dependent manner [40]. BI-97C1 also displays efficacy in a transgenic mouse model, as a single agent in a prostate cancer mouse xenograft model, and sensitizes human prostate cancer (PC3) cells to docetaxel [41] and melanoma differentiation-associated gene-7/interleukin-24 (*mda*-7/IL-24) toxicity [42].

14.3.3 Antiviral

Both gossypol and apogossypol **6** have been shown to inactivate the enveloped parainfluenza-3 and herpes simplex viruses *in vitro*, but not the nonenveloped poliovirus [43]. Apogossypol was shown to inactivate the virus at concentrations greater than 29 μM (with no toxicity to HEp-2 cells), while gossypol decreased the number of plaque-forming units only at concentrations that were toxic to Hep-2 cells [43]. The molecular mechanism of viral inactivation was not determined, but the incubation of infected cells with either agent did not alter subsequent plaque formation, indicating that the antiviral effect was not an intracellular process and was mostly probably due to interactions with the viral coat.

A subsequent study showed that gossypol inhibited viral multiplication and that the anti-HSV-2 effect of gossypol was dose dependent, with no cytotoxicity toward human amniotic epithelial cells being observed [19].

As the toxicity of gossypol has been proposed to be due to the formation of Schiff's bases with cellular proteins, Radloff et al. prepared a series of analogues lacking the aldehyde groups (and so incapable of forming these adducts) and tested them for activity against HSV-2 [44]. All of the periacylated gossylic nitriles **8** prepared exhibited lower cytotoxicity to Vero cells than gossypol itself and had significant anti-HSV-2 activity at 0.5–2 μM, and two analogues **8a** and **b** inhibited viral multiplication in Vero cells.

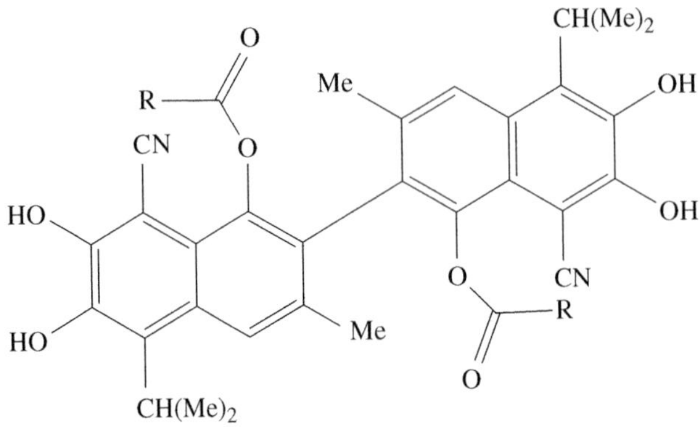

a R = Me
b R = Et
c R = ^{n}Pr
d R = ^{n}Bu

8

The antiviral activity of the two gossypol atropisomers was examined by Lin et al., who showed that, in accordance with its greater biological activity in most biological systems, (−)-gossypol (−)-**1** was a factor of 10× more potent than (+)-gossypol (+)-**1** and inhibited the replication of HIV-1 in peripheral blood monocytes (PBM) at a concentration (EC_{50} = 5.2 μM) 20-fold lower than that required for cytotoxicity [45]. In addition to being less potent, (+)-gossypol exhibited increased cytotoxicity in PBM cells (IC_{50} = 52 μM).

This group then synthesized a range of gossypol derivatives and analogues and tested them as inhibitors of HIV-1 replication in PBM and for their cytotoxicity against PBM and Vero cells. Interestingly, of the four compounds that resulted in greater inhibition of HIV-1 replication than (−)-gossypol, only gossypolone **9**, the major mammalian metabolite, contains an aldehyde group (Table 14.3) [46].

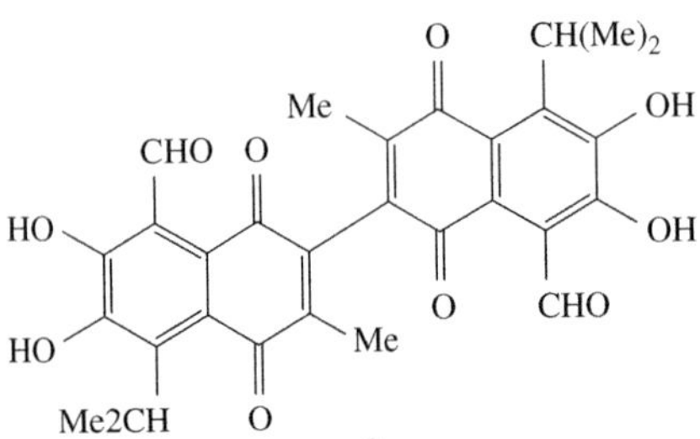

9

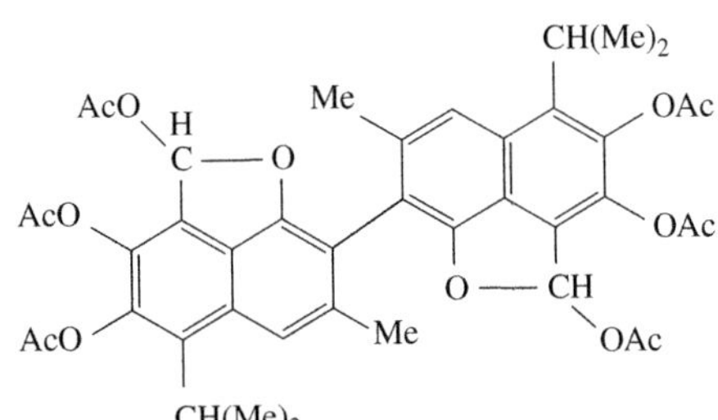

10

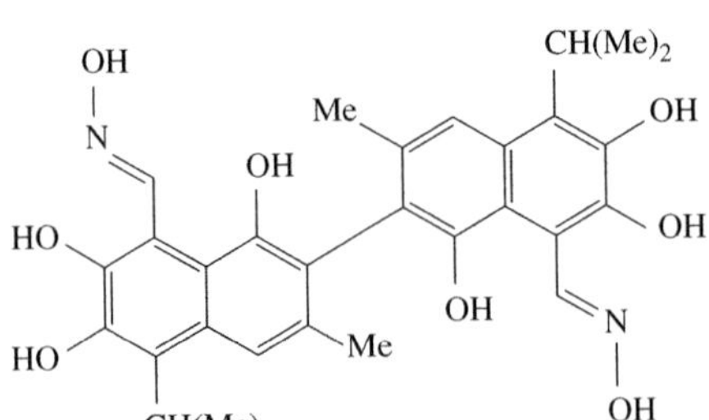

11

TABLE 14.3 Effects of Gossypol and Analogues on HIV-1 Replication in PBM Cells [46]

Compounds	Inhibition of HIV-1 Replication EC_{50} (μM)	Cytotoxicity in PBM Cells IC_{50} (μM)
(−)-**1**	1.7	>100
9	0.86	>100
10	0.93	>100
11	0.41	>100
6	0.82	>100

The antiviral activity of gossypol and its derivatives, for example, **12**, has been suggested to arise from their interferon-inducing ability—with free hydroxyl groups as prerequisite for this activity [47].

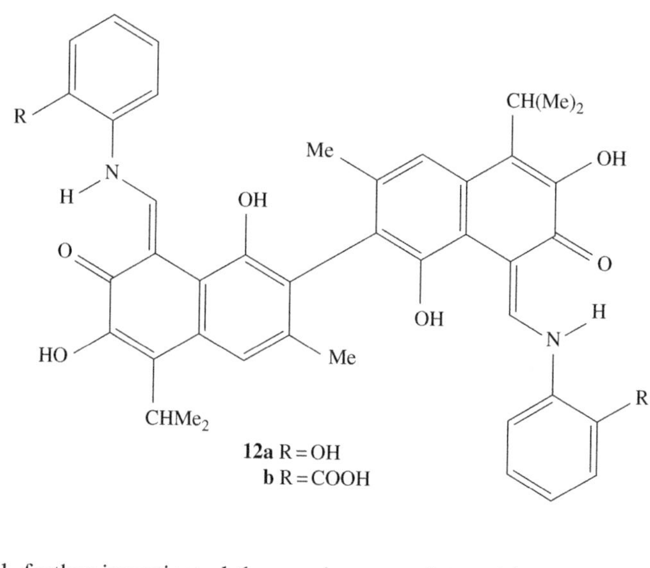

12a R = OH
b R = COOH

Royer et al. further investigated the requirements for *peri*-hydroxyl groups and/or an aldehyde group for anti-HIV activity when they synthesized 1,1′-deoxygossypol **14a** and gossylic acids **14b** and their hemigossypol analogues **15a** and **b** for comparison with gossylic iminolactone **13**, the activity of which had previously been compared to AZT in clinical HIV isolates [48]. Gossylic iminolactone **13** and 1,1′-deoxygossylic acid (DDGA) **14b** were the most potent inhibitors of HIV replication in HeLaT^{4+} cells, with IC_{50}s of 0.4 and 0.7 μM, respectively. No toxicity was observed for either compound at 5 μM, but DDGA exhibited some signs of toxicity at 20 μM [18].

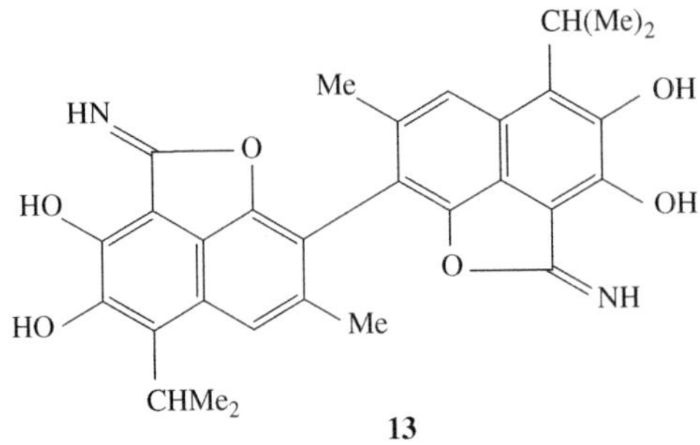

13

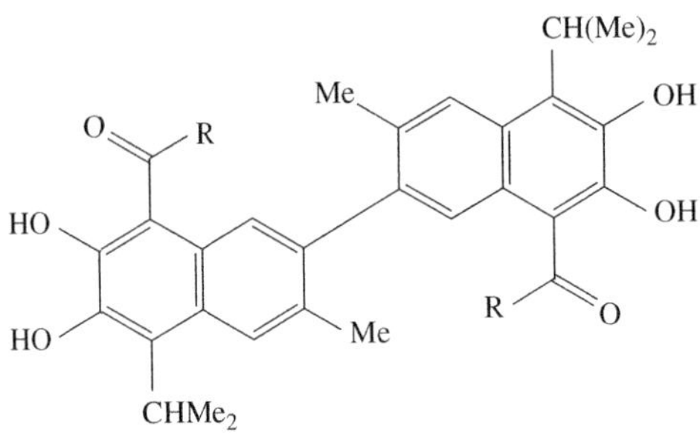

14a R = H
b R = OH

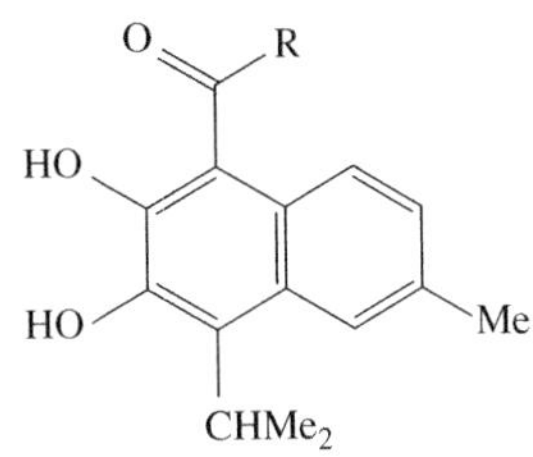

15a R = H
b R = OH

Yang and coworkers have investigated the antiviral activity of gossypol Schiff's bases **3** with the sodium salts of amino acids. The Schiff's bases were less potent than AZT in inhibiting the replication of HIV-1_{IIIB} in TZM-bl cells, with IC_{50}s in the 0.99–3.73 μM range, but some were more potent than 1-adamantylamine in inhibiting H_5N_1 replication in MDCK cells, albeit with slightly increased toxicity (Table 14.4) [49].

In a further study, this group suggested that the anti-H_5N_1 activity of a series of gossypol derivatives was due to their inhibition of the viral entry step as a result of targeting HA2 protein. Interestingly, the more active series in this case was that derived from the usually less active (+)-gossypol, with all derivatives showing increased activity over the diastereoisomeric counterparts (derived from (−)-gossypol **1**) and 1-adamantylamine. Although all compounds were more cytotoxic than 1-adamantylamine, some had improved selectivity indices (CC_{50}/IC_{50})[1] as a result of their enhanced inhibition of H_5N_1 replication [50].

This group then demonstrated that a representative of these gossypol bis-enamines (bis-Schiff's bases), (−)-gossypol-L-Ala **3d**, inhibits the cell fusion-activated gp41

[1] CC_{50} is the concentration required to cause 50% cell death in uninfected MDCK cells; IC_{50} is the concentration required for 50% inhibition of H_5N_1 replication in MDCK cells, as determined by plaque reduction assays.

TABLE 14.4 Effects of Gossypol Schiff Bases 3 on HIV-1_{IIIB} Replication in TZM-bl Cells and H_5N_1 Replication in MDCK Cells [49]

Compound	IC_{50} (μM) HIV-1_{IIIB} in TZM-bl Cells	LD_{50} (μM) in TZM-bl Cells	IC_{50} (μM) H_5N_1 in MDCK Cells	LD_{50} (μM) in MDCK Cells
1-Adamantylamine	ND	ND	2.38	>220.36
AZT	0.0028	ND	ND	ND
3c(L) R=3-indolyl; X=(*S*)-$CHCO_2^-$ Na^+	0.99	47.95	0.28	>35.69
3d R=H; X=(*S*)-$CHCO_2^-$ Na^+	1.22	56.23	0.87	>47.34
3e R=Me; X=(*S*)-CH((*S*)-$CHCH_3$)CO_2^- Na^+	1.12	64.48	0.34	>42.30

ND, not determined.

core domain, possibly by binding to the hydrophobic pocket, and is an effective HIV-1 entry inhibitor. All of these amino acid derivatives exhibited 100-fold enhanced selectivity over (−)-gossypol alone [51].

14.3.4 Antimalarial

In addition to their studies on the antiviral activity of *peri*acylated gossylic nitriles **8**, Vander Jagt and coworkers also investigated the antimalarial activity of these derivatives and found that the activity against both chloroquine-sensitive and chloroquine-resistant *Plasmodium falciparum* strains increased with the length of the *peri*-acyl group alkyl chain [52]. These nitrilo analogues were, once again, selected for evaluation as they lack the formyl (aldehyde) groups that are responsible for toxicity. Later studies suggested that the selective inhibition of the malarial enzyme (*Pf*LDH) over its mammalian counterpart by the acetate analogue **8a** is a result of binding to the pyruvate/nicotinamide-ribose site [53, 54]. In this binding mode, the naphthalene core of gossypol interacts with amino acids that are only found in the *Plasmodium* form of lactate dehydrogenase (LDH), providing a rationale for the specific inhibition of the parasite enzyme [55].

Like L-LDH (anaerobic metabolism), L-malate dehydrogenase (MDH) (oxidative metabolism) is an important enzyme for the survival of *P. falciparum*. During the asexual reproduction and growth phases within the mammalian host, infected erythrocytes have approximately 100× greater demand for glucose than normal, due to an increase in the glycolytic energy production pathway by the parasite [56, 57]. Gossypol inhibits both *Pf*LDH and *Pf*MDH, but, unlike the selective inhibition of *Pf*LDH (IC_{50} = 3.2 μM) in comparison to porcine heart LDH (IC_{50} = 84.5 μM), there is no selectivity between *Pf*MDH (IC_{50} = 1.5 μM) and porcine MDH (IC_{50} = 2.4 μM [mitochondrial]; IC_{50} = 2.9 μM [cytosolic]) [57].

14.3.5 Other Biological Activity

Gossypol has a broad spectrum of biological activity as a result of the many cellular targets on which it acts [20, 58]. Gossypol inhibits a number of enzymes, particularly oxidoreductases (as discussed in Section 14.3.4) and transferases, and also exerts effects on cellular regulatory proteins, as well as cellular and mitochondrial membranes [20]. For example, 6,7′-gossypol diglucoside tetraacetate **16** is a potent inhibitor of the growth of the parasite *Trypanosoma brucei*, with an LD_{50} of 2.44 μM (in the same assay, gossypol has an LD_{50} of 36.90 μM), and has thus been suggested for development as an antitrypanosomal agent [59].

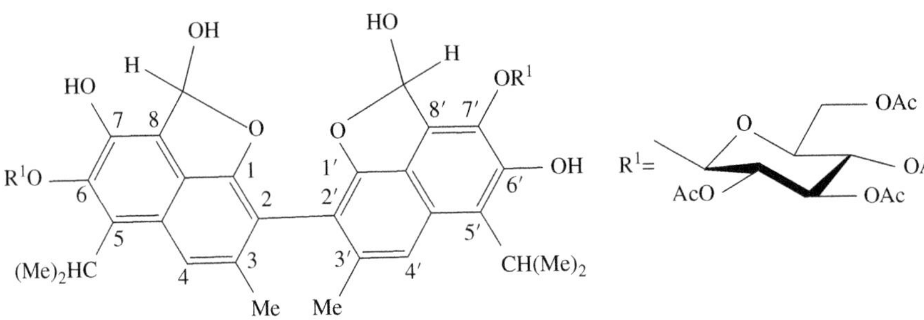

16

As a result of its inhibitory activity on cytoplasmic phospholipase-2, activity against endothelial cell growth, and apoptotic effect on human lymphocytes, in addition to its antioxidant (IC_{50} = 13.1 μM in an iron/ascorbate-dependent lipid peroxidation assay) and antiproliferative activity (GI_{50} 4.8 μM in MTT cell viability assay in an HPV-16 immortalized human keratinocyte cell line [CCD-1106 KERTr]), we have suggested that (−)-gossypol **1** is a candidate for the development as a topical treatment for psoriasis [13].

14.4 BIOLOGICAL ACTIVITY OF CURCUMIN AND ITS ANALOGUES

14.4.1 Introduction

The clinical use of curcumin is limited as a result of its instability at physiological pH [60], low bioavailability [61, 62], limited tissue distribution [63], and rapid metabolism [64, 65].

Structurally modifying curcumin to create new analogues or derivatives is one way of overcoming these drawbacks. The major challenge in designing curcumin analogues is to produce compounds with more desirable pharmacological profiles while retaining the very low toxicity of the parent compound.

As mentioned previously, the general structural features of curcumin include two substituted (3-methoxy-4-hydroxy) aromatic rings joined by a conjugated heptadiene linker containing an α,β-unsaturated diketone **2**.

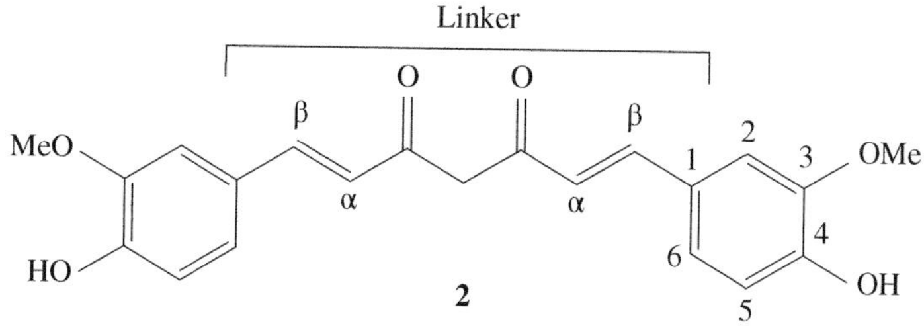

2

Curcumin derivatives can be readily synthesized through the modification of the existing functional groups of curcumin. For example, the aromatic 4-hydroxyl groups can be methylated **17**, acylated **18**, or glycosylated **19** [66], while the diketone moiety of the linker region can be cyclized to give pyrazole **20** and isoxazole analogues **21**.

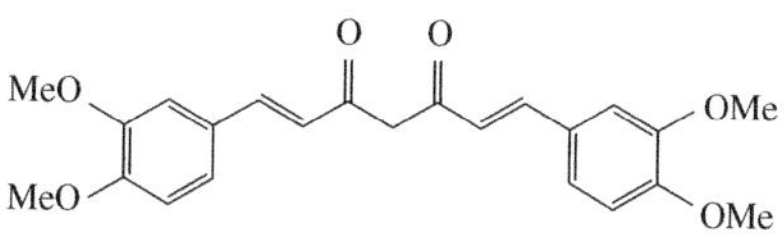

17

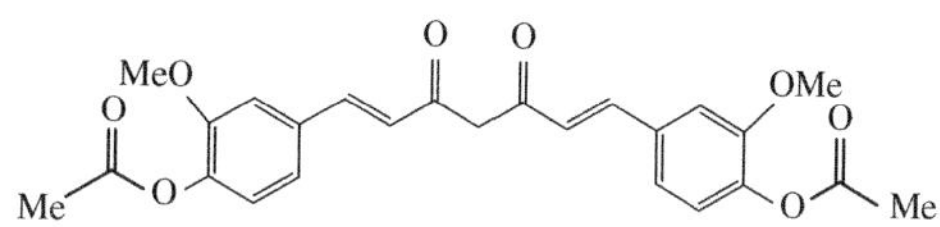

18

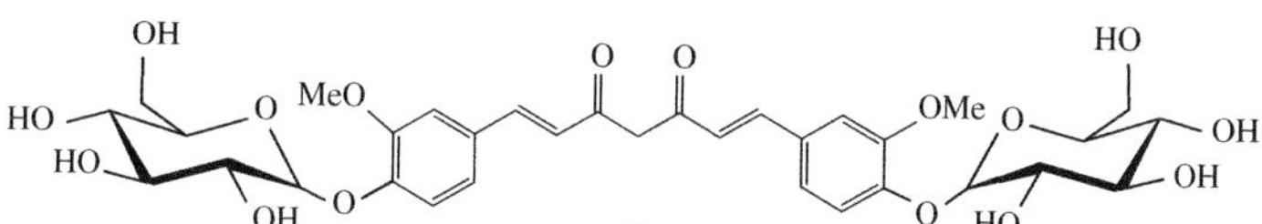

19

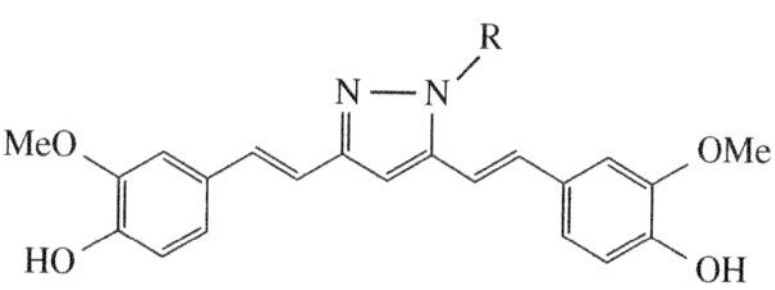

20

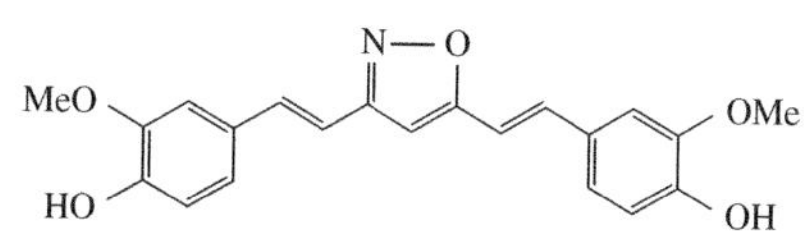

21

Most of the simple analogues of curcumin are synthesized using the Pabon method [67, 68] involving the aldol condensation of two equivalents of a 3,4-substituted benzaldehyde **22** and one equivalent of acetylacetone **23** (Scheme 14.7). This method has been further explored using combinations of benzaldehydes of various substitution patterns (different benzaldehydes can also be used in the same reaction to generate asymmetrical analogues) and diketones other than acetylacetone. One example is the use of acetylcycloalkanones, instead of acetylacetone, to produce analogues that are conformationally restricted. Analogues synthesized via this method still resemble the basic structure of curcumin, and once again, the reactive diketone moiety of these analogues may also be cyclized to give pyrazole and isoxazole analogues.

In the past decade, there has been a more extensive approach to curcumin analogue design, with the structure of the analogues having little resemblance to curcumin, for

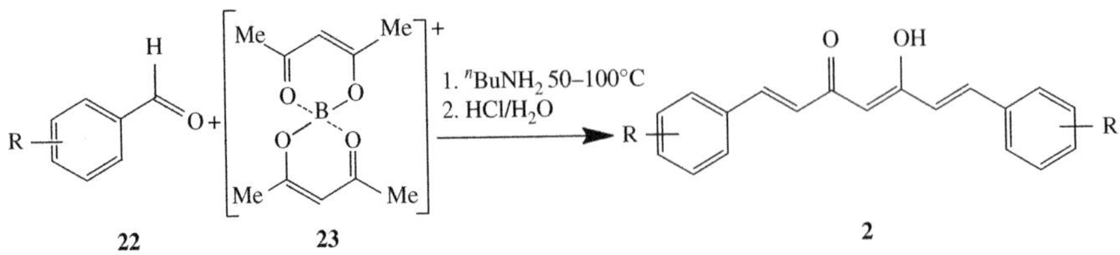

SCHEME 14.7 Pabon synthesis of curcumin.

example, replacing the aromatic moiety with heterocycles and/or the removal of one carbonyl group of the linker to give monocarbonyl [69] and piperidinone analogues [70].

This section aims to give a general overview of the wide range of therapeutic effects and biological activities of curcumin and its analogues, and given the vast amount of published material in this area, the focus will be on a few prominent curcumin analogues from the recent literature.

14.4.2 Anticancer

The process of carcinogenesis is a multistep development of accumulated genetic mutations that ultimately lead to abnormal cell proliferation [71, 72], and recent studies have indicated that 300–500 genes are altered in a given cancer [72]. As curcumin interferes with a diverse range of molecular targets in multiple biochemical pathways, it may be considered to have greater potential as an anticancer agent than single target therapies [73].

Curcumin binds directly to at least 33 different proteins that are overexpressed in cancer [74], including thioredoxin reductase [75], cyclooxygenase-2 (COX-2) [76], and protein kinase C [77]. Curcumin also modulates transcription factors (e.g., NF-κB, STAT3, and AP1), kinases (e.g., EGFR and MAP kinase), and enzymes (e.g., MMP), which are deregulated in cancer (for a more detailed discussion, refer to review articles by Anand et al. [73], Goel et al. [78], and Aggarwal and Harikumar [79]). Curcumin has been shown to inhibit cell growth and cause apoptosis in almost all types of cancer cells, both *in vitro* and *in vivo* [74], but it does not affect the growth of normal cells such as keratinocytes [80], hepatocytes [81], fibroblasts [82], mammary epithelial cells [83], and astrocytes [84].

Curcumin's wide range of biological activity arises from its ability to interact with various biological targets, and diverse libraries of curcumin analogues have been synthesized in order to develop a better understanding of, and to identify the structural features of curcumin responsible for, any particular biological activity. This section will focus on the structure–activity relationship associated with curcumin's anticancer properties. The design and anticancer activities of some highly potent curcumin analogues will also be discussed. For an in-depth description of anticancer analogues of curcumin, see the review article by Agrawal and Mishra [85].

An insight into the basic pharmacophore for curcumins was provided by a study that correlated the biological activity of curcumin **2** and its naturally occurring analogues (demethoxycurcumin **24**, bisdemethoxycurcumin **25**, and tetrahydrocurcumin (THC) **26**) [86]. In this study, curcumin **2** showed greater activity than demethoxycurcumin **24** and bisdemethoxycurcumin **25** in inhibiting NF-κB activation, a transcription factor that is constitutively activated in cancer and inflammation. Curcumin, demethoxycurcumin, and bisdemethoxycurcumin all displayed similar antiproliferative activities. In contrast, THC **26** elicited no suppression of NF-κB activity and had low antiproliferative activity. From these observations, this group concluded that the methoxyphenyl group is important in suppressing NF-κB activation, along with an unsaturated linker, which is also essential for antiproliferative activity. These findings agree with an earlier study by Lin et al. [87] in which an advanced structure–activity relationship was established based on the cytotoxic activities of 50 structurally diverse (symmetrical and asymmetrical) curcumin analogues in prostate cancer (PC3 and LNCaP) cell lines. These workers concluded that a conjugated linker and two phenyl rings are required for cytotoxicity, that elongation of the linker results in the loss of cytotoxic activity, and that 3-methoxy-4-hydroxy is the most favorable aromatic substitution pattern. This study also suggested that an analogue with a monocarbonyl pentadiene linker is more potent than one with a heptadiene diketone linker and the same aromatic substitution pattern. Lin et al. further concluded that the keto–enol tautomer is the biologically active one [87].

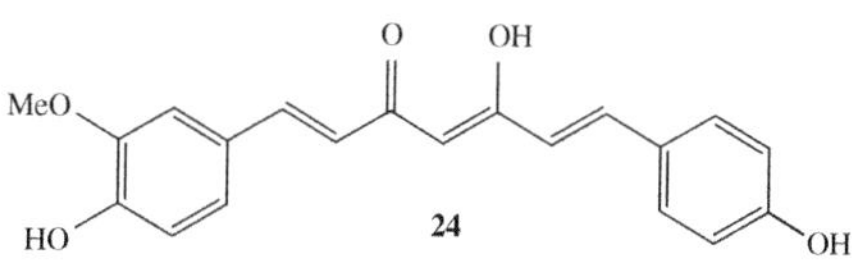

24

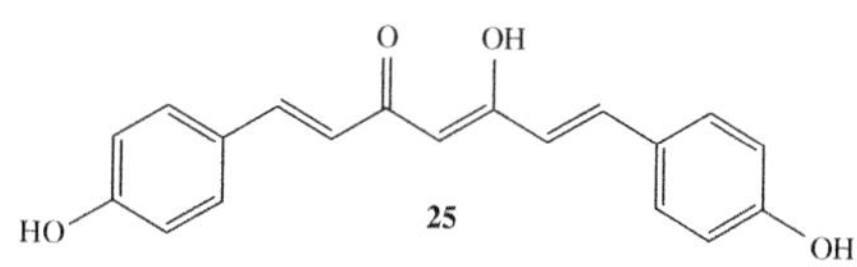

25

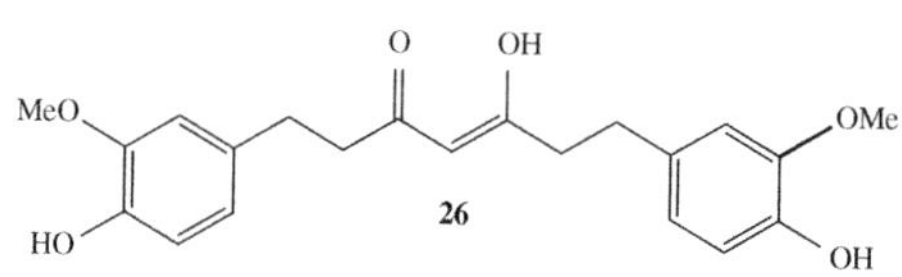

26

As previously discussed, curcumin can form Michael adducts with cellular targets, and Ahn and Sok [88] proposed the Michael acceptor to be the key pharmacophore in curcumin anticancer drug design. Most curcumin analogues retain their Michael acceptor nature, but Amolins et al. [89] assessed the electrophilicity of electron-rich (pyrazole and isoxazole) derivatives of curcumin analogues and concluded that the antiproliferative activity of curcumin is not due to its Michael acceptor functionality [89].

It is known that the instability of curcumin is due to its α,β-unsaturated dicarbonyl moiety [69]. The most prominent class of anticancer curcumin analogues is obtained by omitting the active methylene group in the linker to form more compact 5-carbon monocarbonyl analogues. There are two series of analogues in this class: the open linker bis-benzylidene acetone analogues and the more conformationally restricted analogues (with cyclized linkers, such as piperidinone or tetrahydropyranone).

The anticancer structure–activity relationship of this class of compounds has been thoroughly explored—Yamakoshi et al. [90] synthesized three series of curcumin analogues, compounds with a heptadienyl dicarbonyl linker (series 1), a pentadienyl monocarbonyl linker (series 2), and a pentadienyl cyclohexanone linker (series 3). For analogues with the same aromatic substitution pattern, series 2 compounds have greater anticancer activity than series 1 compounds, and a dramatic loss of activity was observed for series 3 (cyclohexanone-containing) compounds. These findings are exemplified by three compounds: dimethyltetrahydrocurcumin (DMTC; GO-Y25) **27**, GO-035 **28**, and GO-Y032 **29**, with GI_{50} values of 2.0, 1.5, and >50 μM, respectively, in the HCT116 colon cancer cell line [90]. Interestingly, Adams et al. showed that the incorporation of a heteroatom into the four position of the cyclohexanone ring structure (e.g., N or O; forming piperidinone and tetrahydropyranone analogues) increases the anticancer activity of this class of analogues. This trend in anticancer activity is demonstrated by four 2-hydroxy-substituted monocarbonyl compounds. The GI_{50} values, in the melanoma RPMI-7951 cell line, for the open monocarbonyl **30**, cyclohexanone **31**, tetrahydropyranone **32**, and *N*-methylpiperidinone **33** analogues are 1.0, 3.0, 0.8, and 0.9 μM, respectively [70]. It appears that the there are no differences in potency between the tetrahydropyranone and piperidinone series, but the piperidinone series can be further expanded by means of *N*-substitution.

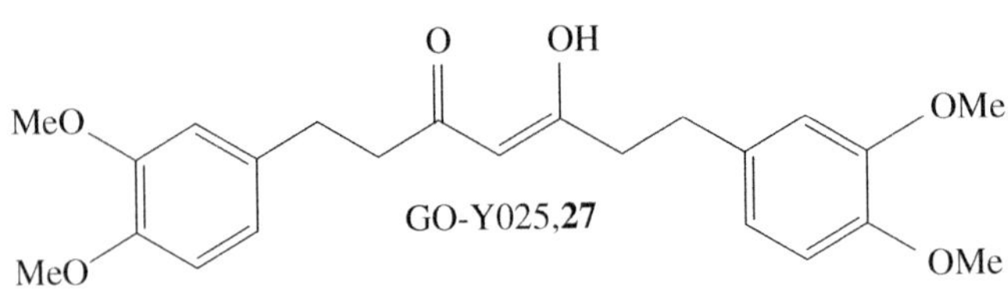

GO-Y025,**27**

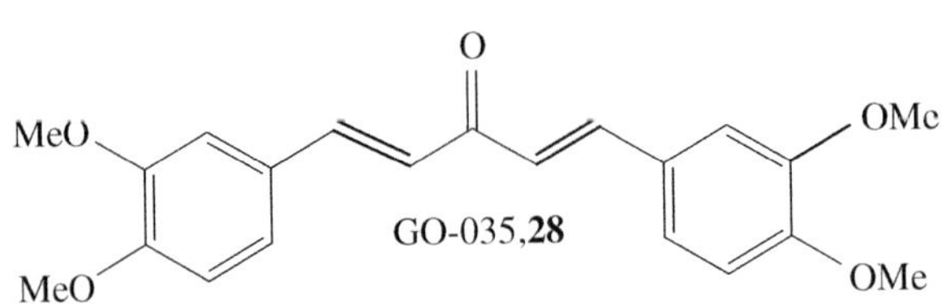

GO-035,**28**

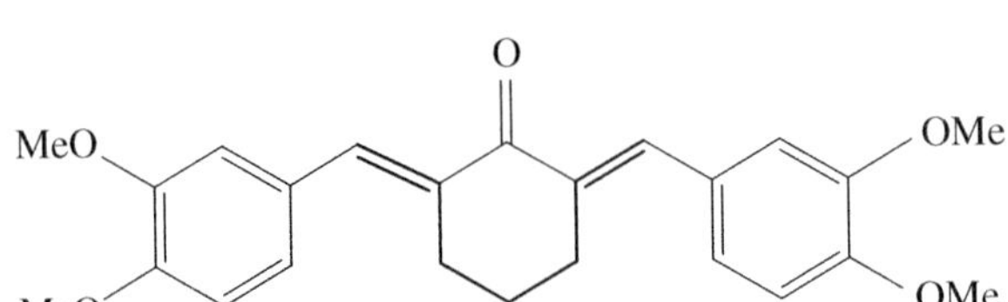

GO-Y032,**29**

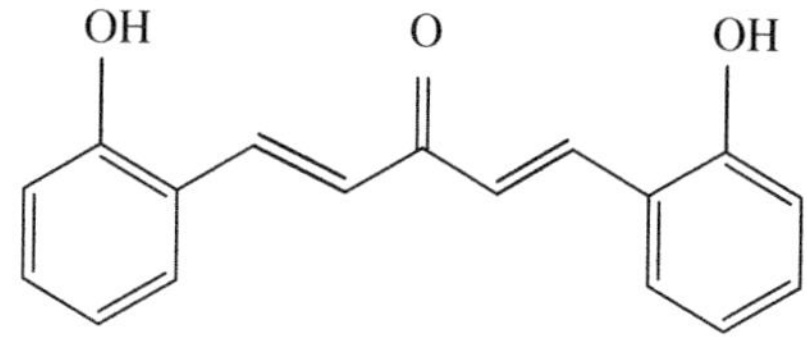

30

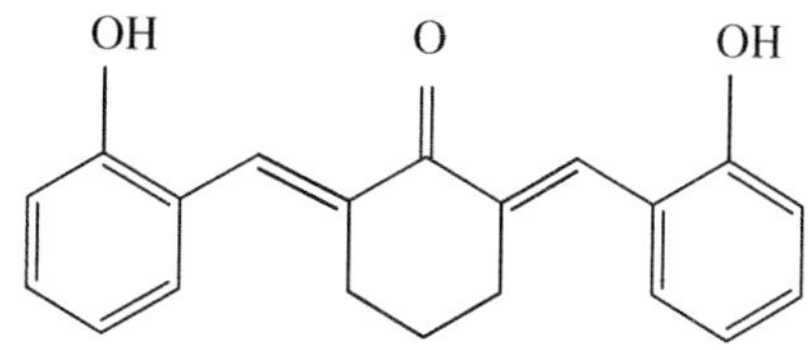

31

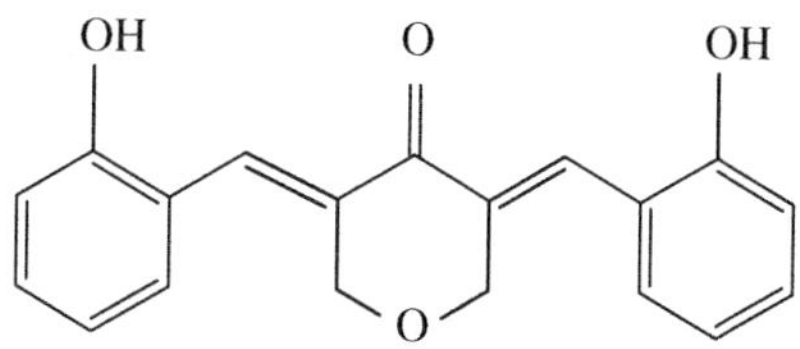

32

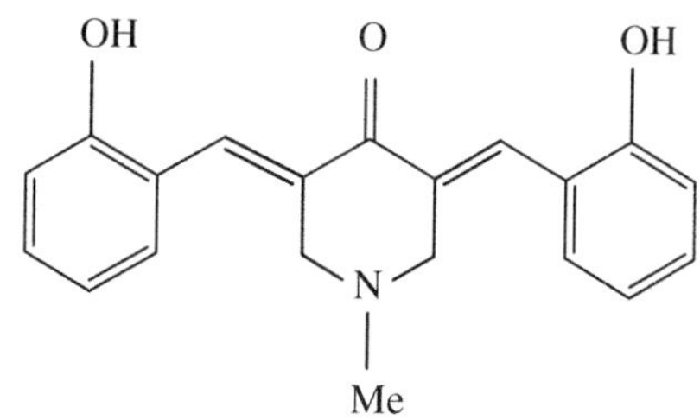

33

In a similar manner to the analogues with a heptadienyl dicarbonyl linker, a complete loss of activity is observed when the linker is saturated [70, 90]. Two analogues are particularly notable due to their potent anticancer activity: one with an open acetone linker (GO-Y030) and the other with a cyclized piperidinone linker (EF-24).

Ohori et al. [91] synthesized three series of symmetrical curcumin analogues; of the 50 compounds that were synthesized, one compound GO-Y030 **34**, a 3,5-bis(methoxymethoxy)-substituted monocarbonyl analogue of curcumin, was identified to be the most potent compound, with a GI_{50} of 0.3 μM in a colon cancer HCT116 cell line (which is 26 times more potent than curcumin; GI_{50} 8 μM). GO-Y030 **34** was also screened against several cancer cell lines and shown to inhibit cell growth at concentrations 8–270 times lower than curcumin (Table 14.5) [91–93].

TABLE 14.5 Growth Inhibition of Human Cancer Cell Lines by Curcumin 2 and GO-Y030 34

	GI_{50} (μM)	
Cell Line	2	34
MDA-MB-231 (**breast**) [92]	19.3	1.2
HPAC (**pancreatic**) [92]	16.7	2.2
PANC-1 (**pancreatic**) [92]	27.8	0.1
8505c (**thyroid**) [93]	10.1	0.79
HuCCT-1 (**bile duct**) [93]	8.6	0.6
HT-29 (**colon**) [93]	13.3	0.96
SW480 (**colon**) [93]	10.3	0.51
KMS12-BM (**myeloma**) [94]	10.3	1.5

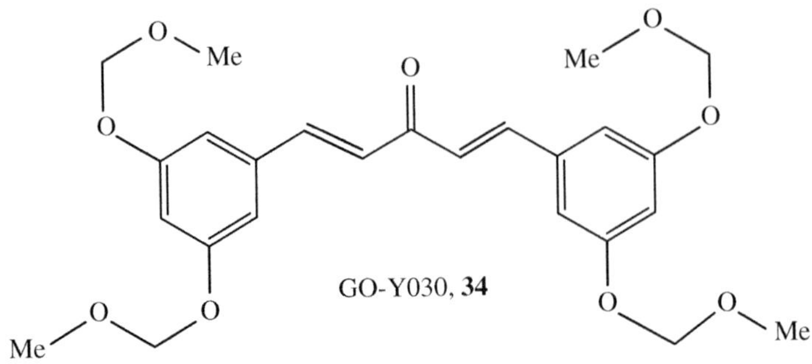

GO-Y030, **34**

The molecular mechanism of GO-Y030 **34** closely resembles that of curcumin; microarray analysis of GO-Y030-treated cells showed similar gene expression profiles to those treated with curcumin [91]. GO-Y030 downregulates oncogenes such as *c*-Myc and cyclin D1 and inhibits the activation of NF-κB and STAT3, a regulator of apoptosis, at concentrations at least one tenth of the curcumin concentrations required for similar effects [91–93]. GO-Y030 is also a potent inducer of apoptosis—it has been shown to promote PARP cleavages, a reliable indicator of apoptosis, at 2 μM (myeloma cell line, KMS12-BM), 2.5 μM (colon cancer cell line, SW480), and 5 μM (colon cancer cell line, HT-29 [93–95]. Curcumin, on the other hand, was shown to induce PARP cleavage at 20 μM concentration in all three of these cell lines [93–95].

Like curcumin, GO-Y030 **34** is shown to be relatively nontoxic with no proliferation suppression in primary human hepatocytes (hNHeps) at 100 μM [91]. No liver or kidney toxicity was observed in mice treated with GO-Y030 0.1% w/w for 2 months [96]. The maximum tolerable dose in mice was 474.5 mg/kg over a month via ip administration [97]. At a concentration that caused apoptosis in a cancer cell line (5 μM), apoptosis did not occur in normal human lung fibroblasts (WI-38), normal bladder smooth muscle cells, and immortalized, nonmalignant human mammary epithelial cells (MCF-10A) [95]. Evidence of apoptosis was observed in normal colonic smooth muscle cells at 5 μM [95], and further experiments will be needed in order to fully elucidate its toxicity.

Despite its potency in cancer cells, GO-Y030 **34** has poor solubility (9.26 mg/l) in PBS [97], so minor structural alterations may be required to improve its solubility. GO-Y030 is a good lead compound for target optimization for the preparation of anticancer agents, but its pharmacokinetic profile will also need to be determined.

Adams et al. [70] synthesized series of curcumin analogues exploring the anticancer activity of compounds with various pentadiene linkers. The most potent compound identified in this study was EF-24 **35**, 3,5-bis((*E*)-2-fluorobenzylidene)piperidin-4-one, a difluorinated pentadiene piperidinone analogue of curcumin.

TABLE 14.6 Data from NCI-60 DTP Cell Line Screen for Curcumin 2 and EF-24 35

Average (Mean Graph Midpoint) over All 60 Lines (μM)	2	35
GI_{50}	7.3	0.7
Total growth inhibition (TGI)	23.5	2.4
LD_{50}	67.2	16

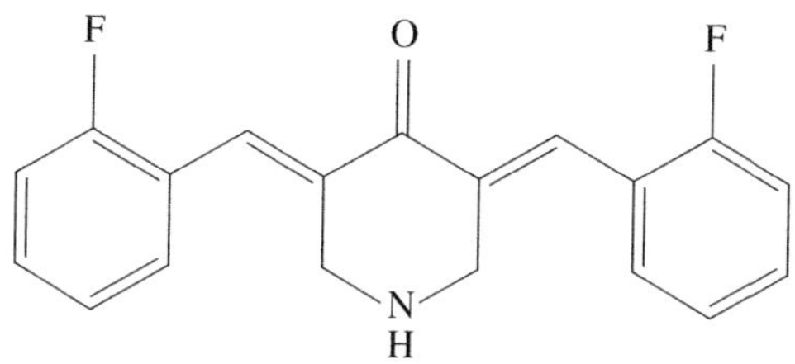

35

EF-24 displayed a broad spectrum of potent anticancer activity; in the NCI-60 DTP human cell line screen (Table 14.6), EF-24 inhibited cell proliferation and angiogenesis at concentrations that were at least a fifth of those required for curcumin [70, 98]. EF-24 also displays potent anticancer activity *in vivo*; mice inoculated with a solid breast tumor showed a 55% decrease in tumor weight compared to the control when treated with 100 mg/kg of EF-24 [70]. Like curcumin, EF-24 also interacts with many molecular targets that are associated with cancer, but usually at far lower concentrations; EF-24 suppresses NF-κB activation in Hep3B and A549 cells at 2 and 1.3 μM, respectively, whereas concentration of 20 and 13 μM, respectively, of curcumin were required for the same effect to be achieved [99].

Several studies have shown that EF-24 **35** induces G_2/M cell cycle arrest and apoptosis at about a tenth of the concentration required for curcumin. Adams et al. [98] demonstrated that in the 1–20 μM concentration range, EF-24 induced G_2/M arrest, followed by apoptosis, in human breast (MDA-MB-231) and prostate (DU-145) cancer cells via redox-dependent mechanisms. Treatment with EF-24 (2 μM concentration) induced G_2/M cell arrest in a time-dependent manner in a cisplatin-resistant ovarian cancer cell line (A2780 cDDP) [100]. In this study, the increase in the G_2/M cell population peaked after 12 h of treatment, and apoptosis was observed after 24 h of treatment. As with curcumin, G_2/M cell arrest and apoptosis induction by EF-24 was the result of increasing phosphatase and tensin (PTEN) homologue expression [100].

Although EF-24 is multitargeting, like curcumin, Thomas et al. [101] showed that the anticancer action of EF-24 is distinctive from curcumin as it inhibits hypoxia-inducible factor (HIF-1), an important mediator for angiogenesis and tumor survival. Curcumin inhibits HIF-1α at the level of transcription, whereas EF-24 suppresses

HIF activity posttranscriptionally, similar to known tubulin inhibitors such as paclitaxel. In contrast to curcumin, EF-24 disrupts the cell cytoskeleton by inducing microtubule stabilization, which curcumin does not affect. The potency of EF-24 is, perhaps, due to this distinctive anticancer mechanism [101].

Like curcumin, EF-24 is relatively nontoxic; 100 mg/kg demonstrated no harmful side effects in athymic nude mice; normal healthy weight gain was seen, with no observed toxicity in the kidneys, the liver, and the spleen in all of the treated mice. The maximum tolerated dose of EF-24 was 200 mg/kg iv and 400 mg/kg ip in mice [70].

Poor water solubility, photosensitivity, low bioavailability, and fast metabolism are factors that impeded the further development of EF-24. Sun et al. proposed a solution to this problem by utilizing the Michael acceptor properties of EF-24; the water solubility and stability of EF-24 improved upon conjugation with the soluble peptide glutathione, forming a bis-Michael adduct (EF-24-$(GSH)_2$). EF-24-$(GSH)_2$ is a prodrug that undergoes a facile retro-Michael addition to release EF-24 in an aqueous environment. The nearly identical antiproliferative dose–response of EF-24 and EF-24-$(GSH)_2$ in a breast cancer cell line (MDA-MB-435) further confirms the release of the active form [102].

There are a few promising highly potent curcumin analogues; however, the physicochemical profiles of these compounds show that they are far from drug-like. More research is required to improve the properties of these new leads without compromising safety profiles and efficacy.

14.4.3 Anti-inflammatory and Antioxidant

As mentioned previously, curcumin **2** is a molecule that interacts with multiple cellular targets, and it modulates multiple targets in the inflammatory pathways, for example, NF-κB, tumor necrosis factor α (TNF-α), IL-6, IL-8, phospholipase A2, and COX [103, 104]. TNF-α and NF-κB are both major mediators of inflammation; TNF-α exerts its inflammatory effects by activating NF-κB, which is associated with the regulation of the expression of COX-2 and cyclin D1, which regulate inflammation and cell proliferation [105]. The anti-inflammatory properties of curcumin have also been attributed to its ability to directly inhibit COX-2 [76].

Curcumin is also a potent antioxidant at low concentrations, and this enhances its anti-inflammatory activity [106]. For an in-depth review of curcumin and its molecular targets in inflammation, refer to review articles by Aggarwal and Harikumar [79] and Strimpakos and Sharma [107].

As mentioned previously, the methoxy group in the phenolic rings and conjugation in the linker moiety of curcumin are important prerequisites for the inhibition of NF-κB [86]. This finding agrees with that of an earlier study by Pan et al. [108], which explored the role of the double bond in the linker moiety of curcumin. NF-κB inhibition by curcumin **2**, THC **26**, hexahydrocurcumin **36**, and octahydrocurcumin **37** was assessed. Curcumin inhibited NF-κB activity in a dose-dependent fashion, while at similar concentrations to those of curcumin, THC **26**, hexahydrocurcumin **36**, and octahydrocurcumin **37** all exhibited similar, but minimal, NF-κB inhibition, thereby indicating the importance of a conjugated linker in suppressing NF-κB activity.

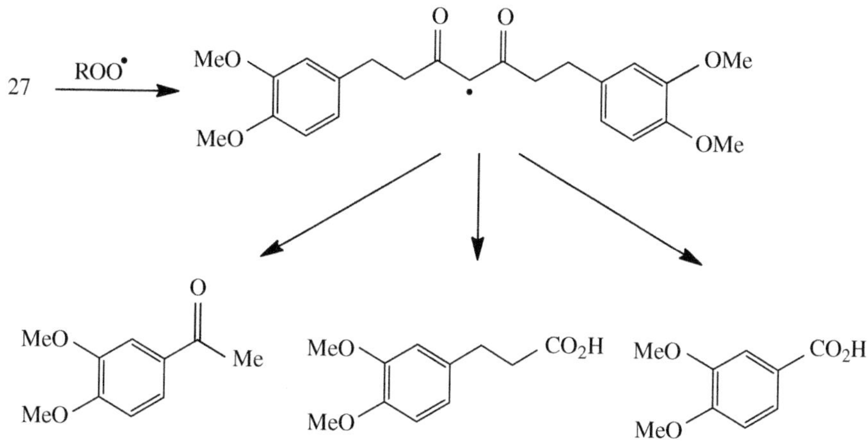

38 39 40

SCHEME 14.8 Antioxidant mechanism of DMTC **27**, leading to C–C bond breaking [110].

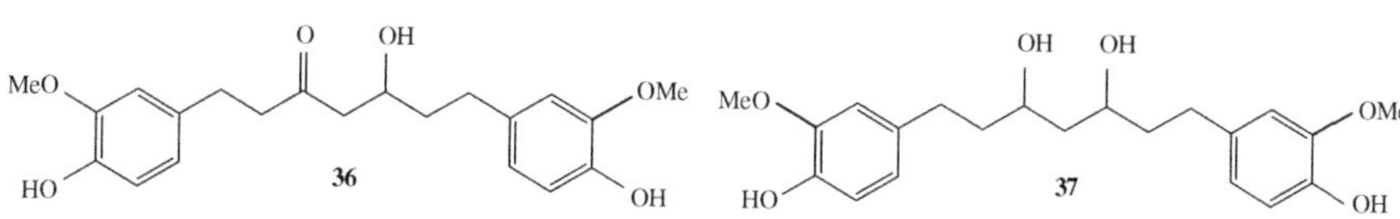

36 37

Although THC **26**, a metabolite of curcumin [109], is not a potent inhibitor of NF-κB, it is a stronger antioxidant than curcumin [110, 111], suggesting that conjugation in the linker section of curcumin is not crucial for its antioxidative activity.

A number of studies have focused on the elucidation of the antioxidant mechanism of curcumin. As a result of a study of the antioxidative activity of curcumin and dimethylcurcumin **17**, Priyadarsini et al. [112] suggested that the phenolic hydroxyl groups are essential for curcumin's antioxidant activity. Curcumin is a more potent inhibitor of lipid peroxidation than dimethylcurcumin **17**, and this observation agrees with earlier studies [113–115]. In addition, Venkateswarlu et al. [116] showed that by increasing the number of hydroxyl substituents on the aromatic rings, the antioxidant activity of curcumin analogues can be increased.

However, Sugiyama et al. [110] suggested that the antioxidant properties of curcumin are mediated through its β-diketone moiety. In this study, the antioxidant activity of curcumin **2**, THC **26**, and DMTC **27** were explored, and the results suggested that the β-diketone moiety of the THC is responsible for their antioxidative activity, resulting in the cleavage of the C–C between the active methylene group (CH_2) and one of the two carbonyl groups of the β-diketone moiety to give, for example, 3,4-dimethoxy-substituted acetophenone **38**, as well as phenylpropionic **39** and benzoic acids **40** (Scheme 14.8).

Litwinienko and Ingold [117] demonstrated that the antioxidant activity of curcumin could be mediated by both the central methylene hydrogens and the aromatic hydroxyl groups, depending on reaction conditions, while the antioxidant property of curcumin has also been attributed to the action of its degradation products, ferulic acid and vanillin [118, 119].

In a similar manner to its anticancer effects, shortening the linker section from a 7-carbon dicarbonyl to a 5-carbon monocarbonyl linker improves its anti-inflammatory activity. Zhao et al. [120] synthesized curcumin analogues focusing on three different 5-carbon unsaturated monocarbonyl linkers—those derived from acetone, cyclopentanone, and cyclohexanone. The trend in the anti-inflammatory activity between these three series of analogues is not pronounced, but of the 16 highly active curcumin analogues, more than half contain the acetone linker. The most active compound from this study, however, is the 3-(dimethylamino)-substituted cyclopentanone analogue AN1 **41**, treatment with which resulted in the dose-dependent inhibition of LPS-induced TNF-α and IL-6 (IC_{50} 1.07 μM) release in LPS-stimulated RAW 264.7 macrophages. In an earlier study by the same research group, an analogue of curcumin that showed high anti-inflammatory activity also contained dimethylamino substituents [121].

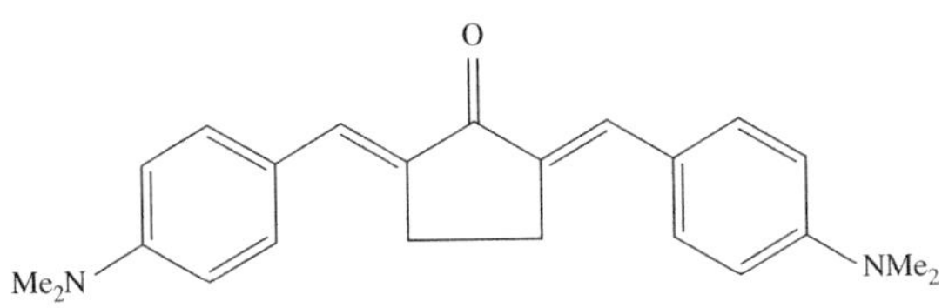

41

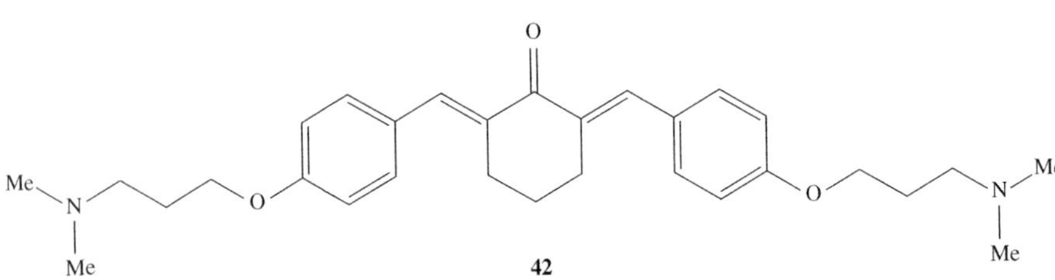

42

Wu et al. synthesized 34 piperidinone curcumin analogues and assessed their anti-inflammatory activity; the most potent compound from this study was F35 **43**, the 3,4,5-methoxy-substituted analogue **43** [122].

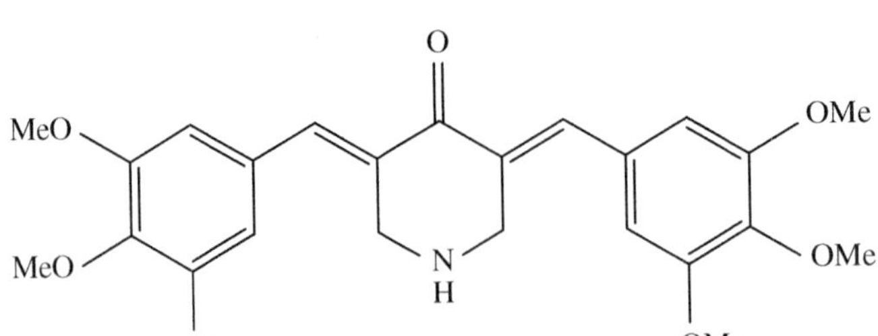

43

Compound **43** showed very promising anti-inflammatory activity, with an IC_{50} of less than 1 μM for LPS-induced release of IL-6 in RAW264.7 macrophage cells. Mechanistic studies suggest that F35 may exert its anti-inflammatory affect partly via the inhibition of NF-κB and ERK pathways. Moreover, the survival rate in mice pretreated with 20 mg/kg of F35 **43** increased by 80% in LPS-induced sepsis.

In contrast, the antioxidative activity of curcumin analogues decreases upon shortening the linker. Shang et al. [123] synthesized four series of symmetrical curcumin analogues varying in the linker region: the 7-carbon diketone, 5-carbon acetone, cyclopentanone, and cyclohexanone linkers. Analogues with a diketone linker were generally more active as DPPH scavengers compared to analogues with a 5-carbon linker. Fang et al. synthesized a small library of dimethylaminomethyl-substituted monocarbonyl curcumin analogues that displayed high antioxidative activities. Most of the analogues had IC_{50} values below 2.5 μM in DPPH free radical scavenging activity, which is 20 times more active than curcumin [124].

As can be deduced from the data presented earlier, there is a need for a more systemic approach in the design of both anti-inflammatory and antioxidative curcumin analogues. Further studies are required in order to build a more complete picture of the relevant structure–activity relationships.

14.4.4 Curcumin in Neurodegenerative Diseases

The accumulation of protein aggregates, inflammation, and oxidative stress are implicated in many neurodegenerative diseases [125]. As previously discussed, curcumin **2** is a pleiotropic molecule; it has the ability to interact with many targets in various diseases, including neurodegenerative. Curcumin limits the formation of protein aggregates associated with various neurodegenerative diseases, via multiple mechanisms, at different levels of the pathogenesis process. Another advantage of curcumin in the treatment of neurodegenerative disease is that it readily crosses the blood–brain barrier [126]. This section will focus on the antiamyloid properties of curcumin and its analogues in the treatment of Alzheimer's disease.

Various *in vitro* and *in vivo* studies have shown that curcumin reduces amyloid plaques through a number of mechanisms. Curcumin **2** binds and inhibits the formation of Aβ fibrils from $A\beta_{40}$ and $A\beta_{42}$, the proteins that comprise Aβ protein [127]. It also decreases the level of mature amyloid precursor protein (APP), the larger transmembrane protein that is cleaved to form Aβ protein [126]. Curcumin destabilizes preformed Aβ fibrils [128] and downregulates the expression of BACE 1, the β-secretase enzyme responsible for the cleavage of the exoplasmic N-terminus of APP [129].

Studies have proposed that iron (III) and Cu (II) influence the formation of $A\beta_{42}$ aggregates [130, 131], and as curcumin can bind to copper(II) and iron (III), this suggests another possible mechanism for its antiamyloid activity [130].

There is limited knowledge of the pharmacophore responsible for curcumin's antiamyloid activity, but it is known that the conjugated linker in curcumin is essential for its prevention of Aβ oligomerization [132]. Furthermore, changing the linker length from a 7-carbon diketone to a 5-carbon monoketone linker does not seem to affect the antiamyloid activity [133]. The design of many antiamyloid curcumin analogues is based on the structure of Congo red and Chrysamine G, known amyloid-binding dyes and inhibitors of Aβ formation [134, 135]. Chen et al. [135] discovered a potent antiamyloid piperidinone **44** analogue of curcumin **2** that has an IC_{50} value of 2.5 μM in inhibiting Aβ aggregation, which is approximately fivefold more potent than curcumin (IC_{50} = 12.1 μM).

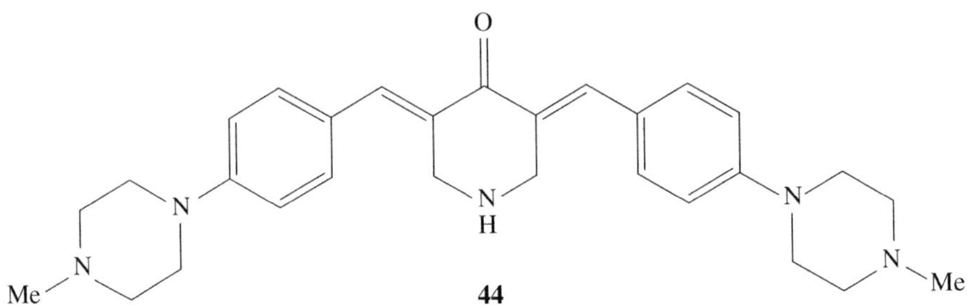

44

Narlawar et al. [136] synthesized a library of *para*-substituted *N*-phenylpyrazole curcumin derivatives as antiamyloid agents; all of the synthesized derivatives showed greater activity toward inhibition of $A\beta_{38}$, $A\beta_{40}$, and $A\beta_{42}$ secretion than curcumin. Further studies showed these derivatives are very potent inhibitors of Tau aggregation and [134] and depolymerized Tau protein aggregates at low micromolar concentrations. In contrast to curcumin, the pyrazole derivatives did not display any affinity toward $A\beta_{42}$ but reduced the secretion of $A\beta_{38}$, $A\beta_{40}$, and $A\beta_{42}$ via the inhibition of γ-secretase [134, 136]. The curcumin isoxazole **21** retained affinity toward $A\beta_{42}$ fibrils and was also found to be a potent inhibitor of Aβ secretion. The most active compounds were analyzed further for their potential to alter Aβ production in primary telencephalic chicken neuronal cultures. However, they displayed a significant decrease in β-APP levels in primary telencephalic chicken neuronal cultures.

Liu et al. combined elements of the structures of cyclohexyl bisphenol A, an antiamyloid compound, and curcumin and synthesized an *N*-phenylpyrazole curcumin derivative, CNB-001 **45**. CNB-001 **45** is a potent antiamyloid compound, both *in vivo* and *in vitro*; it inhibits amyloid-induced cell death in the MC65 system, a cell line that conditionally produces Aβ, with an EC_{50} value of approximately 300 nM [137]. CNB-001 **45** does not directly bind to and inhibit Aβ aggregation, but inhibits γ-secretase. More recently, CNB-001 has been shown to stimulate the removal of Aβ by increasing proteasome activity [138]. CNB-001 **45** crosses the blood–brain barrier and is rapidly distributed to the brain following oral administration, and with maximal concentration achieved 1 h after an oral dose, it has a better pharmacokinetic profile than curcumin [138].

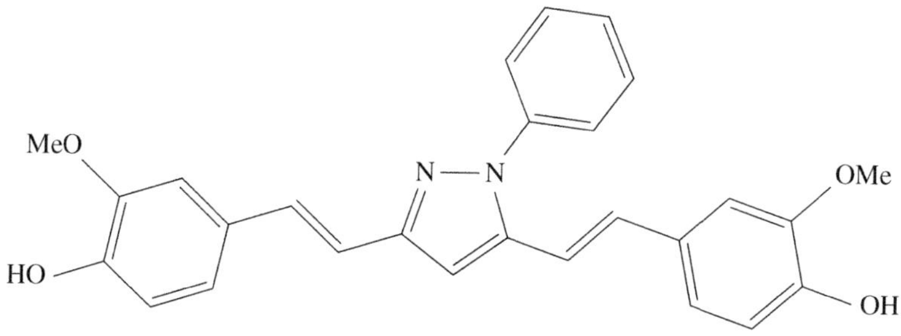

CNB-001, **45**

14.4.5 Antimalarial

Curcumin **2** possess antimalarial activity as a result of its inhibition of the growth of a number of *Plasmodium* species. Reddy et al. [139] demonstrated that curcumin effectively inhibited the growth of *P. falciparum* and a chloroquine-resistant (carrying the PfCRT-K76T mutation) strain in a dose-dependent manner with IC_{50} values of approximately 5 μM. They also showed that the oral administration of curcumin (100 mg/kg/day) resulted in an increase in the survival rate of mice infected with *Plasmodium berghei* (nonhuman species). The antimalarial activity of curcumin is believed to be due to its ability to interfere with many targets, which are necessary for the growth and survival of the parasite.

There are many reported mechanisms for the parasiticidal effect of curcumin on *P. falciparum* strains. Cui et al. [140] reported that the prooxidant activity of curcumin induces DNA damage in *P. falciparum*. This study also demonstrated that curcumin interferes with the histone acetylation status in *P. falciparum* by inhibiting the activity of PfGCN5, a *P. falciparum* homologue of histone acetyltransferase. It has also been suggested, from *in silico* docking studies, that curcumin may also exert its antimalarial activity by binding to PfATP6, a homologue of sarcoplasmic/endoplasmic reticulum Ca^{2+}-ATPase (SERCA), which is a target of artemisinin, a current antimalarial drug [141]. Furthermore, curcumin has been reported to disrupt parasite microtubules [142] and upregulate CD36 expression, which enhances the phagocytosis of *P. falciparum*-parasitized erythrocytes by monocytes and macrophages [143].

There is not much current literature on antimalarial curcumin analogues/derivatives, but these mostly inhibit the growth of both chloroquine-sensitive and chloroquine-resistant strains of *P. falciparum*.

Mishra et al. [144] synthesized a library of curcumin derivatives including cyclized isoxazole, substituted *N*-phenylpyrazole, and substituted benzylidene derivatives, as well as analogues containing substitution by a benzyl group at the active C4 carbon. The most active compound from this study was the unsubstituted pyrazole derivative **20** (R = H) of curcumin, with an IC_{50} of 0.48 μM, which is 10× more potent than curcumin, but 25× less active than artemisinin (IC_{50} 0.019 μM).

Manohar et al. [145] synthesized 5-carbon linker, symmetrical cyclopentanone curcumin analogues, differing in the 3,4-substitution pattern. Of the 76 compounds prepared, 6 analogues showed moderate antimalarial activity against both chloroquine-sensitive and chloroquine-resistant strains of *P. falciparum*, with IC_{50} values ranging from 0.37 to 0.63 μM. These analogues had similar inhibitory activity to chloroquine (IC_{50} = 0.43 μM) in chloroquine-resistant strains of *P. falciparum* but were much less active than artemisinin.

14.4.6 Other Biological Activity

Curcumin **2** also possesses antifungal, antiviral, and antibacterial activity. Its antifungal activity was particularly notable in *Paracoccidioides brasiliensis*, and it has greater activity than fluconazole in certain strains of this fungus [146].

It has recently been shown that curcumin has the ability to abolish the infectivity of enveloped viruses by inhibiting hemagglutination, but this does not occur in nonenveloped viruses [147].

Curcumin has been shown to have antibacterial activity against a number of Gram-positive and Gram-negative bacteria [148]. The antibacterial mechanism of curcumin has been suggested to be due to its ability to bind and inhibit FtsZ assembly [149]. FtsZ is the prokaryotic homologue of tubulin, a cytoskeleton protein, and the assembly of FtsZ protofilaments is crucial in bacterial cytokinesis. Curcumin was found to inhibit cytokinesis in *Bacillus subtilis* 168 by increasing the GTPase activity of FtsZ, thereby inhibiting FtsZ assembly [149].

Mishra et al. [148] synthesized several bioconjugates of curcumin in order to increase the bioavailability and cellular uptake and so enhance curcumin's antibacterial and antifungal activities. Several curcumin derivatives in this study showed better antibacterial activity than the fourth-generation cephalosporin, such as cefepime. Curcumin derivatives containing sulfonamide groups have recently been shown to possess greater activity than curcumin, ciprofloxacin, and fluconazole against a range of pathogenic fungal and Gram-positive and Gram-negative bacterial strains [150].

REFERENCES

[1] Campbell KN, Morris RC, Adams R (1937) The structure of gossypol. I. *Journal of the American Chemical Society* 59: 1723–1728.

[2] Cass QB, Tiritan E, Matlin SA, Freire EC (1991) Gossypol enantiomer ratios in cotton seeds. *Phytochemistry* 30: 2655–2657.

[3] Meyers AI, Willemsen JJ (1997) The synthesis of (S)-(+)-gossypol via an asymmetric Ullmann coupling. *Chemical Communications* 1573–1574.

[4] Aggarwal BB, Sundaram C, Malani N, Ichikawa H (2007) Curcumin: The Indian solid gold. In Aggarwal BB, Surh Y-J, Shishodia S, editors. *Molecular Targets and Therapeutic Uses of Curcumin in Health and Disease*. Springer, New York, 595: 1–75.

[5] Nagabhushan M, Amonkar AJ, Bhide SV (1987) In vitro antimutagenicity of curcumin against environmental mutagens. *Food and Chemical Toxicology* 25: 545–547.

[6] Ammon HP, Wahl MA (1991) Pharmacology of Curcuma longa. *Planta Medica* 57: 1–7.

[7] Singh S (2007) From exotic spice to modern drug? *Cell* 130: 765–768.

[8] Lao CD, Ruffin MT, Normolle D, Heath DD, Murray SI, et al. (2006) Dose escalation of a curcuminoid formulation. *BMC Complementary and Alternative Medicine* 6: 10.

[9] Freedman TB, Cao XL, Oliveira RV, Cass QB, Nafie LA (2003) Determination of the absolute configuration and solution conformation of gossypol by vibrational circular dichroism. *Chirality* 15: 196–200.

[10] Brzezinski B, Olejnik J, Paszyc S (1990) Fourier-transform infrared study on the identification of gossypol tautomers. *Journal of Molecular Structure* 239: 23–31.

[11] Marciniak B, Schroeder G, Kozubek H, Brzezinski B (1991) Spectroscopic and kinetic-studies of the aldehyde lactol tautomerization of gossypol in solution. *Journal of the Chemical Society-Perkin Transactions* 2: 1359–1362.

[12] Stipanov RD, Bell AA, Howell CR (1973) Spectral identification of ketol tautomer of gossypol. *Journal of the American Oil Chemists' Society* 50: 462–463.

[13] Dodou K, Anderson RJ, Lough WJ, Small DAP, Shelley MD, et al. (2005) Synthesis of gossypol atropisomers and derivatives and evaluation of their anti-proliferative and anti-oxidant activity. *Bioorganic & Medicinal Chemistry* 13: 4228–4237.

[14] Payton F, Sandusky P, Alworth WL (2007) NMR study of the solution structure of curcumin. *Journal of Natural Products* 70: 143–146.

[15] Przybylski P, Pyta K, Czupryniak J, Wicher B, Gdaniec M, et al. (2010) The influence of protonation on molecular structure and physico-chemical properties of gossypol Schiff bases. *Organic & Biomolecular Chemistry* 8: 5511–5518.

[16] Fish RG, Groundwater PW, Morgan JJG (1995) The photo-epimerization of gossypol Schiff's bases. *Tetrahedron-Asymmetry* 6: 873–876.

[17] Mathews S, Rao MNA (1991) Interaction of curcumin with glutathione. *International Journal of Pharmaceutics* 76: 257–259.

[18] Royer RE, Deck LM, Vanderjagt TJ, Martinez FJ, Mills RG, et al. (1995) Synthesis and anti-HIV activity of 1,1′-dideoxygossypol and related-compounds. *Journal of Medicinal Chemistry* 38: 2427–2432.

[19] Wichmann K, Vaheri A, Luukkainen T (1982) Inhibiting herpes simplex virus type 2 infection in human epithelial cells by gossypol, a potent spermicidal and contraceptive agent. *American Journal of Obstetrics and Gynecology* 142: 593–594.

[20] Dodou K, Anderson RJ, Small DA, Groundwater PW (2005) Investigations on gossypol: past and present developments. *Expert Opinion on Investigational Drugs* 14: 1419–1434.

[21] Csillag C (1996) Male contraceptive pill to start trial in Brazil. *Lancet* 348: 608–608.

[22] Qian SZ, Wang ZG (1984) Gossypol—a potent antifertility agent for males. *Annual Review of Pharmacology and Toxicology* 24: 329–360.

[23] Hoesl CE, Saad F, Poppel M, Altwein JE (2005) Reversible, non-barrier male contraception: status and prospects. *European Urology* 48: 712–723.

[24] Liang XS, Rogers AJ, Webber CL, Ormsby TJ, Tiritan ME, et al. (1995) Developing gossypol derivatives with enhanced antitumor activity. *Investigational New Drugs* 13: 181–186.

[25] Band V, Hoffer AP, Band H, Rhinehardt AE, Knapp RC, et al. (1989) Antiproliferative effect of gossypol and its optical isomers on human reproductive cancer cell-lines. *Gynecologic Oncology* 32: 273–277.

[26] Blackstaffe L, Shelley MD, Fish RG (1997) Cytotoxicity of gossypol enantiomers and its quinone metabolite gossypolone in melanoma cell lines. *Melanoma Research* 7: 364–372.

[27] Zhang L, Jiang HX, Cao XX, Zhao HY, Wang F, et al. (2009) Chiral gossypol derivatives: evaluation of their anticancer activity and molecular modeling. *European Journal of Medicinal Chemistry* 44: 3961–3972.

[28] Stein RC, Joseph AEA, Matlin SA, Cunningham DC, Ford HT, et al. (1992) A preliminary clinical study of gossypol in advanced human cancer. *Cancer Chemotherapy and Pharmacology* 30: 480–482.

[29] Van Poznak C, Seidman AD, Reidenberg MM, Moasser MM, Sklarin N, et al. (2001) Oral gossypol in the treatment of patients with refractory metastatic breast cancer: a phase I/II clinical trial. *Breast Cancer Research and Treatment* 66: 239–248.

[30] Dao VT, Gaspard C, Mayer M, Werner GH, Nguyen SN, et al. (2000) Synthesis and cytotoxicity of gossypol related compounds. *European Journal of Medicinal Chemistry* 35: 805–813.

[31] Matin A, Doddareddy MR, Gavande N, Nammi S, Groundwater PW, et al. (2013) The discovery of novel isoflavone pan peroxisome proliferator-activated receptor agonists. *Bioorganic & Medicinal Chemistry* 21: 766–778.

[32] Shelley MD, Hartley L, Groundwater PW, Fish RG (2000) Structure-activity studies on gossypol in tumor cell lines. *Anti-Cancer Drugs* 11: 209–216.

[33] Hall C, Troutman SM, Price DK, Figg WD, Kang MH (2013) Bcl-2 family of proteins as therapeutic targets in genitourinary neoplasms. *Clinical Genitourinary Cancer* 11: 10–19.

[34] Sonpavde G, Matveev V, Burke JM, Caton JR, Fleming MT, et al. (2012) Randomized phase II trial of docetaxel plus prednisone in combination with placebo or AT-101, an oral small molecule Bcl-2 family antagonist, as first-line therapy for metastatic castration-resistant prostate cancer. *Annals of Oncology* 23: 1803–1808.

[35] Liu G, Kelly WK, Wilding G, Leopold L, Brill K, et al. (2009) An open-label, multicenter, phase I/II study of single-agent AT-101 in men with castrate-resistant prostate cancer. *Clinical Cancer Research* 15: 3172–3176.

[36] Oliver CL, Miranda MB, Shangary S, Land S, Wang SM, et al. (2005) (-)-Gossypol acts directly on the mitochondria to overcome Bcl-2-mediated and Bcl-X-L-mediated apoptosis resistance. *Molecular Cancer Therapeutics* 4: 23–31.

[37] Becattini B, Kitada S, Leone M, Monosov E, Chandler S, et al. (2004) Rational design and real time, in-cell detection of the proapoptotic activity of a novel compound targeting Bcl-X-L. *Chemistry & Biology* 11: 389–395.

[38] Wei J, Kitada S, Stebbins JL, Placzek W, Zhai DY, et al. (2010) Synthesis and biological evaluation of apogossypolone derivatives as pan-active inhibitors of antiapoptotic B-cell lymphoma/leukemia-2 (Bcl-2) family proteins. *Journal of Medicinal Chemistry* 53: 8000–8011.

[39] Amundson SA, Myers TG, Scudiero D, Kitada S, Reed JC, et al. (2000) An informatics approach identifying markers of chemosensitivity in human cancer cell lines. *Cancer Research* 60: 6101–6110.

[40] Wei J, Stebbins JL, Kitada S, Dash R, Placzek W, et al. (2010) BI-97C1, an optically pure apogossypol derivative as pan-active inhibitor of antiapoptotic B-cell lymphoma/leukemia-2 (Bcl-2) family proteins. *Journal of Medicinal Chemistry* 53: 4166–4176.

[41] Jackson RS, Placzek W, Fernandez A, Ziaee S, Chu CY, et al. (2012) Sabutoclax, a Mcl-1 antagonist, inhibits tumorigenesis in transgenic mouse and human xenograft models of prostate cancer. *Neoplasia* 14: 656–665.

[42] Dash R, Azab B, Quinn BA, Shen XN, Wang XY, et al. (2011) Apogossypol derivative BI-97C1 (Sabutoclax) targeting Mcl-1 sensitizes prostate cancer cells to mda-7/IL-24-mediated toxicity. *Proceedings of the National Academy of Sciences of the United States of America* 108: 8785–8790.

[43] Dorsett PH, Kerstine EE, Powers LJ (1975) Antiviral activity of gossypol and apogossypol. *Journal of Pharmaceutical Sciences* 64: 1073–1075.

[44] Radloff RJ, Deck LM, Royer RE, Vanderjagt DL (1986) Antiviral activities of gossypol and its derivatives against herpes-simplex type-II. *Pharmacological Research Communications* 18: 1063–1073.

[45] Lin TS, Schinazi R, Griffith BP, August EM, Eriksson BFH, et al. (1989) Selective-inhibition of human immunodeficiency virus type-1 replication by the (-) but not the (+) enantiomer of gossypol. *Antimicrobial Agents and Chemotherapy* 33: 2149–2151.

[46] Lin TS, Schinazi RF, Zhu JL, Birks E, Carbone R, et al. (1993) Anti-HIV activity and cellular pharmacology of various analogs of gossypol. *Biochemical Pharmacology* 46: 251–255.

[47] Baram NI, Biktimirov L, Ziyaev KL, Paizieva RZ, Ismailov AI (1995) Antiviral and interferon-inducing activities of gossypol and its derivatives. *Chemistry of Natural Compounds* 31: 299–303.

[48] Royer RE, Mills RG, Young SA, Jagt DLV (1995) Comparison of the antiviral activities of 3′-azido-3′-deoxythymidine (AZT) and gossylic iminolactone (GIL) against clinical isolates of HIV-1. *Pharmacological Research* 31: 49–52.

[49] Yang J, Zhang F, Li JR, Chen G, Wu SW, et al. (2012) Synthesis and antiviral activities of novel gossypol derivatives. *Bioorganic & Medicinal Chemistry Letters* 22: 1415–1420.

[50] Yang J, Chen G, Li LL, Pan W, Zhang F, et al. (2013) Synthesis and anti-H5N1 activity of chiral gossypol derivatives and its analogs implicated by a viral entry blocking mechanism. *Bioorganic and Medicinal Chemistry Letters* 23: 2619–2623.

[51] An T, Ouyang WJ, Pan W, Guo DY, Li JR, et al. (2012) Amino acid derivatives of the (-) enantiomer of gossypol are effective fusion inhibitors of human immunodeficiency virus type 1. *Antiviral Research* 94: 276–287.

[52] Royer RE, Deck LM, Campos NM, Hunsaker LA, Vanderjagt DL (1986) Biologically-active derivatives of gossypol—synthesis and antimalarial activities of peri-acylated gossylic nitriles. *Journal of Medicinal Chemistry* 29: 1799–1801.

[53] Dunn CR, Banfield MJ, Barker JJ, Higham CW, Moreton KM, et al. (1996) The structure of lactate dehydrogenase from Plasmodium falciparum reveals a new target for anti-malarial design. *Nature Structural Biology* 3: 912–915.

[54] Sessions RB, Dewar V, Clarke AR, Holbrook JJ (1997) A model of Plasmodium falciparum lactate dehydrogenase and its implications for the design of improved antimalarials and the enhanced detection of parasitaemia. *Protein Engineering* 10: 301–306.

[55] Conners R, Schambach F, Read J, Cameron A, Sessions RB, et al. (2005) Mapping the binding site for gossypol-like inhibitors of Plasmodium falciparum lactate dehydrogenase. *Molecular and Biochemical Parasitology* 142: 137–148.

[56] Roth E (1990) Plasmodium-falciparum carbohydrate metabolism—a connection between host cell and parasite. *Blood Cells* 16: 453–460.

[57] Tripathi AK, Desai PV, Pradhan A, Khan SI, Avery MA, et al. (2004) An alpha-proteobacterial type malate dehydrogenase may complement LDH function in Plasmodium falciparum—cloning and biochemical characterization of the enzyme. *European Journal of Biochemistry* 271: 3488–3502.

[58] Vander Jagt DL, Deck LM, Royer RE (2000) Gossypol: prototype of inhibitors targeted to dinucleotide folds. *Current Medicinal Chemistry* 7: 479–498.

[59] Yin JJ, Jin LM, Chen F, Wang X, Kitaygorodskiy A, et al. (2011) Novel O-glycosidic gossypol isomers and their bioactivities. *Carbohydrate Research* 346: 2070–2074.

[60] Wang YJ, Pan MH, Cheng AL, Lin LI, Ho YS, et al. (1997) Stability of curcumin in buffer solutions and characterization of its degradation products. *Journal of Pharmaceutical and Biomedical Analysis* 15: 1867–1876.

[61] Wahlstrom B, Blennow G (1978) A study on the fate of curcumin in the rat. *Acta Pharmacologica et Toxicologica* 43: 86–92.

[62] Dempe JS, Scheerle RK, Pfeiffer E, Metzler M (2013) Metabolism and permeability of curcumin in cultured Caco-2 cells. *Molecular Nutrition & Food Research* 57: 1543–1549.

[63] Ravindranath V, Chandrasekhara N (1980) Absorption and tissue distribution of curcumin in rats. *Toxicology* 16: 259–265.

[64] Pan MH, Huang TM, Lin JK (1999) Biotransformation of curcumin through reduction and glucuronidation in mice. *Drug Metabolism and Disposition* 27: 486–494.

[65] Perkins S, Verschoyle RD, Hill K, Parveen I, Threadgill MD, et al. (2002) Chemopreventive efficacy and pharmacokinetics of curcumin in the min/+ mouse, a model of familial adenomatous polyposis. *Cancer Epidemiology, Biomarkers & Prevention* 11: 535–540.

[66] Vijayakumar GR, Divakar S (2007) Amyloglucosidase-catalyzed synthesis of eugenyl and curcuminyl glycosides. *Biotechnology Letters* 29: 575–584.

[67] Pabon HJJ (1964) A synthesis of curcumin and related compounds. *Recueil des Travaux Chimiques des Pays-Bas* 83: 379–386.

[68] Esatbeyoglu T, Huebbe P, Ernst IMA, Chin D, Wagner AE, et al. (2012) Curcumin From molecule to biological function. *Angewandte Chemie-International Edition* 51: 5308–5332.

[69] Sardjiman SS, Reksohadiprodjo MS, Hakim L, van der Goot H, Timmerman H (1997) 1,5-Diphenyl-1,4-pentadiene-3-ones and cyclic analogues as antioxidative agents. Synthesis and structure-activity relationship. *European Journal of Medicinal Chemistry* 32: 625–630.

[70] Adams BK, Ferstl EM, Davis MC, Herold M, Kurtkaya S, et al. (2004) Synthesis and biological evaluation of novel curcumin analogs as anti-cancer and anti-angiogenesis agents. *Bioorganic & Medicinal Chemistry* 12: 3871–3883.

[71] Hanahan D, Weinberg RA (2011) Hallmarks of cancer: the next generation. *Cell* 144: 646–674.

[72] Sa G, Das T (2008) Anti cancer effects of curcumin: cycle of life and death. *Cell Div* 3: 14.

[73] Anand P, Sundaram C, Jhurani S, Kunnumakkara AB, Aggarwal BB (2008) Curcumin and cancer: an "old-age" disease with an "age-old" solution. *Cancer Letters* 267: 133–164.

[74] Ravindran J, Prasad S, Aggarwal BB (2009) Curcumin and cancer cells: how many ways can curry kill tumor cells selectively? *AAPS J* 11: 495–510.

[75] Fang J, Lu J, Holmgren A (2005) Thioredoxin reductase is irreversibly modified by curcumin: a novel molecular mechanism for its anticancer activity. *The Journal of Biological Chemistry* 280: 25284–25290.

[76] Zhang F, Altorki NK, Mestre JR, Subbaramaiah K, Dannenberg AJ (1999) Curcumin inhibits cyclooxygenase-2 transcription in bile acid- and phorbol ester-treated human gastrointestinal epithelial cells. *Carcinogenesis* 20: 445–451.

[77] Mahmmoud YA (2007) Modulation of protein kinase C by curcumin; inhibition and activation switched by calcium ions. *British Journal of Pharmacology* 150: 200–208.

[78] Goel A, Kunnumakkara AB, Aggarwal BB (2008) Curcumin as "Curecumin": from kitchen to clinic. *Biochemical Pharmacology* 75: 787–809.

[79] Aggarwal BB, Harikumar KB (2009) Potential therapeutic effects of curcumin, the anti-inflammatory agent, against neurodegenerative, cardiovascular, pulmonary, metabolic, autoimmune and neoplastic diseases. *International Journal of Biochemistry & Cell Biology* 41: 40–59.

[80] Chakravarti N, Kadara H, Yoon DJ, Shay JW, Myers JN, et al. (2010) Differential inhibition of protein translation machinery by curcumin in normal, immortalized, and malignant oral epithelial cells. *Cancer Prevention Research* 3: 331–338.

[81] Syng-Ai C, Kumari AL, Khar A (2004) Effect of curcumin on normal and tumor cells: role of glutathione and bcl-2. *Molecular Cancer Therapeutics* 3: 1101–1108.

[82] Watson JL, Hill R, Yaffe PB, Greenshields A, Walsh M, et al. (2010) Curcumin causes superoxide anion production and p53-independent apoptosis in human colon cancer cells. *Cancer Letters* 297: 1–8.

[83] Yoon MJ, Kim EH, Kwon TK, Park SA, Choi KS (2012) Simultaneous mitochondrial Ca2+ overload and proteasomal inhibition are responsible for the induction of paraptosis in malignant breast cancer cells. *Cancer Letters* 324: 197–209.

[84] Zanotto-Filho A, Braganhol E, Edelweiss MI, Behr GA, Zanin R, et al. (2012) The curry spice curcumin selectively inhibits cancer cells growth in vitro and in preclinical model of glioblastoma. *The Journal of Nutritional Biochemistry* 23: 591–601.

[85] Agrawal DK, Mishra PK (2010) Curcumin and its analogues: potential anticancer agents. *Medicinal Research Reviews* 30: 818–860.

[86] Sandur SK, Pandey MK, Sung B, Ahn KS, Murakami A, et al. (2007) Curcumin, demethoxycurcumin, bisdemethoxycurcumin, tetrahydrocurcumin and turmerones differentially regulate anti-inflammatory and anti-proliferative responses through a ROS-independent mechanism. *Carcinogenesis* 28: 1765–1773.

[87] Lin L, Shi Q, Nyarko AK, Bastow KF, Wu CC, et al. (2006) Antitumor agents. 250. Design and synthesis of new curcumin analogues as potential anti-prostate cancer agents. *Journal of Medicinal Chemistry* 49: 3963–3972.

[88] Ahn B-Z, Sok S-E (1996) Michael acceptors as a tool for anticancer drug design. *Current Pharmaceutical Design* 2: 247–262.

[89] Amolins MW, Peterson LB, Blagg BSJ (2009) Synthesis and evaluation of electron-rich curcumin analogues. *Bioorganic & Medicinal Chemistry* 17: 360–367.

[90] Yamakoshi H, Ohori H, Kudo C, Sato A, Kanoh N, et al. (2010) Structure–activity relationship of C5-curcuminoids and synthesis of their molecular probes thereof. *Bioorganic & Medicinal Chemistry* 18: 1083–1092.

[91] Ohori H, Yamakoshi H, Tomizawa M, Shibuya M, Kakudo Y, et al. (2006) Synthesis and biological analysis of new curcumin analogues bearing an enhanced potential for the medicinal treatment of cancer. *Molecular Cancer Therapeutics* 5: 2563–2571.

[92] Hutzen B, Friedman L, Sobo M, Lin L, Cen L, et al. (2009) Curcumin analogue GO-Y030 inhibits STAT3 activity and cell growth in breast and pancreatic carcinomas. *International Journal of Oncology* 35: 867–872.

[93] Sato A, Kudo C, Yamakoshi H, Uehara Y, Ohori H, et al. (2011) Curcumin analog GO-Y030 is a novel inhibitor of IKKbeta that suppresses NF-kappaB signaling and induces apoptosis. *Cancer Science* 102: 1045–1051.

[94] Kudo C, Yamakoshi H, Sato A, Ohori H, Ishioka C, et al. (2011) Novel curcumin analogs, GO-Y030 and GO-Y078, are multi-targeted agents with enhanced abilities for multiple myeloma. *Anticancer Research* 31: 3719–3726.

[95] Cen L, Hutzen B, Ball S, DeAngelis S, Chen C-L, et al. (2009) New structural analogues of curcumin exhibit potent growth suppressive activity in human colorectal carcinoma cells. *BMC Cancer* 9: 99.

[96] Shibata H, Yamakoshi H, Sato A, Ohori H, Kakudo Y, et al. (2009) Newly synthesized curcumin analog has improved potential to prevent colorectal carcinogenesis in vivo. *Cancer Science* 100: 956–960.

[97] Kudo C, Yamakoshi H, Sato A, Nanjo H, Ohori H, et al. (2011) Synthesis of 86 species of 1,5-diaryl-3-oxo-1,4-pentadienes analogs of curcumin can yield a good lead in vivo. *BMC Pharmacology* 11: 4.

[98] Adams BK, Cai J, Armstrong J, Herold M, Lu YJ, et al. (2005) EF24, a novel synthetic curcumin analog, induces apoptosis in cancer cells via a redox-dependent mechanism. *Anti-Cancer Drugs* 16: 263–275.

[99] Kasinski AL, Du Y, Thomas SL, Zhao J, Sun SY, et al. (2008) Inhibition of IkappaB kinase-nuclear factor-kappaB signaling pathway by 3,5-bis(2-flurobenzylidene)piperidin-4-one (EF24), a novel monoketone analog of curcumin. *Molecular Pharmacology* 74: 654–661.

[100] Selvendiran K, Tong L, Vishwanath S, Bratasz A, Trigg NJ, et al. (2007) EF24 induces G2/M arrest and apoptosis in cisplatin-resistant human ovarian cancer cells by increasing PTEN expression. *The Journal of Biological Chemistry* 282: 28609–28618.

[101] Thomas SL, Zhong D, Zhou W, Malik S, Liotta D, et al. (2008) EF24, a novel curcumin analog, disrupts the microtubule cytoskeleton and inhibits HIF-1. *Cell Cycle* 7: 2409–2417.

[102] Sun A, Lu YJ, Hu H, Shoji M, Liotta DC, et al. (2009) Curcumin analog cytotoxicity against breast cancer cells: exploitation of a redox-dependent mechanism. *Bioorganic & Medicinal Chemistry Letters* 19: 6627–6631.

[103] Ravindran J, Subbaraju GV, Ramani MV, Sung B, Aggarwal BB (2010) Bisdemethylcurcumin and structurally related hispolon analogues of curcumin exhibit enhanced prooxidant, anti-proliferative and anti-inflammatory activities in vitro. *Biochemical Pharmacology* 79: 1658–1666.

[104] Taylor RA, Leonard MC (2011) Curcumin for inflammatory bowel disease: a review of human studies. *Alternative Medicine Review* 16: 152–156.

[105] Aggarwal S, Ichikawa H, Takada Y, Sandur SK, Shishodia S, et al. (2006) Curcumin (diferuloylmethane) down-regulates expression of cell proliferation and antiapoptotic and metastatic gene products through suppression of IkappaBalpha kinase and Akt activation. *Molecular Pharmacology* 69: 195–206.

[106] Sandur SK, Ichikawa H, Pandey MK, Kunnumakkara AB, Sung B, et al. (2007) Role of pro-oxidants and antioxidants in the anti-inflammatory and apoptotic effects of curcumin (diferuloylmethane). *Free Radical Biology & Medicine* 43: 568–580.

[107] Strimpakos AS, Sharma RA (2008) Curcumin: preventive and therapeutic properties in laboratory studies and clinical trials. *Antioxidants & Redox Signaling* 10: 511–545.

[108] Pan M-H, Lin-Shiau S-Y, Lin J-K (2000) Comparative studies on the suppression of nitric oxide synthase by curcumin and its hydrogenated metabolites through down-regulation of IκB kinase and NFκB activation in macrophages. *Biochemical Pharmacology* 60: 1665–1676.

[109] Hassaninasab A, Hashimoto Y, Tomita-Yokotani K, Kobayashi M (2011) Discovery of the curcumin metabolic pathway involving a unique enzyme in an intestinal microorganism. *Proceedings of the National Academy of Sciences of the United States of America* 108: 6615–6620.

[110] Sugiyama Y, Kawakishi S, Osawa T (1996) Involvement of the beta-diketone moiety in the antioxidative mechanism of tetrahydrocurcumin. *Biochemical Pharmacology* 52: 519–525.

[111] Osawa T, Sugiyama Y, Inayoshi M, Kawakishi S (1995) Antioxidative activity of tetrahydrocurcuminoids. *Bioscience, Biotechnology, and Biochemistry* 59: 1609–1612.

[112] Priyadarsini KI, Maity DK, Naik GH, Kumar MS, Unnikrishnan MK, et al. (2003) Role of phenolic O-H and methylene hydrogen on the free radical reactions and antioxidant activity of curcumin. *Free Radical Biology and Medicine* 35: 475–484.

[113] Sun Y-M, Zhang H-Y, Chen D-Z, Liu C-B (2002) Theoretical elucidation on the antioxidant mechanism of curcumin: a DFT study. *Organic Letters* 4: 2909–2911.

[114] Barclay LR, Vinqvist MR, Mukai K, Goto H, Hashimoto Y, et al. (2000) On the antioxidant mechanism of curcumin: classical methods are needed to determine antioxidant mechanism and activity. *Organic Letters* 2: 2841–2843.

[115] Venkatesan P, Rao MN (2000) Structure-activity relationships for the inhibition of lipid peroxidation and the scavenging of free radicals by synthetic symmetrical curcumin analogues. *The Journal of Pharmacy and Pharmacology* 52: 1123–1128.

[116] Venkateswarlu S, Ramachandra MS, Subbaraju GV (2005) Synthesis and biological evaluation of polyhydroxycurcuminoids. *Bioorganic & Medicinal Chemistry* 13: 6374–6380.

[117] Litwinienko G, Ingold KU (2004) Abnormal solvent effects on hydrogen atom abstraction. 2. Resolution of the curcumin antioxidant controversy. The role of sequential proton loss electron transfer. *Journal of Organic Chemistry* 69: 5888–5896.

[118] Graf E (1992) Antioxidant potential of ferulic acid. *Free Radical Biology and Medicine* 13: 435–448.

[119] Tai A, Sawano T, Yazama F, Ito H (2011) Evaluation of antioxidant activity of vanillin by using multiple antioxidant assays. *Biochimica et Biophysica Acta* 2: 170–177.

[120] Zhao CG, Cai YP, He XZ, Li JL, Zhang L, et al. (2010) Synthesis and anti-inflammatory evaluation of novel mono-carbonyl analogues of curcumin in LPS-stimulated RAW 264.7 macrophages. *European Journal of Medicinal Chemistry* 45: 5773–5780.

[121] Liang G, Zhou H, Wang Y, Gurley EC, Feng B, et al. (2009) Inhibition of LPS-induced production of inflammatory factors in the macrophages by mono-carbonyl analogues of curcumin. *Journal of Cellular and Molecular Medicine* 13: 3370–3379.

[122] Wu JZ, Zhang YL, Cai YP, Wang J, Weng BX, et al. (2013) Discovery and evaluation of piperid-4-one-containing mono-carbonyl analogs of curcumin as anti-inflammatory agents. *Bioorganic & Medicinal Chemistry* 21: 3058–3065.

[123] Shang Y-J, Jin X-L, Shang X-L, Tang J-J, Liu G-Y, et al. (2010) Antioxidant capacity of curcumin-directed analogues: structure–activity relationship and influence of microenvironment. *Food Chemistry* 119: 1435–1442.

[124] Fang XB, Fang L, Gou SH, Cheng L (2013) Design and synthesis of dimethylamino-methyl-substituted curcumin derivatives/analogues: potent antitumor and antioxidant activity, improved stability and aqueous solubility compared with curcumin. *Bioorganic & Medicinal Chemistry Letters* 23: 1297–1301.

[125] Cole GM, Teter B, Frautschy SA (2007) Neuroprotective effects of curcumin. *Advances in Experimental Medicine and Biology* 595: 197–212.

[126] Zhang C, Browne A, Child D, Tanzi RE (2010) Curcumin decreases amyloid-beta peptide levels by attenuating the maturation of amyloid-beta precursor protein. *The Journal of Biological Chemistry* 285: 28472–28480.

[127] Yang F, Lim GP, Begum AN, Ubeda OJ, Simmons MR, et al. (2005) Curcumin inhibits formation of amyloid beta oligomers and fibrils, binds plaques, and reduces amyloid in vivo. *The Journal of Biological Chemistry* 280: 5892–5901.

[128] Ono K, Hasegawa K, Naiki H, Yamada M (2004) Curcumin has potent anti-amyloidogenic effects for Alzheimer's beta-amyloid fibrils in vitro. *Journal of Neuroscience Research* 75: 742–750.

[129] Shimmyo Y, Kihara T, Akaike A, Niidome T, Sugimoto H (2008) Epigallocatechin-3-gallate and curcumin suppress amyloid beta-induced beta-site APP cleaving enzyme-1 upregulation. *Neuroreport* 19: 1329–1333.

[130] House E, Collingwood J, Khan A, Korchazkina O, Berthon G, et al. (2004) Aluminium, iron, zinc and copper influence the in vitro formation of amyloid fibrils of Abeta42 in a manner which may have consequences for metal chelation therapy in Alzheimer's disease. *Journal of Alzheimer's Disease* 6: 291–301.

[131] Tougu V, Karafin A, Zovo K, Chung RS, Howells C, et al. (2009) Zn(II)- and Cu(II)-induced non-fibrillar aggregates of amyloid-beta (1-42) peptide are transformed to amyloid fibrils, both spontaneously and under the influence of metal chelators. *Journal of Neurochemistry* 110: 1784–1795.

[132] Begum AN, Jones MR, Lim GP, Morihara T, Kim P, et al. (2008) Curcumin structure-function, bioavailability, and efficacy in models of neuroinflammation and Alzheimer's disease. *The Journal of Pharmacology and Experimental Therapeutics* 326: 196–208.

[133] Orlando RA, Gonzales AM, Royer RE, Deck LM, Vander Jagt DL (2012) A chemical analog of curcumin as an improved inhibitor of amyloid Abeta oligomerization. *PLoS ONE* 7: 19.

[134] Narlawar R, Pickhardt M, Leuchtenberger S, Baumann K, Krause S, et al. (2008) Curcumin-derived pyrazoles and isoxazoles: Swiss army knives or blunt tools for Alzheimer's disease? *ChemMedChem* 3: 165–172.

[135] Chen SY, Chen Y, Li YP, Chen SH, Tan JH, et al. (2011) Design, synthesis, and biological evaluation of curcumin analogues as multifunctional agents for the treatment of Alzheimer's disease. *Bioorganic & Medicinal Chemistry* 19: 5596–5604.

[136] Narlawar R, Baumann K, Schubenel R, Schmidt B (2007) Curcumin derivatives inhibit or modulate beta-amyloid precursor protein metabolism. *Neuro-Degenerative Diseases* 4: 88–93.

[137] Liu Y, Dargusch R, Maher P, Schubert D (2008) A broadly neuroprotective derivative of curcumin. *Journal of Neurochemistry* 105: 1336–1345.

[138] Valera E, Dargusch R, Maher PA, Schubert D (2013) Modulation of 5-lipoxygenase in proteotoxicity and Alzheimer's disease. *The Journal of Neuroscience* 33: 10512–10525.

[139] Reddy RC, Vatsala PG, Keshamouni VG, Padmanaban G, Rangarajan PN (2005) Curcumin for malaria therapy. *Biochemical and Biophysical Research Communications* 326: 472–474.

[140] Cui L, Miao J, Cui LW (2007) Cytotoxic effect of curcumin on malaria parasite Plasmodium falciparum: inhibition of histone acetylation and generation of reactive oxygen species. *Antimicrobial Agents and Chemotherapy* 51: 488–494.

[141] Ji HF, Shen L (2009) Interactions of curcumin with the PfATP6 model and the implications for its antimalarial mechanism. *Bioorganic & Medicinal Chemistry Letters* 19: 2453–2455.

[142] Chakrabarti R, Rawat PS, Cooke BM, Coppel RL, Patankar S (2013) Cellular effects of curcumin on Plasmodium falciparum include disruption of microtubules. *PLoS ONE* 8: 7.

[143] Mimche PN, Thompson E, Taramelli D, Vivas L (2012) Curcumin enhances non-opsonic phagocytosis of Plasmodium falciparum through up-regulation of CD36 surface expression on monocytes/macrophages. *The Journal of Antimicrobial Chemotherapy* 67: 1895–1904.

[144] Mishra S, Karmodiya K, Surolia N, Surolia A (2008) Synthesis and exploration of novel curcumin analogues as anti-malarial agents. *Bioorganic & Medicinal Chemistry* 16: 2894–2902.

[145] Manohar S, Khan SI, Kandi SK, Raj K, Sun G, et al. (2013) Synthesis, antimalarial activity and cytotoxic potential of new monocarbonyl analogues of curcumin. *Bioorganic & Medicinal Chemistry Letters* 23: 112–116.

[146] Martins CV, da Silva DL, Neres AT, Magalhaes TF, Watanabe GA, et al. (2009) Curcumin as a promising antifungal of clinical interest. *The Journal of Antimicrobial Chemotherapy* 63: 337–339.

[147] Chen TY, Chen DY, Wen HW, Ou JL, Chiou SS, et al. (2013) Inhibition of enveloped viruses infectivity by curcumin. *PLoS ONE* 8: e62482.

[148] Mishra S, Narain U, Mishra R, Misra K (2005) Design, development and synthesis of mixed bioconjugates of piperic acid–glycine, curcumin–glycine/alanine and curcumin–glycine–piperic acid and their antibacterial and antifungal properties. *Bioorganic & Medicinal Chemistry* 13: 1477–1486.

[149] Rai D, Singh JK, Roy N, Panda D (2008) Curcumin inhibits FtsZ assembly: an attractive mechanism for its antibacterial activity. *The Biochemical Journal* 410: 147–155.

[150] Lal J, Gupta SK, Thavaselvam D, Agarwal DD (2013) Biological activity, design, synthesis and structure activity relationship of some novel derivatives of curcumin containing sulfonamides. *European Journal of Medicinal Chemistry* 64: 579–588.

15

PHYTOTHERAPIES FOR THE MANAGEMENT OF OBESITY AND DIABETES

MICHEL RAPINSKI AND ALAIN CUERRIER
Institut de recherche en biologie végétale, l'Université de Montréal, Montréal, Canada

15.1 INTRODUCTION

WHO has cited obesity as a global epidemic [1] and its prevalence is indeed increasing at an alarming rate worldwide [2]. In 2005, 33% of the world population was considered overweight or facing obesity, representing 937 million and 396 million people, respectively [2]. By 2030, these numbers are expected to rise to 1.35 billion and 573 million individuals, respectively, representing 57.8% of the world's projected population [2]. There are important health implications associated with this trend as epidemiological studies suggest that overweight and obesity are important risk factors for noncommunicable chronic diseases such as diabetes and cardiovascular diseases [1, 2]. Not surprisingly, a similar trend is observed in diabetes, which, like its predecessor, has reached pandemic proportions. In 2011, the number of people with diabetes was estimated at 366 million, representing 8.3% of the global population, while this incidence is expected to increase to 552 million people by 2030 [3].

Although the obesity pandemic is thought to have originated in the world's wealthier nations, most notably in North America across to Europe, it is also emerging in developing countries, primarily through the improvement of socioeconomic status and changing lifestyles [4, 5]. This is happening through vectors of subsidized agriculture, the increasing availability of cheap, highly refined fats, oils, and

Phytotherapies: Efficacy, Safety, and Regulation, First Edition. Edited by Iqbal Ramzan.

carbohydrates, labor-saving mechanized devices, affordable motorized transport, rapid urbanization, and the seductions of sedentary pastimes [4, 5]. In fact, a disconcerting dichotomy is beginning to exist between the economically developed and developing world countries. Although the prevalence of obesity is higher in the former, larger populations in the latter result in a higher absolute number: 326 million individuals are overweight and 188 million obese versus 611 and 209 million, respectively [2]. Furthermore, the increase is expected to be relatively larger in economically developing countries due to growth in population, aging population, urbanization, and other lifestyle changes already cited earlier [2].

Diabetes, on the other hand, is a chronic affliction characterized by hyperglycemia, an elevated blood glucose concentration, that is caused by a decreased insulin secretion, which is generally related to problems with pancreatic ß cells [6, 7]. Type 2 diabetes (T2D), accounting for 90% of cases, is the most common form of diabetes [6, 8]. Also, designated non-insulin-dependent diabetes mellitus (NIDDM) generally appears after the development of insulin resistance in target tissues [6, 9]. Excessive accumulation of lipids in abdominal fat or ectopic sites, like the liver or skeletal muscles, plays a significant role in the pathophysiology of insulin resistance [10], and rising cases of obesity will naturally be associated with rising cases of T2D. Hence, it is expected that a reduction in the global burden of overweight and obesity will translate into worldwide decreases in diabetes and cardiovascular diseases [2].

Factors associated with the rise of diabetes, notably urbanization, population growth and aging, as well as lifestyles changes such as decreased physical activity and changing diets [3, 11], are not any different to those of obesity. Nonetheless, certain ethnic groups, for example, indigenous peoples of the Americas and Australasia, are thought to be more at risk [12, 13]. Indeed, the Pima of Arizona, United States, are widely thought to have the highest known prevalence of T2D [14]. Among the James Bay Cree of Eeyou Istchee (CEI), Canada, the age-adjusted incidence of T2D was 29% of CEI adults in 2009 [15]. Although T2D was once rare among Canada's First Nations and other indigenous populations of North America [6, 16], for just over half a century, cases of the disease have risen considerably [17, 18]. Having reached epidemic proportions in some communities, the Canadian aboriginal population with diabetes is three to five times higher than the national mean [6, 18].

Although there is evidence for a genetic predisposition as a contributing factor for the high rate of T2D in indigenous population [13, 19–22], rapid changes in lifestyle and nutrition are still thought to be major causes. Exacerbated further by low compliance to current drug treatment, the inaccessibility of specific healthcare services in remote communities is an important constraining factor that can lead to an increase in complications associated with T2D as it often appears prematurely coupled, with late detection.

Overall, obesity and diabetes will continue to rise. With 60% of the global burden of increase in noncommunicable chronic diseases expected to occur in developing countries [23], governments and healthcare service providers from these countries are not expected to have the resources to tackle the problem [4]. As endorsed by the World Health Organization in its resolutions for traditional medicines [24], phytotherapies from local pharmacopoeias should be investigated and employed, that is, in primary health care to alleviate the burden of obesity and diabetes. This chapter provides support for the rightful place of phytotherapies in the management of these problems by demonstrating how developing rigorous platforms for the pharmacological screening of botanical extracts provide empirical evidence to their efficacy in chronic disease treatment. Finally, it is demonstrated how the necessity of community participation in integrating phytotherapies into culturally adapted treatments can lead to greater compliance.

15.2 PLANTS FROM THE NORTH AMERICAN PHARMACOPOEIA

We consulted Arnason and colleagues [25], Marles and colleagues [26], Moerman [27], and Uprety and colleagues [28] for plants that have been used traditionally in native, North American pharmacopoeias to treat diabetes (Table 15.1). In order to widen the scope of potentially antidiabetic phytotherapies, species that have been used to treat some of the most common symptoms of T2D, notably (a) blurred vision, (b) increased urination, (c) numbness in the extremities of the limbs and feet sores, and finally (d) skin infections and slow healing wounds, which have been ranked by doctors and healthcare professionals to be some of the most relevant [29], were included. To narrow the selection, keywords and uses for treatments of problems indicative of (a) a decreasing state of vision (blind, blindness, vision, and eyesight), (b) a sense of incontinence or frequent and increasing urination (urination, urine, incontinence, and bedwetting), (c) the loss of sensation or feeling in the limbs (numb, numbness, feeling, and sensation) and more specifically feet sores (feet, foot, sores, and ulcers), and (d) problems associated with the skin (skin and skin infections) or slow healing and stubborn wounds (heal, healing, stubborn, and old) were searched.

Of a total of 152 plants listed in Table 15.1, 39 are used traditionally to treat diabetes specifically. These include all the species cited in Cichewicz and Clifford [30], except for *Acer glabrum* Torr., *Equisetum arvense* L., and *Opuntia* spp., which did not appear in the current search. Of the species listed to treat diabetes, five are used to treat at least one other symptom related to T2D. Out of those that were not reported to treat diabetes specifically, six are used to treat two of these symptoms. Using a traditional knowledge (TK) targeted approach to selecting phytotherapies for an in-depth assessment of their pharmacological activity, one may prioritize plant species that have not only been cited for diabetes, but for one or more of its symptoms as well. Likewise, even if a species has not been cited specifically for diabetes, one may still want to consider the pharmacological assessment of those that have a history of traditional use in treating multiple symptoms and complications related to T2D.

Naturally, very few of these species have been assessed for their antidiabetic or antiobesity activity. *Populus tremuloides* and *H. maximum* are the only species in Table 15.1 that have been assessed in humans. Based on a preparation from a Sioux healer who had been using it to treat diabetes in his community, a randomized,

TABLE 15.1 List of North American Plant Species Documented in the Literature for Their Traditional Uses in Treating Diabetes and Symptoms Associated with Diabetes

Plant Species	Family	Traditional Uses
Acer saccharinum L.	Sapindaceae	Skin infections/slow healing wounds
Acer saccharum Marsh.	Sapindaceae	Blurred vision
Achillea millefolium L.	Asteraceae	Skin infections/slow healing wounds, **diabetes**
Acorus americanus (Raf.) Raf.	Acoraceae	**Diabetes**
Acorus calamus L.	Acoraceae	**Diabetes**
Alisma subcordatum Raf.	Alismataceae	Skin infections/slow healing wounds
Alnus viridis ssp. crispa (Ait.) Turrill	Betulaceae	Skin infections/slow healing wounds
Anaphalis margaritacea (L.) Benth. & Hook. f.	Asteraceae	Foot numbness/sores; blurred vision
Andropogon floridanus Scribn.	Poaceae	Increased urination
Anemopsis californica (Nutt.) Hook. & Arn.	Saururaceae	**Diabetes**
Aralia nudicaulis L.	Araliaceae	**Diabetes**
Aralia racemosa L.	Araliaceae	**Diabetes**
Aralia spinosa L.	Araliaceae	Skin infections/slow healing wounds
Arbutus menziesii Pursh	Ericaceae	**Diabetes**
Armoracia rusticana P.G. Gaertn.	Brassicaceae	**Diabetes**
Artemisia alaskana Rydb.	Asteraceae	**Diabetes**
Artemisia norvegica ssp. saxatilis (Besser) H.M. Hall & Clem.	Asteraceae	**Diabetes**
Artemisia frigida Willd.	Asteraceae	Blurred vision; **diabetes**
Artemisia ludoviciana Nutt.	Asteraceae	Skin infections/slow healing wounds
Artemisia tilesii Ledeb.	Asteraceae	Skin infections/slow healing wounds
Aruncus dioicus (Walt.) Fern.	Rosaceae	Increased urination
Asclepias incarnata L.	Apocynaceae	Increased urination
Asclepias speciosa Torr.	Apocynaceae	Blurred vision
Asclepias tuberosa L.	Apocynaceae	Skin infections/slow healing wounds
Athyrium filix-femina (L.) Roth	Woodsiaceae	Increased urination
Baptisia lactea var. lactea (Raf.) Thieret	Fabaceae	Skin infections/slow healing wounds
Brickellia eupatorioides var. eupatorioides (L.) Shinners	Asteraceae	Skin infections/slow healing wounds
Calla palustris L.	Araceae	Blurred vision
Caltha palustris L.	Ranunculaceae	Skin infections/slow healing wounds

(*Continued*)

TABLE 15.1 (*Continued*)

Plant Species	Family	Traditional Uses
Calycanthus floridus L.	Calycanthaceae	Blurred vision
Ceanothus americanus L.	Rhamnaceae	**Diabetes**
Celastrus scandens L.	Celastraceae	Skin infections/slow healing wounds
Chaptalia tomentosa Vent.	Asteraceae	Foot numbness/sores
Chenopodium californicum (S. Wats.) S. Wats.	Chenopodiaceae	Foot numbness/sores
Cicuta maculata L.	Apiaceae	Skin infections/slow healing wounds
Cinna arundinacea L.	Poaceae	**Diabetes**
Cirsium ochrocentrum A. Gray	Asteraceae	**Diabetes**
Citrullus lanatus var. lanatus (Thunb.) Matsum. & Nakai	Curcubitaceae	Increased urination
Clintonia borealis (Ait.) Raf.	Liliaceae	**Diabetes**
Cornus canadensis L.	Cornaceae	Increased urination
Cornus stolonifera Michx.	Cornaceae	Increased urination; blurred vision
Cucurbita pepo L.	Cucurbitaceae	Increased urination
Cypripedium acaule Ait.	Orchidaceae	**Diabetes**
Cypripedium parviflorum Salisb.	Orchidaceae	**Diabetes**
Daucus carota L.	Apiaceae	**Diabetes**
Diervilla lonicera P. Mill.	Caprifoliaceae	Increased urination
Elymus repens (L.) Gould	Poaceae	Increased urination
Chamerion angustifolium (L.) Holub ssp. angustifolium	Onagraceae	Skin infections/slow healing wounds
Equisetum hyemale ssp. affine (Engelm.) Calder & R.L. Taylor	Equisetaceae	Increased urination
Equisetum laevigatum A. Braun	Equisetaceae	Blurred vision
Equisetum sp.	Equisetaceae	Blurred vision
Ericameria arborescens (Gray) Greene	Asteraceae	Foot numbness/sores
Erysimum cheiranthoides L.	Brassicaceae	Skin infections/slow healing wounds
Erythrina herbacea L.	Fabaceae	Foot numbness/sores
Erythronium americanum Ker-Gawl.	Liliaceae	Skin infections/slow healing wounds
Euonymus atropurpureus Jacq.	Celastraceae	Skin infections/slow healing wounds
Euphorbia corollata L.	Euphorbiaceae	**Diabetes**
Euphorbia sp.	Euphorbiaceae	**Diabetes**
Galium aparine L.	Rubiaceae	Skin infections/slow healing wounds
Galium sp.	Rubiaceae	Blurred vision
Galium trifidum L.	Rubiaceae	Skin infections/slow healing wounds

TABLE 15.1 (*Continued*)

Plant Species	Family	Traditional Uses
Geum calthifolium Menzies ex Sm.	Rosaceae	Skin infections/slow healing wounds
Hamamelis virginiana L.	Hamamelidaceae	Skin infections/slow healing wounds
Heracleum maximum Bartram	Apiaceae	Skin infections/slow healing wounds
Heteromeles arbutifolia (Lindl.) M. Roem.	Rosaceae	Skin infections/slow healing wounds
Houstonia caerulea L.	Rubiaceae	Increased urination
Hylotelephium telephioides (Michx.) H. Ohba	Crassulaceae	Skin infections/slow healing wounds
Hyptis pectinata (L.) Poit.	Lamiaceae	Foot numbness/sores
Impatiens capensis Meerb.	Balsaminaceae	Skin infections/slow healing wounds
Impatiens pallida Nutt.	Balsaminaceae	Skin infections/slow healing wounds
Juniperus scopulorum Sarg.	Cupressaceae	**Diabetes**
Kalmia polifolia Wangenh.	Ericaceae	Skin infections/slow healing wounds
Laportea canadensis (L.) Weddell	Urticaceae	Increased urination
Lewisia rediviva Pursh	Montiaceae	**Diabetes**
Licania michauxii Prance	Chrysobalanaceae	Increased urination
Lithospermum incisum Lehm.	Boraginaceae	Skin infections/slow healing wounds
Lobelia cardinalis L.	Campanulaceae	Skin infections/slow healing wounds
Lobelia siphilitica L.	Campanulaceae	Skin infections/slow healing wounds
Lycopodium obscurum L.	Lycopodiaceae	Blurred vision
Lysichiton americanus Hultén & H. St. John	Araceae	**Diabetes**
Malus pumila P. Mill.	Rosaceae	Blurred vision
Mentha arvensis L.	Lamiaceae	**Diabetes**
Monarda fistulosa L.	Lamiaceae	Skin infections/slow healing wounds
Monotropa uniflora L.	Ericaceae	Skin infections/slow healing wounds
Nicotiana attenuata Torr. ex. S. Wats.	Solanaceae	Skin infections/slow healing wounds
Nuphar lutea (L.) Sm	Nymphaceae	Skin infections/slow healing wounds; **diabetes**
Nuphar variegata Dur.	Nymphaceae	Skin infections/slow healing wounds
Onoclea sensibilis L.	Onocleaceae	Increased urination

(*Continued*)

TABLE 15.1 (*Continued*)

Plant Species	Family	Traditional Uses
Lithospermum occidentale (Mack.) Weak., Wits. & D. Estes	Boraginaceae	Foot numbness/sores
Oplopanax horridus (Sm.) Miq.	Araliaceae	**Diabetes**
Peniocereus greggii var. greggii (Engelm.) Britton & Rose	Cactaceae	**Diabetes**
Persea borbonia (L.) Spreng.	Lauraceae	Foot numbness/sores; increased urination
Phoradendron serotinum ssp. serotinum (Raf.) M.C. Johnst.	Santalaceae	Foot numbness/sores
Phytolacca americana L.	Phytolaccaceae	Skin infections/slow healing wounds
Picea glauca (Moench) Voss	Pinaceae	Skin infections/slow healing wounds; increased urination
Picea mariana (Mill.) BSP	Pinaceae	Skin infections/slow healing wounds
Picea sitchensis (Bong.) Carr.	Pinaceae	Blurred vision
Piloblephis rigida (Bartr. ex Benth.) Raf.	Lamiaceae	Foot numbness/sores
Pinus edulis Engelm.	Pinaceae	Skin infections/slow healing wounds
Pinus lambertiana Dougl.	Pinaceae	Blurred vision
Populus balsamifera L.	Salicaceae	Skin infections/slow healing wounds; **diabetes**
Populus tremuloides Michx.	Salicaceae	Foot numbness/sores; **diabetes**
Prunella vulgaris L.	Lamiaceae	**Diabetes**
Prunus cerasus L.	Rosaceae	Skin infections/slow healing wounds
Prunus pensylvanica L. f.	Rosaceae	Skin infections/slow healing wounds
Prunus serotina Ehrh.	Rosaceae	Skin infections/slow healing wounds
Prunus virginiana L.	Rosaceae	Skin infections/slow healing wounds
Psorothamnus polydenius var. polydenius (Torr. ex. S. Watson) Rydb.	Fabaceae	Increased urination
Pteridium aquilinum (L.) Kuhn	Dennstaedtiaceae	Increased urination
Pyrularia pubera Michx.	Santalaceae	Skin infections/slow healing wounds
Rhododendron groenlandicum (Oeder) Kron & Judd	Ericaceae	Blurred vision
Rhus copallinum L.	Anacardiaceae	Increased urination
Rhus glabra L.	Anacardiaceae	Skin infections/slow healing wounds; increased urination
Rhus sp.	Anacardiaceae	**Diabetes**
Rhus typhina L.	Anacardiaceae	Increased urination

TABLE 15.1 (*Continued*)

Plant Species	Family	Traditional Uses
Ribes americanum Mill.	Grossulariaceae	**Diabetes**
Rosa acicularis Lindl.	Rosaceae	Blurred vision
Rubus idaeus L.	Rosaceae	Skin infections/slow healing wounds
Rubus occidentalis L.	Rosaceae	Skin infections/slow healing wounds
Rubus odoratus L.	Rosaceae	Skin infections/slow healing wounds
Rubus spectabilis Pursh	Rosaceae	Skin infections/slow healing wounds
Rumex acetosella L.	Polygonaceae	Skin infections/slow healing wounds
Rumex crispus L.	Polygonaceae	Skin infections/slow healing wounds
Rumex obtusifolius L.	Polygonaceae	Skin infections/slow healing wounds
Sanguinaria canadensis L.	Papaveraceae	Skin infections/slow healing wounds
Sarracenia purpurea L.	Sarraceniacea	Skin infections/slow healing wounds
Sassafras albidum (Nutt.) Nees	Lauraceae	Increased urination
Scrophularia californica Cham. & Schlecht.	Scrophulariaceae	Blurred vision
Sedum sp.	Crassulaceae	Increased urination
Shepherdia canadensis (L.) Nutt.	Elaeagnaceae	**Diabetes**
Solanum douglasii Dunal	Solanaceae	Blurred vision
Sorbus sitchensis M. Roemer	Rosaceae	Increased urination
Sphagnum fuscum (Schimp.) Klinggr.	Sphagnaceae	Skin infections/slow healing wounds
Symphoricarpos albus (L.) Blake	Caprifoliaceae	Blurred vision
Taraxacum officinale F.H. Wigg.	Asteraceae	**Diabetes**
Taxus canadensis Marsh.	Taxaceae	Foot numbness/sores
Thermopsis macrophylla Hook. & Arn.	Fabaceae	Blurred vision
Thuja occidentalis L.	Cupressaceae	Foot numbness/sores; skin infections/slow healing wounds
Toxicodendron diversilobum (Torr. & Gray) Greene	Anacardiaceae	Blurred vision
Toxicodendron vernix (L.) Kuntze	Anacardiaceae	Skin infections/slow healing wounds
Triosteum perfoliatum L.	Caprifolicaceae	Skin infections/slow healing wounds
Tsuga canadensis (L.) Carr.	Pinaceae	Skin infections/slow healing wounds

(*Continued*)

TABLE 15.1 *(Continued)*

Plant Species	Family	Traditional Uses
Typha latifolia L.	Typhaceae	Skin infections/slow healing wounds
Ulmus rubra Muhl.	Ulmaceae	Skin infections/slow healing wounds
Urtica dioica ssp. gracilis (Ait.) Seland.	Urticaceae	Foot numbness/sores
Vaccinium ovatum Pursh	Ericaceae	**Diabetes**
Vaccinium uliginosum L.	Ericaceae	**Diabetes**
Verbascum thapsus L.	Scrophulariaceae	Skin infections/slow healing wounds
Vitis vulpina L.	Vitaceae	**Diabetes**
Ximenia americana L.	Ximeniaceae	Foot numbness/sores
Yucca filamentosa L.	Asparagaceae	**Diabetes**
Zizania aquatica L.	Poaceae	**Diabetes**

placebo-controlled, single-blind study suggested that a herbal tea prepared with both species may have short-term benefits for type 2 diabetics who have poor glycemic control [31]. Nonetheless, a select number have undergone *in vitro* or *in vivo* screening.

The antidiabetic and antiobesity activity of plants such as *Rubus idaeus* [32], *Prunella vulgaris* [33, 34], *Cucurbita pepo* [35, 36], *Acorus calamus* [37–42], *Taraxacum officinale* [43–45], *Prunus cerasus* [46, 47], *Thuja occidentalis* [48], *Phytolacca americana* [49, 50], *Acer saccharum* [51], and *Chamerion angustifolium* [52] have already been evaluated *in vivo*. Their potential to reduce or modulate markers of antidiabetic or antiobesity activity, such as glucose, insulin, and lipid levels in plasma, serum, or blood have been measured in various animal models, mostly genetic or induced diabetic rats or mice. Not surprisingly, most of these have also been tested and shown to be active *in vitro* [53–60]. Although other plants, such as *Malus pumila* [61], *Prunus virginiana* [62], and *Urtica dioica* [63], have only been screened *in vitro*, species with demonstrated pharmacological activities deserve more in-depth studies on their phytotherapeutic use for the treatment of diabetes or obesity.

Species such as *R. idaeus*, raspberry, *C. pepo*, the pumpkin, *T. officinale*, dandelion, and *P. cerasus*, sour cherry, are already part of the common North American diet. Having passed the test of time for safety and palatability, they may already be easily integrated into less intrusive phytotherapies. In basic and simple dietary interventions, for example, these species may be incorporated as functional foods to help control blood glucose. Others may be tailored to specific regions, where they are culturally relevant to their local communities. This is particularly true to locations where some traditional activities still play an important role in people's lives, and where these species are already part of the traditional pharmacopoeia.

In one such Cree community of Eeyou Istchee (CEI), plants from the traditional pharmacopoeia were selected based on their use for 15 symptoms and complications that may be associated with diabetes [29]. *Typha latifolia*, *Sarracenia purpurea*, *Rhododendron groenlandicum*, *Picea mariana*, *Picea glauca*, and *Populus* spp. were species cited by CEI elders and healers that are also included in Table 15.1. Their antidiabetic potential, and others mentioned in the study, has since been assessed using a vigorous platform of pharmacological bioassays (see Tables 15.2 and 15.3). Like other traditional medicinal plants cited earlier, they have shown an extensive range of *in vivo* and *in vitro* biological activities, ranging from their capacity to modulate glucose uptake and insulin sensitivity, to protecting against glucose toxicity or deprivation.

15.3 PHARMACOLOGICAL SCREENING: PROVIDING EMPIRICAL EVIDENCE FOR PHYTOTHERAPIES

The botanical origins of galegin, the precursor to one of the world's most prescribed antidiabetic drug, metformin, brings to light the impact and importance of traditional phytotherapies and medicine in treating such chronic diseases and health problems. In fact, over 42 million prescriptions for metformin were written in the United States in 2009, ranking it tenth among the country's most dispensed drugs [64]. This oral hypoglycemic drug [65] is derived from guanidine, isolated from *Galega officinalis* L. (Fabaceae) at the end of the nineteenth century. The aerial parts of this plant were used medicinally in medieval Europe to treat a range of symptoms and illnesses, some of which were related to T2D [66]. Just as pharmaceutical drugs undergo rigorous screening to test for efficacy, so can phytotherapies, whose activity can be assessed through *in vitro* and *in vivo* experiments.

15.3.1 Diabetes

Due to the complex and multifaceted nature of T2D, Haddad and colleagues [10] suggested a wide array of *in vitro* bioassays and *in vivo* animal models to assess the antidiabetic potential of plant-based medicinal preparations (Table 15.2). Prioritization of these models within this platform gives more importance to *in vivo* studies, where markers such as blood glucose, body weight, and fatty liver can be quantified. Primary antidiabetic biological activities are observed, *in vitro*, on cells involved in the production (i.e., pancreatic) and response (i.e., muscular, hepatic, and adipose tissue) to insulin or in the absorption of glucose (i.e., intestinal). Secondary antidiabetic activities, on the other hand, refer to their interaction with prescribed drugs and effects on complications of diabetes such as oxidative stress, inflammation, and neuropathy.

In vivo models used to evaluate the potential of medicinal plants to reduce blood glucose rely on genetically predisposed, such as the db/db mice and Zucker diabetic fatty (ZDF) rats [10, 33, 42, 46], or exogenously diabetic or obese animals. Diabetes can be induced using chemicals like alloxan and streptozotocin [36, 40, 49], or through such dietary interventions as feeding high fat diets. The male C57BL/6J mouse used in diet-induced obesity (DIO) models, for example, has been found to

TABLE 15.2 Description of *In Vivo* Animal Models and *In Vitro* Bioassays Used by the CIHR-TAAM to Evaluate the Antiobesity and Antidiabetic Potential of Medicinal Plants

ID	Model	Description
1	*In vivo*	DIO C57BL/6 mice (hypoglycemic and antiobesity)
2	*In vivo*	STZ rats (hypoglycemic activity; type 1 diabetes)
3	*In vivo*	KK-A^y mice (hypoglycemic activity; type 2 diabetes)
4	*In vivo*	Prediabetic insulin-resistant rats (insulin sensitising activity)
5	*In vivo*	Normal rats
6	*In vitro*	Glucose-stimulated insulin secretion (e.g., pancreatic β TC-tet cells)
7	*In vitro*	Proliferation of pancreatic β cells (e.g., pancreatic β TC-tet cells)
8	*In vitro*	Glucose transport in skeletal cells (e.g., C2C12 myotubes)
9	*In vitro*	Glucose transport in adipocytes (e.g., 3T3-L1 adipocytes)
10	*In vitro*	Hepatic glucose metabolism (e.g., H4IIE hepatocytes)
11	*In vitro*	Adipogenesis (e.g., 3T3-L1 adipocytes)
12	*In vitro*	Intestinal glucose absorption (e.g., Caco-2 cells)
13	*In vitro*	Drug interactions (e.g., Cytochrome P450s)
14	*In vitro*	Antioxidant activity (e.g., DPPH)
15	*In vitro*	Advanced glycation end products (no cells)
16	*In vitro*	Antiinflammation (e.g., LPS-activated macrophages)
17	*In vitro*	Glucose toxicity (e.g., PC12-AC neural precursor cells)
18	*In vitro*	Glucose deprivation (e.g., PC12-AC neural precursor cells)

The order of presentation for each model represents their relevance in prioritizing the antiobesity or antidiabetic potential of each phytotherapy according to Haddad et al. [10].

provide stable and reliable results [10]. In addition, being prone to obesity, these mice also develop the first signs associated with the development of T2D-like insulin resistance, impaired glucose tolerance, and mild-to-moderate hyperglycemia, therefore offering an appropriate model to test the efficacy of new antidiabetic preparations in prediabetics and in the early stages of obese T2D [10]. Hence, botanical preparations can be tested and their results compared to the effects of insulin or other commonly used drugs from the biguanide, thiazolidinediones, sulfonylurea derivatives, and meglitinides classes [6]. In making this list more extensive, however, approaches to determining the antiobesity potential of a botanical extract may also be added as they may reduce the risk of not only T2D but also cardiovascular diseases [67].

In vitro screening for primary antidiabetic activity is achieved by employing assays for the potentiation of glucose-stimulated insulin secretion (GSIS), glucose transport and adipogenesis, the modulation of hepatic glucose metabolism, and the inhibition of intestinal glucose absorption. Pancreatic cell lines, such as *beta*-tet cells, release insulin in response to changes in glucose concentrations [68, 69]; the effect of plant preparations in potentiating GSIS in these cells may therefore be compared to known insulin secretagogues like the sulfonylureas [10]. In skeletal muscles, like the C2C12 cell line, the uptake of glucose can be measured and compared to the biguanide, metformin [68, 69]. Various protocols for the glucose transport assay and establishing the pathway by which extracts modulate the translocation of Glut4 transporters then allows identification of their insulinomimetic, insulin-sensitizing, or insulin-independent properties [70, 71]. Glitazone-like activity can be tested through the differentiation of preadipocyte cells. Thiazolidinediones such as rosiglitazone acts as an insulin-sensitizer by activating peroxisome proliferator-activated receptor gamma (PPARγ) nuclear receptors [72]. Its effect on the transcription of a number of genes can be observed as an enhanced accumulation of intracellular triglycerides in differentiating adipocytes [68, 69, 72]. On the other hand, measuring hepatic glucose production or storage into glycogen using hepatocyte cell lines, such as murine H4IIE, may assist in the assessment of the insulin-dependent and insulin-independent activity of botanical extracts on hepatic glucose metabolism [70, 73]. Finally, intestinal cell lines such as CaCo-2 cells may be used to evaluate an extract's capacity to decrease intestinal glucose transport, thus reducing blood glucose by inhibiting the absorption and digestion of carbohydrates [74].

In vitro screening for secondary antidiabetic activity is achieved by using assays for the inhibition of cytochrome P450 (CYP) and measuring neuroprotective, antioxidant, antiglycation, as well as antiinflammatory activities of plant extracts. The potential for interaction between these botanical preparations and current therapeutic drugs is assessed by evaluating their inhibitory activity in the CYP enzyme systems that are responsible for the metabolism of these drugs [75, 76]. Furthermore, the protective effect of these preparations on preneuronal and neuronal cells is observed under hyperglycemic and hypoglycemic conditions to assess their potential in mediating diabetic neuropathy [77, 78].

It is known that multiple symptoms and complications associated with T2D are the result of, or are exacerbated by, the oxidative stress related to hyperglycemia on micro- and macrovascular systems [8, 79–81]. Multiple cell-free assays exist to test the antioxidant activity of botanical preparations in stabilizing free radicals such as reactive oxygen species (ROS) produced from glucose autoxidation [8, 80, 81]. Such assays also exist to evaluate an extract's capacity to inhibit the glycation reaction between glucose and proteins. These reactions, more common during hyperglycemia, lead to the production of advanced glycation end-products (AGEs), which also contribute to the development of micro and macrovascular complications in diabetics [82, 83]. Finally, the increased production of tumor necrosis factor alpha (TNF-α) and other proinflammatory cytokines associated with proportionally larger adipose tissue mass are in part related to chronic low-grade inflammation associated with T2D. The lipopolysaccharide (LPS)-activated macrophage assay may also be used to determine the antiinflammatory activity of plant extracts [10].

In a case study on antidiabetic plants from the CEI traditional pharmacopoeia, multiple ethnobotanical studies from various CEI communities were conducted on plants used to treat symptoms associated with diabetes [29, 84–86]. Selected plants were then pharmacologically investigated for their antidiabetic potential based on the platform presented by Haddad and colleagues [10] (Table 15.2). The results of these experiments clearly show this potential among CEI medicinal plants, which, in some cases, possess a range of activities *in vitro* and *in vivo* (Table 15.3). Furthermore, some species have been shown to act like antidiabetic drugs such as metformin

TABLE 15.3 Antidiabetic Activity of Medicinal Plants From the CEI Pharmacopoeia Evaluated Using the *In Vitro* and *In Vivo* Models of Table 15.2

Species	Activity	ID	References
Rhododendron groenlandicum	Hypoglycemic	1, 8, 9	Ouchfoun (2011), Spoor et al. [69]
	Adipogenic	11	Ouchfoun (2011), Spoor et al. [69]
	Antiglycation	15	Harris et al. (2011)
	Antiobesity	1	Ouchfoun (2011)
Larix laricina	Hypoglycemic	1, 8	Harbilas et al. [88], Spoor et al. [69]
	Insulino-sensitizing	1	Harbilas et al. [88]
	Antiobesity	1	Harbilas et al. [88]
	Adipogenic	11	Shang et al. (2012), Spoor et al. [69]
Rhododendron tomentosum	Hypoglycemic	5, 9, 12	Harbilas et al. [68], Nistor Baldea et al. [74]
	Adipogenic	11	Harbilas et al. [68]
	Neuroprotective	17, 18	Harbilas et al. [68]
	Antiglycation	15	Harris et al. (2011)
Picea mariana	Hypoglycemic	5, 9, 12	Nistor Baldea et al. [74], Spoor et al. [69]
	Adipogenic	11	Spoor et al. [69]
	Neuroprotective	17	Downing [84], Spoor et al. [69]
	Antiglycation	15	Harris et al. (2011)
Picea glauca	Hypoglycemic	12	Nistor Baldea et al. [74]
	Neuroprotective	17, 18	Harbilas et al. [68], Harris et al. [77]
	Antiglycation	15	Harris et al. (2011)
Kalmia angustifolia	Adipogenic	11	Harbilas et al. [68]
	Antiglycation	15	Harris et al. (2011)
Sorbus decora	Hypoglycemic	2, 3, 4, 8, 12	Guerrero-Analco et al. (2010), Nistor Baldea et al. [74], Spoor et al. [69], Vianna et al. [87]
	Insulino-sensitizing	4	Vianna et al. [87]
	Neuroprotective	17	Spoor et al. [69]
	Antiglycation	15	Harris et al. (2014)
Abies balsamea	Hypoglycemic	8, 9	Spoor et al. [69]
	Antiglycation	15	Harris et al. (2011)
Alnus incana subsp. rugosa	Antiobesity	11	Martineau, Hervé et al. [72], Martineau, Muhammad et al. (2010)
	Hypoglycemic	8, 9, 12	Nistor Baldea et al. [74], Spoor et al. [69]
	Neuroprotective	18	Spoor et al. [69]

TABLE 15.3 (*Continued*)

Species	Activity	ID	References
Juniperus communis	Hypoglycemic	12	Nistor Baldea et al. [74]
	Antiglycation	15	Harris et al. (2011, 2014)
Pinus banksiana	Hypoglycaemic	9, 12	Nistor Baldea et al. [74], Spoor et al. [69]
	Adipogenic	11	Spoor et al. [69]
	Neuroprotective	18	Spoor et al. [69]
	Antiglycation	15	Harris et al. (2011)
S. planifolia	Hypoglycaemic	9, 12	Harbilas et al. [68], Nistor Baldea et al. [74]
	Neuroprotective	18	Harbilas et al. [68]
	Antiglycation	15	Harris et al. (2011)
Vaccinium vitis-idaea	Hypoglycemic	8, 9, 12	Eid et al. [70], Harbilas et al. [68], Nistor Baldea et al. [74]
	Adipogenic	11	Harbilas et al. [68]
	Antiglycation	15	Beaulieu et al. (2010), Harris et al. (2011, 2014)
Sarracenia purpurea	Neuroprotective	17, 18	Beaulieu et al. (2010), Harris et al. (2011, 2014)
	Hypoglycemic	8, 9, 10, 12	Muhammad et al. [73], Nistor Baldea et al. [74], Spoor et al. [69]
	Antiglycation	15	Harris et al. (2011)
Populus balsamifera	Antiobesity	1, 11	Harbilas et al. (2013), Harbilas, Brault et al. (2012), Martineau, Hervé et al. [72], Martineau, Muhammad et al. (2010a)
	Hypoglycemic	1, 12	Harbilas et al. (2012b), Nistor Baldea et al. [74]
	Insulino-sensitizing	1	Harbilas et al. (2012b)
	Neuroprotective	17	Harbilas et al. [68]
	Antiglycation	15	Harris et al. (2011)
G. hispidula	Hypoglycemic	9, 12	Harbilas et al. [68], Nistor Baldea et al. [74]
	Neuroprotective	17, 18	Harbilas et al. [68]
	Antiglycation	15	Harris et al. (2011)
L. clavatum	Hypoglycemic	12	Nistor Baldea et al. [74]
	Adipogenic	11	Harbilas et al. [68]
	Neuroprotective	17	Harbilas et al. [68]
V. angustifolium	Hypoglycemic	8, 9	Martineau et al. (2006)
	Adipogenic	11	Martineau et al. (2006)
	Neuroprotective	17	Martineau et al. (2006)
	Insulinotropic	6	Martineau et al. (2006)
	Proliferative	7	Martineau et al. (2006)
	Antiglycation	15	Harris et al. (2014), McIntyre et al. (2008)

(i.e., *Abies balsamea*, *Alnus incana* subsp. *rugosa*, *Larix laricina*, *Picea mariana*, *Pinus banksiana*, *Rhododendron groenlandicum*, *Sarracenia purpurea*, *Sorbus decora*, *Vaccinium vitis-idaea*), and rosiglitazone (i.e., *Larix laricina*) [70, 71, 87, 88].

15.3.2 Obesity

Natural antiobesity preparations can induce weight-loss via several mechanisms affecting lipid homeostasis or metabolism, or even by decreasing appetite and inducing satiety [1, 10, 89]. Although Haddad and colleagues [10] provided a rigorous template for screening plants for their antidiabetic potential, none as extensive as that have been suggested for antiobesity activity. Nonetheless, direct activity can be assessed *in vitro* using assays where the inhibition of pancreatic lipase activity and adipocyte differentiation is assessed, as well as the potential of thermogenesis and lipid metabolism [1, 10, 89]. Indirect activity for alternate complications of obesity, such as chronic inflammation, can also be evaluated *in vitro* [10]. Finally, *in vivo* models used to evaluate the antiobesity potential of botanical preparations may use genetically predisposed animal models, such as the *Ob/Ob* mice, or exogenously induced obese animals [10]. The male C57BL/6J mouse, for example, that become phenotypically obese when given an unrestricted high-fat diet, is frequently used in diet-induced obesity (DIO) mouse models to determine the efficacy of novel therapeutic interventions and antiobesity agents [10]. Naturally, animal models of diabetes, such as those previously described (Table 15.3), can also be used to measure markers of obesity, such as reduced body weight, adipose tissue, and lipid accumulation in organs, blood, or serum [10, 32, 46, 48].

Pancreatic lipase plays an important role in digesting triglycerides into glycerol and fatty acids in the intestinal tract [90]. Hence, preventing the absorption of lipids results in the excretion of nonabsorbed fat via oily feces. Inhibition of the enzymatic reaction can be tested *in vitro* using lipase of porcine pancreas type-2 in cell-free assays [67, 90]. When absorbed, however, fats can be stored in specialized cells, adipocytes, that make up adipose tissue. Inhibiting the differentiation of preadipocyte into functional lipid storing adipocytes may therefore represent a potential antiobesity activity [1, 10]. Equally, botanical preparations can be tested for their effects in increasing the metabolism of lipids, such as the body's capacity to burn excess fat [1]. During thermogenesis, mitochondrial activity is uncoupled from oxidative phosphorylation that conserves energy through ATP formation; instead, energy is lost as heat [91]. This can be evaluated in skeletal muscle cells [91] and brown adipose tissue that are known to be highly thermogenic by using their fat expenditure [92]. Furthermore, adipocytes may also be used to assess lipolysis by stimulating, for example, the inhibition of adenosine 3′,5′-cyclic monophosphate (cAMP)-dependent phosphodiesterase (PDE) [93]. Finally, the production of the pro-inflammatory adipokine TNF-α is proportional to the mass of adipose tissue found in visceral obesity [10]. The activity of botanical extracts in inhibiting the release of TNF-α and other pro-inflammatory cytokines from LPS-activated macrophages may be used to assess their antiinflammatory activity [10].

15.4 COMMUNITY-BASED PARTICIPATION: DEVELOPING PHYTOTHERAPIES FROM TRADITIONAL KNOWLEDGE

Alternative systems integrating traditional approaches to modern health care have already been successfully attempted in Inuit communities of northern Quebec's Nunavik region, Canada [94]. Although limited to midwifery, the formation of the Inuulitsivik midwifery service and education program involved consultations with all members of the communities from elders and traditional midwives to childbearing and young women. Due to the cultural dimension of this initiative, its success is not surprising because it plays a key role in the perception of health, disease, and illness. Community-based participation is at the heart of the diabetes education program such as the Kahnawake Schools Diabetes Prevention Project [95]. Although the prevalence of T2D in this Mohawk community was one and half to two times that of the general Canadian population between 1986 and 2003, such ongoing programs are thought to be a contributing factor in explaining why this prevalence was in fact much lower than other Canadian First Nation communities [96].

Indeed, in Canada, a dichotomy exists between the aboriginal and nonaboriginal worlds with regards to nutrition, sickness, and medicine [97]. For T2D, it is often perceived as a "white man's disease" introduced just like tuberculosis and smallpox in the past [18, 98]. Although 90% of diabetics in an Aboriginal Peoples Survey (APS) reported to having consulted a health-care professional in 1991, over 80% of diabetics in the First Nations and Inuit Regional Health Survey (FNIRHS) believed that their health care needed improvement [18]. Hence, participatory research, among the CEI, for example, have shown that community members still conceptualize diabetes treatment in a Cree perspective [98]. It has since become clear that tackling the high rate of diabetes among aboriginal populations in Canada will require an in-depth understanding of the local perception of contemporary medicine and community participation [99].

Ethnobotany has a successful history in drug discovery, specifically due to its targeted approach using TK [100, 101]. Such an approach is founded on the principle that the TK of a plant's possible therapeutic activity accumulates as the success of its use is communicated onto others [100]. Hence, in communities whose culture is intrinsically linked to its environment, such as the forest, elders are thought to be important sources of knowledge [101, 102]. Consultation with such members of the community, especially those recognized by others as being knowledgeable in medicinal plants, has been shown to be an excellent approach for consolidating TK with contemporary understanding of disease and sickness [10, 103]. For this reason, various quantitative methods based on informant consensus, such as the frequency of citation [104], can be used to assess the medicinal potential of plant preparations, in order to prioritize them for pharmacological analysis. Papers such as those of Araújo and colleagues [105] and Phillips and Gentry [106, 107] develop, compare, and discuss some of these methods.

Although North American traditional medicinal plant knowledge is abundant, understanding of T2D and its complications is nonetheless limited in some regions, especially among elders for whom the illness is a relatively new problem [98].

Employing simple informant consensus in these situations may therefore not be the most appropriate approach for developing culturally adapted phytotherapies for T2D. To address this, Leduc and colleagues [29] decoded T2D into a communicable form by breaking it down into its associated symptoms and complications, after which they were ranked by professional health-care practitioners in diabetes according to their importance and relevance to the illness. An index for measuring the importance of a traditional phytotherapy with regards to diabetes, the syndromic importance value (SIV), can then be calculated by taking into account (a) consensus, (b) the number of symptoms cited, and (c) the importance of each of these symptoms to T2D [29].

In an ethnobotanical approach similar to what is advocated by Cox and Balick [101], prioritization first begins with the compilation of TK through consultations and interviews. It is then followed by pharmacological screening, which prioritizes the ranking of plants scientifically. In the comprehensive platform for testing for antidiabetic activities proposed by Haddad and colleagues [10], priority in assessing the antidiabetic potential of traditional phytotherapies is given to positive results obtained via *in vivo* models than those obtained *in vitro*. Second, results pertaining to the regulation of blood glucose are given more importance than achieving homeostasis as the primary goal of T2D therapy [10]. In the hierarchy of relevance, importance was then given to phytotherapies capable of reducing body weight, while assays testing for secondary activity, or those related to complications of T2D, were given lower priority. Ranking of these various methods, based on Haddad and colleagues [10], is presented in Table 15.2.

Consensus between the scientific realm of knowledge and traditional knowledge is important for creating integrated and culturally adapted approaches to treating T2D. It is therefore easy to understand that efficient phytotherapies will be more successful in the management of diabetes if these are thought to be culturally important and relevant. Indeed, many popular plant-derived pharmaceutical drugs, that is, digitoxin, quinine, aspirin, and morphine, were known to Western science from ethnobotanical leads for decades before entering Western health-care systems [100]. Likewise, when CEI elders were asked to rank potential antidiabetic plants based on their traditional medicinal knowledge, highly ranked species were well known by community members and coincided with those that performed best according to the scheme of pharmacological experiments used to screen for antidiabetic potential [10]. Similarly, SIVs appear to have efficiently translated TK as plants ranked according to this algorithm also reflect the pharmacological prioritization [29, 84, 85]. Consequently, plant species that show a greater level of antidiabetic potential and are considered culturally important, as suggested by SIV results, make excellent candidates for in-depth clinical studies.

Safe and efficient use of these phytotherapies will have to involve a well thought out structure, which takes into account all the players involved in all areas of local health care. The development of standardized and quality natural health products will require extensive phytochemical analysis. Finally, studies on the effect of harvesting and wildcrafting on wild populations of medicinal plants, like that of Tendland and colleagues [108] on *R. groenlandicum*, are necessary to insure sustainable use.

However, increasing community participation and integrating traditional medicines in order to develop culturally adapted approaches to treating diabetes may in fact increase compliance and awareness.

15.5 CONCLUSIONS

Adopting a rigorous platform of screening, it is possible to assess the antidiabetic potential of botanical preparations using a series of *in vitro* and *in vivo* experiments. With extensive pharmacological investigation, their mechanisms of action can be elucidated; in some cases, these have been shown to be analogous to current antidiabetic drugs like metformin. Furthermore, with the help of community-based research, culturally appropriate phytotherapies can be developed by integrating traditional medicinal knowledge. Consultations with elders and healers have been shown to be essential to prioritizing locally used plants for pharmacological screening. Although extensive pharmacological analyses have been made on a selected number of CEI medicinal plants, the traditional pharmacopoeia from the flora of North America offer many more plants with antidiabetic potential. Future avenues for developing phytotherapies include (i) clinical studies, (ii) extensive research on phytotherapies to appropriately translate their medicinal potential to healthcare professionals, and (iii) standardization and quality control via phytochemical analysis. Obesity and diabetes are lifestyle problems that require lifestyle changes. Using phytotherapies has a place in their management as they also imply adopting healthier lifestyle changes.

REFERENCES

[1] Kazemipoor M, Radzi C, Cordell GA, Yaze I (2012) Potential of traditional medicinal plants for treating obesity: a review. *International Conference on Nutrition and Food Sciences*, 39: 5. Retrieved from http://arxiv.org/abs/1208.1923.

[2] Kelly T, Yang W, Chen CS, Reynolds K, He J (2008) Global burden of obesity in 2005 and projections to 2030. *Int J Obes (Lond)* 32: 1431–1437.

[3] Whiting DR, Guariguata L, Weil C, Shaw J (2011) IDF diabetes atlas: global estimates of the prevalence of diabetes for 2011 and 2030. *Diabetes Res Clin Pract* 94: 311–321.

[4] Prentice AM (2006) The emerging epidemic of obesity in developing countries. *Int J Epidemiol* 35: 93–99.

[5] WHO (2000) Obesity: preventing and managing the global epidemic. Report of a WHO consultation. World Health Organ Tech Rep Ser 894: i–xii, 1–253.

[6] Agence de la santé publique du Canada (2011) *Le diabete au Canada: Perspective de sante publique sur les faits et chiffres*. Ottawa: Agence de la santé publique du Canada, p. 120. Retrieved from http://www.phac-aspc.gc.ca/cd-mc/publications/diabetes-diabete/facts-figures-faits-chiffres-2011/pdf/facts-figures-faits-chiffres-fra.pdf (Accessed November 19, 2014).

[7] Prentki M, Nolan CJ (2006) Islet beta cell failure in type 2 diabetes. *J Clin Invest* 116: 1802–1812.

[8] Tiwari AK, Rao JM (2002) Diabetes mellitus and multiple therapeutic approaches of phytochemicals: present status and future prospects. *Curr Sci* 83: 30–38.

[9] Bennett PH, Bogardus C, Tuomilehto J, Zimmet P (1992) Epidemiology and natural history of NIDDM: non-obese and obese. In: *Alberti KGMM*, DeFronzo RA, Keen H, Zimmet P (Eds.). *International Textbook of Diabetes Mellitus*. Chichester: John Wiley & Sons, Ltd, pp. 148–176.

[10] Haddad PS, Musallam L, Martineau LC, Harris C, Lavoie L, et al. (2012) Comprehensive evidence-based assessment and prioritization of potential anti-diabetic medicinal plants: a case study from Canadian Eastern James Bay Cree Traditional medicine. *Evid Based Complement Alternat Med* 2012: 893426.

[11] Wild S, Roglic G, Green A, Sicree R, King H (2004) Global prevalence of diabetes: estimates for the year 2000 and projections for 2030. *Diabetes Care* 27: 1047–1053.

[12] WHO, IDF (2004) *Diabetes Action Now: An Initiative of the World Health Organization and the International Diabetes Federation*. Geneva: WHO, p. 17.

[13] Yu CH, Zinman B (2007) Type 2 diabetes and impaired glucose tolerance in aboriginal populations: a global perspective. *Diabetes Res Clin Pract* 78: 159–170.

[14] Lindsay RS, Funahashi T, Hanson RL, Matsuzawa Y, Tanaka S, et al. (2002) Adiponectin and development of type 2 diabetes in the Pima Indian population. *Lancet* 360: 57–58.

[15] Kuzmina E, Lejeune P, Dannenbaum D, Torrie J (2010) *Cree Diabetes Information System: 2009 Annual Update*. Chisasibi: Cree Board of Health and Social Services of James Bay.

[16] Thouez JP, Ekoe JM, Foggin PM, Verdy M, Nadeau M, et al. (1990) Obesity, hypertension, hyperuricemia and diabetes mellitus among the Cree and Inuit of northern Quebec. *Arctic Med Res* 49: 180–188.

[17] Brassard P, Robinson E (1995) Factors associated with glycemia and microvascular complications among James Bay Cree Indian diabetics of Quebec. *Arctic Med Res* 54: 116–124.

[18] Young TK, Reading J, Elias B, O'Neil JD (2000) Type 2 diabetes mellitus in Canada's first nations: status of an epidemic in progress. *CMAJ* 163: 561–566.

[19] Chakravarthy MV, Booth FW (2004) Eating, exercise, and "thrifty" genotypes: connecting the dots toward an evolutionary understanding of modern chronic diseases. *J Appl Physiol* 96: 3–10.

[20] Hegele RA, Zinman B, Hanley AJ, Harris SB, Barrett PH, et al. (2003) Genes, environment and Oji-Cree type 2 diabetes. *Clin Biochem* 36: 163–170.

[21] Neel JV (1999) The "thrifty genotype" in 1998. *Nutr Rev* 57: S2–S9.

[22] Ritenbaugh C, Goodby CS (1989) Beyond the thrifty gene: metabolic implications of prehistoric migration into the New World. *Med Anthropol* 11: 227–236.

[23] Misra A, Singhal N, Khurana L (2010) Obesity, the metabolic syndrome, and type 2 diabetes in developing countries: role of dietary fats and oils. *J Am Coll Nutr* 29: 289S–301S.

[24] WHO (2009) *Sixty-second World Health Assembly*. Geneva: WHO. Retrieved from http://appswhoint/gb/ebwha/pdf_files/WHA62-REC1/WHA62_REC1-enpdf (Accessed November 19, 2014).

[25] Arnason T, Hebda RJ, Johns T (1981) Use of plants for food and medicine by Native Peoples of eastern Canada. *Can J Bot* 59: 2189–2325.

[26] Marles RJ, Clavelle C, Monteleone L, Tays N, Burns D (2008) *Aboriginal Plant Use in Canada's Northwest Boreal Forest*. Edmonton: Natural Resources Canada, p. 368.

[27] Moerman DE (2009) *Native American Medicinal Plants: An Ethnobotanical Dictionary Portanl.* Oregon: Timber Press, Inc, p. 799.

[28] Uprety Y, Asselin H, Dhakal A, Julien N (2012) Traditional use of medicinal plants in the boreal forest of Canada: review and perspectives. *J Ethnobiol Ethnomed* 8: 7.

[29] Leduc C, Coonishish J, Haddad P, Cuerrier A (2006) Plants used by the Cree Nation of Eeyou Istchee (Quebec, Canada) for the treatment of diabetes: a novel approach in quantitative ethnobotany. *J Ethnopharmacol* 105: 55–63.

[30] Cichewicz RH, Clifford LJ (2006) Native American medicine. In: Soumyanath A (Ed.). *Traditional Medicines for Modern Times: Antidiabetic Plants.* Boca Raton: CRC Press, pp. 169–177.

[31] Ryan EA, Imes S, Wallace C, Jones S (2000) Herbal tea in the treatment of diabetes mellitus. *Clin Invest Med* 23: 311–317.

[32] Morimoto C, Satoh Y, Hara M, Inoue S, Tsujita T, et al. (2005) Anti-obese action of raspberry ketone. *Life Sci* 77: 194–204.

[33] Hwang SM, Kim JS, Lee YJ, Yoon JJ, Lee SM, et al. (2012) Anti-diabetic atherosclerosis effect of *Prunella vulgaris* in db/db mice with type 2 diabetes. *Am J Chin Med* 40: 937–951.

[34] Skottova N, Kazdova L, Oliyarnyk O, Vecera R, Sobolova L, et al. (2004) Phenolics-rich extracts from *Silybum marianum* and *Prunella vulgaris* reduce a high-sucrose diet induced oxidative stress in hereditary hypertriglyceridemic rats. *Pharmacol Res* 50: 123–130.

[35] Caili F, Huan S, Quanhong L (2006) A review on pharmacological activities and utilization technologies of pumpkin. *Plant Foods Hum Nutr* 61: 73–80.

[36] Dixit Y, Kar A (2010) Protective role of three vegetable peels in alloxan induced diabetes mellitus in male mice. *Plant Foods Hum Nutr* 65: 284–289.

[37] Arun KS, Augustine A (2013) Hypolipidemic effect of methanol fraction of *Acorus calamus* Linn. in diet-induced obese rats. In: Sabu A, Augustine A (Eds.). *Prospects in Bioscience: Addressing the Issues.* New Delhi: Springer, pp. 399–404.

[38] Souza TD, Mengi, SA, Hassarajani S, Chattopadhayay S (2007) Efficacy study of the bioactive fraction (F-3) of *Acorus calamus* in hyperlipidemia. *Indian J Pharmacol* 39: 196–200.

[39] Parab RS, Mengi SA (2002) Hypolipidemic activity of *Acorus calamus* L. in rats. *Fitoterapia* 73: 451–455.

[40] Prisilla DH, Balamurugan R, Shah HR (2012) Antidiabetic activity of methanol extract of *Acorus calamus* in STZ induced diabetic rats. *Asian Pac J Trop Biomed* 2: S941–S946.

[41] Si MM, Lou JS, Zhou CX, Shen JN, Wu HH, et al. (2010) Insulin releasing and alpha-glucosidase inhibitory activity of ethyl acetate fraction of *Acorus calamus in vitro* and *in vivo*. *J Ethnopharmacol* 128: 154–159.

[42] Wu HS, Zhu DF, Zhou CX, Feng CR, Lou YJ, et al. (2009) Insulin sensitizing activity of ethyl acetate fraction of *Acorus calamus* L. *in vitro* and *in vivo*. *J Ethnopharmacol* 123: 288–292.

[43] Choi UK, Lee OH, Yim JH, Cho CW, Rhee YK, et al. (2010) Hypolipidemic and antioxidant effects of dandelion (*Taraxacum officinale*) root and leaf on cholesterol-fed rabbits. *Int J Mol Sci* 11: 67–78.

[44] Davaatseren M, Hur HJ, Yang HJ, Hwang JT, Park JH, et al. (2013) Taraxacum official (dandelion) leaf extract alleviates high-fat diet-induced nonalcoholic fatty liver. *Food Chem Toxicol* 58: 30–36.

[45] Zhang J, Kang MJ, Kim MJ, Kim ME, Song JH, et al. (2008) Pancreatic lipase inhibitory activity of *Taraxacum officinale in vitro* and *in vivo*. *Nutr Res Pract* 2: 200–203.

[46] Seymour EM, Singer AA, Kirakosyan A, Urcuyo-Llanes DE, Kaufman PB, et al. (2008) Altered hyperlipidemia, hepatic steatosis, and hepatic peroxisome proliferator-activated receptors in rats with intake of tart cherry. *J Med Food* 11: 252–259.

[47] Seymour EM, Lewis SK, Urcuyo-Llanes DE, Tanone II, Kirakosyan A, et al. (2009) Regular tart cherry intake alters abdominal adiposity, adipose gene transcription, and inflammation in obesity-prone rats fed a high fat diet. *J Med Food* 12: 935–942.

[48] Dubey SK, Batra A (2008) Anti diabetic activity of *Thuja occidentalis* Linn. *Res J Pharm Technol* 1: 362–365.

[49] Jeong SI, Kim KJ, Choi MK, Keum KS, Lee S, et al. (2004) α-Spinasterol isolated from the root of *Phytolacca americana* and its pharmacological property on diabetic nephropathy. *Planta Med* 70: 736–739.

[50] Lazarus DD, Trimble LA, Moldawer LL (1998) The metabolic effects of pokeweed mitogen in mice. *Metabolism* 47: 75–82.

[51] Honma A, Koyama T, Yazawa K (2010) Anti-hyperglycemic effects of sugar maple *Acer saccharum* and its constituent acertannin. *Food Chem* 123: 390–394.

[52] Khookhor O, Sato Y (2009) Mongolian plant extracts with potential glucose absorption inhibiting effects in rats. *J Tradit Med* 26: 74–79.

[53] Apostolidis E, Li L, Lee C, Seeram NP (2011) *In vitro* evaluation of phenolic-enriched maple syrup extracts for inhibition of carbohydrate hydrolyzing enzymes relevant to type 2 diabetes management. *J Funct Foods* 3: 100–106.

[54] Gonzalez-Castejon M, Garcia-Carrasco B, Fernandez-Dacosta R, Davalos A, Rodriguez-Casado A (2014) Reduction of adipogenesis and lipid accumulation by *Taraxacum officinale* (Dandelion) extracts in 3T3L1 adipocytes: an *in vitro* study. *Phytother Res* 28: 745–752.

[55] Jeong SI, Kim KJ, Choo YK, Keum KS, Choi BK, et al. (2004) *Phytolacca americana* inhibits the high glucose-induced mesangial proliferation via suppressing extracellular matrix accumulation and TGF-beta production. *Phytomedicine* 11: 175–181.

[56] Lee M-H, Chen YY, Tsai JW, Wang SC, Watanabe T, et al. (2011) Inhibitory effect of β-asarone, a component of *Acorus calamus* essential oil, on inhibition of adipogenesis in 3T3-L1 cells. *Food Chem* 126: 1–7.

[57] McDougall GJ, Kulkarni NN, Stewart D (2009) Berry polyphenols inhibit pancreatic lipase activity *in vitro*. *Food Chem* 115: 193–199.

[58] Sharma N, Sharma VK, Seo SY (2005) Screening of some medicinal plants for anti-lipase activity. *J Ethnopharmacol* 97: 453–456.

[59] Wu HS, Li YY, Weng LJ, Zhou CX, He QJ, et al. (2007) A fraction of *Acorus calamus* L. extract devoid of beta-asarone enhances adipocyte differentiation in 3T3-L1 cells. *Phytother Res* 21: 562–564.

[60] Zheng CD, Duan YQ, Gao JM, Ruan ZG (2010) Screening for anti-lipase properties of 37 traditional Chinese medicinal herbs. *J Chin Med Assoc* 73: 319–324.

[61] Kumar A, Chauhan GS (2010) Extraction and characterization of pectin from apple pomace and its evaluation as lipase (steapsin) inhibitor. *Carbohydr Polym* 82: 454–459.

[62] Burns Kraft TF, Dey M, Rogers RB, Ribnicky DM, Gipp DM, et al. (2008) Phytochemical composition and metabolic performance-enhancing activity of dietary berries traditionally used by Native North Americans. *J Agric Food Chem* 56: 654–660.

[63] Ghedira K, Goetz P, Jeune R (2009) *Urtica dioica* L. Urtica urens et/ou hybrides (Urticaceae). *Phytothérapie* 7: 279–285.

[64] SDI/Verispan (2010) 2009 top 200 generic drugs by total prescriptions. *Drug Topics* June: 1–3.

[65] Marles RJ, Farnsworth NR (1995) Antidiabetic plants and their active constituents. *Phytomedicine* 2: 137–189.

[66] Bailey CJ, Day C (2004) Metformin: its botanical background. *Pract Diabet Int* 21: 115–117.

[67] Almoosawi S, McDougall GJ, Fyfe L, Al-Dujaili EAS (2010) Investigating the inhibitory activity of green coffee and cacao bean extracts on pancreatic lipase. *Nutr Bull* 35: 207–212.

[68] Harbilas D, Martineau LC, Harris CS, Adeyiwola-Spoor DC, Saleem A, et al. (2009) Evaluation of the antidiabetic potential of selected medicinal plant extracts from the Canadian boreal forest used to treat symptoms of diabetes: part II. *Can J Physiol Pharmacol* 87: 479–492.

[69] Spoor DC, Martineau LC, Leduc C, Benhaddou-Andaloussi A, Meddah B, et al. (2006) Selected plant species from the Cree pharmacopoeia of northern Quebec possess anti-diabetic potential. *Can J Physiol Pharmacol* 84: 847–858.

[70] Eid HM, Martineau LC, Saleem A, Muhammad A, Vallerand D, et al. (2010) Stimulation of AMP-activated protein kinase and enhancement of basal glucose uptake in muscle cells by quercetin and quercetin glycosides, active principles of the antidiabetic medicinal plant *Vaccinium vitis-idaea*. *Mol Nutr Food Res* 54: 991–1003.

[71] Martineau LC, Adeyiwola-Spoor DC, Vallerand D, Afshar A, Arnason JT, et al. (2010) Enhancement of muscle cell glucose uptake by medicinal plant species of Canada's native populations is mediated by a common, metformin-like mechanism. *J Ethnopharmacol* 127: 396–406.

[72] Martineau LC, Herve J, Muhamad A, Saleem A, Harris CS, et al. (2010) Anti-adipogenic activities of *Alnus incana* and *Populus balsamifera* bark extracts, part I: sites and mechanisms of action. *Planta Med* 76: 1439–1446.

[73] Muhammad A, Guerrero-Analco JA, Martineau LC, Musallam L, Madiraju P, et al. (2012) Antidiabetic compounds from *Sarracenia purpurea* used traditionally by the Eeyou Istchee Cree First Nation. *J Nat Prod* 75: 1284–1288.

[74] Nistor Baldea LA, Martineau LC, Benhaddou-Andaloussi A, Arnason JT, Levy E, et al. (2010) Inhibition of intestinal glucose absorption by anti-diabetic medicinal plants derived from the James Bay Cree traditional pharmacopeia. *J Ethnopharmacol* 132: 473–482.

[75] Cieniak C, Liu R, Fottinger A, Smiley SA, Guerrero-Analco JA, et al. (2013) *In vitro* inhibition of metabolism but not transport of gliclazide and repaglinide by Cree medicinal plant extracts. *J Ethnopharmacol* 150: 1087–1095.

[76] Tam TW, Liu R, Arnason JT, Krantis A, Staines WA, et al. (2009) Actions of ethnobotanically selected Cree anti-diabetic plants on human cytochrome P450 isoforms and flavin-containing monooxygenase 3. *J Ethnopharmacol* 126: 119–126.

[77] Harris CS, Lambert J, Saleem A, Coonishish J, Martineau LC, et al. (2008) Anti-diabetic activity of extracts from needle, bark, and cone of *Picea glauca*: organ-specific protection from glucose toxicity and glucose deprivation. *Pharm Biol* 46: 126–134.

[78] Harris CS, Asim M, Saleem A, Haddad PS, Arnason JT, et al. (2012) Characterizing the cytoprotective activity of *Sarracenia purpurea* L, a medicinal plant that inhibits glucotoxicity in PC12 cells. *BMC Complement Alternat Med* 12: 245.

[79] Giugliano D, Ceriello A, Paolisso G (1996) Oxidative stress and diabetic vascular complications. *Diabetes Care* 19: 257–267.

[80] Johansen JS, Harris AK, Rychly DJ, Ergul A (2005) Oxidative stress and the use of antioxidants in diabetes: linking basic science to clinical practice. *Cardiovasc Diabetol* 4: 5.

[81] Rahimi R, Nikfar S, Larijani B, Abdollahi M (2005) A review on the role of antioxidants in the management of diabetes and its complications. *Biomed Pharmacother* 59: 365–373.

[82] Hegab Z, Gibbons S, Neyses L, Mamas MA (2012) Role of advanced glycation end products in cardiovascular disease. *World J Cardiol* 4: 90–102.

[83] Yamagishi S, Maeda S, Matsui T, Ueda S, Fukami K, et al. (2012) Role of advanced glycation end products (AGEs) and oxidative stress in vascular complications in diabetes. *Biochim Biophys Acta* 1820: 663–671.

[84] Downing A (2010) Inter and intra-specific differences in medicinal plant use for the treatment of type II diabetes symptoms by the Cree Elders of Eeyou Istchee (QC). Université de Montréal.

[85] Fraser MH (2006) Ethnobotanical investigation of plants used for the treatment of type 2 diabetes by two Cree communities in Québec: quantitative comparisons and antioxidant evaluation. McGill University.

[86] Rapinski M (2012) Ethnobotanique de la Nation crie d'Eeyou Istchee et variation géographique des plantes médicinales antidiabétiques. Université de Montréal.

[87] Vianna R, Brault A, Martineau LC, Couture R, Arnason JT, et al. (2011) *In vivo* anti-*diabetic* activity of the ethanolic crude extract of *Sorbus decora* C.K. Schneid (Rosacea): a medicinal plant used by Canadian James Bay Cree nations to treat symptoms related to diabetes. *Evid Based Complement Alternat Med* 2011: 237941.

[88] Harbilas D, Vallerand D, Brault A, Saleem A, Arnason JT, et al. (2012) *Larix laricina*, an antidiabetic alternative treatment from the Cree of Northern Quebec Pharmacopoeia, decreases glycemia and improves insulin sensitivity *in vivo*. *Evid Based Complement Alternat Med* 2012: 296432.

[89] Hasani-Ranjbar S, Nayebi N, Larijani B, Abdollahi M (2009) A systematic review of the efficacy and safety of herbal medicines used in the treatment of obesity. *World J Gastroenterol* 15: 3073–3085.

[90] Cai S, Wang O, Wang M, He J, Wang Y, et al. (2012) *In vitro* inhibitory effect on pancreatic lipase activity of subfractions from ethanol extracts of fermented Oats (*Avena sativa* L.) and synergistic effect of three phenolic acids. *J Agric Food Chem* 60: 7245–7251.

[91] Riedel A, Pignitter M, Hochkogler CM, Rohm B, Walker J, et al. (2012) Caffeine dose-dependently induces thermogenesis but restores ATP in HepG2 cells in culture. *Food Funct* 3: 955–964.

[92] Wu J, Bostrom P, Sparks LM, Ye L, Choi JH, et al. (2012) Beige adipocytes are a distinct type of thermogenic fat cell in mouse and human. *Cell* 150: 366–376.

[93] Kim J, Lee YS, Kim CS, Kim JS (2012) Betulinic acid has an inhibitory effect on pancreatic lipase and induces adipocyte lipolysis. *Phytother Res* 26: 1103–1106.

[94] Van Wagner V, Epoo B, Nastapoka J, Harney E (2007) Reclaiming birth, health, and community: midwifery in the Inuit villages of Nunavik, Canada. *J Midwifery Womens Health* 52: 384–391.

[95] Potvin L, Cargo M, McComber AM, Delormier T, Macaulay AC (2003) Implementing participatory intervention and research in communities: lessons from the Kahnawake Schools Diabetes Prevention Project in Canada. *Soc Sci Med* 56: 1295–1305.

[96] Horn OK, Jacobs-Whyte H, Ing A, Bruegl A, Paradis G, et al. (2007) Incidence and prevalence of type 2 diabetes in the First Nation community of Kahnawake, Quebec, Canada, 1986–2003. *Can J Public Health* 98: 438–443.

[97] Gittelsohn J, Harris SB, Burris KL, Kakegamic L, Landman LT, et al. (1996) Use of ethnographic methods for applied research on diabetes among the Ojibway-Cree in northern Ontario. *Health Educ Q* 23: 365–382.

[98] Boston P, Jordan S, MacNamara E, Kozolanka K, Bobbish-Rondeau E, et al. (1997) Using participatory action research to understand the meanings aboriginal Canadians attribute to the rising incidence of diabetes. *Chronic Dis Can* 18: 5–12.

[99] Gray-Donald K, Robinson E, Collier A, David K, Renaud L, et al. (2000) Intervening to reduce weight gain in pregnancy and gestational diabetes mellitus in Cree communities: an evaluation. *CMAJ* 163: 1247–1251.

[100] Balick MJ, Cox PA (1996) *Plants that Heal: Plants, People, and Culture the Science of Ethnobotany*. New York: Scientific American Library, pp. 24–61.

[101] Cox PA, Balick MJ (1994) The ethnobotanical approach to drug discovery. *Sci Am* 270: 82–87.

[102] Höft M, Barik SK, Lykke AM (1999) *Quantitative Ethnobotany: Applications of Multivariate and Statistical Analyses in Ethnobotany*. Paris: UNESCO, p. 46.

[103] Cuerrier A, Downing A, Patterson E, Haddad P (2012) Aboriginal antidiabetic plant project with the James Bay Cree of Quebec: an insightful collaboration. *J Enterp Commun People Places Global Econ* 6: 251–270.

[104] Ladio AH, Lozada M (2004) Patterns of use and knowledge of wild edible plants in distinct ecological environments: a case study of a Mapuche community from northwestern Patagonia. *Biodivers Conserv* 13: 1153–1173.

[105] de Sousa Araújo TA, Alencar NL, de Amorim ELC, de Albuquerque UP (2008) A new approach to study medicinal plants with tannins and flavonoids contents from the local knowledge. *J Ethnopharmacol* 120: 72–80.

[106] Phillips O, Gentry AH (1993a) The useful plants of Tambopata, Peru: I. Statistical hypotheses tests with a new quantitative technique. *Econ Bot* 47: 15–32.

[107] Phillips O, Gentry AH (1993b) The useful plants of Tambopata, Peru: II. Additional hypothesis testing in quantitative ethnobotany. *Econ Bot* 47: 33–43.

[108] Tendland Y, Pellerin S, Haddad PS, Cuerrier A (2012) Impacts of experimental leaf harvesting on a North American medicinal shrub, *Rhododendron groenlandicum*. *Botany* 90: 247–251.

16

PHYTOTHERAPEUTICS FOR CANCER THERAPY

Daniel M.-Y. Sze[1], Hao Liu[2], Maureen V. Boost[2], Raimond Wong[3], and Stephen Sagar[3]

[1] *School of Medical Sciences and Health Innovations Research Institute (HiRi), RMIT University, Australia*
[2] *Faculty of Health and Social Sciences, The Hong Kong Polytechnic University, Hong Kong*
[3] *Departments of Oncology and Medicine, McMaster University, Hamilton, Ontario, Canada*

16.1 INTRODUCTION

This chapter aims to examine some of the recent evidence-based scientific information about the contribution of phytotherapeutics for effective cancer management with particular emphasis on results of human clinical trials. It is well recognized that effective cancer therapy is related to multidimensional mechanisms that involve direct killing of cancer cells; enhancement of natural immunity and antiangiogenesis. Conventional pharmaceutical agents usually provide a unique but single mechanism for cancer control, and treatment with a single agent is not as effective as treatment with a combination of conventional agents that exert anticancer effects through different mechanisms. Thus, current effective cancer treatment strategies are often achieved through "multicompound multitarget" approaches.

Phytotherapeutics, which consist of multiple chemical compounds derived from natural products, are also known to interact with and specifically target different receptors on various cell types. The use of phytotherapeutics may provide the advantage of "multicompound multitarget" action in cancer treatment.

It is well accepted that immune deficiency is a critical factor related to the high cancer incidence rate. It has also been shown that medicines that enhance natural

Phytotherapies: Efficacy, Safety, and Regulation, First Edition. Edited by Iqbal Ramzan.

anticancer immunity may prolong overall survival of cancer patients. This chapter discusses the clinical evidence of effectiveness of phytotherapeutics mainly in relation to the anticancer mechanisms through natural killer (NK) cell enhancement.

16.2 ANTICANCER PHYTOTHERAPEUTICS WITH NK ENHANCEMENT

16.2.1 Effects of Clinically Useful Phytocompounds on Cancer Patients' NK Cell Immunity, Quality of Life (QoL), and Overall Survival

The overall goal of cancer management is to provide a quality outcome to improve the whole health status of the patient. While measurement of patients' overall survival is the gold standard of assessing the clinical efficacy of the treatment, it is also important to ensure that patients' QoL has also improved. It is proposed that enhancing patients' natural immunity to cancer and reducing the angiogenesis support to the tumor may result in reduction in the size of the tumor, leading to improvement in the overall well-being of the patients, resulting in longer survival.

It is now known that heterogeneity exists in cancer cells at different stages from primary local proliferation to secondary metastasis [1]. However, the body's innate immunity, in particular through the NK cells, provides a strong protection or fighting power to constantly monitor and remove any developing "problematic" cells such as virus-infected cells or cancer cells because of the low major histocompatibility complex (MHC) class I expression on their cell surfaces. Thus, NK cells play an important role in anticancer immunomodulation. Some receptors are exclusively expressed by NK cells, such as natural cytotoxicity receptors (NCRs). NCRs contain three specific molecules (NKp46, NKp30, and NKp44) for several host ligands of tumor cells [2, 3]. NK cells not only kill tumor cells directly but also perform an important helper role by their stimulation of dendritic cells and improvement of the cytotoxic effects or function of T cells [4, 5].

16.2.2 Commonly Used Phytotherapeutics in Cancer Management

16.2.2.1 Search Methodology A systematic review was conducted; a literature search was performed for articles published between 1990 and July 2013 that were listed in the following four electronic databases: PubMed, Embase, CNKI (one of the biggest Full-Text Database for Journals in China), and CBMdisc (China Biological Medicine database).The English keywords for the search included the following: "Chinese herbal medicine," "Chinese herb," "Herbal medicine," "CHM," "Traditional Chinese medicine," "TCM," "China herb," "China extract," "China fraction," "China formula," "China prescription," "cancer," "carcinoma," "tumour," " NK cell," and "natural killer cell." A search using keywords in Chinese language has not been conducted. The search scope setting was under full text. There was no other special limitation.

A flowchart showing the screening process of the systematic review is presented in Figure 16.1. The initial screen of titles or abstracts was followed by full reading of

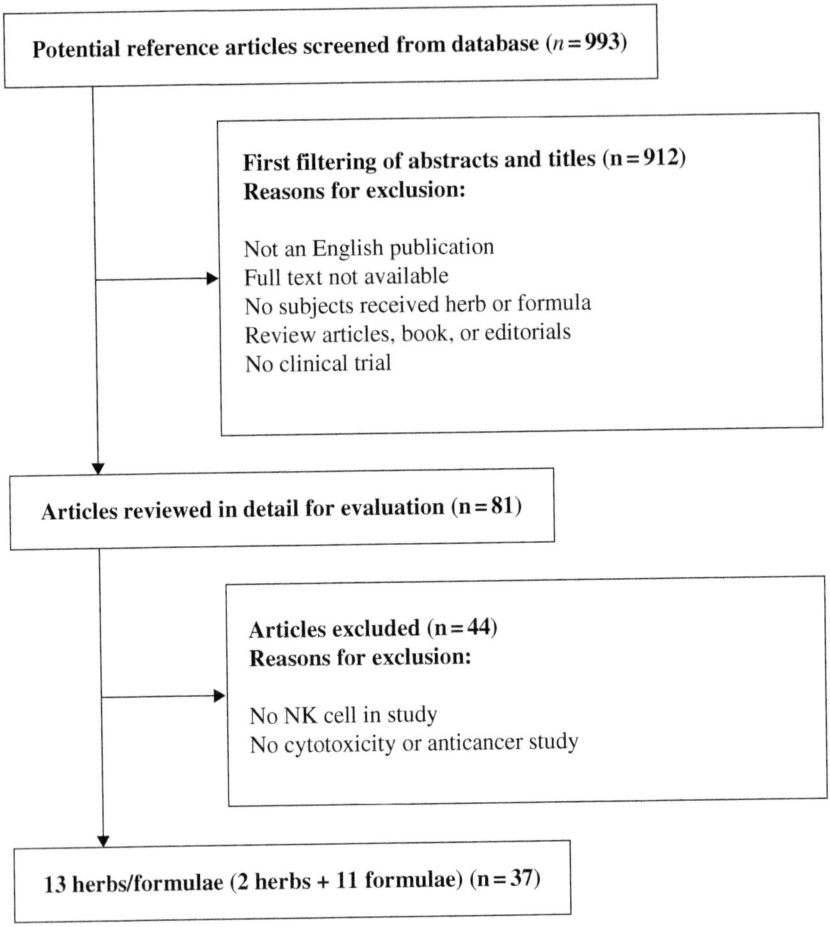

FIGURE 16.1 Flowchart of screening process for publications.

the text. The articles were classified as eligible if they matched with the following five criteria: article in English, full text, clinical trial, cancer phytotherapy, and NK cell. A total of 37 articles were included and 13 herbs/formulae (2 herbs and 11 formulae) were identified in these articles. A detailed study of these articles was performed. NK cell number/percentage and the action of the herbs/formulae on NK cell activities were first tabulated, followed by the description of patients' QoL and overall survival.

16.2.2.2 Two Common Herbs for Cancer Phytotherapy via NK Modulation Two commonly used phytotherapeutics, the medicinal mushrooms *Lentinula edodes* mycelia and *Ganoderma lucidum*, have been found to elicit their anticancer functions by regulating NK cells. The preclinical animal studies and very limited human studies have been reviewed by Shah et al. [6] regarding the use of *L. edodes* mushroom-derived extract (active hexose correlated compound) as a complementary therapy. A recent report summarized very clearly the clinical usefulness of lentinan, which is also a purified *L. edodes* mushroom–derived extract, describing its desirable survival benefits in advanced gastric cancer patients [7]. However, there are additional publications only indexed by some databases in China. These databases were not included in this particular review. For *G. lucidum*, Yuen and Gohel [8] summarized the possible mechanisms of this medicinal mushroom in cancer management, such as direct cytotoxicity, immune activation, and antiangiogenesis, but again literature from the databases in China in relation to cancer phytotherapeutics was omitted. This review has taken a more comprehensive search in English language in databases established in both Western countries and China to provide the readers with a thorough understanding of phytotherapeutics against cancer.

16.2.2.3* Lentinula edodes *Mycelia *L. edodes* mycelia (LEM, also called shiitake) is a medicinal and edible fungus. A hot water extract or lignin derivatives or isolated lentinan/polysaccharides are usually applied as Chinese herbal medicines (CHMs). As it is easy to cultivate, it is widely used. The effects of LEM on NK cell modulation, QoL, and survival benefit in cancer intervention are shown in Tables 16.1 and 16.2, respectively.

Activities on NK There were seven clinical trials involving cancer patients treated with LEM, which are listed in Table 16.1. However, six of these studies involved very few patients (5–16) with a wide range of cancers, and thus their interpretation warrants special consideration.

Three of the seven studies in Table 16.1 reported the effects of lentinan/LEM extract on NK cell number [11, 14, 15]. Only one of these three clinical case studies indicated that LEM extract promoted the percentage of relative NK cells based on CD3-CD161+ subsets ($p<0.1$) [14]. The other two case studies suggested that lentinan had no effect on NK cell number [11, 15].

Five of the seven studies showed that both lentinan and LEM extract improved the cytotoxicity (16.7–66.5% increase compared with baseline) of peripheral blood mononucleated cells (PBMCs), which maybe partly attributable to an increase in NK cell subsets.

The only clinical trial in these seven studies was a randomized control trial (RCT) comparing 42 advanced nonsmall cell lung cancer patients (18 adenocarcinoma, 23 squamous cells, and 1 large cell lung cancer) treated with 1 mg lentinan twice a week plus chemotherapy for 8 weeks and 39 patients in the control group without lentinan supplement [13]. There was a 16.7% increase in cytotoxic effects on PBMC after lentinan treatment. However, all these articles only examined the overall PBMC, but not individual NK populations. Thus, it is not clear whether there was an increase in the cytotoxicity targeting NK after the phytotherapeutic treatment.

Effects on QoL and Survival Benefit Treatment with both lentinan and LEM extract resulted in a decrease in adverse events caused by either the cancers themselves or traditional remedies (chemo, radiotherapy, or resection), or improvements in

TABLE 16.1 Effects of *L. Edodes* Mycelia on NK Cells

No.	Publication Year	Type of Cancer (Age)	Type of Study (n), Treatment	NK Number/Percentage (Subsets, Methods): Results of BA	NK Activity (Cytolysis/%, Methods, Target Cell Line): Results of BA	Reference
LEM-NK-1	1984	Liver, bile duct, lung, gastrointestinal cancer (all adults mainly under chemotherapy, no stages mentioned)	Case reports ($n=9$), lentinan	ND	Only PBMC cytotoxicity NOT specific about NK: individual $p<0.05$ for 7 of 9; total 43.5% increase for 9 (51Cr-release assay, K562)	[9]
LEM-NK-2	1992	Gastric, pancreatic, esophageal, anal, unknown cancer (main metastases, all adults mainly under chemotherapy, main stage III–V)	Case reports ($n=16$), lentinan	ND	PEC cytotoxicity: for responders: ~66.5% increase ($p<0.02$); for nonresponders: no change; PBMC cytotoxicity: ~15.0% decrease ($p<0.05$) (*not* specific about NK) (51Cr-release assay, K562)	[10]
LEM-NK-3	1992	Gastric cancer (adults without other remedies mentioned, no stages mentioned)	Case reports ($n=11$), lentinan	No effect on NK number (identified using CD16, CD56, CD57 markers) by flow cytometry	Only PBMC cytotoxicity NOT specific about NK: 31.2% increase ($p<0.05$) (51Cr-release assay, K562)	[11]

LEM-NK-4	1992	Gastric cancer (adults under gastrectomy, stage I–IV)	Case reports ($n=10$), lentinan	ND	Only PBMC *not* specific about NK: ~44.1% increase ($p<0.05$) (51Cr-release assay, K562)	[12]
LEM-NK-5	2006	Advanced nonsmall cell lung cancer (adults under chemotherapy, stage III–IV)	Randomized control trial with lentinan ($n=42$); control ($n=39$), lentinan	ND	Only PBMC cytotoxicity *not* specific about NK: 16.7% increase ($p<0.05$) TC: for treatment group, only PBMC *not* specific about NK: 20% more than control with no comparison (APAAP assay)	[13]
LEM-NK-6	2011	Breast cancer, gastrointestinal cancer (mainly under postoperative chemotherapy, all adults, stage I–III)	Case reports ($n=5$), LEM extract	NK percentage (CD3-CD161$^+$): 12.3% increase ($p<0.1$) by flow cytometry	Only PBMC cytotoxicity *not* specific about NK: 18.3% increase ($p<0.05$) (51Cr-release assay, K562)	[14]
LEM-NK-7	2013	Breast cancer (postoperative metastasis, adults under chemotherapy, stage II–IIIA)	Case reports ($n=10$), LEM extract	No effect on percentage of NK(CD3-CD161$^+$) by flow cytometry	No effect on the whole PBMC cytotoxicity (51Cr-release assay, K562)	[15]

APAAP, alkaline phosphatase antialkaline phosphatase assay; BA, before treatment versus after treatment; ND, not done; PBMC(s), human peripheral blood mononuclear cells; PEC, peritoneal or pleural exudate mononuclear cells; TC, treatment group versus control group.

TABLE 16.2 Effects of *L. Edodes* Mycelia on QoL and Survival Condition of Cancer Patients

No.	Publication Year	Type of Cancer (Age)	Type of Study (*n*), Treatment	QoL (Indicators): Results of BA/TC	Survival Benefit: Results of BA/TC	Reference
LEM-Q/S-1	1992	Gastric, pancreatic, esophageal, anal, unknown cancer (main metastases, all adults mainly under chemotherapy, main stage III–V)	Case reports ($n=16$), lentinan	ND	Not significant in survival time after treatment	[10]
LEM-Q/S-2	2006	Advanced nonsmall cell lung cancer (adults under chemotherapy, stage III–IV)	RCT with lentinan ($n=42$); control ($n=39$), lentinan	TC: Mean scores of Karnofsky in treatment group before treatment were about 64.5. Rising rate of Karnofsky scores in treatment group were 1.3 times higher than that of control ($p<0.01$, 52% (22/42) vs. 23% (9/39)). Declining rate of Karnofsky scores in treatment group was 61.4% lower than that of control ($p<0.01$, 17% (7/42) vs. 44% (17/39)). According to the total rise and decline after treatment, mean scores in treatment group were almost 70, an average increase for each patient was about 3.6 scores; improved rate of body weight was 1.1 times higher ($p<0.01$); adverse events were 33.7% less ($p<0.05$, leucopenia, and nausea/vomiting)	ND	[13]

LEM-Q/S-3	2008	Liver cancer (not suitable for surgical treatment, adults, main stage III–IV)	RCT with radiotherapy ($n=23$); chemotherapy ($n=24$); radiotherapy-combined with chemotherapy ($n=31$); lentinan + combination ($n=29$), lentinan	ND		TC: survival time of lentinan combination group had 30% longer duration than standard combination (however, no clear mention about $p<0.05$, 28.2 m vs. 21.9 m)	[16]
LEM-Q/S-4	2009	Liver cancer (mainly unresectable, adults receiving chemotherapy, no stages mentioned)	Case reports ($n=36$), lentinan	ND		Survival time of patients who ingested test food for a mean period of 47 weeks was 70% longer than that of patients who ingested for 7–12 weeks ($p<0.05$, 20.9 m vs. 12.4 m)	[17]

(Continued)

TABLE 16.2 (*Continued*)

No.	Publication Year	Type of Cancer (Age)	Type of Study (n), Treatment	QoL (Indicators): Results of BA/TC	Survival Benefit: Results of BA/TC	Reference
LEM-Q/S-5	2009	Advanced pancreatic cancer (main unresectable, adults mainly receiving chemotherapy, no stages mentioned)	Case reports ($n=15$), lentinan	No information on the change of total QoL-ACD scores was provided. However, face scale, activity aspect, physical aspect, physiological aspect, and social aspect of QoL were significantly related to survival duration ($p<0.05$)	Survival time for QoL improved group was 4.2 times longer than QoL deterioration group ($p<0.01$, 33.9 m vs. 6.5 m)	[18]
LEM-Q/S-6	2011	Gastric cancer (not suitable for surgical treatment, adults under chemotherapy, no stages mentioned)	Case–control reports with lentinan ($n=31$); control ($n=37$), lentinan	ND	TC: for treatment group, overall survival was 21.9% longer than control ($p<0.05$, 689 days vs. 565 days)	[19]
LEM-Q/S-7	2011	Advanced gastrointestinal cancer (all metastases, adults under chemotherapy, no stages mentioned)	Case reports ($n=8$), LEM extract	BA: the adverse events of nausea reduced from 75 (6/8) to 0% with no statistical analysis	ND	[20]

LEM-Q/S-8	2011	Breast cancer, gastrointestinal cancer (mainly under postoperative chemotherapy, all adults, stage I–III)	Case reports ($n = 5$), LEM extract	BA: the total QoL-ACD scores in treatment group showed 7.8% increase ($p < 0.05$); functional 9.1% increase ($p < 0.05$); physical 17.4% increase ($p < 0.05$)	ND	[14]
LEM-Q/S-9	2012	Advanced gastric cancer (unresectable, adults receiving chemotherapy, main stage IV)	Case–control study with lentinan ($n = 19$); control ($n = 20$), lentinan	TC: for treatment group, adverse events were 58.6% lower ($p < 0.1$)	No effect on survival rate, comparing treatment and control groups	[21]
LEM-Q/S-10	2013	Breast cancer (postoperative metastases, adults under chemotherapy, stage II–IIIA)	Case reports ($n = 10$), LEM extract	BA: the total QoL-ACD scores were 73.3% lower after first week treatment ($p < 0.05$)	ND	[15]

BA, before treatment versus after treatment; ND, not done; QoL, quality of life; QoL-ACD, QoL questionnaire for cancer patients treated with anticancer drugs; RCT, randomized control trial; TC, treatment group versus control group.

Karnofsky scores, total QoL, and body weight. Table 16.2 shows the effects of *L. edodes* mycelia on QoL and survival condition of cancer patients.

Six of the ten studies in Table 16.2 reported the effects of lentinan/LEM extract on QoL. Three reported enhancement of total QoL after lentinan treatment [14, 15, 18], while three reported reductions in the occurrence of adverse events during lentinan treatment, all reported more than a 30% reduction (33.7–75% reduction compared with baseline or control group) [13, 20, 21]. Only one study showed that the increase of Karnofsky scores in the treatment group was 1.3 times higher than that of the control group [13].

Six of the ten studies reported the effects of lentinan/LEM extract on overall survival as well. Four studies demonstrated that lentinan was effective in prolonging early-stage cancer patient survival time (21.9–70% higher compared with baseline or control group) [16–19], but there were no obvious effects in two studies with cancer patients in advanced stages of the disease [10, 21]. Specifically in one study, patients ($n=36$) receiving lentinan for nearly a year were shown to have a 70% increase in duration of survival [17]. This supports the claim that lentinan appeared to enhance patients' QoL, as well as overall survival.

Next, we examined another commonly used phytotherapeutic, mushroom, *G. lucidum*, for its effects on cancer patients' NK cells, QoL and related survival duration.

16.2.2.4 Ganoderma lucidum *G. lucidum* is also known as Lingzi in China and Reishi in Japan. Traditionally, this herb has been used for the treatment of symptoms of consumptive diseases, cough, shortness of breath, sleeplessness, and dyspepsia. It has properties for invigorating the liver and removing toxins, cardiovascular protection, and enhancement of general well-being. *G. lucidum* has also been used widely in China as a complementary herbal medicine for cancer treatment. For instance in a study of breast cancer patients, 58% of surveyed participants used *G. lucidum* [22]. This section examines the evidence of *G. lucidum* for clinical outcomes (survival and QoL) together with the NK immunomodulation activity (Tables 16.3 and 16.4).

Effects of G. lucidum on NK Cell Numbers and Activities The literature search yielded only five clinical trials focusing on immune modulation of a *G. lucidum* polysaccharide (product name: Ganopoly®) in advanced cancer patients with standardized administration strategy (1800 mg oral Ganopoly three times daily before meals for 12 weeks). Four of the five studies reported the effects of Ganopoly on NK number in advanced cancer patients. Of these four, only one study showed an increase in the number of $CD56^+$ NK [24]. On the other hand, two studies confirmed that Ganopoly had a positive effect on the cytotoxicity of PBMC used as NK anticancer agent [23, 24].

In the only randomized, placebo-controlled, multicenter study of GL-NK-1, patients in the Ganopoly group had 77.6% higher NK activity compared with baseline ($42.8 \pm 19.7\%$ vs. $24.1 \pm 12.3\%$, $p<0.05$) in comparison with the control group ($24.5 \pm 8.7\%$) [23]. In the case report GL-NK-2, 30 advanced cancer patients (including lung, colon, breast, liver, prostate, bladder, brain, and unknown cancer)

TABLE 16.3 Effects of *Ganoderma Lucidum* on NK Cells

No.	Publication Year	Type of Cancer (Age)	Type of Study (n), Treatment	NK Number/ Percentage (Subsets, Methods): Results of BA	NK Activity (Cytolysis/%, Methods, Target Cell Line): Results of BA	Reference
GL-NK-1	2003	Advanced lung cancer (adults without other remedies, mainly stage IV)	Randomized, placebo-controlled, multicenter with Ganopoly ($n=32$); placebo ($n=28$), Ganopoly	ND	Whole PBMC cytotoxicity of treatment group increased 77.6% ($p<0.05$) TC: Treatment group was 74.7% more cytotoxic than placebo group without statistical analysis (51Cr-release assay, K562)	[23]
GL-NK-2	2003	Several types of advanced-stage cancers (all metastases, adults without other remedies, no stages mentioned)	Case reports ($n=30$), Ganopoly	NK number ($CD56^+$): 16.4% increase after treatment ($p<0.05$)	Whole PBMC cytotoxicity of treatment group increased 29.7% ($p<0.05$) (51Cr-release assay, K562)	[24]

(*Continued*)

TABLE 16.3 *(Continued)*

No.	Publication Year	Type of Cancer (Age)	Type of Study (n), Treatment	NK Number/ Percentage (Subsets, Methods): Results of BA	NK Activity (Cytolysis/%, Methods, Target Cell Line): Results of BA	Reference
GL-NK-3	2005	Advanced lung cancer (adults without other remedies, no stages mentioned)	Case reports ($n=30$), Ganopoly	Not significant on NK number ($CD56^+$) after GL treatment	Not significant on cytotoxicity of the whole PBMC population	[25]
GL-NK-4	2005	Advanced colorectal cancer (adults without other remedies, no stages mentioned)	Case reports ($n=41$), Ganopoly	Not significant on NK number ($CD56^+$) after GL treatment	Not significant on cytotoxicity of the whole PBMC population	[26]
GL-NK-5	2006	Advanced colorectal cancer (adults without other remedies, no stages mentioned)	Case reports ($n=41$), Ganopoly	Not significant on NK number ($CD56^+$) after GL treatment	Not significant on cytotoxicity of the whole PBMC population	[27]

BA, before treatment versus after treatment; ND, not done; PBMC(s), human peripheral blood mononuclear cells; TC, treatment group versus control group.

TABLE 16.4 Effects of *Ganoderma Lucidum* on QoL and Survival Condition of Cancer Patients

No.	Publication Year	Type of Cancer (Age)	Type of Study (n), Treatment	QoL (Indicators): Results of BA/TC	Survival Benefit: Results of BA/TC	Reference
GL-Q/S-1	2003	Advanced lung cancer (adults without other remedies, main stage IV)	Randomized, placebo-controlled, multicenter with Ganopoly ($n=32$); placebo ($n=28$), Ganopoly	TC: Karnofsky scores of all patients screened were ≥60, but without detailed information. Improvement rate of Karnofsky score in treatment group was 2.5 times higher than placebo ($p<0.05$, 50.0% (16/32) vs. 14.3% (4/28)). Declining rate of Karnofsky scores was 61.4% lower than that of placebo ($p<0.05$, 21.9% (7/32) vs. 39.3% (11/28)). According to the total condition of rise and decline after treatment, average increase for each patient was about 2.8 scores; total cancer-related symptoms of treatment group were 1.9 times less ($p<0.05$, fever, sweating, body weight loss, weakness, insomnia, and anorexia)	ND	[23]
GL-Q/S-2	2012	Breast cancer (adults under endocrine therapy, stage I–III)	RCT with *G. lucidum* extract ($n=25$); control ($n=23$), *G. lucidum* extract	TC and BA: treatment group showed multiple significant improvements in patients' QoL based on the scores of HADS, FACT-F, and EORTC QLQ-C30 ($p<0.05$, 17.4–35.3% increase)	ND	[28]
GL-Q/S-3	2012	Breast cancer (adults, main stage I–II)	Survey with *G. lucidum* extract ($n=1501$); control ($n=2440$), *G. lucidum* extract	TC: total QoL scores of treatment group showed no change; social well-being was 1.8% higher ($p<0.05$, social support, interpersonal relationships, and total)	ND	[22]

BA, before treatment versus after treatment; EORTC QLQ-C30, European Organization for Research and Treatment of Cancer Core Quality of Life Questionnaire C30; FACT-F, functional assessment of cancer therapy: Fatigue; HADS, Hospital Anxiety and Depression Scale; ND, not done; PBMC(s), human peripheral blood mononuclear cells; QoL, quality of life; RCT, randomized control trial; TC, treatment group versus control group.

who completed the 12-week treatment had changes in lymphocyte subsets, such as a 16.4% increase of NK cell ($CD56^+$) proliferation (baseline vs. 12 weeks treatment: 238.3±54.1 vs. 277.4+/−56.8, $p<0.05$) and 29.7% NK activity increase compared to baseline (34.5±11.8% vs. 26.6+/−8.3%, $p<0.05$) [24]. In the case report GL-NK-3 of advanced lung cancer, neither NK activity nor NK cell numbers were significantly improved after 12 weeks [25]. Results from another two case reports (GL-NK-4 and GL-NK-5) in advanced colorectal cancer patients were similar with no effects on NK activity and NK cell ($CD56^+$) number [26, 27].

The few available clinical trials are all based on the use of *G. lucidum* polysaccharide (product name: Ganopoly) in cancer patients with standardized administration strategy (1800 mg oral Ganopoly three times daily before meals for 12 weeks). Of the five clinical trials, only GL-NK-2 showed a relative increase in $CD56^+$ NK number, 16.4% increase after treatment ($p<0.05$) [24]. There was also a 29.7% increase in cytotoxicity for the overall PBMC population, not mentioning the individual NK cells, after the Ganopoly treatment ($p<0.05$). The other three studies, GL-NK-3, 4, and 5, all with similar number of patients (~30–40), showed no changes in the NK cell number after the Ganopoly treatment.

Effects of G. lucidum on QoL and Survival Benefit Only two clinical trials and one survey focused on QoL associated with the use of *G. lucidum* (Ganopoly or *Ganoderma lucidum* extract) in cancer patients. None of the studies reported patient survival data and thus no discussion on this aspect is possible.

In the randomized, placebo-controlled, multicenter study, GL-Q/S-1, advanced lung cancer patients in the Ganopoly group had an increased proportion of patients with improved Karnofsky performance status scores (50.0% (16/32) vs. 14.4% (4/28), $p<0.05$) and improvements of cancer-related symptoms (fever: 65.6 vs. 25.0%; sweating: 81.3 vs. 28.6%; body weight loss: 43.8 vs. 10.7%; weakness: 75.0 vs. 25.0%; insomnia: 84.4 vs. 42.9%; anorexia: 71.9 vs. 35.7%, all $p<0.05$) compared with those in the control group [23].

In the study by Zhao et al. [28] examining the effects of *G. lucidum* on the improvement of cancer-related fatigue, several related questionnaires were used. The Hospital Anxiety and Depression Scale (HADS) composed of two parts: anxiety subscale (HADS-A) and depression subscale (HADS-D) each with seven items were used. The European Organization for Research and Treatment of Cancer Core Quality of Life Questionnaire C30 (EORTC QLQ-C30) consisting of 30 items to test the multidimensionality of patients' QoL condition was also used. In the RCT GL-Q/S-2 on breast cancer patients receiving endocrine treatment, spore powder of *G. lucidum* extract (1000 mg three times a day for 4 weeks) improved the QoL assessments by both HADS and EORTC QLQ-C30 compared with control group (HADS: anxiety: 4.1±2.9 vs. 6.1±3.2, $p<0.05$; depression: 3.1±2.8 vs. 4.6±2.9, $p<0.01$; total: 7.1±3.1 vs. 9.8±3.4, $p<0.010$. Functioning and global QoL scores on EORTC QLQ-C30: physical: 78.2±26.1 vs. 64.5±28.7, $p<0.01$; emotional: 79.5±31.5 vs. 64.3±31.8, $p<0.01$; cognitive: 75.1±26.5 vs. 68.3±26.3, $p<0.05$; global QoL: 68.9±21.4 vs. 57.7±24.2, $p<0.01$. Symptoms scores on EORTC QLQ-C30 were fatigue: 31.1±18.1 vs. 40.2±16.8, $p<0.01$; sleep disturbance: 42.3±26.2 vs.

53.9 ± 24.8, $p<0.01$; appetite loss: 24.3 ± 18.4 vs. 30.3 ± 16.5, $p<0.05$) [28]. This study also highlights the lack of a universally accepted questionnaire for the measurement of QoL. While some research groups continue to report Karnofsky scores, others have started to design specific questionnaires with emphasis on particular areas. Thus, reports with different questionnaires make comparison of clinical effects on QoL difficult.

The postdiagnosis population-based, longitudinal, prospective survey of breast cancer management (GL-Q/S-3) deserves some discussion because this is a major QoL study of more than 4000 breast cancer patients or survivors who had either taken or not taken *G. lucidum* extract at 6 and 36 months after the diagnosis of cancer. This study employed two QoL instruments, the General Quality of Life Inventory-74 and the Short-Form Health Survey (SF-36, Chinese version). Six months after cancer diagnosis, 59% of the 4149 breast cancer patients reported taking the extract as complementary treatment. At 36 months after cancer diagnosis, 36% of these breast cancer patients continued to take *G. lucidum* extract as part of their cancer management therapy. Importantly, consideration of the QoL outcome at 36 months after cancer diagnosis shows that while there is clear improvement in the social well-being (in terms of social support and interpersonal relationships), unexpectedly those patients that consumed *G. lucidum* extract reported significantly poorer physical well-being in comparison with the group who did not take the extract. The physical well-being parameter includes aspects of sleep and energy, physical discomfort, and eating function. The researchers carefully analyzed the multiple linear regression results by adjusting for various parameters such as age at diagnosis, education level, income, marital status, exercise participation, tea consumption, menopausal status, symptoms, comorbidity, body mass index, vitamin supplement use, TCM use, TNM stage, type of surgery, chemotherapy, radiotherapy, tamoxifen use, ER/PR status, recurrence/metastasis, and the baseline total QoL obtained at the 6-month postdiagnosis period. The researchers suggest that this poor physical well-being result may be due to either *G. lucidum* use negatively influencing patients' physical well-being or that patients with low physical well-being were more likely to seek *G. lucidum* as a complementary treatment. This outcome warrants further in-depth research investigation.

16.2.3 Phytotherapeutic Formulae for Cancer via NK Modulation

16.2.3.1 Effects of Phytotherapeutic Formulae on NK Cell Numbers

The effects of 11 phytotherapeutic formulae on NK cell numbers are summarized in Table 16.5.

Nine formulae in Table 16.5 were tested for their effects on NK cell proliferation. Of these nine formulae (in total including 10 clinical studies), only six formulae (7 clinical studies) showed obvious effects on NK number (mainly about 50% increase compared with baseline or control group) including three studies based on NK subset of $CD16^+$, $CD56^+$, or $CD16^+/CD56^+$, respectively [34, 37, 42], two based on $CD16^+CD56^+$ [30, 31] and two with no mention of the NK subsets [29, 41].

It is particularly worth noting the apparent differences in NK cell numbers between two studies examining the effects of Shenqi Fuzheng Injection (SFI) on NK

TABLE 16.5 Effects of Phytotherapeutic Formulae on NK Cells in Clinical Cancer Treatment

No.	Phytotherapy Drugs	Herb(s) Included	Type of Cancer (Age)	Type of Study (n)	NK Number/Percentage (Subsets, Methods): Results of BA/TC	NK Activity (Cytolysis/%, Methods, Target Cell Line): Results of BA/TC	Reference
PHYTO-NK-1	Shenyi capsule	*Panax ginseng*	Nonsmall cell lung cancer (adults, stage II–III)	RCT with Shenyi alone ($n=43$); Shenyi and chemo ($n=46$); chemo alone ($n=44$)	BA: NK percentage (no mention of NK markers) Shenyi alone group showed 50% increase ($p<0.05$) by flow cytometry	ND	[29]
			Mammary cancer (adults under chemotherapy, no stages mentioned)	RCT with treatment ($n=58$); control ($n=52$)	TC: NK number ($CD16^+56^+$) of treatment group was 57.3% more than control ($p<0.05$, 118±95/μl vs. 75±89/μl)	TC: the whole PBMC cytotoxicity of treatment group was 18.8% higher ($p<0.05$) (MTT assay, mammary tumor cell)	[30]
PHYTO-NK-2	Shenqi Fuzheng Injection (SFI)	*Radix codonopsis*, *Radix astragali*, and *Radix hedysari*	Advanced breast cancer (adults under chemotherapy, stage II–III)	RCT with treatment ($n=65$); control ($n=61$)	BA and TC: NK percentage ($CD16^+/56^+$) of treatment group had 6.1% increase after treatment ($p<0.05$), 17.7% higher than control ($p<0.05$) by 3H-TdR assay	ND	[31]
			Gastric cancer (adults under operation, main stage III–IV)	Case–control study with treatment ($n=32$); control ($n=31$)	ND	TC: the whole PBMC cytotoxicity of treatment group was 13.5% higher ($p<0.05$) (MTT assay)	[32]

PHYTO-NK-3	CHM complex-1	Rose geranium, *Ganoderma tsugae*, *Codonopsis pilosula*, and *Angelica sinensis*	Breast cancer (adults under chemo- or radiotherapy, main stage I–III)	Double-blind, placebo-controlled randomized trial with treatment ($n = 31$); placebo ($n = 27$)	Not significant comparing treatment and control groups	ND	[33]
PHYTO-NK-4	CHM complex-2	*Ganoderma lucidum*, *Codonopsis pilosula*, *Angelica sinensis*, and citronellol powder	Breast, colorectal, nasopharyngeal, and lung cancer (all adults under chemo- or radiotherapy, main stage II–IV)	Double-blind, placebo-controlled randomized study with treatment ($n = 55$); placebo ($n = 50$)	TC: NK number ($CD16^{+}56^{+}$ with no specific NK subsets identified) of treatment group was ~30.7% higher than placebo group with no comparison	ND	[34]
PHYTO-NK-5	Aidi Injection	*Radix ginseng*, *Radix astragalus membranaceus bge*, *Radix acanthopanacis*, and mylabris	Nonsmall cell lung cancer III A (adults under chemotherapy and resection, stage IIIA)	RCT with treatment ($n = 30$); control ($n = 30$)	ND	TC: cytotoxicity of treatment group was 68.6% higher than control ($p < 0.05$) (radionuclide release method)	[35]

(Continued)

TABLE 16.5 *(Continued)*

No.	Phytotherapy Drugs	Herb(s) Included	Type of Cancer (Age)	Type of Study (*n*)	NK Number/Percentage (Subsets, Methods): Results of BA/TC	NK Activity (Cytolysis/%, Methods, Target Cell Line): Results of BA/TC	Reference
PHYTO-NK-6	Danggui Buxue Decoction NO.1	Milkvetch root, Chinese angelica root, spatholobus stem, forsythia fruit, globethistle root, chuanxiong, and white-stiff silkworm	Nonsmall cell lung cancer (adults under conventional treatment, stage I–III)	RCT with treatment ($n=41$); control ($n=41$)	ND	BA and TC: cytotoxicity of treatment group was 33.3% higher after treatment ($p<0.05$), 31.8% higher than control ($p<0.05$) (radionuclide release method)	[36]
PHYTO-NK-7	Hochu-ekki-to	Astragalus root, atractylodes lancea rhizome, ginseng root, Japanese angelica root, bupleurum root, jujube fruit, citrus unshiu peel, glycyrrhiza root, Cimicifuga rhizome, and Ginger rhizome	Gastrointestinal malignancy (adults under surgery, no stages mentioned)	Case–control study with treatment ($n=20$); control ($n=27$)	TC: NK number ($CD57^+$) of post-surgery treatment group was 56.4% lower ($p<0.05$) by flow cytometry	TC: the whole PBMC cytotoxicity of postsurgery treatment group was 38.5% lower ($p<0.05$) (51Cr-release assay, K562)	[37]

PHYTO-NK-8	Juzen-Taiho-to	*Radix astragali*, *Cortex cinnamomi*, *Radix angelicae*, *Radix paeoniae*, *Rhizoma cnidii*, *Radix rehmanniae*, *Radix ginseng*, *Rhizoma atractylodis lanceae*, *Poria*, and *Radix glycyrrhizae*	Gastric cancer (adults under gastrectomy, stage I–III)	Case reports ($n=13$)	ND	Not significant on cytotoxicity after treatment	[38]
			Advanced pancreatic cancer (adults without other remedies mentioned, stage IV)	Case reports ($n=30$)	Not significant on NK percentage ($CD57^+$ with no specific NK subsets identified) after treatment	ND	[39]
PHYTO-NK-9	CHM complex-3	Poria cocos, codonopsis root, stir-fried ovate Atractylodes rhizome, umbrellate pore fungus, eriobotruee folium, semen coicis, tangerine peel, *pinelliae preparatum rhizome*, fructus oryzae germinates, and frutus hordei germinates	Ovarian cancer (adults under chemotherapy, stage I–IV)	Double-blind placebo-controlled RCT with treatment ($n=31$); placebo ($n=28$)	Not significant on NK number (no mention on NK defining) comparing treatment and control groups	Not significant on cytotoxicity of the whole PBMC comparing treatment and control groups	[40]

(Continued)

TABLE 16.5 *(Continued)*

No.	Phytotherapy Drugs	Herb(s) Included	Type of Cancer (Age)	Type of Study (n)	NK Number/Percentage (Subsets, Methods): Results of BA/TC	NK Activity (Cytolysis/%, Methods, Target Cell Line): Results of BA/TC	Reference
PHYTO-NK-10	Shenqi mixture	Pseudo starwort root, milkvetch root, poria, white atractylodes tuber, rehmannia root, dendrobium, white peony root, Chinese angelica root, oldenlandia, Chinese lobelia, wild ginger and licorice root	Primary hepatocellular cancer (adults under microwave coagulation, stage II–III)	RCT with treatment ($n=36$); control ($n=36$)	TC and BA: NK percentage (no mention on NK subset), treatment group had 48.2% increase after treatment ($p<0.01$), 40.3% higher than control ($p<0.01$) by flow cytometry	ND	[41]

PHYTO-NK-11	CHM complex-4	Astragalus membranaceus, red ginseng, ophiopogonis radix, asiabell root, milkvetch root, indian bread, bighead atractylodes, angelica root, chuanxiong rhizome, prepared rehmannia root, white peony root, licorice, barbary wolfberry fruit, Chinese dates, and dodder seed	Advanced nonsmall cell lung cancer (adults under CT regimen, stage III–IV)	RCT with treatment ($n=50$); control ($n=50$)	TC: NK percentage (CD16$^+$ with no specific NK subsets identified) of treatment group was 55.2% higher ($p<0.05$) by high-power microscopy	ND	[42]

3H-TdR, 3H-thymidine riboside; BA, before treatment versus after treatment; ND, not done; PBMC(s), human peripheral blood mononuclear cells; RCT, randomized controlled trial; TC, treatment group versus control group.

cells with one study reporting a 60% increase [30] in contrast to the second finding a much smaller change of only 6% [31]. SFI is a CHM injection preparation only used in China composed of two Chinese medicinal herbs: *Radix astragali* (root of astragalus; Chinese name: huangqi) and *Radix codonopsis* (root of *Codonopsis pilosula*; Chinese name: dangshen). Based on the Chinese Medicine theory, this mixture is used for tonifying Qi, protecting haemogram, enhancing immune function, and improving Qi deficiency symptoms and QoL for cardiovascular and cancer patients along with Qi deficiency.

Three clinical trials have reported using SFI, of which two were RCTs using SFI in mammary cancer patients receiving chemotherapy, In one, combined SFI (250 ml/day for 2 weeks) markedly increased the absolute number of $CD16^{+}/56^{+}$ NK cells by 118±95/μl compared with the control group with chemotherapy alone, 75±89/μl ($p<0.05$) [30]. This greater than 50% increase in the absolute number of NK cells is in sharp contrast to that of the second study in which the relative % of NK cells increased only 6.1% after treatment [31]. Even when another more sensitive proliferation test of 3H-TdR assay was employed, there was only a 17.7% increase when compared with the control group.

The fundamental reason for this discrepancy lies in the fact that phytoceutical compounds are likely to upregulate other immunological cell populations such as T-cell subsets of CD4 or CD8 cells [43], or monocytes [44]. Thus, if expansion in other populations occur, the relative percentage increase of NK cells in overall PBMC may be relatively reduced because the NK population represents only a small component of about 5–10% of PBMC (T cells contribute 50–60% of PBMC). This emphasizes the importance of measuring the absolute number of each immune cell population and subset by flow cytometry to have an accurate understanding of the genuine effects or functions of immunomodulation medicines.

16.2.3.2 Effects of Phytotherapeutic Formulae on NK Cytotoxicity Function

The effects of 11 phytotherapeutic formulae on NK cell cytotoxicity are summarized in Table 16.5.

Six of the eleven formulae were tested for their effects on NK cell activity (cytotoxicity). Of these six formulae, five out of the seven published studies indicated a positive effect on the PBMC cytotoxicity (13.5–68.6% higher than control group).

Notably, only 3 of the 14 studies examined the changes in both NK cell number and NK cytotoxicity functionality: SFI in PHYTO-NK-2 (57.3% increase in absolute NK number compared with control: ($p<0.05$), 18.8% higher on PBMC cytotoxicity ($p<0.05$)); Hochu-ekki-toin PHYTO-NK-7 (56.4% increase in absolute NK number compared with control ($p<0.05$), 38.5% reduction in PBMC cytotoxicity ($p<0.05$)); CHM complex-3 in PHYTO-NK-9 (not significant on either NK number or PBMC cytotoxicity compared with control). Of the 14 studies, only one reported the absolute number of NK cells instead of the relative NK percentage that is likely to be affected by simultaneous effects of phytotherapeutics on other important immune populations in PBMC. Given that both proliferation and cytotoxicity of NK cells are important in final effects of phytotherapies on cancer management, they must be considered as a whole. Based on the limited data on the 11 phytotherapy formulae, although the 68.6% higher PBMC cytotoxicity produced by Aidi Injection compared with control group was the highest need, there were no data on its effect on NK proliferation, so Hochu-ekki-to remains the most promising therapy and appears superior to SFI.

16.2.3.3 Effects of Phytotherapeutic Formulae on Cancer Patients' QoL and Survival Period A total of 11 formulae examined the effects of other phytotherapeutics for cancer intervention on QoL and benefit to survival (Table 16.6).

The effects of the formulae on QoL will be discussed here. Seven formulae reported the effects on QoL. Of these seven, five formulae (which represented 5 of the 10 clinical trials) showed improvements in adverse events in the treatment group (mainly above 50%, range 29.7–100% times higher than the control group). Four of the seven formulae (6 of 10 clinical trials) indicated improvement of Karnofsky scores in the treatment group (three of six studies reported an elevation in Karnofsky scores: 1.5–1.6 times higher than the control group with three of six studies showing approximately 20% improvement (range of 5.9–25.5%) compared with baseline or control group). One formula (2/10 clinical trials) reported an increase in body weight in the SFI group, which was 86.6–110% times higher than in the control group [30, 45]. In two formulae involving Hochu-ekki-to and CHM complex-4 (2/10 clinical trials), other QoL scores were improved (5.6–23.2%, VAS-F, FACT-G, FACT-F, TOI-F, and FACT-L) [42, 48].

Five formulae reported effects on survival differences. Among these five formulae (five studies), two studies implied that SFI and Shenqi mixture markedly improved survival rates compared with control groups (SFI: 1.4 times higher over a 3-year period; Shenqi mixture: 24.8–40.0% higher over a 1–2-year period).

Overall only 8 out of the 11 phytotherapeutic formulae reported the effects on QoL or survival differences or both. These parameters were only assessed for four formulae: Shenyi capsule, SFI, Aidi Injection, and Shenqi mixture. However, only SFI and Shenqi mixture reported improvements in both QoL and survival benefit. As the improvements of SFI on Karnofsky scores, body weight, and survival rate were much better than Shenqi mixture, SFI would appear to be the most superior of these 11 phytotherapeutic formulae.

Considering the effects of all 11 phytotherapeutic formulae on cancer patients' QoL and survival benefit, only two mixtures appeared to have clinically desirable effects. Therefore, detailed information on these is provided next.

SFI: In the RCT of 110 mammary cancer patients receiving chemotherapy, combined with SFI (250 ml/day for 2 weeks) marked increased Karnofsky QoL scores compared with the control group administered chemotherapy alone (increase: 34.5 vs. 13.5%, $p<0.05$; stable: 46.6 vs. 38.5%, $p<0.05$; decrease: 19.0 vs. 48.1%, $p<0.05$, respectively) were observed [30]. In another clinical study with 126 advanced breast cancer patients receiving chemotherapy, combination with SFI (250 ml/day for 56 days) was related to less severe side effects and shorter recovery periods compared with the control group receiving chemotherapy alone (neutropenia: 55.4% (17/36) vs. 77.1% (47/61), $p<0.01$; anemia: 44.6% (29/65) vs. 62.3% (38/61),

TABLE 16.6 Effects of Phytotherapeutic Formulae on QoL and Survival Rate in Clinical Cancer Treatment

No.	Phytotherapy Drugs	Type of Cancer (Age)	Type of Study (n)	QoL (Indicators): Results of BA/TC	Survival benefit: results of BA/TC	Reference
PHYTO-Q/S-1	Shenyi capsule	Nonsmall cell lung cancer (adults, stage II–III)	RCT with Shenyi alone ($n=43$); Shenyi and chemo ($n=46$); chemo alone ($n=44$)	TC: adverse events in Shenyi only group were 94.7% less than in combined group, 94.6% less than chemo alone ($p<0.05$, nausea and vomiting, leucopenia)	Not significant among three groups	[29]
PHYTO-Q/S-2	Shenqi Fuzheng Injection	Mammary cancer (adults under chemotherapy, no stage mentioned)	RCT with treatment ($n=58$); control ($n=52$)	TC: the information about Karnofsky scores before treatment was not mentioned at all. Karnofsky score in treatment group was 1.6 times higher ($p<0.05$, 34.5% (20/58) vs. 13.5% (7/52)). Karnofsky scores in treatment group were 60.5% lower than that of control ($p<0.05$,19.0% (11/58) vs. 48.1% (25/52)). According to the total condition of rise and decline after treatment, average increase for each patient was about 1.6 scores; elevated rate of body weight in treatment group was 86.6% higher ($p<0.05$)	ND	[30]

		Advanced breast cancer (adults under chemotherapy, stage II–III)	RCT with treatment ($n=65$); control ($n=61$)	TC: Adverse events of treatment group were 29.7% less ($p<0.05$, neutropenia, anemia, and gastrointestinal reactions)	ND	[31]
		Nonsmall cell lung cancer (adults under chemotherapy, main stage III–IV)	Case–control study with treatment ($n=60$); control ($n=60$)	BA: Karnofsky scores of all patients screened were ≥60, but without detailed information. Karnofsky scores in treatment group showed 25.5% increase after treatment ($p<0.01$, 86.2 ± 9.3 vs. 68.7 ± 8.6). According to the data in treatment group before and after treatment, average increase for each patient was about 17.5 scores; TC: elevation rate of body weight in treatment group was 1.1 times higher ($p<0.01$)	TC: survival rate over 3 years of treatment group was 1.4 times higher ($p<0.05$, 45.7% vs. 19.4%)	[45]
PHYTO-Q/S-3	CHM complex-1	ND	ND	ND	ND	—
PHYTO-Q/S-4	CHM complex-2	ND	ND	ND	ND	—
PHYTO-Q/S-5	Aidi Injection	Advanced colorectal cancer (adults under chemotherapy, main stage I–III)	Case–control study with treatment ($n=50$); control ($n=50$)	TC: Karnofsky scores of all patients screened were ≥70, but without information detailed. Elevated rate of Karnofsky scores in treatment group was 1.5 times higher ($p<0.05$, 60% (30/50) vs. 24% (12/50)). Declining rate of Karnofsky scores in treatment group were 61.4% lower than that of control ($p<0.05$, 14% (7/50) vs. 22% (11/50)). According to the total condition of rise and decline after treatment, average increase for each patient was about 4.6 scores	ND	[46]

(Continued)

TABLE 16.6 *(Continued)*

No.	Phytotherapy Drugs	Type of Cancer (Age)	Type of Study (n)	QoL (Indicators): Results of BA/TC	Survival benefit: results of BA/TC	Reference
		Advanced nonsmall-cell lung cancer (adults under chemotherapy, stage III–IV)	RCT with treatment ($n=53$); control ($n=51$)	TC: mean Karnofsky scores in treatment group before treatment was about 77.5. Karnofsky score of treatment group was 1.5 times higher ($p<0.01$, 58.5% (31/53) vs. 23.5% (12/51)). Karnofsky scores in treatment group were 58.0% lower than that of control ($p<0.01$, 13.2% (7/53) vs. 31.4% (16/51)). According to the total condition of rise and decline after treatment, mean scores in treatment group were about 82, as average increase for each patient was about 4.5 scores; adverse events in treatment group were 56.5% lower ($p<0.05$, leucopenia, anemia, nausea, etc.)	Not significant comparing treatment and control groups	[47]
PHYTO-Q/S-6	Danggui Buxue Decoction NO.1	ND	ND	ND	ND	—
PHYTO-Q/S-7	Hochu-ekki-to	Breast, stomach, colorectal, lung and other cancer (all adults, no stages mentioned)	RCT with treatment ($n=20$); control ($n=20$)	BA: VAS-F of treatment group showed 18.5% decrease ($p<0.05$); FACT-G: 5.6% decrease ($p<0.05$); FACT-F: 7.8% decrease ($p<0.05$); TOI-F: 10.6% decrease ($p<0.05$)	ND	[48]
PHYTO-Q/S-8	Juzen-Taiho-to	Advanced pancreatic cancer (adults under chemotherapy, stage III)	Case–control study with treatment ($n=10$); control ($n=10$)	ND	Not significant comparing treatment and control groups	[39]

PHYTO-Q/S-9	CHM complex-3	Ovarian cancer (adults under chemotherapy, stage I–IV)	Double-blind placebo-controlled RCT with treatment ($n=31$); placebo ($n=28$)	Not significant comparing treatment and control groups on EORTC LQL-C30, GHS scores and adverse events	ND	[40]
PHYTO-Q/S-10	Shenqi mixture	Primary hepatocellular cancer (adults under microwave coagulation, stage II–III)	RCT with treatment ($n=36$), control ($n=36$)	TC and BA: Karnofsky scores of all patients screened were ≥50, but without detailed information. Karnofsky scores in treatment group showed 22.5% increase after treatment ($p<0.01$, 64.5 ± 11.5 vs. 79.0 ± 13.8), 18.4% higher than control ($p<0.01$, 79.0 ± 13.8 vs. 66.7 ± 15.0). According to the data of treatment group before and after treatment, average increase for each patient was about 14.5 scores; adverse events reduction was about 2 times as effective as control ($p<0.01$, pain in hepatic region, fever, fatigue, etc.)	TC: survival rate of treatment group in 12 months was 24.8% higher ($p<0.05$); at 18 months was 40.0% higher ($p<0.05$); in 24 months was 33.3% higher ($p<0.05$); recurrent rate in treatment group over 12 months was 50.0% lower ($p<0.05$); over 18 months was 52.9% lower ($p<0.05$); over 24 months was 40.0% lower ($p<0.05$)	[41]

(*Continued*)

TABLE 16.6 (*Continued*)

No.	Phytotherapy Drugs	Type of Cancer (Age)	Type of Study (n)	QoL (Indicators): Results of BA/TC	Survival benefit: results of BA/TC	Reference
PHYTO-Q/S-11	CHM complex-4	Advanced nonsmall cell lung cancer (adults under CT regimen, stage III–IV)	RCT with treatment ($n=50$); control ($n=50$)	TC: Karnofsky scores of all patients screened were ≥60, but without detailed information. Karnofsky scores in treatment group were 5.9% higher ($p<0.05$, 86.0 ± 10.62 vs. 81.2 ± 9.5). However, the difference before and after treatment in treatment group was not mentioned; FACT-L was 23.2% higher ($p<0.01$, total score, body, emotion, and function); adverse reaction was 70.9% less ($p<0.01$)	ND	[42]

BA, before treatment versus after treatment; EORTC QLQ-C30, European Organization for Research and Treatment of Cancer Core Quality of Life Questionnaire C30; FACT-F, functional assessment of cancer therapy–fatigue; FACT-G, functional assessment of cancer therapy–general; FACT-L, evaluation of cancer treatment function; GHS, global health status; ND, not done; PBMC(s), human peripheral blood mononuclear cells; QoL, quality of life; RCT, randomized control trial; TC, treatment group versus control group; TOI-F, trial outcome index–fatigue; VAS-F, visual analogue scale of global fatigue.

$p<0.05$; gastrointestinal reactions: 47.7% (31/65) vs. 70.5% (43/61), $p<0.01$, respectively) [31].

In the control study of PHYTO-Q/S-3 with nonsmall cell lung cancer patients (9 in stage II, 32 in stage III, and 19 in stage IV with average age of 69.5 years), the combination therapy (250 ml/day once a day for 63 days) resulted in improved QoL (compared with previous condition, Karnofsky scores: 86.15±9.33 vs. 68.66±8.56, $p<0.01$; compared with control group, body weight: $\chi^2=4.97$, $p<0.01$), and higher survival rate compared with control group (10 in stage II, 34 in stage III, and 16 in stage IV with average age of 68.0 years) of chemotherapy alone (survival rate in 3 years: 45.7% (16/35) vs. 19.4% (7/36), $\chi^2=4.46$, $p<0.05$) [45].

Shenqi Mixture: For primary hepatoma patients in a RCT of T6-10, 20 ml SQM (three times a day for 6 months) combined with microwave coagulation resulted in significant increase in Karnofsky QoL score compared with the control group receiving microwave therapy alone (79.0±13.8 vs. 66.7±15.0, $p<0.01$, respectively). It also markedly improved major clinical complications compared with the control group such as pain in the hepatic region (80.76 vs. 36.36%, $p<0.01$), fever (87.50 vs. 33.33%, $p<0.01$), fatigue (83.33 vs. 53.85%, $p<0.01$), abdominal distention (45.83 vs. 43.75%, $p<0.01$), poor appetite (86.67 vs. 40.00%, $p<0.01$), and jaundice (83.33 vs. 45.16%, $p<0.01$) [41]. Combination therapy resulted in obvious survival benefit and a reduced recurrence rate compared with the control group (survival rate: 12 months: 83.33% (30/36) vs. 66.77% (24/36), $p<0.05$; 18 months: 77.78% (28/36) vs. 55.56% (20/36), $p<0.05$; 24 months: 55.56% (20/36) vs. 41.67% (15/36), $p<0.05$. Recurrence rate: 12 months: 13.89% (5/36) vs. 27.78% (10/36), $p<0.05$; 18 months: 22.22% (8/36) vs. 47.22% (17/36), $p<0.05$; 24 months: 41.67% (15/36) vs. 69.44% (25/36), $p<0.05$).

In summary, considering that there have been more clinical trials on SFI, this represents an important phytotherapeutic that warrants further extensive clinical trials in the future with a larger sample size and detailed measurements of patients' QoL and overall survival.

16.3 CONCLUSIONS

According to the current state of knowledge from clinical studies on phytotherapies, several key issues can be summarized:

1. The current knowledge on the effects of phytotherapeutics on NK cells remains very limited. Most studies suffer from very unclear or vague definitions of NK subsets and in some cases no details of the methods used were described. This makes it difficult to determine whether phytotherapies are really effective specifically on NK cells or on lymphocytes in general such as T/B cells as similar markers of the relevant receptors. If the phytotherapy only affects some specific subsets, it may be ignored if the test was based on total NK cells. Determining the effects on specific NK subsets may make the mechanism of

action of phytotherapy on NK cells clearer. On the other hand, more RCT studies on NK with herbal drugs are needed.

2. The total sample size of patients used to study the effects of phytotherapies on NK were low (only 103 for lentinan, 170 for Ganopoly, 535 for all other herbal formulae). Based on the available data, further RCT with larger sample sizes, which should also be double-blinded and multicentered, should be performed.
3. Studies of effects of the phytotherapeutics on survival benefit were very limited. None of the RCTs involving Ganoderma included survival benefit as an indicator of its activity. In addition, 6 of the 11 phytotherapeutic formulae studies did not report the effects on survival.
4. The reported effects of Lentinula and Ganoderma on NK number were very weak, in contrast to obvious effects on NK number noted in 7 of the 10 clinical studies involving six of the nine other phytotherapeutic agents, which showed increases of around 50% compared with baseline or control groups. The therapies having the most potent effects could not be determined but the following appear to be the most promising: SFI, Hochu-ekki-to, CHM complex-4, Shenyi capsule, and Shenqi mixture.
5. As to the activities of Lentinula on NK cytotoxicity (16.7–66.5% increase compared with baseline) was better than Ganoderma, and four of the six clinical trials involving five of the phytotherapy products indicated positive effect on PBMC cytotoxicity (13.5–68.6% higher than control group). Therefore, the five recommended phytotherapy strategies for NK cytotoxicity improvement may be listed as Aidi Injection, Hochu-ekki-to, Danggui Buxue Decoction NO.1, Lentinula, and SFI.
6. If effects on both NK numbers and toxicity are taken into account, the recommended phytotherapy strategies appear to be Hochu-ekki-to, SFI, and Aidi Injection.
7. This review has shown that only Lentinula, SFI, and Shenqi exhibit activities on both QoL and survival benefit. Therefore, mainly based on Karnofsky scores and survival difference, the recommendations for QoL and survival management were SFI, Lentinula, and Shenqi mixture.
8. There is lack of a universally accepted QoL questionnaire that is reflected by the observation that various surveys have been used in the different reports. The development of such a QoL questionnaire should be one of the priorities in the research of phytotherapeutics in cancer management.
9. Based on the current results from all the parameters investigated and reported in the literature, SFI may provide the best outcomes in cancer management and it warrants a more thorough and robust multicentre clinical trial investigation.

This chapter highlights the importance of careful documentation of the clinical effects for both overall survival and quality of life, in relation to potential mechanisms of action for further evidence-based use of selected cancer phytotherapeutics. In the future, more detailed reviews on different mechanism(s) of action are needed. It is known that complex conditions such as cancer will benefit by use of "multicompound multitarget" approaches. Recognition of cancer phytotherapies that target more than one anticancer mechanism of action and with proven effects on QoL may ultimately lead to improved management of cancer.

REFERENCES

[1] Jamal-Hanjani M, Thanopoulou E, Peggs KS, Quezada SA, Swanton, C (2013) Tumour heterogeneity and immune-modulation. *Curr Opin Pharmacol.* 13: 497–503.

[2] Bottino C, Moretta L, Pende D, Vitale M, Moretta A (2004) Learning how to discriminate between friends and enemies, a lesson from Natural Killer cells. *Mol Immunol.* 41: 569–575.

[3] Moretta L, Bottino C, Pende D, Vitale M, Mingari C, et al. (2004) Different checkpoints in human NK-cell activation. *Trends Immunol.* 25: 670–676.

[4] Kalinski P, Giermasz A, Nakamura Y, Basse P, Storkus WJ, et al. (2005) Helper role of NK cells during the induction of anticancer responses by dendritic cells. *Mol Immunol.* 42: 535–539.

[5] Kalinski P, Nakamura Y, Watchmaker P, Giermasz A, Muthuswamy R, et al. (2006) Helper roles of NK and CD8+ T cells in the induction of tumour immunity. Polarized dendritic cells as cancer vaccines. *Immunol Res.* 36: 137–146.

[6] Shah SK, Walker PA, Moore-Olufemi SD, Sundaresan A, Kulkarni AD, et al. (2011) An evidence-based review of a *Lentinula edodes* mushroom extract as complementary therapy in the surgical oncology patient. *JPEN J Parenter Enteral Nutr.* 35: 449–458.

[7] Ina K, Kataoka T, Ando T (2013) The use of lentinan for treating gastric cancer. *Anticancer Agents Med Chem.* 13: 681–688.

[8] Yuen JW, Gohel MD (2005) Anticancer effects of *Ganoderma lucidum*: a review of scientific evidence. *Nutr Cancer.* 53: 11–17.

[9] Miyakoshi H, Aoki T, Mizukoshi M (1984) Acting mechanisms of Lentinan in human—II. Enhancement of non-specific cell-mediated cytotoxicity as an interferon inducer. *Int J Immunopharmacol.* 6: 373–379.

[10] Oka M, Yoshino S, Hazama S, Shimoda K, Suzuki T (1992) Immunological analysis and clinical effects of intraabdominal and intrapleural injection of lentinan for malignant ascites and pleural effusion. *Biotherapy.* 5: 107–112.

[11] Arinaga S, Karimine N, Takamuku K, Nanbara S, Inoue H, et al. (1992) Enhanced induction of lymphokine-activated killer activity after lentinan administration in patients with gastric carcinoma. *Int J Immunopharmacol.* 14: 535–539.

[12] Fujimoto T, Omote K, Mai M, Natsuume-Sakai S (1992) Evaluation of basic procedures for adoptive immunotherapy for gastric cancer. *Biotherapy.* 5: 153–163.

[13] Wang W, Dai X, Ouyang X (2006) Efficacy of Lentinan combined with chemotherapy in advanced non-small cell lung cancer (in Chinese). *Zhongguo Fei Ai Za Zhi* 9: 78–81.

[14] Yamaguchi Y, Miyahara E, Hihara J (2011) Efficacy and safety of orally administered *Lentinula edodes* mycelia extract for patients undergoing cancer chemotherapy: a pilot study. *Am J Chin Med.* 39: 451–459.

[15] Nagashima Y, Maeda N, Yamamoto S, Yoshino S, Oka M (2013) Evaluation of host quality of life and immune function in breast cancer patients treated with combination of adjuvant chemotherapy and oral administration of *Lentinula edodes* mycelia extract. *Onco Targets Ther.* 6: 853–859.

[16] Yang P, Liang M, Zhang Y, Shen B (2008) Clinical application of a combination therapy of lentinan, multi-electrode RFA and TACE in HCC. *Adv Ther.* 25: 787–794.

[17] Isoda N, Eguchi Y, Nukaya H, Hosho K, Suga K, et al. (2009) Clinical efficacy of superfine dispersed lentinan (beta-1,3-glucan) in patients with hepatocellular carcinoma. *Hepatogastroenterology.* 56: 437–441.

[18] Shimizu K, Watanabe S, Watanabe S, Matsuda K, Suga T, et al. (2009) Efficacy of oral administered superfine dispersed lentinan for advanced pancreatic cancer. *Hepatogastroenterology.* 56: 240–244.

[19] Ina K, Furuta R, Kataoka T, Kayukawa S, Yoshida T, et al. (2011) Lentinan prolonged survival in patients with gastric cancer receiving S-1-based chemotherapy. *World J Clin Oncol.* 2: 339–343.

[20] Okuno K, Uno K (2011) Efficacy of orally administered *Lentinula edodes* mycelia extract for advanced gastrointestinal cancer patients undergoing cancer chemotherapy: a pilot study. *Asian Pac J Cancer Prev.* 12: 1671–1674.

[21] Higashi D, Seki K, Ishibashi Y, Egawa Y, Koga M, et al. (2012) The effect of lentinan combination therapy for unresectable advanced gastric cancer. *Anticancer Res.* 32: 2365–2368.

[22] Bao PP, Lu W, Cui Y, Zheng Y, Gu K, et al. (2012) Ginseng and *Ganoderma lucidum* use after breast cancer diagnosis and quality of life: a report from the Shanghai Breast Cancer Survival Study. *PLoS One.* 7: e39343.

[23] Gao XH, Dai XH, Chen GL, Ye JX, Zhou SF (2003) A randomized, placebo-controlled, multicenter study of *Ganoderma lucidum* (W. Curt. Fr.) LLoyd (Aphyllophoromycetideae) polysaccharides (Ganopoly(R)) in patients with advanced lung cancer. *Int J Med Mushrooms.* 5: 369–381.

[24] Gao Y, Zhou S, Jiang W, Huang M, Dai X (2003) Effects of ganopoly (a *Ganoderma lucidum* polysaccharide extract) on the immune functions in advanced-stage cancer patients. *Immunol Invest.* 32: 201–215.

[25] Gao Y, Tang W, Dai W, Gao H, Chen G, et al. (2005) Effects of water-soluble *Ganoderma lucidum* polysaccharides on the immune functions of patients with advanced lung cancer. *J Med Food.* 8: 159–168.

[26] Huang M, Gao YH, Tang WB, Dai XH, Gao H, et al. (2005) Immune responses to water-soluble Ling Zhi Mushroom *Ganoderma lucidum* (W. Curt. Fr.) P. Karst. Polysaccharides in patients with advanced colorectal cancer. *Int J Med Mushrooms.* 7: 525–538.

[27] Chen X, Hu ZP, Yang XX, Huang M, Gao Y (2006) Monitoring of immune responses to a herbal immuno-modulator in patients with advanced colorectal cancer. *Int Immunopharmacol.* 6: 499–508.

[28] Zhao H, Zhang Q, Zhao L, Huang X, Wang J, et al. (2012) Spore powder of *Ganoderma lucidum* improves cancer-related fatigue in breast cancer patients undergoing endocrine therapy: a pilot clinical trial. *Evid Based Complement Alternat Med.* 2012: 809614.

[29] Lu P, Su W, Miao ZH, Niu HR, Liu J, et al. (2008) Effect and mechanism of ginsenoside Rg3 on postoperative life span of patients with non-small cell lung cancer. *Chin J Integr Med.* 14: 33–36.

[30] Bo Y, Li HS, Qi YC, Lu MY (2007) Clinical study on treatment of mammary cancer by shenqi fuzheng injection in cooperation with chemotherapy. *Chin J Integr Med.* 13: 37–40.

[31] Dai Z, Wan X, Kang H, Ji Z, Liu L, et al. (2008) Clinical effects of shenqi fuzheng injection in the neoadjuvant chemotherapy for local advanced breast cancer and the effects on T-lymphocyte subsets. *J Tradit Chin Med.* 28: 34–38.

[32] Zhao Q, Li Y, Yang JQ, Chen SX, Jiao ZK, et al.(2004) Effects of Chinese medicine strengthening the spleen and replenishing qi on spleen deficiency syndrome and biological behavior of patients with gastric carcinoma. *Chin J Clin Rehabil.* 8: 8148–8151.

[33] Zhuang SR, Chiu HF, Chen SL, Tsai JH, Lee MY, et al. (2012) Effects of a Chinese medical herbs complex on cellular immunity and toxicity-related conditions of breast cancer patients. *Br J Nutr.* 107: 712–718.

[34] Zhuang SR, Chen SL, Tsai JH, Huang CC, Wu TC, et al. (2009) Effect of citronellol and the Chinese medical herb complex on cellular immunity of cancer patients receiving chemotherapy/radiotherapy. *Phytother Res.* 23: 785–790.

[35] Sun XF, Pei YT, Yin QW, Wu MS, Yang GT (2010) Application of Aidi injection in the bronchial artery infused neo-adjuvant chemotherapy for stage IIIA non-small cell lung cancer before surgical operation. *Chin J Integr Med.* 16: 537–541.

[36] Du QC, Yang KZ, Sun XF (2009) Efficacy of auxiliary therapy with Danggui Buxue Decoction No.1 in treating patients of non-small cell lung cancer at peri-operational stage. *Chin J Integr Med.* 15: 184–188.

[37] Kimura M, Sasada T, Kanai M, Kawai Y, Yoshida Y, et al. (2008) Preventive effect of a traditional herbal medicine, Hochu-ekki-to, on immunosuppression induced by surgical stress. *Surg Today.* 38: 316–322.

[38] Horie Y, Kato K, Kameoka S, Hamano K (1994) Bu ji (hozai) for treatment of postoperative gastric cancer patients. *Am J Chin Med.* 22: 309–319.

[39] Ikemoto T, Shimada M, Iwahashi S, Saito Y, Kanamoto M, et al. (2013) Changes of immunological parameters with administration of Japanese Kampo medicine (Juzen-Taihoto/TJ-48) in patients with advanced pancreatic cancer. *Int J Clin Oncol.* 19: 81–86.

[40] Chan KK, Yao TJ, Jones B, Zhao JF, Ma FK, et al. (2011) The use of Chinese herbal medicine to improve quality of life in women undergoing chemotherapy for ovarian cancer: a double-blind placebo-controlled randomized trial with immunological monitoring. *Ann Oncol.* 22: 2241–2249.

[41] Lin JJ, Jin CN, Zheng ML, Ouyang XN, Zeng JX, et al. (2005) Clinical study on treatment of primary hepatocellular carcinoma by Shenqi mixture combined with microwave coagulation. *Chin J Integr Med.* 11: 104–110.

[42] Yang ZY, Wu XM, Ou YL, Yu P, Luo J, et al. (2004) Effect of TCM combined with chemotherapy on immune function and quality of life of patients with non-small cell lung cancer in stage III–IV. *Chin J Integr Med.* 10: 181–186.

[43] Sze DM, Giesajtis G, Brown RD, Raitakari M, Gibson J, et al. (2001) Clonal cytotoxic T cells are expanded in myeloma and reside in the CD8(+)CD57(+)CD28(−) compartment. *Blood.* 98: 2817–2827.

[44] Sekhon BK, Sze DM, Chan WK, Fan K, Li GL, et al. (2013) PSP activates monocytes in resting human peripheral blood mononuclear cells: immunomodulatory implications for cancer treatment. *Food Chem.* 138: 2201–2209.

[45] Liu CL, Chen WP, Cui SZ, Deng G, Liu L, et al. (2004) Effect of shenqi fuzheng injection for assistance of chemotherapy in treating senile patients with non-small cell lung cancer. *Zhongguo Zhong Xi Yi Jie He Za Zhi.* 24: 901–903.

[46] Xu HX, Huang XE, Li Y, Li CG, Tang JH (2011) A clinical study on safety and efficacy of Aidi injection combined with chemotherapy. *Asian Pac J Cancer Prev.* 12: 2233–2236.

[47] Li H, Dong L, Li Y, Fu S (2008) A randomized clinical trial of combination of Aidi injection with Gemcitabine and Oxaliplatin regimen or Go regimen only in the treatment of advanced non-small-cell lung cancer. *Zhongguo Fei Ai Za Zhi.* 11: 570–573.

[48] Jeong JS, Ryu BH, Kim JS, Park JW, Choi WC, et al. (2010) Bojungikki-tang for cancer-related fatigue: a pilot randomized clinical trial. *Integr Cancer Ther.* 9: 331–338.

17

PHYTOMEDICINES FOR FATTY LIVER DISEASE AND FUNCTIONAL GASTROINTESTINAL CONDITIONS

George Q. Li[1], Moon-Sun Kim[1], Fangming Jin[2,3], and Jun-Lae Cho[1]

[1] *Faculty of Pharmacy, The University of Sydney, Sydney, New South Wales, Australia*
[2] *School of Pharmacy, Shaanxi University of Chinese Medicine, Xi'an, Shaanxi, China*
[3] *Global Therapeutics Pty Ltd, Byron Bay, New South Wales, Australia*

17.1 INTRODUCTION

Fatty liver disease (FLD), irritable bowel syndrome (IBS), and constipation are common gastrointestinal (GI) conditions worldwide. For instance, the prevalence of FLD ranged from 10 to 20% in the general population with or without alcohol consumption all over the world [1]. Complementary and alternative medicine (CAM) and phytotherapies have been used in many countries in the management of these conditions, originating from traditional Chinese medicine (TCM), Ayurvedic medicine, Japanese medicine, and Western herbal medicine [2].

There is a rich body of literature on clinical management of these conditions in TCM. Among 18 Chinese herbal categories, 9 of them are related to liver and GI conditions [3, 4]. They are heat-clearing herbs (antipyretics); laxative herbs; diuretics and dampness-removing herbs; aromatic herbs for resolving dampness; internal warming herbs; herbs for regulating Qi (carminatives); herbs for removing food stagnation; tonics; and astringents. Many of the herbs covered in this chapter fall within these categories. Fatty liver is characterized in TCM as deficiency of three TCM organs, liver, spleen, and kidney, resulting in phlegm and dampness retaining, Qi stagnation, and

Phytotherapies: Efficacy, Safety, and Regulation, First Edition. Edited by Iqbal Ramzan.

blood stasis. The treatment principles are to tonify Qi, strengthen liver and spleen, clear heat in kidney, discharge phlegm, and invigorate the blood. Therefore, formulae are prescribed to treat the symptoms [5], and the ingredients will include a number of categories of TCM herbs with the functions mentioned above.

The objectives of modern research on herbal medicines are to understand the active ingredients, pharmacological actions, safety, and clinical efficacy. The studies are also part of drug discovery and development in pharmaceutical sciences. Therefore, we have searched databases mainly PubMed and textbooks to identify the most popular herbal medicines. The preclinical and clinical data of the selected herbs are summarized to provide an evidence basis on their clinical usage.

17.2 PHYTOMEDICINES FOR FLD

17.2.1 Introduction and Pharmacotherapy

The liver is not only the largest organ that manages toxicity and decomposes food nutrients but also plays a pivotal role in the regulation of glucose and lipid energy homeostasis. Long-term chronic liver diseases are initially manifested with increased fatty depots in the liver, known as fatty liver. Many researchers define this condition wherein there is excess fat accumulation in the form of lipid droplet (LD) in the cytoplasm of hepatocytes that is associated with an enlargement of the liver via the process of hepatic steatosis (HS) [6]. HS is a set of benign condition but may lead to severe secondary condition with inflammation and lipotoxicity, named steatohepatitis (SH) that may progress to fibrosis.

FLD is clinically categorized into two broad entities: alcoholic fatty liver disease (AFLD) and nonalcoholic fatty liver disease (NAFLD). The initially proposed "two-hit" model by James and Day provides a pathophysiologic rationale for the progression to SH [7]. The first hit of FLD is the process of HS, the excess accumulation of triglyceride (TG) [2] by esterification of overload free fatty acids (FFAs), in the liver [8]. Uptake of large quantity of FFAs by liver is caused from many factors. Adipocyte malfunction to insulin disrupts the regulation of the lipase in the adipose tissue and then excess FFAs are released [2, 9, 10]. Excess production of exogenous and endogenous glucose in obesity and diabetes, together with hyperinsulinemia, increases the synthesis of FFAs in the liver [11]. Lipotoxicity of FFAs has been suggested to be a key factor to progress SH as a second hit through the oxidative stress and release of proinflammatory cytokines such as TNF-α, IL-6, and IL-8 [12]. In other proposals, HS and its progression to SH may result from improper fatty acid (FA) oxidation, as observed in patients carrying variant alleles of mitochondrial acyl-CoA dehydrogenases and on the basis of studies with transgenic mice with deficiencies in mitochondrial FA oxidation [13–15]. Hepatic inflammation and apoptosis are important in the pathogenesis of end-stage liver disease by leading the activation of hepatic stellate cells, which play a pivotal role in hepatic fibrosis [16]. This aggressive fibrogenesis as well as sustained hepatocellular proliferation contributes to the development of liver cirrhosis. The end stage is progression toward the development of hepatocellular carcinoma.

Hepatic regulation of lipid homeostasis is influenced by a complex system of hormones. Initially, dysfunction of FA oxidation-related signaling often leads to HS followed by other rapidly progressive and severe complications [17]. Peroxisome proliferator-activated receptors (PPAR) [18] that belong to a superfamily of nuclear receptor play a significant role in the transcriptional regulation of genes responsible for the management of lipid utilization and storage [19]. PPARα is responsible for lipid catabolism such as FA uptake, FA binding, and FA oxidation to control lipid and glucose metabolism in the liver and muscle [20]. Deficiency of this factor displays severe HS and failure to upregulate the FA β-oxidation system [21, 22]. PPARα also has beneficial effect to improve SH through decreasing of proinflammatory factors [23–25]. AMP-activated protein kinase (AMPK), which is a fuel-sensing enzyme activated by physiological and pathological stresses that deplete cellular ATP, also gives rise to beneficial effect of lipid homeostasis in the liver. It has been reported that activation of AMPK contributed the improvement of NAFLD in high-sucrose diet-induced hepatic SH mice [26]. The AMPK pathway, cross-talking with two enzymes, acetyl-CoA carboxylase (ACC) [27] and carnitine palmitoyltransferase 1 (CPT-1), is important in the improvement of HS and SH through the activation of FA oxidation and inhibition of FA synthesis [28–30]. Fatty liver with insulin resistance (IR) is controlled by sterol regulatory element-binding protein-1c (SREBP-1c), which is increased in response to high insulin levels. Thus, SREBP-1c may play a crucial role in the regulation of hepatic glucose production and TG (Fig. 17.1).

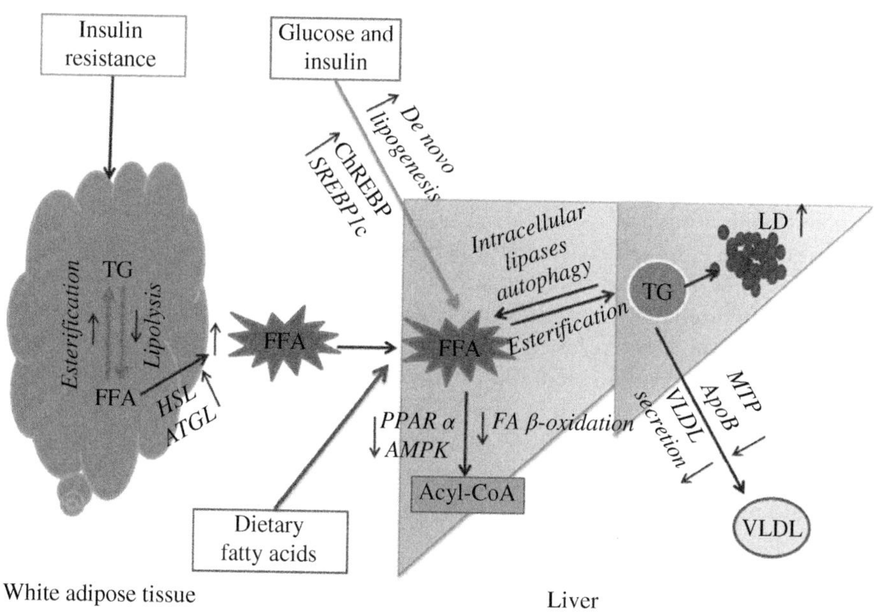

FIGURE 17.1 Manifestation of fatty liver disease (adapted from [31, 32]). AMPK, AMP-activated protein kinase; ApoB, apolipoprotein B; ATGL, adipose triglyceride lipase; ChREBP, carbohydrate-responsive element-binding protein; FFA, free fatty acid; HSL, hormone-sensitive lipase; LD, lipid droplet; MTP, microsomal triglyceride transfer protein; PPAR**α**, peroxisome proliferator-activated receptor alpha; SREBP-1c, sterol regulatory element-binding protein-1c; TG, triglyceride; VLDL, very-low-density lipoprotein.

Pharmacological intervention for FLD also has been carried out, and now many targets are available for therapies that may rescue the liver from lipotoxicity by restoring adipose tissue insulin sensitivity and reduction of excess hepatic TG accumulation or reverse intracellular inflammation and oxidative stress to determine their long-term safety and efficacy.

Metformin is the molecule being tested currently, as it can reduce the steatosis [33]. Metformin, which is used to treat type 2 diabetes (T2D) in overweight people, reduces blood glucose levels and IR, improves HS, and decreases TNF-α levels, an inflammatory cytokine highly expressed in NAFLD patients [34–36]. Experimental evidences showed that a reduction of intrahepatic lipid content with metformin treatment reduced the activities of ACC and lipogenic transcription factor SREBP-1c in steatotic models [37]. It was also reported that metformin was shown to inhibit hepatic gluconeogenesis, such as cAMP response element-binding protein (CREB) activation [38] and liver kinase B1 (LKB1)/AMPK signal [39] involved in the suppression of genes encoding gluconeogenic and lipogenic liver enzymes, by modulating the AMPK-dependent regulation [40]. However, there was only a modest improvement in HS and inflammation. In a 48-week open-label study in 26 patients, metformin improved NASH activity in only 30% of patients, although interpretation of study was confounded by a significant weight loss in the responders (19% lost > 10 kg) [41].

Thiazolidinediones (TZDs) such as pioglitazone and rosiglitazone are potent PPARγ activators, which increase insulin sensitivity, reduce hepatic lipid content, and lower resistance in plasma levels in patients with T2D [42, 43]. Other beneficial effects of TZDs include suppression of inflammation and cell proliferation and their proapoptotic and antifibrotic effects. Rosiglitazone improved NAFLD by decreasing TNF-α in the Otsuka-Long Evans-Tokushima fatty (OLETF) rat [44], further highlighting the potential of TZDs to reduce hepatic inflammation. However, there has been considerable debate about the long-term safety of TZDs regarding congestive heart failure, cardiovascular disease (CVD), and bladder cancer. It was reported that pioglitazone or rosiglitazone for the therapeutic treatment of NASH after 48 weeks resulted in increased liver fat, inflammatory activity, and aggravation of histopathological scores. Therefore, TZD treatment should be critically evaluated [45, 46]. Beneficial effects of TZDs may also be limited due to treatment-associated weight gain [46, 47]. In a randomized placebo-controlled study, 50% of patients responded to rosiglitazone therapy with improved steatosis and transaminase levels, despite an additional gain of 0.5 kg body weight compared with the placebo group [47].

Fibrates are ligands of PPARα, affecting FA oxidation and lowering serum TG levels [48]. Studies in animal models of HS and SH have demonstrated that fibrates ameliorated IR, stimulated FA oxidation, and suppressed inflammation [49–53]. The potent effects of fibrates on macrovesicular steatosis were reported in a study, in which pretreatment of 11 obese living liver transplant donors with bezafibrate (400 mg/day) significantly improved HS and was able to normalize liver function and lipid metabolism [54, 55]. Dietary supplementation with unsaturated FAs including omega-n-3 fatty acid [56] is thought to improve NAFLD by activating PPARα and normalizing peripheral IR but also preventing lipid peroxidation and suppressing the lipogenic transcription factor SREBP-1 [57–60]. Recent clinical trials from 16 consecutive patients with NAFLD treated with 200 mg/day of fenofibrate, noted significant decreases in TG, alkaline phosphatase, and gamma-glutamyl transpeptidase was observed.

17.2.2 Treatment of Fatty Liver with Herbal Medicines

The research literature is now growing and providing the improved knowledge on the population of complementary medicines through the possible mechanism of pathogenesis of FLD, although more clinical research evidence is required. Many herbal medicines from TCMs, Ayurvedic medicines, Japanese medicines, and Western herbal medicines are used in the treatment of chronic liver diseases [2]. Chinese formulae for FLD can include the categories of heat-clearing herbs (antipyretics), diuretics and dampness-removing herbs, tonifying herbs, and herbs for removing food stagnation. A popular Chinese formula, Qushi Huayu Decoction, containing *Artemisia capillaries*, *Polygonum cuspidata*, *Hypericum japonicum*, *Curcuma longa*, and *Gardenia jasminoides*, inhibits hepatic lipid accumulation by activation AMPK both *in vitro* and *in vivo* [61]. It was previously reported to ameliorate HS in rat models and HepG2 cells [62] and inhibited TNF-α expression in rats with NAFLD and steatotic HepG2 cells [63]. The mixture of natural extracts (*Astragalus membranaceus*, *Salvia miltiorrhiza*, and *Pueraria lobata*) also improved fatty liver in rats induced by alcohol diet by recovery of hepatic lipase activity including decrease of plasma aspartate aminotransferase (AST) and alanine transaminase (ALT) [64]. Polyphenolic compounds and polysaccharides have also been revealed to improve inflammation in the liver [65, 66].

Pharmacologically, these herbs have lipid lowering effects, restoring the AST/ALT levels, anti-inflammation, and antioxidative activities. Based on their availability of literature and popularity in Western countries, the following herbs—*Silybum marianum* (milk thistle), *Panax ginseng and Panax notoginseng (ginseng)*, *Momordica charantia* (bitter melon), *Lycium barbarum* (barberry), *Allium sativum* (Garlic), *Camellia sinensis* (green tea), *Alisma orientalis* (water plantain), *Trigonella foenum-graecum* (fenugreek), and *Glycyrrhiza glabra* (licorice)—are selected and summarized in the next section, Table 17.1, and Figure 17.2.

17.2.3 Common Herbs Used in Fatty Liver Management

17.2.3.1 **S. marianum** *S. marianum* (St. Mary's Thistle, milk thistle) is often used as a therapy by patients who have liver disease. It has been shown to be safe and well tolerated and to improve liver chemistry and symptoms in a range of conditions including acute and chronic viral hepatitis, alcoholic liver disease, and drug-related hepatitis [11].

Traditionally, milk thistle extract is made from the seeds [83]. The crude extract termed silymarin comprises of 20–30% FAs, including linoleic acid, and 65–80% of a complex of 7 flavonolignans (silybin A, silybin B, isosilybin A, isosilybin B,

TABLE 17.1 Summary of Activities of Herbal Medicines for Fatty Liver Disease

Herbal Medicines	Experimental Model	Parts/Active Constituents	Effects and Proposed Mechanisms	References
Silybum marianum	Alcohol-induced rats	Silymarin (a flavonoid complex extract)	Reduction of lipid peroxidation in the liver Enhancement of glutathione antioxidant system	[67]
	Methionine and choline deficient (MCD) induced rats	Extract with polyphenolic content	Reduction of serum ALT and AST Decrease of hepatic TNF-α and TGF-β mRNA expression and lowering of activation of procaspase-3	[66]
Panax ginseng	Sixty-six percent hepatectomized rats	Extract at 125 or 250 mg/kg/day	Reduction of TG and total cholesterol (TC) in the liver 3 days after hepatectomy	[68]
	HepG2 cell and CCl_4-induced mice	Red ginseng essential oil	Reduction of H_2O_2-mediated oxidative stress and inhibition of the phosphorylation of upstream mitogen-activated protein kinases (MAPKs)	[69]
Panax notoginseng	THP-1 cell and a collagen-induced mice	*n*-Butanol extract	Anti-inflammatory effect by decreasing of TNF-α and NF-κB and JNK pathway	[70]
Lycium barbarum	High-fat diet mice	Polysaccharide	Increase of blood and hepatic antioxidant levels such as ROS and decrease of MDA formation	[71]
Allium sativum	High-fat diet in C57BL/6J mice	Five percent whole garlic	Reduced weight gain and activation of AMPK pathway	[72]
	Ethanol-induced rats	Aged black garlic	Reduction of fat accumulation in the liver and possession of antioxidant properties	[73]

Camellia sinensis	*ob/ob* mice	Total extract	Reduction of hepatic lipid accumulation and ALT level by decreasing hepatic TNF-α	[74]
	Diet-induced obese rats with NASH	Powder extract containing 30% of total catechins	Reduction of inflammation by suppressing proinflammatory cytokines	[75]
	Leptin-deficient obese (*ob/ob*) mice	AIN-93G containing green tea extract at 1% (wt/wt)	Reduction of lipid accumulation, inflammation, oxidative stress, and nitrative stress	[76]
Momordica charantia	Fat fed diet rats	Methanol fraction	Reduction of hepatic TG levels and TG synthesis while activation of FA oxidation	[77]
	Diabetic rats induced by sucrose	Total extract	Improvement of antioxidant status, indicated by low levels of thiobarbituric acid-reactive substances	[78]
Trigonella foenum-graecum	Alcohol-fed rats	Polyphenol extract	Improvement of lipid profile and decrease of collagen content	[79]
	Obese diabetic mice	Diosgenin (aglycone form)	Decrease of plasma and hepatic TG	[80]
Alisma orientalis	High-fat diet rats	Methanol extract	Reduction of serum and hepatic lipids and histological improvement of HS	[10]
Glycyrrhiza glabra	Hepatocytes	Glycyrrhizin	Inhibition of production of TNF-α and hepatic fibrogenesis	[81]
	Chronic hepatitis patients	Glycyrrhizin (80 mg for 1 month)	Decrease of transaminases and no significant side effect	[82]

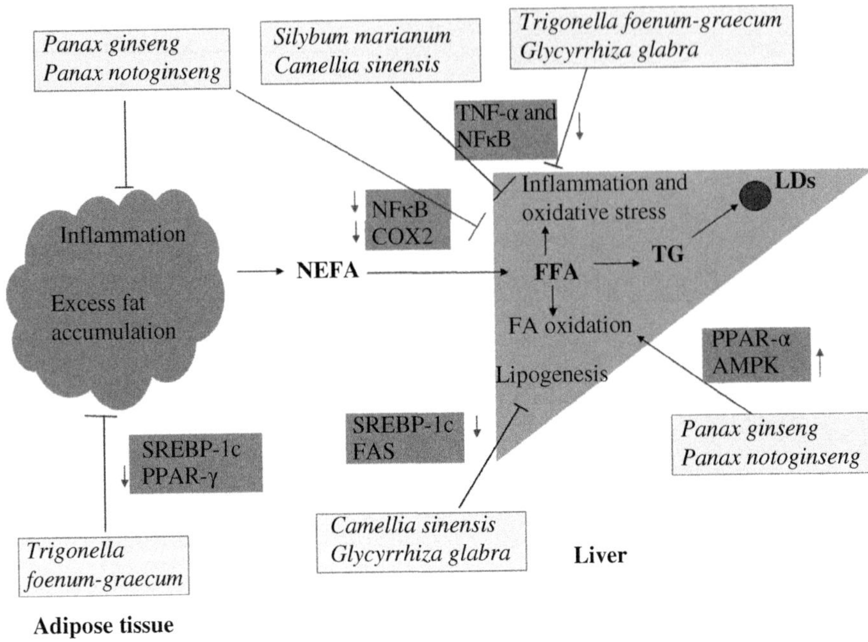

FIGURE 17.2 The pleiotropic effects and their molecular mechanisms of herbal medicines in amelioration of FLD. AMPK, AMP-activated protein kinase; COX-2, cyclooxygenase 2; FAS, fatty acid synthase; FFA, free fatty acid; LD, lipid droplet; NEFA, nonesterified fatty acid; NF-κB, nuclear factor-kappa B; PPARα, peroxisome proliferator-activated receptor alpha; PPARγ, peroxisome proliferator-activated receptor gamma; SREBP-1c, sterol regulatory element-binding protein-1c; TG, triglyceride.

silychristin, isosilychristin, silydianin) and 1 flavonoid (taxifolin) [84]. Silibinin is a semipurified fraction of silymarin that is a mixture of 2 diastereoisomers, silybin A and silybin B with 1:1 ratio [84, 85].

The crude extract has antiapoptotic and anti-inflammatory activities [66]. Silymarin is one of the most investigated plant extracts with known mechanisms of action for oral treatment of toxic liver damage [86] and protective treatment in acute and chronic liver diseases [87]. Silymarin constituent protects animals against multiple types of experimental liver injury, such as carbon tetrachloride, acetaminophen, and iron overload [87, 88]. Silymarin also acts as an antioxidant by scavenging reactive oxygen species (ROS) and increasing glutathione (GSH) content in the liver [89, 90]. Silymarin has been reported to suppress TNF-α-induced NF-κB activation and protected the liver against T-cell-induced injury with increasing the expression of IL-10 [91, 92].

In randomized control trials, silymarin reduced the mortality of patients with cirrhosis of the liver without side effects [6]. It was reported from pooled analysis of clinical trials, a 7% reduction in liver mortality in cirrhotic patients treated with silymarin [93]. A meta-analysis of milk thistle, however, found no improvements in histology at liver biopsy and no reduction in mortality or in biochemical markers of liver function among patients with chronic liver disease [94].

17.2.3.2 **Ginseng** Ginseng, including both *P. ginseng* and *P. notoginseng* species, is one of the popular and widely used herbal medicines in the world. Ginseng has also been used as a traditional antidiabetic remedy for many years in many countries such as Korea and China. Ginseng intake possibly controls hyperglycemia both in health and diabetic individuals [95–97]. Evidence from clinical and experimental studies also suggest that ginseng may have beneficial effects as an antihyperlipidemic agent on reducing serum total cholesterol level and enhancing antioxidant status [98, 99]. Ginseng extracts were reported to suppress the formation of fatty liver by reducing serum TG and TC and elevating of HDL cholesterol level in animal and hyperlipidemia patients [68, 100]. Another study showed that red ginseng extract suppressed the progression of AFLD in a rat model induced by ethanol and inhibited fibrotic initiation [101]. It was reported that *P. notoginseng* extract lowered fibrinogen and lipid plasma level of TG and TC in rats fed a high-fat diet [102, 103] and had anti-inflammatory effects in mice and human neutrophils [104].

Ginsenoside Rb1, which is one of the active compounds of *P. ginseng*, has recently been tested and shown to improve FLD by activation of AMPK [9] and compound K, metabolite form of ginsenoside Rb1, recently also demonstrated the same activity in high FA-induced human hepatocytes [105]. Recently, ginsenoside Rb1 was shown to inhibit TNF-α-mediated NF-κB transcriptional activity in HepG2 cells with IC_{50} of 27.45 μM and gene expression of inducible nitric oxide synthase (iNOS) and COX-2 inducible inflammatory enzymes [106]. Ginsenoside Rg1 prevented hepatic fibrosis induced by thioacetamide in rats [107]. Ginsenoside R1, the active form of *P. notoginseng*, was shown to have an anti-inflammatory effect by reducing TNF-α-induced plasminogen activator inhibitor-1 (PAI-1) mRNA and protein expression [108].

17.2.3.3 **C. sinensis** Green tea (*C. sinensis*) is one of the popular herbs and drinks receiving much attention in the last few decades [109, 110]. The constituents of green tea include polyphenols (catechins and flavonoids), alkaloids (caffeine and theobromine), polysaccharides, and vitamins (vitamin C). The polyphenols are primarily responsible for the beneficial properties of tea. Catechins consist of six primary compounds, namely, catechin, gallocatechin, epicatechin (EC), epigallocatechin (EGC), epicatechin gallate (ECG), and epigallocatechin gallate (EGCG) [111]. EGCG is the predominant catechin ranging from 50 to 75% of the total catechins [112].

The hepatoprotective mechanisms of green tea extract in FLD have been tested in various *in vitro* and *in vivo* models. Catechins are under biotransformation by colonic microflora, including hydrogenation and oxidation. These metabolites have antioxidant, anti-inflammatory, and enzyme inhibition activities [113, 114]. Green tea extract containing polyphenol and caffeine has been shown to stimulate fat oxidation, boosting the metabolic rate without increasing the heart rate [115]. It has been shown that green tea polyphenol could prevent oxygen free radical-induced hepatocyte lethality, prevent lipopolysaccharide-induced liver injury through inhibition of iNOS and TNF-α expression, and inhibit carcinogen or toxin-induced liver oxidative DNA damage [116, 117]. Some animal studies also demonstrated that green tea reduced hepatic TG level by suppression of enzyme activities for FA synthesis [118] and protected against hepatic injury [119] in obese mice. EGCG protected against NAFLD in high fat [120, 121] and

SREBP-1c overexpression models [122]. Green tea also possesses abilities to decrease circulating nonesterified FAs and de novo lipogenesis including FA synthase and sterol-CoA desaturate-1 in various FLD models [74, 123]. The anti-inflammatory effect of green tea extract was also reported through downregulation of Toll-like receptor 4 and blocking of degradation of inhibitor of NF-κB [124].

A meta-analysis of 10 clinical trials showed that regular intake of green tea can significantly reduce the risk of various manifestations of NAFLD as well as other liver diseases [125]. One large cross-sectional study (1371 men, over 40 years old) also reported that intake of green tea decreased serum TG and TC, AST, and ALT [126]. It is still elusive whether green tea protects against FLD directly or indirectly through its positive effects from obesity and diabetes.

17.2.3.4 **T. foenum-graecum** Fenugreek (*T. foenum-graecum*) is one of the oldest medicinal plants originating in India and Northern Africa [127] and has a long history as a traditional medicine in Ayurvedic and Chinese medicine. The seeds have been initially used for antidiabetic effects. Fenugreek seeds contain lysine- and L-tryptophan-rich proteins, saponins, coumarin, fenugreekine, nicotinic acid, sapogenins, scopoletin, and trigonelline [128]. The seeds also contain polyphenolic flavonoids.

An *in vitro* system was used to assess the molecular mechanism of hypolipidemic effect of fenugreek. It inhibited accumulation of fat in 3T3-L1 cells via decreased expression of adipogenic factors such as PPARγ, SREBP-1, and CAAT element-binding proteins-alpha (c/EBP-α). The extract also gave rise to the reduction of cellular TG and cholesterol concentrations in HepG2 cells via reduced expression of SREBP-1 [129]. The methanol extract led to anti-inflammatory effect through inhibition of TNF-α in THP-1 cells [130]. In an obese rat model, fenugreek seed extract was demonstrated to decrease lipid accumulation in the liver of Zucker obese rat [131]. It also showed its potential effects for alcoholic-induced fatty changes in the liver [79].

Clinical trial of small scale has shown some hypocholesterolemic effects. Fifteen nonobese, asymptomatic, hyperlipidemic adults ingested 100 g defatted fenugreek powder per day for 3 weeks, and the result showed lower level of TG and LDL cholesterol than baseline values [132]. Similarly, in hypercholesterolemic type 2 diabetic patients treated with fenugreek for 3 and 6 weeks, serum lipid profile was improved with reduction of serum TC and TG [133].

17.2.3.5 **G. glabra** Licorice is the root of *G. glabra*. Glycyrrhizin is an active component of a triterpenoid saponin glycoside. Upon hydrolysis, the glycoside loses its sweet taste and is converted to the aglycone glycyrrhetinic acid and two molecules of glucuronic acid [134].

The main mechanisms for the hepatoprotective effects of glycyrrhizin include anti-lipid peroxidation, antioxidant, and anti-inflammatory effect [81, 135]. Glycyrrhizin also suppressed the phospholipase A2 activation, decreasing ALT release by hepatocytes [136]. Glycyrrhizin also inhibited of the production of TNF-α and hepatic fibrogenesis in a mouse model [81] and increased production of IL-10 in rat liver dendritic cells [137]. The aqueous extract had hepatoprotective effects in rabbit model of acute liver injury induced by carbon tetrachloride [138]. Carbenoxolone, which is the

3-hemisuccinate of glycyrrhetinic acid, the active principal of licorice, also showed the prevention of the development of fatty liver in C57BL/6-Lep *ob/ob* mice through the inhibition of SREBP-1c activity and apoptosis [139].

Clinical trials using glycyrrhizin have mostly been associated with the treatment of hepatitis C. One clinical trial in hepatitis C patients given glycyrrhizin showed that ALT level dropped and the overall rate of hepatocellular carcinoma was lower [82]. The clinical test for the direct effect of this herb for FLD has not been demonstrated yet.

17.3 PHYTOMEDICINES FOR IBS

17.3.1 Introduction and Pharmacotherapy

IBS is one of the most common intestinal diseases. It is a chronic functional GI disorder characterized by the presence of episodic abdominal pain or discomfort in association with altered bowel habits (diarrhea and/or constipation) and other GI symptoms such as bloating and distension unexplained by structural or biochemical abnormalities. There are three major categories of IBS designated as IBS with predominate diarrhea (IBS-D), IBS with predominate constipation (IBS-C), and IBS with alternating constipation and diarrhea (IBS-M) [140]. Epidemiologic studies indicated that IBS affected 10–20% of the general population depending on the diagnostic criteria used to define the condition [141–145] and generally a higher rate in women [145, 146]. Annually, IBS results in 3.5 million office visits and an estimated total burden of $45 billion on the healthcare system in the United States [147].

The pathophysiological mechanisms of IBS relate to altered bowel motility, increased visceral hypersensitivity, psychosocial abnormalities, altered brain–gut interactions, low-grade inflammation, alterations in the intestinal microflora, and genetic factors [148–152]. Accelerated and delayed rectal, colonic, and small bowel motility may be associated with characteristic IBS symptoms in some IBS-D and IBS-C patients, respectively [153–155]. Visceral hypersensitivity at different sites in the gut can often be identified in IBS patients [156, 157], and it may be the main mechanism inducing abdominal pain [158]. Possible mediators in the visceral hyperalgesia response relate to serotonin, calcitonin gene-related peptide, substance P, bradykinin, tachykinins, and neurotrophins [150]. Psychological abnormality is a common comorbidity in IBS patients and a factor that amplifies IBS symptoms. There is a growing body of evidence to suggest that alteration in the microbiome may play a role in the development of IBS. It is found that around 33% of patients with IBS reported a family history. The role of genetic factors is reported concerning genetic polymorphisms involving genes targeting proinflammatory cytokines, the serotonin reuptake transporter, tryptophan hydroxylase, sodium ion channel proteins, and the alpha 2A adrenergic receptor [159].

Treatments of IBS are largely based upon the frequency and severity of constipation, diarrhea, bloating, or pain. Dietary fiber supplements (psyllium, ispaghula husk, wheat and corn bran, methylcellulose, calcium polycarbophil, and partially hydrolyzed guar gum), laxatives (polyethylene glycol), prokinetic agents (mosapride, prucalopride, pumosetrag), prosecretory agents (lubiprostone), and bile acid modulators (chenodeoxycholic acid) are used for IBS-C. Antidiarrheals (loperamide), serotonergic agents (alosetron, ramosetron), and antibiotics (rifaximin) are used in patients with IBS-D. Alosetron also improved quality of life (QOL) in women with IBC-D and IBC-M. Antispasmodics and psychotropic agents are used to relieve abdominal pain and discomfort. Besides the medicines, probiotics (*Bifidobacterium infantis*) offer benefits to IBS patients [147, 159].

17.3.2 Treatment of IBS in Traditional Medicine

CAM therapies have been used in many countries in the management of IBS. A population-based survey in 1409 subjects from the United Kingdom indicated that more than 50% of IBS patients used CAM treatment [160], and in the United States, 35% of patients with functional GI disease used CAM [161].

In the management of IBS, TCM use Zang-Fu in diagnosis and treatment system with substantial herbal strategy. TCM principle considers that IBS is associated with liver and spleen disharmony, in particular liver Qi stagnation (liver excess) and spleen deficiency. It may also involve stomach weakness and intestinal disharmony. Emotional distress, anxiety, mood changes, frustration, and irritability can lead to liver Qi stagnation, which transforms into heat drying fluid in the intestines and disrupts spleen and stomach function. Improper diets also disrupt spleen and stomach function leading to symptoms of abdominal pain, dry or loose stool, and sluggish or frequent bowel movement. TCM treatment principles for IBS focus on improving the flow of liver Qi, strengthening spleen and stomach, and harmonizing stomach and intestine.

There are a number of herbs and formulae used in the treatment of IBS in traditional medicine. Tong-Xie-Yao-Fang (formula for painful diarrhea) has been used to relieve symptoms associated with IBS in TCM since the fifteenth century [162]. This formula is composed of four Chinese herbs—*Atractylodes macrocephala*, *Paeonia lactiflora*, *Citrus reticulata*, and *Saposhnikovia divaricata*—having the property of regulating the functions of liver and spleen and has been reported to be effective from preclinical and clinical studies in improving IBS-associated disorders of digestive system, alleviating abdomen pain and diarrhea [163–165]. In a Cochrane systematic review of herbal medicines for the treatment of IBS, formulae of Tong-Xie-Yao-Fang, Padma Lax, STW 5, and an Indian Ayurvedic formula improved global IBS symptoms compared with placebo [166]. Other alternative treatments frequently used for IBS patients are peppermint oil [167–169], massage, and acupuncture [170, 171]. Pharmacological and clinical evidence for some of the most commonly used herbs in IBS—*Cynara scolymus*, *Mentha×piperita*, *C. longa*, and *Citrus reticulate*—and their active compounds are reviewed in the following section and in Fig. 17.3.

17.3.3 Common Herbs Used in the Management of IBS

17.3.3.1 **C. scolymus** *C. scolymus* L., known as artichoke, is widely cultivated in Mediterranean, American, and African countries and commonly eaten as a vegetable. Extracts from artichoke leaves are traditionally used in the treatment of dyspeptic and hepatic disorders, obesity, and hyperlipidemia. *C. scolymus* leaves contain several

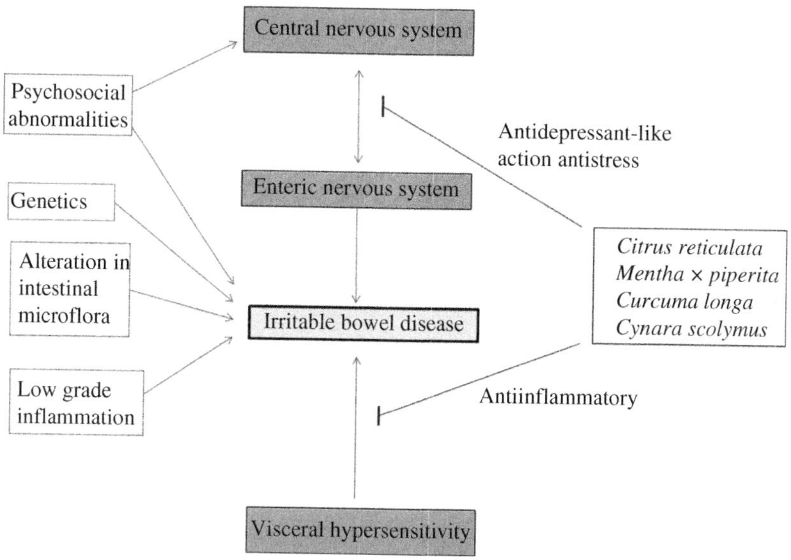

FIGURE 17.3 The pleiotropic effects of herbal medicines in amelioration of IBS. Brain–gut axis mediators involved: serotonin, calcitonin gene-related peptide, substance P, calcitonin gene-related peptide, histamine, and glucagon-like peptide 1.

phenolic compounds and flavonoids whose main chemical constituents include cynarin, caffeic acid (CA), chlorogenic acid, apigenin-7-*O*-*β*-D-glucopyranoside, isorhoifolin, cynaroside, scolymoside, and luteolin. Sesquiterpenes, such as cynaropicrin and lupeol, and volatile oils are detected in *C. scolymus* [172–174].

The antispasmodic activity of *C. scolymus* extract against guinea pig ileum contracted by acetylcholine was reported [175], and the dichloromethane fraction showed the most promising biological effects. Its main active component, cynaropicrin, exhibits about 14-fold more activity than dichloromethane fraction and has similar potency to that of papaverine, a well-known antispasmodic agent [176]. Flavonoids from *C. scolymus* upregulate endothelial-type nitric oxide synthase gene expression in human endothelial cells [177]. A study revealed that *C. scolymus* prevent gastric mucosal injury induced by ethanol and stress [178].

Clinical studies reported that *C. scolymus* leaf extract reduces symptoms of IBS and improves health-related QOL. Significant fall in IBS incidence and a significant shift in self-reported usual bowel pattern away from "alternating constipation/diarrhea" toward "normal" were observed. Nepean Dyspepsia Index (NDI) total symptom score significantly decreased by 41% ($p<0.001$) after treatment. Similarly, there was a significant 20% improvement in the NDI total QOL score in the subset after treatment [179]. In a postmarketing surveillance study of *C. scolymus* extract for 6 weeks, analysis of the data from the IBS subgroup revealed significant reductions in the severity of symptoms and favorable evaluations of overall effectiveness by both physicians and patients. Furthermore, 96% of patients rated *C. scolymus* extract as better than or at least equal to previous therapies administered for their symptoms, and *C. scolymus* extract had good tolerability [180].

Some work revealed adverse event incidence of *C. scolymus* leaf extract to be low [181] and not found in the clinical study [179]. Orally, artichoke extract might increase flatulence in some patients [180]. People sensitive to the Asteraceae/Compositae family may be at the greatest risk of the allergic reaction from *C. scolymus* leaf extract. Allergic contact dermatitis, which would be attributed to the constituent of cynaropicrin, may occur with the use of *C. scolymus* leaf extract [182]. Theoretically, *C. scolymus* leaf extract might worsen bile duct obstruction and gallstones by increasing bile flow [181].

After administration of *C. scolymus* leaf extracts standardized by its main caffeoylquinic acids and flavonoids (caffeoylquinic acids, CA, and luteolin glycosides), none of the genuine target extract constituents were detected in human plasma and urine. However, CA, its methylated derivatives (ferulic acid (FA) and isoferulic acid) [183], and the hydrogenation products (dihydrocaffeic acid (DHCA) and dihydroferulic acid (DHFA)) were identified as metabolites derived from caffeoylquinic acids. Except for DHFA, all of these compounds were present as sulfates or glucuronides. Peak plasma concentrations of total CA, FA, and IFA were reached within 1 h and declined over 24 h showing almost biphasic profiles. In contrast, maximum concentrations of total DHCA and DHFA were observed only after 6–7 h, indicating two different metabolic pathways for caffeoylquinic acids. Luteolin administered as a glycoside was recovered from plasma and urine only as sulfate or glucuronide but neither in the form of genuine glycosides nor as free luteolin. Peak plasma concentrations were reached rapidly within 0.5 h. The elimination showed a biphasic profile [184].

17.3.3.2 **M. × piperita** *M.×piperita*, commonly known as peppermint, contains phenolic constituents (including rosmarinic acid), several flavonoids (primarily eriocitrin, luteolin, and hesperidin), and volatile components (such as menthol and menthone). Pharmacological studies revealed that peppermint oil had a relaxing effect on GI tissue, analgesic and anesthetic effects in the central and peripheral nervous systems, immunomodulating actions, and chemopreventive potential [185].

Several clinical trials demonstrate the effects of peppermint oil on IBS, especially effective in relieving abdominal pain in diarrhea-predominant IBS [167, 186]. In a double-blind placebo-controlled randomized trial, 57 patients with IBS according to the Rome II criteria were treated with peppermint oil (two enteric-coated capsules twice per day or placebo) for 4 weeks. Peppermint oil was able to significantly reduce the total IBS score and relive the symptoms of abdominal bloating, abdominal pain or discomfort, diarrhea, constipation, feeling of incomplete evacuation, pain at defecation, passage of gas or mucus, and urgency at defecation [168]. A review of 16 clinical trials investigating 180–200 mg enteric-coated peppermint oil in IBS or recurrent abdominal pain in children (1 study) showed statistically significant effects in favor of peppermint oil. Average response rates in terms of "overall success" are 58% (range 39–79%) for peppermint oil and 29% (range 10–52%) for placebo. Adverse events reported were generally mild and transient, but very specific. Peppermint oil caused the typical GI effects like heartburn and anal/perianal burning or discomfort sensations [187].

Peppermint oil is relatively rapidly absorbed after oral administration and eliminated mainly via the bile. The major biliary metabolite is menthol glucuronide, which undergoes enterohepatic circulation. The urinary metabolites result from hydroxylation at the C-7 methyl group at C-8 and C-9 of the isopropyl moiety, forming a series of mono- and dihydroxymenthols and carboxylic acids, some of which are excreted in part as glucuronic acid conjugates. Studies with tritiated I-menthol in rats indicated about equal excretion in feces and urine. The main metabolite identified was menthol glucuronide. Additional metabolites are mono- or dihydroxylated menthol derivatives [188].

On the herb–drug interactions, peppermint oil may raise serum levels of simvastatin and felodipine, which are used to treat high cholesterol and high blood pressure, respectively [189]. Peppermint oil has choleretic effects and may be contraindicated in patients with hiatus hernia, severe gastroesophageal reflux disease, gallstones, cholecystitis (gallbladder inflammation), and severe liver disease [189]. Peppermint oil may trigger menstruation and should be avoided during pregnancy [189].

17.3.3.3 **C. longa** *C. longa* (turmeric) is also known as one of the most important spice and food additives. It is normally standardized to curcuminoids (particularly curcumin but also rich in demethoxycurcumin and bidemethoxycurcumin), which are its main constituents [190]. Since the time of Ayurveda (1900 B.C.), numerous therapeutic activities have been assigned to turmeric for a wide variety of diseases and conditions, including those of the skin, pulmonary, and GI systems, aches, pains, wounds, sprains, and liver disorders [191]. *C. longa* has antioxidant, antihistamine, anti-inflammatory (inhibits NF-kappa B), antimicrobial (antibacterial, antiviral, antifungal), antiproliferative, chemopreventive, anticarcinogenic, hepatoprotective, antiulcer, and antispasmodic properties [191–193].

A clinical study revealed that *C. longa* extract may improve IBS symptomology. In a partially blinded, randomized and two-dose pilot study, 207 IBS patients were given one or two tablets of standardized turmeric extract for 8 weeks. Results revealed that abdominal pain/discomfort score reduced significantly by 22% and 25% in the one- and two-tablet group, respectively. There were significant improvements in all IBS Quality of Life scales of between 5 and 36% in both groups, and approximately two thirds of all subjects reported an improvement in symptoms after treatment [194].

In a phase I clinical study, no clinical, hematological, renal, or hepatic toxicity of turmeric oil at 1 and 3 months was observed [195]. *C. longa* has been shown *in vitro* to inhibit platelet activity [196] and to have anticoagulant effects *in vivo* [197]. On the herb–drug interaction, curcumin has been shown to enhance the antitumor effect of cisplatin when used in combination against fibrosarcoma in rats by decreasing tumor marker enzymes, including aminotransferases, lactate dehydrogenase, gamma-glutamyl transpeptidase, alkaline phosphatase, and 5′-nucleotidase to near-normal levels [198]. *In vitro* studies suggest that curcumin has a synergistic effect with cisplatin in decreasing the expression of certain cancer genes [199]. Curcumin has been shown to protect against cyclophosphamide-induced lung injury *in vivo* [200]. Six days of treatment with 300 mg of curcumin was associated with a decrease in the bioavailability of talinolol in 12 healthy volunteers [201]. A pharmacokinetic study has found that pretreatment with curcumin increased the plasma concentrations of losartan and its metabolite in rats [202].

17.3.3.4 **C. reticulata** *C. reticulate*, commonly called Ripe Tangerine Peel and Chen Pi in Chinese, is rich in essential oils, such as limonene, 1,8-cineole, and γ-terpinene [203]. Apigenin is a bioflavonoid widely found in citrus and possesses a variety of pharmacological actions on the central nervous system. Research results suggest that the antidepressant-like actions of oral apigenin treatment could be related to a combination of multiple biochemical effects and might help to elucidate its mechanisms of action that are involved in normalization of stress-induced changes in brain monoamine levels, the HPA axis, and the platelet adenylyl cyclase activity [204]. *C. reticulata* extract also possesses antibacterial action against *Helicobacter pylori* and its essential oil rich in limonene provides gastroprotective action [205, 206].

C. reticulate is an ingredient in the popular IBS formula, Tong-Xie-Yao-Fang. Clinical studies showed that the formula quickly reduced abdominal pain and distention, improved the property of stool, and mental tension and depression in patients [165]. Experimental studies also revealed that it attenuated behavioral hyperalgesia by regulating substances associated with the brain–gut axis, including decreasing the expression of 5-HT and substance P in the periphery and that of corticotrophin-releasing factor in the brain [163].

17.4 PHYTOMEDICINES FOR CONSTIPATION

Constipation is a syndrome that is generally defined by the symptoms of straining, infrequent and difficult evacuation of feces, hardness of stool, and feeling of incomplete evacuation [207]. The estimated prevalence of constipation is up to 28%, representing 56 million adults in the United States alone, and is commonly reported in Western societies, especially in female and elder individuals and residents of chronic care facilities and patients with concurrent psychiatric illness, even though it affects people of all ages and both genders [208, 209]. The Rome III symptom criteria broadly separates constipation into two categories: functional constipation and constipation-predominant IBS [210].

Pharmacological agents are beneficial to treat patients with chronic constipation. There are numerous options that are effective against chronic constipation. First of all, several over-the-counter laxatives are adopted to increase stool frequency or ease stool passage via direct stimulation on the gut or gut contents: bulk laxatives, stool softener or wetting agents, stimulant laxatives, and osmotic laxative [211, 212]. Secondly, chloride channel activators allow secretion of fluid, and lubiprostone has been approved by the US Food and Drug Administration (FDA) [213]. Linaclotide is a novel guanylate cyclase activator that induces chloride ion and water secretion. Thus, it helps to prevent dehydration and obstruction of the intestine and to increase intestinal transit [214]. Next, serotonergic enterokinetic agents are widely used pharmacological agents to treat constipation. Among them, 5-HT_4 receptor agonists such as indoles, substituted benzamides, and dihydrobenzofurancarboxamide (Prucalopride) are the most extensively studied and tested for their function to enhance proximal smooth muscle contraction and to relax distal smooth muscles [212]. Neurotrophin-3 is a neurotrophic factor, being explored to ameliorate the softness of stool inconsistency and the frequency of bowel movement [215]. Furthermore, opioid antagonists are widely used to treat opioid-induced constipation [216].

17.4.1 Treatment of Constipation with Herbal Medicines

CAM suggests many strategies to treat chronic constipation: acupuncture, herbal medicines, auriculotherapy, massage, aromatherapy, and reflexology. Epidemiological research demonstrates that treatments using CAM methods are effective in improving chronic constipation in Korea, Japan, and Turkey [217–219].

In fact, traditional folk medicine has been adopted to treat constipation in many countries based on ancient philosophy and long-term clinical practice [220]. According to etiology of TCM, constipation is usually caused by the imbalance of homeostasis of Qi, blood, yin, and yang. It is largely divided into two different types: excessive and deficient patterns [220, 221]. Most constipation cases are triggered by the deficiency: yin deficiency type, blood and yin deficiency type, and Qi deficiency type. On the contrary, pathological excess of Qi is considered to be the other cause [222]. The excess of Qi is also called stagnation and is characterized by the symptom of stiffness, heaviness, and tightness in the intestine, while the deficiency types are characterized by dryness, failure of lubrication, and consequently accumulation of stool in the intestine [221, 222]. For example, a classic TCM formula for constipation, Wen-Pi Tang, contains *P. ginseng*, *Zingiber officinale*, *Glycyrrhiza uralensis*, *Aconitum carmichaelii*, and *Rheum palmatum*, which are laxatives, simultaneously promoting the energy of yang [221] (Fig. 17.4).

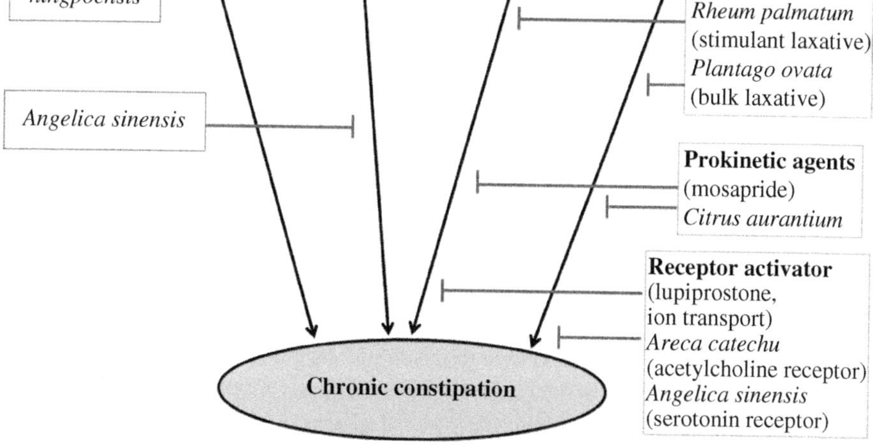

FIGURE 17.4 The pleiotropic effects of herbal medicines in amelioration of constipation.

17.4.2 Common Herbs Used in the Management of Constipation

Herbs can also be broadly classified as laxatives and prokinetic agents. Common laxatives are *Plantago ovata*, *R. palmatum*, *Cassia senna*, and *Aloe vera*, which will be covered in detail in the following sections; and some foods are weak laxatives: oriental sesame seed, walnut kernel, mulberry fruit, honey, and bitter apricot seed. Prokinetic agents are *Areca catechu*, *Magnolia officinale*, *Citrus aurantium*, *C. reticulate*, *Angelica sinensis*, and *A. membranaceus* [212, 222]. Other Chinese herbs commonly used in constipation often have tonifying functions, for instance, *P. ginseng*, *Z. officinale*, *A. macrocephala*, and *A. carmichaelii* [221, 223]. An epidemiological research reveals that *R. palmatum*, *M. officinale*, *Scrophularia ningpoensis*, *C. aurantium*, and *Rehmannia glutinosa* are most popular herbs used for constipation in Taiwan in 2004 [224]. Pharmacologically, *A. catechu* and *A. sinensis* are found to react with acetylcholine receptor and serotonin receptor in the intestine, respectively [224].

17.4.2.1 **A. vera** *A. vera* has a long history of traditional medical usage to treat many human diseases including constipation [225]. The extract of aloe demonstrated beneficial effects in increasing intestinal motility and fecal volume and normalizing body weight in loperamide-induced constipation rats [226]. Furthermore, studies have been demonstrating that *A. vera* reduced fluid absorption and simultaneously inhibited chloride channel and NA+/K+-adenosine triphophatase in colonic mucosa [227–229]. Previous studies reported that aloe had revealed considerable strong benefits to improve constipation symptom of the adult even though, like any other laxatives, overdose or abuse was reported to show gripping and severe diarrhea [230, 231]. In terms of drug and herb interactions, the consumption of *A. vera* reduced absorption of orally administered drug and, on the other hand, induced hypokalemia [232] (Fig. 17.5).

17.4.2.2 **P. ovata** Psyllium seed husk of *P. ovata* is commonly used for chronic constipation. Its active compounds include polysaccharides and arabinoxylans with high gel-forming property [233]. It was reported that patients with constipation, aged 18–75 years old of both genders, who received psyllium husk seed treatment had considerably increased defecations per week [234]. Other human trials have also demonstrated that it significantly decreased the time necessary to pass bowel movements and increased the number of bowel movements per day and the amount of stool passed [235–237]. In terms of side effects, even though this is usually the safest laxative, it might be contraindicated in metabolic syndrome patients [238]. In addition, it has the potential to decrease drug absorption by decreasing GI transit time [239].

17.4.2.3 **R. palmatum** *R. palmatum*, or rhubarb, is one of the extremely famous ingredients of TCM as a purgative and heat-clearing agent. It has been used for clearing toxic heat and purging knotted heat and stool from the colon and many other symptoms [240]. Rhubarb was shown to have obvious purgative action in the rodent [241]. Oral administration is identified to induce colonic mobility and inhibition of N+/K+-exchanging ATPase or chloride channel in the colon [227, 242]. Its cathartic action is limited to the large intestine [243, 244]. Like psyllium seed husk, rhubarb has the potential to decrease drug absorption in the intestine due to shorter GI transit

FIGURE 17.5 *Aloe vera.*

time [239]. It was shown that processed rhubarbs can either induce or inhibit activities of CYP1A2, CYP2C6, CYP2E1, and CYP3A1 and modify the metabolism of saxagliptin, which can impact on the drug if coadministered with processed rhubarbs [245]. In addition, rhubarb has the potential to induce electrolyte imbalance, caused by potassium loss [231].

17.4.2.4 **C. senna** Senna leaves originated from *C. senna* (*Senna alata, or Senna alexandrina*) have been used to treat skin rashes, constipation, and fungal infectious like athlete's foot in traditional medicine. It contains many chemical constituents and the anthraquinones are likely to be the active compounds [246]. Its main action is to stimulate nerve ending of colonic mucosa. In a study of 104 women undergoing pelvic reconstructive surgery, senna leaves significantly decreased the first bowel movement time and reduced the use of magnesium citrate [247]. It is contraindicated in intestinal obstructions, acute abdominal conditions, and inflammatory bowel conditions. Long term usage of senna leaves should be cautious due to potential side effect of anthraquinones [246]. Decreased intestinal transit time may reduce absorption of orally administered drugs. Simultaneous use with other drugs or herbs that induce hypokalemia, such as thiazide diuretics, adrenocorticosteroids, or liquorice root, may exacerbate electrolyte imbalance, which may potentiate the effects of cardiotonic glycosides (digitalis) and antiarrhythmic drugs [248, 249].

17.5 SUMMARY AND FUTURE PERSPECTIVES

FLD, IBS, and constipation are very common GI conditions worldwide. These chronic conditions are dealt in traditional medicine systems such as Western herbal medicine and TCM. For example, 9 out of the 18 Chinese herbal medicine categories are used for these conditions. Pharmaceuticals used in these conditions are of limited success in terms of efficacy. For FLD, only some drugs for treatment of diabetes and hyperlipidemia are used for prevention. *S. marianum*, *P. ginseng* and *P. notoginseng*, *C. sinensis*, *T. foenum-graecum*, and *G. glabra* have hepatoprotective effects and improve lipid profile in animal studies.

IBS is often associated with constipation and managed with medicines similar to constipation, using dietary fiber supplements, laxatives, and prokinetic and prosecretory agents for IBS-C; antidiarrheals, serotonergic agents, and antibiotics are applied in patients with IBS-D. Antispasmodic activity is a common property of herbs used in diarrheal type IBS. They contain similar chemicals particularly essential oils as active compounds: sesquiterpenes, phenolic compounds, and flavonoids in *C. scolymus*; monoterpene, phenolic constituents, and flavonoids in peppermint oil; phenolic curcuminoids and essential oils in *C. longa*; and monoterpenes and flavonoids in *C. reticulata*.

Stimulant laxative and bulk-forming agents are the most commonly used remedies in the management of constipation, together with saline laxatives or osmotic laxatives and stool softeners, chloride channel activators, guanylate cyclase activators, and serotonergic enterokinetic agents. *Plantago ovate* is the most commonly used bulk-forming agent. Stimulant laxatives including *A. vera*, *C. senna*, and *R. palmatum* all contain anthraquinones that stimulate nerve ending of colonic mucosa and augment propulsion and accelerate colonic transit. These actions are described as function on Qi, blood, Yin, and Yang in TCM. The anthraquinones are contraindicated with intestinal obstructions, acute abdominal conditions, and inflammatory bowel conditions; can interact with drugs, and cause side effects.

Clinical trials have generally reported positive outcomes and showed that herbal medicines are promising. However, they are not conclusive in their efficacy due to small sample size, weak methodology, lack of defined outcomes, and nonhomogeneity in diagnosis. Nevertheless, preclinical research supports traditional usage of herbal medicines in FLD, constipation, and IBS. Future studies linking the active compounds, standardized extracts, multiple actions, and molecular mechanisms, are warranted to support well-designed clinical studies.

REFERENCES

[1] Angulo P (2002) Nonalcoholic fatty liver disease. *N Engl J Med* 346: 1221–1231.

[2] Batey R, Salmond S, Bensoussan A (2005) Complementary and alternative medicine in the treatment of chronic liver disease. *Curr Gastroenterol Rep* 7: 63–70.

[3] Maciocia G (2005) *The Foundations of Chinese Medicine*. Philadelphia: Elsevier Churchill Livingstone.

[4] Bensky D (1993) *Chinese Herbal Medicine: Materia Medica*. Seattle: Eastland Press.

[5] Liu ZL, Xie LZ, Zhu J, Li GQ, Grant SJ, Liu JP (2013) Herbal medicines for fatty liver diseases. *Cochrane Database Syst Rev* August 24 (8): CD009059.

[6] Ferenci P, Dragosics B, Dittrich H, Frank H, Benda L, Lochs H, Meryn S, Base W, Schneider B (1989) Randomized controlled trial of silymarin treatment in patients with cirrhosis of the liver. *J Hepatol* 9: 105–113.

[7] James OF, Day CP (1998) Non-alcoholic steatohepatitis (NASH): a disease of emerging identity and importance. *J Hepatol* 29: 495–501.

[8] Sanyal AJ, Campbell-Sargent C, Mirshahi F, Rizzo WB, Contos MJ, Sterling RK, Luketic VA, Shiffman ML, Clore JN (2001) Nonalcoholic steatohepatitis: association of insulin resistance and mitochondrial abnormalities. *Gastroenterology* 120: 1183–1192.

[9] Shen L, Xiong Y, Wang DQ-H, Howles P, Basford JE, Wang J, Xiong YQ, Hui DY, Woods SC, Liu M (2013) Ginsenoside Rb1 reduces fatty liver by activating AMP-activated protein kinase in obese rats. *J Lipid Res* 54: 1430–1438.

[10] Hong X, Tang H, Wu L, Li L (2006) Protective effects of the Alisma orientalis extract on the experimental nonalcoholic fatty liver disease. *J Pharm Pharmacol* 58: 1391–1398.

[11] Gordon A, Hobbs DA, Bowden DS, Bailey MJ, Mitchell J, Francis AJP, Roberts SK (2006) Effects of Silybum marianum on serum hepatitis C virus RNA, alanine aminotransferase levels and well-being in patients with chronic hepatitis C. *J Gastroenterol Hepatol* 21: 275–280.

[12] de Almeida IT, Cortez-Pinto H, Fidalgo G, Rodrigues D, Camilo ME (2002) Plasma total and free fatty acids composition in human non-alcoholic steatohepatitis. *Clin Nutr* 21: 219–223.

[13] Tolwani RJ, Hamm DA, Tian L, Sharer JD, Vockley J, Rinaldo P, Matern D, Schoeb TR, Wood PA (2005) Medium-chain acyl-CoA dehydrogenase deficiency in gene-targeted mice. *PLoS Genet* 1: e23.

[14] Grosse SD, Khoury MJ, Greene CL, Crider KS, Pollitt RJ (2006) The epidemiology of medium chain acyl-CoA dehydrogenase deficiency: an update. *Genet Med* 8: 205–212.

[15] Zhang D, Liu Z-X, Choi CS, Tian L, Kibbey R, Dong J, Cline GW, Wood PA, Shulman GI (2007) Mitochondrial dysfunction due to long-chain Acyl-CoA dehydrogenase deficiency causes hepatic steatosis and hepatic insulin resistance. *Proc Natl Acad Sci U S A* 104: 17075–17080.

[16] Malhi H, Gores GJ (2008) Cellular and molecular mechanisms of liver injury. *Gastroenterology* 134: 1641–1654.

[17] Reddy JK, Rao MS (2006) Lipid metabolism and liver inflammation. II. Fatty liver disease and fatty acid oxidation. *Am J Physiol Gastrointest Liver Physiol* 290: G852–858.

[18] Caldwell SH, Hespenheide EE, Redick JA, Iezzoni JC, Battle EH, Sheppard BL (2001) A pilot study of a thiazolidinedione, troglitazone, in nonalcoholic steatohepatitis. *Am J Gastroenterol* 96: 519–525.

[19] Torra IP, Chinetti G, Duval C, Fruchart JC, Staels B (2001) Peroxisome proliferator-activated receptors: from transcriptional control to clinical practice. *Curr Opin Lipidol* 12: 245–254.

[20] Kersten S, Desvergne B, Wahli W (2000) Roles of PPARs in health and disease. *Nature* 405: 421–424.

[21] Ip E, Farrell GC, Robertson G, Hall P, Kirsch R, Leclercq I (2003) Central role of PPARalpha-dependent hepatic lipid turnover in dietary steatohepatitis in mice. *Hepatology* 38: 123–132.

[22] Harano Y, Yasui K, Toyama T, Nakajima T, Mitsuyoshi H, Mimani M, Hirasawa T, Itoh Y, Okanoue T (2006) Fenofibrate, a peroxisome proliferator-activated receptor alpha agonist, reduces hepatic steatosis and lipid peroxidation in fatty liver Shionogi mice with hereditary fatty liver. *Liver Int* 26: 613–620.

[23] Matsusue K, Haluzik M, Lambert G, Yim SH, Gavrilova O, Ward JM, Brewer B, Jr., Reitman ML, Gonzalez FJ (2003) Liver-specific disruption of PPARgamma in leptin-deficient mice improves fatty liver but aggravates diabetic phenotypes. *J Clin Invest* 111: 737–747.

[24] Gavrilova O, Haluzik M, Matsusue K, Cutson JJ, Johnson L, Dietz KR, Nicol CJ, Vinson C, Gonzalez FJ, Reitman ML (2003) Liver peroxisome proliferator-activated receptor gamma contributes to hepatic steatosis, triglyceride clearance, and regulation of body fat mass. *J Biol Chem* 278: 34268–34276.

[25] Hofmann C, Lorenz K, Braithwaite SS, Colca JR, Palazuk BJ, Hotamisligil GS, Spiegelman BM (1994) Altered gene expression for tumor necrosis factor-alpha and its receptors during drug and dietary modulation of insulin resistance. *Endocrinology* 134: 264–270.

[26] Song Z, Deaciuc I, Zhou Z, Song M, Chen T, Hill D, McClain CJ (2007) Involvement of AMP-activated protein kinase in beneficial effects of betaine on high-sucrose diet-induced hepatic steatosis. *Am J Physiol Gastrointest Liver Physiol* 293: G894–902.

[27] Bellentani S, Saccoccio G, Masutti F, Crocè LS, Brandi G, Sasso F, Cristanini G, Tiribelli C (2000) Prevalence of and risk factors for hepatic steatosis in Northern Italy. *Ann Intern Med* 132: 112–117.

[28] Koteish A, Diehl AM (2001) Animal models of steatosis. *Semin Liver Dis* 21: 89–104.

[29] Lelliott CJ, Lopez M, Curtis RK, Parker N, Laudes M, Yeo G, Jimenez-Linan M, Grosse J, Saha AK, Wiggins D, Hauton D, Brand MD, O'Rahilly S, Griffin JL, Gibbons GF, Vidal-Puig A (2005) Transcript and metabolite analysis of the effects of tamoxifen in rat liver reveals inhibition of fatty acid synthesis in the presence of hepatic steatosis. *FASEB J* 19: 1108–1119.

[30] Bandyopadhyay GK, Yu JG, Ofrecio J, Olefsky JM (2006) Increased malonyl-CoA levels in muscle from obese and type 2 diabetic subjects lead to decreased fatty acid oxidation and increased lipogenesis; thiazolidinedione treatment reverses these defects. *Diabetes* 55: 2277–2285.

[31] Postic C, Girard J (2008) Contribution of de novo fatty acid synthesis to hepatic steatosis and insulin resistance: lessons from genetically engineered mice. *J Clin Invest* 118: 829–838.

[32] Adams LA, Angulo P, Lindor KD (2005) Nonalcoholic fatty liver disease. *CMAJ* 172: 899–905.

[33] Haukeland JW, Konopski Z, Eggesbø HB, von Volkmann HL, Raschpichler G, Bjøro K, Haaland T, Løberg EM, Birkeland K (2009) Metformin in patients with non-alcoholic fatty liver disease: a randomized, controlled trial. *Scand J Gastroenterol* 44: 853–860.

[34] Solomon SS, Mishra SK, Cwik C, Rajanna B, Postlethwaite AE (1997) Pioglitazone and metformin reverse insulin resistance induced by tumor necrosis factor-alpha in liver cells. *Horm Metab Res* 29: 379–382.

[35] Marchesini G, Brizi M, Bianchi G, Tomassetti S, Bugianesi E, Lenzi M, McCullough AJ, Natale S, Forlani G, Melchionda N (2001) Nonalcoholic fatty liver disease: a feature of the metabolic syndrome. *Diabetes* 50: 1844–1850.

[36] Duseja A, Das A, Das R, Dhiman RK, Chawla Y, Bhansali A, Kalra N (2007) The clinicopathological profile of Indian patients with nonalcoholic fatty liver disease (NAFLD) is different from that in the west. *Dig Dis Sci* 52: 2368–2374.

[37] Zhou G, Myers R, Li Y, Chen Y, Shen X, Fenyk-Melody J, Wu M, Ventre J, Doebber T, Fujii N, Musi N, Hirshman MF, Goodyear LJ, Moller DE (2001) Role of AMP-activated protein kinase in mechanism of metformin action. *J Clin Invest* 108: 1167–1174.

[38] Koo SH, Flechner L, Qi L, Zhang X, Screaton RA, Jeffries S, Hedrick S, Xu W, Boussouar F, Brindle P, Takemori H, Montminy M (2005) The CREB coactivator TORC2 is a key regulator of fasting glucose metabolism. *Nature* 437: 1109–1111.

[39] Shaw RJ, Lamia KA, Vasquez D, Koo SH, Bardeesy N, Depinho RA, Montminy M, Cantley LC (2005) The kinase LKB1 mediates glucose homeostasis in liver and therapeutic effects of metformin. *Science* 310: 1642–1646.

[40] Kim YD, Park K-G, Lee Y-S, Park Y-Y, Kim D-K, Nedumaran B, Jang WG, Cho W-J, Ha J, Lee I-K, Lee C-H, Choi H-S (2008) Metformin inhibits hepatic gluconeogenesis through AMP-activated protein kinase–dependent regulation of the orphan nuclear receptor SHP. *Diabetes* 57: 306–314.

[41] Loomba R, Lutchman G, Kleiner DE, Ricks M, Feld JJ, Borg BB, Modi A, Nagabhyru P, Sumner AE, Liang TJ, Hoofnagle JH (2009) Clinical trial: pilot study of metformin for the treatment of non-alcoholic steatohepatitis. *Aliment Pharmacol Ther* 29: 172–182.

[42] Jung HS, Youn BS, Cho YM, Yu KY, Park HJ, Shin CS, Kim SY, Lee HK, Park KS (2005) The effects of rosiglitazone and metformin on the plasma concentrations of resistin in patients with type 2 diabetes mellitus. *Metabolism* 54: 314–320.

[43] Bajaj M, Suraamornkul S, Hardies LJ, Pratipanawatr T, DeFronzo RA (2004) Plasma resistin concentration, hepatic fat content, and hepatic and peripheral insulin resistance in pioglitazone-treated type II diabetic patients. *Int J Obes Relat Metab Disord* 28: 783–789.

[44] Seo YS, Kim JH, Jo NY, Choi KM, Baik SH, Park JJ, Kim JS, Byun KS, Bak YT, Lee CH, Kim A, Yeon JE (2008) PPAR agonists treatment is effective in a nonalcoholic fatty liver disease animal model by modulating fatty-acid metabolic enzymes. *J Gastroenterol Hepatol* 23: 102–109.

[45] Neuschwander-Tetri BA, Caldwell SH (2003) Nonalcoholic steatohepatitis: summary of an AASLD Single Topic Conference. *Hepatology* 37: 1202–1219.

[46] Lutchman G, Modi A, Kleiner DE, Promrat K, Heller T, Ghany M, Borg B, Loomba R, Liang TJ, Premkumar A, Hoofnagle JH (2007) The effects of discontinuing pioglitazone in patients with nonalcoholic steatohepatitis. *Hepatology* 46: 424–429.

[47] Ratziu V, Giral P, Jacqueminet S, Charlotte F, Hartemann-Heurtier A, Serfaty L, Podevin P, Lacorte JM, Bernhardt C, Bruckert E, Grimaldi A, Poynard T (2008) Rosiglitazone for nonalcoholic steatohepatitis: one-year results of the randomized placebo-controlled Fatty Liver Improvement with Rosiglitazone Therapy (FLIRT) Trial. *Gastroenterology* 135: 100–110.

[48] Cook WS, Yeldandi AV, Rao MS, Hashimoto T, Reddy JK (2000) Less extrahepatic induction of fatty acid beta-oxidation enzymes by PPAR alpha. *Biochem Biophys Res Commun* 278: 250–257.

[49] Chou CJ, Haluzik M, Gregory C, Dietz KR, Vinson C, Gavrilova O, Reitman ML (2002) WY14,643, a peroxisome proliferator-activated receptor alpha (PPARalpha) agonist, improves hepatic and muscle steatosis and reverses insulin resistance in lipoatrophic A-ZIP/F-1 mice. *J Biol Chem* 277: 24484–24489.

[50] Akbiyik F, Cinar K, Demirpence E, Ozsullu T, Tunca R, Haziroglu R, Yurdaydin C, Uzunalimoglu O, Bozkaya H (2004) Ligand-induced expression of peroxisome proliferator-activated receptor alpha and activation of fatty acid oxidation enzymes in fatty liver. *Eur J Clin Invest* 34: 429–435.

[51] van Raalte DH, Li M, Pritchard PH, Wasan KM (2004) Peroxisome proliferator-activated receptor (PPAR)-alpha: a pharmacological target with a promising future. *Pharm Res* 21: 1531–1538.

[52] Haluzik MM, Lacinova Z, Dolinkova M, Haluzikova D, Housa D, Horinek A, Vernerova Z, Kumstyrova T, Haluzik M (2006) Improvement of insulin sensitivity after peroxisome proliferator-activated receptor-alpha agonist treatment is accompanied by paradoxical increase of circulating resistin levels. *Endocrinology* 147: 4517–4524.

[53] Nagasawa T, Inada Y, Nakano S, Tamura T, Takahashi T, Maruyama K, Yamazaki Y, Kuroda J, Shibata N (2006) Effects of bezafibrate, PPAR pan-agonist, and GW501516, PPARdelta agonist, on development of steatohepatitis in mice fed a methionine- and choline-deficient diet. *Eur J Pharmacol* 536: 182–191.

[54] Nakamuta M, Morizono S, Soejima Y, Yoshizumi T, Aishima S, Takasugi S, Yoshimitsu K, Enjoji M, Kotoh K, Taketomi A, Uchiyama H, Shimada M, Nawata H, Maehara Y (2005) Short-term intensive treatment for donors with hepatic steatosis in living-donor liver transplantation. *Transplantation* 80: 608–612.

[55] Perkins JD (2006) Saying "Yes" to obese living liver donors: short-term intensive treatment for donors with hepatic steatosis in living-donor liver transplantation. *Liver Transpl* 12: 1012–1013.

[56] Li ZZ, Berk M, McIntyre TM, Feldstein AE (2009) Hepatic lipid partitioning and liver damage in nonalcoholic fatty liver disease: role of stearoyl-CoA desaturase. *J Biol Chem* 284: 5637–5644.

[57] Keller H, Dreyer C, Medin J, Mahfoudi A, Ozato K, Wahli W (1993) Fatty acids and retinoids control lipid metabolism through activation of peroxisome proliferator-activated receptor-retinoid X receptor heterodimers. *Proc Natl Acad Sci U S A* 90: 2160–2164.

[58] Sekiya M, Yahagi N, Matsuzaka T, Najima Y, Nakakuki M, Nagai R, Ishibashi S, Osuga J, Yamada N, Shimano H (2003) Polyunsaturated fatty acids ameliorate hepatic steatosis in obese mice by SREBP-1 suppression. *Hepatology* 38: 1529–1539.

[59] Gutiérrez AM, Reboredo GR, Mosca SM, Catalá A (2007) Non-enzymatic lipid peroxidation of microsomes and mitochondria from liver, heart and brain of the bird Lonchura striata: relationship with fatty acid composition. *Comp Biochem Physiol A Mol Integr Physiol* 146: 415–421.

[60] Le Foll C, Corporeau C, Le Guen V, Gouygou J-P, Bergé J-P, Delarue J (2007) Long-chain n-3 polyunsaturated fatty acids dissociate phosphorylation of Akt from phosphatidylinositol 3′-kinase activity in rats. *Am J Physiol Endocrinol Metab* 292: E1223–E1230.

[61] Feng Q, Gou XJ, Meng SX, Huang C, Zhang YQ, Tang YJ, Wang WJ, Xu L, Peng JH, Hu YY (2013) Qushi Huayu Decoction inhibits hepatic lipid accumulation by activating AMP-activated protein kinase in vivo and in vitro. *Evid Based Complement Alternat Med* 2013: 184358–.

[62] Zhang H, Feng Q, Li HS, Chen SD, Wang XN, Peng JH, Zhang N, Hu YY (2008) Effects of Qushi Huayu Decoction on cathepsin B and tumor necrosis factor-alpha expression in rats with non-alcoholic steatohepatitis. *Zhong Xi Yi Jie He Xue Bao* 6: 928–933.

[63] Feng Q, Cheng Y, Hu YY, Zhang H, Peng JH, Zhang N (2010) Qushi Huayu Decoction inhibits protein and gene expression of cathepsin B in HepG2 cells induced by free fatty acids. *Chin J Integr Med* 16: 518–524.

[64] Kwon HJ, Kim Y-Y, Chung SY (2004) Effects of natural product extract on the fatty liver induced by alcohol diet in rats. *J Health Sci* 50: 466–473.

[65] Park CM, Youn HJ, Chang HK, Song YS (2010) TOP1 and 2, polysaccharides from Taraxacum officinale, attenuate CCl(4)-induced hepatic damage through the modulation of NF-kappaB and its regulatory mediators. *Food Chem Toxicol* 48: 1255–1261.

[66] Aghazadeh S, Amini R, Yazdanparast R, Ghaffari SH (2011) Anti-apoptotic and anti-inflammatory effects of Silybum marianum in treatment of experimental steatohepatitis. *Exp Toxicol Pathol* 63: 569–574.

[67] González-Correa JA, de la Cruz JP, Gordillo J, Ureña I, Redondo L, Sánchez de la Cuesta F (2002) Effects of silymarin MZ-80 on hepatic oxidative stress in rats with biliary obstruction. *Pharmacology* 64: 18–27.

[68] Cui X, Sakaguchi T, Ishizuka D, Tsukada K, Hatakeyama K (1998) Orally administered ginseng extract reduces serum total cholesterol and triglycerides that induce fatty liver in 66% hepatectomized rats. *J Int Med Res* 26: 181–187.

[69] Bak MJ, Jun M, Jeong WS (2012) Antioxidant and hepatoprotective effects of the red ginseng essential oil in H(2)O(2)-treated HepG2 cells and CCl(4)-treated mice. *Int J Mol Sci* 13: 2314–2330.

[70] Chang SH, Choi Y, Park JA, Jung DS, Shin J, Yang JH, Ko SY, Kim SW, Kim JK (2007) Anti-inflammatory effects of BT-201, an n-butanol extract of Panax notoginseng, observed in vitro and in a collagen-induced arthritis model. *Clin Nutr* 26: 785–791.

[71] Wu H-T, He X-J, Hong Y-K, Ma T, Xu Y-P, Li H-H (2010) Chemical characterization of Lycium barbarum polysaccharides and its inhibition against liver oxidative injury of high-fat mice. *Int J Biol Macromol* 46: 540–543.

[72] Lee M-S, Kim I-H, Kim C-T, Kim Y (2011) Reduction of body weight by dietary garlic is associated with an increase in uncoupling protein mRNA expression and activation of AMP-activated protein kinase in diet-induced obese mice. *J Nutr* 141: 1947–1953.

[73] Kim MH, Kim MJ, Lee JH, Han JI, Kim JH, Sok DE, Kim MR (2011) Hepatoprotective effect of aged black garlic on chronic alcohol-induced liver injury in rats. *J Med Food* 14: 732–738.

[74] Park HJ, DiNatale DA, Chung M-Y, Park Y-K, Lee J-Y, Koo SI, O'Connor M, Manautou JE, Bruno RS (2011) Green tea extract attenuates hepatic steatosis by decreasing adipose lipogenesis and enhancing hepatic antioxidant defenses in ob/ob mice. *J Nutr Biochem* 22: 393–400.

[75] Park HJ, Lee J-Y, Chung M-Y, Park Y-K, Bower AM, Koo SI, Giardina C, Bruno RS (2012) Green tea extract suppresses NFκB activation and inflammatory responses in diet-induced obese rats with nonalcoholic steatohepatitis. *J Nutr* 142: 57–63.

[76] Chung M-Y, Park HJ, Manautou JE, Koo SI, Bruno RS (2012) Green tea extract protects against nonalcoholic steatohepatitis in ob/ob mice by decreasing oxidative and nitrative stress responses induced by proinflammatory enzymes. *J Nutr Biochem* 23: 361–367.

[77] Senanayake GK, Fukuda N, Nshizono S, Wang Y-M, Nagao K, Yanagita T, Iwamoto M, Ohta H (2012) Mechanisms underlying decreased hepatic triacylglycerol and cholesterol by dietary bitter melon extract in the rat. *Lipids* 47: 495–503.

[78] Chaturvedi P, George S (2010) Momordica charantia maintains normal glucose levels and lipid profiles and prevents oxidative stress in diabetic rats subjected to chronic sucrose load. *J Med Food* 13: 520–527.

[79] Kaviarasan S, Viswanathan P, Anuradha CV (2007) Fenugreek seed (Trigonella foenum graecum) polyphenols inhibit ethanol-induced collagen and lipid accumulation in rat liver. *Cell Biol Toxicol* 23: 373–383.

[80] Uemura T, Goto T, Kang MS, Mizoguchi N, Hirai S, Lee JY, Nakano Y, Shono J, Hoshino S, Taketani K, Tsuge N, Narukami T, Makishima M, Takahashi N, Kawada T (2011) Diosgenin, the main aglycon of fenugreek, inhibits LXRalpha activity in HepG2 cells and decreases plasma and hepatic triglycerides in obese diabetic mice. *J Nutr* 141: 17–23.

[81] Yoshikawa M, Matsui Y, Kawamoto H, Umemoto N, Oku K, Koizumi M, Yamao J, Kuriyama S, Nakano H, Hozumi N, Ishizaka S, Fukui H (1997) Effects of glycyrrhizin on immune-mediated cytotoxicity. *J Gastroenterol Hepatol* 12: 243–248.

[82] Arase Y, Ikeda K, Murashima N, Chayama K, Tsubota A, Koida I, Suzuki Y, Saitoh S, Kobayashi M, Kumada H (1997) The long term efficacy of glycyrrhizin in chronic hepatitis C patients. *Cancer* 79: 1494–1500.

[83] Greenlee H, Abascal K, Yarnell E, Ladas E (2007) Clinical applications of Silybum marianum in oncology. *Integr Cancer Ther* 6: 158–165.

[84] Kroll DJ, Shaw HS, Oberlies NH (2007) Milk thistle nomenclature: why it matters in cancer research and pharmacokinetic studies. *Integr Cancer Ther* 6: 110–119.

[85] Hogan FS, Krishnegowda NK, Mikhailova M, Kahlenberg MS (2007) Flavonoid, silibinin, inhibits proliferation and promotes cell-cycle arrest of human colon cancer. *J Surg Res* 143: 58–65.

[86] Hikino H, Kiso Y, Wagner H, Fiebig M (1984) Antihepatotoxic actions of flavonolignans from Silybum marianum fruits. *Planta Med* 50: 248–250.

[87] Flora K, Hahn M, Rosen H, Benner K (1998) Milk thistle (Silybum marianum) for the therapy of liver disease. *Am J Gastroenterol* 93: 139–143.

[88] Luper S (1998) A review of plants used in the treatment of liver disease: part 1. *Altern Med Rev* 3: 410–421.

[89] Mira L, Silva M, Manso CF (1994) Scavenging of reactive oxygen species by silibinin dihemisuccinate. *Biochem Pharmacol* 48: 753–759.

[90] Valenzuela A, Aspillaga M, Vial S, Guerra R (1989) Selectivity of silymarin on the increase of the glutathione content in different tissues of the rat. *Planta Med* 55: 420–422.

[91] Saliou C, Rihn B, Cillard J, Okamoto T, Packer L (1998) Selective inhibition of NF-κB activation by the flavonoid hepatoprotector silymarin in HepG2: evidence for different activating pathways. *FEBS Lett* 440: 8–12.

[92] Schümann J, Prockl J, Kiemer AK, Vollmar AM, Bang R, Tiegs G (2003) Silibinin protects mice from T cell-dependent liver injury. *J Hepatol* 39: 333–340.

[93] Saller R, Meier R, Brignoli R (2001) The use of silymarin in the treatment of liver diseases. *Drugs* 61: 2035–2063.

[94] Jacobs BP, Dennehy C, Ramirez G, Sapp J, Lawrence VA (2002) Milk thistle for the treatment of liver disease: a systematic review and meta-analysis. *Am J Med* 113: 506–515.

[95] Vuksan V, Stavro MP, Sievenpiper JL, Beljan-Zdravkovic U, Leiter LA, Josse RG, Xu Z (2000) Similar postprandial glycemic reductions with escalation of dose and administration time of American ginseng in type 2 diabetes. *Diabetes Care* 23: 1221–1226.

[96] Soonthornpun S, Rattarasarn C, Leelawattana R, Setasuban W (1999) Postprandial plasma glucose: a good index of glycemic control in type 2 diabetic patients having near-normal fasting glucose levels. *Diabetes Res Clin Pract* 46: 23–27.

[97] Turner RC (1998) The U.K. prospective diabetes study. A review. *Diabetes Care* 21: C35–C38.

[98] Muwalla MM, Abuirmeileh NM (1990) Suppression of avian hepatic cholesterogenesis by dietary ginseng. *J Nutr Biochem* 1: 518–521.

[99] Zheng X, Yan Y (1991) The effect of ginsenosides of ginseng stem and leaf (GSL) on the lipid regulation and lipid peroxidation in chronic hyperlipidemic rabbits. *Zhongguo Yaolixue Tonbao* 7: 110–116.

[100] Yamamoto M, Uemura T, Nakama S, Uemiya M, Kumagai A (1983) Serum HDL-cholesterol-increasing and fatty liver-improving actions of Panax ginseng in high cholesterol diet-fed rats with clinical effect on hyperlipidemia in man. *Am J Chin Med* 11: 96–101.

[101] Park Y-S, Kang J-S (2011) Korean red ginseng extract suppresses the progression of alcoholic fatty liver in a rat model. *J Health Sci* 57: 512–520.

[102] Cicero AF, Vitale G, Savino G, Arletti R (2003) Panax notoginseng (Burk.) effects on fibrinogen and lipid plasma level in rats fed on a high-fat diet. *Phytother Res* 17: 174–178.

[103] Ji W, Gong BQ (2007) Hypolipidemic effects and mechanisms of Panax notoginseng on lipid profile in hyperlipidemic rats. *J Ethnopharmacol* 113: 318–324.

[104] Jin UH, Park SG, Suh SJ, Kim JK, Kim DS, Moon SK, Lee YC, Park WH, Kim CH (2007) Inhibitory effect of Panax notoginseng on nitric oxide synthase, cyclo-oxygenase-2 and neutrophil functions. *Phytother Res* 21: 142–148.

[105] Kim MS, Lee KT, Iseli TJ, Hoy AJ, George J, Grewal T, Roufogalis BD (2013) Compound K modulates fatty acid-induced lipid droplet formation and expression of proteins involved in lipid metabolism in hepatocytes. *Liver Int* 33: 1583–1593.

[106] Song SB, Tung NH, Quang TH, Ngan NTT, Kim KE, Kim YH (2012) Inhibition of TNF-α-mediated NF-κB transcriptional activity in HepG2 cells by dammarane-type saponins from Panax ginseng leaves. *J Ginseng Res* 36: 146–152.

[107] Geng J, Peng W, Huang Y, Fan H, Li S (2010) Ginsenoside-Rg1 from Panax notoginseng prevents hepatic fibrosis induced by thioacetamide in rats. *Eur J Pharmacol* 634: 162–169.

[108] Zhang HS, Wang SQ (2006) Notoginsenoside R1 from Panax notoginseng inhibits TNF-alpha-induced PAI-1 production in human aortic smooth muscle cells. *Vascul Pharmacol* 44: 224–230.

[109] Xiao J, So KF, Liong EC, Tipoe GL (2013) Recent advances in the herbal treatment of non-alcoholic fatty liver disease. *J Tradit Complement Med* 3: 88–94.

[110] Namita P, Mukesh R, Vijay KJ (2012) Camellia sinensis (green tea): a review. *Global J Pharmacol* 6: 52–59.

[111] Sharangi AB (2009) Medicinal and therapeutic potentialities of tea (Camellia sinensis L.)—a review. *Food Res Int* 42: 529–535.

[112] Lee M-J, Maliakal P, Chen L, Meng X, Bondoc FY, Prabhu S, Lambert G, Mohr S, Yang CS (2002) Pharmacokinetics of tea catechins after ingestion of green tea and (–)-epigallocatechin-3-gallate by humans: formation of different metabolites and individual variability. *Cancer Epidemiol Biomarkers Prev* 11: 1025–1032.

[113] Feng WY (2006) Metabolism of green tea catechins: an overview. *Curr Drug Metab* 7: 755–809.

[114] Lambert JD, Sang S, Yang CS (2007) Biotransformation of green tea polyphenols and the biological activities of those metabolites. *Mol Pharm* 4: 819–825.

[115] Dulloo AG, Duret C, Rohrer D, Girardier L, Mensi N, Fathi M, Chantre P, Vandermander J (1999) Efficacy of a green tea extract rich in catechin polyphenols and caffeine in increasing 24-h energy expenditure and fat oxidation in humans. *Am J Clin Nutr* 70: 1040–1045.

[116] Hasegawa R, Chujo T, Sai-Kato K, Umemura T, Tanimura A, Kurokawa Y (1995) Preventive effects of green tea against liver oxidative DNA damage and hepatotoxicity in rats treated with 2-nitropropane. *Food Chem Toxicol* 33: 961–970.

[117] Cai Y-J, Ma L-P, Hou L-F, Zhou B, Yang L, Liu Z-L (2002) Antioxidant effects of green tea polyphenols on free radical initiated peroxidation of rat liver microsomes. *Chem Phys Lipids* 120: 109–117.

[118] Kim H-J, Jeon S-M, Lee M-K, Jung UJ, Shin S-K, Choi M-S (2009) Antilipogenic effect of green tea extract in C57BL/6J-Lepob/ob mice. *Phytother Res* 23: 467–471.

[119] Bruno RS, Dugan CE, Smyth JA, DiNatale DA, Koo SI (2008) Green tea extract protects leptin-deficient, spontaneously obese mice from hepatic steatosis and injury. *J Nutr* 138: 323–331.

[120] Kuzu N, Bahcecioglu IH, Dagli AF, Ozercan IH, Ustündag B, Sahin K (2008) Epigallocatechin gallate attenuates experimental non-alcoholic steatohepatitis induced by high fat diet. *J Gastroenterol Hepatol* 23: e465–e470.

[121] Bose M, Lambert JD, Ju J, Reuhl KR, Shapses SA, Yang CS (2008) The major green tea polyphenol, (-)-epigallocatechin-3-gallate, inhibits obesity, metabolic syndrome, and fatty liver disease in high-fat–fed mice. *J Nutr* 138: 1677–1683.

[122] Ueno T, Torimura T, Nakamura T, Sivakumar R, Nakayama H, Otabe S, Yuan X, Yamada K, Hashimoto O, Inoue K, Koga H, Sata M (2009) Epigallocatechin-3-gallate improves nonalcoholic steatohepatitis model mice expressing nuclear sterol regulatory element binding protein-1c in adipose tissue. *Int J Mol Med* 24: 17–22.

[123] Shrestha S, Ehlers SJ, Lee J-Y, Fernandez M-L, Koo SI (2009) Dietary green tea extract lowers plasma and hepatic triglycerides and decreases the expression of sterol regulatory element-binding protein-1c mRNA and its responsive genes in fructose-fed, ovariectomized rats. *J Nutr* 139: 640–645.

[124] Hong Byun E, Fujimura Y, Yamada K, Tachibana H (2010) TLR4 signaling inhibitory pathway induced by green tea polyphenol epigallocatechin-3-gallate through 67-kDa laminin receptor. *J Immunol* 185: 33–45.

[125] Jin X, Zheng R-H, Li Y-M (2008) Green tea consumption and liver disease: a systematic review. *Liver Int* 28: 990–996.

[126] Imai K, Nakachi K (1995) Cross sectional study of effects of drinking green tea on cardiovascular and liver diseases. *BMJ* 310: 693–696.

[127] Basch E, Kuo G, Smith M (2003) Therapeutic applications of fenugreek. *Altern Med Rev* 8: 20–27.

[128] Bukhari SB, Bhanger MI, Memon S (2008) Antioxidative activity of extracts from fenugreek seeds (Trigonella foenum-graecum). *Pak J Anal Environ Chem* 9: 78–83.

[129] Vijayakumar MV, Pandey V, Mishra GC, Bhat MK (2010) Hypolipidemic effect of fenugreek seeds is mediated through inhibition of fat accumulation and upregulation of LDL receptor. *Obesity* 18: 667–674.

[130] Kawabata T, Cui M-Y, Hasegawa T, Takano F, Ohta T (2011) Anti-inflammatory and anti-melanogenic steroidal saponin glycosides from fenugreek (Trigonella foenum-graecum L.) seeds. *Planta Med* 77: 705–710.

[131] Raju J, Bird RP (2006) Alleviation of hepatic steatosis accompanied by modulation of plasma and liver TNF-α levels by Trigonella foenum graecum (fenugreek) seeds in Zucker obese (fa/fa) rats. *Int J Obes* 30: 1298–1307.

[132] Sharma RD, Raghuram TC, Dayasagar RV (1991) Hypolipidaemic effect of fenugreek seeds. A clinical study. *Phytother Res* 3: 145–147.

[133] Moosa ASM, Rashid MU, Asadi AZS, Ara N, Uddin MM, Ferdaus A (2006) Hypolipidemic effects of fenugreek seed powder. *Bangladesh J Pharmacol* 1: 64–67.

[134] Kilgore KS, Tanhehco EJ, Park JL, Naylor KB, Anderson MB, Lucchesi BR (1998) Reduction of myocardial infarct size in vivo by carbohydrate-based glycomimetics. *J Pharmacol Exp Ther* 284: 427–435.

[135] Shiki Y, Shirai K, Saito Y, Yoshida SHO, Mori Y, Wakashin M (1992) Effect of glycyrrhizin on lysis of hepatocyte membranes induced by anti-liver cell membrane antibody. *J Gastroenterol Hepatol* 7: 12–16.

[136] Rossum TGJV, Vulto AG, Man RAD, Brouwer JT, Schalm SW (1998) Review article: glycyrrhizin as a potential treatment for chronic hepatitis C. *Aliment Pharmacol Ther* 12: 199–205.

[137] Abe M, Akbar F, Hasebe A, Horiike N, Onji M (2003) Glycyrrhizin enhances interleukin-10 production by liver dendritic cells in mice with hepatitis. *J Gastroenterol Hepatol* 38: 962–967.

[138] Al-Razzuqi RAM, Al-Jawad FH, Hussaini JAA, Al-Jeboori AA (2012) Hepatoprotective effect of Glycyrrhiza glabra in carbon tetrachloride-induced model of acute liver injury. *J Phys Pharm Adv* 2: 259–263.

[139] Dal Rhee S, Kim C-H, Seon Park J, Hoon Jung W, Bum Park S, Youn Kim H, Hwan Bae G, Jan Kim T, Young Kim K (2012) Carbenoxolone prevents the development of fatty liver in C57BL/6-Lep ob/ob mice via the inhibition of sterol regulatory element binding protein-1c activity and apoptosis. *Eur J Pharmacol* 691: 9–18.

[140] Emmanuel A, EQuigley E (2013) *Irritable Bowel Syndrome Diagnosis and Clinic Management*. West Sussex: Wiley-Blackwell.

[141] Longstreth GF, Wilson A, Knight K, Wong J, Chiou CF, Barghout V, Frech F, Ofman JJ (2003) Irritable bowel syndrome, health care use, and costs: a U.S. managed care perspective. *Am J Gastroenterol* 98: 600–607.

[142] Krogsgaard LR, Engsbro AL, Bytzer P (2013) The epidemiology of irritable bowel syndrome in Denmark. A population-based survey in adults </=50 years of age. *Scand J Gastroenterol* 48: 523–529.

[143] Gwee KA, Bak YT, Ghoshal UC, Gonlachanvit S, Lee OY, Fock KM, Chua AS, Lu CL, Goh KL, Kositchaiwat C, Makharia G, Park HJ, Chang FY, Fukudo S, Choi MG, Bhatia S, Ke M, Hou X, Hongo M (2010) Asian consensus on irritable bowel syndrome. *J Gastroenterol Hepatol* 25: 1189–1205.

[144] Gwee KA, Lu CL, Ghoshal UC (2009) Epidemiology of irritable bowel syndrome in Asia: something old, something new, something borrowed. *J Gastroenterol Hepatol* 24: 1601–1607.

[145] Saito YA, Schoenfeld P, Locke GR, 3rd (2002) The epidemiology of irritable bowel syndrome in North America: a systematic review. *Am J Gastroenterol* 97: 1910–1915.

[146] Thompson WG, Irvine EJ, Pare P, Ferrazzi S, Rance L (2002) Functional gastrointestinal disorders in Canada: first population-based survey using Rome II criteria with suggestions for improving the questionnaire. *Dig Dis Sci* 47: 225–235.

[147] Trinkley KE, Nahata MC (2011) Treatment of irritable bowel syndrome. *J Clin Pharm Ther* 36: 275–282.

[148] Clark C, DeLegge M (2008) Irritable bowel syndrome: a practical approach. *Nutr Clin Pract* 23: 263–267.

[149] Crowell MD, Harris L, Jones MP, Chang L (2005) New insights into the pathophysiology of irritable bowel syndrome: implications for future treatments. *Curr Gastroenterol Rep* 7: 272–279.

[150] Karantanos T, Markoutsaki T, Gazouli M, Anagnou NP, Karamanolis DG (2010) Current insights in to the pathophysiology of Irritable Bowel Syndrome. *Gut Pathog* 2: 3.

[151] Hasler WL (2011) Traditional thoughts on the pathophysiology of irritable bowel syndrome. *Gastroenterol Clin North Am* 40: 21–43.

[152] Ohman L, Simren M (2010) Pathogenesis of IBS: role of inflammation, immunity and neuroimmune interactions. *Nat Rev Gastroenterol Hepatol* 7: 163–173.

[153] Manabe N, Wong BS, Camilleri M, Burton D, McKinzie S, Zinsmeister AR (2010) Lower functional gastrointestinal disorders: evidence of abnormal colonic transit in a 287 patient cohort. *Neurogastroenterol Motil* 22: 293–e282.

[154] Deiteren A, Camilleri M, Burton D, McKinzie S, Rao A, Zinsmeister AR (2010) Effect of meal ingestion on ileocolonic and colonic transit in health and irritable bowel syndrome. *Dig Dis Sci* 55: 384–391.

[155] Lee OY (2010) Asian motility studies in irritable bowel syndrome. *J Neurogastroenterol Motil* 16: 120–130.

[156] Mertz H, Naliboff B, Munakata J, Niazi N, Mayer EA (1995) Altered rectal perception is a biological marker of patients with irritable bowel syndrome. *Gastroenterology* 109: 40–52.

[157] Nozu T, Kudaira M, Kitamori S, Uehara A (2006) Repetitive rectal painful distention induces rectal hypersensitivity in patients with irritable bowel syndrome. *J Gastroenterol* 41: 217–222.

[158] Bueno L, Fioramonti J (2002) Visceral perception: inflammatory and non-inflammatory mediators. *Gut* 51 Suppl 1: i19–i23.

[159] Chey WD, Maneerattaporn M, Saad R (2011) Pharmacologic and complementary and alternative medicine therapies for irritable bowel syndrome. *Gut Liver* 5: 253–266.

[160] Kong SC, Hurlstone DP, Pocock CY, Walkington LA, Farquharson NR, Bramble MG, McAlindon ME, Sanders DS (2005) The incidence of self-prescribed oral complementary and alternative medicine use by patients with gastrointestinal diseases. *J Clin Gastroenterol* 39: 138–141.

[161] van Tilburg MA, Palsson OS, Levy RL, Feld AD, Turner MJ, Drossman DA, Whitehead WE (2008) Complementary and alternative medicine use and cost in functional bowel disorders: a six month prospective study in a large HMO. *BMC Complement Altern Med* 8: 46.

[162] Bensky D, Barolet R (1990) *Chinese Herbal Medicine: Formulas and Strategies*. Seattle: Eastland Press.

[163] Hu XG, Xu D, Zhao Y, Yang XB, Meng J, Shen H, Guo J (2009) The alleviating pain effect of aqueous extract from Tong-Xie-Yao-Fang, on experimental visceral hypersensitivity and its mechanism. *Biol Pharm Bull* 32: 1075–1079.

[164] Bian Z, Wu T, Liu L, Miao J, Wong H, Song L, Sung JJ (2006) Effectiveness of the Chinese herbal formula TongXieYaoFang for irritable bowel syndrome: a systematic review. *J Altern Complement Med* 12: 401–407.

[165] Pan F, Zhang T, Zhang YH, Xu JJ, Chen FM (2009) Effect of Tongxie Yaofang Granule in treating diarrhea-predominate irritable bowel syndrome. *Chin J Integr Med* 15: 216–219.

[166] Liu JP, Yang M, Liu YX, Wei M, Grimsgaard S (2006) Herbal medicines for treatment of irritable bowel syndrome. *Cochrane Database Syst Rev* CD004116.

[167] Alam MS, Roy PK, Miah AR, Mollick SH, Khan MR, Mahmud MC, Khatun S (2013) Efficacy of Peppermint oil in diarrhea predominant IBS - a double blind randomized placebo - controlled study. *Mymensingh Med J* 22: 27–30.

[168] Cappello G, Spezzaferro M, Grossi L, Manzoli L, Marzio L (2007) Peppermint oil (Mintoil) in the treatment of irritable bowel syndrome: a prospective double blind placebo-controlled randomized trial. *Dig Liver Dis* 39: 530–536.

[169] Kline RM, Kline JJ, Di Palma J, Barbero GJ (2001) Enteric-coated, pH-dependent peppermint oil capsules for the treatment of irritable bowel syndrome in children. *J Pediatr* 138: 125–128.

[170] Tian FJ, Hu CJ, Chen Y (2013) Research progress of Traditional Chinese non-drug therapy in the treatment of irritable bowel syndrome. *Chin J Chin Med* 28: 118–120.

[171] Lim B, Manheimer E, Lao L, Ziea E, Wisniewski J, Liu J, Berman B (2006) Acupuncture for treatment of irritable bowel syndrome. *Cochrane Database Syst Rev* May 16 (5): CD005111.

[172] Noldin VF, Cechinel V, Delle Monache F, Benassi JC, Christmann IL, Pedrosa RC, Yunes RA (2003) Chemical composition and biological activities of the leaves of Cynara scolymus L. (artichoke) cultivated in Brazil. *Quím Nova* 26: 331–334.

[173] Schutz K, Kammerer D, Carle R, Schieber A (2004) Identification and quantification of caffeoylquinic acids and flavonoids from artichoke (Cynara scolymus L.) heads, juice, and pomace by HPLC-DAD-ESI/MS(n). *J Agric Food Chem* 52: 4090–4096.

[174] Nassar MI, Mohamed TK, Elshamy AI, El-Toumy SA, Lateef AM, Farrag AR (2013) Chemical constituents and anti-ulcerogenic potential of the scales of Cynara scolymus (artichoke) heads. *J Sci Food Agric* 93: 2494–2501.

[175] Emendorfer F, Emendorfer F, Bellato F, Noldin VF, Niero R, Cechinel-Filho V, Cardozo AM (2005) Evaluation of the relaxant action of some Brazilian medicinal plants in isolated guinea-pig ileum and rat duodenum. *J Pharm Pharm Sci* 8: 63–68.

[176] Emendorfer F, Emendorfer F, Bellato F, Noldin VF, Cechinel-Filho V, Yunes RA, Delle Monache F, Cardozo AM (2005) Antispasmodic activity of fractions and cynaropicrin from Cynara scolymus on guinea-pig ileum. *Biol Pharm Bull* 28: 902–904.

[177] Li H, Xia N, Brausch I, Yao Y, Forstermann U (2004) Flavonoids from artichoke (Cynara scolymus L.) up-regulate endothelial-type nitric-oxide synthase gene expression in human endothelial cells. *J Pharmacol Exp Ther* 310: 926–932.

[178] Ishida K, Kojima R, Tsuboi M, Tsuda Y, Ito M (2010) Effects of artichoke leaf extract on acute gastric mucosal injury in rats. *Biol Pharm Bull* 33: 223–229.

[179] Bundy R, Walker AF, Middleton RW, Marakis G, Booth JC (2004) Artichoke leaf extract reduces symptoms of irritable bowel syndrome and improves quality of life in otherwise healthy volunteers suffering from concomitant dyspepsia: a subset analysis. *J Altern Complement Med* 10: 667–669.

[180] Walker AF, Middleton RW, Petrowicz O (2001) Artichoke leaf extract reduces symptoms of irritable bowel syndrome in a post-marketing surveillance study. *Phytother Res* 15: 58–61.

[181] Kraft K (1997) Artichoke leaf extract—recent findings reflecting effects on lipid metabolism, liver and gastrointestinal tracts. *Phytomedicine* 4: 369–378.

[182] Leung AY, Foster S (1996) *Encyclopedia of Common Natural Ingredients Used in Food, Drugs and Cosmetics*. 2nd ed. New York: John Wiley & Sons.

[183] Menghini L, Genovese S, Epifano F, Tirillini B, Ferrante C, Leporini L (2010) Antiproliferative, protective and antioxidant effects of artichoke, dandelion, turmeric and rosemary extracts and their formulation. *Int J Immunopathol Pharmacol* 23: 601–610.

[184] Wittemer SM, Ploch M, Windeck T, Muller SC, Drewelow B, Derendorf H, Veit M (2005) Bioavailability and pharmacokinetics of caffeoylquinic acids and flavonoids after oral administration of Artichoke leaf extracts in humans. *Phytomedicine* 12: 28–38.

[185] McKay DL, Blumberg JB (2006) A review of the bioactivity and potential health benefits of peppermint tea (Mentha piperita L.). *Phytother Res* 20: 619–633.

[186] Merat S, Khalili S, Mostajabi P, Ghorbani A, Ansari R, Malekzadeh R (2010) The effect of enteric-coated, delayed-release peppermint oil on irritable bowel syndrome. *Dig Dis Sci* 55: 1385–1390.

[187] Grigoleit HG, Grigoleit P (2005) Peppermint oil in irritable bowel syndrome. *Phytomedicine* 12: 601–606.

[188] Grigoleit HG, Grigoleit P (2005) Pharmacology and preclinical pharmacokinetics of peppermint oil. *Phytomedicine* 12: 612–616.

[189] Kligler B, Chaudhary S (2007) Peppermint oil. *Am Fam Physician* 75: 1027–1030.

[190] Pistelli L, Bertoli A, Gelli F, Bedini L, Ruffoni B, Pistelli L (2012) Production of Curcuminoids in different in vitro organs of Curcuma longa. *Nat Prod Commun* 7: 1037–1042.

[191] Aggarwal BB, Sundaram C, Malani N, Ichikawa H (2007) Curcumin: the Indian solid gold. *Adv Exp Med Biol* 595: 1–75.

[192] Singh S, Aggarwal BB (1995) Activation of transcription factor NF-kappa B is suppressed by curcumin (diferuloylmethane). *J Biol Chem* 270: 24995–25000.

[193] Araujo CC, Leon LL (2001) Biological activities of Curcuma longa L. *Mem Inst Oswaldo Cruz* 96: 723–728.

[194] Bundy R, Walker AF, Middleton RW, Booth J (2004) Turmeric extract may improve irritable bowel syndrome symptomology in otherwise healthy adults: a pilot study. *J Altern Complement Med* 10: 1015–1018.

[195] Joshi J, Ghaisas S, Vaidya A, Vaidya R, Kamat DV, Bhagwat AN, Bhide S (2003) Early human safety study of turmeric oil (Curcuma longa oil) administered orally in healthy volunteers. *J Assoc Physicians India* 51: 1055–1060.

[196] Mayanglambam A, Dangelmaier CA, Thomas D, Damodar Reddy C, Daniel JL, Kunapuli SP (2010) Curcumin inhibits GPVI-mediated platelet activation by interfering with the kinase activity of Syk and the subsequent activation of PLCgamma2. *Platelets* 21: 211–220.

[197] Chen HW, Kuo HT, Chai CY, Ou JL, Yang RC (2007) Pretreatment of curcumin attenuates coagulopathy and renal injury in LPS-induced endotoxemia. *J Endotoxin Res* 13: 15–23.

[198] Navis I, Sriganth P, Premalatha B (1999) Dietary curcumin with cisplatin administration modulates tumour marker indices in experimental fibrosarcoma. *Pharmacol Res* 39: 175–179.

[199] Notarbartolo M, Poma P, Perri D, Dusonchet L, Cervello M, D'Alessandro N (2005) Antitumor effects of curcumin, alone or in combination with cisplatin or doxorubicin, on human hepatic cancer cells. Analysis of their possible relationship to changes in NF-kB activation levels and in IAP gene expression. *Cancer Lett* 224: 53–65.

[200] Venkatesan N, Chandrakasan G (1995) Modulation of cyclophosphamide-induced early lung injury by curcumin, an anti-inflammatory antioxidant. *Mol Cell Biochem* 142: 79–87.

[201] Juan H, Terhaag B, Cong Z, Bi-Kui Z, Rong-Hua Z, Feng W, Fen-Li S, Juan S, Jing T, Wen-Xing P (2007) Unexpected effect of concomitantly administered curcumin on the pharmacokinetics of talinolol in healthy Chinese volunteers. *Eur J Clin Pharmacol* 63: 663–668.

[202] Liu AC, Zhao LX, Xing J, Liu T, Du FY, Lou HX (2012) Pre-treatment with curcumin enhances plasma concentrations of losartan and its metabolite EXP3174 in rats. *Biol Pharm Bull* 35: 145–150.

[203] Bourgou S, Rahali FZ, Ourghemmi I, Saidani Tounsi M (2012) Changes of peel essential oil composition of four Tunisian citrus during fruit maturation. *ScientificWorldJournal* 2012: 528593–.

[204] Yi LT, Li JM, Li YC, Pan Y, Xu Q, Kong LD (2008) Antidepressant-like behavioral and neurochemical effects of the citrus-associated chemical apigenin. *Life Sci* 82: 741–751.

[205] Li Y, Xu C, Zhang Q, Liu JY, Tan RX (2005) In vitro anti-Helicobacter pylori action of 30 Chinese herbal medicines used to treat ulcer diseases. *J Ethnopharmacol* 98: 329–333.

[206] Moraes TM, Kushima H, Moleiro FC, Santos RC, Rocha LR, Marques MO, Vilegas W, Hiruma-Lima CA (2009) Effects of limonene and essential oil from Citrus aurantium on gastric mucosa: role of prostaglandins and gastric mucus secretion. *Chem Biol Interact* 180: 499–505.

[207] Bharucha AE, Pemberton JH, Locke GR, 3rd (2013) American Gastroenterological Association technical review on constipation. *Gastroenterology* 144: 218–238.

[208] Higgins PD, Johanson JF (2004) Epidemiology of constipation in North America: a systematic review. *Am J Gastroenterol* 99: 750–759.

[209] Irvine EJ, Ferrazzi S, Pare P, Thompson WG, Rance L (2002) Health-related quality of life in functional GI disorders: focus on constipation and resource utilization. *Am J Gastroenterol* 97: 1986–1993.

[210] Bharucha AE (2007) Constipation. *Best Pract Res Clin Gastroenterol* 21: 709–731.

[211] Ford AC, Talley NJ (2012) Laxatives for chronic constipation in adults. *BMJ* 345: e6168–.

[212] Singh S, Rao SS (2010) Pharmacologic management of chronic constipation. *Gastroenterol Clin North Am* 39: 509–527.

[213] Ambizas EM, Ginzburg R (2007) Lubiprostone: a chloride channel activator for treatment of chronic constipation. *Ann Pharmacother* 41: 957–964.

[214] Potter LR (2011) Regulation and therapeutic targeting of peptide-activated receptor guanylyl cyclases. *Pharmacol Ther* 130: 71–82.

[215] Parkman HP, Rao SS, Reynolds JC, Schiller LR, Wald A, Miner PB, Lembo AJ, Gordon JM, Drossman DA, Waltzman L, Stambler N, Cedarbaum JM (2003) Neurotrophin-3 improves functional constipation. *Am J Gastroenterol* 98: 1338–1347.

[216] Diego L, Atayee R, Helmons P, Hsiao G, von Gunten CF (2011) Novel opioid antagonists for opioid-induced bowel dysfunction. *Expert Opin Investig Drugs* 20: 1047–1056.

[217] Koc Z, Topatan S, Saglam Z (2012) Use of and attitudes toward complementary and alternative medicine among midwives in Turkey. *Eur J Obstet Gynecol Reprod Biol* 160: 131–136.

[218] Kim SR, Lee TY, Kim MS, Lee MC, Chung SJ (2009) Use of complementary and alternative medicine by Korean patients with Parkinson's disease. *Clin Neurol Neurosurg* 111: 156–160.

[219] Iwase S, Yamaguchi T, Miyaji T, Terawaki K, Inui A, Uezono Y (2012) The clinical use of Kampo medicines (traditional Japanese herbal treatments) for controlling cancer patients' symptoms in Japan: a national cross-sectional survey. *BMC Complement Altern Med* 12: 222–.

[220] Cheng CW, Kwok AO, Bian ZX, Tse DM (2012) The quintessence of traditional Chinese medicine: syndrome and its distribution among advanced cancer patients with constipation. *Evid Based Complement Alternat Med* 2012: 739642.

[221] Huang CH, Lin JS, Li TC, Lee SC, Wang HP, Lue HC, Su YC (2012) Comparison of a Chinese Herbal Medicine (CCH1) and lactulose as first-line treatment of constipation in long-term care: a randomized, double-blind, double-dummy, and placebo-controlled trial. *Evid Based Complement Alternat Med* 2012: 923190.

[222] Shen H (2009) Discussion on certain issues of the diagnosis and treatment of functional constipation. *Chin J Integr Med* 15: 89–92.

[223] Jia G, Meng MB, Huang ZW, Qing X, Lei W, Yang XN, Liu SS, Diao JC, Hu SY, Lin BH, Zhang RM (2010) Treatment of functional constipation with the Yun-chang capsule: a double-blind, randomized, placebo-controlled, dose-escalation trial. *J Gastroenterol Hepatol* 25: 487–493.

[224] Jong MS, Hwang SJ, Chen YC, Chen TJ, Chen FJ, Chen FP (2010) Prescriptions of Chinese herbal medicine for constipation under the national health insurance in Taiwan. *J Chin Med Assoc* 73: 375–383.

[225] Foster M, Hunter D, Samman S (2011) Evaluation of the nutritional and metabolic effects of Aloe vera. In: Benzie IFF, Wachtel-Galor S, editors. *Herbal Medicine: Biomolecular and Clinical Aspects*. Boca Raton: CRC Press.

[226] Wintola OA, Sunmonu TO, Afolayan AJ (2010) The effect of Aloe ferox Mill. in the treatment of loperamide-induced constipation in Wistar rats. *BMC Gastroenterol* 10: 95.

[227] Bradley PR (1992) *British Herbal Compendium*. Bournemouth: British Herbal Medicine Association.

[228] de Witte P (1993) Metabolism and pharmacokinetics of anthranoids. *Pharmacology* 47 Suppl 1: 86–97.

[229] Ishii Y, Tanizawa H, Takina Y (1990) Studies of Aloe. III.: mechanism of cathartic effect. (2). *Chem Pharm Bull* 38: 197–200.

[230] Vogler BK, Ernst E (1999) Aloe vera: a systematic review of its clinical effectiveness. *Br J Gen Pract* 49: 823–828.

[231] Brunton LL, Chabner BA, Knollmann BC (1990) *Goodman and Gilman's The Pharmacological Basis of Therapeutics*. New York: McGraw Hill.

[232] American Society of Hospital Pharmacists (1990) *American Hospital Formulary Service*. Bethesda: Board of Directors of the American Society of Hospital Pharmacists.

[233] Mehmood MH, Aziz N, Ghayur MN, Gilani AH (2011) Pharmacological basis for the medicinal use of psyllium husk (Ispaghula) in constipation and diarrhea. *Dig Dis Sci* 56: 1460–1471.

[234] Wang HJ, Liang XM, Yu ZL, Zhou LY, Lin SR, Geraint M (2004) A randomised, controlled comparison of low-dose polyethylene glycol 3350 plus electrolytes with ispaghula husk in the treatment of adults with chronic functional constipation. *Clin Drug Investig* 24: 569–576.

[235] Singh B (2007) Psyllium as therapeutic and drug delivery agent. *Int J Pharm* 334: 1–14.

[236] Qvitzau S, Matzen P, Madsen P (1988) Treatment of chronic diarrhoea: loperamide versus ispaghula husk and calcium. *Scand J Gastroenterol* 23: 1237–1240.

[237] Dettmar PW, Sykes J (1998) A multi-centre, general practice comparison of ispaghula husk with lactulose and other laxatives in the treatment of simple constipation. *Curr Med Res Opin* 14: 227–233.

[238] Pal S, Radavelli-Bagatini S (2012) Effects of psyllium on metabolic syndrome risk factors. *Obes Rev* 13: 1034–1047.

[239] Izzo AA (2005) Herb-drug interactions: an overview of the clinical evidence. *Fundam Clin Pharmacol* 19: 1–16.

[240] Sui F, Huo HR, Zhang CB, Yang N, Guo JY, Du XL, Zhao BS, Liu HB, Li LF, Guo SY, Jiang TL (2010) Emodin down-regulates expression of TRPV1 mRNA and its function in DRG neurons in vitro. *Am J Chin Med* 38: 789–800.

[241] Takayama K, Tsutsumi H, Ishizu T, Okamura N (2012) The influence of rhein 8-O-beta-D-glucopyranoside on the purgative action of sennoside A from rhubarb in mice. *Biol Pharm Bull* 35: 2204–2208.

[242] Waller SL, Misiewicz JJ (1969) Prognosis in the irritable-bowel syndrome. A prospective study. *Lancet* 2: 754–756.

[243] Pierre Duez MV, Vanhaelen-Fastre R, Hanocq M, Molle L (1984) Comparison between high-performance thin-layer chromatography-flourometry and high-performance liquid chromatography for the determination of sennosides A and B in Senna (Cassia spp.) pods and leaves. *J Chromatogr A* 303: 391–395.

[244] Leng-Peschlow E (1986) Dual effect of orally administered sennosides on large intestine transit and fluid absorption in the rat. *J Pharm Pharmacol* 38: 606–610.

[245] Gao J, Shi Z, Zhu S, Li GQ, Yan R, Yao M (2013) Influences of processed rhubarbs on the activities of four CYP isozymes and the metabolism of saxagliptin in rats based on probe cocktail and pharmacokinetics approaches. *J Ethnopharmacol* 145: 566–572.

[246] Hennebelle T, Weniger B, Joseph H, Sahpaz S, Bailleul F (2009) Senna alata. *Fitoterapia* 80: 385–393.

[247] Patel M, Schimpf MO, O'Sullivan DM, LaSala CA (2010) The use of senna with docusate for postoperative constipation after pelvic reconstructive surgery: a randomized, double-blind, placebo-controlled trial. *Am J Obstet Gynecol* 202: 479 e471–e475.

[248] Biradar YS, Jagatap S, Khandelwal KR, Singhania SS (2008) Exploring of antimicrobial activity of Triphala Mashi-an ayurvedic formulation. *Evid Based Complement Alternat Med* 5: 107–113.

[249] World Health Organisation (1999) *WHO monographs on selected medicinal plants.* Geneva: World Health Organization.

18

PHYTOMEDICINES FOR INFLAMMATORY CONDITIONS

SIGRUN CHRUBASIK-HAUSMANN
Institute of Forensic Medicine, University of Freiburg, Freiburg, Germany

18.1 TRADITIONAL MEDICINES FOR INFLAMMATORY CONDITIONS IN EUROPE

Since ancient times until the era when chemotherapy was introduced, phytomedicines were the only treatment available for inflammatory conditions and pain. The most popular treatment was the oral use of the leaf or bark of *Salix* species or of the aspen tree, the herb of meadowsweet, goldenrod, or wintergreen. In case of severe pain, unprocessed opium (poppy tears), the dried latex or juice of the seed pod of *Papaver somniferum*, was used, a remedy in use since the Neolithic Age. But inflammatory conditions were also treated topically with the herb of chamomile, peppermint, *Arnica*, comfrey, and others. With the advancement of the practical and theoretical knowledge of chemistry initiated by the Arabs and further developed by the Europeans, isolated plant compounds gained popularity in the treatment of inflammatory conditions: salicin was extracted from willow bark, salicylate from meadowsweet and wintergreen, colchicin from autumn crocus, and morphine from opium. Shortly afterward, compounds could be synthesized that were much cheaper to produce than extracted directly from plants. At the end of the past century, however, the increasing occurrence of severe adverse events (e.g., gastrointestinal, hepatic, and/or renal) during treatment with synthetic nonsteroidal anti-inflammatory drugs (NSAIDs) warranted reconsideration of the use of phyto-anti-inflammatory drugs (PAIDs). In the meantime, research has shed light into the mechanism(s) of action of the PAIDs, and many clinical studies have investigated their efficacy.

Phytotherapies: Efficacy, Safety, and Regulation, First Edition. Edited by Iqbal Ramzan.

18.2 TWENTY-FIRST-CENTURY UPDATE ON PAIDs

Phytomedicines contain multiple compounds with different actions. Identification of the chemical structures is steadily increasing for all PAIDs, but for none of these the composition of the active principle (the sum of all compounds contributing to the PAID's overall action) has been fully characterized. Table 18.1 shows that PAIDs have a broader mechanism of action than synthetic NSAIDs.

The mechanism of action of the latter has been explained on the basis of their inhibition of the enzymes that synthesize prostaglandins. Inhibition of COX-2 (cyclo-oxygenase-2) activity represents the most likely mechanism of action for NSAID-mediated analgesia, while the ratio of inhibition of COX-1 to COX-2 by NSAIDs reflects the likelihood of adverse effects (e.g., gastrointestinal, hepatic, renal). In addition, some NSAIDs inhibit the lipoxygenase pathway that is also involved in the production of algogenic metabolites. Interference with G-protein-mediated signal transduction by NSAIDs is another target independent of other mechanisms and is a central mechanism of action that augments the peripheral mechanism [2]. PAIDs act on the same and/or other similar targets of NSAID action or at targets of synthetic biologicals (chondroprotective drugs). Animal studies indicate a chondroprotective effect for some of the PAIDs (e.g., avocado–soybean unsaponifiables, Devil's claw, *boswellia*); however, clinical studies so far have not proven a demonstrable chondroprotective effect [1, 3]. Topical PAIDs may, in addition to the conventional pathways, have a skin-irritating component or may contain toxic compounds so they cannot be applied orally.

18.3 ORAL EXTRACTS FROM *SALIX* SPECIES

In 1763, an observation was published, "An account of the success of the bark of the willow in the cure of agues," after patients suffering from high temperature were treated with an aqueous extract of dried and powdered willow bark [4]. In the yeast-induced hyperthermia model in rodents, the antipyretic effect of willow bark extract was weaker than that of aspirin in equipotent doses [5]. However, on a milligram-per-kilogram basis, the proprietary aqueous extract, STW 33-I, was at least as effective as aspirin (ASS) in reducing inflammatory exudates. In rats, willow bark extract (in 30% ethanol) inhibited carrageenan-induced paw edema, adjuvant-induced arthritis, heat-induced inflammation, and dextran-induced hind paw edema. The anti-inflammatory effect was dose-dependent; a dose of 120 mg/kg of the ethanolic extract was equivalent to 600 mg/kg ASS. Willow bark extract analgesia was also demonstrated in the hot plate test in mice at doses of 60–120 mg/kg (equivalent to ASS dose of 600 mg/kg) [5].

The empirically chosen daily dose (up to 24 g) by Stone in the eighteenth century might have contained up to 1000 mg of salicin (crude plant material generally contains about 4% salicin [6]). Unfortunately, regulatory authorities have restricted the daily dose recommendation of willow bark preparations to 120 mg [7] or at most 240 mg of salicin [8] One dose-finding study indicated that an

TABLE 18.1 Mechanisms of Action of PAIDs Suggested by *In Vitro* Studies

		Inhibition of					
Oral PAIDs		COX-1	COX-2	LOX	Cytokines	Elastase (E) Hyaluronidase (H)	Antioxidative Effect
Salix species	Willow bark	Y	Y	Y	Y	H	Y
Harpagophytum procumbens	Devil's claw	No activity	Y	Y	Y	E	Y
Persea americana + *Glycine max*	Avocado-soybean Unsaponifiables	Not investigated	Y	Not investigated	X	H	Y
Tripterygium wilfordii	Thunder God vine	Y	Y	Y	Y	Not investigated	Y
Seed oils with GLA	Evening primrose, borage, blackcurrant	Y	Not investigated	Y	Y	Not investigated	Y
Boswellia serrata	Salai guggal	Y	No activity	Y	Y	H	Y pro-oxidative
Rosa canina	Rose hip and seed	Y	Y	No activity	Y	E	Y
Uncaria species	Cat's claw	Y	Y	Not investigated	Y	Not investigated	Y
Zingiber officinalis	Ginger	Y	Y	Y	Y	E	Y
Urtica dioica	Nettle herb	Y	ni	Y	Y	E	Y
Curcuma species	Tumeric root	Y	Y	Y	Y	H	Y
Tanacetum parthenium	Feverfew	Y	Y	Y	Y	E	Y
Pinus pinaster	Polyphenols	Y	Y	Y	Y	E	Y
Topical PAIDs							
Capsicum species	Capsaicin	Not investigated	Y	Y	Not investigated	Not investigated	Y
Mentha piperita	Peppermint	Weak	Y	Y	Y	Not investigated	Y
Arnica montana	Arnica	Y	Y	Y	Y	E	Y
Symphytum officinale	Comfrey	Y	Y	Not investigated	Not investigated	Not investigated	Y

Compared to NSAIDs, phytomedicines act via different targets reflecting their multicomponent active principle. The overall action of phytomedicines is the sum of the coactive compounds' actions.

COX-1, cyclooxygenase-1; COX-2, cyclooxygenase-2; LOX, lipoxygenase; Y, yes (modified after Cameron et al., 2009 [1]).

extract dose with 240 mg salicin/day was superior to an extract dose with 120 mg salicin/day in treating acute exacerbations of chronic low-back pain [9, 10]. However, the putative beneficial effects of even higher doses of willow bark extracts in alleviating inflammation and pain have not been substantiated by clinical studies.

18.3.1 Efficacy

Five high-quality trials investigated ethanolic extracts from willow bark over up to 6 weeks of treatment. Forty percent of the patients with acute exacerbations of chronic low-back pain who consumed an extract with 240 mg salicin/day were pain-free after a 4-week treatment compared with 20% of those taking half that dose. One study indicated that the high-dose ethanolic extract was not inferior to rofecoxib in low-back pain sufferers. However, in one study, comparing this dose against 100 mg diclofenac, the NSAID was superior for pain caused by gonarthrosis and was ineffective in patients suffering from rheumatoid arthritis [11]. This systematic review concluded that for extract doses up to 240 mg salicin/day, only moderate evidence of effectiveness was found in the treatment of musculoskeletal pain [11]. There is no doubt that other pain-like migraine or headache may also respond to willow bark extract [6]. Treatment of pain after dental extraction is a promising indication (see later section), but further studies with a confirmatory study design need to provide evidence of this.

18.3.2 Safety

It is generally accepted that aspirin originates from *Salix* species. However, willow bark contains only a small amount of salicin that is the prodrug for various salicylic acid derivatives. A willow bark extract dose with 240 mg salicin/day corresponds to 100 mg salicylic acid derivatives [12], an insufficient dose to treat pain. Aspirin, which was synthesized in 1897, is also metabolized to salicylic acid. The aspirin acetyl group is responsible for preventing blood clotting even at a dose of 100 mg/day (cardioprotective dose). Treatment with willow bark preparations do not exhibit a major impact on coagulation and can therefore be safely used perioperatively [13]. In contrast to aspirin, willow bark extract (while inhibiting COX-1 *in vitro*) does not damage the gastrointestinal mucosa due to gastroprotective constituents in the extract [14, 15]. Whereas 9 of 10 aspirin-treated rats (100 mg/kg) showed stomach lesions, ethanolic willow bark extract containing 12% salicin, in a dose of up to 120 mg/kg, did not elicit any adverse gastric effects [16].

Severe adverse events were not observed in the five clinical studies up to 6 weeks mentioned earlier or in three surveys over 8 weeks including a total of 5871 patients [17–19] or in one survey over 6 months including 436 patients suffering from chronic back pain, fibromyalgia, and/or osteoarthritis [20]. The overall adverse event rate was below 3%. It was observed that it was implausible that the regulatory authority, European Medicines Agency (EMA), had restricted the use of willow bark preparations to 4 weeks [21] in light of the observation that NSAIDs, in current use, with a higher risk to benefit ratio than willow extract are used for longer treatment periods, for example, up to 138 weeks [22].

Acute toxicity studies in rats could not determine a lethal dose of willow bark extract even in doses 200 times the experimental level [16]. However, chronic toxicity data are still lacking [6, 8]. Possible interactions with natural or synthetic blood thinners (wafarin-like drugs) need to be elucidated, especially if higher doses of willow bark extract (with 360 or 480mg salicin/day) are used. A life-threatening anaphylactic reaction was observed in a patient with a history of allergy to salicylates [23]. Known salicylate allergy is therefore a contraindication for willow bark preparations.

18.4 ORAL EXTRACTS FROM *HARPAGOPHYTUM PROCUMBENS*

During the First World War, soldiers stationed in Africa brought the local traditional medicine Devil's claw (*Harpagophytum procumbens*) back to Germany. In the first medical report published in 1958, Zorn presented results on the anti-inflammatory and antiarthritic effects of aqueous *Harpagophytum* extract in rats and also preliminary observations in patients. He observed "that the inflammatory process either ceased or that the healing process continued after the treatment had been stopped" [24]. A majority of the animal studies were carried out thereafter to indicate that *Harpagophytum procumbens* is an effective anti-inflammatory and analgesic preparation for the treatment of inflammation and pain [25]. *Harpagophytum* extract inhibits the induction of proinflammatory gene expression, possibly by blocking the AP-1 pathway [26]. The harpagoside fraction is probably responsible for the effect of Devils claw on enzyme activities.

However, other components from Devil's claw crude extract may antagonize or increase the synthesis of inflammatory mediators [27]. It has been suggested that a ratio based on the amount and relative proportions of anti- and proinflammatory compounds may be used to predict the relative anti-inflammatory properties of *Harpagophytum* preparations in order to minimize risk at the expense of maximizing therapeutic response [28]. It seems advisable to rely on *Harpagophytum* preparations prepared according to good manufacturing practice (GMP) that have demonstrated efficacy in clinical studies [29].

For relief of minor articular pain, the EMA regulatory authority recommends a daily dose of 4.5g dried root of *Harpagophytum procumbens* or *Harpagophytum zeyheri* in 500ml water as herbal tea or extract (water or ethanol as solvent) [30]. The EMA is not acknowledging that commercially available aqueous extracts contain more active principles (50–60mg of the leading compound, harpagoside) than ethanolic extracts (at most 30mg harpagoside). This was demonstrated in a study investigating commercially available extracts [31] (Table 18.1). One dose-finding study indicated that aqueous extracts with 100mg of harpagoside in a daily dose (corresponding to extract based on 9g of crude plant material) were superior to half that dose [32]. This higher dose has mainly been used in European countries other than Germany [33]. It seems likely that higher doses of *Harpagophytum* preparations than the empirically chosen dose based on 4.5 g of crude plant material are likely to be more clinically effective.

18.4.1 Efficacy

The quality of 20 clinical trials investigating the effectiveness of *Harpagophytum* preparations in the treatment of musculoskeletal pain has been reviewed in 2003 [29]; three of these studies had a confirmatory study design. A systematic Cochrane review concluded that two high-quality trials utilizing *Harpagophytum procumbens* demonstrated short-term improvements in pain and consumption of rescue medication for daily doses standardized to 50 or 100 mg harpagoside with another high-quality trial demonstrating relative equivalent efficacy of an aqueous *Harpagophytum* extract with 60 mg harpagoside with 12.5 mg/day of rofecoxib in the treatment of chronic back pain [34]. Two high-quality trials using ethanolic *Harpagophytum* extract (<30 mg harpagoside/day) for the treatment of knee and hip osteoarthritis failed to show any improvement in the established WOMAC outcome measure [35]. Use of *Harpagophytum procumbens* powder containing 60 mg harpagoside, in a daily dose, demonstrated moderate evidence of effectiveness in the treatment of osteoarthritis of the spine, hip, and knee [35]. It is therefore difficult to understand the rationale for the regulatory authority, EMA, arriving at the conclusion that there is insufficient scientific basis for recommending the use of *Harpagophytum* products [25].

18.4.2 Safety

In a systematic review on the safety of *Harpagophytum* preparations, 28 clinical trials were identified, of which 20 identified adverse events. In none of the double-blind studies was there any incidence of adverse events during treatment with *Harpagophytum procumbens* that were higher than during placebo treatment. Minor adverse events occurred in approximately 3% of the patients, mainly gastrointestinal adverse events [36]. *Harpagophytum* preparations are contraindicated in case of gastric and duodenal ulcers [25]. Since the dosage used in most of the studies is at the lower end and since long-term treatment with *Harpagophytum* products is advisable, more safety data are warranted. Safety has not been established in children or pregnant and lactating women. Preclinical safety data indicate very low acute and chronic toxicity [33]. No case of overdose has been reported [30]. Two studies over 1 year indicated a favorable outcome in terms of OMERACT-OARSI responders (Fig. 18.1) and a high benefit-to-risk ratio when compared to long-term use of NSAIDs [42].

18.5 ORAL AVOCADO–SOYBEAN UNSAPONIFIABLES

About 20 years ago, the French company "Laboratoires EXPANSCIENCE" discovered, developed, and produced a pharmaceutical product, PIASCLEDINE® 300, for the *symptomatic, delayed-effect treatment of arthrosis of the hip and knee*. The active principle (ASU EXPANSCIENCE) comprises a mixture of soybean unsaponifiables

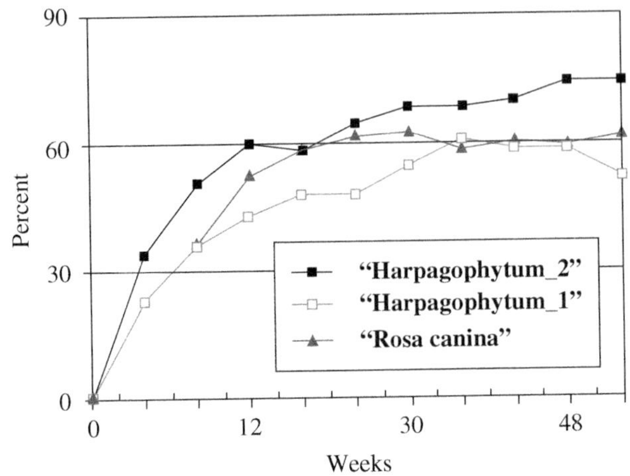

FIGURE 18.1 OMERACT-OARSI response considering pain and function [37, 38] during treatment with *Harpagophytum* extract (60 mg harpagoside/day) in two studies over 1 year [39, 40] and during treatment with a rose hip and seed powder (*Rosa canina*) over 1 year [41].

with more than 10 identified compounds. The product specifications and elements of the process are protected by patents [43] and are not declared (Table 18.2). A dose-finding study revealed that 600 mg dose of PIASCLEDINE per day was no more effective than half that dose (ceiling effect); a dose of 450 mg has not been tested.

18.5.1 Efficacy

The ASU product, PIASCLEDINE 300, was investigated in four high-quality confirmatory studies [21]. Pooled data from two studies (326 participants) showed that pain improved at 3 months, one study (163 participants) found pain improved at 6 months, but a further study demonstrated no effects on pain, or structural joint changes at 12 months. In the latter study, a significant reduction in joint space narrowing in most of the radiologically severe patients (baseline joint width below 2.45 mm) was identified in a *post hoc* analysis. In order to get further information on the structure modification of PIASCLEDINE 300, another 399 patients suffering from hip osteoarthritis were randomized in a 3-year study duration [3]; unfortunately, no significant difference in mean joint width space was observed, but there were 20% less progressors in the PIASCLEDINE 300 group than in the placebo group (40 versus 50%). A patient was defined as a progressor when the joint width space was more than −0.5 mm at 3 years. The clinical relevance of this result however requires further assessment. In this study, there was no effect on pain and the established validated outcome measure WOMAC was not different between the PIASCLEDINE 300 and the placebo groups.

A direct comparison between PIASCLEDINE 300 once daily and chondroitin sulfate 400 mg three times daily, in 364 patients revealed no difference in efficacy in a noninferior study [45]. Since the clinical relevance of the efficacy of chondroitin

TABLE 18.2 Characteristics of Some Phytomedicines Used for the Treatment of Osteoarthritis and Rheumatoid Arthritis[a]

Plant					Marker	
Name	Part	Preparation	Drug/extract ratio	Milligram per day	Constituent	Milligram per day
Harpagophytum procumbens	Root	Aqueous extract	1.5–2.5:1	2400	Harpagoside	30
		Ethanolic (60%) extract	4.5–5.5:1	960	Harpagoside	<30
		Cryoground powder		2610	Harpagoside	60
Salix pupurea + daphnoides	Bark	Ethanolic (70%) extract	10–20:1	1360	Salicin	240
Rosa canina	Hip + seed	Powder		5000	Galactolipid	1.5
Curcuma longa	Root	Extract, solvent not stated	Not stated	200	Curcuminoids	20%
Cucuma domestica	Root	Ethanolic extract	Not stated	Not stated	Curcuminoids	500 mg
Zingiber officinale	Root	Acetone extract Eurovita 33	20:1[b]	510	Not stated	—
		CO_2 extract	Not stated	1000	Gingerol	40
Zingiber officinale + Alpinia officinale		Acetone extract EXT 77	20:1[b]	Not stated	Not stated	—
		Not stated	Not stated	Not stated	Not stated	—
Persea gratissima (P)	Oil	Unsaponifiable fraction 1/3 P; 2/3 G		300	Not stated	—
Glycine max (G)				600		
Boswellia serrata	Gum resin	Extract, solvent not stated	Not stated	999	Boswellic acid	40%
					Organic acid	65%
		Extract, solvent not stated	Not stated	1200–3600	Boswellic acid	—

(*Continued*)

TABLE 18.2 *(Continued)*

Plant					Marker	
Name	Part	Preparation	Drug/extract ratio	Milligram per day	Constituent	Milligram per day
Boswellia carteri + *Curcuma longa*	Gum resin	Extract, solvent not stated +	Not stated	Not stated	Boswellic acid	37.5%
	Root	*Curcuma longa*	Not stated	Not stated	Not stated	Not stated
	Root	Extract, solvent not stated	Not stated	Not stated	Not stated	—
Tripterygium wilfordii	Root	Ethanol/ethyl subsequent Acetate extract	45:1	180	Triptolide	0.09
				360	Tripdiolide	0.108
				180		—
		Chloroform/methanol extract	Not stated	60	Tripdiolide	0.021
					Tripdiolide	0.041
					Tripdiolide	0.002
					Alcaloides[c]	0.88
Uncaria tomentosa	Bark	Aqueous extract	Not stated	60		
Uncaria guianensis	Bark	Aqueous extract	Not stated	100	Not stated	—
Symphytum officinale	Radix	Ethanolic (60%) extract	2:1	3×6 cm	Allantoin	0.2–0.5%
Arnica montana	Herb	Tincture, 50% ethanol	20:1	3×4 cm	Not stated	—
Petiveria alliacea	Herb	Aqueous extract	9 g/600 ml	600 ml	Not stated	—

[a]Modified after Refs. [1, 44].
[b]Information obtained from the company
[c]Pentacyclic oxindole alcaloides drug extract ratio: how many parts of crude plant material are required to produce one part of extract.

sulfate in slowing or arresting progression of osteoarthritis has been questioned [46], further data are necessary to prove the effectiveness of PIASCLEDINE 300 in the treatment of osteoarthritis.

18.5.2 Safety

When the data on adverse events from the clinical studies mentioned earlier were assessed using meta-analysis, there was only a negligible difference in the odds ratio in either a placebo group or intervention group. A postmarketing safety profile of the product since its commercialization in France until 2008 revealed that cutaneous, hepatic, and gastrointestinal disorders were the most frequently reported adverse drug reactions associated with PIASCLEDINE 300. In a recent 3-year study, safety was excellent. Since the product is largely prescribed in France, the incidence of adverse drug reactions seems to be "very rare." However, this assumption is based on the belief that all adverse reports are reported (and recorded) and this is probably unlikely [47]. Therefore, safety pharmacological studies according to US or European published guidelines (www.fda.gov/downloads/Drugs/GuidanceComplianceRegulatoryInformation/Guidances/ucm074957.pdf, www.ema.europa.eu/docs/en_GB/document_library/Scientific_guideline/2009/09/WC500002720.pdf) should confirm safe use of this herbal product.

18.6 ORAL EXTRACTS FROM *TRIPTERYGIUM WILFORDII*

Tripterygium wilfordii Hook F (TWHF), also known as thunder god vine, is part of the Traditional Chinese Medicine that recommends crude water extracts (decoctions) from the root to treat inflammatory conditions. A large number of uncontrolled studies reported beneficial effects of the decoctions. However, the therapeutic benefit was associated with the occurrence of adverse events (see later section). A variety of extraction procedures were therefore used to minimize toxicity. Two of these extracts are now widely used in China: an ethanol/ethyl acetate extract and a chloroform/methanol extract, both standardized for diterpenes (like triptolide and tripdiolide) and developed in the 1970s [44] (Table 18.2).

Numerous preclinical studies have demonstrated that root extracts of TWHF inhibit the expression of proinflammatory cytokines, proinflammatory mediators, adhesion molecules and matrix metalloproteinases by macrophages, lymphocytes, synovial fibroblasts, and chondrocytes and provide potent anti-inflammatory and immunosuppressive effects in animals, indicating that this PAID is useful for the treatment of rheumatoid arthritis [48]. Such effects are also observed with isolated coactive triptolide in adjuvant-induced arthritis [48] or when the total alkaloid fraction was administered to rats with collagen-induced arthritis [49].

18.6.1 Efficacy

TWHF extract (solvent chloroform/methanol, 60 mg daily) was studied in a cross-over trial versus placebo, the first arm of which lasted for 12 weeks. Improvements were reported with the extract with statistically significant decreases in joint tenderness and swollen joint count. Nonsignificant decreases were reported in morning stiffness and walking time. An increase in grip strength was also demonstrated, although this was not statistically significant [48].

Larger doses of extract were investigated with the ethanol/ethyl acetate TWHF extract, 180 mg ($n = 10$) and 360 mg ($n = 10$) per day. Eight high-dose and four low-dose-per-day patients satisfied the ACR20 improvement criteria at the end of the intervention period compared to none of the patients in the placebo group. These dichotomous data convert to an odds ratio of 17 and 85 for the high- and the low-extract doses, respectively. The 360 mg dose was more effective than the 180 mg dose and the 180 mg dose was more effective than placebo. The mean times to reach the ACR20 were 7 and 12 weeks, respectively [48].

In a randomized, double-blind study at 11 US centers, the effectiveness of the ethanol/ethyl acetate TWHF extract 180 mg/day was compared with sulfasalazine 2 g/day. Six-month treatment with the extract resulted in rapid improvement in clinical signs, including joint pain, joint swelling, and measures of overall well-being and markers of inflammation (C-reactive protein (CRP), erythrocyte sedimentation rate (ESR), pro-inflammatory cytokine, interleukin (IL)-6). Significant differences from baseline and significantly larger improvements compared to sulfasalazine were apparent at 2 weeks of therapy and persisted throughout the study for the number of swollen and tender joints, quality of life, disability, pain, global assessments of health, ESR, and CRP. Moreover, radiographic progression was lower in the TWHF group [48].

18.6.2 Safety

More adverse events were seen among patients receiving TWHF extract than patients receiving placebo. Fever and aplastic anemia occurred in one patient following an overdose of TWHF extract. One death occurred also, although not thought to be related to the intervention. The most common adverse events associated with TWHF extract include gastrointestinal tract disturbances (diarrhea), headache, hair loss, rash, skin pigmentation, dysmenorrhea, decrease of male fertility, renal insufficiency, hematotoxicity, embryotoxicity, and immune suppression as demonstrated by increased rates of infection [44, 48, 50]. Most of the adverse events observed during long-term treatment for up to 5.8 years could be resolved by dose adjustment except amenorrhea. LD_{50} in rodents for the ethyl alcohol and chloroform extracts were 608–858 and 160 mg/kg, respectively, depending on the source of the plant and the time of harvest. Subacute toxicity manifested as pathological changes mainly in the lymphatic and reproductive systems. The risk–benefit trade-off of the decoctions and extracts of TWHF may therefore be judged as unfavorable [50]. However, recently solid lipid nanoparticles loaded with *Tripterygium wilfordii* extract were shown to have protective effects on male reproductive toxicity and hepatotoxicity in rats [51] while enhancing anti-inflammatory effects. Future studies are required to prove the superiority of this preparation in humans; nanotechnology may resolve some of these problems, but future studies are needed to prove the utility of TWHF beyond doubt.

18.7 ORAL PAIDs CONTAINING UNSATURATED FATTY ACIDS

Unsaturated fatty acids contribute to the anti-inflammatory action of rose hip and the seed [52]. Seed oils of evening primrose (*Oenothera biennis*), blackcurrant (*Ribes nigrum*), and borage (*Borago officinalis*) are standardized to gamma-linolenic acid (GLA). Preclinical studies demonstrate that the GLA mechanism of action is similar to the other PAIDs (inhibition of cyclooxygenase, lipoxygenase, cytokines, antioxidative effect).

18.7.1 Efficacy

The efficacy of a powder from rose hip and the seed in the treatment of osteoarthritis has been summarized in a systematic review [53]. None of the studies had a confirmatory study design. Two further exploratory studies supported the claims of effectiveness in the treatment of osteoarthritis [54] and in the treatment of low-back pain [41]. Further studies are required to confirm the effectiveness of rose hip and the seed beyond any doubt.

Seven studies investigated the effects of plant-sourced GLA in 286 participants. Three of these investigated evening primrose seed oils (EPOs), two blackcurrant seed oils, and two borage seed oils. None of the studies had a confirmatory study design. The approximate daily intake of GLA varied from 525 to 2800 mg. Although the active principle of the seed oils may be slightly different due to the composition of the oil, studies were considered collectively based on the daily consumption of GLA. Placebo oils included olive oil, sunflower oil, liquid paraffin, cottonseed oil, and soybean oil.

The studies investigating GLA doses between 525 and 540 mg did not provide sufficient evidence of effectiveness. Four studies investigated daily GLA doses between 1400 and 2800 mg and were reported appropriately to allow data extraction and some data mining/pooling. Three studies show that 6 months of treatment with oils containing at least 1400 mg GLA improved self-reported pain assessed on a visual analog scale in rheumatoid arthritis patients [44]. It is unclear why the results from a further study differed from those of the other three high-dose studies, but possible explanations include a shorter intervention period (12 weeks), use of a noninert oil (olive oil) in the placebo group, and a small sample size that may have contributed to type II error. Duration of morning stiffness measured in minutes showed improvement in the GLA versus the placebo group after 6 months. Likewise, joint tenderness, swollen joints, and global evaluation of disease activity were improved [44]. One problem identified in studies where larger doses of GLA were administered was related to the large quantity (and size) of capsules required to administer the higher doses.

18.7.2 Safety

Neither the 3- to 4-week-long studies in patients suffering from osteoarthritis nor the 1-year long study in patients suffering from low-back pain identified any specific adverse events for rose hip and seed powder [41, 53, 54]. Since the powder formulation absorbs the gastrointestinal fluid if patients consume insufficient liquid together with the powder, constipation may occur. It is also not advisable to consume the powder within 2h of intake of other medication since the powder may interfere with the absorption of lipophilic drugs.

In the clinical studies discussed earlier, investigating herbal medicines containing GLA, the relative risk of adverse events was higher among patients using the GLA oil than among patients using the placebo oil [44]. The US National Library of Medicine and the National Institutes of Health summarized safety data for EPO [55]. Reports on allergy or hypersensitivity are rare. Seizures may occur in individuals taking EPO, particularly in people with a history of seizure disorders and among individuals taking EPO in combination with anesthetics or other centrally acting drugs such as chlorpromazine, thioridazine, trifluoperazine, or fluphenazine. Doses of antiseizure medications may therefore require to be increased. Patients who undergo surgery requiring general anesthesia are advised to stop taking EPO 2 weeks prior to surgery. Other adverse events include occasional headache, abdominal pain, nausea, and loose stools. In animal studies, GLA decreased blood pressure. Early results from human studies do not show consistent changes in blood pressure, but patients on blood pressure medications should closely monitor their blood pressure. Since borage and blackcurrant seed oils also contain GLA, the adverse event profiles may be similar.

18.8 OTHER ORAL PAIDs

A total of 30 and 14 different PAID products have been investigated in patients suffering from osteoarthritis and rheumatoid arthritis. The characteristics of the single oral PAIDs are summarized in Table 18.2. For oral PAID mixtures from China, Europe, and India, the evidence of effectiveness in osteoarthritis or rheumatoid arthritis has been insufficient [44, 54]. The available data provide trends of their effectiveness. Future rigorous studies need to identify the optimum daily dose corresponding to clinically relevant effectiveness without adverse events. In no studies that investigated PAIDs so far severe adverse events were observed except for TWHF extracts developed in the 1970s. In light of the observation that greater effects may be achieved with higher PAID doses, safety profiles are warranted for all PAIDs [56]. In most of the studies so far, the study medications were insufficiently characterized. Preparation of the study medications should consider the World Health Organization (WHO) recommendations with detailed description of the manufacturing procedure (e.g., "if other substances are added during manufacture in order to adjust the plant preparation to a certain level of active or characteristic constituents or for any other purpose") and the method for identification (e.g., detailed assay description). If identification of an active principle is not possible, it should be sufficient to identify a characteristic or marker substance or mixture of substances (e.g., "chromatographic fingerprint") to ensure consistent quality of the preparation in order to be able to replicate the study with an essentially similar product [57].

The minimum information provided for a PAID should include the plant part, the brand or commercial name (if the preparation has not been solely prepared for the study), the excipient added in case of extracts, and the drug extract ratio if the crude plant material is not used. The daily dose of the "native" preparation should be stated (otherwise the extract dose may also contain additives) [58]. Although not required by regulatory authorities, it is desirable to know the content of at least one characteristic marker substance (if possible a coactive ingredient). Results of studies with insufficient declared characteristics are only attributable to a particular product used in a particular study and cannot be translated to other products of the same plant material unless bioequivalence of the products have been demonstrated.

The WHO definition of herbal medicines should also be considered, that is, "… finished, labeled, medicinal products that contain as active ingredient, aerial or underground parts of plants, or other plant material, or combinations thereof, whether in the crude state or as plant preparations (e.g., comminuted or powdered plant materials, extracts, tinctures, fatty or essential oils)." Medicines containing plant material combined with chemically defined active substances, including chemically defined, isolated constituents of plants or organic or inorganic active ingredients that are not of plant origin, are not considered to be herbal medicines [59]. Thus, studies investigating extracted or synthetic capsaicin cannot be considered among the studies on herbal medicines.

Recent Western monographs provide useful reviews on the mechanism of action and experimental and clinical studies of various plants. Whereas the monographs of the European Scientific Cooperative on Phytotherapy [60, 61], the monographs of the "American Herbal Pharmacopeia" [62], and the WHO monographs on selected medicinal plants [63], while not being official, do provide scientific information on safety, efficacy, and quality of medicinal plants and provide recommendations for their use in clinical practice (e.g., doses, types of preparation). By contrast, the EMA [64] serves as a guide for application dossiers to obtain marketing authorization from the regulatory authorities of the individual countries in the European Union. Unfortunately, only few EMA monographs have used an evidence-based approach.

18.9 TOPICAL PAIDs

The mechanism of action of topical PAIDs is different from that of oral products (Table 18.1) in that they also act as counter-irritants via the skin (nettle, capsaicin, peppermint oil) or they are toxic when taken orally (*Arnica*, comfrey) [65]. For example, nettle leaf is covered with needle-like hairs that pierce the skin on contact injecting irritant substances (formic acid, acetic acid, serotonin, histamine, acetylcholine, 5-hydroxytryptamine), which cause an irritant skin reaction. In the Middle Ages, urtication (beating with nettle) belonged to the armentarium of treatments for osteoarthritic pain. The peppermint oil coactive compound, menthol, triggers cold-sensitive receptors in skin sensory neurons, for example, transient receptor potential melastatine 8 (TRPM8). *In vitro* studies demonstrate that menthol inhibits the arachidonic acid cascade and cytokine release and exerts local anesthetic, spasmolytic, and antioxidative effects. The analgesic action of menthol is based on a weak kappa opioid receptor agonist effect and cumulative inactivation of voltage-gated sodium channels [65].

The *capsicum* active principle, the capsaicinoids, triggers the heat-sensitive transient receptor potential vanilloid-1 (TRPV1) receptors, which results in a decrease in membrane resistance, depolarization, and activation of synaptosomal neurotransmitter release. Following the initial activation (which is often associated with heat sensation), desensitation and depletion of neurotransmitters produce the capsaicin analgesic effect. If capsaicin exposure persists, nerve terminals will degenerate (dysfunctionalization), which causes the prolonged analgesic effect after the treatment has ended. Other capsaicin effects include the inhibition of inducible COX-2 mRNA expression, LOX, and a free radical scavenging activity (Table 18.2). Clinical studies demonstrate that neuropathic pain, rheumatoid arthritis, osteoarthritis, and soft tissue pain may all respond well to topical capsaicin. In addition, patients suffering from psoriasis, pruritus, cluster headache, postmastectomy, and other pain may benefit from topical capsaicin treatment [66]. However, most of the studies applied isolated pure capsaicin, which is not PAID.

18.9.1 Efficacy

Six studies with six topical PAIDs included in the recent Cochrane review were not suitable for data pooling; it was not possible to draw firm conclusions from the single studies and the two studies that used topical nettle leaf. However, no serious side effects were reported.

18.9.2 Safety

In general, adverse events from local capsaicin application occurred in one-third of the patients and would not have occurred if these patients were treated with placebo [67]. However, in a few patients, the skin irritation led to treatment withdrawal.

Preclinical studies demonstrate that capsaicin has a genotoxic, carcinogenic, and tumor promotion potential. Skin irritation and tumor-promoting effects appear to be mediated via interaction with the same vanilloid receptor. Thus, a limit on the capsaicin content that would significantly reduce its skin irritation potential is expected to lessen concerns around its tumor promotion potential [68]. Further studies are required to rule out concerns over long-term capsaicinoid use and to confirm the safe use of the other topical PAIDs.

REFERENCES

[1] Cameron M, Blumle A, Gagnier JJ, Little CV, Parsons T, Chrubasik S (2009) Evidence of effectiveness of herbal medicinal products in the treatment of arthritis. Part 1: osteoarthritis. *Phytother Res* 23: 1497–1515.

[2] Cashman JN (1996) The mechanisms of action of NSAIDs in analgesia. *Drugs* 52 (Suppl 5): 13–23.

[3] Maheu E, Cadet C, Marty M, Moyse D, Kerloch I, Coste P, Dougados M, Mazières B, Spector TD, Halhol H, Grouin JM, Lequesne M (2013) Randomised, controlled trial of avocado-soybean unsaponifiable (Piascledine) effect on structure modification in hip osteoarthritis: the ERADIAS study. *Ann Rheum Dis* 73: 376–384.

[4] Stone E (1763) An account of the success of the bark of the willow in the cure of agues. *Philos Trans R Soc Lond* 53: 195–200.

[5] Vlachojannis J, Magora F, Chrubasik S (2011) Willow species and aspirin: different mechanism of actions. *Phytother Res* 25: 1102–1104.

[6] Anonymous (2003a) *Salicis cortex*. In *ESCOP Monographs. Hrgs. European Scientific Cooperative on Phytotherapy*. Thieme-Verlag, Stuttgart/New York: pp. 445–451.

[7] Blumenthal, M (1998) *The Complete German Commission E Monographs*. American Botanical Council, Austin.

[8] EMA (2009a) www.ema.europa.eu/docs/en_GB/document_library/Herbal_-_HMPC_assessment_report/2009/12/WC500018258.pdf

[9] Chrubasik S, Eisenberg E, Balan E, Weinberger T, Luzzati R, Conradt C (2000) Treatment of low back pain exacerbations with willow bark extract: a randomized double-blind study. *Am J Med* 109: 9–14.

[10] Chrubasik S, Künzel O, Acker A, Conradt C, Kerschbaumer F (2001) Potential economic impact of using a proprietary willow bark extract in outpatients treatment of low back pain: an open non-randomized study. *Phytomedicine* 8: 241–251.

[11] Vlachojannis JE, Cameron M, Chrubasik S (2009) A systematic review on the effectiveness of willow bark for musculoskeletal pain. *Phytother Res* 23: 897–900.

[12] Schmid B, Kötter I, Heide L (2001) Pharmacokinetics of salicin after oral administration of a standardised willow bark extract. *Eur J Clin Pharmacol* 57: 387–391.

[13] Krivoy N, Pavlotzky E, Chrubasik S, Eisenberg E, Brooks G (2001) Effect of salicis cortex extract on human platelet aggregation. *Planta Med* 67: 209–213.

[14] Pobłocka-Olech L, Krauze-Baranowska M (2008) SPE-HPTLC of procyanidins from the barks of different species and clones of *Salix*. *J Pharm Biomed Anal* 48: 965–968.

[15] Pobłocka-Olech L, Krauze-Baranowska M, Głód D, Kawiak A, Łojkowska E (2010) Chromatographic analysis of simple phenols in some species from the genus *Salix*. *Phytochem Anal* 21: 463–469.

[16] Glinko A (1998) Pharmacological properties of a standardized extract from Willow Bark (Cortex salicis). *Chair of Pharmacology and Toxicology*. Pomeranian Academy of Medicine, Szczecin.

[17] Werner G, Scheithe K (2004) *Willow bark extract (AssalixR) for chronic back pain and arthralgia, a post-authorization surveillance study*. Congress Phytopharmaka and Phytotherapy, February 26–28, Berlin.

[18] Zenner-Weber MA (2004) *Successful treatment of chronic rheumatic diseases (ICD M-codes) with willow bark extract (Assalix), a seeding trial. Gemeinsamer Kongress der Schweizer Gesellschaften für Rheumatologie und für Physikalische Medizin*, September 16–17, Locarno.

[19] Müller-Faßbender H, Müller J, Bach G (2007) Wirksamkeit und Verträglichkeit eines Weidenrindenpräparates bei Erkrankungen des Bewegungsapparates. *Arthritis und Rheuma* 27: 267–271.

[20] Uehleke B, Preis S, Mueller J, Stange R, Kelber O, Melzer J (2013) Willow bark extract STW 33–1 is safe and effective in the long-term treatment of outpatients with rheumatic pain, esp. osteoarthritis or back pain. A subgroup analysis. *Planta Med* 79: 1273, PN103.

[21] Vlachojannis C, Magora F, Chrubasik S (2013) Pro and contra duration restriction of willow bark treatment. *Phytother Res* 28: 148–149.

[22] Reginster JY, Malmstrom K, Mehta A, Bergman G, Ko AT, Curtis SP, Reicin AS (2007) Evaluation of the efficacy and safety of etoricoxib compared with naproxen in two, 138-week randomised studies of patients with osteoarthritis. *Ann Rheum Dis* 66: 945–951.

[23] Boullata JI, McDonnell PJ, Oliva CD (2003) Anaphylactic reaction to a dietary supplement containing willow bark. *Ann Pharmacother* 37: 832–835.

[24] Zorn B (1958) Über die antiarthritische Wirkung der Harpagophytum-Wurzel. *Dtsch Rheumaforsch* 17: 134–138.

[25] EMA (2009b) http://www.ema.europa.eu/docs/en_GB/document_library/Herbal_-_HMPC_assessment_report/2010/01/WC500059019.pdf

[26] Fiebich BL, Muñoz E, Rose T, Weiss G, McGregor GP (2012) Molecular targets of the antiinflammatory *Harpagophytum procumbens* (Devil's claw): inhibition of TNFα and COX-2 gene expression by preventing activation of AP-1. *Phytother Res* 26 (6): 806–811.

[27] Anauate MC, Torres LM, de Mello SB (2010) Effect of isolated fractions of *Harpagophytum procumbens* D.C. (Devil's claw) on COX-1, COX-2 activity and nitric oxide production on whole-blood assay. *Phytother Res* 24: 1365–1369.

[28] Ouitas NA, Heard C (2010) Estimation of the relative antiinflammatory efficacies of six commercial preparations of *Harpagophytum procumbens* (Devil's claw). *Phytother Res* 24: 333–338.

[29] Chrubasik S, Conradt C, Black A (2003) The quality of clinical trials with *Harpagophytum procumbens*. *Phytomedicine* 10: 613–623.

[30] EMEA (2008) www.ema.europa.eu/docs/en_GB/document_library/Herbal_-_Community_herbal_monograph/2010/01/WC500059018.pdf

[31] Sporer F, Chrubasik S (1999) Präparate aus der Teufelskralle (*Harpagophytum procumbens*) Zschr. *Phytotherapie* 20: 235–236.

[32] Chrubasik S, Junck H, Breitschwerdt H, Conradt C, Zappe H (1999) Effectiveness of *Harpagophytum* extract WS 1531 in the treatment of exacerbation of low back pain: a randomized placebo controlled double blind study. *Eur J Anaesthesiol* 16: 118–129.

[33] Anonymous (2003b) *Harpagophyti radix*. In ESCOP Monographs. Hrgs. European Scientific Cooperative on Phytotherapy. Thieme-Verlag, Stuttgart/New York: 233–240.

[34] Gagnier JJ, van Tulder M, Berman B, Bombardier C (2006) Herbal medicine for low back pain. *Cochrane Database Syst Rev* April 19 (2): CD004504.

[35] Gagnier JJ, Chrubasik S, Manheimer E (2004) *Harpgophytum procumbens* for osteoarthritis and low back pain: a systematic review. *BMC Complement Altern Med* 4: 13.

[36] Vlachojannis J, Roufogalis BD, Chrubasik S (2008) Systematic review on the safety of *Harpagophytum* preparations for osteoarthritic and low back pain. *Phytother Res* 22: 149–152.

[37] Pham T, van der Heijde D, Lassere M, Altman RD Anderson JJ, Bellamy N, Hochberg M, Simon L, Strand V, Woodworth T, Dougados M (2003) OMERACT-OARSI. *Outcome variables for osteoarthritis clinical trials: the OMERACT-OARSI set of responder criteria. J Rheumatol* 30: 1648–1654.

[38] Pham T, van der Heijde D, Altman RD, Anderson JJ, Bellamy N, Hochberg M, Simon L, Strand V, Woodworth T, Dougados M (2004) OMERACT-OARSI initiative: Osteoarthritis Research Society International set of responder criteria for osteoarthritis clinical trials revisited. *Osteoarthritis Cartilage* 12: 389–399.

[39] Chrubasik S, Künzel O, Thanner J, Conradt C, Black A (2005) A 1-year follow-up after a pilot study with Doloteffin for low back pain. *Phytomedicine* 12: 1–9.

[40] Chrubasik S, Chrubasik C, Künzel O, Black A (2007) Patient-perceived benefit during one year of treatment with Doloteffin. *Phytomedicine* 14: 371–376.

[41] Chrubasik C, Wiesner L, Black A, Müller-Ladner U, Chrubasik S (2008b) A one-year survey on the use of a powder from *Rosa canina* lito in acute exacerbations of chronic pain. *Phytother Res* 22: 1141–1148.

[42] Dubois RW, Melmed GY, Henning JM, Bernal M (2004) Risk of upper gastrointestinal injury and events in patients treated with cyclooxygenase (COX)-1/COX-2 nonsteroidal antiinflammatory drugs (NSAIDs), COX-2 selective NSAIDs and gastroprotective cotherapy: an appraisal of the literature. *J Clin Rheumatol* 10: 178–189.

[43] Msika P, Baudouin C, Saunois A, Bauer T (2008) Avocado/soybean unsaponifiables, ASU EXPANSCIENCE, are strictly different from the nutraceutical products claiming ASU appellation. *Osteoarthritis Cartilage* 16: 1275–1276.

[44] Cameron M, Gagnier JJ, Chrubasik S (2011) Herbal therapy for treating rheumatoid arthritis. *Cochrane Database Syst Rev* February 16 (2): CD002948.

[45] Pavelka K, Coste P, Géher P, Krejci G (2010) Efficacy and safety of piascledine 300 versus chondroitin sulfate in a 6 months treatment plus 2 months observation in patients with osteoarthritis of the knee. *Clin Rheumatol* 29: 659–670.

[46] Black C, Clar C, Henderson R, MacEachern C, McNamee P, Quayyum Z, Royle P, Thomas S (2009) The clinical effectiveness of glucosamine and chondroitin supplements in slowing or arresting progression of osteoarthritis of the knee: a systematic review and economic evaluation. *Health Technol Assess* 13: 1–148.

[47] Olivier P, Montastruc JL (2010) Réseau français des centres régionaux de pharmacovigilance. Post-marketing safety profile of avocado-soybean unsaponifiables. *Presse Med* 39: e211–e216.

[48] Bao J, Dai SM (2011) A Chinese herb *Tripterygium wilfordii* Hook F in the treatment of rheumatoid arthritis: mechanism, efficacy and safety. *Rheumatol Int* 31: 1123–1129.

[49] Zhang Y, Xu W, Li H, Zhang X, Xia Y, Chu K, Chen L (2013) Therapeutic effects of total alkaloids of *Tripterygium wilfordii* Hook F. on collagen-induced arthritis in rats. *J Ethnopharmacol* 145: 699–705.

[50] Canter PH, Lee HS, Ernst E (2006) A systematic review of randomised clinical trials of *Tripterygium wilfordii* for rheumatoid arthritis. *Phytomedicine* 13: 371–377.

[51] Xue M, Jiang ZZ, Wu T, Li J, Zhang L, Zhao Y, Li XJ, Zhang LY, Yang SY (2012) Anti-inflammatory effects and hepatotoxicity of *Tripterygium*-loaded solid lipid nanoparticles on adjuvant-induced arthritis in rats. *Phytomedicine* 19: 998–1006.

[52] Wenzig EM, Widowitz U, Kunert O, Chrubasik S, Bucar F, Knauder E, Bauer R (2008) Phytochemical composition and *in vitro* pharmacological activity of two rose hip (*Rosa canina* L.) preparations. *Phytomedicine* 15: 826–835.

[53] Chrubasik C, Roufogalis BD, Müller-Ladner U, Chrubasik S (2008a) A systematic review on the *Rosa canina* effect and efficacy profiles. *Phytother Res* 22: 725–733.

[54] Cameron M, Chrubasik S (2014) Oral herbal therapies for treating osteoarthritis. *Cochrane Database Syst Rev* May 22 (5): CD002947.

[55] US National Library Website (2012) www.nlm.nih.gov/medlineplus/druginfo/natural/patient-primrose.html

[56] APIC Website (2004) http://www.api-conference.org/pa4.cgi?src=eca_news_data.htm&nr=488&show=daten/news/GMP_News_488.htm&id=S11510781142

[57] WHO Website (2012) http://apps.who.int/medicinedocs/en/d/Jh2984e/2.html

[58] Chrubasik S, Sporer F, Wink M (1996) Zum Wirkstoffgehalt in Arzneimitteln aus *Harpagophytum procumbens*. *Forsch Komplementärmed* 3: 57–63.

[59] WHO Website (2013a) www.who.int/medicines/areas/traditional/definitions/en/index.html

[60] Anonymous (2003c) *ESCOP Monographs*. 2nd Edition. Thieme-Verlag, Stuttgart/New York.

[61] Anonymous (2009) *ESCOP Monographs*. Supplement. Thieme-Verlag, Stuttgart/New York.

[62] AHP Website (2011) www.herbal-ahp.org

[63] WHO Website (2013b) http://apps.who.int/medicinedocs/en/d/Js2200e/

[64] EMA Website (2013) www.ema.europa.eu/ema/index.jsp?curl=search.jsp&q=Herbal+monographs&btnG=Search&mid=WC0b01ac05800240cf

[65] Cameron M, Chrubasik S (2013) Topical herbal therapies for treating osteoarthritis. *Cochrane Database Syst Rev* May 31 (5): CD010538.

[66] Weiser T, Roufogalis B, Chrubasik S (2013) Comparison of the effects of pelargonic acid vanillylamide and capsaicin on human vanilloid receptors. *Phytother Res* 27: 1048–1053.

[67] Mason L, Moore RA, Derry S, Edwards JE, McQuay HJ (2004) Systematic review of topical capsaicin for the treatment of chronic pain. *BMJ* 328 (7446): 991.

[68] Anonymous (2007) Final report on the safety assessment of capsicum annuum extract, capsicum annuum fruit extract, capsicum annuum resin, capsicum annuum fruit powder, capsicum frutescens fruit, capsicum frutescens fruit extract, capsicum frutescens resin and capsaicin. *Int J Toxicol* 26 (Suppl 1): 103–106.

19

PHYTOTHERAPIES FOR INFECTIOUS DISEASES: ARE THESE REALLY USEFUL?

Gail B. Mahady[1], Gabrielle Escalante[1], Pooja Mikkilineni[1], Laura J. Mahady[2], Temitope O. Lawal[3], and Bolanle A. Adeniyi[3]

[1] *Department of Pharmacy Practice, College of Pharmacy, PAHO/WHO Collaborating Center for Traditional Medicine, University of Illinois at Chicago, Chicago, Illinois, USA*
[2] *The Barrow Neurological Institute and Arizona State University, Phoenix, Arizona, USA*
[3] *Department of Pharmaceutical Microbiology, University of Ibadan, Ibadan, Nigeria*

THE HISTORY OF MEDICINE

2000 B.C.—Here, eat this root.
1000 A.D.—That root is heathen. Here say this prayer.
1850 A.D.—That prayer is superstition. Here, drink this potion.
1920 A.D.—That potion is snake oil. Here swallow this pill.
1945 A.D.—That pill is ineffective. Here take this penicillin.
1955 A.D.—Oops … bugs have mutated. Here take this tetracycline.
1960–1999 A.D.—39 more "oops" … Here, take this more powerful antibiotic.
2013 A.D.—The bugs are still winning! Here, eat this root.
Anonymous

Phytotherapies: Efficacy, Safety, and Regulation, First Edition. Edited by Iqbal Ramzan.

19.1 INTRODUCTION

Infectious diseases have plagued mankind since the beginning of time, and to date, they are still a primary cause of morbidity and mortality, responsible for approximately one-third of all deaths worldwide [1]. Infectious and parasitic diseases continue to be devastating for children under the age of 5 years in developing countries due at least in part to the HIV/AIDS epidemic [2]. According to the World Health Organization, over 8 million children (<5 years of age) died in 2009, of which 98% lived in developing countries [1]. Although one might expect to see high death rates from infectious diseases in developing countries, mortality caused by infectious diseases has also been increasing in the United States since 1980 [3]. In the United States, mortality due to infectious disease was ranked fifth in 1981; but by 1991, mortality due to infectious diseases has risen to the third leading cause of death, an increase of 58% [3]. While there can be no doubt that increases in the HIV/AIDS pandemic and resultant secondary infections play a role in increasing mortality due to infections, new, re-emerging and resistant pathogens are also playing an increased role in mortality from infections. Since 1973, more than 20 well-recognized infectious diseases, such as tuberculosis (TB), malaria, and cholera, have re-emerged or spread geographically and many of these diseases are more virulent and are presenting in drug-resistant forms. Also since 1973, at least 30 previously unrecognized infectious agents have been identified, including HIV, Ebola, hepatitis C, West Nile, and Nipah viruses, most having no cure [3]. Other factors that contribute to the threat of infectious diseases include a significant increase in antibiotic resistance for serious diseases such as multidrug-resistant TB, malaria, HIV, and *Staphylococcus aureus* [3]. Thus, despite the scientific progress that has been made in microbiology, serious incidents of epidemics caused by drug-resistant microorganisms, along with the emergence of previously unknown disease-causing pathogens, still pose an enormous threat to public health and safety. These serious threats in healthcare require a focused global initiative for the development of new strategies for the prevention and treatment of infectious disease [4].

Since 1992, increased alarm over emerging and re-emerging infectious diseases has resulted in a number of national and international initiatives to restore and improve surveillance and control of communicable diseases [5]. The World Health Organization (WHO) has the largest health directive within the United Nations, including establishing health priorities, coordinating global health surveillance, and emergency assistance in the event of disease outbreaks. WHO has been credited for success in the eradication of smallpox, near eradication of polio, and they have made substantial progress in controlling childhood diseases, and in the expansion of primary health-care systems in developing countries. Other international organizations such as the World Bank, and several institutions in developed countries such as the US Centers for Disease Control, and nongovernmental organizations (NGOs) play an important role in strengthening both international and national surveillance and response systems for infectious diseases. However, due to the magnitude of the problem, and the lack of funds, capacity, and commitment, progress toward a truly global surveillance system has been slow.

Where Are the New Antimicrobial Drugs in the Pharmaceutical Pipeline? Since the 1970s, the number of new antibiotics entering the pharmaceutical pipeline has declined. In fact, the majority of antibiotic drug classes were discovered between

1930 and 1970, and many of the newly approved antibiotics have been based on these templates [6]. Since the 1970s, only three new antibacterial classes, the topical antibiotic mupirocin (Fig. 19.1) in 1985, the oxazolidinone linezolid (Fig. 19.2) in 2000, and the lipopeptide, daptomycin (Fig. 19.3) in 2003, have been approved for use [7]. In addition, over the past 20 years, there has been a 56% decline in the number of antibiotics approved annually by the Food and Drug Administration (FDA), and only 22 new antibacterial drugs have been launched in the past 15 years [7]. Twelve of these are natural product-derived drugs belonging to five different structural classes (β-lactam, streptogramin, macrolide, tetracycline (Fig. 19.4) and daptomycin (Fig. 19.3)). The other 10 are synthetic drugs belonging to only two antibacterial classes, with the quinolone class accounting for nine of these drugs [7]. A recent review of the antibiotics in the current pharmaceutical pipeline has shown that there is a significant lack of drugs that address the need for treatments of multidrug-resistant bacteria, particularly Gram-negative bacteria, which have become a public health hazard [8]. The lack of emerging new antibiotics and the development of multidrug-resistant bacteria, as well as the economic and regulatory challenges of antibiotic research, have been reviewed previously [6, 9–11]. Most scientists agree that in terms of infectious diseases, a major health-care crisis is eminent, as there are very few potential drugs in current clinical development that offer significant benefits over existing drugs and that target Gram-negative, hospital-based infections [6]. A review by Bourlioux published in 2013 provides a list of practical alternatives and investigative options at our disposal for the prevention and treatment of multidrug-resistant bacteria in light of the growing need for alternative treatments, and this list includes herbal medicines [12].

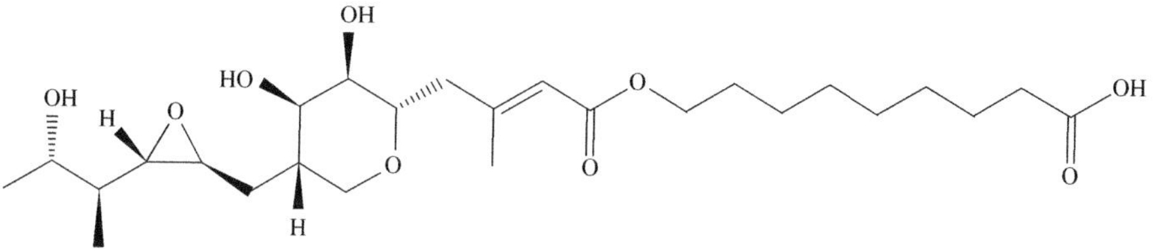

FIGURE 19.1 Structure of mupirocin, a topical antibiotic.

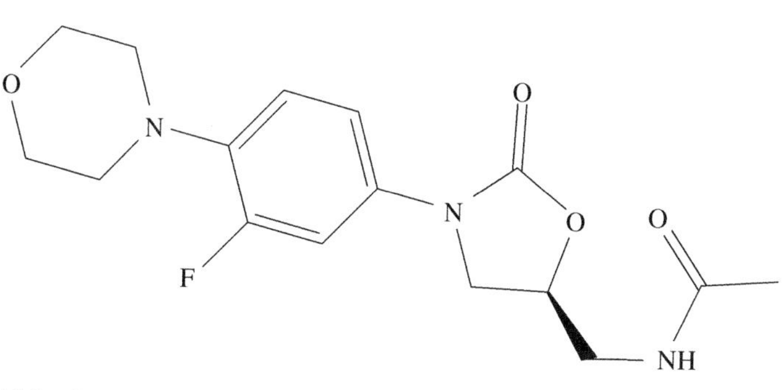

FIGURE 19.2 Structure of one of the oxazolidinone classes of antibiotics, linezolid, a novel class of antibiotics that was approved by the US FDA in 2000.

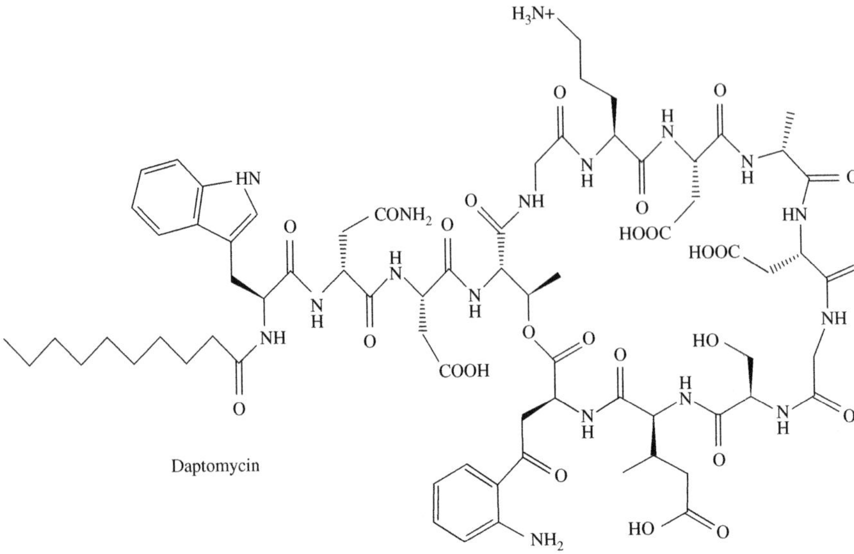

FIGURE 19.3 Chemical structure of the lipopeptide antibiotic, daptomycin.

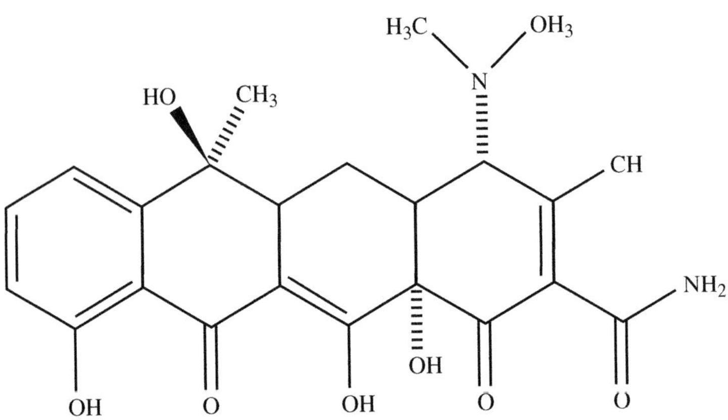

FIGURE 19.4 Structure of tetracycline.

19.2 HISTORICAL PRECEDENT FOR NATURAL PRODUCTS AS ANTIMICROBIAL DRUGS

Since the 1930s, natural products have played a significant role in the discovery of new classes and novel antibiotic drugs; and in fact, most antibacterial drugs were derived directly from natural products or discovered using a natural product lead [7].

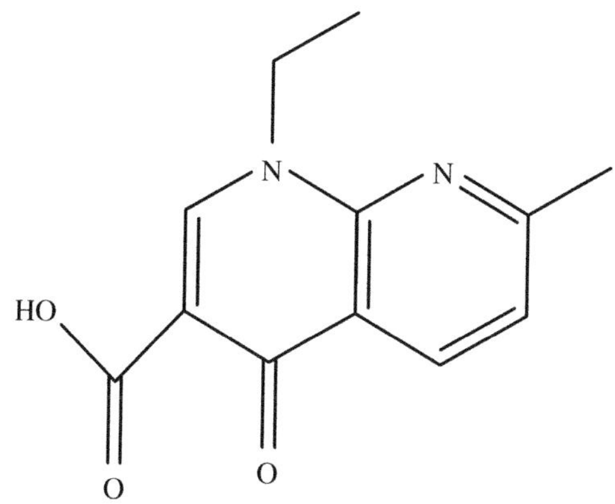

FIGURE 19.5 Structure of nalidixic acid.

It is estimated that 70 of the 90 antibiotics marketed during the period 1982–2002 originated or were derived from natural products [13]. However, many of the other 20 compounds belonged to the fluoroquinolones class of antibiotics, which are actually derivatives of another naturally occurring compound, nalidixic acid (Fig. 19.5) [13]. Despite these successes, over the past 30 years there has been a significant decline in antibiotic research and natural products drug discovery in the pharmaceutical industry. Surprisingly, both were abandoned in the late 1990s due to a perceived lack of success, poor return on investment, and the introduction of new screening tools and methodologies such as combinatorial chemistry, which were expected to be highly successful in generating new chemical diversity, making natural products obsolete [13]. In addition, another new method initiated was high throughput (HT) screening, which allows for the rapid screening of high numbers of samples. However, natural products do not lend themselves well to HT screening due to the complexity of the extracts, the labor intensiveness involved in isolating a new compound, and the time required to obtain a hit. Interestingly, neither HT screening nor combinatorial chemistry have increased the development of new antimicrobial drugs over the past 10–15 years. What we do know is that new and emerging microbes, as well as the development of resistance by some major human pathogens, have created a critical need for new antimicrobial agents, a need that has not been met by the use of new technologies or the discovery of new therapeutic agents.

19.3 ARE PHYTOTHERAPIES USEFUL FOR THE TREATMENT OF INFECTIOUS DISEASES?

In addition to bacterial and fungal sources of natural antimicrobial agents, traditional plant-based medicines have been used for centuries in most countries worldwide for the treatment of all diseases, including infectious diseases. This fact is reflected in the enormous amount of published scientific data that describe the antimicrobial activities of thousands of plant extracts against numerous bacteria, viruses, and fungi, including Gram-negative bacilli and multidrug-resistant organisms [14–21]. Searches of the PubMed database (from 1975 to 2012) produced over 1200 reports in the scientific and medical literature that described the antimicrobial activities of various plant species and their chemical constituents. Furthermore, a search of the Napralert database, a natural products database housed within the University of Illinois at Chicago, has shown that of the almost 60,000 plant species listed in the database, over 6,000 species have experimental antimicrobial activity, of which approximately 4,000 species have ethnomedical data supporting the use of these plants by various cultures to treat infectious diseases [16–22]. The majority of these plant species have *in vitro* activity against a broad range of bacteria, fungi, mycobacterium, and viruses.

But Are Phytotherapies Really Useful for the Prevention and Treatment of Infectious diseases? Absolute answers to this question are difficult for many reasons, most having to do with the lack of clinical trials, funding for natural products research, and the perceived notion that antibiotics must be bactericidal and totally eradicate the bacteria in question at very low concentrations of the drug. Furthermore, of the thousands of reports of *in vitro* antimicrobial activities for herbal medicines, very few animal and human studies have been published due in part to a lack of funding for this type of research. Unfortunately, antimicrobial activity *in vitro* does not always translate into activity *in vivo* or in human studies. This can become even more complicated when complex herbal formulae are used that combine multiple plant extracts. Moreover, *in vitro* data do not provide reliable information on dosing, bioavailability, and other clinically relevant issues. Thus, of the thousands of medicinal plants with antimicrobial effects *in vitro*, very few have actually been tested in animals and even less in humans. However, a few medicinal plants have been tested extensively in clinical trials for antimicrobial effects such as tea tree oil (extensively reviewed in 19, 20) and garlic, *Allium sativum* L. reviewed in Refs. [15, 23, 24], and these data have already been reviewed, so they will not be included here. Cranberry (*Vaccinium macrocarpon* Ait) has also been extensively reviewed [25]; however, there are some more recent and interesting clinical trials data since 2010 that are included in this review.

19.3.1 Cranberry (*Vaccinium macrocarpon* Ait)

Cranberry, known scientifically as *Vaccinium macrocarpon* Ait. (Ericaceae, heath family), is a native North American plant [26]. Historically, the small, edible red-black berries were used by Native Americans as a dye, food, and for the treatment of wounds [26]. During the eighteenth century, cranberry juice was commonly employed throughout Europe and New England for the treatment of urinary tract infections (UTIs). In an early clinical trial in the 1960s, Pappas and coworkers published a clinical trial involving 60 patients who had the symptoms of UTIs (frequency, dysuria, urgency, nocturia) associated with *Escherichia coli* infections and were treated with 16 oz. of cranberry juice daily for 21 days [27]. Following therapy, 53% of the patients reported a reduction in urine bacterial counts and were symptom free, while another 20% were reported to show a moderate response to the cranberry therapy. However, the effects were short-lived, with 61% of patients experiencing a relapse of symptoms and bacteriuria 6 weeks after discontinuing the cranberry juice [27]. In a 1994 clinical trial, Avorn and coworkers published a randomized, double-blind, placebo-controlled trial involving 153 elderly women, and reported a reduction in bacteriuria and pyuria following 6 months of therapy with 300 ml daily of cranberry juice cocktail [28]. The incidence of urogenital symptoms in patients with bacteriuria and pyuria was 4% in the cranberry treated group versus 7% in the placebo group [28].

In 2008, the Cochrane group published a systematic review of the clinical trials for cranberry to assess the efficacy of cranberry products in preventing UTIs in susceptible populations [25]. Ten clinical trials ($n = 1049$, five cross-over, five parallel group) were included in the analysis. In seven studies, the authors analyzed a cranberry or cranberry–lingonberry juice combination versus placebo, juice, or water. In the other studies, cranberry tablets versus placebo were used. The results of the analysis showed that when compared with placebo or control, cranberry products significantly reduced the incidence of UTIs at 12 months (RR 0.65, 95% CI 0.46–0.90). Cranberry products more effectively reduced the incidence of UTIs in women with recurrent UTIs than elderly men or in people requiring catheterization. Thus, there is evidence that regular consumption of cranberry juice may decrease the number of symptomatic UTIs over a 12-month period, particularly for women with recurrent UTIs [25]. Since this review was published, at least five more randomized-controlled clinical trials have been published that investigated the effects of cranberry on UTIs and *Helicobacter pylori* infections [29–34]. In 2012, Afshar and coworkers assessed the effectiveness of cranberry juice for the prevention of UTIs in a randomized double-blind clinical trial involving 40 children [29]. The children were randomized to receive daily doses (2 ml/kg body weight) of cranberry juice with high concentrations (37%) of proanthocyanidin (Fig. 19.6) versus a cranberry juice with no proanthocyanidin for a 1-year treatment period. Children up to the age of 18 years were eligible if they had at least two culture documented nonfebrile UTIs in the calendar year before enrollment. Urinary infection was defined as a positive culture of a midstream sample with an uropathogenic bacterium of 100 million cfu/l (105 ml^{-1}) in symptomatic children, with or without fever. After 12 months of follow-up, the average incidence of UTI in the treatment group was 0.4 per patient per year and 1.15 in the placebo group ($p < 0.045$), representing a 65% reduction in the risk of UTI [29]. The study concluded that cranberry juice with high concentrations of proanthocyanidin is effective for the prevention of pediatric nonfebrile UTIs [29].

Another randomized-controlled clinical trial compared the time to UTI and the rates of asymptomatic bacteriuria and urinary P-fimbriated *Escherichia coli* over 6 months in premenopausal women with history of UTI [33]. The volunteers were randomized to 1 of 3 arms: 4 oz of cranberry juice daily, 8 oz of cranberry juice daily, or placebo juice, with the time to UTI (symptoms plus pyuria) being the primary outcome. A total of 176 participants were randomized (120 to cranberry juice and 56 to placebo) and followed up for a median of 168 days. The cumulative rate of UTI was 0.29 in the cranberry juice group and 0.37 in the placebo group ($P = 0.82$). The adjusted hazard ratio for UTI in the cranberry juice group versus the placebo group was 0.68 (95% confidence interval, 0.33–1.39; $P = 0.29$). The proportion of women with P-fimbriated urinary *E. coli* isolates during the intervention phase was 10 of 23

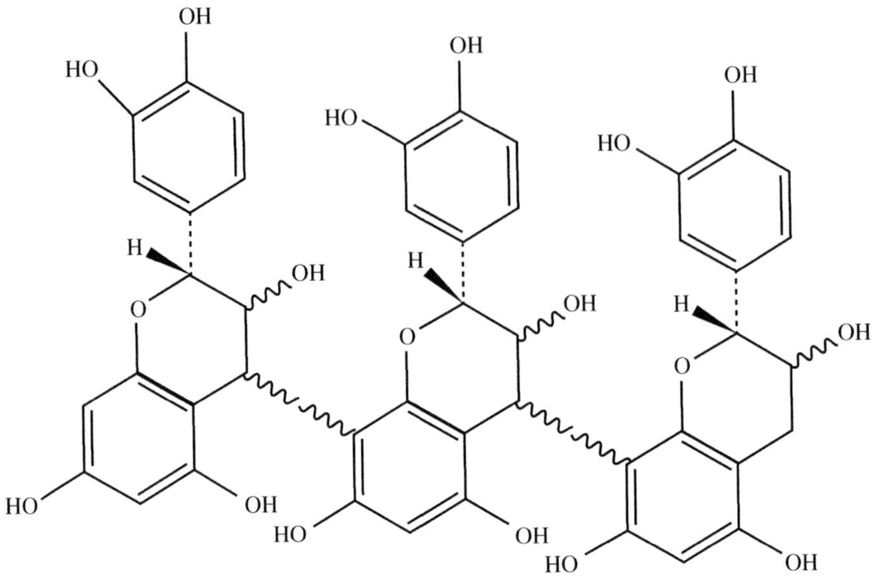

FIGURE 19.6 Example of a proanthocyanidin (proanthocyanidin trimer), compound from cranberry reported to have antibacterial effects.

(43.5%) in the cranberry juice group and 8 of 10 (80.0%) in the placebo group ($p = 0.07$). While cranberry juice did not significantly reduce UTI risk compared with placebo, a reduction in urinary P-fimbriated *E. coli* strains was observed and supports the biological potential of cranberry activity [33].

In 2011, a double-blind, double-dummy comparison trial involving 221 premenopausal women with recurrent UTIs assessed the efficacy of 12-month prophylactic use of trimethoprim-sulfamethoxazole (TMP-SMX), 480 mg once daily, and cranberry capsules, 500 mg twice a day [30]. Primary outcomes measured were the mean number of symptomatic UTIs over 12 months, the proportion of patients with at least 1 symptomatic UTI, the median time to first UTI, and development of antibiotic resistance in indigenous *E. coli*. After 12 months, the results showed that the mean number of patients with at least 1 symptomatic UTI was higher in the cranberry than in the TMP-SMX group (4.0 vs. 1.8; $P = 0.02$), and the proportion of patients with at least 1 symptomatic UTI was slightly higher in the cranberry than in the TMP-SMX group (78.2% vs. 71.1%). Median time to the first symptomatic UTI was 4 months for the cranberry and 8 months for the TMP-SMX group. After 1 month, in the cranberry group, 23.7% of fecal and 28.1% of asymptomatic bacteriuria *E. coli* isolates were TMP-SMX resistant, whereas in the TMP-SMX group, 86.3% of fecal and 90.5% of asymptomatic bacteriuria *E. coli* isolates were TMP-SMX resistant. Also, there were increased resistance rates for trimethoprim, amoxicillin, and ciprofloxacin in these *E. coli* isolates after only 1 month in the TMP-SMX group, whereas antibiotic resistance did not increase in the cranberry group [30]. Thus, while TMP-SMX was more effective in premenopausal women than cranberry capsules, cranberry did not increase bacterial resistance in these women, suggesting that long-term use may be feasible [30].

Along with its potential use for the prevention of uropathogenic *E. coli*, cranberry juice and its chemical constituents have been reported to exert antiadhesion activities on *H. pylori* in human studies. Thus, cranberry has been used as adjunct therapy with standard antibiotics. One clinical trial assessed the potential additive effects of cranberry with triple therapy (omeprazole, amoxicillin, and clarithromycin (OAC)), in a double-blind randomized study of *H. pylori* infections [32]. The 177 subjects were treated with OAC for 1 week and then were randomly allocated to receive 250 ml of either cranberry juice (cranberry-OAC, $n=89$) or placebo beverage (placebo-OAC, $n=88$) twice daily and only cranberry juice or placebo beverage for the next 2 weeks. Treatment outcome was determined with the 13C urea breath test (13C-UBT). An additional control group consisted of patients referred to the same center during the same period who were treated with OAC alone for 1 week (nonplacebo-OAC, $n=712$). While the overall rate of *H. pylori* eradication (13C-UBT *a 3.5*) was 82.5%, with no statistically significant difference among the 3 arms, gender analysis revealed that for female subjects only, the eradication rate was higher in the cranberry-OAC arm ($n=42$, 95.2%) than in the placebo-OAC arm ($n=53$, 86.8%) and significantly higher than in the nonplacebo-OAC group ($n=425$, 80%; $p=0.03$). The results suggest that the addition of cranberry juice to OAC therapy improves the rate of *H. pylori* eradication in females [32]. Thus, cranberry may be useful for the management of both pathogenic Gram-negative *E. coli* and *H. pylori*.

In a 2012 randomized trial, 32 volunteers were administered a proanthocyanidin standardized cranberry, and urine samples were collected and tested in an antiadhesion assay [31]. The results showed significant bacterial antiadhesion activity in urine samples collected from volunteers that consumed cranberry powder as compared with placebo ($p<0.001$). This inhibition was dose dependent. The study concluded that administration of PAC-standardized cranberry powder at dosages containing 72 mg of PAC per day offered some protection against bacterial adhesion and virulence in the urinary tract [31].

Vaccinium species contain high concentrations of quinic acid, which is aromatized to benzoic acid, and with the addition of glycine, is changed to hippuric acid [34]. Hippuric acid is found in the urine following consumption of cranberries and inhibits bacterial adhesion [34]. Fructose, found in all fruit, also prevents adhesion of type 1 fimbriated bacteria. Proanthocyanidins found in cranberries have been shown to prevent attachment of P-fimbriated *E. coli* to cellular surfaces [35]. Proanthocyanidins appear to be easily absorbed and metabolized, and there is evidence that they have high bioavailability in humans [34]. Thus, for cranberry proanthocyanidins, there are both *in vitro* and human trials data that support the antibacterial effects of these compounds.

In 2012, a Cochrane updated review of the randomized clinical trials for cranberry was published that included new clinical data from the new trials published since 2008 [35]. The review assessed the effectiveness of cranberry products in the prevention of UTIs in susceptible populations [35]. The review incorporated 10 studies included in the 2008 update, and 14 new studies. Thirteen studies ($n=2380$) investigated only cranberry juice/concentrate, nine studies ($n = 1032$) assessed only cranberry tablets/capsules, one study compared cranberry juice and tablets; and one study compared cranberry capsules and tablets [35]. Comparison/control arms included placebo, no treatment, water, methenamine hippurate, antibiotics, or lactobacillus. Data included in the meta-analyses showed that when compared with placebo, water or no treatment, cranberry products did not significantly reduce the occurrence of symptomatic UTI overall (RR 0.86, 95% CI 0.71–1.04) or for any of the subgroups: women with recurrent UTIs (RR 0.74, 95% CI 0.42–1.31); older people (RR 0.75, 95% CI 0.39–1.44); pregnant women (RR 1.04, 95% CI 0.97–1.17); children with recurrent UTI (RR 0.48, 95% CI 0.19–1.22); cancer patients (RR 1.15, 95% CI 0.75–1.77); or people with neuropathic bladder or spinal injury (RR 0.95, 95% CI: 0.75–1.20) [35]. The effectiveness of cranberry was not significantly different to antibiotics for women (RR 1.31, 95% CI 0.85, 2.02) and children (RR 0.69, 95% CI 0.32–1.51). There was no significant difference between gastrointestinal adverse effects from cranberry product compared to those of placebo/no treatment (RR 0.83, 95% CI 0.31–2.27) [35]. The authors concluded that cranberry products are not associated with prevention of UTIs. However, the lack of association of cranberry with a reduced incidence of UTIs in clinical trials may be due to lack of participant adherence, lack of sufficient active ingredient in the cranberry product, or lack of sufficient statistical power [36, 37].

19.3.2 Turmeric (*Curcuma longa* L.) as an Antimicrobial Agent

Turmeric (*Curcuma longa* L.) and ginger (*Zingiber officinale* L.), both from the family Zingiberaceae, have long been reported to have antimicrobial activities against many bacteria [26, 38, 39]. Dried *Curcuma longa* is the source of the spice turmeric, the ingredient that gives curry powder its characteristic yellow color. Turmeric has a long tradition of use in the Chinese and Ayurvedic systems of medicine, particularly as an anti-inflammatory agent, anticancer treatment, and also as an antimicrobial agent. Turmeric is reported to have antibacterial effects against many Gram-positive bacteria [39] and also Gram-negative bacteria such as *H. pylori* [40]. A methanol extract of the dried powdered turmeric rhizome and curcumin were tested against 19 strains of *H. pylori*, including five *CagA*+ strains (cancer causing strains of *H. pylori*). The results of this study showed that the methanol extract inhibited the growth of all strains of *H. pylori* and that curcumin was the active constituent and inhibited the growth of all strains of *H. pylori in vitro* with a minimum inhibitory concentration range of 6.25–12.5 μg/ml (positive control amoxicillin MICs 0.0039–0.25 μg/ml) and was more active against the five *CagA*+ strains [40]. One *in vivo* study investigated the mechanism of curcumin's action on *H. pylori*-infections in C57BL/6 mice [38]. In addition to eradication of *H. pylori* strains from infected mice, curcumin was also shown to regulate the expression and activities of MMP-3 and -9 in gastric tissues [41]. In addition, curcumin was highly effective in restoring the denudation of epithelial region, disruption in gastric mucosal layer, and infiltration of inflammatory cells that occurred due to *H. pylori* infection in mouse gastric tissues [41]. When curcumin treatments were compared with the effect of triple antibiotic (TT) therapy in Sydney Strain 1 (*H. pylori* strain)-infected gastric tissues, curcumin was more effective in reducing the infiltration of inflammatory cells as compared with TT. This study demonstrated that alterations in the balance between the MMPs and TIMPs during *H. pylori* infections were normalized more efficiently by curcumin treatments than TT treatments, suggesting a protective effect, as well as an antibacterial effect [41].

The antimicrobial activities of turmeric are thought to be due to the various chemical constituents of the essential oil (5.8%) that are obtained by steam distillation of rhizomes, namely α-phellandrene (1%), sabinene (0.6%), cineol (1%), borneol (0.5%), zingiberene (25%), and sesquiterpines (53%). However, the phenolic constituents of the root, namely curcumin (diferuloylmethane) (3–4%), also have activity and these comprised curcumin I (94%), curcumin II (6%), and curcumin III (0.3%) (Fig. 19.7) [39].

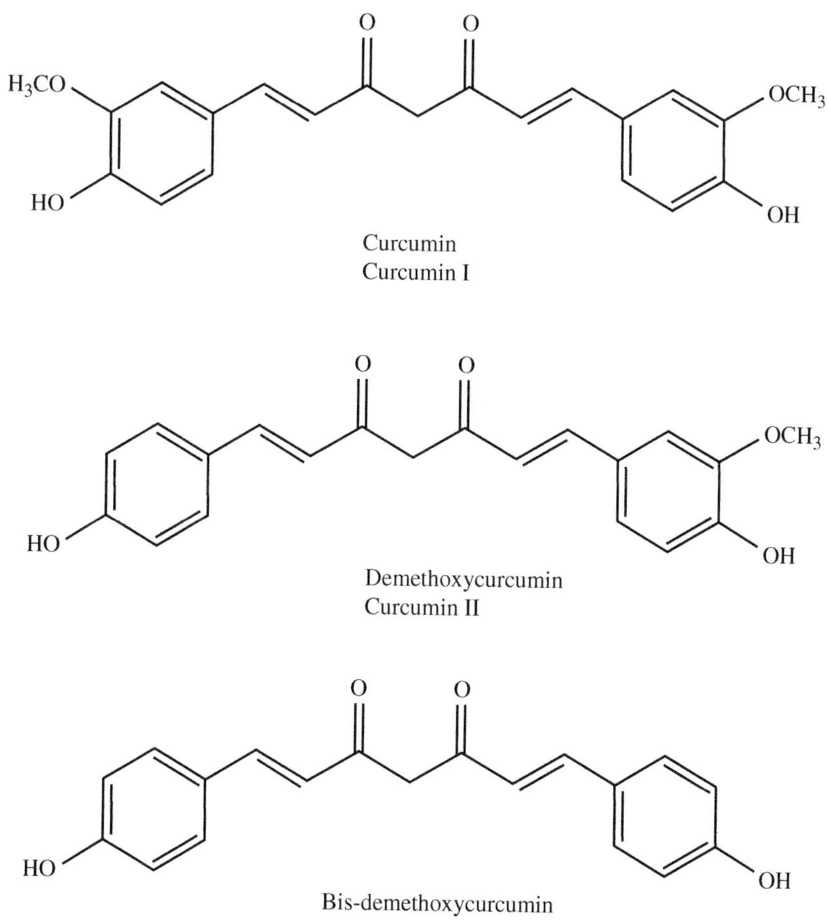

FIGURE 19.7 Chemical structures of the phenolic constituents of turmeric.

19.3.3 Ginger (*Zingiber officinale* L.) as an Antimicrobial Agent

Ginger root is well known as a condiment and food plant that is used traditionally for the treatment of a wide range of disorders in traditional systems of medicine, including the treatment of gastrointestinal ailments such as motion sickness, dyspepsia, and hyperemesis gravidarum [26, 39, 42–44]. Ginger root has also been reported to have antimicrobial activities against a wide range of bacteria and fungi [26, 39]. Methanol extracts of the dried powdered ginger rhizome, as well as the isolated constituents, 6-,8-,10-gingerol and 6-shogoal, were tested against 19 strains of *H. pylori*, including five *CagA*+ strains [42]. The methanol extract inhibited the growth of all *H. pylori* strains *in vitro* with a minimum inhibitory concentration range of 6.25–50 μg/ml [42, 44]. One fraction of the crude extract, containing the gingerols, was active and inhibited the growth of all HP strains with an MIC range of 0.78–12.5 μg/ml and with significant activity against the *CagA*+ strains [42]. In a later *in vivo* study, a standardized methanol extract of ginger root was tested in a rodent model of *H. pylori*-induced disease, the Mongolian gerbil, to examine the effects of the extract on both prevention and eradication of infection [43]. Oral administration of the extract to Mongolian gerbils at a daily dose of 100 mg/kg body weight in rations for either 3 weeks prior to infection or 6 weeks postinfection, reduced bacterial load by 50%, and significantly ($P<0.05$) reduced both acute and chronic muscosal and submucosal inflammation, cryptitis, as well as epithelial cell degeneration and erosion induced by *H. pylori*. The results showed that ginger has anti-*H. pylori* activities but also has significant anti-inflammatory effects against *H. pylori*-induced inflammation [43].

The gingerols (Fig. 19.8) are a group of structurally related polyphenolic compounds isolated from ginger and have demonstrated anti-*H. pylori* activities, and are thought to be the active antimicrobial constituents of ginger. 10-Gingerol had an MIC of 1.56 μg/ml against *CagA*+ strains of *Helicobacter* [42].

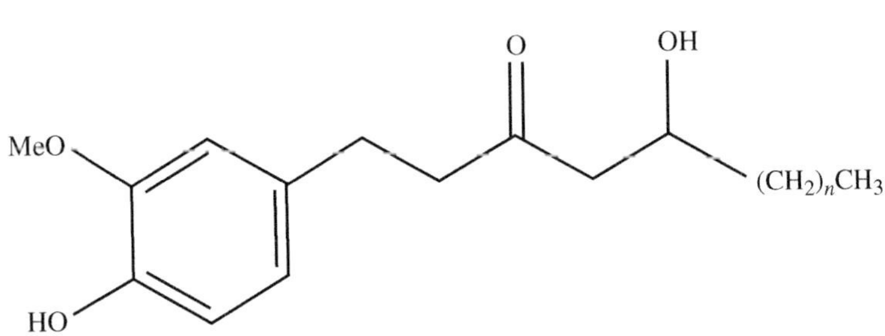

Compound	*n*
6-Gingerol	4
8-Gingerol	6
10-Gingerol	8

FIGURE 19.8 Chemical structure of the gingerols.

19.4 NATURALLY OCCURRING COMPOUNDS THAT MAY REDUCE ZOONOSIS

Any infection that can be transmitted from vertebrate animals into humans and vice versa is classified as a zoonosis according to the WHO [5]. Currently, there are over 200 diseases that are classified as zoonoses, many of which have been recognized for many centuries. These infectious diseases are caused by various types of pathogenic infectious agents, including bacteria, parasites, fungi, and viruses. Reducing pathogen carriage in foods to prevent zoonosis transmission to humans is an important part of reducing the rate of infectious diseases, thus there is much interest in naturally occurring substances that could be added to food or animal feed to reduce the bacterial carriage and thus reduce transmission to humans.

Salmonella enteritidis and *Campylobacter jejuni* are the two major food-borne pathogens transmitted through poultry products, a common source of zoonosis [45]. Chickens are the reservoir hosts for these Gram-negative bacteria, with their intestinal colonization being one of the most significant factors responsible for the contamination of meat and eggs. Effective strategies for reducing bacterial colonization of chickens would be a major step to improve microbiological safety of poultry products, and thus reduce transmission to humans. A natural, safe antimicrobial treatment that can be added to feed presents the most practical and cheap method for most farms, and would also be acceptable due to no or low toxicity. Thus, there is a potential for using plant-derived, generally recognized as safe (GRAS) status molecules, such as caprylic acid, trans-cinnamaldehyde, eugenol, carvacrol, and thymol as feed supplements for reducing caecal populations of *Salmonella enteritidis* and *C. jejuni* in chickens [45, 46].

One example that has been tested *in vivo* is caprylic acid (Fig. 19.9), which is a medium-chain fatty acid reported to be effective in killing a variety of bacterial pathogens, including *Campylobacter jejuni*. The therapeutic effect of caprylic acid on *C. jejuni* counts in the caecal contents of 42-day-old chickens was investigated in four trials [45]. In the first two trials, day-of-hatch chicks ($n=60$ per trial) were assigned to six treatment groups ($n=10$ birds per treatment group): positive controls (Campylobacter, no caprylic acid), 0.7 or 1.4% of caprylic acid in feed for the last 3 days of the trial with or without a 12-h feed withdrawal. Treatments were similar for trials three and four with the exception that the doses used were lower (0.35 or 0.7% caprylic acid supplementation) for the last 7 days of the trial. On day 42, the caeca were excised and Campylobacter cultured and counts determined. The supplementation of caprylic acid at 0.35 and 0.7% consistently decreased ($P<0.05$) the colonization of

FIGURE 19.9 Structure of caprylic acid. A naturally occurring compound having antibacterial effects in chickens.

C. jejuni in the chicken caeca compared with positive control treatment [45]. When these treatments were evaluated after a 12-h feed withdrawal period, 0.7% caprylic acid decreased Campylobacter colonization in the 3-day treatment. Body weight and feed consumption did not differ between the caprylic acid and control groups. The results suggest that therapeutic supplementation of caprylic acid in the feed can effectively decrease Campylobacter in market-aged chickens and may be a potential treatment for decreasing pathogen carriage in poultry [45].

19.5 SYNERGISTIC AND ADDITIVE EFFECTS WITH ANTIBIOTICS

In 2009, the results of a prospective randomized study to evaluate the combined therapeutic effect of *Serenoa repens*, *Urtica dioica* (ProstaMEV), as well as quercitin and curcumin (FlogMEV) extracts as adjunct therapy with prulifloxacin in patients affected by chronic bacterial prostatitis (CBP) [47], were published. The study involved 143 patients affected by CBP (National Institutes of Health (NIH) class II prostatitis). All patients received prulifloxacin 600 mg daily for 14 days, but were split into two groups: group A received prulifloxacin along with ProstaMEV and FlogMEV ($n=107$); group B received only prulifloxacin ($n=37$). Microbiological and clinical efficacies were tested by two follow-up visits at 1 month and 6 months, respectively. Quality of life (QoL) was measured using the NIH Chronic Prostatitis Symptom Index (CPSI) and International Prostatic Symptom Score (IPSS) questionnaires. One month after treatment, 89.6% of patients who had received prulifloxacin associated with ProstaMEV and FlogMEV did not report any symptoms related to CBP compared with only 27% of patients who received antibiotic therapy alone were recurrence free ($P<0.0001$). Significant differences were also found between groups in terms of symptoms and QoL ($P<0.0001$ for both). Six months after treatment, no patients in group A had recurrence of the disease while two patients in group B did. The combination of ProstaMEV, FlogMEV extracts improved the clinical efficacy of prulifloxacin in patients affected by CBP [47].

Ginger extracts and 10-Gingerol have been shown to have both additive and synergistic effects with antibiotics against Gram-negative bacteria. An extract from ginger reduced the minimum inhibitory concentrations (MICs) of aminoglycosides in vancomycin-resistant enterococci (VRE) [48]. 10-Gingerol was identified as the active compound, and reduced the MIC of arbekacin, and other aminoglycosides, as well as of bacitracin and polymixin B [48].

19.6 NEW EMERGING INFECTIOUS DISEASES AND THOSE WITH NO KNOWN TREATMENTS

Dengue is a serious major mosquito-borne infectious viral disease with very little effective antiviral therapies and no approved vaccines. However, very recently Zandi et al. [49, 50] have shown that aqueous extracts of the roots of *Scutellaria baicalensis* had virucidal activity against four dengue virus serotypes. The authors used quantitative

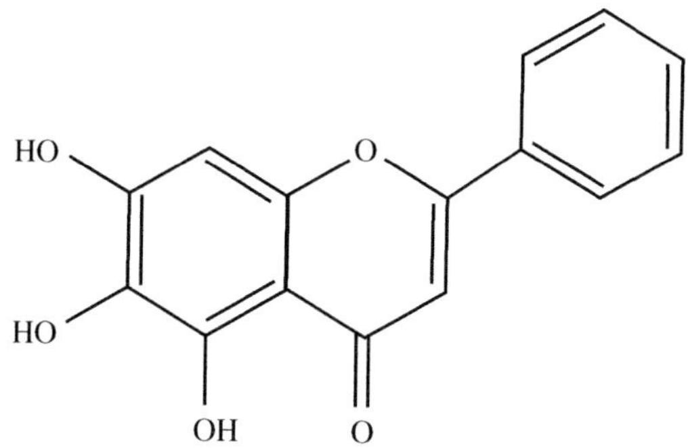

FIGURE 19.10 Structure of baicalein.

real-time polymerase chain reaction (qRT-PCR) assays to measure the actual dengue virus RNA copy number. The median inhibitory concentration (IC_{50}) for the extract in Vero cells following virus infection was 86.59–95.19 μg/ml for the different dengue serotypes. However, when the Vero cells were treated with the extract at the time of virus infection, the IC50 concentrations were reduced to 56.02–77.41 μg/ml [50]. The concentration of baicalein (5,6,7-trihydroxyflavone) (Fig. 19.10), a flavone in the *S. baicalensis* extract, was approximately 1%. The same authors also investigated the effects of baicalein against different stages of dengue virus type 2 (DENV-2) replication in Vero cells using focus forming unit reduction assay and quantitative RT-PCR [49]. The results of this investigation showed that baicalein strongly inhibited DENV-2 replication in Vero cells (IC_{50} = 6.46 μg/ml and selectivity index SI = 17.8) when added to the cells after viral infection. However, again the IC_{50} decreased to 5.39 μg/ml (SI = 21.3) when cells were treated 5 h before virus infection and continuously up to 4 days postinfection. Baicalein showed direct virucidal effect against DENV-2 *in vitro* (IC_{50} = 1.55 μg/ml) and an antiadsorption effect (IC_{50} = 7.14 μg/ml) [49]. This demonstrates that specific phytotherapies and purified naturally occurring compounds are potential candidates as new antiviral drugs against viruses that currently have little or no effective drug treatments.

19.7 SARS

In 2003, outbreaks of severe acute respiratory syndrome (SARS and a novel coronavirus associated with the disease) occurred in China, Hong Kong, and Taiwan. The cumulative SARS cases reported to WHO from 32 countries were 8098, including 774 deaths [51]. Since the pathogenesis of SARS is not well understood, currently there is no effective treatment for it. For thousands of years, traditional Chinese medicine (TCM) has been used to control infectious diseases, and due to the possibility of a beneficial effect of TCM on SARS, TCM has been investigated as a potential treatment of SARS. In one clinical trial, the efficacy of Integrative Chinese and Western medicine (ICWM) was assessed for treating SARS. In this trial, 49 patients with SARS were studied; they were divided into the control group (n = 29) and the

ICWM group ($n=20$). The former was treated according to the "Recommended Program for Treatment of SARS" provided by Health Ministry, by administering such drugs as ribavirin, levofloxacin, thymopentin, azithromycin, and so on, the latter was treated with the ICWM protocol for SARS that included traditional Chinese medicines. The time to symptom improvement in the ICWM group was 5.10 days and that in the control group was 7.62 days, the difference being significant ($P<0.05$), indicating that treatment with the ICWM protocol was superior to the treatment with Western medicine alone with respect to improving clinical symptom, promoting recovery of immune function and reducing lung inflammation, decreasing the dosage of hormone used, and shortening the therapeutic course [51].

19.8 REDUCING MRSA CARRIAGE

Methicillin-resistant *Staphylococcus aureus* (MRSA) is a significant pathogenic bacteria and an important cause of morbidity and mortality worldwide. It is estimated that approximately 2.5 million persons (1.4% of the population) are MRSA nasal carriers [52]. Since both tea and coffee have antimicrobial properties, a study was carried out to determine if the consumption of tea, coffee, or both would reduce the nasal carriage of MRSA [52]. In an adjusted logistic regression analysis controlling for age, race, sex, poverty–income ratio, current health status, hospitalization over 12 months, and use of antibiotics in the past month, individuals who reported consuming hot tea were only 50% as likely to have MRSA nasal carriage relative to individuals who drank no hot tea (odds ratio=0.47; 95% confidence interval, 0.31–0.71). Similarly, individuals who reported consuming coffee also had a 50% reduction in the risk of MRSA nasal carriage relative to individuals who drank no coffee (odds ratio=0.47; 95% confidence interval, 0.24–0.93) [52]. The study concluded that the consumption of hot tea or coffee is a promising new method to decrease MRSA nasal carriage that is safe, inexpensive, and easily accessible to the global population [52].

In 2006, a randomized placebo-controlled clinical trial was performed to determine the effects of tea catechin inhalation on MRSA carriage in disabled elderly patients [53]. Seventy-two patients aged 78 ± 11 years with cerebrovascular diseases, who were classified as disabled according to daily living activity, and were either bedridden or required assistance for standing, and also diagnosed with the presence of MRSA in sputum participated in the study. Subjects in the treatment group received inhalation of 2 ml of tea catechin extract solution along with saline (3.7 mg/ml catechins, 43% of catechins comprise epigallocatechin gallate) (Fig. 19.11), or inhalation of saline alone, thrice daily using a handheld nebulizer for 7 days. The reduction rates calculated as the summation of decrease and disappearance of MRSA in sputum at 7 days were 47% (17 of 36 patients) in the catechin group and 15% (5 of 33 patients) in the control group; the difference in the reduction rates between the two groups was statistically significant ($P=0.014$). The results suggest that tea catechins, including EGG, can reduce MRSA nasal carriage [53]. In a previous study, the same group investigated the effects of inhalation of tea catechin on MRSA in 24 elderly in-patients with confirmed MRSA in their sputum [54]. The patients in the

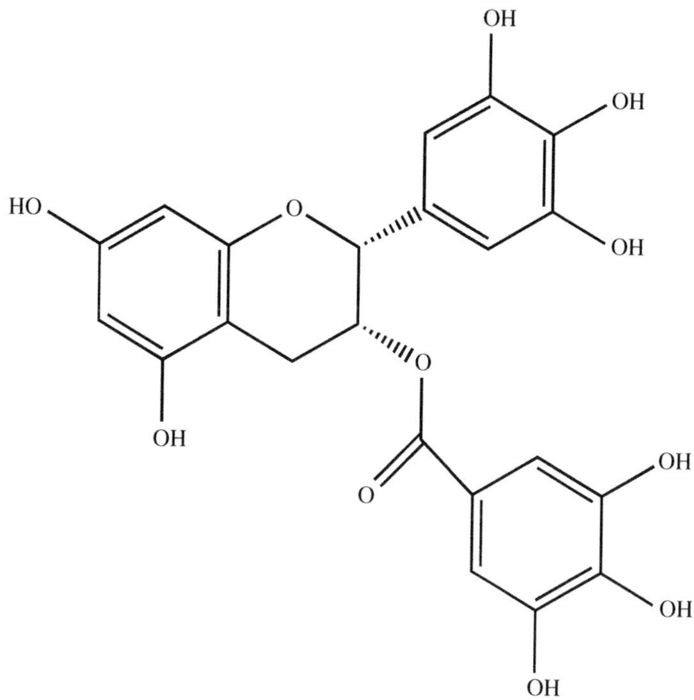

FIGURE 19.11 Structure of epigallocatechin gallate (EGCG), a compound present in a tea catechin extract that as an inhalant, can reduce nasal MRSA carriage.

catechin group ($N=12$) were administered an inhalation of tea catechin extracts (in saline/bromhexine) (3.7 g/l catechins, 43% were composed of epigallocatechin gallate), thrice daily with hand nebulizer for 4 weeks. The clinical effects were compared with the control group ($N=12$) who were given an inhalation of saline/bromhexine alone. After a week of treatment, the number of patients with decreased or disappearance of MRSA in their sputum was significantly higher in the catechin group compared with that in the control group (seven vs. no patients; $P<0.05$). The number of patients discharged during the study was significantly increased, and the days of hospital stay were significantly decreased in the catechin-treated group compared with the control group (six vs. one patient; $P<0.05$, 51 ± 22 vs. 85 ± 0 days, mean ± S.D.; $P<0.05$, respectively). No adverse effects were observed in any patient during the study [53]. Catechin inhalation appears to be a safe and effective method for the reduction of MRSA and shortening of hospitalization.

19.9 CONCLUSIONS

For thousands of years, plant-based medicines (phytotherapies) have been used to treat infectious diseases in every country and culture worldwide as part of a traditional system of medicine. There are thousands of publications indicating that medicinal plant extracts inhibit the growth of many bacteria, fungi, and viruses, including MRSA, dengue, SARS, and Gram-negative bacteria. Some of these infections have no known treatment. In addition, extracts of cranberry, garlic, tea tree, and others have clinical data supporting their use for the treatment of Gram-negative *E. coli* and *H. pylori* infections, thereby demonstrating their usefulness. Furthermore, there is *in vitro* evidence that combining plant extracts and pure natural compounds with common antibiotics may reduce resistance and enhance the life span of specific antibiotics, as well as clinical data showing that combinations can indeed improve overall outcomes and reduce reinfection. Over the past 10 years, there have been many reviews on the topic of phytotherapies and their effects on MRSA, *H. pylori*, and many other Gram-negative and positive bacilli. However, beyond garlic, cranberry and a few others, there is currently no focused effort to test these active phytotherapies *in vivo* or in human studies due to a lack of funding for such work and a reduced focus on natural products research by the majority of pharmaceutical companies. Considering the overwhelming need for new antimicrobial therapies for infectious diseases, the reduced number of new antibiotics in the pharmaceutical pipeline, and the fact that natural products hold the most promise for the development of novel antimicrobial therapies, research and development of phytotherapies should be a global priority.

REFERENCES

[1] World Health Organization (2011) *World Health Statistics 2011*, WHO Press, Geneva, Switzerland, pp. 1–15.

[2] World Health Organization (2008) *World Health Report*, World Health Organization, Geneva, Switzerland, pp. 1–10.

[3] Pinner R, Teutsch S, Simonsen L, et al. (1996) Trends in infectious diseases mortality in the United States. *JAMA* 275: 189–193.

[4] Mahady GB (2005) Medicinal plants for the treatment and prevention of bacterial infections. *Curr Pharm Des* 19: 2405–2427.

[5] World Health Organization (2013) Global infectious disease surveillance, fact sheet (www.who.int/mediacentre/factsheets/fs200/en/) accessed June 14, 2013.

[6] Butler MS, Cooper MA (2011) Antibiotics in the clinical pipeline in 2011. *J Antibiot (Tokyo)* 64: 413–425.

[7] Butler MS, Buss AD (2006) Natural products-the future scaffolds for novel antibiotics? *Biochem Pharmacol* 71: 919–929.

[8] Freire-Moran L, Aronsson B, Manz C, et al. (2011) Critical shortage of new antibiotics in development against multidrug-resistant bacteria-time to react is now. *Drug Resist Update* 14: 118–124.

[9] Boucher HW (2010) Challenges in anti-infective development in the era of bad bugs, no drugs: a regulatory perspective using the example of bloodstream infection as an indication. *Clin Infect Dis* 50: S4–S9.

[10] Projan SJ (2008) Whither antibacterial drug discovery? *Drug Dis Today* 13: 279–280.

[11] Trémolières F, Cohen R, Gauzit R, et al. (2010) Save antibiotics! What can be done to prevent a forecasted disaster? Suggestions to promote the development of new antibiotics. *Réanimation* 19: 354–360.

[12] Bourlioux P (2013) Which alternatives are at our disposal in the anti-infectious therapeutics face to multi-drug resistant bacteria? (Article in French). *Ann Pharm Franc* 7: 150–158.

[13] Peláez F (2006) The historical delivery of antibiotics from microbial natural products-can history repeat? *Biochem Pharmacol* 71: 981–990.

[14] Kuete V, Alibert-Franco S, Eyong KO (2011) Antibacterial activity of some natural products against bacteria expressing a multidrug-resistant phenotype. *Int J Antimicro Agents* 37: 156–161.

[15] Lee SY, Shin YW, Hahm KB (2008) Phytoceuticals: mighty but ignored weapons against *Helicobacter pylori* infection. *J Digest Dis* 9: 129–139.

[16] Mahady GB (2009) Medicinal plants for the treatment and prevention of bacterial infections. In: Atta-Ur-Rahman (ed.) *Frontiers in Medicinal Chemistry*, Vol. 4, Elsevier Science Publishers, Amsterdam, the Netherlands, pp. 248–284.

[17] Mahady GB, Huang Y, Doyle BJ, et al. (2009) Natural products as antibacterial agents. In: Atta-Ur-Rahman (ed.) *Studies in Natural Products*, Vol. 35, Elsevier Science Publishers, Amsterdam, the Netherlands, pp. 423–444.

[18] Lawal TO, Soni KK, Saxena RC, et al. (2012) Anti-*Helicobacter pylori* activities of compounds of natural origin. In: B Brahmachari (ed.) *Bioactive Natural Products: Opportunities and Challenges in Medicinal Chemistry*, World Scientific Press, New Jersey, pp. 475–497.

[19] Slover C, Danziger L, Mahady GB (2009) Recent advances in natural products for methicillin resistant *Staphylococcus aureus Staphylococcus aureus* (MRSA). In: I Ahmed, F Aqil (eds.) *New Strategies Combating Bacterial Infection*, Wiley-Blackwell Publishers, Weinheim, Germany, pp. 127–134.

[20] Slover C, Locklear T, Doyle BJ, et al. (2008) Efficacy of herbal medicines for *Helicobacter pylori* infections. In: P Houghton, P Murkerje (eds.) *Health Promotion through Herbal Medicine (Approach for Evaluation of Natural Health Products)*, Pharmaceutical Press, London, pp. 149–159.

[21] Mahady GB, Locklear TD, Doyle BJ, et al. (2007) Botanicals as treatment for infectious diseases. In: RR Watson, VR Preedy (eds.) *The Encyclopedia of Herbal Medicine in Clinical Practice*, CABI Publishing, Wallingford, pp. 1–10.

[22] Lawal TO, Soni KK, Adeniyi BA, et al. (2013) Botanicals as adjunct therapy and treatment of multi-drug resistant Staphylococcal infections. In: M Rai, K Kon (eds.) *Fighting Multidrug Resistance with Herbal Extracts*, Essential Oils and their Components, Elsevier Publishers, Chennai, India, pp. 135–145.

[23] Dini C, Fabbri A, Geraci A (2011) The potential role of garlic (*Allium sativum*) against the multi-drug resistant tuberculosis pandemic: a review. *Ann Ist Sup San* 47: 465–473.

[24] Goncagul G, Ayaz E (2010) Antimicrobial effect of garlic (*Allium sativum*). *Recent Patents in Anti-Infective Drug Discovery* 5: 91–93.

[25] Jepson RG, Craig JC (2008) Cranberries for preventing urinary tract infections. *Cochrane Database of Systematic Reviews* 1: CD001321.

[26] Mahady GB, Fong HHS, Farnsworth NR (eds.) (2001) Cranberry. In: *Botanical Dietary Supplements: Quality, Safety and Efficacy*, Swets & Zeilinger, Lisse, the Netherlands, pp. 45–53.

[27] Papas PN, Brusch CA, Ceresia GC (1966) Cranberry juice in the treatment of urinary tract infections. *Southwest Med J* 47: 17–20.

[28] Avorn J, Monane M, Gurwitz JH, et al. (1994) Reduction of bacteriuria and pyuria after ingestion of cranberry juice. *JAMA* 271: 751–754.

[29] Afshar K, Scott SH, MacNeily AE (2012) Cranberry juice for the prevention of pediatric urinary tract infection: a randomized controlled trial. *J Urol* 188: 1584–1587.

[30] Beerepoot M, ter Reit G, Nys S, et al. (2011) Cranberries vs. antibiotics to prevent urinary tract infections: a randomized double-blind non-inferiority trial in premenopausal women. *Arc Int Med* 171: 1270–1278.

[31] Howell AB, Botto H, Combescure C, et al. (2010) Dosage effect on uropathogenic *Escherichia coli* anti-adhesion activity in urine following consumption of cranberry powder standardized for proanthocyanidin content: a multicentric randomized double blind study. *MC Infect Dis* 10: 94.

[32] Shmuely H, Yahav J, Samra Z, et al. (2007) Effect of cranberry juice on eradication of *Helicobacter pylori* in patients treated with antibiotics and a proton pump inhibitor. *Mol Nutr and Food Res* 51: 746–751.

[33] Stapleton A, Dziura R, Hooton DM, et al. (2012) Recurrent urinary tract infection and urinary *Escherichia coli* in women ingesting cranberry juice daily: a randomized controlled trial. *Mayo Clin Proceed* 87: 143–150.

[34] Stothers L (2002) A randomized trial to evaluate effectiveness and cost effectiveness of naturopathic cranberry products as prophylaxis against urinary tract infection in women. *Canadian J of Urol* 9: 1558–1562.

[35] Howell AB, Vorsa N, Der Marderosian A, et al. (1998) Inhibition of adherence of P-fmbriated *Escherichia coli* to uroepithelial-cell surfaces by proanthocyanidin extracts from cranberries. *NEJM* 339: 108–1086.

[36] Jepson RG, Williams G, Craig JC (2012) Cranberries for preventing urinary tract infections. *Cochrane Database Systematic Reviews* 10: CD001321.

[37] Jepson RG, Craig J, Williams G (2013) Cranberry products and prevention of urinary tract infections. *JAMA* 310: 1395–1396.

[38] De R, Kundu P, Swarnakar S, et al. (2009) Antimicrobial activity of curcumin against *Helicobacter pylori* isolates from India and during infections in mice. *Antimicrob Agents Chemother* 53: 1592–1597.

[39] Farnsworth NR, Fong HHS, Mahady GB (1999) Rhizoma curcumin. In: *WHO Monographs of Selected Medicinal Plants*, Vol. 1, World Health Organization, Traditional Medicine Programme, Geneva, Switzerland, pp. 45–51.

[40] Mahady GB, Pendland SL, Yun G, et al. (2002) Turmeric (*Curcuma longa*) and curcumin inhibit the growth of *Helicobacter pylori*, a group 1 carcinogen. *Anticancer Res* 22: 4179–4181.

[41] Kundu P, De R, Pal I, et al. (2011) Curcumin alleviates matrix metalloproteinase-3 and -9 activities during eradication of *Helicobacter pylori* infection in cultured cells and mice. *PLoS One* 6: e16306.

[42] Mahady GB, Pendland SL, Yun GS, et al. (2003) Ginger (*Zingiber officinale* Roscoe) and the gingerols inhibit the growth of *CagA*+ strains of Helicobacter pylori. *Anticancer Res* 23: 3699–3702.

[43] Gaus K, Huang Y, Israel DA, et al. (2009) Standardized ginger (*Zingiber officinale*) extract reduces bacterial load and suppresses acute and chronic inflammation in Mongolian gerbils infected with *CagA*+ Helicobacter pylori. *Pharm Biol* 47: 92–98.

[44] Mahady GB, Pendland SL, Stoia A, et al. (2005) *In vitro* susceptibility of *Helicobacter pylori* to botanical extracts used traditionally for the treatment of gastrointestinal disorders. *Phytother Res* 19: 988–991.

[45] Solis de los Santos F, Donoghue AM, Venkitanarayanan K, et al. (2009) The natural feed additive caprylic acid decreases *Campylobacter jejuni* colonization in market-aged broiler chickens. *Poul Sci* 88: 61–64.

[46] Venkitanarayanan K, Kollanoor-Johny A, Darre MJ, et al. (2013) Use of plant-derived antimicrobials for improving the safety of poultry products. *Poul Sci* 92: 493–501.

[47] Cai T, Mazzoli S, Bechi A, et al. (2009) *Serenoa repens* associated with *Urtica dioica* (ProstaMEV) and curcumin and quercitin (FlogMEV) extracts are able to improve the efficacy of prulifloxacin in bacterial prostatitis patients: results from a prospective randomized study. *Int J Antimicrob Agents* 33: 549–553.

[48] Nagoshi C, Shiota S, Kuroda T, et al. (2006) Synergistic effect of (10)-gingerol and aminoglycosides against vancomycin-resistant enterococci (VRE). *Biol Pharm Bull* 29: 443–447.

[49] Zandi K, Teoh BT, Sam SS, et al. (2012) Novel antiviral activity of baicalein against dengue virus. *BMC CAM* 12: 214–216.

[50] Zandi K, Lim TH, Rahim NA, et al. (2013) Extract of *Scutellaria baicalensis* inhibits dengue virus replication. *BMC CAM* 13: 91–95.

[51] Zhang RL, Jiao Q, Wang BG (2003) Controlled clinical study on 49 patients of SARS treated by integrative Chinese and Western medicine (Article in Chinese). *Zhongguo Zhong Xi Yi Jie He Za Zhi* 23: 654–657.

[52] Matheson EM, Mainous AG 3rd, Everett CJ, et al. (2011) Tea and coffee consumption and MRSA nasal carriage. *Ann Fam Med* 9: 299–304.

[53] Yamada H, Tateishi M, Harada K, et al. (2006) A randomized clinical study of tea catechin inhalation effects on methicillin-resistant *Staphylococcus aureus Staphylococcus aureus* in disabled elderly patients. *J Am Med Dir Assoc* 7: 79–83.

[54] Yamada H, Ohashi K, Atsumi T, et al. (2003) Effects of tea catechin inhalation on methicillin-resistant *Staphylococcus aureus* in elderly patients in a hospital ward. *J Hosp Infect* 53: 229–231.

20

PHYTOMEDICINES FOR CNS DISORDERS: SAFETY ISSUES FOR USE WITH ANTIEPILEPTIC DRUGS

Sophia Yui Kau Fong, Rosina Yau Mok, Qiong Gao, Yin Cheong Wong, and Zhong Zuo

School of Pharmacy, The Chinese University of Hong Kong, Shatin, Hong Kong

20.1 INTRODUCTION

Epilepsy is one of the most common neurologic diseases affecting more than 50 million people worldwide [1] and antiepileptic drugs (AEDs) are the mainstay of treatment. Many patients require more than one AED and most of these patients are on life-long therapy. Moreover, AEDs are regarded as "broad-spectrum" drugs that have efficacy not only in epilepsy but also in various neurological conditions such as neuropathic pain, and psychiatric and neuromuscular disorders [2]. The widespread use of AEDs makes it essential for clinicians to identify potential interactions between AEDs and other coadministered substances, including phytomedicines.

Phytomedicines, or herbal medicines, usually contain numerous pharmacologically active constituents, including alkaloids, anthraquinones, anthocyanins, coumarins, essential oils, flavonoids, glycosides, tannins, and saponins, all of which may potentially interact with AEDs and may lead to potential clinical consequences. One of the contributing factors toward increasing incidence of herb–drug interactions is the increased popularity of herbal medicines [3]. According to pharmacoepidemiologic surveys, the percentage of epileptic patients concurrently taking complementary and alternative medicines and AEDs is relatively high in both developed and developing regions: United States (39%), Cambodia (36%), United Kingdom (34%), Taiwan

Phytotherapies: Efficacy, Safety, and Regulation, First Edition. Edited by Iqbal Ramzan.

(16%), Nigeria (15%), and India (12%); while more than 60% of patients did not inform their physicians [4–9]. In China, integrated medicine is common practice where Western and traditional Chinese medicines are prescribed concurrently for the treatment of epilepsy [10]. Therefore, the frequency of patients taking AEDs with herbal medicines is high, and it is necessary to address the safety issues of such combined use.

Herbal medicines may interact with AEDs in two ways: pharmacokinetically and/or pharmacodynamically (Fig. 20.1). Pharmacokinetic interactions relate to the alteration of the processes of absorption, distribution, metabolism, and excretion. Pharmacodynamic interactions relate to interactions at the site of action of a drug and may be additive, synergistic, or antagonistic in nature. Interactions between herbal medicines and AEDs may lead to potential clinical consequences including lack of efficacy, toxic reactions, unexpected effects, unforeseen side effects, and noncompliance and are therefore of major importance for patient outcome. In this chapter, existing information from preclinical (animal and *in vitro*) and clinical studies for pharmacokinetic and pharmacodynamics interactions between herbal medicines and AEDs are systematically reviewed and summarized. The mechanisms underlying such interactions will also be discussed.

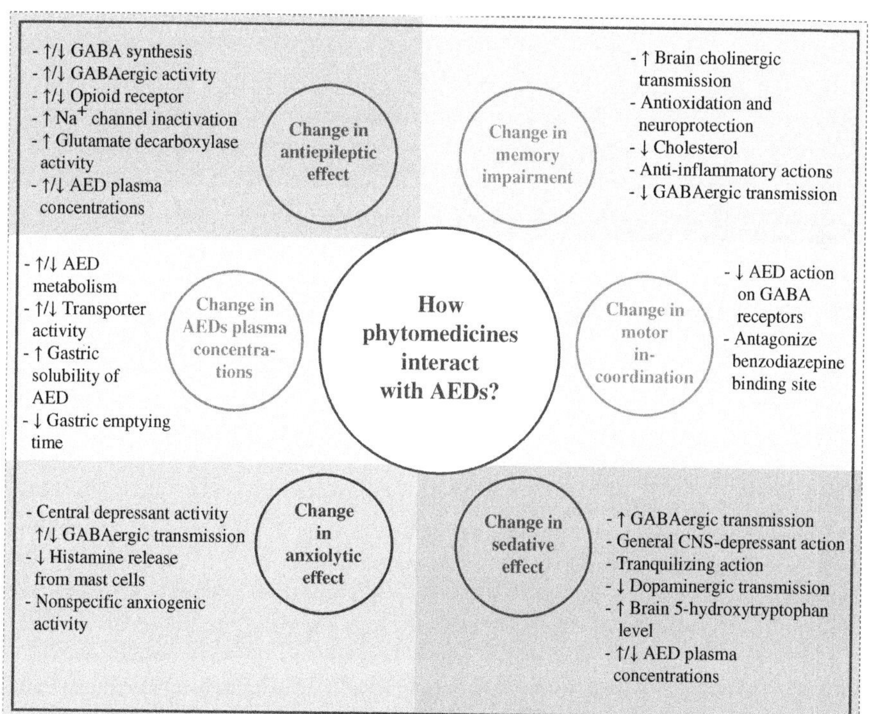

FIGURE 20.1 How phytomedicines interact with AEDs.

20.2 METHODOLOGY OF SYSTEMATIC LITERATURE SEARCH

A computer-based search of the following English databases was conducted: AMED (1985–Mar 2013), CINAHL Plus (1937–Oct 2012), Cochrane Database of Systematic Reviews (2005–Dec 2012), CENTRAL (Mar 2013), Embase (1947–Mar 2013), Medline (1946–Mar 2013), and SciFinder Scholar (1907–Mar 2013). Sixteen commonly prescribed AEDs were included in the search: carbamazepine, clobazam, clonazepam, diazepam, ethosuximide, gabapentin, lamotrigine, levetiracetam, oxcarbazepine, paraldehyde, phenobarbital, phenytoin, pregabalin, topiramate, valproate, and vigabatrin. Drug names and the brand names/common names were all searched as keywords and they were combined, using the combination term and with a comprehensive list of keyword and MeSH search terms for herbal medicines listed in the published review [11]. No language restriction was imposed during the search, but non-English articles were included only if they contained an English abstract with sufficient information. In addition to the English databases, four Chinese databases were searched, including Chinese BioMedical Literature Database (1978–Oct 2012), China Journal Net (1915–Oct 2012), Traditional Chinese Medical Database System (1984–Oct 2012), and Chinese Medical Academic Conference Database (1994–Oct 2012). The MeSH headings and keywords used for the search were the Chinese names of AEDs, in combination with the Chinese equivalent terms of "interaction," "Chinese herbal medicines," and "Chinese and Western medicines" ("jie he," "xiang hu zuo yong," "zhong yao," "zhong cao yao," and "zhong xi yi"). Bibliographies of every retrieved article were checked for any additional pertinent studies. Articles were considered eligible for evaluation if they contained original data involving herbal medicine interactions with AEDs without restriction on the study type (*in vitro*, animal and human studies) or the report type (original and review articles). All relevant literature fulfilling these inclusion criteria was extracted and compiled, except for the interacting pairs that have beneficial effects.

20.3 PHARMACOKINETIC INTERACTIONS

Herbal medicines may alter the pharmacokinetics of AEDs via a number of mechanisms. Affecting the hepatic metabolism is the most common mechanism since majority of the AEDs are extensively metabolized by cytochrome P450s (CYP) and/or Uridine Glucuronyl Transferases (UGTs). Known substrates of CYP3A4 include carbamazepine, diazepam, ethosuximide, and tiagabine. Phenytoin, phenobarbitone, diazepam, primidone, and valproate are substrates for CYP2C9, while phenytoin, phenobarbitone, and primidone are also metabolized by CYP2C19. AEDs that undergo glucuronidation include carbamazepine, valproate, phenytoin, oxcarbazepine, and lamotrigine [12]. Herbal medicines may alter the plasma levels of AEDs via inhibition or induction of the enzymes involved in the metabolism of AEDs. Inhibition of hepatic enzymes may result in accumulation of AEDs, which can cause toxicity within hours to days. This is particularly the case for AEDs having concentration-dependent adverse effects [13]. On the other hand, herbal medicines may increase metabolism of AEDs by inducing enzymes. The induction effect may take a few days to a few weeks to manifest after coadministration of herbal medicines and AEDs, which may result in reduced level and therefore reduced efficacy of AEDs. Some of the AEDs such as phenytoin, valproate, diazepam, and tiagabine are highly bound to proteins (>90%). Constituents in herbal medicines may competitively shift these AEDs from the serum protein and increase their free levels, which may lead to toxicity or increased clearance. The displacement is of clinical importance for AEDs with narrow therapeutic indices, such as phenytoin. Furthermore, some of the AEDs have poor solubility and their rate and extent of absorption are easily affected by coadministration of herbal medicines that may alter gastric conditions, for example, by delaying/increasing the gastric emptying time or by decreasing/increasing acid secretion. In the following sections, recent information on pharmacokinetic interactions between individual AEDs and herbal medicines from the literature will be discussed. Herbal medicines that lead to potential alteration in the plasma concentrations of AEDs are summarized in Table 20.1.

20.3.1 Carbamazepine

In animal studies, the plasma level or oral bioavailability of carbamazepine was increased when coadministered with four Chinese herbal medicines, namely *Cassia auriculata* Linn., piperine (an active compound in *Piper longum* Linn.), Platycodonis Radix, and Polygoni Cuspidati Rhizoma et Radix, possibly through decreasing the metabolism of carbamazepine or improving gastric solubility of carbamazepine [22, 25, 28, 32].

In contrast, *Ginkgo biloba* Linn., Jia-Wei-Xiao-Yao-San and Xiao-Yao-San decreased the plasma level/oral bioavailability of carbamazepine by increasing the metabolism of carbamazepine via CYP3A4 induction in animals [26, 29, 30, 33]. Gensenoside (an active constituent of ginseng) also activates CYP3A4 activity *in vitro* and thereby increases carbamazepine metabolism [23].

Xiao-Qing-Long-Tang and Xiao-Chai-Hu-Tang delayed the time for carbamazepine to reach peak plasma concentration (T_{max}) by decreasing gastric emptying rate [19, 21], whereas Paeoniae Radix decreased the T_{max} of carbamazepine possibly by improving dissolution of carbamazepine [35].

Melatonin, berberine (an active compound in Coptidis rhizome), *Cardiospermum halicacabum* Linn., Chai-Hu-Jia-Long-Gu-Mu-Li-Tang, and *Hypericum perforatum* Linn., did not alter plasma levels or other pharmacokinetic parameters of carbamazepine in animal or human studies [16–18, 24, 27, 31, 34]. For a more detailed description of the pharmacokinetic interaction between carbamazepine and herbal medicines, readers are referred to the systematic review by Fong et al. [13].

20.3.2 Phenytoin

The effect of piperine, an active compound in *Piper longum* Linn., on the pharmacokinetics of phenytoin was studied in healthy volunteers and patients with uncontrolled epilepsy. A single dose (20 mg) and multiple doses (20 mg for 7 days) of

TABLE 20.1 Effects of Phytomedicines on Plasma Concentrations of Antiepileptic Drugs[a]

Antiepileptic Drugs (AEDs)	Changes in Plasma Concentrations of AEDs		
	Increase	Decrease	No Change
Carbamazepine	Mentat[b] [14] Xiao-Chai-Hu-Tang[c] [19] **Piperine** (an active compound in *Piper longum* Linn.) [22] *Cassia auriculata* Linn.[25]	Septilin [15] **Ispaghula husk** [20] Gensenoside (an active compound in *Panax ginseng*) [23] *Ginkgo biloba* Linn. [26]	**Melatonin**[f] [16–18] Xiao-Qing-Long-Tang[g] [21] *Cardiospermum halicacabum* Linn. [24] Berberine (an active compound in Coptidis rhizome) [27]
	Polygoni Cuspidati Rhizoma et Radix [28] Platycodonis Radix [32]	Jia-Wei-Xiao-Yao-San[d] [29, 30] Xiao-Yao-San[e] [33]	*Hypericum perforatum* Linn. [31] Chai-Hu-Jia-Long-Gu-Mu-Li-Tang[h] [34] Paeoniae Radix [35]
Diazepam	Angelicae dahuricae Radix (oral administration) [36]	Salvia miltiorrhizae Radix et Rhizoma [37]	**Ginkgo biloba Linn**. [38] Xiao-Chai-Hu-Tang[b] [39] Angelicae dahuricae Radix (intravenous administration) [36]
Phenobarbitone		*Ginkgo biloba* Linn. [40]	**Melatonin**[e] [16] Phyllanthus amarus Schum & Thonn [41]
Phenytoin	Mentat[a] [14, 42] *Ginkgo biloba* Linn. [44] **Piperine** [47–49] *Nardostachys jatamansi* DC. [50] Shankhapushpi[i] [51]	Rhei Radix et Rhizoma [43] Paeoniae Radix [45]	Melatonin[e] [18] Jujubae Fructus [46] Paeoniae Radix [45]
Valproate		Gegen-Qinlian-Tang[j] [52] *Ginkgo biloba* Linn. [54]	**Paenoiae Radix** [53] Osthole (a coumarin-like derivative isolated from *Peucedanum ostruthium* (Linn.) Koch.) [55] **Melatonin** [16]

[a] The herb names highlighted with bold are evidenced by clinical studies.

[b] **Mentat**: Also known as BR-16A, containing 13 herbs: *Bacopa monnieri* Linn., *Centella asiatica* Linn., *Withania somnifera* (Linn.) Dunal, *Evolvulus alsinoides* Linn., *Nardostachys jatamansi* Linn., *Acorus calamus* Linn., *Celastrus paniculatus* Linn., *Zingiber officinale* Linn., *Valeriana wallachii*, Prunus amygdalus, *Orchis mascula* Linn., *Syzgium aromaticum* Linn., and Mukta pishti.

[c] **Xiao-Chai-Hu-Tang**: Known as Sho-saiko-to in Japanese, containing seven herbs: Bupleuri Radix, Pinelliae Tuber, Scutellariae Radix, Zizyphi Fructus, Ginseng Radix, Glycyrrhizae Radix, and Zingiberis Rhizoma.

[d] **Jia-Wei-Xiao-Yao-San**: Also known as free and easy wanderer plus, which contains 11 herbs: Bupleuri Radix, Scutellariae Radix, Zingiberis Rhizoma, Angelicae Sinensis Radix, Zizyphi Fructust, Moutan Cortex, Paeoniae Radix Alba, Atractylodis Macrocephalae Rhizoma, Poria, Menthae Haplocalycis Herba, and Glycyrrhizae Radix

[e] **Xiao-Yao-San**: Also known as free and easy wanderer, which contains eight herbs: Bupleuri Radix, Angelicae Sinensis Radix, Paeoniae Radix Alba, Atractylodis Macrocephalae Rhizoma, Poria, Zingiberis Rhizoma, Glycyrrhizae Radix, Menthae, and Haplocalycis Herba.

[f] **Melatonin**: The major secretory product of pinealocytes (*N*-acetyl-5-methoxytryptamine).

[g] **Xiao-Qing-Long-Tang**: Known as Sho-seiryu (to extract) in Japanese, containing eight herbs: Pinelliae Tuber, Glycyrrhizae Radix, Cinnamomi Cortex, Schisandrae Fructus, Asiasari Radix, Paeoniae Radix, Ephedrae Hebra, and Zingiberis Siccatum Rhizoma.

[h] **Chai-Hu-Jia-Long-Gu-Mu-Li-Tang**: Known as Saiko-ka-ryukostsu-borei-to in Japanese, containing 10 herbs: Bupleuri Radix, Pinelliae Tuber, Cinnamomi Cortex, Hoelen, Scutellariae Radix, Zizyphi Fructus, Ginseng Radix, Ostreae testa, Fossilia Ossis Mastodi, and Zingiberis Rhizoma.

[i] **Shankhapushpi**: An Ayurvedic syrup which contains *Convolvulus pluricaulis* Chois., *Centella asiatica* Urban., *Nardostachys jatamansi* DC., *Nepeta hindostana* Haines, *Nepeta elliptica* Royle., and *Onosma bracteatum* Wall.

[j] **Gegen-Qinlian-Tang**: A Chinese herbal medicine containing four herbs: Puerariae Radix, Scutellariae Radix, Coptidis Rhizoma, and Glycyrrhizae Radix.

piperine increased the oral bioavailability ($AUC_{0\to t}$), peak plasma concentrations (C_{max}), and the rate of absorption of phenytoin under both steady-state and non-steady-state conditions. While piperine did not affect the elimination of phenytoin at steady state, piperine may enhance the absorption of phenytoin by delaying gastric emptying time, increasing splanchnic blood flow, decreasing acid secretion, and altering membrane dynamics that aids efficient membrane permeability [47–49].

In animal studies, pretreatment for more than 7 days with Gingko biloba and an Indian herbal medicine Mentat increased the oral bioavailability of phenytoin possibly via inhibition of hepatic CYP2C9 enzymes [14, 42, 44].

Coadministration of Ayurvedic syrup Shankhapushpi and phenytoin for 5 days significantly lowered plasma phenytoin levels by 50% in rats. Shankhapushpi contains six herbs which might induce hepatic microsomal enzymes responsible for the metabolism of phenytoin [51].

In another animal study, coadministration of a single dose or multiple doses of Rhei Radix et Rhizoma decoction (2 g/kg) significantly decreased the C_{max} and $AUC_{0\to t}$ as well as the elimination rate of phenytoin and its metabolites, phenytoin glucuronide, 4-hydroxyphenytoin, and 4-hydroxyphenytoin glucuronide, possibly via alteration of the transporter systems. Rhei Radix et Rhizoma was demonstrated to induce the P-glycoprotein-mediated efflux of phenytoin and inhibited the multiple resistance protein 2-medicated transport of phenytoin and 4-hydroxyphenytoin *in vitro* [43].

The administration of phenytoin in combination with Paenoniae Radix increased the T_{max} of phenytoin threefold in rats and this was attributed to a delay in phenytoin absorption [45]. Paenoniae Radix did not affect other phenytoin pharmacokinetic parameters except for a 44% reduction in the apparent volume of distribution.

20.3.3 Valproate

A randomized, open-label, two-way crossover study was conducted in six healthy volunteers to examine the effect of multiple oral doses of Paenoiae Radix (1.2 g powder of Paeoniae Radix extract for 7 days) on a single oral dose of valproate. No changes in pharmacokinetic parameters including C_{max}, T_{max}, oral bioavailability and clearance, apparent volume of distribution, or clinically relevant adverse events were observed [53].

The effect of Gegen-Qinlian-Tang, a Chinese herbal medicine containing four herbs on the pharmacokinetics of valproate was investigated in rats. C_{max} and $AUC_{0\to t}$ of valproate were significantly decreased when coadministered with Gegen-Qinlian-Tang (2 and 4 g/kg). The four herbs in Gegen-Qinlian-Tang synergistically inhibited the monocarboxylate transporter-mediated absorption of valproate *in vitro* and this was attributed to the increased oral bioavailability *in vivo* [52]. In an animal study, intraperitoneal injection of osthole (a coumarin-like derivative isolated from *Peucedanum ostruthium* (Linn.) Koch.) at 150 mg/kg did not affect the total brain concentration of valproate administered at a dose of 173.8 mg/kg. In parallel, it did not alter the antiepileptic effect, motor performance, long-term memory, or skeletal muscular effects of valproate [55].

In an *in vitro* study, *Gingko biloba* Linn. extract decreased the hepatic microsomal formation of 3-OH-valproate, 4-OH-valproate, 5-OH-valproate, and 4-ene-valproate by inhibiting CYP2C9. Quercetin, kaempferol, and isorhamnetin but not other terpene trilactones or flavonol glycosides, isolated from *Gingko biloba* Linn. produced such inhibitory effects [54].

20.3.4 Diazepam

In an open-label study, the effect of *Ginkgo biloba* Linn. extract on diazepam metabolism was investigated in 12 healthy male volunteers. Ginkgo biloba Linn. extract (120 mg twice daily) taken for 28 days did not affect C_{max}, $AUC_{0\rightarrow t}$, elimination half-life ($t_{1/2}$), and the time to reach C_{max} (T_{max}) of diazepam and its metabolite N-desmethyldiazepam. The authors concluded that *Ginkgo biloba* Linn. did not affect the pharmacokinetics of diazepam via CYP2C19 and the excretion of N-desmethyldiazepam in healthy volunteers [38].

In an animal study, the Chinese herb Angelicae dahuricae Radix (1 g/kg) caused a fourfold increase in the C_{max} of diazepam after oral but not intravenous administration of diazepam, possibly by decreasing the first-pass metabolism via inhibition of CYP2C, CYP3A, and CYP2D1 as demonstrated *in vitro*. As a result of this effect on the presystemic clearance of diazepam, Angelicae dahuricae Radix had little effect on the pharmacokinetics of diazepam after intravenous administration [36].

In another animal study, pretreatment of Salvia miltiorrhizae Radix et Rhizoma water extract (100 mg/kg per day, po) for 15 days reduced the C_{max} and $AUC_{0\rightarrow t}$ of diazepam (15 mg/kg) by 72.7 and 44.4%, respectively, while the total body clearance was markedly increased twofold. The effect was attributed to the induction of the metabolic activity of cytochrome P450 by Salvia miltiorrhizae Radix et Rhizoma [37].

Pretreatment of Xiao-Cha-Hu-Tang (2 g/kg) in rats for 7 days did not affect the plasma concentration and protein binding of diazepam, nor the activities of hepatic drug-metabolizing enzymes attributed to diazepam metabolism (CYP1A2, 2C6, 2C11, 2D1, and 3A2) [39].

20.3.5 Phenobarbitone

Phenobarbitone is a well-known inducer of various CYP enzymes [56]. The effect of a Chinese herbal medicine, *Phyllanthus amarus* SCHUM & THONN, on the phenobarbitone-induced P450 enzyme activities was investigated *in vivo* and *in vitro*.

Phyllanthus amarus SCHUM & THONN (250 and 750 mg/kg) was orally administered to rats once daily for 15 days. Administration of phenobarbitone was started on day 12 and continued for 4 days (60 mg/kg, intraperitoneal, once daily). Livers were collected and liver microsomes were prepared for the determination of CYP enzymes activities. Both doses of *Phyllanthus amarus* SCHUM & THONN inhibited the elevated enzyme activities of CYP1A1, CYP1A2, and CYP 2B1/2 produced by phenobarbitone by 50%. *In vitro* results also indicated that *Phyllanthus amarus* SCHUM & THONN potently inhibited the activities of these CYP enzymes [41].

In another animal study, the C_{max} and $AUC_{0\rightarrow t}$ of phenobarbitone were decreased by 40% and 20% respectively in rats that received 2-week pretreatment with *Gingko biloba* Linn. (0.1, 0.5, and 1% *Gingko biloba* extract). These changes were possibly due to enhancement of CYP2B expression, since liver weight was increased by 35% with the highest dose of *Gingko biloba* extract [40].

20.3.6 Newer Generations of Antiepileptic Drugs

Newer AEDs have been developed in the past 20 years in attempts to improve the tolerability and side-effect profiles, namely vigabatrin, gabapentin, lamotrigine, pregabalin, topiramate, and levetiracetam. Compared to the older AEDs, these AEDs are less susceptible to metabolic interactions since most do not undergo hepatic metabolism but are eliminated unchanged via the kidneys. Consequently, potential interactions between newer AEDs and coadministered substances are rarely reported. A retrospective observational study found that approximately 45% of epileptic patients treated with monotherapy using older AEDs had a potential interaction with another drug, whereas only 3.9% of patients taking a newer AED had a potential drug interaction [57]. There are no studies or case reports currently on the pharmacokinetic interactions between newer AEDs and herbal medicines. Nevertheless, it should be remembered that some of the newer AEDs are still prone to pharmacokinetic drug interactions. Oxcarbazepine and topiramate are CYP3A4 inducers and have clinically significant pharmacokinetic interactions with oral contraceptives [58]. Lamotrigine undergoes extensive glucuronidation and is susceptible to both UGT inhibitors and inducers [59]. It is expected that more studies in the coming decades will reveal the potential interactions between these apparently "safe" AEDs and herbal medicines.

20.4 PHARMACODYNAMIC INTERACTIONS

AEDs have multiple pharmacodynamic actions that make them valuable in treating not only epilepsy but also many other neurological diseases. For instance, benzodiazepines and phenobarbitone possess sedative and hypnotic effects for treatment of insomnia. Benzodiazepines, gabapentin, and several other AEDs have also been frequently indicated for anxiety since they have proven anxiolytic effects [2]. However, AEDs, particularly the older ones, are well known for their side effects including impairment of memory and other cognitive functions.

Mechanistically, AEDs mainly target proteins involved in neuronal excitation, via modulation of voltage-gated sodium channels, calcium channels, or ligand-gated receptors for γ-amino butyric acid (GABA) and glutamate; or by affecting intracellular pathways. Most AEDs have several mechanisms of action and, consequently, there are many sites at which a pharmacodynamic interaction may be elicited [60].

In recent years, a growing body of research has explored the activity of herbal medicines in the central nervous system (CNS). Herbal medicines were found to possess a variety of CNS activities: antiepileptic, proconvulsive, sedative, anxiolytic, and cognitive improvement effects, to name but a few [61, 62]. All of these may potentially alter the pharmacodynamic activities of AEDs. In the following sections, up-to-date information from the literature on the pharmacodynamic interactions between individual AEDs and herbal medicines on (i) antiepileptic, (ii) sedative, (iii) anxiolytic, (iv) memory impairment, and (v) miscellaneous effects will be discussed. Preclinical and clinical studies will both be included but the majority of the results are from animal studies. Readers are recommended to review the publication by Fong et al. [11] for a more detailed discussion.

20.4.1 Antiepileptic Effects

AEDs exert their antiepileptic effects via multiple mechanisms of action. One of the major mechanisms is the enhancement of the inhibitory events mediated by gamma-aminobutyric acid (GABA) (benzodiazepines, phenobarbitone, gabapentin, topiramate, vigabatrin, valproate). Another mechanism for AEDs to reduce the seizure frequency is to inhibit voltage-dependent sodium or calcium channels, thereby blocking sustained repetitive firing in individual neurons (carbamazepine, gabapentin, lamotrigine, oxcarbazepine, phenobarbitone, phenytoin, topiramate, valproate). The third type of action is to block T-type calcium channels and this effect is known to be possessed by ethosuximide only. Some AEDs may also reduce events mediated by excitatory amino acids like glutamate (phenobarbitone and topiramate) [63].

The effects of herbal medicines on the antiepileptic activities of AEDs are summarized in Table 20.2. The affected AEDs included are carbamazepine, clonazepam, diazepam, gabapentin, phenobarbitone, phenytoin, and valproate. Most of the herbal medicines identified from the literature potentiated the antiepileptic effects of AEDs in animal models using stimuli such as maximal electroshock, pentylenetetrazole, kainic acid, and so on, via several mechanisms proposed by the investigators: (i) *Nardostachys jatamansi* DC. and Moutan Cortex increased GABA level possibly through increasing GABA synthesis [50, 77]; (ii) *Acorus calamus* Linn., *Annona squamosa* Linn., *Vitex negundo* Linn., essential oils, and *Zizyphus jujube* increased GABAergic activity, possibly through direct binding to GABA-benzodiazepine receptor [46, 67, 74, 81]; (iii) *Withania somnifera* (Linn.) Dunal potentiated the acute and chronic antiepileptic effects of subtherapeutic doses of clonazepam and diazepam possibly via modulation of $GABA_A$ receptor, as it was also shown to enhance the binding of a radiolabeled benzodiazepine agonist, flunitrazepam, while inhibiting the binding of radiolabeled GABA to its putative receptor sites *in vitro* [65, 73]; (iv) thymoquinone, an active component of *Nigella sativa* Linn., increased the potency of valproate against maximal electroshock- and pentylenetetrazole-induced convulsions in mice probably via an increase in opioid receptor-mediated GABAergic tone [84]; (v) *Zizyphus jujube* increased the antiepileptic effect of phenytoin and phenobarbitone via prolongation of Na^+ channel inactivation [46]; (vi) Moutan Cortex stimulated glutamate decarboxylase activity and thus decreased brain glutamate level [77]; (vii) melatonin potentiated the subtherapeutic antiepileptic effects of carbamazepine, phenobarbitone, and phenytoin but not valproate. Melatoin exerted its CNS activity via multiple possible mechanisms, including direct inhibitory action

TABLE 20.2 Effect of Phytomedicines on the Anti-epileptic Effects of Antiepileptic Drugs[a]

Antiepileptic Drugs (AEDs)	Alteration in Antiepileptic Effect of AEDs		
	Potentiate Increase	Decrease	No Change
Carbamazepine	Melatonin [16, 18] **Chai-Hu-Gui-Zhi-Tang**[b] [64] *Zizyphus jujube* [46]		
Clonazepam	*Withania somnifera* [65]	*Hypericum perforatum* Linn. [66]	
Diazepam	*Annona squamosa* Linn. [67] *Panax gingseng* C.A. Mey [68] Siotone granule[c] [70] **Chai-Hu-Gui-Zhi-Tang**[d] [64, 71] *Valeriana officinalis* Linn. [72] *Withania somnifera* [73] *Vitex negundo* Linn.[74]		Xiao-Chai-Hu-Tang[f] [39] Oroxylin A (a component commonly found in Scutellariae Radix) [69]
Ethosuximide	**Chai-Hu-Gui-Zhi-Tang**[d] [64]		
Gabapentin	*Centella asiatica* Linn. [75]		
Phenobarbitone	Melatonin[e] [16] Moutan Cortex [77] *Zizyphus jujube* [46]	*Bacopa monnieri* Linn. [76] *Centella asiatica* Linn. [76]	
Phenytoin	△9-tetrahydro-cannabinol(a cannabis extract) [78] *Bacopa monnieri* Linn. [76] Cannabidiol (a cannabis extract) [78, 80] *Zizyphus jujube* [46] Melatonin[e] [18] *Nardostachys jatamansi* DC. [50]		***Ginkgo biloba* Linn**. [79] Hypericum perforatum Linn. [66] Shankhapushpi[g] [51]

Valproate	*Acorus Calamus* Linn. [81]	*Ginkgo biloba* Linn. [79, 82]	Osthole (a coumarin-like derivative isolated from *Peucedanum ostruthium* (Linn.) Koch.) [55]
	Centella asiatica Linn. [75]		Melatonin[e] [16]
	Ginkgo biloba Linn.[83]		
	Chai-Hu-Gui-Zhi-Tang[d] [64]		
	Thymoquinone (an active component of *Nigella sativa* Linn.) [84]		

[a] The herb names highlighted in bold are evidenced by clinical studies.
[b] **Chai-Hu-Gui-Zhi-Tang**: Known as Saiko-Keishi-To in Japanese, contains nine herbs: Bulpeuri Radix, Cinnamomi Cortex, Paeoniae Radix, Zingiberis Rhizoma, Glycyrrhizae Radix, Gingseng Radix et Rhizoma, Scutellariae Radix, Pinelliae Rhizoma, Zizyphi Fructus.
[c] **Siotone granule**: A herbal psychotropic preparation.
[d] **Gegen-Qinlian-Tang**: A Chinese herbal medicine containing four herbs: Puerariae Radix, Scutellariae Radix, Coptidis Rhizoma, and Glycyrrhizae Radix.
[e] **Xiao-Yao-San**: Also known as free and easy wanderer, which contains eight herbs: Bupleuri Radix, Angelicae Sinensis Radix, Paeoniae Radix Alba, Atractylodis Macrocephalae Rhizoma, Poria, Zingiberis Rhizoma, Glycyrrhizae Radix, Menthae, and Haplocalycis Herba.
[f] **Mentat**: Also known as BR-16A, containing 13 herbs: *Bacopa monnieri* Linn., *Centella asiatica* Linn., *Withania somnifera* (Linn.) Dunal, *Evolvulus alsinoides* Linn., *Nardostachys jatamansi* Linn., *Acorus calamus* Linn., *Celastrus paniculatus* Linn., *Zingiber officinale* Linn., *Valeriana wallachii*, Prunus amygdalus, *Orchis mascula* Linn., *Syzgium aromaticum* Linn., and Mukta pishti.
[g] **Chai-Hu-Jia-Long-Gu-Mu-Li-Tang**: Known as Saiko-ka-ryukostsu-borei-to in Japanese, containing 10 herbs: Bupleuri Radix, Pinelliae Tuber, Cinnamomi Cortex, Hoelen, Scutellariae Radix, Zizyphi Fructus, Ginseng Radix, Ostreae testa, Fossilia Ossis Mastodi, and Zingiberis Rhizoma.

on neural activity, conversion of melatonin to an anticonvulsant compound that resembles the kynurenines, enhancement of GABA-ergic transmission in the CNS, and increasing GABA level in cerebral cortex [16, 18]. Besides pharmacodynamic interactions, CHMs may also enhance the antiepileptic activities of AEDs through pharmacokinetic interaction. For example, *Nardostachys jatamansi* DC., significantly increased not only the serum levels of phenytoin but also the seizure-protective effect of phenytoin [50]. Cannabidiol and $\triangle^9$-tetrahydro-cannabinol, two active components of cannabis, potentiated the antiepileptic effects of phenytoin possibly through metabolic interaction in the liver [78].

Fewer studies have reported that herbal medicines reduce the antiepileptic effect of AEDs. *Hypericum perforatum* Linn., commonly known as St John's wort, decreased the anticonvulsant activity of clonazepam and phenytoin in rat seizure models. It was suggested that the interaction is more likely of a pharmacodynamic rather than pharmacokinetic nature and that the possible mechanism is by altering central synaptic neurotransmitter levels [66]. On the other hand, the Ayurvedic syrup, Shankhapushpi, reduced the anticonvulsant activity of phenytoin via both pharmacodynamic interaction as well as reduction of plasma phenytoin [51]. *Centella asiatica* Linn. potentiated the antiepileptic activity of gabapentin, phenytoin, and valproate in pentylenetetrazole-induced seizure in ICR mice but decreased the antiepileptic effect of phenobarbitone in maximal electroshock-induced seizure model in Wistar rats [75, 76]. Similarly, *Gingko biloba* Linn., increased the antiepileptic effects of valproate in kainic-acid-induced seizure in Swiss-Webster mice but reduced the antiepileptic effects of carbamazepine and valproate in picrotoxin- and strychnine-induced seizure model in mice [82, 83]. While the mechanisms remain uncertain, the aforementioned examples highlight that the same herbal medicines may exert different effects on different AEDs. In addition, the interaction outcome may vary with the animal species (mice vs. rat), seizure model, as well as the doses of herbal medicines (relative to that of AED) that are used.

Compared to animal studies, the available evidence in humans is limited since only two human studies/reports were identified, which were both of poor quality. The first one was a case report: a 55-year-old epileptic patient taking phenytoin and valproate self-medicated with a cornucopia of herbal supplements and nutraceuticals, prominent among which was *Gingko biloba* Linn., suffered a fatal breakthrough seizure. Pharmacokinetic interaction was suggested since the autopsy report revealed subtherapeutic phenytoin and valproate serum levels [79]. However, it should be noted that the serum concentrations of AEDs in this patient prior to the intake of *Gingko biloba* Linn. were not reported, making it difficult to conclude whether the subtherapeutic effect was due to poor compliance or herb–drug interaction. Another study is an open-label study in 32 patients with generalized major motor seizures refractory to their existing standard AEDs treatment. When *Cynanchum otophyllum* Schneid (Qingyangsen, or Cynanchi Otophylli Radix) was used as an adjunctive treatment to AEDs for a course of 2–9 months, nine patients achieved complete seizure control without side effects. The investigator concluded that Cynanchum otophyllum Schneid would improve the antiepileptic effects of AEDs [85]. However, it should be noted that the study is more of a clinical observation rather than a well-designed controlled clinical trial.

20.4.2 Sedative Effects

One of the oldest AEDs, phenobarbitone, was used as a hypnotic and sedative before its antiepileptic effect was discovered serendipitously and it is still used for insomnia although it has been largely replaced by much safer benzodiazepines. The sedative effects of benzodiazepines, similar to their antiepileptic effects, are due to positive allosteric modulation of $GABA_A$ receptor, predominantly those containing the α1 subunit [86]. Phenobarbitone also acts via effects on $GABA_A$ receptor, but it has additional actions on Ca^+ and Na^+ channels [87]. A newer AED, gabapentin, also possesses sedative activity.

The sedative effects of diazepam and phenobarbitone were potentiated by majority of the herbal medicines identified from the literature; was reflected by the potentiation of diazepam/phenobarbitone-induced sleeping time in animals (Table 20.3). Herbal medicines may increase AED-induced sleeping time by different mechanisms: (i) enhancement of GABAergic transmission (*Acanthus montanus* Nees [90], *Bauhinia tomentosa* Linn. [99], *Cassia fistula* Linn. [112], *Cecropia pachystachya* Mart. (ambay) [107], *Coccos nucifera* Linn. [91], *Cuscuta reflexa* Roxb. [110], *Cynodon dactylon* Pers [96], *Cyperus rotundus* Linn. [94], *Hyptis spicigera* Lam [90], Mentat [106], *Microglossa pyrifolia* Kuntze [90], *Nauclea latifolia* Smith [104], *Phyllanthus discoideus* Baill [103], *Piliostigma reticulatum* Hochst [90], *Russelia equisetiformis* [132], *Senna spectabilis* DC. [97], *Vernonia condensata* baker [98], and *Voacanga africana* Stapf [90]); (ii) general CNS depressant or tranquilizing action (*Acorus calamus* Linn. [95], *Annona senegalensis* Pers. [122], *Barleria lupulina* Linn. [125], *Cassia fistula* Linn. [112], *Cissus cornifolia* (Bak.) Planch [105], *Clerodendrum phlomidis* Linn. [123], *Clerodendron colebrookianum* Walp. [134], *Coccos nucifera* Linn. [91], *Cynodon dactylon* Pers. [96], *Cyperus rotundus* Linn. [94], *Dodonaea viscosa* (Linn.) Jacq [135], *Drynaria quercifolia* J. Smith [92], *Hybanthus Enneaspermus* Muell [121], *Hydrilla verticillata* Linn. [108], *Hygrophila difformis* Linn. [101], *Jussiaea suffruticosa* Linn. [127], *Mallotus peltatus* (Geist) Muell Arg. var *acuminatus* [128], *Mikania scandens* (Linn.) Wild. [117], *Nauclea latifolia* Smith [119], *Ocimum sanctum* Linn. [130], *Parthenium hysterophorus* Linn. [113], *Psidium guajava* Linn. [133], *Rumex nepalensis* Spreng. [131], *Tecoma stans* (Linn.) Juss. ex Kunth. [115], and *Vitex negundo* Linn. [88, 89]); (iii) inducing actions similar to that induced by benzodiazepines (*Balanites aegyptiaca* Linn. [100], *Coptis chinensis* Franch [93], and *Gargenia jasminoides* Ellis [93]); (iv) decreasing dopaminergic transmission and, hence, increasing the sensitivity of the CNS to the depressant actions of diazepam or phenobarbitone (*Cuscuta reflexa* Roxb. [110] and *Morus alba* Linn. [129]); (v) increasing brain 5-hydroxytryptophan (*Moringa oleifera* Lamk. [111]); and (vi) inhibiting AED metabolism (*Nauclea latifolia* Smith [119]).

Gingko biloba Linn. is the only herbal medicine reported to shorten phenobarbitone-induced sleeping time. It appeared to interact pharmacokinetically with phenobarbitone via induction of the hepatic cytochrome 2B enzymes and thereby decreased the maximal plasma concentration and area under the plasma concentration–time curve of phenobarbitone in rats [40].

TABLE 20.3 Effect of Phytomedicines on the Sedative Effects of Antiepileptic Drugs

Antiepileptic Drugs (AEDs)	Alteration in Sedative Effect of AEDs		
	Potentiate Increase	Decrease	No Change
Diazepam	*Vitex negundo* Linn. [88, 89]		*Alchornea laxiflora* Pax and Hoffman [90]
	Coccos nucifera Linn. [91]		Oroxylin A (an active component of *Scutellaria baicalensis*) [69]
	Drynaria quercifolia J. Smith [92]		Huang-Lian-Jie-Du-Tang[c] [93]
	Cyperus rotundus Linn. [94]		
	Acorus calamus Linn. [95]		
	Cynodon dactylon Linn. [96]		
	Senna spectabilis DC. [97]		
	Vernonia condensata baker [98]		
	Bauhinia tomentosa Linn. [99]		
	Balanites aegyptiaca Linn. [100]		
	Hygrophila difformis Linn. [101]		
	Arq Gulab[a] [102]		
	Phyllanthus discoideus Baill [103]		
Diazepam	*Nauclea latifolia* Smith [104]		
	Acanthus montanus Nees [90]		
	Hyptis spicigera Lam [90]		
	Microglossa pyrifolia Kuntze [90]		
	Piliostigma reticulatum Hochst [90]		
	Voacanga africana Stapf [90]		
	Cissus cornifolia (Bak.) Planch [105]		
	Mentat[b] [106]		
	Cecropia pachystachya Mart. (ambay) [107]		
	Hydrilla verticillata Linn. [108]		
	Abies webbiana Lindl. [109]		
	Cuscuta reflexa Roxb. [110]		
	Moringa oleifera Lamk. [111]		
	Cassia fistula Linn.[112]		

Phenobarbitone	*Parthenium hysterophorus* Linn. [113]	*Ginkgo biloba* Linn. [40]	*Spathodea campanulata* P. Beauv [114]
	Tecoma stans (L.) Juss. ex Kunth. [115]		*Clausena lansium* (Lour.) Skeels [116]
	Mikania scandens (Linn.) Wild. [117]		*Ichnocarpus frutescens* Linn. [118]
	Nauclea latifolia Smith [119]		
	Alstonia macrophylla Wall ex A. DC. [120]		
	Hybanthus Enneaspermus Muell [121]		
Phenobarbitone	*Annona senegalensis* Pers. [122]		
	Clerodendrum phlomidis Linn. [123]		
	Acori Calami Rhizoma [124]		
	Barleria lupulina Linn. [125]		
	Cleome viscosa Linn. [126]		
	Jussiaea suffruticosa Linn. [127]		
	Mallotus peltatus (Geist) Muell Arg var *acuminatus* [128]		
	Morus alba Linn. [129]		
	Ocimum sanctum Linn. [130]		
	Rumex nepalensis Spreng. [131]		
	Russelia equisetiformis [132]		
	Psidium guajava Linn. [133]		
	Clerodendron colebrookianum Walp. [134]		
	Dodonaea viscosa (Linn.) Jacq [135]		

[a] **Arq Gulab**: A clear, nonviscous liquid preparation obtained by the aqueous distillation of duly macerated flowers of the plant Rosa damascene Mill.
[b] The herb names highlighted in bold are evidenced by clinical studies.
[c] **Huang-Lian-Jie-Du-Tang**: Known as Oren-geduko-to in Japanese, contained four herbs: Scutellariae, Coptidis Rhizoma, Gardeniae Fructus, Phellodendri Chinensis Cortex.

20.4.3 Anxiolytic Effects

Benzodiazepines are a class of AEDs that are frequently used for anxiety disorders. The anxiolytic effects of benzodiazepines, similar to their antiepileptic effects, are due to allosteric positive modulation of $GABA_A$ receptors but the subunit involved is α2 subtype [86]. Apart from benzodiazepines, some other AEDs also possess anxiolytic effects, including valproate, gabapentin, pregabalin, and vigabatrin (by increasing brain GABA levels or neurotransmission) and tiagabine (by selectively increasing synaptic GABA availability via blockage of the reuptake of GABA through transporter inhibition) and are prescribed for anxiety disorders [136].

The effects of herbal medicines on the anxiolytic effect of diazepam are summarized in Table 20.4. Most of the herbal medicines identified from the literature potentiated the anxiolytic effects of AEDs in animal models including elevated plus maze, light–dark model and social interaction test, via different mechanisms proposed by the investigators: (i) central depressant activity (*Cymbopogon citratus* (DC) Stapf [137]), (ii) modulation of GABAergic transmission (*Aloysia citriodora* Palau [145], *Gingko biloba* Linn. [139, 140], *Nelumbo nucifera* Gaertn. [142]), (iii) inhibition of histamine release from mast cells (Xiao-Chai-Hu-Tang [39]); (iv) involvement of benzodiazepine receptors but exact mechanism(s) are not known (Ban-Xia-Hou-Pu-Tang and Jia-Wei-Gui-Pi-Tang [143]).

Trimyristin, an active component of Myristicae Semen, antagonized the anxiolytic effects of diazepam in elevated plus maze model via induction of a nonspecific anxiogenic activity [141]. Two active components in *Scutellaria baicalensis* Georgi, baicalin and oroxylin A, increased and decreased the anxiolyitc effects of diazepam, respectively [69, 144]. The investigators proposed that oroxylin A might selectively produce antagonistic activity at the benzodiazepine site, while the mechanism of baicalin remains uncertain. The overall effect of *Scutellaria baicalensis* Georgi on the anxiolytic effect of diazepam needs further investigation.

20.4.4 Memory Impairment Effects

Impaired memory is among the most common complaints by patients with epilepsy, where consistent data show that AEDs were contributing to the decline in memory function. AEDs induce various effects on memory but the exact mechanism is yet to be identified [149, 150]. Benzodiazepines reportedly cause dose-related anterograde amnesia, that is, forgetfulness for events that occur following drug intake, and which persists for several hours [151]. The anterograde amnesic effects of benzodiazepines, similar to their sedative effects, are due to allosteric positive modulation of $GABA_A$ receptor containing the α1 subunit [152]. The reduction of cerebral glucose metabolism by phenobarbitone was one of the proposed contributing factors to the adverse cognitive effects [153].

The effect of herbal medicines on AEDs-induced memory impairment was studied using various animal models such as Morris water maze, elevated plus-maze, Hebb–Williams maze, hexagonal swimming pool, and passive avoidance test. All of the studied herbal medicines reduced diazepam or phenytoin-induced memory impairment, with diverse proposed mechanisms (Table 20.5): (i) enhancement of brain cholin-

TABLE 20.4 Effect of Phytomedicines on the Anxiolytic Effects of Antiepileptic Drugs[a]

Antiepileptic Drugs (AEDs)	Alteration in Anxiolytic Effect of AEDs		
	Increase	Decrease	No Change
Diazepam	*Cymbopogon citratus* (DC.) Stapf [137]	Oroxylin A [69]	Ding-Zhi-Wan[h] [138]
	Ginkgo biloba Linn. [139, 140]	Myristicae Semen [141]	
	Nelumbo nucifera Gaertn [142]		
	Xiao-Chai-Hu-Tang[b] [39]		
	Jia-Wei-Gui-Pi-Tang[c] [143]		
	Ban-Xia-Hou-Pu-Tang[d] [143]		
	Yi-Gan-San[e] [143]		
	Chai-Pu-Tang[f] [143]		
	Baicalin (an active component of *Scutellaria baicalensis*) [144]		
	Aloysia citriodora Palau [145]		
	Xue-Fu-Zhu-Yu-Tang[g] [146]		
	Panax ginseng C. A. Mey [147]		
	Hypericum perforatum Linn. [147]		
Phenytoin	Curcumin (an active component of *Curcuma longa* Linn.) [148]		

[a] The herb names highlighted with underlines are evidenced by clinical studies.

[b] **Mentat**: Also known as BR-16A, containing 13 herbs: *Bacopa monnieri* Linn., *Centella asiatica* Linn., *Withania somnifera* (Linn.) Dunal, *Evolvulus alsinoides* Linn., *Nardostachys jatamansi* Linn., *Acorus calamus* Linn., *Celastrus paniculatus* Linn., *Zingiber officinale* Linn., *Valeriana wallachii*, Prunus amygdalus, *Orchis mascula* Linn., *Syzgium aromaticum* Linn., and Mukta pishti.

[c] **Jia-Wei-Gui-Pi-Tang**: Known as Kami-Kihi-To in Japanese, contains 11 herbs: Ginseng Radix et Rhizoma, Astragali Radix, Poria, Glycyrrhizae Radix et Rhizoma, Atractylodis Macrocephalae Rhizoma, Aucklandiae Radix, Polygalae Radix, Ziziphi Spinosae Semen, Longan Arillus, Angelicae Sinensis Radix, Moutan Cortex.

[d] **Ban-Xia-Hou-Pu-Tang**: Known as Hange-Koboku-To in Japanese, contains five herbs: Pinelliae Rhizoma, Magnoliae Officinalis Cortex, Poria, Zingiberis Rhizoma Recens, Perillae Folium.

[e] **Yi-Gan-San**: Known as Yoku-Kan-San in Japanese, contains seven herbs: Angelicae Sinensis Radi, Atractylodis Macrocephalae Rhizoma, Poria, Uncariae Ramulus Cum Uncis, Chuanxiong Rhizoma, Bupleuri Radix, Glycyrrhizae Radix et Rhizoma.

[f] **Chai-Pu-Tang**: Known as Saiboku-To in Japanese, contains 10 herbs: Jujubae Fructus, Pinelliae Rhizoma, Glycyrrhizae Radix et Rhizoma Praeparata cum Melle, Zingiberis Rhizoma Recens, Bupleuri Radix, Poria, Magnoliae Officinalis Cortex, Ginseng Radix et Rhizoma, Perillae, Scutellariae Radix.

[g] **Xue-Fu-Zhu-Yu-Tang**: A Chinese herbal medicine containing 11 herbs: Bupleuri Radix, Angelicae Sinensis Radix , Rehmanniae Radix, Paeoniae Radix Rubra , Chuanxiong Rhizoma, Aurantii Fructus Immaturus, Persicae Ramulus, Carthami Flos, Achyranthis Bidentatae Radix, Platycodonis Radix, Glycyrrhizae Radix et Rhizoma.

[h] **Ding-Zhi-Wan**: Also known as Ting-Chih-Wan, contains five herbs: Ginseng Radix et Rhizoma, Polygalae Radix, Jujubae Fructus, Platycladi Semen, Poria.

TABLE 20.5 Effect of Phytomedicines on the Memory Impairment Effects of Antiepileptic Drugs

Antiepileptic Drugs (AEDs)	Alteration in Memory Impairment Effect of AEDs	
	Decrease	No Change
Diazepam	Salvia miltiorrhizae Radix et Rhizoma [154, 155]	*Bacopa monnieri* Linn. [156]
	Coriandrum sativum Linn. [157, 158]	*Semecarpus anacardium* Linn. [159]
	Bacopa monnieri Linn. [160, 161]	Uncaria tomentosa total alkaloid [162]
	Oroxylin A [163, 164]	Osthole [55]
	Granati pericarpium [165]	
	Daucus carota Linn. [166, 167]	
	Thespesia populnea Linn. [168]	
	Ocimum sanctum Linn. [169]	
	Nardostachys jatamansi [170]	
	Zingiber officinale Roscoe [171]	
	Glycyrrhiza glabra Linn. [172]	
	Myristica fragrans Houtt [173]	
	Ocimum tenuiflorum [174]	
	Clerodendron phlomidis Linn. [175, 176]	
	L-33[a] [177]	
	Calotropis procera (Ait.) R. Br. [178]	
	Rose alba [179]	
	Phyllanthus amarus [180]	
	Emblica officinalis Gaertn. [181, 182]	
	Argyreia speciosa Linn. [183]	
	Murraya koenigii (Linn.) Spreng. [184]	
	Thespesia populnea Linn. [168]	
	Zingiber officinale Roscoe [171]	
Phenytoin	Osthole [55]	

[a] **L-33**: A polyherbal formulation containing: *Bacopa monniera* Linn., *Glycyrrhiza glabra* Linn., *Valeriana wallechii* DC., *Withania somnifera* Linn.

ergic transmission from increased acetylcholine synthesis or inhibition of acetylcholinesterase (*Argyreia speciosa* Linn. [183], *Calotropis procera* (Ait.) R. Br. [178], *Clerodendron phlomidis* Linn. [175, 176], *Coriandrum sativum* Linn. [157], *Daucus carota* Linn. [166, 167], *Emblica officinalis* Gaertn. [181, 182], *Murraya koenigii* (Linn.) Spreng [184], *Myristica fragrans* Houtt [173], *Ocimum sanctum* Linn. [169], *Phyllanthus amarus* [180], *Thespesia populnea* Linn. [168], and *Zingiber officinale* Roscoe [171]); (ii) antioxidative and neuroprotective actions (*Bacopa monniera* Linn. [156, 160, 161], *Coriandrum sativum* Linn. [158], curcumin in *Curcuma longa* Linn. [148], *Glycyrrhiza glabra* Linn. [172], *Nardostachys jatamansi* [170], *Ocimum sanctum* Linn. [169], *Semecarpus anacardium* Linn. [159], *Thespesia populnea* Linn. [168], and *Zingiber officinale* Roscoe [171]); (iii) anti-inflammatory actions (*Coriandrum sativum* Linn. [158], *Glycyrrhiza glabra* Linn. [172], *Semecarpus anacardium* Linn. [159], *Thespesia populnea* Linn. [168], and *Zingiber officinale* Roscoe [171]); (iv) cholesterol-lowering activity, which may prevent the accumulation of amyloid plaques and intra-neuronal neurofibrillary tangles (*Coriandrum sativum* Linn. [158], *Daucus carota* Linn. [166, 167], *Murraya koenigii* (Linn.) Spreng. [184], and *Thespesia populnea* Linn. [168]); and (v)decreasing brain GABAergic transmission (*Emblica officinalis* Gaertn. [181, 182], L-33 (a polyherbal formulation consisting of *Bacopa monniera* Linn., *Glycyrrhiza glabra* Linn., *Valeriana wallechii* DC., and *Withania somnifera* (Linn.) Dunal), *Myristica fragrans* Houtt [173], oroxylin A from Scutellariae Radix [163, 164], and tanshinone I from Salvia miltiorrhizae Radix et Rhizoma [154]).

20.4.5 Motor Incoordination Effects

One of the common cognitive adverse effects induced by AEDs is motor incoordination [185]. AEDs such as diazepam, valproate, and phenyotin influence the coordination of movement as is often evaluated by the ability of animals (usually mice) to stay on a rotating rod [186, 187].

The effects of herbal medicines on AEDs-induced motor incoordination are summarized in Table 20.6. Magnum hop and Aroma hop reduced diazepam-induced motor incoordination by acting on GABA receptors and significantly reducing the

TABLE 20.6 Effect of Phytomedicines on the Motor Incoordination Effects of Antiepileptic Drugs

Antiepileptic Drugs (AEDs)	Alteration in Motor Incoordination Effect of AEDs		
	Increase	Decrease	No Change
Diazepam	Angelicae dahuricae Radix [36]	Oroxylin A [69]	*Dodonaea viscosa* Linn. Jacq [135]
	Zingiberis Rhizoma [188]	Magnum hop [189]	Baicalin [144]
		Aroma hop [189]	
Phenytoin			Osthole [55]
Valproate			Osthole [55]

activity of diazepam [189]. Oroxylin A, the active component of Scutellariae Radix, also reduced the motor incoordination caused by diazepam possibly via its antagonistic action mediated by the $\alpha_{2,3,5}$-containing benzodiazepine binding site [69].

In contrast, Angelicae dahuricae Radix [36] and Zingiberis Rhizoma [188] potentiated the motor impairment caused by diazepam as they both significantly decreased the time that mice spent on the rotating rod following intravenous diazepam. However, the exact mechanism was not proposed by the investigators.

20.5 CONCLUSIONS

This chapter has demonstrated that phytomedicines may alter the pharmacokinetics and pharmacodynamics of AEDs, which may lead to changes in therapeutic and/or adverse outcomes. From the systematic literature search, majority (>90%) of the evidence of interactions are from animal studies, while there are only seven human reports. Since most of the existing evidence is based on preclinical studies, physicians should be cautious when extrapolating these results to their clinical practice. In order to address the question "are phytomedicines safe to be used in combination with AEDs?," more extensive clinical data are needed to provide concrete answers. Until then, it is prudent to note that caution should be exercised for the combined use of phytomedicines and AEDs.

REFERENCES

[1] Leonardi M, Ustun TB (2002) The global burden of epilepsy. *Epilepsia* 43: 21–25.

[2] Rogawski MA, Loscher W (2004) The neurobiology of antiepileptic drugs for the treatment of nonepileptic conditions. *Nature Medicine* 10: 685–692.

[3] Mohamed ME, Frye RF (2010) Effects of herbal supplements on drug glucuronidation. Review of clinical, animal and *in vitro* studies. *Planta Medica* 77: 311–321.

[4] Chen LC, Chen YF, Yang LL, Chou MH, Lin MF (2000) Drug utilization pattern of antiepileptic drugs and traditional Chinese medicines in a general hospital in Taiwan—a pharmacoepidemiologic study. *Journal of Clinical Pharmacy and Therapeutics* 25: 125–129.

[5] Easterford K, Clough P, Comish S, Lawton L, Duncan S (2005) The use of complementary medicines and alternative practitioners in a cohort of patients with epilepsy. *Epilepsy & Behavior* 6: 59–62.

[6] Liow K, Ablah E, Nguyen JC, Sadler T, Wolfe D, et al. (2007) Pattern and frequency of use of complementary and alternative medicine among patients with epilepsy in the midwestern United States. *Epilepsy & Behavior* 10: 576–582.

[7] Bhalla D, Chea K, Hun C, Vannareth M, Huc P, et al. (2012) Population-based study of epilepsy in Cambodia associated factors, measures of impact, stigma, quality of life, knowledge-attitude-practice and treatment gap. *PLoS ONE* 7: e46296.

[8] Tandon M, Prabhakar S, Pandhi P (2002) Pattern of use of complementary/alternative medicine (CAM) in epileptic patients in a tertiary care hospital in India. *Pharmacoepidemiology and Drug Safety* 11: 457–463.

[9] Danesi MA, Adetunji JB (1994) Use of alternative medicine by patients with epilepsy: a survey of 265 epileptic patients in a developing country. *Epilepsia* 35: 344–351.

[10] White SD (1999) Deciphering "integrated Chinese and Western medicine" in the rural Lijiang basin: state policy and local practice(s) in socialist China. *Social Science & Medicine* 49: 1333–1347.

[11] Fong SYK, Wong YC, Zuo Z (2014) Alterations in the CNS effects of anti-epileptic drugs by Chinese herbal medicines. *Expert Opinion on Drug Metabolism & Toxicology* 10: 249–267.

[12] Mani R, Pollard J (2009) Antiepileptic drugs and other medications: what interactions may arise? *Current Treatment Options in Neurology* 11: 253–261.

[13] Fong SYK, Gao Q, Zuo Z (2013) Interaction of carbamazepine with herbs, dietary supplements and food: a systematic review. *Journal of Evidence-Based Complementary & Alternative Medicine* 15: 898261.

[14] Tripathi M, Sundaram R, Rafiq M, Venkataranganna MV, Gopumadhavan S, Mitra SK (2000) Pharmacokinetic interactions of Mentat with carbamazepine and phenytoin. *European Journal of Drug Metabolism and Pharmacokinetics* 25: 223–226.

[15] Garg SK, Islam Afm S, Kumar N (1998) Effect of septilin—a herbal preparation on pharmacokinetics of carbamazepine in rabbits. *Indian Journal of Physiology and Pharmacology* 42: 527–532.

[16] Borowicz KK, Kaminski R, Gasior M, Kleinrok Z, Czuczwar SJ (1999) Influence of melatonin upon the protective action of conventional anti-epileptic drugs against maximal electroshock in mice. *European Neuropsychopharmacology* 9: 185–190.

[17] Gupta M, Gupta YK, Agarwal S, Aneja S, Kalaivani M, Kohli K (2004) Effects of add-on melatonin administration on antioxidant enzymes in children with epilepsy taking carbamazepine monotherapy: a randomized, double-blind, placebo-controlled trial. *Epilepsia* 45: 1636–1639.

[18] Gupta YK, Gupta M, Chaudhary G, Kohli K (2004) Modulation of antiepileptic effect of phenytoin and carbamazepine by melatonin in mice. *Methods and Findings in Experimental and Clinical Pharmacology* 26: 99–102.

[19] Ohnishi N, Okada K, Yoshioka M, Kuroda K, Nagasawa K, Takara K, et al. (2002) Studies on interactions between traditional herbal and western medicines. V. effects of Sho-saiko-to (Xiao-Cai-hu-Tang) on the pharmacokinetics of carbamazepine in rats. *Biological and Pharmaceutical Bulletin* 25: 1461–1466.

[20] Etman MA (1995). Effect of a bulk forming laxative on the bioavailability of carbamazepine in man. *Drug Development and Industrial Pharmacy* 21: 1901–1906.

[21] Ohnishi N, Yonekawa Y, Nakasako S, Nagasawa K, Yokoyama T, Yoshioka M, et al. (1999) Studies on interactions between traditional herbal and western medicines. I. Effects of Sho-seiryu-to on the pharmacokinetics of carbamazepine in rats. *Biological and Pharmaceutical Bulletin* 22: 527–531.

[22] Pattanaik S, Hota D, Prabhakar S, Kharbanda P, Pandhi P (2009) Pharmacokinetic interaction of single dose of piperine with steady-state carbamazepine in epilepsy patients. *Phytotherapy Research* 23: 1281–1286.

[23] Hao M, Zhao Y, Chen P, Huang H, Liu H, Jiang H, et al. (2008) Structure-activity relationship and substrate-depedent phenomena in effects of ginsenosides on activities of drug-metabolizing P450 enzymes. *PLoS ONE* 3: e2697.

[24] Thabrew I, Munasinghe J, Chackrewarthi S, Senarath S (2004) The effects of Cassia auriculata and Cardiospermum halicacabum teas on the steady state blood level and toxicity of carbamazepine. *Journal of Ethnopharmacology* 90: 145–150.

[25] Thabrew MI, Munasinghe TMJ, Chackrewarthi S, Senarath S (2003) Possible interaction of herbal tea and carbamazepine. *Drug Metabolism and Drug Interactions* 19: 177–187.

[26] Harish Chandra R, Rajkumar M, Veeresham C (2009) Pharmacokinetic interaction of ginkgo biloba with carbamazepine. Planta Medica; Conference: 8th Annual Oxford International Conference on the Science of Botanicals University, MS United States: 454.

[27] Qiu W, Jiang XH, Liu CX, Ju Y, Jin JX (2009) Effect of berberine on the pharmacokinetics of substrates of CYP3A and P-gp. *Phytotherapy Research: PTR* 23: 1553–1558.

[28] Chi YC, Chao PDL, Tsai SY, Hou YC (2011) Coadministration of rhizoma polygoni cuspidati affected carbamazepine pharmacokinetics through inhibiting CYP3A4 and MRP2. Drug Metabolism Reviews; Conference: 4th Asia Pacific Regional Meeting of the International Society for the Study of Xenobiotics, ISSX T'ainan Taiwan (Republic of China). Conference Start: 20110422 Conference End: 20110425. Conference Publication: (var.pagings). 43: 53–54.

[29] Zhang ZJ, Kang WH, Li Q, Tan QR (2007) The beneficial effects of the herbal medicine Free and Easy Wanderer Plus (FEWP) for mood disorders: double-blind, placebo-controlled studies. *Journal of Psychiatric Research* 41: 828–836.

[30] Zhang ZJ, Kang WH, Tan QR, Li Q, Gao CG, Zhang FG, et al. (2007) Adjunctive herbal medicine with carbamazepine for bipolar disorders: a double-blind, randomized, placebo-controlled study. *Journal of Psychiatric Research* 41: 360–369.

[31] Burstein AH, Horton RL, Dunn T, Alfaro RM, Piscitelli SC, Theodore W (2000) Lack of effect of St John's wort on carbamazepine pharmacokinetics in healthy volunteers. *Clinical Pharmacology & Therapeutics* 68: 605–612.

[32] Liu P, Wei L (2008) Effect of platycodon grandiflorum on the blood concentration of carbamazepine in rabbits. *Evaluation and Analysis of Drug-Use in Hospitals of China* 8: 366–368.

[33] Li Q, Tan Q, Zhang Z, Gao C, Xing Y, et al. (2005) Randomized double-blind controlled clinical trial of the combination of Xiao-yao-san and carbamazepine in the treatment of bipolar disorders. *Chinese Journal of Clinical Pharmacology* 21: 336–340.

[34] Ohnishi N, Nakasako S, Okada K, Umehara S, Takara K, Nagasawa K, et al. (2001) Studies on interactions between traditional herbal and Western medicines. IV: lack of pharmacokinetic interactions between Saiko-ka-ryukotsu-borei-to and carbamazepine in rats *European Journal of Drug Metabolism and Pharmacokinetics* 26: 129–135.

[35] Chen LC, Chen YF, Chou MH, Lin MF, Yang LL, Yen KY (2002) Pharmacokinetic interactions between carbamazepine and the traditional Chinese medicine Paeoniae Radix. *Biological and Pharmaceutical Bulletin* 25: 532–535.

[36] Ishihara K, Kushida H, Yuzurihara M, Wakui Y, Yanagisawa T, Kamei H, et al. (2000) Interaction of drugs and Chinese herbs: pharmacokinetic changes of tolbutamide and diazepam caused by extract of Angelica dahurica. *Journal of Pharmacy and Pharmacolog* 52: 1023–1029.

[37] Jinping Q, Peiling H, Yawei L, Abliz Z (2003) Effects of the aqueous extract from Salvia miltiorrhiza Bge on the pharmacokinetics of diazepam and on liver microsomal cytochrome P450 enzyme activity in rats. *The Journal of Pharmacy and Pharmacology* 55: 1163–1167.

[38] Zuo XC, Zhang BK, Jia SJ, Liu SK, Zhou LY, Li J, et al. (2010) Effects of Ginkgo biloba extracts on diazepam metabolism: a pharmacokinetic study in healthy Chinese male subjects. *European Journal of Clinical Pharmacology* 66: 503–509.

[39] Yuzurihara M, Ikarashi Y, Ishihara K, Kushida H, Ishige A, Sasaki H, et al. (2000) Effects of subacutely administered saiboku-to, an oriental herbal medicine, on pharmacodynamics and pharmacokinetics of diazepam in rodents. *European Journal of Drug Metabolism and Pharmacokinetics* 25: 127–136.

[40] Kubota Y, Kobayashi K, Tanaka N, Nakamura K, Kunitomo M, Shinozuka K, et al. (2004) Pretreatment with Ginkgo biloba extract weakens the hypnosis action of phenobarbital and its plasma concentration in rats. *The Journal of Pharmacy and Pharmacology* 56: 401–405.

[41] Harikumar KB, Kuttan R (2006) Inhibition of drug metabolizing enzymes (cytochrome P450) *in vitro* as well as *in vivo* by Phyllanthus amarus SCHUM & THONN. *Biological & Pharmaceutical Bulletin* 29: 1310–1313.

[42] Garg SK, Islam AS, Kumar N, Sehgal M, Bhargava VK (1999) Effect of 'Mentat' on the pharmacokinetics of single and multiple doses o phenytoin in rabbits. *Neurology India* 47: 104–107.

[43] Chi YC, Juang SH, Chui WK, Hou YC, Chao PD (2012) Acute and chronic administrations of rheum palmatum reduced the bioavailability of phenytoin in rats: a new herb-drug interaction. *Evidence-Based Complementary and Alternative Medicine* 2012: 701205.

[44] Harish Chandra R, Veeresham C (2011) Herb—drug interaction of noni juice and Ginkgo biloba with phenytoin. *Pharmacognosy Journal* 2: 33–41.

[45] Chen LC, Chou MH, Lin MF, Yang LL (2001) Effects of Paeoniae Radix, a traditional Chinese medicine, on the pharmacokinetics of phenytoin. *Journal of Clinical Pharmacy and Therapeutics* 26: 271–278.

[46] Pahuja M, Kleekal T, Reeta KH, Tripathi M, Gupta YK (2012) Interaction profile of Zizyphus jujuba with phenytoin, phenobarbitone and carbamazepine in maximal electroshock-induced seizures in rats. *Epilepsy & Behavior* 25: 368–373.

[47] Bano G, Amla V, Raina RK, Zutshi U, Chopra CL (1987) The effect of piperine on pharmacokinetics of phenytoin in healthy volunteers. *Planta Medica* 53: 568–569.

[48] Pattanaik S, Hota D, Prabhakar S, Kharbanda P, Pandhi P (2006) Effect of piperine on the steady-state pharmacokinetics of phenytoin in patients with epilepsy. *Phytotherapy Research* 20: 683–686.

[49] Velpandian T, Jasuja R, Bhardwaj RK, Jaiswal J, Gupta SK (2001) Piperine in food: interference in the pharmacokinetics of phenytoin. *European Journal of Drug Metabolism and Pharmacokinetics* 26: 241–247.

[50] Rao VS, Rao A, Karanth KS (2005) Anticonvulsant and neurotoxicity profile of Nardostachys jatamansi in rats. *Journal of Ethnopharmacology* 102: 351–356.

[51] Dandekar UP, Chandra RS, Dalvi SS, Joshi MV, Gokhale PC, Sharma AV, et al. (1992) Analysis of a clinically important interaction between phenytoin and Shankhapushpi, an Ayurvedic preparation. *Journal of Ethnopharmacology* 35: 285–288.

[52] Liu H-J, Yu C-P, Hsieh Y-W, Tsai S-Y, Hou Y-C (2013) Inhibition of monocarboxylate transporter-mediated absorption of valproic acid by Gegen-Qinlian-Tang. *The American Journal of Chinese Medicine* 41: 369–378.

[53] Chen LC, Chou MH, Lin MF, Yang LL (2000) Lack of pharmacokinetic interaction between valproic acid and a traditional Chinese medicine, Paeoniae Radix, in healthy volunteers. *Journal of Clinical Pharmacy and Therapeutics* 25: 453–459.

[54] Numa AM, Abbott FS, Chang TKH (2007) Effect of Ginkgo biloba extract on oxidative metabolism of valproic acid in hepatic microsomes from donors with the CYP2C9*1/*1

genotype. This article is one of a selection of papers published in this special issue (part 1 of 2) on the Safety and Efficacy of Natural Health Products. *Canadian Journal of Physiology and Pharmacology* 85: 848–855.

[55] Luszczki JJ, Marczewski T, Mazurkiewicz LP, Karwan S, Teresinska M, Florek-Luszczki M, et al. (2011) Influence of osthole on the anticonvulsant activity of phenytoin and valproate in the maximal electroshock-induced seizures in mice. *Annales Universitatis Mariae Curie-Sklodowska, Sectio DDD: Pharmacia* 24: 33–44.

[56] Waxman DJ, Azaroff L (1992) Phenobarbital induction of cytochrome P-450 gene expression. *The Biochemical Journal* 281: 577–592.

[57] Dickson M, Bramley TJ, Kozma C, Doshi D, Rupnow MFT (2008) Potential drug–drug interactions with antiepileptic drugs in Medicaid recipients. *American Journal of Health-System Pharmacy* 65: 1720–1726.

[58] Sabers A (2008) Pharmacokinetic interactions between contraceptives and antiepileptic drugs. *Seizure* 17: 141–144.

[59] Johannessen Landmark C, Patsalos PN (2009) Drug interactions involving the new second- and third-generation antiepileptic drugs. *Expert Review of Neurotherapeutics* 10: 119–140.

[60] Landmark CJ (2007) Targets for antiepileptic drugs in the synapse. *Medical science monitor: International Medical Journal of Experimental and Clinical Research* 13: RA1–7.

[61] Spinella M (2001) Herbal medicines and epilepsy: the potential for benefit and adverse effects. *Epilepsy & Behavior* 2: 524–532.

[62] Tyagi A, Delanty N (2003) Herbal remedies, dietary supplements and seizures. *Epilepsia* 44: 228–235.

[63] Czapinski P, Blaszczyk B, Czuczwar SJ (2005) Mechanisms of action of antiepileptic drugs. *Current Topics in Medicinal Chemistry* 5: 3–14.

[64] Narita Y, Satowa H, Kokubu T, Sugaya E (1982) Treatment of epileptic patients with the Chinese herbal medicine 'Saiko-Keishi-To' (SK). *IRCS Medical Science* 10: 88–89.

[65] Kulkarni SK, George B, Mathur R (1998) Protective effect of Withania somnifera root extract on electrographic activity in a lithium-pilocarpine model of status epilepticus. *Phytotherapy Research* 12: 451–453.

[66] Radhika MS, Patil PA, Patil MI, Mayur SS (2009) Proconvusant activity of Hypericum perforatum L. extract powder and its interaction with phenytoin and clonazepam in wistar rats. *Pharmacologyonline* 3: 240–246.

[67] Porwal M, Sharma K, Malik P (2011) Anticonvulsant effect of Annona squamosa Linn. leaves in mice. *Pharmacologyonline* 2: 44–52.

[68] Mitra SK, Chakraborti A, Bhattacharya SK (1996) Neuropharmacological studies on Panax ginseng. *Indian Journal of Experimental Biology* 34: 41–47.

[69] Huen MSY, Leung JWC, Ng W, Lui WS, Chan MNS, Wong JTF, et al. (2003) 5,7-Dihydroxy-6-methoxyflavone, a benzodiazepine site ligand isolated from Scutellaria baicalensis Georgi, with selective antagonistic properties. *Biochemical Pharmacology* 66: 125–132.

[70] Kulkarni SK, Joseph P (1998) Anticonvulsant profile of Siotone granules, a herbal preparation. *Indian Journal of Experimental Biology* 36: 658–662.

[71] Sugiyama K, Kano T, Muteki T (1997) Intravenous anesthetics, acting on the -amino butric acid (GABA)(A) receptor, potentiate the herbal medicine 'Saiko-Keishi-To'-induced chloride current. [Japanese]. *Japanese Journal of Anesthesiology* 46: 1197–1203.

[72] Nouri MHK, Abad ANA (2011) Gabaergic system role in aqueous extract of Valeriana officinalis L. root on PTZ-induced clonic seizure threshold in mice. *African Journal of Pharmacy and Pharmacology* 5: 1212–1217.

[73] Kulkarni SK, Akula KK, Dhir A (2008) Effect of Withania somnifera Dunal root extract against pentylenetetrazol seizure threshold in mice: possible involvement of GABAergic system. *Indian Journal of Experimental Biology* 46: 465–469.

[74] Khokra SL, Jain S, Prakash O (2011) Anticonvulsant activity of essential oils isolated from Vitex negundo Linn. *Pharmaceutical Chemistry Journal* 44: 646–650.

[75] Vattanajun A, Watanabe H, Tantisira MH, Tantisira B (2005) Isobolographically additive anticonvulsant activity between Centella asiatica's ethyl acetate fraction and some antiepileptic drugs. *Journal of the Medical Association of Thailand* 88 Suppl 3: S131–S140.

[76] Sudha S, Bindu R, Joyce G, Amit A, Venkataraman BV (2005) Pharmacological interaction of Centella asiatica and Bacopa monnieri with antiepileptic drugs—an experimental study in ratsPharmacological interaction of Centella asiatica and Bacopa monnieri with antiepileptic drugs—an experimental study in rats. *Journal of Natural Remedies* 5: 63–69.

[77] Wang Y, Ming L, Cen DY, Zhang JS, Guang WH, Xu SY (1997) Antiepileptic action of total glucosides of moutan cortex. *Chinese Pharmacological Bulletin* 13: 268–270.

[78] Chesher GB, Jackson DM (1974) Anticonvulsant effects of cannabinoids in mice: drug interactions within cannabinoids and cannabinoid interactions with phenytoin. *Psychopharmacologia* 37: 255–264.

[79] Kupiec T, Raj V (2005) Fatal seizures due to potential herb-drug interactions with Ginkgo biloba. *Journal of Analytical Toxicology* 29: 755–758.

[80] Consroe P, Wolkin A (1977) Cannabidiol—antiepileptic drug comparisons and interactions in experimentally induced seizures in rats. *Journal of Pharmacology and Experimental Therapeutics* 201: 26–32.

[81] Katyal J, Sarangal V, Gupta YK (2012) Interaction of hydroalcoholic extract of Acorus calamus Linn. with sodium valproate and carbamazepine. *Indian Journal of Experimental Biology* 50: 51–55.

[82] Manocha A, Pillai K, Husain Z (1996) Influence of Ginkgo biloba on the effect of anticonvulsants. *Indian Journal of Pharmacology* 28: 84–87

[83] Abdel-Wahab BA, Metwally ME (2011) Ginkgo biloba enhances the anticonvulsant and neuroprotective effects of sodium valproate against kainic acid-induced seizures in mice. *Journal of Pharmacology and Toxicology* 6: 679–690.

[84] Raza M, Alghasham AA, Alorainy MS, El-Hadiyah TM (2006) Beneficial interaction of thymoquinone and sodium valproate in experimental models of epilepsy: reduction in hepatotoxicity of valproate. *Scientia Pharmaceutica* 74: 159–173.

[85] Ding YJ, He XX (1986) Traditional Chinese herbs in treatment of neurological and neurosurgical disorders. *The Canadian Journal of Neurological Sciences* 13: 210–213.

[86] Rudolph U, Möhler H (2004) Analysis of GABAA receptor function and dissection of the pharmacology of benzodiazepines and general anesthetics through mouse genetics. *Annual Review of Pharmacology and Toxicology* 44: 475–498.

[87] French-Mullen J, Barker J, Rogawski M (1993) Calcium current block by (-)-pentobarbital, phenobarbital and CHEB but not (+)-pentobarbital in acutely isolated hippocampal CA1 neurons: comparison with effects on GABA-activated Cl- current. *The Journal of Neuroscience* 13: 3211–3221.

[88] Gupta M, Mazumder UK, Bhawal SR (1999) CNS activity of Vitex negundo Linn. in mice. *Indian Journal of Experimental Biology* 37: 143–146.

[89] Gupta M, Mazumder UK, Bhawal SR, Swamy SMK (1997) CNS activity of petroleum ether extract of Vitex negundo Linn in mice. *Indian Journal of Pharmaceutical Sciences* 59: 240–245.

[90] Bum EN, Taiwe GS, Nkainsa LA, Moto FCO, Seke Etet PF, Hiana IR, et al. (2009) Validation of anticonvulsant and sedative activity of six medicinal plants. *Epilepsy and Behavior* 14: 454–458.

[91] Pal D, Abhijit S, Sumanta G, Sandip J, Mandal S (2011) CNS depressant activities of roots of Coccos nucifera in mice. *Acta Poloniae Pharmaceutica* 68: 249–254.

[92] Khan A, Haque E, Rahman BM, Rahman M (2009) Neuropharmacological effect of the rhizome of Drynaria quercifolia in mice. *Iranian Journal of Pharmacology and Therapeutics* 8: 23–27.

[93] Sasaki K (1996) Effect of Oren-geduko-to on change of the hexobarbital-induced sleep time in diazepam-treated mice. *Nature Medicine* 50: 300–302.

[94] Pal D, Dutta S, Sarkar A (2009) Evaluation of CNS activities of ethanol extract of roots and rhizomes of Cyperus rotundus in mice. *Acta Poloniae Pharmaceutica* 66: 535–541.

[95] Pandy V, Jose N, Subhash H (2009) CNS activity of methanol and acetone extracts of Acorus calamus leaves in mice. *Journal of Pharmacology and Toxicology* 4: 79–86.

[96] Pal D (2008) Evaluation of CNS activities of aerial parts of Cynodon dactylon Pers. in mice. *Acta Poloniae Pharmaceutica* 65: 37–43.

[97] Bum EN, Nkantchoua GN, Njikam N, Taiwe GS, Ngoupaye GT, Pelanken MM, et al. (2010) Anticonvulsant and sedative activity of leaves of Senna spectabilis in mice. *International Journal of Pharmacology* 6: 123–128.

[98] da Silva JB, Temponi VS, Fernandes FV, Alves GAD, de Matos DM, Gasparetto CM, et al. (2011) New approaches to clarify antinociceptive and anti-inflammatory effects of the ethanol extract from Vernonia condensata leaves. *International Journal of Molecular Sciences* 12: 8993–9008.

[99] Sathya B, Ariharasivakumar G, Vimalson DC, Subramani M, Magesh M (2011) Psychopharmacological evaluation of ethanolic extract of leaves of bauhinia tomentosa L. In mice. *International Journal of Pharmacy and Technology* 3: 3693–3709.

[100] Yau J, Abdulmalik UN, Yaro AH, Chindo BA, Anuka JA, Hussaini IM (2011) Behavioral properties of Balanites aegyptiaca in rodents. *Journal of Ethnopharmacology* 135: 725–729.

[101] Pal D, Samanta K (2011) CNS activities of ethanol extract of aerial parts of Hygrophila difformis in mice. *Acta Poloniae Pharmaceutica—Drug Research* 68: 75–81.

[102] Golla U, Kumar AK, Solomon Suder Raj B (2011) Assessment of antioxidant and hypnotic activity of unani formulation arq gulab. *Pharmacologyonline* 1: 930–941.

[103] Bum Ngo E, Pelanken MM, Njikam N, Talla E, Taiwe GS, Nkantchoua GCN, et al. (2009) The decoction of leaves of Phyllanthus discoideus possesses anticonvulsant and sedative properties in mice. *International Journal of Pharmacology* 5: 168–172.

[104] Ngo Bum E, Taiwe GS, Moto FCO, Ngoupaye GT, Nkantchoua GCN, Pelanken MM, et al. (2009) Anticonvulsant, anxiolytic and sedative properties of the roots of Nauclea latifolia Smith in mice. *Epilepsy and Behavior* 15: 434–440.

[105] Musa AM, Yaro AH, Usman H, Magaji MG, Habu M (2008) Phytochemical and some neuropharmacological studies on the methanolic leaf extracts of Cissus cornifolia [Vitaceae] in mice. *International Journal of Pharmacology* 4: 145–148.

[106] Kumar A, Kulkarni SK (2006) Protective effect of BR-16A, a polyherbal preparation against social isolation stress: possible GABAergic mechanism. *Phytotherapy Research* 20: 538–541.

[107] Consolini AE, Ragone MI, Migliori GN, Conforti P, Volonte MG (2006) Cardiotonic and sedative effects of Cecropia pachystachya Mart. (ambay) on isolated rat hearts and conscious mice. *Journal of Ethnopharmacology* 106: 90–96.

[108] Pal D, Balasaheb NS, Khatun S, Bandyopadhyay PK (2006) CNS activities of the aqueous extract of Hydrilla verticillata in mice. *Natural Product Sciences* 12: 44–49.

[109] Nayak SS, Ghosh AK, Debnath B, Vishnoi SP, Jha T (2004) Synergistic effect of methanol extract of Abies webbiana leaves on sleeping time induced by standard sedatives in mice and anti-inflammatory activity of extracts in rats. *Journal of Ethnopharmacology* 93: 397–402.

[110] Pal D, Panda C, Sinhababu S, Dutta A, Bhattacharya S (2003) Evaluation of psychopharmacological effects of petroleum ether extract of Cuscuta reflexa Roxb. stem in mice. *Acta Poloniae Pharmaceutica* 60: 481–486.

[111] Gupta M, Mazumder UK, Chakrabarti S (1999) CNS activities of methanolic extract of Moringa oleifera root in mice. *Fitoterapia* 70: 244–250.

[112] Mazumder UK, Gupta M, Rath N (1998) CNS activities of Cassia fistula in mice. *Phytotherapy Research* 12: 520–522.

[113] Jha U, Chhajed PM, Shelke TT, Oswal RJ, Adkar PP (2011) CNS activity of methanol extract of Parthenium hysterophorus L. in experimental animals. *Der Pharmacia Lettre* 3: 335–341.

[114] Ilodigwe EE, Akah PA, Okoye TC, Omeje EO (2010) Anticonvulsant effects of a glycoside isolated from the leaf of Spathodea campanulata P. Beauv. *Journal of Medicinal Plants Research* 4: 1895–1900.

[115] Borse LB, Kottai Muthu A, Thangatripathi A, Borse SL (2011) CNS activity of the methanol extracts of heartwood of tecoma stans in experimental animal model. *Pharmacologyonline* 3: 959–968.

[116] Adebajo A, Iwalewa E, Obuotor E, Ibikunle G, Omisore N, Adewunmi C, et al. (2009) Pharmacological properties of the extract and some isolated compounds of *Clausena lansium* stem bark: anti-trichomonal, antidiabetic, anti-inflammatory, hepatoprotective and antioxidant effects. *Journal of Ethnopharmacology* 122: 10–19.

[117] Dey P, Chandra S, Chatterjee P, Bhattacharya S (2011) Neuropharmacological properties of Mikania scandens (L.) Willd. (Asteraceae). *Journal of Advanced Pharmaceutical Technology and Research* 2: 255–259.

[118] Kumarappan C, Vijayakumar M, Thilagam E, Balamurugan M, Thiagarajan M, Senthil S, et al. (2011) Protective and curative effects of polyphenolic extracts from Ichnocarpus frutescense leaves on experimental hepatotoxicity by carbon tretrachloride and tamoxifen. *Annals of Hepatology* 10: 63–72.

[119] Amos S, Abbah J, Chindo B, Edmond I, Binda L, Adzu B, et al. (2005) Neuropharmacological effects of the aqueous extract of Nauclea latifolia root bark in rats and mice. *Journal of Ethnopharmacology* 97: 53–57.

[120] Chattopadhyay D, Arunachalam G, Ghosh L, Mandal AB (2004) CNS activity of Alstonia macrophylla leaf extracts: an ethnomedicine of Onge of Bay Islands. *Fitoterapia* 75: 673–682.

[121] Kar DM, Maharana L, Rout SP (2010) CNS activity of aerial parts of Hybanthus Enneaspermus Mull. *Pharmacologyonline* 3: 959–981.

[122] Okoye TC, Akah PA, Omeke CP (2010) Evaluation of the anticonvulsant and muscle relaxant effects of the methanol root bark extracts of Annona senegalensis. *Asian Pacific Journal of Tropical Medicine* 3: 25–28.

[123] Murugesan T, Saravanan KS, Lakshmi S, Ramya G, Thenmozhi K (2001) Evaluation of psychopharmacological effects of Clerodendrum phlomidis Linn. extract. *Phytomedicine* 8: 472–476.

[124] Rajangam J, Thiyagarajan A, Joshi VD, Palei N (2010) Psychopharmacological activities of Acorus calamus roots in rat and mice models. *Latin American Journal of Pharmacy* 29: 1009–1013.

[125] Suba V, Murugesan T, Rao RB, Pal M, Mandal SC, Saha BP (2002) Neuropharmacological profile of Barleria lupulina Lindl. Extract in animal models. *Journal of ethnopharmacology* 81: 251–255.

[126] Parimala Devi B, Boominathan R, Mandal SC (2004) Studies on psychopharmacological effects of Cleome viscosa Linn. extract in rats and mice. *Phytotherapy Research* 18: 169–172.

[127] Murugesan T, Ghosh L, Das J, Pal M, Saha BP (1999) CNS activity of Jussiaea suffruticosa Linn. extract in rats and mice. *Pharmacy and Pharmacology Communications* 5: 663–666.

[128] Chattopadhyay D, Arunachalam G, Mandal SC, Bhadra R, Mandal AB (2003) CNS activity of the methanol extract of Mallotus peltatus (Geist) Muell Arg. leaf: an ethnomedicine of Onge. *Journal of Ethnopharmacology* 85: 99–105.

[129] Yadav AV, Nade VS (2008) Anti-dopaminergic effect of the methanolic extract of Morus alba L. leaves. *Indian Journal of Pharmacology* 40: 221–226.

[130] Mukherjee J, Bhaumik U, Mukherjee PK, Saha BP (2009) CNS activity of the methanol extract obtained from the roots of Ocimum sanctum linn. *Pharmacologyonline* 2: 673–685.

[131] Ghosh L, Arunachalam G, Murugesan T, Pal M, Saha BP (2002) Studies on the psychopharmacological activities of Rumex nepalensis Spreng. root extract in rats and mice. *Phytomedicine* 9: 202–206.

[132] Kolawole OT, Makinde JM, Olajide OA (2007) Central nervous system depressant activity of Russelia equisetiformis. *Nigerian Journal of Physiological Sciences* 22: 59–63.

[133] Olajide OA, Awe SO, Makinde JM (1999) Pharmacological studies on the leaf of Psidium guajava. *Fitoterapia* 70: 25–31.

[134] Gupta M, Mazumder UK, Das S (1998) Effect of leaf extract from Clerodendron colebrookianum on CNS function in mice. *Indian Journal of Experimental Biology* 36: 171–174.

[135] Krupanidhi AM, Vagdevi HM, Shreedhara CS, Vaidya VP, Muralikrishna KS (2007) Investigation of neuropharmacological activities of ethanolic extract of Dodonaea viscosa seeds. *Journal of Natural Remedies* 7: 263–268.

[136] Nemeroff CB (2003) The role of GABA in the pathophysiology and treatment of anxiety disorders. *Psychopharmacology Bulletin* 37: 133–146.

[137] Costa CARDA, Kohn DO, De Lima VM, Gargano AC, Florio JC, Costa M (2011) The GABAergic system contributes to the anxiolytic-like effect of essential oil from Cymbopogon citratus (lemongrass). *Journal of Ethnopharmacology* 137: 828–836.

[138] Lin YC, Hsieh MT, Chen CF, Cheng HY, Peng WH (2003) Anxiolytic effect of ting-chih-wan in mouse behavior models of anxiety. *The American Journal of Chinese Medicine* 31: 47–59.

[139] Kuribara H, Weintraub ST, Yoshihama T, Maruyama Y (2003) An anxiolytic-like effect of Ginkgo biloba extract and its constituent, ginkgolide-A, in mice. *Journal of Natural Products* 66: 1333–1337.

[140] Chermat R, Brochet D, DeFeudis FV, Drieu K (1997) Interactions of Ginkgo biloba extract (EGb 761), diazepam and ethyl beta-carboline-3-carboxylate on social behavior of the rat. *Pharmacology, Biochemistry, and Behavior* 56: 333–339.

[141] Sonavane GS, Sarveiya VP, Kasture VS, Kasture SB (2002) Anxiogenic activity of Myristica fragrans seeds. *Pharmacology, Biochemistry, and Behavior* 71: 239–244.

[142] Kulkarni MP, Juvekar AR (2009) Anti-anxiety effects of leaves of Nelumbo nucifera gaertn. in mice. *Pharmacologyonline* 2: 292–299.

[143] Kuribara H, Maruyama Y (1996) The anxiolytic effect of oriental herbal medicines by an improved plus maze test in mice—involvement of benzodiazepine receptors. *Japanese Journal of Neuropsychopharmacol* 18: 179–190.

[144] Xu Z, Wang F, Tsang SY, Ho KH, Zheng H, Yuen CT, et al. (2006) Anxiolytic-like effect of baicalin and its additivity with other anxiolytics. *Planta Medica* 72: 189–192.

[145] Ragone MI, Sella M, Pastore A, Consolini AE (2010) Sedative and cardiovascular effects of Aloysia citriodora Palau, on mice and rats. *Latin American Journal of Pharmacy* 29: 79–86.

[146] Sun M, Zhan XP, Jin CY, Shan JZ, Xu S, Wang YL (2008) Clinical observation on treatment of post-craniocerebral traumatic mental disorder by integrative medicine. *Chinese Journal of Integrative Medicine* 14: 137–141.

[147] Genkova-Papazova M, Petkova B, Lazarova-Bakarova M (2010) Influence of phyto-adaptogens on diazepam withdrawal in rats. *Autonomic and Autacoid Pharmacology* 30: 19.

[148] Reeta KH, Mehla J, Gupta YK (2009) Curcumin is protective against phenytoin-induced cognitive impairment and oxidative stress in rats. *Brain Research* 1301: 52–60.

[149] Motamedi GK, Meador KJ (2004) Antiepileptic drugs and memory. *Epilepsy & Behavior* 5: 435–439.

[150] Kwan P, Brodie MJ (2001) Neuropsychological effects of epilepsy and antiepileptic drugs. *The Lancet* 357: 216–222.

[151] Bentué-Ferrer D, Akwa Y (2008) Benzodiazepines: effects on memory functioning. In: Pandi-Perumal S, Verster JC, Monti JM, Lader M, Langer SZ, editors. *Sleep Disorders: Diagnosis and Therapeutics*. First edition. Informa Healthcare, London; pp. 105–114.

[152] Beracochea D (2006) Anterograde and retrograde effects of benzodiazepines on memory. *Scientific World of Journal* 6: 1460–1465.

[153] Theodore WH (1988) Antiepileptic drugs and cerebral glucose metabolism. *Epilepsia* 29: S48–S55.

[154] Kim DH, Kim S, Jeon SJ, Son KH, Lee S, Yoon BH, et al. (2009) Tanshinone I enhances learning and memory and ameliorates memory impairment in mice via the extracellular signal-regulated kinase signalling pathway. *British Journal of Pharmacology* 158: 1131–1142.

[155] Kim DH, Park SJ, Kim JM, Jeon SJ, Cho YW, Son KH, et al. (2011) Cognitive dysfunctions induced by a cholinergic blockade and Abeta25-35 peptide are attenuated by salvianolic acid B. *Neuropharmacology* 61: 1432–1440.

[156] Prabhakar S, Saraf MK, Banik A, Anand A (2011) Bacopa monniera selectively attenuates suppressed superoxide dismutase activity in diazepam induced amnesic mice. *Annals of Neurosciences* 18: 8–13.

[157] Mani V, Parle M, Ramasamy K, Abdul Majeed AB (2011) Reversal of memory deficits by Coriandrum sativum leaves in mice. *Journal of the Science of Food and Agriculture* 91: 186–192.

[158] Mani V, Parle M (2009) Memory-enhancing activity of Coriandrum sativum in rats. *Pharmacologyonline* 2: 827–839.

[159] Farooq SM, Alla TR, Venkat Rao N, Prasad K, Shalam Nandakumar, K, et al. (2007) A study on CNS effects of milk extract of nuts of Semecarpus anacardium. Linn, (Anacardiaceae). *Pharmacologyonline* 1: 49–63.

[160] Prabhakar S, Saraf MK, Pandhi P, Anand A (2008) Bacopa monniera exerts antiamnesic effect on diazepam-induced anterograde amnesia in mice. *Psychopharmacology* 200: 27–37.

[161] Saraf MK, Prabhakar S, Pandhi P, Anand A (2008) Bacopa monniera ameliorates amnesic effects of diazepam qualifying behavioral-molecular partitioning. *Neuroscience* 155: 476–484.

[162] Mohamed AF, Matsumoto K, Tabata K, Takayama H, Kitajima M, Watanabe H. (2000) Effects of Uncaria tomentosa total alkaloid and its components on experimental amnesia in mice: elucidation using the passive avoidance test. *Journal of Pharmacy & Pharmacology* 52: 1553–1561.

[163] Kim DH, Kim S, Jeon SJ, Son KH, Lee S, Yoon BH, et al. (2008) The effects of acute and repeated oroxylin A treatments on Abeta25-35-induced memory impairment in mice. *Neuropharmacology* 55: 639–647.

[164] Kim DH, Jeon SJ, Son KH, Jung JW, Lee S, Yoon BH, et al. (2007) The ameliorating effect of oroxylin A on scopolamine-induced memory impairment in mice. *Neurobiology of Learning & Memory* 87: 536–546.

[165] Adiga S, Trivedi P, Ravichandra V, Deb D, Mehta F (2010) Effect of Punica granatum peel extract on learning and memory in rats. *Asian Pacific Journal of Tropical Medicine* 3: 687–690.

[166] Mani V, Parle M, Ramasamy K, Abdul Majeed AB (2010) Anti-dementia potential of Daucus carota seed extract in rats. *Pharmacologyonline* 1: 552–565.

[167] Vasudevan M, Parle M (2006) Pharmacological evidence for the potential of Daucus carota in the management of cognitive dysfunctions. *Biological & Pharmaceutical Bulletin* 29: 1154–1161.

[168] Vasudevan M, Parle M (2007) Memory-enhancing activity of Thespesia populnea in rats. *Pharmaceutical Biology* 45: 267–273.

[169] Joshi H, Parle M (2006) Evaluation of nootropic potential of Ocimum sanctum Linn. in mice. *Indian Journal of Experimental Biology* 44: 133–136.

[170] Joshi H, Parle M (2006) Nardostachys jatamansi improves learning and memory in mice. *Journal of Medicinal Food* 9: 113–118.

[171] Joshi H, Parle M (2006) Zingiber officinale: evaluation of its nootropic effect in mice. *The African Journal of Traditional, Complementary and Alternative Medicines* 3: 64–74.

[172] Parle M, Dhingra D, Kulkarni SK (2004) Memory-strengthening activity of Glycyrrhiza glabra in exteroceptive and interoceptive behavioral models. *Journal of Medicinal Food* 7: 462–466.

[173] Parle M, Dhingra D, Kulkarni SK (2009) Improvement of mouse memory by Myristica fragrans seeds. *Journal of Medicinal Food* 7: 157–161.

[174] Karichedu Joshi H, Parle M (2011) Evaluation of neuroprotective and cognition improving potentials of Ocimum tenuiflorum in mice. Neurodegenerative Diseases Conference: 10th International Conference AD/PD Alzheimer's and Parkinson's Diseases: Advances, Concepts and New Challenges Barcelona Spain Conference Start, Barcelona, Spain; 20110309.

[175] Joshi H, Megeri K (2011) Cognition improving effects of Clerodendron phlomidis linn. Bark extract in mice. European Psychiatry Conference: 19th European Congress of Psychiatry, EPA, Vienna Austria Conference Start: 20110312 Conference End: 20110315 Conference Publication: (varpagings) 26, Date of Publication: March 2011.

[176] Joshi H, Megeri K (2008) Antiamnesic evaluation of C. phlomidis Linn. bark extract in mice. *Revista Brasileira de Ciencias Farmaceuticas/Brazilian Journal of Pharmaceutical Sciences* 44: 717–725.

[177] Yadav P, Uplanchiwar V, Gahane A, Modi A, Telrandhe U, Bheemachari (2010) Nootropic activity of L-33—a polyherbal formulation. *Pharmacologyonline* 2: 818–827.

[178] Gavimath CC, Havannavar V, Hulekal P, Pattar P, Joshi H (2009) Evaluation of antiamnesic potentials of Calotropis procera in mice. *Pharmacologyonline* 3: 457–462.

[179] Naikwade N, Mule S, Adnaik R, Magdum C (2009) Memory-enhancing activity of Rose alba in mice. *International Journal of Green Pharmacy* 3: 239–242.

[180] Joshi H, Parle M (2007) Evaluation of the antiamnesic effects of Phyllanthus amarus in mice. *Colombia Médica* 38: 132–139.

[181] Vasudevan M, Parle M (2007) Effect of Anwala churna (Emblica offcinalis GAERTN.): an ayurvedic preparation on memory deficit rats. *Yakugaku Zasshi* 127: 1701–1707.

[182] Vasudevan M, Parle M (2007) Memory enhancing activity of Anwala churna (Emblica officinalis Gaertn.): an Ayurvedic preparation. *Physiology & Behavior* 91: 46–54.

[183] Hanumanthachar J, Navneet K, Jyotibala C (2007) Evaluation of nootropic effect of Argyreia speciosa in mice. *Journal of Health Science* 53: 382–388.

[184] Vasudevan M, Parle M (2009) Antiamnesic potential of Murraya koenigii leaves. *Phytotherapy Research* 23: 308–316.

[185] Mula M, Trimble M (2009) Antiepileptic drug-induced cognitive adverse effects. *CNS Drugs* 23: 121–137.

[186] Tsudaa M, Suzuki T, Misawaa M, Nagase H (1996) Involvement of the opioid system in the anxiolytic effect of diazepam in mice. *European Journal of Pharmacology* 307: 7–14.

[187] Freed WJ (1989) Impaired motor coordination in mice induced by 2-amino-7-phosphonoheptanoic acid (APH), glutamic acid diethyl ester (GDEE) and other compounds. *Pharmacology, Biochemistry, and Behavior* 32: 733–736.

[188] Vishwakarma SL, Pal SC, Kasture VS, Kasture SB (2002) Anxiolytic and antiemetic activity of Zingiber officinale. *Phytotherapy Research* 16: 621–626.

[189] Raskovic A, Horvat O, Jakovljevic V, Sabo J, Vasic R (2007) Interaction of alcoholic extracts of hops with pentobarbital and diazepam in mice. *European Journal of Drug Metabolism & Pharmacokinetics* 32: 81–85.

21

PHYTOTHERAPIES: DRUG INTERACTIONS IN CANCER

ANDREW J. MCLACHLAN[1] AND STEPHEN J. CLARKE[2]

[1] *Faculty of Pharmacy and Centre for Education and Research on Ageing, The University of Sydney and Concord Hospital, Sydney, New South Wales, Australia*
[2] *Sydney School of Medicine and Northern Clinical School, Kolling Institute of Medical Research, Royal North Shore Hospital, The University of Sydney, Sydney, New South Wales, Australia*

21.1 INTRODUCTION

Phytotherapy medicines are included within the broader definition of herbal and complementary medicines encompassing a diverse group of treatments ranging from traditional Chinese medicines, herbal medicines, plant extracts, plant/fruit juices, and so-called functional foods [1]. The past two decades have seen the increasing use of complementary and alternative medicines in developed countries [2–5]. We will discuss the potential for clinically significant interactions between commonly used phytotherapy medicines and conventional medicines [6] used in people with cancer. The rationale for this chapter is based on the high use of complementary and alternative therapies in people living with cancer. We will also report on the prevalence of use of phytotherapy medicines in people with cancer and consider the most commonly used phytotherapies in this clinical setting. The chapter will focus on understanding the mechanisms of potential interactions between phytotherapies and conventional medicines used in the treatment and prevention of cancer and its consequences. We have included evidence for and examples of selected herbal medicine interactions with drugs used in people with cancer. The authors' acknowledge a number of previous excellent reviews and commentaries on this topic [7–10] and

Phytotherapies: Efficacy, Safety, and Regulation, First Edition. Edited by Iqbal Ramzan.

hope that this chapter provides further insight and evidence to inform the safe use of phytotherapies in cancer.

It is acknowledged that the use of phytotherapy medicines is highly prevalent around the world and varies significantly across different countries based on the availability of conventional medicines, influenced by cultural practices and the nature of the health systems in both developed and developing countries. A 1990 survey of the general public in the United States estimated that 34% of the respondents used at least one form of complementary therapy in the previous 12 months [2]. This figure had increased to 42% by 1997 [3]. The popularity of the use of complementary and alternative medicines, including phytotherapy medicines, has also been observed in Australia [4]. In 2004, a South Australian survey reported 52% of respondents had used at least one nonmedically prescribed complementary therapy in the previous year. More than 57% of respondents reported using complementary therapies without their health practitioner's knowledge and 50% took conventional medicines on the same day, creating the potential for interactions between conventional medicines and complementary and alternative medicines [4]. A number of studies [5, 11–13] have identified the most commonly used herbal medicines, which are presented in Table 21.1. Taken together, these data demonstrate the high prevalence of the use of complementary and alternative medicines in the general community of developed countries, a phenomenon that is set to continue. The focus of this discussion will be limited to commonly used herbal medicines (Table 21.1) and especially those with a potential to cause significant phytotherapy–drug interactions with conventional medicines.

TABLE 21.1 Most Commonly used Herbal Medicines in Developed Countries Identified from a Range of Studies[a]

Aloe vera	Echinacea	Kava
Bilberry	Evening primrose	Liquorice
Black cohosh	Fenugreek	Milk thistle
Chamomile	Garlic	Peppermint
Chasteberry	Ginger	Saw palmetto
Cranberry	Ginkgo biloba	Senna
Dandelion	Ginseng	Slippery elm
Dong quai	Green tea	St. John's wort
	Hawthorne	Valerian

[a]From Refs. [5, 11–13].

21.2 USE OF HERBAL AND COMPLEMENTARY MEDICINES BY PEOPLE LIVING WITH CANCER

A systematic review of the literature in 1998 estimated that 31% of people with cancer reported using complementary and alternative medicines [14]. Recent studies suggest that this figure may now exceed 80%, although there is variability in the use of phytotherapies depending on cancer type and ethnic group studied;

complementary medicines use being more common in breast cancer patients and individuals from East Asian geographical ancestry [15, 16]. A more recent systematic review has specifically focused on the studies investigating the prevalence of herbal medicine use by people living with cancer in the United Kingdom [17]. This review reported that the prevalence of phytotherapy medicines use ranged from 3.1 to 24.9% [17]. The increased use of complementary and alternative medicines, particularly phytotherapy medicines, in people with cancer is highly relevant as even under optimal circumstances medicines used in cancer treatment have a low therapeutic index, which may be further lowered by adverse interactions between complementary and the conventional cancer medicines [8, 9].

The reason people living with cancer take phytotherapies is complex. A recent systematic review has investigated the reasons why people living with cancer use complementary and alternative medicines [18]. Although there was a wide range of responses, the most frequent was a perceived beneficial response (38%), wanting "control" (17%), complementary medicine as a "last resort" (10%), and "finding hope" (10%) [19]. Not surprisingly complementary medicines, including phytotherapies, is big business. It has been estimated that cancer patients spend over US$30 billion in out-of-pocket expenses on complementary medicines in the United States, even though there are limited data to indicate the cost-effectiveness of complementary medicines [20]. This increased use by patients and expense of complementary medicines has highlighted issues in regard to the safety and efficacy of these treatments. This is particularly the case for systemically administered complementary medicines including phytotherapies, where there is the potential for clinically significant interactions with conventional treatments [6].

21.3 MECHANISMS OF PHYTOTHERAPY–DRUG INTERACTIONS

For many conventional medicines, phytotherapy–drug interactions may not always lead to any serious adverse clinical consequences; however, for medicines with a low therapeutic index (where there is a close relationship between a safe/effective dose and dose associated with adverse effects), such as anticancer drugs, even minor changes in drug clearance could produce dramatic effects on patient outcomes [6, 8]. Phytotherapy–drug interactions occur via several broad mechanisms, including pharmaceutical (or direct physicochemical interactions), pharmacokinetic (PK), and pharmacodynamic (PD) interactions. Elucidating the mechanism of phytotherapy–drug interactions is a key focus of many of the controlled pharmacological studies. Understanding the mechanism of such interactions allows the prediction of other clinically important interactions but more importantly can guide strategies to manage and prevent clinical consequences of potential phytotherapy–drug interactions [6].

PK interactions can result when common pathways of absorption, metabolism, distribution, or elimination exist between the constituents of phytotherapy medicines and conventional therapeutic agents. These interactions most commonly involve intestinal and hepatic drug metabolizing enzymes (cytochrome P450, or "CYP" enzymes), which have been the focus of much of the research in the area of phytotherapy–drug interaction studies [6]. Of growing interest, and an expanding research literature, are interactions involving drug transporters [21] such as the ABC transporters (including p-glycoprotein) [22], breast cancer resistance protein (BCRP), organic anion transporting polypeptides (OATPs), and multidrug resistance proteins (MRPs), which are found in numerous healthy tissues including the gut epithelium, liver, and central nervous system, as well as chemotherapy-resistant tumor cells. Two of the most important CYP enzymes for metabolism of xenobiotics in humans are CYP3A4 and CYP2D6. CYP3A4 is involved in the metabolism of numerous therapeutic agents used in cancer including the taxanes (docetaxel and paclitaxel), vinca alkaloids (vincristine, vinblastine, vindesine, and vinorelbine), camptothecins (irinotecan), the hormones exemestane, tamoxifen, and letrozole, and the epidermal growth factor receptor inhibitors (gefitinib and erlotinib) [23, 24]. Substrates for the drug transporter, p-glycoprotein, among cancer drugs include many of the naturally derived anticancer drugs including the taxanes, vinca alkaloids, epipodophyllotoxins, and anthracyclines [22].

The constituents of phytotherapy medicines have the potential to induce or inhibit these drug metabolizing or transport pathways leading to changes in bioavailability or the systemic clearance of the conventional cancer chemotherapy agents. If bioavailability is increased, this might lead to increased risk of drug-related adverse effects, while a reduction in bioavailability might lead to compromised therapeutic efficacy due to reduced systemic exposure. It is clear that some phytotherapy constituents mediate PK interactions through activation of the pregnane X receptor (PXR), a ligand activated nuclear receptor that is part of the superfamily of nuclear receptors [25]. PXR regulates induction of CYP3A gene expression by xenobiotics, but may also regulate the induction of other genes involved in drug metabolizing pathways, and drug transporters, including CYP2B, CYP2C, CYP24, glutathione S-transferases, sulfotransferases, glucuronosyltransferases, OATP 1A4, p-glycoprotein (MDR1), and multidrug-resistance-associated proteins 2 and 3 [25, 26]. It has been shown that PXR is activated by a number of phytotherapies including *gingko biloba* (higher doses), St. John's wort, and some traditional Chinese medicines, demonstrating that phytoconstituents have the potential to have a major impact on drug metabolism and systemic exposure of interacting drugs [27, 28].

PD interactions may occur when the bioavailable constituents of a phytotherapy medicine demonstrate additive, synergistic, or antagonistic effects with a coadministered therapeutic agent. It is worth noting that disease states themselves can change the PK or PD of a drug and extrapolating data generated in controlled clinical studies in healthy participants to patients with cancer is not always possible [29, 30]. For example, CYP3A-mediated drug metabolism may be impaired in patients with an acute phase response, as occurs in numerous illnesses including comorbid rheumatological conditions, acute infections, and patients with advanced cancer and probably contributes to the marked variability in drug pharmacokinetics and toxicity that has been noted in these circumstances [31].

21.4 SELECTED EXAMPLES OF PHYTOTHERAPY MEDICINES THAT HAVE THE POTENTIAL TO CAUSE DRUG INTERACTIONS IN CANCER

In this chapter, we have chosen to provide representative examples of the types of interactions with commonly used phytotherapies to demonstrate that phytotherapy–drug interactions do occur and may lead to adverse outcomes. Often the potential for an interaction with anticancer drugs has to be extrapolated from preclinical studies or interactions with drugs from other therapeutic classes based on a rigorous understanding of the mechanism of such interactions and the pharmacological properties of a conventional drug and the constituents of a phytotherapy [6]. These examples emphasize the need to perform well-designed pharmacological (PK/PD) studies with phytotherapies and anticancer treatments to improve our knowledge of the mechanism and clinical significance of concomitant phytotherapy medicine use in cancer.

21.4.1 Black Cohosh (*Cimicifuga racemosa*)

Black cohosh is used and promoted in the setting of cancer to minimize the vasomotor side effects of antiestrogen therapies, such as tamoxifen [32]. Black cohosh is not a phytoestrogen, but its modest effect on vasomotor symptoms may be due to serotonergic or dopaminergic properties [33], or the result of constituents that have selective estrogen receptor modulator (SERM) activity [34]. Based on this theoretical interaction, caution in regard to administration of black cohosh in patients with estrogen-dependent tumors seems warranted.

Whilst there have been no direct *in vivo* studies, an *in vitro* study suggests that constituents of black cohosh may influence the efficacy of selected chemotherapeutic agents used in the treatment of breast cancer [35]. Constituents in black cohosh have been found to enhance the sensitivity of mouse mammary cancer cells to doxorubicin and docetaxel, but reduced sensitivity to cisplatin. Whilst the mechanisms of interaction and clinical relevance of this study are not yet clear, caution may be warranted in cancer patients receiving black cohosh in conjunction with chemotherapy. An *in vivo* preclinical study in rats also investigated the use of black cohosh and tamoxifen on implanted endometrial adenocarcinoma cells and showed that black cohosh did not enhance or reduce the inductive effect of tamoxifen on tumor growth, but may have reduced the metastasizing potential of the tumor potentiated by tamoxifen [36].

A controlled clinical phytotherapy–drug interaction study has shown that administration of black cohosh may have an inhibitory effect on CYP2D6 activity, but had no significant effect on the activities of CYP3A4, CYP1A2, and CYP2E1, in healthy participants [37]. Caution may be warranted therefore in patients receiving cancer therapeutic agents or supportive treatments primarily metabolized by CYP2D6. A further study, again in healthy participants, has shown that black cohosh has no effect on the drug disposition of digoxin, which may be indicative of a lack of effect of the herb on the activity of p-glycoprotein [38].

In summary, evidence regarding the potential interaction between black cohosh and therapeutic agents used in cancer is suggestive, but limited, and further clinical and PK studies are required.

21.4.2 Echinacea (*Echinacea purpurea*)

Echinacea is traditionally used as an immunostimulant, antibacterial and antiviral agent, and is commonly included in cough and cold preparations to help alleviate symptoms [39]. Evidence from clinical studies regarding its efficacy remains conflicting largely as a result of the investigation of varying preparations, dosages, and species [40]. Echinacea is commonly used in the cancer setting for its claimed benefits of stimulating the immune response, which could provide potential benefits in protecting against infection and recovery from chemotherapy-related side effects [41].

Echinacea has been shown *in vitro* to stimulate B-lymphocytes, targeted by monoclonal antibodies such as rituximab used in cancer treatment [42]. This provides the theoretical possibility of a PD phytotherapy–drug interaction, suggesting that Echinacea supplements should be used with caution.

There is conflicting *in vivo* evidence regarding the effect of Echinacea on the activity of drug metabolizing enzymes. Gorski et al. [43] concluded that Echinacea was an inhibitor of CYP1A2 and causes induction of the CYP3A with no effect on CYP2C9 or CYP2D6. However, a clinical study by Gurley et al. [44] observed no effect on the activity of CYP3A4, CYP2D6, CYP1A2, or CYP2E1. It is likely that the phytotherapy–drug interactions with Echinacea are dependent on the preparation and/or dosage of Echinacea used. However, caution is still warranted with the use of anticancer agents that are metabolized by CYP3A4 which have the potential to interact with coadministered Echinacea products. A recent case report in a patient receiving etoposide has implicated coadministration of Echinacea as contributing to profound adverse effects (thrombocytopenia) [45].

21.4.3 Fenugreek (*Trigonella foenum graecum*)

Fenugreek and its constituents have been actively investigated for their possible benefits in cancer prevention and treatment [46, 47]. Although no phytotherapy–drug interactions have been reported for fenugreek, it has several constituents that have the potential to interact with some medicines. Fenugreek contains several flavonoids, including quercetin, which has been implicated in p-glycoprotein and CYP3A4 inhibition. A preclinical study in rodents found that coadministration of quercetin significantly increased the oral bioavailability of etoposide which is a substrate for p-glycoprotein and CYP3A [48]. The authors postulated that this is related to inhibition of transport and first-pass intestinal metabolism. A clinical study showed that the systemic exposure of cyclosporin (a CYP3A4 substrate) was increased when it was coadministered with quercetin to healthy volunteers ($n=8$), the highest increase occurring when participants received quercetin for 3 days prior to commencement of cyclosporin [49]. While previous *in vitro* studies have demonstrated an inhibitory effect of quercetin on p-glycoprotein [50], information regarding plasma concentrations and bioavailability of quercetin following oral administration of recommended doses of fenugreek is largely unknown. Thus, there is the potential for interaction between fenugreek and conventional therapeutic agents as a result of the quercetin content and caution is warranted in coadministering fenugreek together with agents that are CYP3A4 substrates and/or substrates for p-glycoprotein.

21.4.4 Ginkgo Biloba

Ginkgo is a herbal medicine that is commonly used in the setting of cancer in combination with anticancer medicines [9]. Ginkgo remains an actively researched herbal medicine in the area of cancer [51]. The major flavonoids found in ginkgo (particularly, kaempferol and quercetin) have been found to modulate p-glycoprotein [52] and cytochrome P450 drug metabolizing enzymes [53, 54] with data primarily derived from *in vitro* experiments. However, controlled clinical studies [55] failed to demonstrate that such interactions were clinically significant [9].

Ginkgo has been implicated as a possible phytotherapy that can militate cognitive changes after chemotherapy [56]. Vardy et al. [57] conducted a prospective open-label crossover phytotherapy–drug interaction study in women with early-stage breast cancer taking either tamoxifen, anastrozole, or letrozole, but also receiving *ginkgo biloba* (EGb 761, 120 mg twice daily for 3 weeks). A PK assessment of trough concentrations for each of the drugs before and after ginkgo treatment found that steady-state trough concentrations of tamoxifen, anastrozole, or letrozole before and after treatment with *ginkgo biloba* were not significantly different. This study provides relevant data on the safety of ginkgo coadministration with hormone therapies in women with early-stage breast cancer.

21.4.5 Asian Ginseng (*Panax ginseng*)

Several different types of ginseng are used in phytotherapy medicine products (Asian, Siberian, American, and Japanese ginseng varieties) although Asian Ginseng is the most commonly used, especially in people with cancer [9]. Asian ginseng is a complex mixture of phytoconstituents including saponin glycosides (i.e., ginsenosides), antioxidants, volatile oils, fatty acids, vitamins, and polysaccharides, which have been investigated for their effects in cancer as a treatment and adjunct [58–60]. Numerous studies have investigated potential interactions between ginseng and conventional medicines [10]. Data from *in vitro* studies suggest that the constituents of *panax ginseng* inhibit a range of drug metabolizing enzymes [6, 10], while ginsenosides have moderate inhibitory effects on p-glycoprotein [61] and BCRP [62] activity.

A number of controlled phytotherapy–drug interaction studies in healthy participants have provided insights into potentially clinically important interactions, but the results have been conflicting. In a comprehensive study by Gurley et al. [53] in older people (mean age 67 years), *panax ginseng* (500 mg, three times daily, standardized to 5% ginsenosides) was administered for 28 days. This study used a probe drug cocktail of midazolam (CYP3A4), caffeine (CYP1A2), chlorzoxazone (CYP2E1), and debrisoquine (CYP2D6) to investigate possible metabolic interactions and found *panax ginseng* led to a small, but not clinically significant, reduction in CYP2D6 activity, while CYP3A4, CYP1A2, and CYP2E1 activity was unaffected. This was confirmed in a clinical trial in healthy young participants taking recommended doses of Asian ginseng in which no effect was observed on the phenotypic ratios for CYP3A4, CYP1A2, CYP2D6, and CYP2E1 enzymes [63]. However, a recent well-controlled clinical study in 12 healthy participants by Malati et al. [64] found that *panax ginseng* (500 mg twice daily) administered for 28 days led to a significant induction of the apparent clearance of the CYP3A substrate midazolam (50% increase) after oral administration; however, fexafenadine (p-glycoprotein substrate) pharmacokinetics were unchanged.

Ginseng has also been implicated in an interaction with imatinib, which resulted in hepatotoxicity in a case report of a person with chronic myelogenous leukemia [65]. The authors attributed the effects of ginseng on CYP3A4 as contributing to the development of imatinib-induced hepatic injury and recommended that the combination be avoided.

While there have been no specific reports of interaction between Asian ginseng and anticancer agents, it is reasonable to carefully monitor people with cancer who are taking ginseng, while further research is needed to establish the clinical significance of potential phytotherapy–drug interactions in people with cancer [10].

21.4.6 Green Tea (*Camellia sinensis*)

Green tea is one of the most commonly ingested phytotherapies [5, 66] which has the potential to interact with a number of medicines used in the setting of cancer. A number of studies have demonstrated that constituents of green tea, especially catechins, including (–)-epigallocatechin-3-gallate (EGCG), can inhibit p-glycoprotein [67, 68] and organic anion transporting polypeptides (OATP1A2 and OATP2B1) [69]. A very recent study by Misaka et al. [70] demonstrated that coadministration of green tea (700 mL/day) significantly reduced the bioavailability of the OATP1A2 (SLCO1A2) substrate nadolol in healthy volunteers ($n = 10$) by 85.0% ($P < 0.01$). The authors concluded that green tea reduces plasma concentrations of nadolol possibly in part by inhibition of OATP1A2-mediated uptake of nadolol in the intestine. This has implications for a number of orally administered anticancer medicines, including methotrexate and paclitaxel, which are known to be OATP substrates [71]. However, again, evidence from clinical studies is lacking at this time.

Green tea extract has also been implicated in the inhibition of selected drug metabolism pathways [72]. *In vitro* studies using pooled human liver and intestinal microsomes suggest that green tea catechins have the potential to cause clinically relevant interactions with substrates for CYP3A, CYP2B6, and CYP2C8, common drug metabolizing pathways for a number of medicines used in the setting of cancer. At this time, direct evidence of metabolic inhibition from clinical studies with medicines used in cancer is lacking.

One of the more interesting PD phytotherapy–drug interactions has been reported between green tea and the clinically used proteasome inhibitor bortezomib [73]. This study investigated the interaction with green tea polyphenols using multiple myeloma and glioblastoma cell lines *in vitro* and via *in vivo* preclinical experiments. The study showed that (−)-epigallocatechin gallate and other polyphenols prevented bortezomib-induced tumor cell death both *in vitro* and *in vivo*. The antagonistic effects of green tea polyphenols was evident for boronic-acid-based proteasome inhibitors (such as bortezomib) but not with non-boronic acid proteasome inhibitors (such as nelfinavir). The authors proposed that polyphenols undergo a direct physicochemical interaction with bortezomib and concluded that green tea polyphenols have the potential to reduce the efficacy of bortezomib such that ingestion of green tea should be avoided during cancer therapy with bortezomib [73].

21.4.7 Kava Kava (*Piper methysticum* Forst. f.)

Kava is a commonly used herbal medicine for anxiety and as such is used by people living with cancer. The kavalactones found in kava have been implicated in causing heptatotoxicity [74] and also influencing the activity of drug metabolizing enzymes [75]. Despite suggestions of metabolic interactions *in vitro*, a series of rigorously conducted controlled clinical studies have confirmed that coadministration of kava does not have a significant effect on metabolic substrates of CYP1A2 [37], CYP3A4 [76], or CYP2D6 [77] in healthy participants. As a phytotherapy with sedative properties, kava has the potential to contribute to additive PD interactions with other sedatives, anxiolytics, or medicines with sedative properties (i.e., opioid analgesics) commonly used in the clinical management of people with cancer [75].

21.4.8 Liquorice (*Glycyrrhiza uralensis*)

Liquorice root is used for a variety of possible effects in people with cancer. It has a number of active phytoconstituents, such has glycyrrhizin and its more potent metabolite glycyrrhetinic acid, which are inhibitors of cortisol metabolism [78]. Some constituents of liquorice have been evaluated for their effects on biological pathways related to cancer [79]. One recent study found that coadministration of liquorice led to significantly higher bioavailability of cortisone after its oral administration [80]. Prednisone and prednisolone are commonly used agents in many chemotherapy protocols and thus patients receiving these agents should be cautioned against administration of liquorice root to reduce the risk of severe mineralocorticoid side effects.

Repeated dosing studies in rodents found that liquorice extracts or purified glycyrrhizin (at high doses) significantly induced CYP3A4 and CYP1A2 production [81]. This finding was replicated in an *in vivo* study in healthy participants who received repeated doses of liquorice for 14 days and found a clinically significant induction of CYP3A4 activity assessed using changes in the pharmacokinetics of midazolam (CYP3A4 substrate) [82]. Caution seems warranted regarding the use of liquorice, especially at higher doses, in people with cancer due to the risk of potentially serious phytotherapy–drug interactions.

21.4.9 Milk Thistle (*Silybum marianum*)

Milk thistle (also called St Mary's Thistle) is commonly used for its hepatoprotective properties and silymarin, a mixture of closely related flavonoids, is the principal constituent believed to be responsible for this effect [83–85]. Although clinical evidence to support its efficacy remains elusive, milk thistle continues to be one of the most commonly used phytotherapies with a range of potential uses in people with cancer [84, 85]. The results of *in vitro* studies suggest that some phytoconstituents may have direct effects in cancer or favorable PD interactions with selected chemotherapeutic agents [86, 87]; however, the clinical relevance of findings from *in vitro* studies remains unclear.

There have been numerous *in vitro* studies that have investigated potential phytotherapy–drug interactions and shown that extracts of milk thistle inhibit the activity of CYP3A4 and CYP2C9 as well as UGT isoforms and p-glycoprotein and MRP1 transporter proteins [88–92]. However, *in vivo* studies have shown lack of effect of milk thistle on the pharmacokinetics of indinavir (CYP3A4 and MRP1 substrate) [93] or midazolam (CYP3A4 substrate) [44]. van Erp et al. [94] studied the short-term and long-term effects of the administration of milk thistle on the pharmacokinetics of irinotecan in a small number ($n=6$) of patients with cancer. This well-designed (and adequately powered) clinical study found that the use of milk thistle (200 mg three times a day for 14 consecutive days) did not affect the pharmacokinetics of irinotecan (nor its active metabolite, SN-38) when taken concurrently [94]. The enzymes and transporters responsible for irinotecan metabolism are also important in the metabolism of a number of other anticancer drugs. Interestingly, the results demonstrated that when recommended doses of the herb have been administered orally, plasma concentrations of silybin did not reach the levels needed in *in vitro* studies to affect CYP3A4 or UGT1A1 function [94].

The balance of available evidence suggests that milk thistle can be safely combined with cancer chemotherapy agents that are CYP3A- or p-glycoprotein substrates but close monitoring of changes in drug effects remains an appropriate strategy.

21.4.10 St. John's Wort (*Hypericum perforatum*)

St. John's wort is commonly used for the treatment of mild-to-moderate depression as well as other mood disorders and mild anxiety, suggesting that it has a clear role as a supportive treatment in people with cancer. St. John's wort has been one of the most widely studied phytotherapies with respect to potential interactions [6] but also for the potential effects of its phytoconstituents in cancer [95]. St. John's wort and its phytoconstituent hyperforin have been shown to be potent modulators of several cytochrome P450 enzymes and transporters [10]. *In vivo* studies have shown that St. John's wort derivatives produce significant induction of hepatic and intestinal CYP3A4 [27, 53] if administered for longer than a 2-week period. In the clinical setting, the predominant effect of coadministration of St. John's wort is induction of metabolism with the associated risk of lack of efficacy due to subtherapeutic concentrations [96].

Two clinical studies have directly investigated clinically significant interactions between St. John's wort and anticancer agents. The first of these examined the effect of St. John's wort on the pharmacokinetics of irinotecan and its active metabolite SN-38 [97]. Coadministration with St. John's wort led to a clinically significant 42% decrease in the systemic exposure of SN-38, which could potentially lead to a therapeutic failure based on clinical observations in the reduction of neutrophils and leukocytes in the patients studied [97]. The second study investigated the effect of St. John's wort on imatinib and found that the apparent clearance of imatinib increased by 43% after 2 weeks of St. John's wort administration [98]. CYP3A4 is the major enzyme responsible for the metabolism of imatinib with CYPs 1A2, 2D6, 2C9, and 2C19 contributing to a lesser extent.

Taken together, numerous studies have now established that administration of St. John's wort for a period of at least 2 weeks leads to significant induction of intestinal and hepatic CYP3A4 as well as induced expression of intestinal p-glycoprotein drug transporters [6, 9]. This has major implications for many anticancer therapies which have been identified as CYP3A4 and p-glycoprotein substrates. This has led to a recommendation to avoid the use of St. John's wort in people with cancer to minimize the major risk of therapeutic failure due to clinically significant phytotherapy–drug interactions.

21.4.11 Valerian (*Valeriana officinalis*)

Valerian is purported to be useful as an antispasmodic, an anxiolytic, and as an antidepressant but it is most often used for insomnia, a common problem in people living with cancer. Studies indicate that valerian does not interact with CYPs 1A2, 2E1, 2D6, or 3A4/5 metabolism pathways at recommended doses *in vivo* [33, 99], although *in vitro* studies do not support this evidence and suggest it not only inhibits CYP3A4 but also p-glycoprotein [10, 100]. While there may be some rationale against the concomitant use of sedatives and drugs with sedative effects (such as opioid analgesics) with valerian, there is no *in vivo* evidence of any interactions between it and anticancer agents or other conventional medicines. Block et al. [101] suggests that while enough evidence exists to recommend valerian for short-term use in cases of mild insomnia in cancer patients, its use over a prolonged period should be cautioned against until its long-term safety is determined.

21.5 FUTURE PERSPECTIVES: NEED FOR EVIDENCE AND ADVICE TO CANCER PATIENTS AND PHYSICIANS

From a consumer's perspective, phytotherapy medicines will continue to have a place as complementary therapies in the treatment and prevention of cancer [18]. It is essential that clinicians and regulators appreciate and respect the importance of cultural and social beliefs of consumers regarding the use of phytotherapy medicines, and their desire to take control and responsibility for their treatment options and care. Clinicians and regulators do have a role in providing and translating evidence and information for consumers in a nonjudgmental manner. This will help consumers appreciate the credibility of information about the safety and efficacy of the phytotherapies and their potential interaction with conventional treatments they are receiving. It is important to appreciate the aspects of phytotherapy medicine quality and the need to identify products produced in accordance with relevant standards and supported by sound evidence related to the claims made about a specific product. A further complication is the risk of contamination and misidentification [102] of phytotherapy products. The nature of the evidence supporting phytotherapy–drug interactions has also been discussed here. Recent analyses continue to highlight limitations in trial design, implementation, and reporting [103], which have important implications for the translation of evidence and advice for consumers with cancer who are taking phytotherapy medicines. A key future perspective for consumers, clinicians, and regulators is the need to gather, analyze, and interpret real-world data on the safety of phytotherapy–drug combinations to identify dangerous or perhaps beneficial combinations in popular use [104].

21.6 CONCLUSIONS

The increasing and highly prevalent use of phytotherapy medicines has led to concerns about the appropriate concomitant use of conventional and phytotherapy medicines in people with cancer. This chapter has highlighted concerns about potential adverse interactions between commonly used phytotherapies and conventional cancer treatments. Despite the available information, there remain considerable gaps in clinically relevant evidence to inform safe and appropriate use of phytotherapies because of the lack of well-conducted clinical and pharmacological studies of phytotherapies and conventional treatments in people with cancer. This must remain a priority for the future. In the meantime, the provision of rigorous advice to well-informed consumers, coupled with open disclosure and careful clinical monitoring remain essential elements for the safe management of potential phytotherapy–drug interactions in people living with cancer.

ACKNOWLEDGMENTS

The authors thank Michael Dolton, Darrin Brown, Stephen Carbonara, and Rachel Kissane who contributed to identifying and evaluating published studies included in this review.

CONFLICT OF INTEREST

Professor McLachlan receives financial research support from UnityHealth Pty Ltd (Kew, Victoria, Australia) to maintain a database of herb–drug interactions.

REFERENCES

[1] Ernst E (2001) A primer of complementary and alternative medicine commonly used by cancer patients. *Med J Aust* 174: 88–92.

[2] Eisenberg DM, Kessler RC, Foster C, Norlock FE, Calkins DR, et al. (1993) Unconventional medicine in the United States. Prevalence, costs, and patterns of use. *N Engl J Med* 328: 246–252.

[3] Eisenberg DM, Davis RB, Ettner SL, S Appel, Wilkey S, et al. (1998) Trends in alternative medicine use in the United States, 1990–1997: results of a follow-up national survey. *JAMA* 280: 1569–1575.

[4] MacLennan AH, Myers SP, Taylor AW (2006) The continuing use of complementary and alternative medicine in South Australia: costs and beliefs in 2004. *Med J Aust* 184: 27–31.

[5] Zhang AL, Story DF, Lin V, Vitetta L, Xue CC (2008) A population survey on the use of 24 common medicinal herbs in Australia. *Pharmacoepidemiol Drug Saf* 17: 1006–1013

[6] Coxeter PD, McLachlan AJ, Duke CC, Roufogalis BD (2004) Herb-drug interactions: an evidence based approach. *Curr Med Chem* 11: 1513–1525.

[7] Tascilar M, de Jong FA, Verweij J, Mathijssen RH (2006) Complementary and alternative medicine during cancer treatment: beyond innocence. *Oncologist* 11: 732–741

[8] Sparreboom A, Cox MC, Acharya MR, Figg WD (2004) Herbal remedies in the United States: potential adverse interactions with anticancer agents. *J Clin Oncol* 22: 2489–2503.

[9] Haefeli WE, Carls A (2014) Drug interactions with phytotherapeutics in oncology. *Expert Opin Drug Metab Toxicol* 10: 359–377.

[10] Goey AK, Mooiman KD, Beijnen JH, Schellens JH, Meijerman I (2013) Relevance of in vitro and clinical data for predicting CYP3A4-mediated herb-drug interactions in cancer patients. *Cancer Treat Rev* 39: 773–783.

[11] Bent S, Ko R (2004) Commonly used herbal medicines in the United States: a review. *Am J Med* 116: 478–485.

[12] Bruno JJ, Ellis JJ (2005) Herbal use among US elderly: 2002 National Health Interview Survey. *Ann Pharmacother* 39: 643–648.

[13] Bent S (2008) Herbal medicine in the United States: review of efficacy, safety, and regulation: grand rounds at University of California, San Francisco Medical Center. *J Gen Intern Med* 23: 854–859.

[14] Ernst E, Cassileth BR (1998) The prevalence of complementary/alternative medicine in cancer: a systematic review. *Cancer* 83: 777–782.

[15] Richardson MA, Sanders T, Palmer JL, Greisinger A, Singletary SE (2000) Complementary/alternative medicine use in a comprehensive cancer center and the implications for oncology. *J Clin Oncol* 18: 2505–2514.

[16] Boon HS, Olatunde F, Zick SM (2007) Trends in complementary/alternative medicine use by breast cancer survivors: comparing survey data from 1998 and 2005. *BMC Womens Health* 7:4.

[17] Gratus C, Damery S, Wilson S, Warmington S, Routledge P, Grieve R, Steven N, Jones J, Greenfield S (2009) The use of herbal medicines by people with cancer in the UK: a systematic review of the literature. *QJM* 102: 831–842.

[18] Oh B, Butow P, Mullan B, Beale P, Pavlakis N, Rosenthal D, Clarke S (2010) The use and perceived benefits resulting from the use of complementary and alternative medicine by cancer patients in Australia. *Asia Pac J Clin Oncol* 6: 342–349.

[19] Verhoef MJ, Balneaves LG, Boon HS, Vroegindewey A (2005) Reasons for and characteristics associated with complementary and alternative medicine use among adult cancer patients: a systematic review. *Integr Cancer Ther* 4: 274–286.

[20] Herman PM, Craig BM, Caspi O (2005) Is complementary and alternative medicine (CAM) cost-effective? A systematic review. *BMC Complement Altern Med* 5: 11.

[21] König J, Müller F, Fromm MF (2013) Transporters and drug-drug interactions: important determinants of drug disposition and effects. *Pharmacol Rev* 65: 944–966.

[22] Marchetti S, Mazzanti R, Beijnen JH, Schellens JH (2007) Concise review: clinical relevance of drug drug and herb drug interactions mediated by the ABC transporter ABCB1 (MDR1, P-glycoprotein) *Oncologist* 12: 927–941.

[23] Meijerman I, Beijnen JH, Schellens JH (2006) Herb-drug interactions in oncology: focus on mechanisms of induction. *Oncologist* 11: 742–752.

[24] Beijnen JH, Schellens JH (2004) Drug interactions in oncology. *Lancet Oncol* 5: 489–496.

[25] Ma X, Idle JR, Gonzalez FJ (2008) The pregnane X receptor: from bench to bedside. *Expert Opin Drug Metab Toxicol* 4: 895–908.

[26] Köhle C, Bock KW (2009) Coordinate regulation of human drug-metabolizing enzymes, and conjugate transporters by the Ah receptor, pregnane X receptor and constitutive androstane receptor. *Biochem Pharmacol* 77: 689–699.

[27] Moore LB, Goodwin B, Jones SA, Wisely GB, Serabjit-Singh CJ, Willson TM, Collins JL, Kliewer SA (2000) St. John's wort induces hepatic drug metabolism through activation of the pregnane X receptor. *Proc Natl Acad Sci U S A* 97: 7500–7502.

[28] Li L, Stanton JD, Tolson AH, Luo Y, Wang H (2009) Bioactive terpenoids and flavonoids from Ginkgo biloba extract induce the expression of hepatic drug-metabolizing enzymes through pregnane X receptor, constitutive androstane receptor, and aryl hydrocarbon receptor-mediated pathways. *Pharm Res* 26: 872–882.

[29] McLachlan AJ, Hilmer SN, Le Couteur DG (2009) Variability in response to medicines in older people: phenotypic and genotypic factors. *Clin Pharmacol Ther* 85: 431–433.

[30] He X, Clarke SJ, McLachlan AJ (2011) Clinical pharmacology of chemotherapy agents in older people with cancer. *Curr Gerontol Geriatr Res* 2011: 628670.

[31] Rivory LP, Slaviero KA, Clarke SJ (2002) Hepatic cytochrome P450 3A drug metabolism is reduced in cancer patients who have an acute-phase response. *Br J Cancer* 87: 277–280.

[32] Villaseca P (2012) Non-estrogen conventional and phytochemical treatments for vasomotor symptoms: what needs to be known for practice. *Climacteric* 15: 115–124.

[33] Mahady GB (2003) Is black cohosh estrogenic? *Nutr Rev* 61: 183–186.

[34] Seidlova-Wuttke D, Hesse O, Jarry H, Christoffel V, Spengler B, Becker T, Wuttke W (2003) Evidence for selective estrogen receptor modulator activity in a black cohosh (Cimicifuga racemosa) extract: comparison with estradiol-17beta. *Eur J Endocrinol* 149: 351–362.

[35] Rockwell S, Liu Y, Higgins SA (2005) Alteration of the effects of cancer therapy agents on breast cancer cells by the herbal medicine black cohosh. *Breast Cancer Res Treat* 90: 233–239.

[36] Nisslein T, Freudenstein J (2004) Concomitant administration of an isopropanolic extract of black cohosh and tamoxifen in the in vivo tumor model of implanted RUCA-I rat endometrial adenocarcinoma cells. *Toxicol Lett* 150: 271–275.

[37] Gurley BJ, Gardner SF, Hubbard MA, Williams DK, Gentry WB, Khan IA, Shah A (2005) In vivo effects of goldenseal, kava kava, black cohosh, and valerian on human cytochrome P450 1A2, 2D6, 2E1, and 3A4/5 phenotypes. *Clin Pharmacol Ther* 77: 415–426.

[38] Gurley BJ, Barone GW, Williams DK, Carrier J, Breen P, et al. (2006) Effect of milk thistle (Silybum marianum) and black cohosh (Cimicifuga racemosa) supplementation on digoxin pharmacokinetics in humans. *Drug Metab Dispos* 34: 69–74.

[39] Barnes J, Anderson LA, Gibbons S, Phillipson JD (2005) Echinacea species (Echinacea angustifolia (DC.) Hell., Echinacea pallida (Nutt.) Nutt., Echinacea purpurea (L.) Moench): a review of their chemistry, pharmacology and clinical properties. *J Pharm Pharmacol* 57: 929–954

[40] Shah SA, Sander S, White CM, Rinaldi M, Coleman CI (2007) Evaluation of echinacea for the prevention and treatment of the common cold: a meta-analysis. *Lancet Infect Dis* 7:473–480.

[41] Block KI, Mead MN (2003) Immune system effects of echinacea, ginseng, and astragalus: a review. *Integr Cancer Ther* 2: 247–267.

[42] Werneke U, Earl J, Seydel C, Horn O, Crichton P, Fannon D (2004) Potential health risks of complementary alternative medicines in cancer patients. *Br J Cancer* 90: 408–413.

[43] Gorski JC, Huang SM, Pinto A, Hamman MA, Hilligoss JK, Zaheer NA, Desai M, Miller M, Hall SD (2004) The effect of echinacea (Echinacea purpurea root) on cytochrome P450 activity in vivo. *Clin Pharmacol Ther* 75: 89–100.

[44] Gurley BJ, Gardner SF, Hubbard MA, Williams DK, Gentry WB, et al. (2004) In vivo assessment of botanical supplementation on human cytochrome P450 phenotypes: Citrus aurantium, Echinacea purpurea, milk thistle, and saw palmetto. *Clin Pharmacol Ther* 76: 428–440.

[45] Bossaer JB, Odle BL (2012) Probable etoposide interaction with Echinacea. *J Diet Suppl* 9: 90–95.

[46] Sung B, Prasad S, Yadav VR, Aggarwal BB (2012) Cancer cell signaling pathways targeted by spice-derived nutraceuticals. *Nutr Cancer* 64: 173–197.

[47] Shabbeer S, Sobolewski M, Anchoori RK, Kachhap S, Hidalgo M, et al. (2009) Fenugreek: a naturally occurring edible spice as an anticancer agent. *Cancer Biol Ther* 8: 272–278.

[48] Li X, Choi JS (2009) Effects of quercetin on the pharmacokinetics of Etoposide after oral or intravenous administration of etoposide in rats. *Anticancer Res* 29: 1411–1415.

[49] Choi JS, Choi BC, Choi KE (2004) Effect of quercetin on the pharmacokinetics of oral cyclosporine. *Am J Health Syst Pharm* 61: 2406–2409.

[50] Scambia G, Ranelletti FO, Panici PB, De Vincenzo R, Bonanno G, et al. (1994) Quercetin potentiates the effect of adriamycin in a multidrug-resistant MCF-7 human breast-cancer cell line: P-glycoprotein as a possible target. *Cancer Chemother Pharmacol* 344: 59–64.

[51] DeFeudis FV, Papadopoulos V, Drieu K (2003) Ginkgo biloba extracts and cancer: a research area in its infancy. *Fundam Clin Pharmacol* 17: 405–417.

[52] Limtrakul P, Khantamat O, Pintha K (2005) Inhibition of P-glycoprotein function and expression by kaempferol and quercetin. *J Chemother* 17: 86–95.

[53] Gurley BJ, Gardner SF, Hubbard MA, Williams DK, Gentry WB, et al. (2005) Clinical assessment of effects of botanical supplementation on cytochrome P450 phenotypes in the elderly: St John's wort, garlic oil, Panax ginseng and Ginkgo biloba. *Drugs Aging* 22: 525–539.

[54] Gaudineau C, Beckerman R, Welbourn S, Auclair K (2004) Inhibition of human P450 enzymes by multiple constituents of the Ginkgo biloba extract. *Biochem Biophys Res Commun* 318: 1072–1078.

[55] Zadoyan G, Rokitta D, Klement S, Dienel A, Hoerr R, et al. (2012) Effect of Ginkgo biloba special extract EGb 761 on human cytochrome P450 activity: a cocktail interaction study in healthy volunteers. *Eur J Clin Pharmacol* 68: 553–560.

[56] Barton DL, Burger K, Novotny PJ, Fitch TR, Kohli S, et al. (2013) The use of Ginkgo biloba for the prevention of chemotherapy-related cognitive dysfunction in women receiving adjuvant treatment for breast cancer, N00C9. *Support Care Cancer* 21: 1185–1192.

[57] Vardy J, Dhillon HM, Clarke SJ, Olesen I, Leslie F, et al. (2013) Investigation of herb-drug interactions with ginkgo biloba in women receiving hormonal treatment for early breast cancer. *Springerplus* 2: 126.

[58] Yun TK (2001) Panax ginseng—a non-organ-specific cancer preventive? *Lancet Oncol* 2 :49–55.

[59] Voglert BK, Pittler MH, Ernst E (1999) The efficacy of ginseng. A systematic review of randomised clinical trials. *Eur J Clin Pharmacol* 55:567–575.

[60] Qi LW, Wang CZ, Du GJ, Zhang ZY, Calway T, Yuan CS (2011) Metabolism of ginseng and its interactions with drugs. *Curr Drug Metab* 12: 818–822.

[61] Kitagawa S, Takahashi T, Nabekura T, Tachikawa E, Hasegawa H (2007) Inhibitory effects of ginsenosides and their hydrolyzed metabolites on daunorubicin transport in KB-C2 cells. *Biol Pharm Bull* 30: 1979–1981.

[62] Jin J, Shahi S, Kang HK, van Veen HW, Fan TP (2006) Metabolites of ginsenosides as novel BCRP inhibitors. *Biochem Biophys Res Commun* 345: 1308–1314.

[63] Gurley BJ, Gardner SF, Hubbard MA, Williams DK, Gentry WB, et al. (2002) Cytochrome P450 phenotypic ratios for predicting herb-drug interactions in humans. *Clin Pharmacol Ther* 72: 276–287.

[64] Malati CY, Robertson SM, Hunt JD, Chairez C, Alfaro RM, et al. (2012) Influence of Panax ginseng on cytochrome P450 (CYP)3A and P-glycoprotein (P-gp) activity in healthy participants. *J Clin Pharmacol* 52: 932–939.

[65] Bilgi N, Bell K, Ananthakrishnan AN, Atallah E (2010) Imatinib and Panax ginseng: a potential interaction resulting in liver toxicity. *Ann Pharmacother* 44: 926–928.

[66] Graham HN (1992) Green tea composition, consumption, and polyphenol chemistry. *Prev Med* 21: 334–350.

[67] Jodoin J, Demeule M, Beliveau R (2002) Inhibition of the multidrug resistance P-glycoprotein activity by green tea polyphenols. *Biochim Biophys Acta* 1542: 149–159.

[68] Qian F, Wei D, Zhang Q, Yang S (2005) Modulation of P-glycoprotein function and reversal of multidrug resistance by (–)-epigallocatechin gallate in human cancer cells. *Biomed Pharmacother* 59: 64–69.

[69] Roth M, Timmermann BN, Hagenbuch, B (2011) Interactions of green tea catechins with organic anion-transporting polypeptides. *Drug Metab Dispos* 39: 920–926.

[70] Misaka S, Yatabe J, Müller F, Takano K, Kawabe K, et al. (2014) Green tea ingestion greatly reduces plasma concentrations of nadolol in healthy subjects. *Clin Pharmacol Ther* 95: 432–438.

[71] van de Steeg E, van Esch A, Wagenaar E, Kenworthy KE, Schinkel AH (2013) Influence of human OATP1B1, OATP1B3, and OATP1A2 on the pharmacokinetics of methotrexate and paclitaxel in humanized transgenic mice. *Clin Cancer Res* 19: 821–832.

[72] Misaka S, Kawabe K, Onoue S, Werba JP, Giroli M, et al. (2013) Effects of green tea catechins on cytochrome P450 2B6, 2C8, 2C19, 2D6 and 3A activities in human liver and intestinal microsomes. *Drug Metab Pharmacokinet* 28: 244–249.

[73] Golden EB, Lam PY, Kardosh A, Gaffney KJ, Cadenas E, et al. (2009) Green tea polyphenols block the anticancer effects of bortezomib and other boronic acid-based proteasome inhibitors. *Blood* 113: 5927–5937.

[74] Rowe A, Zhang LY, Ramzan I (2011) Toxicokinetics of kava. *Adv Pharmacol Sci* 2011: 326724.

[75] Anke J, Ramzan I (2004) Pharmacokinetic and pharmacodynamic drug interactions with Kava (Piper methysticum Forst. f.) *J Ethnopharmacol* 93: 153–160.

[76] Gurley BJ, Swain A, Hubbard MA, Hartsfield F, Thaden J, et al. (2008) Supplementation with goldenseal (Hydrastis canadensis), but not kava kava (Piper methysticum), inhibits human CYP3A activity in vivo. *Clin Pharmacol Ther* 83: 61–69.

[77] Hubbard MA, Williams DK, Barone G, Hartsfield F, Tong Y, et al. (2008) Clinical assessment of CYP2D6-mediated herb-drug interactions in humans: effects of milk thistle, black cohosh, goldenseal, kava kava, St. John's wort, and Echinacea. *Mol Nutr Food Res* 2: 755–763.

[78] Isbrucker RA, Burdock GA (2006) Risk and safety assessment on the consumption of Licorice root (Glycyrrhiza sp.), its extract and powder as a food ingredient, with emphasis on the pharmacology and toxicology of glycyrrhizin. *Regul Toxicol Pharmacol* 46: 167–192.

[79] Hsu YL, Wu LY, Hou MF, Tsai EM, Lee JN, et al. (2011) Glabridin, an isoflavan from licorice root, inhibits migration, invasion and angiogenesis of MDA-MB-231 human breast adenocarcinoma cells by inhibiting focal adhesion kinase/Rho signaling pathway. *Mol Nutr Food Res* 55: 318–327.

[80] Methlie P, Husebye EE, Hustad S, Lien EA, Løvås K (2011) Grapefruit juice and licorice increase cortisol availability in patients with Addison's disease. *Eur J Endocrinol* 165: 761–769.

[81] Paolini M, Barillari J, Broccoli J, Pozetti L, Perocco P, et al. (1999) Effect of Liquorice and glycyrrhizin on rat liver carcinogen metabolizing enzymes. *Cancer Lett* 145: 35–42.

[82] Tu JH, He YJ, Chen Y, Fan L, Zhang W, et al. (2010) Effect of glycyrrhizin on the activity of CYP3A enzyme in humans. *Eur J Clin Pharmacol* 66: 805–810.

[83] Ramasamy K, Agarwal R (2008) Multitargeted therapy of cancer by silymarin. *Cancer Lett* 269: 352–362.

[84] Greenlee H, Abascal K, Yarnell E, Ladas E (2007) Clinical applications of Silybum marianum in oncology. *Integr Cancer Ther* 6: 158–165.

[85] Sagar SM (2007) Future directions for research on Silybum marianum for cancer patients. *Integr Cancer Ther* 6: 166–173.

[86] Deep G, Agarwal R (2010) Antimetastatic efficacy of silibinin: molecular mechanisms and therapeutic potential against cancer. *Cancer Metastasis Rev* 29: 447–463.

[87] Ting H, Deep G, Agarwal R (2013) Molecular mechanisms of silibinin-mediated cancer chemoprevention with major emphasis on prostate cancer. *AAPS J* 15: 707–716.

[88] Zuber R, Modriansky M, Dvorak Z (2002) Effect of silybin and its congeners on human liver microsomal cytochrome P450 activities. *Phytother Res* 16: 632–638.

[89] Venkataramanan R, Ramachandran V, Komoroski BJ, Zhang S, Schiff PL, Strom SC (2000) Milk thistle, a herbal supplement, decreases the activity of CYP3A4 and uridine diphosphoglucuronosyl transferase in human hepatocyte cultures. *Drug Metab Dispos* 28: 1270–1273.

[90] Zhang S, Morris ME (2003) Effects of the flavanoids biochanin A, morin, phloretin, and silymarin on P-glycoprotein-mediated transport. *J Pharmacol Exp Ther* 304: 1258–1267.

[91] Zhang S, Morris ME (2003) Effect of the flavonoids biochanin A and silymarin on the P-glycoprotein-mediated transport of digoxin and vinblastine in human intestinal Caco-2 cells. *Pharm Res* 20: 1184–1191.

[92] Morris ME, Zhang S (2006) Flavonoid-drug interactions: effects of flavonoids on ABC transporters. *Life Sci* 78: 2116–2130.

[93] Mills E, Wilson K, Clarke M, Foster B, Walker S, et al. (2005) Milk thistle and indinavir: a randomized controlled pharmacokinetics study and meta-analysis. *Eur J Clin Pharmacol* 61: 1–7.

[94] van Erp NPH, Baker SD, Zhao M, Rudek MA, Guchelaar H-J, et al. (2005) Effect of milk thistle (Silybum marianum) on the pharmacokinetics of irinotecan. *Clin Cancer Res* 11: 7800–7806.

[95] Billard C, Merhi F, Bauvois B (2013) Mechanistic insights into the antileukemic activity of hyperforin. *Curr Cancer Drug Targets* 13: 1–10.

[96] Mills E, Montori VM, Wu P, Gallicano K, Clarke M, et al. (2004) Interaction of St John's wort with conventional drugs: systematic review of clinical trials. *BMJ* 329: 27–30.

[97] Mathijssen RH, Verweij J, de Bruijn P, Loos WJ, Sparreboom A (2002) Effects of St. John's wort on irinotecan metabolism. *J Natl Cancer Inst* 94: 1247–1249.

[98] Frye RF, Fitzgerald SM, Lagattuta TF, Hruska MW, Egorin MJ (2004) Effect of St John's wort on imatinib mesylate pharmacokinetics. *Clin Pharmacol Ther* 76: 323–329.

[99] Donovan JL, DeVane CL, Chavin KD, Wang JS, Gibson BB, et al. (2004) Multiple night-time doses of valerian (Valeriana officinalis) had minimal effects on CYP3A4 activity and no effect on CYP2D6 activity in healthy volunteers. *Drug Metab Dispos* 32:1333–1336.

[100] Lefebvre T, Foster BC (2004) In vitro activity of commercial valerian root extracts against human cytochrome P450 3A4. *J Pharm Pharm Sci* 7: 265–273.

[101] Block KI, Gyllenhaal C, Mead MN (2004) Safety and efficacy of herbal sedatives in cancer care. *Integr Cancer Ther* 3: 128–148.

[102] Newmaster SG, Grguric M, Shanmughanandhan D, Ramalingam S, Ragupathy S (2013) DNA barcoding detects contamination and substitution in North American herbal products. *BMC Med* 11: 222.

[103] Gagnier JJ, Moher D, Boon H, Beyene J, Bombardier C (2011) Randomized controlled trials of herbal interventions underreport important details of the intervention. *J Clin Epidemiol* 64: 760–769.

[104] Skalli S, Soulaymani Bencheikh R (2012) Safety monitoring of herb-drug interactions: a component of pharmacovigilance. *Drug Saf* 35: 785–791.

22

QUALITY USE OF MEDICINES: CONSIDERATIONS IN PHYTOTHERAPY

LYNN WEEKES
NPS MedicineWise, Surry Hills, New South Wales, Australia

22.1 INTRODUCTION

Phytotherapies or herbal medicines are used in all cultures, sometimes as mainstream medicines and sometimes as adjuncts to traditional Western medicines. They are used to treat conditions, to moderate symptoms, and to promote well-being. Health practitioners may prescribe or recommend their use and they are also used by consumers who self-select and self-manage. Clearly, phytotherapies are used in many of the same and in some unique ways compared with other medicines and the appropriate and safe use of these medicines is as critical as for any other medicines. To achieve the best outcomes from a medicine we need to apply quality use of medicines (QUM) principles and practices.

QUM has been defined as selecting options wisely, choosing the most appropriate medicine when one is deemed necessary and using medicines safely and effectively [1].

22.1.1 Judicious Use

Selecting management options wisely means that consideration should be given to the place of medicines in treating illness and maintaining health while recognizing that there may be better ways than a medicine to manage many disorders. This is especially relevant for herbal medicines, which in Western countries are frequently

Phytotherapies: Efficacy, Safety, and Regulation, First Edition. Edited by Iqbal Ramzan.

used to boost well-being and in fact may be used instead of a balanced diet or other healthy lifestyle management options.

22.1.2 Appropriate Selection

If a medicine is considered necessary, the most suitable medicine should be selected from those available to ensure the right medicine, for the right person, for the right condition, at the right time, and in the right dose. This means taking into account clinical, social, and cultural traits of the person taking the medicine as well as their preferences and beliefs. The comparative harms and benefits of specific medicines, coexisting conditions or therapies, the burden of monitoring treatment, and the quality of the evidence to support it would all be weighed up to tailor the best treatment for a person. The affordability of the medicine for the person, the community, and the health system will be an additional consideration in many situations.

22.1.3 Safe and Effective Use

Medicines should be used safely and effectively to get the best possible results. This requires being clear about the therapeutic outcome, monitoring for its achievement, and making the necessary alterations, including stopping the medicine, if the medicine does not achieve this outcome. Misuse, overuse, and underuse of medicines can all lead to suboptimal outcomes and these should be minimized. The person taking the medicine should be empowered to manage their medicines, problem solving when appropriate, and referring to their health professional as needed. Health literacy is important to support people taking medicines to access and use information and take part in decision making with or without their health professional.

22.1.4 The QUM Paradigm

QUM is consumer-centric. At the same time, it considers various perspectives—individual, community development, and public health—in planning, implementing, and evaluating initiatives to improve use of medicines. This approach involves all stakeholders, includes all stages of learning, and addresses all relevant settings [1].

QUM is sometimes represented as a pyramid with three levels and four faces. The levels of the pyramid relate to stages of learning:

- *Level 1. The Awareness Level*
  Awareness of medicines as a health issue. For an individual, this could mean recognizing a medicine as such, no matter what formulation or form it comes in and no matter if it is sourced via a prescription or bought in a supermarket.
- *Level 2. The Knowledge and Skills Level*
  Knowledge, skills, information, and tools needed to make appropriate decisions. For an individual this could mean knowing where to find reliable information on medicines, being able to read and interpret the medicines label, and knowing what information or questions to share with their health professional.

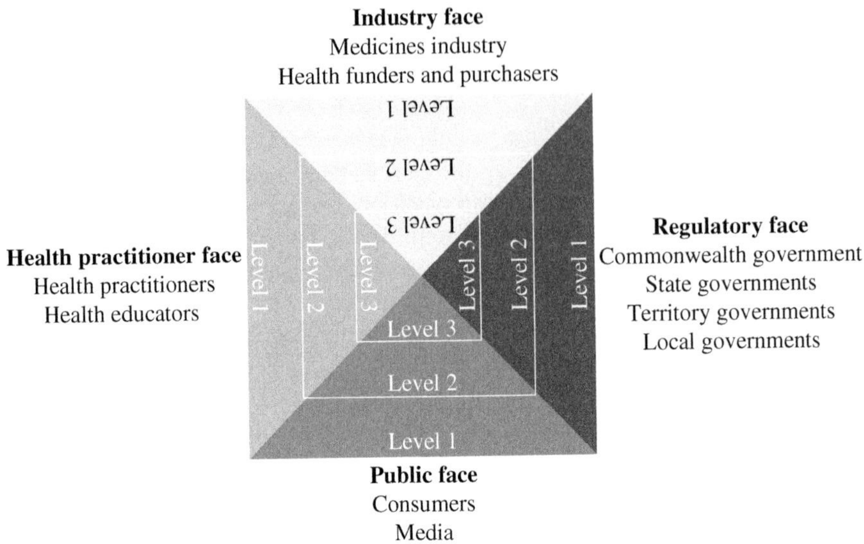

FIGURE 22.1 Faces of the QUM pyramid. Reproduced with permission, Australian Government Department of Health 2013.

- *Level 3. The Action and Evaluation Level*
  This involves monitoring for positive and adverse effects of medicines, quality assurance, and problem solving. For an individual this could include keeping a symptom diary, titrating doses in response to exacerbations of symptoms, and discussing concerns about medicines with their doctor and pharmacist.

The faces of the pyramid identify those most directly able to influence QUM: the consumer, the health professional, the pharmaceutical industry, and government (Fig. 22.1).

Each player has a role to play in QUM and information needs that allow them to reach the best decisions for medicines use.

22.2 RELEVANCE OF QUM FOR HERBAL MEDICINES

22.2.1 Is the QUM Framework Relevant for Herbal Therapies?

There are several reasons to answer, yes, to this question:

- Herbal medicines *are* medicines and all medicines have the potential for benefit as well as harm.
- Herbal medicines are used alone, in combination with and sometimes in preference to conventional Western medicines.
- Herbal medicines are no longer used solely within particular health paradigms such as Chinese medicine or Ayurvedic medicine, complete with the health philosophy and systems that support them.

- Increasingly, herbal medicines from different health philosophies and cultures are being used together by consumers who take advantage of a global market and popularization of herbal remedies.

Consumers have access to herbal medicines via doctors, pharmacists, alternative health practitioners, and directly from retail outlets. A 2007 consumer survey in Australia found that 85% of respondents had used complementary medicines in the past 12 months and a substantial proportion of these were herbal medicines [2]. About half of these people used complementary medicines on the same day as a prescription medicine and more than half had been taking them for more than 5 years. It seems safe to say that complementary medicines, including herbal medicines are a standard part of many consumers' routine medicine regimen and so the principles of QUM should and do apply.

Further, the World Health Organization (WHO) estimates that 70–90% of the rural population in developing countries use herbal medicines to meet or partly meet their health needs. This, together with widespread use in developed countries, has seen herbal medicines recognized as an essential component of primary health care by WHO [3].

While consumers frequently self-initiate and manage herbal medicines, it is also true that practitioners of Western medicines prescribe or recommend herbal remedies. A survey of general practitioners and pharmacists in 2008 found that both groups regularly recommended complementary medicines to their patients. The most commonly recommended herbal medicines were St John's wort, valerian, black cohosh, Echinacea, *gingko biloba*, and traditional Chinese medicines [4].

In 1998, WHO recommended that countries should adopt a regulatory system to manage the appropriate use of herbal medicines to ensure that herbal medicines have acceptable quality, safety, and efficacy [5]. By 2007, 48 countries had a national policy on traditional medicines and over 110 countries had mechanisms in place to regulate traditional medicines [3]. Regulators are increasingly taking a risk-based approach to herbal medicines, requiring safe, uncontaminated products that have consistency in formulation or presentation. Less emphasis is placed on evidence for efficacy in this approach which accepts hundreds of years of traditional use among its considerations [6]. Policy makers now include herbal medicines in their definition of medicines and in some circumstances in the medicines they reimburse as part of government funded or third-party insurance schemes. A survey of Health Maintenance Organizations in 2011 found that of 18 major HMOs and insurance providers in the United States, including Aetna, Medicare, Prudential, and Kaiser Permanente, 14 covered at least 11 of 34 alternative therapies, including some herbal medicines [7]. In Australia, there are no herbal medicines on the Pharmaceutical Benefits Scheme, most likely because of difficulties in presenting a case for cost-effectiveness according to the rigorous guideline requirements. However, insurers such as BUPA and Medibank Private include limited reimbursement for herbal medicines at least at higher levels of coverage.

In Western countries, herbal medicines are often manufactured like other pharmaceuticals in facilities that comply with Good Manufacturing Practice

(GMP). This is not true for most Chinese Medicine, which is prescribed as herbs, teas, and decoctions. Nor is it the case for indigenous or traditional herb use in many parts of Asia, Africa, and South America where whole or parts of plants may be used.

Herbal medicines are pervasive across health systems and closely integrated with other types of care. The QUM framework is a useful way to ensure we use herbal medicines for greatest benefit and least harm.

22.3 USE OF PHYTOTHERAPIES BY CONSUMERS

A national census of medicine use by Australians aged 50 years and older in 2009–2010 reported that complementary medicines were used by 46.3% of participants, representing just over half (53.2%) of all medicines users. In this study, 87.4% of people used both conventional and complementary medicines and women used more than men. The complementary medicines most commonly recorded included fish oil, glucosamine, and herbal medicines and they were used for general health (29.3%), arthritis (20.2%), bone health and disease prevention (6.7%), and joint health and disease prevention (4.7%) [8].

This is consistent with previous surveys that found general health was a common reason for taking Echinacea, *gingko biloba*, and ginseng while people taking St John's wort and black cohosh were more likely to be targeting particular symptoms. Up to 10% of people were unsure why they were taking a particular herbal medicine [2].

The use of complementary medicines is similarly popular in the United Kingdom. A postal survey in 1998 of over 5000 people found 22% had purchased herbal or homeopathic products in the last year with over-the-counter herbal medicines accounting for the vast majority of these [9].

Against this background of common use of herbal medicines for general well-being there is more intensive use for some clinical conditions and symptoms.

A cross-sectional UK study of almost 1500 people with cancer found that 20% used a herbal medicine with users more likely to be affluent, female and younger. Evening Primrose was the most frequently used herb, followed by Echinacea and garlic. Breast cancer patients were more likely to use Agnus castus, Dong quai, red vine leaf, Wild Yam, and Willow. Saw palmetto was used exclusively by men with genital cancers [10].

An earlier European survey found that 36% of cancer patients were either past or present users of complementary medicines. Herbal medicines were the most popularly used group of treatments with use rising from 5.3% prior to cancer diagnosis to 13.9% after diagnosis [11]. A wide range of herbs was used and these tended to vary by country. For example, Turkey most commonly used nettle leaves and thyme, Scotland used more green tea, Switzerland more mistletoe, the Czech republic used Ovosan, selenium, ginseng, *gingko biloba*, and Echinacea, Sweden used more blood salts and ginseng, Serbia and Spain more aloe vera, and Iceland more lupine extracts (angelica).

A systematic review of people with cardiovascular disease found that complementary medicine use was common; users were likely to be using more than one complementary medicine and to be using them in conjunction with conventional medicines [12]. This review included seven studies reporting data on concomitant use of prescription and herbal medicines. On average, cardiovascular patients consumed seven prescription medicines and two herbal, vitamin, or mineral products daily.

Predictors for use of herbal medicines in Western countries include higher educational status, poorer health, a holistic orientation to health, having had a transformational experience, specific health problems (anxiety, back problems, chronic pain, urinary tract problems), and self-identification with certain cultural commitments (environmentalism, feminism, spirituality, person growth). Dissatisfaction with conventional Western medicine is not necessarily a predictor of herbal medicine use [13].

Traditional indigenous herbal medicines are commonly used in many parts of the world alongside or in place of conventional Western medicine. The particular herbs used tend to vary by region and by ethnicity of the population and there can be blurred borders between medicinal herbs and dietary supplements.

22.4 CONSUMER ATTITUDES AND BELIEFS ABOUT HERBAL MEDICINES

There are data to suggest that consumers do not think about herbal medicines in the same way that they regard conventional medicines. Their expectations for both benefit and harms are different as is their estimation of how the medicine fits with their worldview.

22.4.1 Holistic View of Health and Well-Being

A qualitative study of German users of herbal medicines found that people felt a traditional link to herbal medicines which was augmented by the experiences of family and friends. They felt more autonomy using herbal medicines and this correlated with higher health awareness, curiosity, and motivation to try something else. Respondents described their experience of herbal medicines as softer and slower and therefore more sustainable than conventional medicines [14].

A second qualitative study of German seniors found that most people had a life-long experience with medicinal herbs and home remedies due to the unavailability of conventional medicines during a childhood that coincided with postwar poverty. The reasons for using the herbal medicines included dealing with various physical and mental problems, prevention of diseases, and personal control over health. Self-care in the sense of empowerment, self-initiation, and self-responsibility were important factors. Fear of dementia and dependence on medical care were motivators in this study. Additional motivations included enhancing well-being and quality of life and fear of side effects from conventional medicines [15].

TABLE 22.1 Disadvantages of Complementary Medicines As Perceived By Users[a]

Disadvantages	Percentage Respondents (n=612), %
None	29.9
Expensive	18.6
Lack of research/not clinically tested	12.1
Lack of efficacy/placebo/ not sure it will work	10.5
Not enough information/don't know what you are getting	9.8
Take longer to work	7.7
Not prescribed properly	5.2
Problems mixing with other medicines	4.7
Side effects and allergies	2.9
Other, don't know	21.8

[a] From Williamson et al. [2].
Reproduced with permission, NPS Medicinewise 2013.

People who take herbal medicines often want to increase their well-being rather than treat a particular set of symptoms. As such, they are prepared to try something and see if it works for them rather than expect data from clinical trials. They also tend to believe that natural is best and in fact that natural is safe.

The 2007 Australian survey reported differences in attitudes between people who used more than five complementary medicines daily and those who used less. High users were more likely to say that natural medicines were safe for children (57 vs. 37%), not as risky as prescription products (54 vs. 47%), and safe because the ingredients were natural. Only about 40% of people in both groups thought that natural therapies could cause side effects. Furthermore, efficacy and the quality of evidence were seen to be a concern by a minority of consumers [2] (Table 22.1).

Respondents to a European survey of cancer patients said their reasons for using complementary therapies were to increase the body's ability to fight cancer (51%), to improve physical well-being (41%), to counteract ill-effects from the tumor or medical treatments (25%), because it "might help, can't hurt" (23%), and a desire to do something positive to fight the disease (23%) [11].

22.4.2 It Is Natural, So It Must Be Safe

Consumers generally believe that the risk of harm from herbal medicines is low or nonexistent.

An NPS Medicinewise survey [2] of complementary medicine users found that participants reported complementary medicines as having several advantages over conventional therapy. Most of these related to safety, and when asked about disadvantages less than 3% cited safety or side effects specifically, although 4.7% were concerned about possible interactions (Tables 22.1 and 22.2).

TABLE 22.2 Advantages of Complementary Medicines As Perceived By Users[a]

Advantages	Percentage Respondents (n=612), %
Natural	33.0
Fewer chemicals	28.6
Seems to help	24.3
General health and well-being	24.0
Not many side effects/gentle on body	16.8
Alternative to pharmaceuticals	12.7
Safe	7.7
Boosts immune system	7.7
Cheaper	4.6
Complements/supplements prescription medicines	3.8
None, other, don't know	18.5

[a] From Williamson et al. [2].
Reproduced with permission, NPS Medicinewise 2013.

22.5 APPLYING THE QUM FRAMEWORK TO PHYTOTHERAPIES

22.5.1 Judicious Use

22.5.1.1 Is a (Herbal) Medicine the Best Option? As for conventional medicines, herbal medicines are used as a replacement for other interventions that involve changes in lifestyle, such as weight loss, cessation of smoking, reduction in alcohol intake, and more exercise.

The use of herbal medicines and dietary supplements for weight loss is a case study for injudicious use of herbal medicines.

Chitosan has been promoted in various products as an aid to weight loss. Manufacturers claim that chitosan will remove unwanted calories and fats and so you can eat whatever you like provided you titrate the dose of chitosan to mop up the fats. In Australia, the claims for one product ran as: "the undoit pill lets you have the cake without the calories. The snack without the guilt. The taste without the tummy." The pill "binds the fat and carbs from your snack while it's still in your stomach and stops them from being absorbed" [16].

A scientific panel for the Therapeutic Goods Administration reported: "[T]here was no evidence that either ingredient could block all of the fat or carbohydrate content from any given meal (as opposed to blocking some proportion of the fat or carbohydrate content)...[and] there was no clear evidence, other than a simple extrapolation made by the advertiser, to support the specific numeric claims in the advertisement, such as 'each pill undoes 210g of carbs'" [16].

While such promotion of products is deplorable from many perspectives, from a QUM standpoint, the main issue is that consumers are not being supported to take up healthy eating and exercise activity as the preferred means of losing weight. The alternative solution is in fact likely to result in weight gain if the product results in complacency regarding eating habits.

22.5.2 Appropriate Selection

22.5.2.1 Is a Herbal Medicine a Better Option Than a Conventional Medicine? In weighing up the comparative information between any two medicines, whether herb–herb, herb–drug, or drug–drug, it is important to consider the quality of the evidence for safety and efficacy, the preferences of the patient, the affordability of the medicine and how the medicine will fit into the person's overall health plan and medication regimen.

Health professionals sometimes worry that consumers will delay or refuse conventional treatment that has been proven to be effective, even potentially life-saving, and use a herbal medicine instead for which the benefits are less clear.

Steve Jobs, the founder of Apple, was reported in the media to have regretted delaying conventional cancer treatment by 9 months while he explored alternative therapies. Fortunately, Jobs is the exception not the rule.

Surveys suggest that most people with medical conditions use complementary medicines in addition to, rather than as a substitute for, conventional medicines [2].

Breast cancer provides a useful case study for what we know about the impact of treatment refusals and delay, and the effectiveness of herbal medicines as substitute treatment. Conventional treatment for early breast cancer is surgical resection of the tumor, followed by radiation and/or chemotherapy to reduce the risk of recurrence and metastases. Given the potential for excellent treatment outcomes, few women reject conventional treatment in order to substitute complementary medicine as opposed to the many who use herbal medicines as an adjuvant to conventional therapy.

These findings are reassuring because there is evidence that people who preferentially use herbal medicines do less well.

A review of patients who used complementary medicines in preference to conventional medicine found that for 11 patients, who refused surgery, 10 experienced progressive disease and of 10 who refused radiation four developed further disease. Nine patients refused chemotherapy, raising their estimated 10-year mortality from 17 to 25%. Patients who accepted surgery, but refused adjuvant treatments, did better than those who rejected surgery, but even this group had increased 10-year mortality estimates [17].

A subgroup analysis of the Women's Healthy Eating and Living (WHEL) study examined 2562 breast cancer survivors and surveyed for rejection of chemotherapy and use of complementary medicines. Among survivors, 177 declined chemotherapy and 80% of this group reported using complementary medicines. Compared to women that received chemotherapy, women who declined it had a 90% greater risk of an additional breast cancer event, and the risk of death increased by 70%. Complementary medicine use was not statistically correlated with the findings [18].

These and other small studies are not conclusive but they do provide consistent results to suggest that using complementary medicines including herbal medicines in preference to conventional medicine is likely to impact negatively on health [17–19].

Finally, it is reasonable to consider affordability and overall cost when selecting the "best" medicine for an individual. The global market for herbal medicines in 2013 was estimated to be $83 billion and the herbal medicine category has had a high rate of growth over the past 10 years [20, 21].

In Australia, more than four times as much is spent on complementary medicines in out-of-pocket expenses as on pharmaceuticals by consumers [21]. The most common disadvantage of complementary medicines according to Australians is that they are expensive [2].

It is clear that herbal medicines have significant costs associated with their use and most of these costs are borne by consumers as out-of-pocket expenses with a small amount being reimbursed by some insurers. It is important then that herbal medicines are affordable and cost-effective for the individual and the health system and considered to be value for money from both perspectives.

German users of herbal medicines said that they were mostly willing to pay extra for herbal medicines because health is worth the money, although some would only pay for proven therapies. Some people claimed that prices were too high and would like them to be reimbursed by their insurer [14].

Conversely, there is evidence that herbal medicines are relied upon in many communities when conventional medicines and/or access to medical services is too expensive. A US study of people with medical conditions who did not access conventional care found that 25% used some form of alternative medicine. Users of alternative medicine had poorer health and had more barriers to care, including cost in 20% of cases [22].

22.5.3 Safe and Effective Use

The principles for safe and effective use of herbal medicines are consistent with those for all medicines. The likely benefits and risks of treatments need to be well understood, the therapeutic goal should be clear and the progress should be monitored against the treatment goal while always being alert to safety concerns.

Failure to meet the therapeutic goal should result in a change to or cessation of the herbal medicine, while safety concerns should also include reporting the adverse event to the regulator and documenting its occurrence in the patient clinical record.

Safety can be considered in terms of side effects, interactions, allergies, and product safety. Many people use herbal medicines safely but it is useful to know about the types of problems that most commonly arise to ensure that harm is minimized for all patients.

22.5.4 Adverse Reactions

An overview of 50 systematic reviews of 50 herbal medicines found serious side effects noted for four herbal medicines and moderately severe side effects for another 15 [23]. The medicines causing serious side effects were *Belladona*, *Larrea tridentate* (creosote), *Piper methysticum* (kava), and *Cassia senna*. The most common severe adverse effects were liver or kidney damage, colon perforation, carcinoma, coma, and death (Table 22.3).

Hepatotoxicity, one of the more concerning adverse effects of herbal medicines is uncommon but can be catastrophic. The clinical presentation and severity can range from mild hepatitis to fulminant liver failure requiring transplantation [24]. Herbal medicines associated with liver failure have included Ayurvedic and Chinese herbs,

TABLE 22.3 Herbal Medicines Associated with Side Effects in Systemic Reviews[a]

Severe side effects	*Belladona, Larrea tridentate, Piper methysticum, Cassia senna*
Moderately severe side effects	*Pelargonium sidoides, Perna canaliculus,* Aloe vera, *Mentha piperita (peppermint), Medicago sativa, Cimicifuga racemosa* (black cohosh), *Caulophyllum thalictroides, Serenoa repens* (Saw palmetto), *Taraxacum officinale* (dandelion), *Camellia sinensis* (green tea), *Commifora mukul, Hoodia gordonii, Viscum album* (mistletoe), *Trifolium pratense* (Red clover), *Stevia rebaudiana*

[a] From Posadzki et al. [23].

black cohosh, creosote (chaparral), germander, greater celandine, green tea, Herbalife®, Hydroxycut®, kava, pennyroyal, skullcap, pyrrolizidine alkaloids, and usnis acid [24].

High rates of herbal medicine use in clinically vulnerable groups such as those with impaired immune systems, cancer, or in pregnant women led some to question the safety of these treatments. A review of herbal medicine use by cancer patients identified 21 case reports of adverse effects. Some herbal medicines such as laetrile are toxic in their own right, depending on the dose, formulation, and duration of treatment. Others have the potential to interact with chemotherapy although actual cases of harm are not well-documented [25].

A study of Malay women found 34% of women used herbs during pregnancy and 73% during labor. This use was largely unsupervised and products were directly supplied by traditional midwives. The authors note that this use of herbal medicines is likely to be safe but given the potential for harm to the baby and mother during this period more information and research is needed [26].

Likewise, people with cardiovascular disease may be more prone to side effects from some herbal medicines than other people. For example, licorice is known to have hypertensive and hypokalemic effects, aconite can cause cardiac arrythmias and yohimbine is contraindicated in people with hypertension, angina, and renal impairment [27].

22.5.5 Interactions

One of the most relevant safety concerns with herbal medicines is their potential to interact with conventional Western medicine. A 2008 survey of 1818 patients found 107 interactions with potential clinical significance. The five most common natural products with a potential for interaction were garlic, valerian, kava, ginkgo, and St John's wort, which accounted for 68% of the potential clinically significant interactions. The four most common classes of prescription medications with a potential for interaction were antithrombotic medications, sedatives, antidepressant agents, and antidiabetic agents, which accounted for 94% of the potential clinically significant interactions [28].

A literature review reported herb–drug interactions with clinical significance, from case reports and clinical observations. Enhanced anticoagulation and bleeding was reported for warfarinized patients who also took danshen, sativumor, or *gingko biloba*. Ginseng was reported to reduce blood concentrations of alcohol and warfarin, and induce mania when used concomitantly with phenelzine. *Piper methysticum* (kava) apparently increased the "off" periods in patients with Parkinson's disease taking levodopa and induced a semicomatose state when given concomitantly with alprazolam. St John's wort was associated with decreased blood concentrations of cyclosporin, midazolam, tacrolimus, amitriptyline, digoxin, indinavir, warfarin, phenprocoumon, and theophylline. It also was linked with breakthrough bleeding and unplanned pregnancies when used concomitantly with oral contraceptives and serotonin syndrome when used in combination with selective serotonin reuptake inhibitors [29].

22.5.6 Allergy

Like any biological substance, herbal medicines can cause allergy in some individuals. Royal jelly is a well-known example but many others have been reported.

As for any allergic reaction, it is important that it is documented in the patient's medical record and that the patient understands the extent and nature of their allergic reaction. Similarly, a complete list of a patient's medicines, including any herbal medicines is important during any health event especially when the diagnosis is problematic so that unexpected allergies or side effects can be ruled out.

22.5.7 Safe Formulation

Accidental or intentional contamination of herbal material with conventional drugs, toxic substances, or microorganisms has been reported. Examples include reports of Chinese herbal creams containing corticosteroids and Ayurvedic remedies and herbal teas containing heavy metals. There are also reasonably frequent reports of herbal medicines containing ingredients not declared on the label or substituted herbs [20, 30].

Complementary medicines are often less powerful than prescription medicines—but they still need to be used with care. Like all medicines, they can have benefits. However, they can also have side effects, cause allergic reactions, and may interact with prescription medicines. Risks of harm should be weighed against the likelihood of benefit in all cases. Careful clinical monitoring should be undertaken in more vulnerable people and where the potential for interactions is more likely.

22.5.8 Effectiveness

Most of the evidence for the effectiveness of herbal medicines stems from traditional use accumulated over hundreds of years, which is documented in key monographs. Randomized control trials have also been conducted for some herbs but the majority vary in quality, lack methodological rigor, are often of short duration and have small sample sizes.

The main question for QUM with respect to effective use is: Has the consumer achieved the health or well-being effects they intended? It is important therefore at the outset of therapy to be clear about the reasons for using a herbal medicine—be it for general well-being or to treat a specific condition or set of symptoms.

Most consumers are aware that herbal medicines come from a long tradition of use and are less worried that they have not been scientifically tested than they would be for prescription medicines. Nevertheless, there are Cochrane reviews and research that demonstrate the potential effectiveness of some herbs. For example, cranberry tablets for the prevention of recurrent urinary tract infections in young women and St John's wort for depression.

In Western countries, consumers do expect herbal medicines to be safe and of good quality and many assume more intervention by the government regulator than actually occurs [2, 27]. Researchers have also questioned the adequacy of safety information provided with herbal medicines to allow safe and effective use [31].

Additional concerns have been raised that herbal medicine can lead to nonadherence with conventional medicine and so less effective treatment overall.

However, most studies do not support this view particularly where people have chosen to use both types of medicines. For example, a study in older people on multiple medicines found those taking over-the-counter medicines, including herbal medicines, were more likely to be adherent [32] and a study of people with HIV found no correlation between use of herbal medicines and adherence to the overall therapeutic regimens [33].

Conversely, a study of inner-city asthmatic patients in the United States did find an association between use of herbal medicines and lower adherence with inhaled corticosteroids after adjusting for confounders [34]. It could be speculated that in some situations medical beliefs, such as concern about the side effects of corticosteroids, can result in a preference for herbal medicines.

The expectations of consumers, their preferences and beliefs are important considerations in assessing the safe and effective use of herbal medicines for an individual. Essentially, people assume safety and may be prepared to take their chances with efficacy. The latter is acceptable on the proviso of full disclosure and this assumes the consumer is aware that the herbal product may not have rigorous evidence of efficacy, that promotional claims are accurate, and that purchase of the product does not represent an unreasonable financial burden. In the final analysis, the consumer should be achieving more benefits than harms from the medicine and ongoing monitoring should take into account both the original rationale for treatment and the consumer's perception of what is most important.

22.6 BUILDING BLOCKS FOR QUALITY USE OF HERBAL MEDICINES

Quality use of herbal products is dependent on the availability of objective information, consumer and health professional knowledge, information sharing to ensure safety, and supportive systems for shared decision making.

22.6.1 Objective Information and Ethical Promotion

22.6.1.1 Health Literacy As with all other medicines, poor health literacy can impact on safe use of herbal medicines. Although people often self-manage their herbal medication, there is evidence that significant knowledge gaps still exist.

A study of complementary use among cancer patients found that the majority believed herbal supplements would increase the efficacy of standard anticancer treatment and prolong survival. About half advised their doctor of their complementary medicine use and the most common sources of information were the Internet (36%), and books and brochures (25%). Most trusted the information they accessed and 73% would have liked more information, for example, in a specialized consultation (60%), written information (44%), or the Internet (20%) [35].

A Malaysian study of pregnant women attempted to assess their knowledge of herbal medicines. Of 460 participants, 90% were in the low herb knowledge group and among the women with a high knowledge score, 42.6% had used herbs during pregnancy. Only 8.5% said they knew the ingredient(s) in their medication and none could list them. These women were also generally unaware of quality or contamination problems with herbal medicines such as presence of heavy metals [26].

22.6.1.2 Information Sources for Consumers A UK study reported that the majority of herbal medicine users with cancer obtained their medicine from high street stores and supermarkets. Common sources of information included the recommendation of a health professional, the Internet, or a mail order vendor [10]. A similar European survey reported that the most common sources of information were friends (56.5%), family (29%), doctor (18.6%), complementary medicine practitioner (12.9%), and the Internet (9.3%) [25].

Thai outpatients with chronic kidney disease reported that family and friends were an important source of information (52%) about herbal medicines as well as being an important source of the medicines themselves (27%). Most herbal products were bought from pharmacies, herbal or dietary supplement shops (41%), or by direct sales as advertised on television and radio (30%) [36].

German users of herbal medicines reported obtaining information from across all media, health care professionals, family and friends, advertising, and patient information leaflets [14].

22.6.1.3 Promotion and Advertising of Herbal Medicines Herbal medicines are widely promoted through advertising, testimonials, at the point of sale (e.g., in pharmacy) and by conventional, integrative, and alternative health professionals.

Unethical marketing, including false claims regarding safety and efficacy is more common with herbal medicines than for prescription medicines with the latter being much more heavily regulated. One US study found that more than 80% of websites for herbal products made one or more health claim, 50% claimed to treat, prevent, diagnose, or cure a condition despite regulations barring such statements [37].

22.6.1.4 Objective Information Sources NPS MedicineWise commissioned a review of the quality of complementary medicines information resources in 2009 [38].

The review identified 52 information products from reputable sources and rated each according to accessibility, currency of information, content quality, transparency of information sources, coverage of herbal medicines, and accuracy. Nine resources were identified as providing high-quality objective information, although many of

TABLE 22.4 High-Quality Objective Information On Herbal Medicines[a]

Natural Standard Professional Database package. 2008
Natural Medicines Comprehensive Database (Health Professional Edition). 2008
Natural Standard Professional Database—Professional monographs. 2008
Herbal Medicines and Dietary Supplements package through MedicinesComplete. 2008
Natural Standard Professional Database—Bottom line monographs. 2008
MedlinePlus: Drugs, Supplements and Herbal Information. 2008
Herbal Medicines by Barnes et al. 3rd Edition. 2007
Natural and Alternative Treatments. EBSCO. 2008
Herbs and Natural Supplements. An evidence-based guide by Braun and Cohen. 2nd Edition. 2007

[a] From McGuire et al. [38].

these may only be accessible to specialist practitioners or health professionals (Table 22.4).

22.6.2 Education and Training

22.6.2.1 Knowledge of Health Professionals

The following principles have been recommended as a way for prescribers to minimize adverse events related to herb–drug interactions [39]:

- Take care with drugs that have a narrow therapeutic window, such as anticoagulants and anticonvulsants.
- Be aware of which drugs are enzyme inducers or inhibitors.
- Consider drug pharmacology to avoid pharmacodynamic interactions with drugs, herbal medicines, and food supplements.
- Carefully weigh up risks and benefits of adding each additional medicine especially for patients already on multiple medicines.
- Be especially careful in people who are more vulnerable to adverse events such as older people and people with multiple coexisting conditions.

Surveys and qualitative interviews of general practitioners and pharmacists in Australia found significant knowledge gaps regarding herbal medicines [4]. Only 38% of general practitioners and 43% of pharmacists felt confident when discussing herbal medicines with a patient. In terms of specific knowledge: 40% of general practitioners reported minimal or no knowledge about black cohosh and *ginkgo biloba*; less than 40% were aware of possible interactions or side effects from these herbs, and 38% had heard of potential for liver toxicity from black cohosh. Results for pharmacists showed slightly higher awareness although a third had minimal or no knowledge about black cohosh and only 44% knew of its potential for liver toxicity.

Doctors and pharmacists identified a need for additional education and training on herbal medicines both during their undergraduate training and as part of continuing professional development [4].

22.6.3 Systems and Interventions

22.6.3.1 Adverse Drug Reaction Reporting The manufacturers of herbal medicines rarely report adverse reactions to the national regulator. For example, according to one commentator, of the 2500 reports of herbal medicine adverse events made to the FDA between 1990 and 1994, fewer than 10 were made by manufacturers of the product. A 2008 report from the US Health Department revealed that less than 1% adverse reactions caused by dietary supplements were reported to the FDA [27].

Better systems are required internationally to support consumer reporting of potential adverse events and to increase the awareness of sponsors and manufacturers about their responsibility to report adverse events.

22.6.4 Shared Decision Making

Patients want doctors and pharmacists to be involved in how they use herbal medicines provided the health professional is nonjudgmental. A qualitative study among senior Germans found that patients wanted more empathetic physicians who were better trained in herbal medicine use and respectful of the patient's life experiences. Participants in this study also noted the need to build their own confidence to discuss herbal medicine options more openly with their doctor [2, 15].

A second German study highlighted the range of perceptions that consumers have regarding their doctor's attitude to herbal medicines: from unbelieving, dogmatic, unconcern, to sincerity and a positive conviction [14].

The patient should be alerted to possible side effects or interactions (herb–drug or herb–condition) and given information on the safe use of the herb. Formulation, brand, dose, and dosage regimen should be documented in the patient's clinical record, together with any history of allergy or adverse events associated with herbs. Patients on medicines with a narrow therapeutic index such as warfarin, immunosupressants, and epilepsy medicines should be advised of specific herbal medicines to avoid or use with care. For example, patients on warfarin should be advised not to use *gingko biloba*, dong quai, garlic, papaya, or danshen, or to do so only under medical supervision [37].

Health professionals often do not know what their patients are taking and sometimes it is because they fail to ask about herbal medicines as part of the consultation. Studies in cancer patients reported that only 50% of patients undergoing chemotherapy or radiotherapy advise their doctor or oncologist that they were also taking herbal medicines [35, 40]. A systematic review of complementary medicine use in people with cardiovascular disease reported that physician awareness of their patients' use of complementary medicines was lowest in a Turkish study at 8% and ranged from 39 to 65% in five other studies. Reasons for nondisclosure included fear that the doctor would disapprove and that the doctor had not asked about it [12]. This compares with more than 90% of Ugandan patients with HIV who used herbal medicine not telling their doctor about their herbal medicines because most did not think the doctor would find it relevant (68.5%) [33]. Only 28% of Thai people with chronic kidney failure informed their doctor that they were using herbal medicines and the main reason was that the doctor did not ask them [36].

It may be that indigenous herbal medicine use is thought of differently to Western herbal medicine use and this could explain differences in disclosure to health professionals. Cultural differences in how patients and doctors interact together with the quality of the conversation between the health professional and his or her patient are also likely to be important.

22.7 CONCLUSION

Herbal medicines should be thought of as falling within the QUM paradigm especially when being used alongside conventional Western medicines.

Judicious, safe, effective, and affordable medicines use is as critical for herbal medicines as for other prescription and over-the-counter medicines. There are significant gaps in the data we have to guide effective use of herbal medicines and to a lesser extent safe use. In addition, the ready availability of high-quality, objective information is a problem for both health professionals and consumers. Health professionals would like to have more knowledge to better care for their patients, and consumers would appreciate greater positive engagement from their health professionals regarding their herbal medicine use.

REFERENCES

[1] Pharmaceutical Health and Rational use of Medicines Committee (2002) *The national strategy for quality use of medicines*. Plain English edition. Commonwealth of Australia, Canberra.

[2] Williamson M, Tudball J, Toms M, Garden F, Grunseit A (2008) *Information use and needs of complementary medicine users*. National Prescribing Service, Sydney.

[3] World Health Organization medicines strategy December 2008–2013. http://www.who.int/medicines/publications/Medicines_Strategy_draft08-13.pdf. Accessed November 15, 2014.

[4] Brown J, Morgan T, Adams J, et al. (2008) *Complementary medicines information use and needs of health professionals: general practitioners and pharmacists*. National Prescribing Service, Sydney.

[5] World Health Organization (1998) *Guidelines for the appropriate use of herbal medicines*. WHO Regional Publications, Western Pacific Series No. 23. WHO Regional Office for Western Pacific, Manila.

[6] Saraf A, Saraf S (2012) Legal regulations of complementary and alternative medicines in different countries. *Pharmacogn Rev* 6: 154–160.

[7] Wong C (2012) 12 Common questions about health insurance and complementary alternative medicine. http://altmedicine.about.com/od/alternativemedicinebasics/a/Insurance.htm. Accessed November 15, 2014.

[8] Morgan TK, Williamson M, Pirotta M, et al. (2012) A national census of medicines use: a 24-hour snapshot of Australians aged 50 years and older. *Med J Aust* 196: 50–53.

[9] Thomas KJ, Nicholl JP, Coleman P (2001) Use and expenditure on complementary medicines in England. A population based survey. *Complement Ther Med* 9: 2–11.

[10] Damery S, Gratus C, Grieve R, et al. (2011) The use of herbal medicines by people with cancer: a corss-sectional survey. *Br J Cancer* 104: 927–933.

[11] Molassiotis A, Fernandez-Ortega P, Pud D, et al. (2005) Use of complementary and alternative medicine in cancer patients: a European survey. *Ann Oncol* 16: 655–663.

[12] Grant SJ, Bin YS, Kiat H, Chang DHT (2012) The use of complementary and alternative medicine by people with cardiovascular disease: a systematic review. *BMC Public Health* 12: 299.

[13] Astin JA (1998) Why patients use alternative medicines. Results of a national study. *JAMA* 279: 1548–1553.

[14] Joos S, Glassen K, Musselmann B (2012) Herbal medicine in primary healthcare in Germany: the patient's perspective. *Evid Based Complement Alternat Med* 2012: 294638.

[15] Stockigt B, Witt CM (2013) CAM use and suggestions for medical care of senior citizens: a qualitative study using World café method. *Evid Based Complement Alternat Med* 2013: 951245.

[16] Norrie J (2012) Claim that undoit pill blocks all fat and carbs is baseless. *The Conversation.* 3 July 2012. https://theconversation.com/claim-that-undoit-pill-blocks-all-fat-and-carbs-is-baseless-8052. Accessed November 15, 2014.

[17] Chang EY, Glissmeyer M, Tonnes S, Hudson T, Johnson N (2006) Outcomes of breast cancer in patients who use alternative therapies as primary treatment. *Am J Surg* 192: 471–473.

[18] Saquib J, Parker BA, Natarajan L, et al. (2012) Prognosis following the use of complementary and alternative medicine in women diagnosed with breast cancer. *Complement Ther Med* 20: 283–290.

[19] Han E, Johnson N, DelaMelena T, Glissmeyer M, Steinbock K (2011) Alternative therapy used as primary treatment for breast cancer negatively impacts outcomes. *Ann Surg Oncol* 18: 912–916.

[20] Barnes J (2003) Quality, efficacy and safety of complementary medicines: fasions, facts and the future. Part 1 regulation and quality. *Br J Clin Pharmacol* 55: 226–233.

[21] National Institute of Complementary Medicines (October 2013) http://www.nicm.edu.au/understanding-cm/facts-and-statistics. Accessed November 15, 2014.

[22] Nahin RL, Dahlhamer JM, Stussman BJ (2010) Health need and the use of alternative medicine among adults who do not use conventional medicine. *BMC Health Serv Res* 10: 220.

[23] Posadzki P, Watson LK, Ernst E (2013) Adverse effects of herbal medicines: an overview of systematic reviews. *Clin Med* 13: 7–12.

[24] Bunchorntavakul C, Reddy KR (2013) Review article: herbal and dietary supplement hepatotoxicity. *Aliment Pharmacol Ther* 37: 3–17.

[25] Olaku O, White JD (2011) Herbal therapy use by cancer patients. A literature review on case reports. *Eur J Cancer* 47: 508–514.

[26] Sooi LK, Keng SL (2013) Herbal medicines: Malaysian women's knowledge and practice. *Evid Based Complement Alternat Med* 2013: 438139.

[27] Tachijan A, Maria V, Jahangir A (2010) Use of herbal products and potential interactions in patients with cardiovascular diseases. *J Am Coll Cardiol* 55: 515–525.

[28] Sood A, Sood R, Brinker FJ, et al. (2008) Potential for interactions between dietary supplements and prescription medicines. *Am J Med* 121: 207–211.

[29] Hu Z, Yang X, Ho PC, et al. (2005) Herb-drug interactions: a literature review. *Drugs* 65: 1239–1283.

[30] Newmaster SG, GrGuric M, Shanmughanandhan D, et al. (2013) DNA barcoding detects contamination and substitution in North American herbal products. *BMC Med* 11: 222.

[31] Raynor DK, Dickinson R, Knapp P, et al. (2011) Buyer beware? Does the information provided with herbal products available over the counter enable safe use? *BMC Med* 9: 94.

[32] Olesen C, Harbig P, Barat I, Damsgaard EM (2014) Correlation between the use of "over-the counter" medicines and adherence in elderly patients on multiple medications. *Int J Clin Pharm* 36: 92–97.

[33] Lubinga SJ, Kintu A, Atuhaire J, Asiimwe S (2012) Concomitant herbal medicine and antiretroviral therapy (ART) use among HIV patients in Western Uganda: a cross sectional analysis of magnitude and patterns of use, associated factors and impact on ART adherence. *AIDS Care* 24: 1375–1383.

[34] Roy A, Lurslurchachai L, Halm EA, et al. (2010) Use of herbal remedies and adherence to inhaled corticosteroids among inner-city asthmatics. *Ann Allergy Asthma Immunol* 104: 132–138.

[35] Pihlak R, Liivand R, Trelin O, et al. (2014) Complementary medicine use among cancer patients receiving radiotherapy and chemotherapy: methods, sources of information and the need for counselling. *Eur J Cancer Care* 23: 249–254.

[36] Tangkiatkumjai M, Boardman H, Praditpornsilpa K, Walker DM (2013) Prevalence of herbal and dietary supplement usage in Thai outpatients with chronic kidney disease: a cross-sectional survey. *BMC Complement Altern Med* 13: 153.

[37] Hussain S (2011) Patient counselling about herb-drug interactions. *Afr J Tradit Complement Altern Med* 8: 152–163.

[38] McGuire TM, Walters JA, Dean AJ, et al. (2009, March) *Review of the quality of complementary medicines information resources: summary report.* National Prescribing Service, Sydney.

[39] Morris CA, Avorn J (2003) Internet marketing of herbal products. *JAMA* 2003 290: 1505–1509.

[40] Boon H, Stewart M, Kennard MA, et al. (2000) Use of complementary/alternative medicine by breast cancer survivors in Ontario: prevalence and perceptions. *J Clin Oncol* 18: 2515–2521.

23

INTELLECTUAL PROPERTY AND PATENT ISSUES WITH PHYTOTHERAPY PRODUCTS

GINT SILINS[1], JENNIFER TAN[2], AND KELVIN CHAN[3,4]

[1] *Cullens Patent & Trade Mark Attorneys, Brisbane, Queensland, Australia*
[2] *E-TQCM Consultants Limited, Tsuen Wan, Hong Kong*
[3] *Faculty of Pharmacy, The University of Sydney, Sydney, New South Wales, Australia*
[4] *National Institute for Complementary Medicine, University of Western Sydney, Sydney, New South Wales, Australia*

23.1 INTRODUCTION

23.1.1 Historical and Current Aspects of Intellectual Property

Phytotherapy, the medical use of materials from medicinal plants (herbal *materia medica*, HMM), is an important part of traditional medicine that has been practised in developing regions over centuries as a key source of health care, even in the present advanced era of science and technology. The world market for herbal medicines has been estimated at US$ 60 billion in 2003 with an annual growth rate of between 5 and 15% [1]. For traditional Chinese medicines alone, figures are estimated as having exceeded €103.5 billion in 2010; and in 2015 the world market for Chinese medicine will rise to €138.5 billion, and optimistically to €175.9 billion in 2020 and €220.7 billion in 2025 [2]. Researchers and commercial industries who are working toward research and development (R&D) of HHM may attempt to claim intellectual property (IP) rights over the HMM sources and/or traditional knowledge [3].

Historically, IP can be considered as a legal concept, which is regarded as the creation of the mind that should be recognized with exclusive rights as protection [4].

Phytotherapies: Efficacy, Safety, and Regulation, First Edition. Edited by Iqbal Ramzan.

The legal principles governing *intellectual property* rights have evolved over centuries. It was not until the nineteenth century that the term *intellectual property* began to be used, and not until the late twentieth century that it became commonplace in the majority of the world. The British Statute of Anne (1710) and the Statute of Monopolies (1623) are now considered as the origins of copyright and patent law, respectively [5].

The modern use of IP can be traced as far back as 1867 with the founding of the North German Confederation with constitutional details granted for its legislative power over the protection of IP [6]. The administrative secretariats, after being subsequently established by the Paris Convention (1883), then merged with the Berne Convention (1886) in 1893 and located in Berne. The secretariat adopted the term IP in their new combined title to form the United International Bureaux for the Protection of Intellectual Property. The IP secretariat subsequently moved to Geneva in 1960, and was succeeded in 1967 with the establishment of the World Intellectual Property Organization (WIPO) by treaty as an agency of the United Nations. According to Lemley, it was only at this point that the term IP really began to be used in the United States, and it did not enter popular usage until passage of the Bayh–Dole Act in 1980 [7].

23.1.2 Types of Intellectual Property Rights

"Intellectual property" (IP) broadly refers to industrial, scientific, literary, and artistic endeavors for which one has, or one perceives as having, legal rights [8]. For particular types of endeavors, IP rights are available automatically upon creation, whereas for other types of endeavors IP rights are not automatic but first require registration. Certain types of endeavors can enjoy IP protection both automatically and upon registration. IP rights are important in that they enable a person to establish or maintain legal protection for an endeavor such as an innovation/invention, product design, or artistic or literary work, thereby providing that person with a competitive advantage.

Examples of registrable or nonregistrable automatic IP rights include the following:

- **Patents** for protecting innovations/inventions such as products and processes
- **Trademark registrations** for protecting badges or signs of origin or quality used in connection with the promotion or sale of products or services
- **Industrial designs** for protecting product appearances
- **Copyright** for protecting literary and artistic works
- **Trade secrets** for protecting proprietary information, discoveries, products, and processes
- **Plant variety rights** for protecting new plant varieties

IP rights can notionally also include:

- **Regulatory exclusivity** for therapeutic products, medical devices, and veterinary products
- **Restricted third-party access** to commercially sensitive information/data and biological materials deposited with regulatory bodies

23.1.3 Worldwide IP Laws Have Yet to Be Harmonized

IP laws in industrialized jurisdictions tend to be more developed as opposed to in less industrialized jurisdictions. Historically, IP laws have largely evolved in accordance with changing needs and attitudes of the nation, even to the point where otherwise closely allied nations have significant differences in IP laws.

Although jurisdictions of the world have yet to harmonize their IP laws, steps in that direction have been taken in that many jurisdictions are signatories to international agreements, declarations, treaties, and conventions relating to, for example, patents, trademarks, industrial designs, copyright, and plant variety rights [9].

For example, most industrialized jurisdictions are signatories to the TRIPS Agreement (159 signatories), which to date is the most comprehensive multilateral agreement on intellectual property law [10]. Although the TRIPS Agreement specifies general requirements to be met by all member jurisdictions, the Agreement also provides a degree of flexibility such that different jurisdictions can adopt their preferred practice.

23.2 IP RIGHTS—PHYTO-INDUSTRY

In one of the author's experience, a widely held notion within the phyto-industry is that "complementary medicines" (however you wish to define this term) are simply not patentable and so, in so far as it is possible, one must rely on trade secrets for IP protection or accept that no IP protection is possible. Consequently, stemming from this is the notion that IP rights are nonexistent or are quite limited in nature for:

1. Known (including traditional) phytotherapy products and phytotherapies
2. Innovations based on known (including traditional) phytotherapy products and phytotherapies
3. New innovative phytotherapy products and phytotherapies

These notions have probably translated into decreased research interest in, and commercialization of, phytotherapy products and phytotherapies, particularly those that are already known to the public. These notions may have also resulted in poor IP protection by those commercializing known or new phytotherapy products and phytotherapies, thereby providing less of a barrier for competitors to enter the marketplace with competing products and therapies.

23.2.1 IP Protection for Phytotherapy Products and Phytotherapies

With regard to known (including traditional) phytotherapy products and phytotherapies, the notion that patent protection is not available is correct, and trade secrets and trademarks (branding) play more of a role in maintaining a market advantage.

In so far as it is possible, trade secrets may be used in connection with: how the phytotherapy products are formulated and prepared; from where active phyto-ingredients/extracts are sourced; and how phytotherapies are carried out by practitioners. Trademarks will usually play an important role in brand awareness and consumer acceptance. Unique product packaging protected as a trademark or by way of a design registration may also play an important role (trademarks and registered designs will be addressed in greater detail elsewhere in this chapter).

Regarding innovations based on known (including traditional) phytotherapy products and phytotherapies, the notion that patent protection is not available may be incorrect. As will be discussed elsewhere in this chapter, patents may play more of a role than trade secrets in maintaining a market advantage in respect of innovations based on known phytotherapy products and phytotherapies. If registering a new improved phytotherapy product or new phytotherapeutic indication with a regulatory authority for marketing approval, then regulatory exclusivity may also provide IP protection, as will be discussed elsewhere in this chapter. Again, trademarks and design registration may play an important role.

With regard to entirely new innovative phytotherapy products and phytotherapies, the notion that patent protection is not available may also be incorrect. Again, patents should play more of a role than trade secrets in maintaining a market advantage. If registering a new phytotherapy product or new phytotherapeutic indication with a regulatory authority for marketing approval, then regulatory exclusivity may also provide IP protection. In addition, trademarks and design registration may play an important role.

In summary, patents, trade secrets, and regulatory exclusivity will usually be the most important IP rights in respect of phytotherapy products and phytotherapies, and registered trademarks as well as possible design registrations may play a role in the later stages of commercialization.

23.2.2 Patents

Generally speaking, filing of a patent application describing and claiming an invention is usually followed by examination of the application and, provided all legal requirements are met, a patent will be granted (registered) based on the invention as defined in the patent claim or claims [11].

A granted patent provides a patent owner with exclusive rights to a patented invention (as claimed) for a set period of time. Although the set period is usually 20 years, the period will depend on the type of patent granted and whether the patent is eligible for an extension of patent term (due to regulatory delays in obtaining marketing approval for a claimed therapeutic, for example). Twenty-year-term patents are most common, but some jurisdictions also have utility model and innovation patent regimes that provide a shorter term and require a lower threshold of inventiveness—hence the reason for a shorter patent term [12].

23.2.3 Patents as IP Assets

A patent, being personal property, can be instrumental to the successful commercialization of an invention and for maximizing financial returns. For example, a patent can be pivotal in establishing: a research and development collaboration; a joint venture, partnership, or alliance; or, a licensing, supply, or comarketing agreement. Moreover, a patent may entitle the owner to apply for government funding, and research and development tax concessions [13].

Patents are usually must-have IP assets for the pharmaceutical and biotechnology industries [14].

23.2.4 Patents for Protecting Phyto-Inventions

Generally speaking, patents can be used to protect phyto-inventions that are new (novel), involve an inventive step (i.e., are not obvious), and are capable of industrial application. Typically, patentable phyto-inventions include: phytotherapeutic agents, compounds, and compositions, including isolated plant bioactives, bioactive plant extracts/preparations, and pharmaceutical compositions; processes for preparing the phytotherapeutic agents, compounds, and compositions, including isolation and purification procedures, and recombinant and modification procedures; and, uses for those agents, compounds, and compositions, including for diagnosis and therapy. Some jurisdictions (such as Australia) allow novel and nonobvious plant varieties to be patented as well.

In the case of known phytotherapy products and phytotherapies, these are not normally patentable, although new, nonobvious improvements in connection with known phytotherapy products and phytotherapies may be patentable. Examples of patentable phyto-inventions include:

- New, nonobvious therapeutic uses (indications) for known phytotherapy products.
- New dosage regimes utilizing known phytotherapy products for known phytotherapeutic uses, provided that the dosage regimes provide an unexpected benefit.
- New ways of preparing known phytotherapeutic agents, compounds, and compositions, provided that there is an unexpected benefit.
- Improvements to known phytotherapeutic agents, compounds, and compositions provided that there is an unexpected benefit. The improvement could be to the phytotherapeutic agent or compound itself, or to a way of delivering the agent or compound for greater efficacy in therapy.

As will be appreciated, each of these patentable phyto-invention examples could provide an important commercial advantage.

23.2.5 Exclusions to Patentability

Currently, therapeutic products and therapeutic methods, including phytobased products and therapies, are not patentable in some jurisdictions. Most industrialized jurisdictions of the world are signatories to the TRIPS Agreement [15]. With regard to patents, but for three possible exceptions, the TRIPS Agreement requires member nations to make patents available for product and process inventions in all fields of technology.

TRIPS jurisdictions may exclude from patentability: (i) inventions deemed contrary to *ordre public* or morality, including to protect human, animal, or plant life or health, or to avoid serious prejudice to the environment provided that such exclusion is not merely because the exploitation is prohibited by their law; (ii) diagnostic, therapeutic, and surgical methods for the treatment of humans or animals; and (iii) plants and animals other than microorganisms and essentially biological processes for the production of plants or animals other than nonbiological and microbiological processes, provided that protection of plant varieties is made available by other means.

At the time of writing, most TRIPS jurisdictions will allow claims to phytotherapy products, but far fewer jurisdictions will allow claims to therapeutic uses [16]. Australia and the United States of America, for example, allow claims to therapeutic uses. For those jurisdictions that will not allow claims to therapeutic uses, it may yet be possible to claim the phytotherapies in a different way, focusing on the phytotherapy product or manufacture of the product for its intended use, as opposed to the use of the product per se, thereby providing a measure of protection nonetheless.

For example, in the case of a European patent application, a therapeutic use can be redefined as a "for use" claim, that is, "Phytotherapy product X for use in the treatment of disease Y." For example, in the case of a Canadian, Chinese, Israeli, Japanese, or New Zealand patent application, a therapeutic use can be redefined in the form of a "Swiss-style use" claim, that is, "Use of phytotherapy product X in the manufacture of a medicament in the treatment of disease Y." Other definitions may also be acceptable in some jurisdictions.

23.3 BRIEF OVERVIEW OF PATENTS AND THE PATENTING PROCESS

23.3.1 Patent Searching

Prior to commencing research, development, and commercialization of a phyto-invention, it would be prudent to:

1. Conduct a freedom to operate search of live patents and patent applications in each jurisdiction of interest to see whether any third-party patent or application could pose a barrier to research, development, and commercialization of the phyto-invention
2. Conduct a patentability search of worldwide patent and nonpatent literature to determine whether a phyto-invention is likely to be novel and inventive, and to determine the scope of patent protection that may be obtainable

23.3.2 Patent Ownership

Prior to filing a patent application, each inventor of the phyto-invention will need to be identified and each patent applicant will need to have clear legal entitlement to the phyto-invention. If a patent is not granted to the lawful owner and/or inventorship has not been correctly established, then in some jurisdictions the patent could be deemed invalid.

23.3.3 Patent Filing

A (nonprovisional) patent application will include a patent specification describing and claiming an invention. Generally speaking, the specification should describe the invention fully, including each commercial embodiment of the invention, to the extent that the invention as claimed can be put into practice by a skilled person. This does not necessarily mean that the actual mechanism of action leading to the desired result needs to be described or even understood for that matter, provided that the desired result can be reproduced by the skilled person without undue experimentation. The claim or claims of the specification will include one or more definitions of the invention.

23.3.4 Examination and Classification

In some jurisdictions, the patent application will be examined by a patent office for compliance with laws of that jurisdiction. Provided that all legal requirements are met with regard to novelty, inventiveness, patentable subject matter, and so forth, the application will be allowed. For those applications claiming nonallowable subject matter, such as therapies, these types of claims may need to be recast in an allowable format, as mentioned elsewhere in this chapter.

The patent offices of some jurisdictions will classify the invention of a patent application to assist with searching and examination. Many patent offices classify inventions in accordance with the International Patent Classification (IPC) system [17].

For example, IPC A61K36/00 relates to "*Medicinal preparations of undetermined constitution containing material from algae, lichens, fungi or plants, or derivatives thereof, e.g., traditional herbal medicines.*" Various subclasses of IPC A61K36/00 relate only to plants and to particular plant types, but the full listing of subclasses is shown in Table 23.1 [18].

One or more of these A61K36 subclasses may be assigned to patent applications describing phytotherapy products. However, entirely different IPCs may also be assigned to an application if the invention further relates to newly characterized phytotherapeutic products, phytotherapies, or processes for preparing phytotherapeutic products.

23.3.5 Allowance and Grant

Upon allowance, the application may be published for third-party opposition purposes, although publication for opposition purposes could occur at an earlier or later stage. In the absence of a third-party opposition, a patent should be granted for the invention as claimed.

23.3.6 Extension of Patent Term

In some jurisdictions, the term of a patent may be extendable due to regulatory delays in obtaining marketing approval for a claimed phyto-invention (normally for a phytotherapeutic product, but it could be for a new therapeutic indication or method of

TABLE 23.1 International Patent Classifications Assigned To Herbal *Materia Medica*

A61K 36/00	Medicinal preparations of undetermined constitution containing material from algae, lichens, fungi, or plants, or derivatives thereof, e.g., traditional herbal medicines [8]
A61K 36/02	Algae
A61K 36/03	Phaeophycota or phaeophyta (brown algae), e.g., Fucus
A61K 36/04	Rhodophycota or rhodophyta (red algae), e.g., Porphyra
A61K 36/05	Chlorophycota or chlorophyta (green algae), e.g., Chlorella
A61K 36/06	Fungi, e.g., yeasts
A61K 36/062	Ascomycota
A61K 36/064	Saccharomycetales, e.g., baker's yeast
A61K 36/066	Clavicipitaceae
A61K 36/068	Cordyceps
A61K 36/07	Basidiomycota, e.g., Cryptococcus
A61K 36/074	Ganoderma
A61K 36/076	Poria
A61K 36/09	Lichens
A61K 36/10	Bryophyta (mosses)
A61K 36/11	Pteridophyta or Filicophyta (ferns)
A61K 36/12	Filicopsida or Pteridopsida
A61K 36/126	Drynaria
A61K 36/13	Coniferophyta (gymnosperms)
A61K 36/14	Cupressaceae (Cypress family), e.g., juniper or cypress
A61K 36/15	Pinaceae (Pine family), e.g., pine or cedar
A61K 36/16	Ginkgophyta, e.g., Ginkgoaceae (Ginkgo family)
A61K 36/17	Gnetophyta, e.g., Ephedraceae (Mormon tea family)
A61K 36/18	Magnoliophyta (angiosperms)
A61K 36/185	Magnoliopsida (dicotyledons)
A61K 36/19	Acanthaceae (Acanthus family)
A61K 36/195	Strobilanthes
A61K 36/20	Aceraceae (Maple family)
A61K 36/21	Amaranthaceae (Amaranth family), e.g., pigweed, rockwort, or globe amaranth
A61K 36/22	Anacardiaceae (Sumac family), e.g., smoketree, sumac, or poison oak
A61K 36/23	Apiaceae or Umbelliferae (Carrot family), e.g., dill, chervil, coriander, or cumin
A61K 36/232	Angelica
A61K 36/233	Bupleurum
A61K 36/234	Cnidium (snow parsley)
A61K 36/235	Foeniculum (fennel)
A61K 36/236	Ligusticum (licorice root)
A61K 36/237	Notopterygium
A61K 36/238	Saposhnikovia
A61K 36/24	Apocynaceae (Dogbane family), e.g., plumeria or periwinkle
A61K 36/25	Araliaceae (Ginseng family), e.g., ivy, aralia, schefflera, or tetrapanax
A61K 36/254	Acanthopanax or Eleutherococcus
A61K 36/258	Panax (ginseng)

TABLE 23.1 (*Continued*)

A61K 36/26	Aristolochiaceae (Birthwort family), e.g., heartleaf
A61K 36/264	Aristolochia (Dutchman's pipe)
A61K 36/268	Asarum (wild ginger)
A61K 36/27	Asclepiadaceae (Milkweed family), e.g., hoya
A61K 36/28	Asteraceae or Compositae (Aster or Sunflower family), e.g., chamomile, feverfew, yarrow, or echinacea
A61K 36/282	Artemisia, e.g., wormwood or sagebrush
A61K 36/284	Atractylodes
A61K 36/285	Aucklandia
A61K 36/286	Carthamus (distaff thistle)
A61K 36/287	Chrysanthemum, e.g., daisy
A61K 36/288	Taraxacum (dandelion)
A61K 36/289	Vladimiria
A61K 36/29	Berberidaceae (Barberry family), e.g., barberry, cohosh, or mayapple
A61K 36/296	Epimedium
A61K 36/30	Boraginaceae (Borage family), e.g., comfrey, lungwort, or forget-me-not
A61K 36/31	Brassicaceae or Cruciferae (Mustard family), e.g., broccoli, cabbage, or kohlrabi
A61K 36/315	Isatis, e.g., Dyer's woad
A61K 36/32	Burseraceae (Frankincense family)
A61K 36/324	Boswellia, e.g., frankincense
A61K 36/328	Commiphora, e.g., mecca myrrh or balm of Gilead
A61K 36/33	Cactaceae (Cactus family), e.g., prickly pear or Cereus
A61K 36/34	Campanulaceae (Bellflower family)
A61K 36/342	Adenophora
A61K 36/344	Codonopsis
A61K 36/346	Platycodon
A61K 36/35	Caprifoliaceae (Honeysuckle family)
A61K 36/355	Lonicera (honeysuckle)
A61K 36/36	Caryophyllaceae (Pink family), e.g., baby's breath or soapwort
A61K 36/37	Celastraceae (Staff-tree or Bittersweet family), e.g., tripterygium or spindle tree
A61K 36/38	Clusiaceae, Hypericaceae or Guttiferae (Hypericum or Mangosteen family), e.g., common St John's wort
A61K 36/39	Convolvulaceae (Morning-glory family), e.g., bindweed
A61K 36/40	Cornaceae (Dogwood family)
A61K 36/41	Crassulaceae (Stonecrop family)
A61K 36/42	Cucurbitaceae (Cucumber family)
A61K 36/424	Gynostemma
A61K 36/428	Trichosanthes
A61K 36/43	Cuscutaceae (Dodder family), e.g., Cuscuta epithymum or greater dodder
A61K 36/44	Ebenaceae (Ebony family), e.g., persimmon
A61K 36/45	Ericaceae or Vacciniaceae (Heath or Blueberry family), e.g., blueberry, cranberry, or bilberry
A61K 36/46	Eucommiaceae (Eucommia family), e.g., hardy rubber tree

(*Continued*)

TABLE 23.1 (*Continued*)

A61K 36/47	Euphorbiaceae (Spurge family), e.g., Ricinus (castorbean)
A61K 36/48	Fabaceae or Leguminosae (Pea or Legume family); Caesalpiniaceae; Mimosaceae; Papilionaceae
A61K 36/481	Astragalus (milkvetch)
A61K 36/482	Cassia, e.g., golden shower tree
A61K 36/483	Gleditsia (locust)
A61K 36/484	Glycyrrhiza (licorice)
A61K 36/485	Gueldenstaedtia
A61K 36/486	Millettia
A61K 36/487	Psoralea
A61K 36/488	Pueraria (kudzu)
A61K 36/489	Sophora, e.g., necklace pod or mamani
A61K 36/49	Fagaceae (Beech family), e.g., oak or chestnut
A61K 36/50	Fumariaceae (Fumitory family), e.g., bleeding heart
A61K 36/505	Corydalis
A61K 36/51	Gentianaceae (Gentian family)
A61K 36/515	Gentiana
A61K 36/52	Juglandaceae (Walnut family)
A61K 36/53	Lamiaceae or Labiatae (Mint family), e.g., thyme, rosemary, or lavender
A61K 36/532	Agastache, e.g., giant hyssop
A61K 36/533	Leonurus (motherwort)
A61K 36/534	Mentha (mint)
A61K 36/535	Perilla (beefsteak plant)
A61K 36/536	Prunella or Brunella (self heal)
A61K 36/537	Salvia (sage)
A61K 36/538	Schizonepeta
A61K 36/539	Scutellaria (skullcap)
A61K 36/54	Lauraceae (Laurel family), e.g., cinnamon or sassafras
A61K 36/55	Linaceae (Flax family), e.g., Linum
A61K 36/56	Loganiaceae (Logania family), e.g., trumpet flower or pinkroot
A61K 36/57	Magnoliaceae (Magnolia family)
A61K 36/575	Magnolia
A61K 36/58	Meliaceae (Chinaberry or Mahogany family), e.g., Azadirachta (neem)
A61K 36/59	Menispermaceae (Moonseed family), e.g., hyperbaena or coral bead
A61K 36/60	Moraceae (Mulberry family), e.g., breadfruit or fig
A61K 36/605	Morus (mulberry)
A61K 36/61	Myrtaceae (Myrtle family), e.g., tea tree or eucalyptus
A61K 36/62	Nymphaeaceae (Water-lily family)
A61K 36/63	Oleaceae (Olive family), e.g., jasmine, lilac, or ash tree
A61K 36/634	Forsythia
A61K 36/638	Ligustrum, e.g., Chinese privet
A61K 36/64	Orobanchaceae (Broom-rape family)
A61K 36/65	Paeoniaceae (Peony family), e.g., Chinese peony
A61K 36/66	Papaveraceae (Poppy family), e.g., bloodroot

TABLE 23.1 (*Continued*)

A61K 36/67	Piperaceae (Pepper family), e.g., Jamaican pepper or kava
A61K 36/68	Plantaginaceae (Plantain family)
A61K 36/69	Polygalaceae (Milkwort family)
A61K 36/70	Polygonaceae (Buckwheat family), e.g., spine flower or dock
A61K 36/704	Polygonum, e.g., knotweed
A61K 36/708	Rheum (rhubarb)
A61K 36/71	Ranunculaceae (Buttercup family), e.g., larkspur, hepatica, hydrastis, columbine, or goldenseal
A61K 36/714	Aconitum (monkshood)
A61K 36/716	Clematis (leather flower)
A61K 36/718	Coptis (goldthread)
A61K 36/72	Rhamnaceae (Buckthorn family), e.g., buckthorn, chewstick, or umbrella-tree
A61K 36/725	Ziziphus, e.g., jujube
A61K 36/73	Rosaceae (Rose family), e.g., strawberry, chokeberry, blackberry, pear, or firethorn
A61K 36/732	Chaenomeles, e.g., flowering quince
A61K 36/734	Crataegus (hawthorn)
A61K 36/736	Prunus, e.g., plum, cherry, peach, apricot, or almond
A61K 36/738	Rosa (rose)
A61K 36/739	Sanguisorba (burnet)
A61K 36/74	Rubiaceae (Madder family)
A61K 36/744	Gardenia
A61K 36/746	Morinda
A61K 36/748	Oldenlandia or Hedyotis
A61K 36/75	Rutaceae (Rue family)
A61K 36/752	Citrus, e.g., lime, orange, or lemon
A61K 36/754	Evodia
A61K 36/756	Phellodendron, e.g., cork tree
A61K 36/758	Zanthoxylum, e.g., prickly ash
A61K 36/76	Salicaceae (Willow family), e.g., poplar
A61K 36/77	Sapindaceae (Soapberry family), e.g., lychee or soapberry
A61K 36/78	Saururaceae (Lizard's-tail family)
A61K 36/79	Schisandraceae (Schisandra family)
A61K 36/80	Scrophulariaceae (Figwort family)
A61K 36/804	Rehmannia
A61K 36/808	Scrophularia (figwort)
A61K 36/81	Solanaceae (Potato family), e.g., tobacco, nightshade, tomato, belladonna, capsicum, or jimsonweed
A61K 36/815	Lycium (desert-thorn)
A61K 36/82	Theaceae (Tea family), e.g., camellia
A61K 36/83	Thymelaeaceae (Mezereum family), e.g., leatherwood or false ohelo
A61K 36/835	Aquilaria
A61K 36/84	Valerianaceae (Valerian family), e.g., valerian
A61K 36/85	Verbenaceae (Verbena family)
A61K 36/855	Clerodendrum, e.g., glorybower
A61K 36/86	Violaceae (Violet family)

(*Continued*)

TABLE 23.1 *(Continued)*

A61K 36/87	Vitaceae or Ampelidaceae (Vine or Grape family), e.g., wine grapes, muscadine or peppervine
A61K 36/88	Liliopsida (monocotyledons)
A61K 36/882	Acoraceae (Calamus family), e.g., sweetflag or Acorus calamus
A61K 36/884	Alismataceae (Water-plantain family)
A61K 36/886	Aloeaceae (Aloe family), e.g., aloe vera
A61K 36/888	Araceae (Arum family), e.g., caladium, calla lily, or skunk cabbage
A61K 36/8884	Arisaema, e.g., Jack in the pulpit
A61K 36/8888	Pinellia
A61K 36/889	Arecaceae, Palmae or Palmaceae (Palm family), e.g., date or coconut palm or palmetto
A61K 36/8895	Calamus, e.g., rattan
A61K 36/89	Cyperaceae (Sedge family)
A61K 36/8905	Cyperus (flat sedge)
A61K 36/894	Dioscoreaceae (Yam family)
A61K 36/8945	Dioscorea, e.g., yam, Chinese yam, or water yam
A61K 36/896	Liliaceae (Lily family), e.g., daylily, plantain lily, Hyacinth, or narcissus
A61K 36/8962	Allium, e.g., garden onion, leek, garlic, or chives
A61K 36/8964	Anemarrhena
A61K 36/8965	Asparagus, e.g., garden asparagus or asparagus fern
A61K 36/8966	Fritillaria, e.g., checker lily or mission bells
A61K 36/8967	Lilium, e.g., tiger lily or Easter lily
A61K 36/8968	Ophiopogon (Lilyturf)
A61K 36/8969	Polygonatum (Solomon's seal)
A61K 36/898	Orchidaceae (Orchid family)
A61K 36/8984	Dendrobium
A61K 36/8988	Gastrodia
A61K 36/899	Poaceae or Gramineae (Grass family), e.g., bamboo, corn, or sugarcane
A61K 36/8994	Coix (Job's tears)
A61K 36/8998	Hordeum (barley)
A61K 36/90	Smilacaceae (Catbrier family), e.g., greenbrier or sarsaparilla
A61K 36/902	Sparganiaceae (Bur-reed family)
A61K 36/904	Stemonaceae (Stemona family), e.g., croomia
A61K 36/906	Zingiberaceae (Ginger family)
A61K 36/9062	Alpinia, e.g., red ginger or galangal
A61K 36/9064	Amomum, e.g., round cardamom
A61K 36/9066	Curcuma, e.g., common turmeric, East Indian arrowroot, or mango ginger
A61K 36/9068	Zingiber, e.g., garden ginger

manufacture). Patent terms, for example, are potentially extendable in Australia, Europe, and the United States. (For a general summary, albeit an outdated summary with regard to some jurisdictions, see the WIPO publication on "*Supplementary Protection Certificates*") [19].

23.4 OTHER TYPES OF IP RIGHTS

23.4.1 Trade Secrets

Confidential information of a commercial nature, such as innovative products and processes, in essence remains "protected" whilst secrecy can be maintained and unable to be reverse engineered [8]. Although an owner of commercially important confidential information may in some instances have legal rights to prevent unauthorized disclosure or theft of confidential information/trade secrets, often it is a case of too little protection and too late as an irreparable leak may have already occurred. However, trade secrets are appropriate for proprietary products and processes, discoveries, or information for which patent protection is not available.

Trade secrets are usually compromised to some, or a large, degree due to product labeling compliance and other disclosure laws.

As mentioned elsewhere in this chapter, it is common practice for known (including traditional) phytomedicines and therapies of a proprietary nature to be kept as trade secrets in so far as this is possible. Also, many proprietary improvements to traditional phytomedicines and therapies are kept as trade secrets in preference, or possibly ignorance, of patents.

Note that it is possible for innovative products and processes to enjoy protection both as a trade secret and by way of an unpublished patent application for a limited period of time. This is because a patent application is not normally published until about 18 months after its initial filing date (but the time frame will be shorter for innovation patent applications, for example), which means that an owner can mark its products and processes as "patent pending" or "patent applied for" before choosing whether or not to withdraw the application prior to the invention being published, so as to retain confidentiality.

23.4.2 Regulatory Exclusivity and Restricted Third-Party Access

At the time of approval of a new product, such as a phytotherapeutic product, by a regulatory body, such as the Food and Drug Administration in the United States or the Therapeutic Goods Administration of Australia, the regulatory body may grant to the product registrant a period of exclusivity against competitor activity. The granted right will differ depending on the jurisdiction and type of product requiring registration. In some jurisdictions, a period of exclusivity against competitor activity may also be granted for a new therapeutical indication or phytotherapeutic product manufacturing process [20–22].

For example, at the time of approval of a new phytotherapeutic product by the Food and Drug Administration in the United States or Therapeutic Goods Administration of Australia, the regulatory body will grant to the product registrant a period of exclusivity during which competitors are unable to rely on data submitted by the product registrant to obtain regulatory approval of a competitor product.

As another example, at the time of approval of a phytotherapeutic product by the Food and Drug Administration in the United States, the regulatory body will grant to the product registrant a period of marketing exclusivity during which competitors with competing products are unable to enter the market.

These granted exclusions can provide a phytotherapeutic product or phytotherapy innovator with a market advantage.

23.4.3 Plant Variety Protection

Plant Variety Rights (also known as Plant Breeder's Rights) provide plant breeders exclusive rights over propagating material and harvested material of registered plant varieties for a set number of years [23].

Many countries provide for the protection of plant varieties. In Australia, for example, the duration of Plant Breeder's Rights in a plant variety is 25 years for trees and vines, and 20 years for any other variety [24].

To obtain Plant Breeder's Rights in a variety, the plant variety must (i) have a breeder, (ii) be distinct, (iii) be uniform, (iv) be stable, and (v) not have been exploited or must have been only recently exploited. Applying for Plant Breeder's Rights generally involves the steps of: (i) filing and obtaining acceptance of an application; (ii) filing a detailed description of the plant variety; and (iii) requesting examination of the application.

23.4.4 Industrial Designs

A registered industrial design, also known as a design patent, provides an owner with exclusive rights to an industrial design for a set period of time. The set period differs between jurisdictions as does the requirements for registration [8].

Generally speaking, design registrations are used to protect product appearances. In the phyto-industry, designs for unique product packaging may be registered and this may be important when seeking to establish brand distinction and awareness within a marketplace.

23.4.5 Trademarks

A trademark/branding is used as a badge or sign of origin or quality in respect of a product, process, or service. Although a trademark need not be registered, there are advantages in doing so. Generally speaking, a registered trademark may provide the registrant with rights to use the trademark and to stop others from adopting the same or similar trademark in respect of the same or similar products and services. A trademark registration usually lasts for 10 years and is renewable indefinitely [8].

Registrable trademarks can be letters, numerals, words, phrases, devices, logos, colors, shapes, smells, and sounds, for example.

In the phyto-industry, trademarks are important when seeking to establish brand distinction and awareness within a marketplace.

23.5 PATENTING TRENDS FOR PHYTOTHERAPEUTICS

During the patenting process, it is necessary for a patent and/or patent application to be published. Various "worldwide" patent databases and national patent office databases are accessible over the Internet, and can be used to locate and view published patent applications and granted patents.

Once such "worldwide" patent database, maintained by the European Patent Office, is Espacenet [25]. This database contains data from over 90 countries. Other "worldwide" patents databases include Patent Lens [26], which contains data from about 80 countries, and Patentscope [27], which contains PCT application data as well as data from about 30 countries.

According to the Espacenet patent database (containing patent data from ~90 countries), the total number of database entries having the IPC. A61K36 for "Medicinal preparations of undetermined constitution containing material from algae, lichens, fungi, or plants, or derivatives thereof, e.g., traditional herbal medicines" has increased over the years to a total of approximately 17,600 entries in the year 2012. Assuming that the data are correct and devoid of (patent family) duplicates, the trend may indicate the increasing commercial importance of medicinal preparations of undetermined constitution and may also suggest that such preparations constitute patentable subject matter. Based on another search of the ESPACENET patent database, more than half of those entries are Chinese patents or published patent applications.

23.6 TRADITIONAL KNOWLEDGE AND IP RIGHTS

An estimate given by author Kartal (citing a source more than 10 years old) is that between 25,000 and 75,000 plant species are used for traditional medicine and, of these, only a small percentage have to date been scientifically studied and commercialized [3]. Kartal further reports that a significant part of the modern pharmaceutical industry is founded, and most likely will continue to be based, on traditional medicinal plants, yet associated economic benefits largely remain with pharmaceutical companies, with little recognition or compensation being given to traditional knowledge owners.

On the one hand, industry is looking to traditional medicinal plants for new phytoproducts and phytotherapies ("bioprospecting"). On the other hand, the patenting and commercialization by industry of phytotherapy products and phytotherapies stemming from traditional knowledge has been a cause of alarm for traditional knowledge owners (and governments). Hence, governments have sought to develop a legal framework that satisfies both the interests of industry and traditional knowledge owners [28].

Communities and governments have openly expressed concern over the number of patent filings for inventions based on traditional knowledge for which prior consent had not been obtained from the traditional knowledge owners. There has also been resistance by some traditional knowledge owners to the patenting of such inventions in cases where their prior consent had not been obtained [29].

For example, representatives of traditional knowledge holders have opposed patents concerning:

- The use of fungicidal extracts from the Neem tree (in one example), resulting in European Patent Number 0436257 [30] being revoked [31]
- The use of turmeric as a wound-healing agent, resulting in US Patent 5,401,504 [32] being revoked [31]
- The use of Kakadu plum extract as an antioxidant and stimulant for collagen production in cosmetics products, whereby Australian Application Number 2007205838 [33] was formally withdrawn by the applicant following examination [34]

Many communities and governments are currently seeking legal instruments for further protecting traditional knowledge because existing laws are deemed inadequate [29]. In this respect, some jurisdictions are already signatories to international agreements, conventions, and declarations, including the Convention on Biological Diversity 1993, the International Treaty on Plant Genetic Resources for Food and Agriculture 2001, and the United Nations Declaration on the Rights of Indigenous Peoples 2007 [35].

The effects of the Convention on Biological Diversity 1993 have already begun filtering through to the patent system in that for Brazil, Colombia, Denmark, Egypt, Germany, India, Italy, Norway, Philippines, South Africa, Sweden, and Venezuela there is a requirement that the source and/or country of origin of genetic resources and/or traditional knowledge be indicated in patent applications for inventions based on such genetic resources or traditional knowledge. The penalty for noncompliance (if any) differs between the jurisdictions [36]. This requirement may be applicable to patent applications describing phyto-inventions that are based on traditional phytotherapy products and phytotherapies.

The Convention on Biological Diversity 1993 enables signatories to establish regimes regulating access to genetic resources and traditional knowledge while establishing benefit-sharing mechanisms relating to that access [35]. On the one hand, this Convention is likely to be welcomed by traditional knowledge owners due to the benefits-sharing aspect, whereas, on the other hand, nontraditional knowledge owners/industry wishing to commercialize traditional knowledge or phytotherapeutic products based on traditional knowledge may view the Convention as an unwelcome additional (and possibly insurmountable) obstacle in the already difficult path to commercialization.

Instances of patent filings benefiting both traditional knowledge owners and those wishing to exploit inventions based on that knowledge do exist: two examples are the following:

- In South India, the Tropical Botanic Garden and Research Institute filed patent applications and successfully commercialized a medicinal plant drug based on knowledge of the Kani tribes. A trust fund was established to share benefits with the tribes [37].
- In Australia, the University of South Australia together with the Chuulangun Aboriginal Corporation filed patent applications for anti-inflammatories derived from traditional medicinal plants [38].

In the Australian context, the legal IP framework includes the Patents Act 1990, Trade Marks Act 1995, Designs Act 2003, Plant Breeder's Rights Act 1994, Convention on Biological Diversity, Copyright Act, as well as other pieces of legislation [39].

As at the date of writing, traditional phytotherapy products and phytotherapies are unlikely to be validly patented or otherwise protected exclusively in Australia by anyone. If patent protection for a traditional knowledge "invention" has been sought, that protection can be contested either at the pre-grant stage or post-grant stage, in that the invention is not novel or inventive/innovative. Usually, a commercial advantage for an "unprotectable" traditional phytotherapy product or phytotherapy is established or enjoyed by way of a trade secret (in so far as this is possible) or brand recognition.

Concerning "innovative" improvements to traditional phytotherapy products and phytotherapies (as discussed elsewhere in this chapter), "protection" may be sought either by way of a trade secret (in so far as this is possible) and/or a patent (usually for up to 20 years; 25 years maximum for new therapeutic products). Patents usually play a primary role in the protection of IP, whereas trade secrets usually play a secondary role. Patent protection and/or Plant Breeder's Rights may also be sought for new plant varieties (for up to 20 or 25 years). A significant obstacle to commercialization, however, may be that industry may be required to engage with traditional knowledge owners and negotiate commercial terms due to the Convention on Biological Diversity.

Looking forward, based on existing trends, it appears that willing traditional knowledge owners of less industrialized or unindustrialized jurisdictions rich in traditional phytoknowledge may yet benefit from the patent system and the legal instruments currently being put in place for benefit sharing.

DISCLAIMER

This chapter is intended to provide general information on intellectual property in so far as it relates to the phyto-industry. The chapter contents should not be relied upon as detailed legal advice for any specific case. While every effort has been made to ensure that the contents are correct at the time of writing, please note, the relevant laws and practice are subject to change. For specific cases, readers should seek advice from their legal advisor.

REFERENCES

[1] Timmermans K (2003) Intellectual property rights and traditional medicine: policy dilemmas at the interface. *Soc Sci Med* 57: 745–756.

[2] Kaiser H (2012) *Traditional Chinese medicine with overview of worldwide market, development, products and companies 2010–2015–2025*. Helmut Kaiser Consultancy HKC22. http://www.hkc22.com/TraditionalChineseMedicine.html. Accessed December 3, 2014.

[3] Kartal M (2007) Intellectual property protection in the natural product drug discovery, traditional herbal medicine and herbal medicinal products. *Phytother Res* 21: 113–119.

[4] Raysman R, Pisacreta EA, Adler KA, Ostrow SH. (1999) *Intellectual property licensing: forms and analysis*. Law Journal Press, 1998–2008. ISBN 973-58852-086-9.

[5] Sherman B, Bently L (1999) *The making of modern intellectual property law: the British experience, 1760–1911*. Cambridge, England: Cambridge University Press. pp. 242.

[6] Mossoff A (2003) *Rethinking the development of patents: an intellectual history, 1550–1800*. East Lansing, MI: Intellectual Property & Communication Law Program, Michigan State University-DCL College of Law. pp. 22.

[7] Lemley MA (2005) Property as a common descriptor of the field probably traces to the foundation of the World Intellectual Property Organization (WIPO) by the United Nations, Property, Intellectual Property and Free Riding. *Tex Law Rev* 83: 1031–1033.

[8] World Intellectual Property Organization (WIPO) (2004) WIPO intellectual property handbook: policy, law and use. WIPO Publication No. 489 (E). Available at http://www.wipo.int/about-ip/en/iprm/. Accessed January 21, 2014.

[9] World Intellectual Property Organization. WIPO-administered treaties. Available at http://www.wipo.int/treaties/en/. Accessed January 21, 2014.

[10] World Intellectual Property Organization. Uruguay round agreement: TRIPS trade-related aspects of intellectual property rights. Available at http://www.wipo.int/treaties/en/agreement/trips.html. Accessed January 21, 2014.

[11] World Intellectual Property Organization. About patents. General FAQs on patents. Available at www.wipo.int/patentscope/en/patents/. Accessed January 21, 2014.

[12] World Intellectual Property Organization. Protecting innovations by utility models. Available at http://www.wipo.int/sme/en/ip_business/utility_models/utility_models.htm. Accessed January 21, 2014.

[13] IP Australia (2013) Funding and grants. Available at http://www.ipaustralia.gov.au/understanding-intellectual-property/ip-for-business/funding-and-grants/. November 23. Accessed January 21, 2014.

[14] World Intellectual Property Organization. Public health and patents. Available at http://www.wipo.int/patent-law/en/developments/publichealth.html. Accessed January 21, 2014.

[15] World Trade Organization. A more detailed overview of the TRIPS agreement. www.wto.org/english/tratop_e/trips_e/intel2_e.htm. Accessed January 21, 2014.

[16] Basheer S, Purohit S, Reddy P (2010) Patent exclusions that promote public health objectives. SCP/15/3, Annex IV. Available at http://www.wipo.int/edocs/mdocs/scp/en/scp_15/scp_15_3-annex4.pdf. Accessed December 3, 2014.

[17] World Intellectual Property Organization. International patent classification (IPC). Available at http://www.wipo.int/classifications/ipc/en/. Accessed January 21, 2014.

[18] World Intellectual Property Organization. IPC version 2013.01. Available at http://www.wipo.int/classifications/ipc/en/. Accessed January 21, 2014.

[19] World Intellectual Property Organization (2002) WIPO handbook on industrial property information and documentation; survey on the grant and publication of "supplementary protection certificates" for medicinal and phytopharmaceutical products or equivalent industrial property rights (SPCs). Adopted by the PCIPI Executive Coordination Committee at its fourteenth session on May 20, 1994, and further updated by the International Bureau in January 2002. Available at www.wipo.int/standards/en/pdf/07-07-01.pdf. Accessed January 21, 2014.

[20] European Medicines Agency. Available at www.ema.europa.eu. Accessed January 21, 2014.

[21] Australian Therapeutic Goods Administration. Available at www.tga.gov.au. Accessed January 21, 2014.

[22] U.S. Food and Drug Administration. Available at www.fda.gov. Accessed January 21, 2014.

[23] World Intellectual Property Organization. Introduction to plant variety protection under the UPOV convention. Available at http://www.wipo.int/edocs/mdocs/sme/en/wipo_ip_bis_ge_03/wipo_ip_bis_ge_03_11-main1.pdf. Accessed January 21, 2014.

[24] IP Australia. Plant breeder's rights. Available at http://www.ipaustralia.gov.au/get-the-right-ip/plant-breeders-rights/. Accessed January 21, 2014.

[25] European Patent Office. Espacenet database. Available at http://www.epo.org/searching/free/espacenet.html. Accessed January 21, 2014.

[26] Cambia. Patent lens database. Available at http://www.patentlens.net/daisy/patentlens/patentlens.html. Accessed January 21, 2014.

[27] World Intellectual Property Organization. Patentscope database. Available at http://patentscope.wipo.int/search/en/search.jsf. Accessed January 21, 2014.

[28] Davis M (1998) Biological diversity and indigenous knowledge. Parliament of Australia, science, technology, environment and resources group research paper 17 1997–98. Available at http://www.aph.gov.au/About_Parliament/Parliamentary_Departments/Parliamentary_Library/pubs/rp/RP9798/98rp17. Accessed January 21, 2014.

[29] World Intellectual Property Organization. Traditional knowledge. Available at www.wipo.int/tk/en/. Accessed January 21, 2014.

[30] European patent 0436257. December 20, 1990. Available at http://worldwide.espacenet.com/searchResults?submitted=true&locale=en_EP&DB=EPODOC&ST=advanced&TI=&AB=&PN=EP0436257&AP=&PR=&PD=&PA=&IN=&CPC=&IC= Accessed December 31, 2014.

[31] World Intellectual Property Organization (2006) WIPO Intergovernmental committee on intellectual property and genetic resources, traditional knowledge and folklore; ninth session. Geneva. April 24 to 29, 2006. Available at www.wipo.int/edocs/mdocs/tk/en/wipo_grtkf_ic_9/wipo_grtkf_ic_9_13.pdf. Accessed January 21, 2014.

[32] US patent 5,401,504. December 28, 1993. Available at http://www.lens.org/lens/patent/US_5401504_A. Accessed March 14, 2015.

[33] Australian Application Number 2007205838. January 19, 2007. Available at http://pericles.ipaustralia.gov.au/ols/auspat/quickSearch.do;jsessionid=k1j9J2tHnLkpkl3GplyyDJHncnhd3QG2VdNXWJw1GFhT15SrkbBl!-1678157587?queryString=2007205838&resultsPerPage= Accessed December 31, 2014.

[34] IP Australia. Auspat extract for Australian application number 2007205838. Available at http://www.ipaustralia.gov.au/auspat/. Accessed January 20, 2014.

[35] Stoianoff NP. 2012. Navigating the landscape of indigenous knowledge—a legal perspective. *Intellect Property Forum* 90: 23–40.

[36] Becker K (2013) AIPPI report Q166: intellectual property and genetic resources/traditional knowledge and folklore'. Available at https://www.aippi.org/download/commitees/166/Report166Report+Executive+Committee+Meeting+Helsinki+2013English.pdf. Accessed January 21, 2014.

[37] World Intellectual Property Organization. Using traditional knowledge to revive the body and a community. Available at www.wipo.int/ipadvantage/en/details.jsp?id=2599. Accessed January 21, 2014.

[38] IP Australia (2013) *Public consultations*. Chuulangun Aboriginal Corporation and the University of South Australia. Available at www.ipaustralia.gov.au. Accessed January 21, 2014.

[39] World Intellectual Property Organization. Australia. Available at http://www.wipo.int/wipolex/en/profile.jsp?code=AU. Accessed January 21, 2014.

24

INTERNATIONAL REGULATORY STATUS OF PHYTOTHERAPIES

ERNEST V. LINEK

Banner & Witcoff, Ltd., Boston, Massachusetts, USA

24.1 INTRODUCTION

Laws, especially laws regulating health products such as phytotherapies and the ingredients used therein, are intended to protect the public from harm. Laws change often, so consult the appropriate governmental websites to ensure that you are aware of the current laws in a country of interest. None of the laws reviewed in this chapter use the term "phytotherapy." Instead, "traditional medicines" is the commonly used term for this subject matter, except in the United States, where such products are simply designated as "dietary supplements."

Wikipedia defines phytotherapy as follows:

> Phytotherapy is the study of the use of extracts of natural origin as medicines or health-promoting agents. Phytotherapy medicines differ from plant-derived medicines in standard pharmacology. Where standard pharmacology isolates an active compound from a given plant, phytotherapy aims to preserve the complexity of substances from a given plant with relatively less processing.

Phytotherapy is distinct from homeopathy and anthroposophic medicine and avoids mixing plant and synthetic bioactive substances. Traditional phytotherapy is a synonym for herbalism and regarded as alternative medicine by much of Western medicine. Although the medicinal and biological effects of many plant constituents such as alkaloids (morphine, atropine, etc.) have been proven through clinical studies, there

Phytotherapies: Efficacy, Safety, and Regulation, First Edition. Edited by Iqbal Ramzan.

is continued debate about the efficacy and the place of phytotherapy in medical therapies. Modern phytotherapy, following the scientific method, can be considered the study of the effects and clinical use of herbal medicines.

24.1.1 Country Law Sources

Current public information sources for the laws discussed in this chapter are as follows:

1. Australia—The Therapeutic Goods Administration (TGA) website: www.tga.gov.au
2. Canada—The Health Canada website: www.hc-sc.gc.ca
3. China—The State Administration of Traditional Chinese Medicine website: www.satcm.gov.cn
4. European Union—The European Medicines Agency website: www.ema.europa.eu/ema/
5. India—The Central Drugs Standard Control Organization website: www.cdsco.nic.in
6. Japan—The Ministry of Health, Labour and Welfare website: http://www.mhlw.go.jp/english
7. United Kingdom—The Medicines and Healthcare Products Regulatory Agency (MHRA) website: www.mhra.gov.uk
8. United States—The Food and Drug Administration website: www.fda.gov

24.1.2 Common Requirement: Good Manufacturing Practices

The countries reviewed for this chapter all require manufacturers of natural phytotherapeutic products to follow Good Manufacturing Practices (GMP), thereby, ensuring that the products are safe for use by consumers. Good manufacturing practice guidelines provide manufacturing, testing, and quality assurance techniques, all designed to ensure that a product is safe for human consumption. Some of the common GMP guidelines are as follows:

(a) *Hygiene:* The manufacturing facility must maintain a clean and hygienic manufacturing area.
(b) *Process Controls:* Manufacturing processes must be clearly defined and controlled. All critical processes are confirmed to ensure consistency and compliance with specifications.
(c) *Validation:* Manufacturing changes that have an impact on the quality of the product are validated as necessary.
(d) *Procedures:* Instructions and procedures are written in clear and unambiguous language. Operators are trained to carry out and document procedures.
(e) *Records:* Manufacturing records are made, manually or by instruments, during manufacture that demonstrate that all the steps required by the defined

procedures and instructions were in fact taken and that the quantity and quality of the product was as expected. Deviations are investigated and documented.

(f) *Distribution Records:* Distribution records are made such that the complete history of a batch to be traced are retained in a comprehensible and accessible form.

(g) *Recall Systems:* A system is available for recalling any batch of product from sale or supply.

(h) *Records of Complaints:* Complaints about marketed products are examined, the causes of quality defects are investigated, and appropriate measures are taken with respect to the defective products to prevent recurrence.

GMP guidelines are not prescriptive instructions on how to manufacture products. They are instead a series of general principles that must be observed during manufacturing. When a company is setting up its quality program and manufacturing process, there may be many ways it can fulfill GMP requirements. It is the company's responsibility to determine the most effective and efficient manufacturing methods that comply with the local requirements imposed by the regulatory agencies.

GMPs are enforced in the United States by the U.S. Food and Drug Administration (FDA), under Section 501(B) of the 1938 Food, Drug and Cosmetic Act (21 USC § 351). The Regulations use the phrase "current good manufacturing practices" (cGMP) to describe these guidelines—since they are often subject to change. Thus keeping up to date with the current regulations is a mandatory requirement for manufacturers. Note, unlike the other countries discussed in this chapter, the United States does not treat natural health products as drugs. Instead, it treats these products as dietary supplements, regulated under the food protection aspect of the laws. However, cGMP regulations have been enacted by the FDA for these products.

The FDA's final cGMP for dietary supplements was effective on August 24, 2007. To limit any disruption for dietary supplements produced by small businesses, the rule included a 3-year phase-in for small businesses. Companies with more than 500 employees had until June 2008 to comply; companies with less than 500 employees had until June 2009 to comply; and companies with fewer than 20 employees had until June 2010 to comply with the regulations. Since June 2010, the cGMP rule has applied to all companies manufacturing dietary supplements in the United States. In 2010, China's Ministry of Health announced the adoption of Good Manufacturing requirements, effective January 17, 2011.

In the United Kingdom, the *Medicines Act* (1968) covers most aspects of GMP in what is commonly referred to as "The Orange Guide," which is named so because of the color of its cover; it is officially known as Rules and Guidance for Pharmaceutical Manufacturers and Distributors. The World Health Organization (WHO) version of GMP is used by pharmaceutical regulators and the pharmaceutical industry in over 100 countries worldwide, primarily in the developing world. The European Union's GMP (EU-GMP) enforces similar requirements to WHO's GMP, as does the FDA's version in the United States.

24.2 SPECIFIC COUNTRY REGULATIONS

24.2.1 Current Regulations in Australia

References Cited/Additional Reading Resources:

The Therapeutic Goods Administration (TGA) website: www.tga.gov.au.

In Australia, phytotherapies and health products containing ingredients such as herbs, vitamins, minerals, nutritional supplements, homoeopathic, and certain aromatherapy preparations are referred to as "complementary medicines" and are regulated as medicines under the *Therapeutic Goods Act 1989* (the Act).

A complementary medicine is defined in the Therapeutic Goods Regulations 1990 as a therapeutic good consisting principally of one or more designated active ingredients mentioned in Schedule 14 of the Regulations, each of which has a clearly established identity and traditional use. These medicines include herbal medicines, homeopathic medicines, vitamins, mineral supplements, traditional medicines such as Ayurvedic medicines, traditional Chinese medicines, other nutritional supplements, and aroma therapy oils.

Specific designated active ingredients include an amino acid; charcoal; a choline salt; an essential oil; plant or herbal material (or a synthetically produced substitute for material of that kind), including plant fibers, enzymes, algae, fungi, cellulose, and derivatives of cellulose and chlorophyll; a homeopathic preparation; a microorganism, whole or extracted, except a vaccine; a mineral including a mineral salt and a naturally occurring mineral; a mucopolysaccharide; non-human animal material (or a synthetically produced substitute for material of that kind) including dried material, bone and cartilage, fats and oils, and other extracts or concentrates; a lipid, including an essential fatty acid or phospholipid; a substance produced by or obtained from bees, including royal jelly, bee pollen, and propolis; a sugar, polysaccharide, or carbohydrate; and a vitamin or a pro-vitamin.

Australia has a risk-based approach with a two-tiered system for the regulation of all medicines, including natural health products and complementary medicines:

1. Lower risk medicines can be listed on the Australian Register of Therapeutic Goods (ARTG).
2. Higher risk medicines must be registered on the ARTG.

Some complementary medicines are exempt from the requirement to be included on the ARTG, such as certain preparations of homoeopathic medicines. The Australian Regulatory Guidelines for Complementary Medicines (ARGCM) provides detail on the regulation of complementary medicines and assists sponsors to meet their legislative obligations.

TGA post-market regulatory activities relate to the monitoring of the continuing safety, quality, and efficacy of listed, registered, and included therapeutic goods once they are on the market. Information on the TGA's approach to managing compliance risk is available at the TGA website.

The TGA Office of Manufacturing Quality inspects manufacturers on an ongoing basis for compliance with good manufacturing practices. The TGA also undertakes listed complementary medicine compliance reviews. Sometimes medicines, including complementary medicines, have unexpected and undesirable effects. The TGA has an active review program, which involves the assessment of adverse events that are reported to the TGA by consumers, health professionals, the pharmaceutical industry, international medicines regulators, or by the medical and scientific experts on TGA advisory committees. Sponsors of medicines are required to report to the TGA suspected adverse reactions for their medicines that they are aware of.

Under the TGA, the marketing and advertising of therapeutic goods, including complementary medicines, is to be conducted in a manner that promotes the quality use of the product, is socially responsible, and does not mislead or deceive the consumer. The advertising of therapeutic goods in Australia is subject to the advertising requirements of the Therapeutic Goods Act, which adopts the Therapeutic Goods Advertising Code (TGAC) and the supporting Regulations, the Trade Practices Act 1975, and other relevant laws.

24.2.2 Current Regulations in Canada

References Cited/Additional Reading Resources:
The Health Canada website: www.hc-sc.gc.ca.

In Canada, complementary/alternative and traditional medicines are known as natural health products (NHPs) and are subject to food and drug regulations. Natural health products include herbal medicines; traditional Chinese, Ayurvedic, and native North American medicines; homeopathic preparations; and vitamin and mineral supplements. NHPs are products that are used and marketed for a variety of health purposes, such as for the prevention or treatment of an ailment or condition, the reduction of health risks, or the maintenance of good health.

As defined in Canada, a NHP is a substance or a combination of substances described in Schedule 1 of the Natural Health Products Regulations (NHPR), a homeopathic medicine, or a traditional medicine, which is intended to provide a pharmacological activity or other direct effect in diagnosing, treating, mitigating, or preventing a disease, disorder, or abnormal physiological state or its symptoms in humans:

(a) Restoring or correcting organic functions in humans
(b) Modifying organic functions in humans, such as modifying those functions in a manner that maintains or promotes health

In addition, some common consumer products, such as certain toothpastes, antiperspirants, shampoos, facial products, and mouthwashes are also classified as NHPs in Canada because of their medicinal ingredients and intended uses.

In Canada, the commercial sale of natural health products is subject to the licensing provisions of the Natural Health Products Regulations, which came into force on January 1, 2004. The purpose of the Regulations is to help assure that Canadians have access to NHPs that are safe, effective, and of high quality.

To be legally sold in Canada, all NHPs must have a Product License and the Canadian sites that manufacture, package, label, and import these products must have Site Licenses. There are specific labeling and packaging requirements as well as good-manufacturing practice standards and evidence norms that must be met in order to obtain product and site licenses. The licensing requirements of the Natural Health Products Regulations apply to any person or company that manufactures, packages, labels, and/or imports NHPs for commercial sale in Canada. The licensing requirements do not apply to health-care practitioners who compound products on an individual basis for their patients or to retailers of NHPs. Canadian physicians choosing to provide alternative treatments must comply with guidelines set by the relevant province's College of Physicians and Surgeons. The Canadian food and drug laws do not recognize traditional Chinese doctors, naturopaths, homeopaths, or herbalists. Most of Canada's healthcare legislation, such as the Canada Health Act, focuses on traditional medical practitioners. However, the regulation of medical professionals is a provincial matter and some Canadian provinces have become tolerant of non-traditional healthcare providers.

The Natural Health Products Regulations are administered by Health Canada's Natural Health Products Program, which is comprised of three directorates, each with their own specific roles and responsibilities. They are the Natural Health Products Directorate, Marketed Health Products Directorate, and Health Products and Food Branch Inspectorate. The Natural Health Products Directorate (NHPD) is lead directorate of the Natural Health Products Program and is the regulating authority for the commercial sale of NHPs in Canada. The NHPD is responsible for the assessment and issuance of product and site licenses.

The Marketed Health Products Directorate (MHPD) provides a consistent approach to post-approval safety surveillance, assessment of signals, and safety trends and risk communications concerning all regulated marketed health products, including NHPs. The management of adverse reactions involving NHPs is also the responsibility of the MHPD. Individuals having experienced an adverse event are encouraged to report it to the MHPD immediately.

The Health Products and Food Branch Inspectorate (HPFBI) is responsible for the enforcement of the Regulations and carries out any required compliance action, including product recalls and investigations.

Health Canada also consults regularly with two external advisory committees: the Management Advisory Committee (MAC) and Expert Advisory Committee (EAC).

Comprised of industry and consumer representatives, the MAC advises Health Canada on the administration of the Natural Health Products Regulations. Members of professional and scientific communities make up the EAC, which provides expert advice to Health Canada on issues relating to the safety, efficacy, and quality of NHPs.

NHPs that have been licensed for sale in Canada will bear a Natural Product Number (NPN) or Homeopathic Medicine Number (DIN-HM) on the label, followed by an eight-digit number (e.g., NPN 12345678). The presence of a NPN or DIN-HM means that the product has been authorized for sale in Canada and is safe and effective when used in accordance with the instructions on the label. Consumers can learn more about licensed NHPs by accessing the Licensed Natural Health Products

Database (LNHPD), which has the following website: http://www.hc-sc.gc.ca/dhp-mps/prodnatur/applications/licen-prod/lnhpd-bdpsnh-eng.php

As discussed earlier, all NHPs sold in Canada require a product license before being marketed. Obtaining this license requires the submission of a Product License Application (PLA), which must include detailed information on the product (e.g., its medicinal and non-medicinal ingredients, dose, potency, and recommended uses) along with evidence supporting its recommended conditions of use. Once a product has been assessed and deemed safe, effective and of high quality, a product license will be issued along with an eight-digit Natural Product Number (NPN) or Homeopathic Medicines Number (DIN-HM) that must appear on the label. The safety and efficacy of health claims associated with NHPs must be supported by appropriate evidence so that consumers and Health Canada can have assurance that products are indeed safe and effective. Health Canada accepts varying types of evidence to support the safety and efficacy of NHPs, ranging from clinical trial data to references to published studies, journals, pharmacopoeias, and traditional resources. The type and amount of supporting evidence required is dependent on the proposed health claim of the product and its overall risks.

In Canada, all NHPs must adhere to specific labeling requirements that are in place to ensure that consumers can make safe and informed choices about the NHPs they choose to use. Examples of the type of information required on the label of NHPs include product name, dose and route of administration, product license number, recommended use (i.e., its health claim), complete list of ingredients, and any risks associated with the product's use, such as cautions, warnings, contra-indications, and known adverse reactions.

Canadian manufacturers, importers, labelers, and packagers of natural health products must have site licenses. Obtaining this license requires the submission of a Site License Application (SLA) complete with supporting documentation demonstrating that the manufacture, packaging, labeling, and importation of NHPs follows appropriate good manufacturing practice standards.

Good Manufacturing Practices, or GMPs, are ongoing measures designed to ensure an effective overall approach to product quality control and risk management. They do so by setting appropriate standards and practices for product testing, manufacturing, storage, handling, and distribution.

24.2.2.1 Canada: Special Rules for Traditional Medicines Traditional medicine is defined as medicine based on the sum total of knowledge, skills, and practices based on the theories, beliefs, and experiences indigenous to different cultures, used in the maintenance of health, as well as in the prevention, diagnosis, improvement, or treatment of physical and mental illness. Canada's traditional medicine definition is a modified version from the World Health Organization Traditional Medicine Program, recognizing traditional medicines at their core as ancient medical practice that existed in human societies before the application of modern science to health and that have evolved to reflect different philosophical backgrounds and cultural origins.

The level of evidence (type and amount) that can be provided to support the safety and efficacy of a natural health product (NHP) varies depending on the proposed health claim(s) of the product and the overall risk profile of the product or its ingredients. These special rules apply to product license applications for traditional medicines. It does not apply to product license applications for natural health products (NHPs) that do not meet the definition of a traditional medicine, or homeopathic medicines. In addition, these rules do not apply to healthcare practitioners (including, but not limited to, practitioners of Traditional Chinese Medicine (TCM) or Ayurvedic medicine) who compound products for their patients in the context of a practitioner–patient relationship.

In order to obtain authorization to sell a natural health product (NHP) in Canada, a product license application must be submitted to Health Canada. As part of this product license application, evidence supporting the safety and efficacy of the NHP according to its recommended conditions of use must be included. It is the responsibility of the applicant to provide a complete product license application, including evidence demonstrating that safety (risk) has been established and any risks sufficiently mitigated, that efficacy (benefit) has been demonstrated and that quality is supported.

It is the responsibility of the NHPD to review the information provided as part of the product license application in order to assess the safety, efficacy, and quality of a natural health product, to ensure benefits outweigh risks and to clearly document the product licensing decision. An applicant may be able to get a product license quickly by using pre-cleared information, including monographs and labeling standards available in the Natural Health Products Ingredients Database.

The purpose of the assessment is to determine whether the evidence supports the safety and efficacy of the product, including whether there is reasonable assurance that benefits of the product outweigh any risk inherent in the product's ingredients or associated use of the product. The assessment of safety (risk) for a product depends on a variety of factors, including the conditions of use and the physical form and pharmacology of each ingredient in the product as well as the product as a whole. The benefit-to-risk profile of a product is always considered prior to a product licensing decision being made (i.e., license issuance or application rejection).

24.2.2.2 ***Health Claims*** A health claim is a statement that indicates the intended beneficial effect of a product when used in accordance with its recommended conditions of use. The term "recommended use or purpose" is often used interchangeably with "health claim" or "indications for use." Given the definition of a traditional medicine, traditional use claims should reflect the sum total of knowledge, skills, and practices based on theories, beliefs, and experiences indigenous to a specific culture, used in the maintenance of health, as well as prevention, diagnosis, improvement, or treatment of physical and mental illness. For a health claim to be categorized as "traditional use," it should be founded upon the theories, experiences, and beliefs embodying the respective ancient practice of medicine. Products with general health claims include those that have low therapeutic impact and are therefore subject to the appropriate evidence requirements.

Annex I of the Pathway for Licensing Natural Health Products Making Modern Health Claims outlines a regulatory pathway for NHPs with general health claims.

These claims are permitted provided that the health and safety of Canadians would not be at risk; this is consistent with a risk-based product approach where health claims are indexed against the level of evidence provided to support the safe use of the products.

24.2.2.3 Demonstrating a Long History of Use The Canadian law permits a number of methods to demonstrate and support a claim of a long history of use. Demonstrating a timespan representing two generations of safe traditional use of an NHP, for example, by

(a) Describing the use in the context of a particular cultural belief system or a system of traditional medicine that has been in existence for at least two generations
(b) Demonstrating a period that was at least two generations ago with the implication that the ingredient was used from that period onward
(c) Demonstrating a time-specific event even though neither a concrete date nor time frame was given (e.g., "used in the time of King Edward II to alleviate coughs")
(d) Demonstrating that the product meets the European Directive on traditional herbal medicines—Directive 2004/24/EC, Article 16
(e) Demonstrating traditional use as per the Australian Therapeutic Goods Administration, that is, used over three or more generations of recorded use for a specific health related or medicinal purpose

Products with multiple medicinal ingredients with "traditional use" claims are permitted for assessment as traditional medicines when certain conditions are met. Evidence should demonstrate "traditional use" of the medicinal ingredients within a single recognized system of traditional medicine (e.g., Traditional Chinese Medicine (TCM), Ethnomedicines of the First Nations, Ayurvedic Medicine, and Traditional Herbal Medicine) as a whole formulation; modifications of a classic recipe that would still be accepted within the system of traditional medicine; or as individual medicinal ingredients. Efficacy should be based on the belief systems, theories, and/or experiences specific to the relevant traditional healing paradigm, not on modern evidence. All ingredients should be documented medicinal ingredients within the same system of traditional medicine and have been prepared based on a suitable traditional method of preparation utilized for that medicinal ingredient and a clear and logical rationale should be provided to support the presence of each medicinal ingredient. Not all ingredients in the formulation need to contribute to the recommended purpose. However, each ingredient should contribute, in a logical manner, to the overall safety and efficacy of the product within the given traditional system of medicine.

The intent of the Regulations is to capture modifications to classic traditional formulations that are supported by the knowledge and experience of the traditional system of medicine. For example, modifications of the classical formula of a TCM that are still based on TCM principles would fall into this category, but the modifications of ingredients based on other evidence would not. The intent is also to capture formulations in Traditional Herbal Medicine or Traditional Ayurvedic Medicine that are logical within those systems of medicine, but not additions of unrelated herbs, vitamins, minerals, or other ingredients. If the recommended conditions of use (i.e., the recommended use or purpose, dose including the method of preparation, dosage form, or route of administration) for the product are not comparable to the recommended conditions of use contained within the references on the traditional use of the product, the product will not be assessed as a traditional medicine, a traditional use claim may not be permitted, and additional evidence requirements may apply.

Health claims may refer to treatment, prevention, risk reduction, or general health maintenance or promotion. Evidence should support the medicinal formulation or modification of a classic recipe or individual ingredient use within a single system of traditional medicine when appropriate.

24.2.2.4 Efficacy Requirements Two independent references are required supporting efficacy of traditional medicines based on belief/theories/experiences within a single system of traditional medicine: pre-cleared information for traditional use or published peer-reviewed compilations (e.g., monographs and pharmacopoeias).

24.2.2.5 Safety Requirements Two independent references are required supporting safety based on belief, theories, and experiences within a single system of traditional medicine. When scientific evidence suggests a risk, scientific evidence should be provided to substantiate safety. (Note: History of use is not an acceptable measure for mitigating potential risk identified in the scientific literature).

24.2.2.6 Demonstrating Traditional Use Demonstrating the traditional use of a product is required under these special rules. Use may be considered traditional if it is supported by at least one reference describing the method's use within the practice of traditional medicine. Just as there are many ways to demonstrate a long history of use, the same references may help to establish the required proof of traditional medicine use, including, for example, the conditions of use (indications, dose, cautions, contra-indications, etc.) of specific ingredients. References should also refer to a health condition that can be diagnosed in the relevant system of traditional medicine independent of whether the reference indicates "traditionally used" or not. References should also refer to use, dosage form, route of administration, dose, and duration of use, when applicable, which is consistent with traditional use and comparable to that of the recommended product. This includes traditional methods of preparation such as

(a) The use of a whole organism or specific parts (leaf, root, fruiting body, etc.,), whether fresh, dried or freeze-dried, or preserved with alcohol, honey, or sugar
(b) Extracts produced by the application of pressure to the source material
(c) Aqueous extracts such as infusions, decoctions, and syrups
(d) Ethanol-based extracts such as tinctures, fluid extracts, and succi
(e) Glycerine-based extracts

(f) Vinegar-based extracts
(g) Oil, grease, or fat-based infusions
(h) Beeswax salves and ointments

Other methods of preparation may be considered traditional if supported by at least one reference describing the method's use within the practice of traditional medicine and assessed as acceptable by the Natural Health Products Directorate (NHPD).

24.2.2.7 ***Safety Evidence for Traditional Medicines*** Safety may be supported in several ways. History of use is an important consideration. References demonstrating that the ingredient has an extensive history of use should be submitted. Although the indication need not be the same, other conditions of use (dose, duration of use, source material, method of preparation, etc.,) should be comparable to the conditions of use being proposed. Cautions, warnings, and contra-indications found in the accompanying references are a primary source of safety information. In many cases, safety concerns can be mitigated by limiting dose and/or duration of use, by adding risk statements, or by limiting sub-populations, for example, pregnant and breastfeeding women.

Finally, a search should be made of the totality of evidence to ensure that no new, unknown safety concerns have been identified by the findings outside of evidence for traditional use. In many cases, other data will not exist to confirm safety, in which case existing safety data as described earlier will be considered sufficient. Should the available evidence suggest that a medicinal ingredient is unsafe when used according to the product's recommended conditions of use, it will be necessary to submit further evidence of equal or higher validity, causality, and credibility to demonstrate that the balance of evidence supports a favorable benefit to risk ratio for the product.

Theoretical concerns will not be considered in the absence of clinical data. Manufacturers may add substances to their medicinal ingredients to aid stability or manufacturing processes. If these remain in significant quantities in the finished NHP (e.g., including any quantity that still provides a technical effect), they must be declared as non-medicinal ingredients on the product license application form and label. It may be necessary to communicate with the manufacturer directly in order to identify these types of ingredients.

24.2.2.8 ***Efficacy Evidence for Traditional Medicines*** Traditional use claims for traditional medicines are divided into two subcategories according to the evidence provided:

1. Pharmacopoeial evidence alone
2. Other types and/or combinations of references supporting traditional use

24.2.2.9 ***Pharmacopoeial Evidence for Traditional Medicines*** Products providing only pharmacopoeial evidence and answering "yes" to all elements of the Checklist for the Traditional Pharmacopoeial Evidence Category only require one supporting reference. Answering "no" to any of the questions posed on this checklist excludes the product from being assessed within the pharmacopoeial stream.

Applications suitable for assessment within the pharmacopoeial evidence category should provide one of the following as evidence in supporting the claim:

1. A copy of the relevant pages of a monograph from a recognized pharmacopoeia (e.g., The Ayurvedic Pharmacopoeia of India and The Pharmacopoeia of the People's Republic of China (PPRC))
2. A copy of a monograph published by a reputable agency with a definition of traditional medicines comparable to that of the Natural Health Products Directorate (e.g., translated version of the Drug Standard of People's Republic of China [also called the State Drug Standard (SDS)]).

24.2.2.10 ***Qualifying Claims*** When a product meets the Natural Health Products Directorate (NHPD) definition of traditional medicine and relies on traditional evidence to support safety or efficacy, claims should be prefaced with qualifiers indicating the specific traditional system of medicine such as "traditionally used in Ayurvedic medicine," to identify for consumers, that efficacy is based on a specific system of traditional medicine. If the claim uses terminology specific to a particular culture or system of medicine, that specific terminology along with the culture or system of medicine should be specified in the claim (e.g., "Traditional Chinese Medicine (TCM) used to replenish Qi (vital energy)..." or "traditionally used in Ayurvedic medicine to improve agni [digestive fire]"). If both traditional use evidence and modern evidence are available to support a proposed claim, the use of the wording "traditionally used" is optional; however, a product making a traditional use claim based primarily on traditional use evidence is to be assessed as a traditional medicine. If a health claim is solely supported by modern evidence, it should not include the words "traditionally used."

The NHPD recognizes that some systems of traditional medicine may communicate risk information in language that is specific to that healing paradigm or culture. In cases where it is not evident to the consumer that the risk information is based on use within a traditional system of medicine, a traditional qualifier can be included in the risk information as an option (e.g., "do not use in cases of external pathogenic heat [TCM]").

24.2.3 Current Regulations in China

References Cited/Additional Reading Resources

The State Administration of Traditional Chinese Medicine website: www.satcm.gov.cn

The administrative department of public health under the State Council is responsible for the supervision and control of the protection of types of traditional Chinese medicines throughout the country. The State administrative departments for the production and trading of traditional Chinese medicine shall assist the administrative department with the control of the protection of types of traditional Chinese medicine throughout the country.

Regulations in China have been developed for the purposes of improving the quality of traditional Chinese medicine, protecting the legitimate rights and interests of traditional Chinese medicine producing enterprises, and promoting the development of traditional Chinese medicine. The regulations apply to types of traditional Chinese medicine produced and manufactured within the territory of China, including patent traditional Chinese medicines, extracts and preparations from natural medicinal materials, as well as artificial traditional Chinese medicines.

China encourages research and development of traditional Chinese medicine with clinical effects and practices a classification protection system for types of traditional Chinese medicine with reliable quality and accurate curative effects. The types of traditional Chinese medicine under the protection of the Regulations are those listed in the national pharmaceutical standards. Upon the determination of the administrative department of public health under the State Council, protection of types of traditional Chinese medicine listed in the pharmaceutical standards of provinces, autonomous regions, and municipalities directly under the Central Government may also be applied for. Two types of protection of traditional Chinese medicine are available, namely, first-class protection and second-class protection.

With regard to types of traditional Chinese medicine that conform to any of the following conditions, *first-class protection* may be applied for

- Having special curative effects for a certain disease
- Artificial medicines prepared from varieties of wild medicinal materials analogously under first-class protection
- Use for the prevention and cure of special diseases

With regard to types of traditional Chinese medicine that conform to any of the following conditions, *second-class protection* may be applied for

- conforming to the provisions of the Regulations, or having once been listed under first-class protection but now being cancelled
- having outstanding curative effects for a certain disease
- effective substances and special preparations extracted from natural medicinal materials

New medicines approved by the administrative department of public health under the State Council shall be protected within a period of protection as described by the administrative department. For those medicines that conform to the provisions in the Regulations, an application for the renewal of the protection may, 6 months before the expiration date of protection approved by the administrative department of public health under the State Council, be offered according to the provisions of the Regulations.

Procedures for handling applications for the protection of types of traditional Chinese medicine are as follows:

- Any traditional Chinese medicine producing enterprise may, if it thinks that the type of traditional Chinese medicine it produces conforms to provisions in the Regulations, apply for protection with the local administrative department for the production and trading of traditional Chinese medicine in the province, autonomous region, or municipality directly under the Central Government.

- The local administrative department for the production and trading of traditional Chinese medicine shall write down its comments on the application, then transmit it to the administrative department of public health at the same level, which shall make a preliminary examination and write down its comments and submit the application, with comments, to the administrative department of public health under the State Council.
- Under special circumstances, a traditional Chinese medicine producing enterprise may directly apply to the State administrative department for the production and trading of traditional Chinese medicine that shall write down comments on the application and transmit it to the administrative department of public health under the State Council, or may directly apply to the administrative department of public health under the State Council.
- The State examination and evaluation committee for the protection of types of traditional Chinese medicine shall, under the authorization of the administrative department of public health under the State Council, be responsible for the examination and evaluation of the types of traditional Chinese medicine for which the protection is applied for. The committee shall provide an examination and evaluation conclusion within 6 months of the date of receiving an application.
- Based on the conclusion of the State examination and evaluation committee for the protection of types of traditional Chinese medicine, the administrative department of public health under the State Council shall, in consultation with the State administrative department for the production and trading of traditional Chinese medicine, decide whether or not to grant the protection thereto.
- For the types of traditional Chinese medicine of which the protection has been approved, the administrative department of public health under the State Council shall issue a Certificate of Protection of Types of Traditional Chinese Medicine.
- The administrative department of public health under the State Council shall be responsible for the formation of the State examination and evaluation committee for the protection of types of traditional Chinese medicine, members of which shall, in consultation with the State administrative department for the production and trading of traditional Chinese medicine, be appointed from experts in the field of medical service, scientific research, inspection, as well as trading and management of traditional Chinese medicine.
- Any enterprise applying for protection of types of traditional Chinese medicine shall provide the State examination and evaluation committee for the protection of types of traditional Chinese medicine with complete sets of materials required by the administrative department of public health under the State Council.
- The administrative department of public health under the State Council shall make announcements in the designated professional newspapers and periodicals regarding the types of traditional Chinese medicine to which protection has been granted or those for which the period of protection has expired.

24.2.3.1 ***Protection of Protected Types of Traditional Chinese Medicine*** The period of protection for types of traditional Chinese medicine is as follows:

1. The period of *first-class protection* is (Group 1) 30 years, (Group 2) 20 years, and (Group 3) 10 years.
2. For all groups, the period of *second-class protection* is 7 years.

Within the period of protection, the prescriptions and pharmaceutical techniques of types of traditional Chinese medicine under first-class protection shall be kept secret and shall not be published by the producing enterprises having been granted the Certificate of Protection of Types of Traditional Chinese Medicine, the administrative department for the production and trading of traditional Chinese medicine, the administrative departments of public health, and other units or individuals concerned. Departments, enterprises, and units concerned that have the duty to keep secrets shall set up necessary security systems as required by the State. Transfer to any foreign country of prescriptions and pharmaceutical techniques of types of traditional Chinese medicine under first-class protection shall be dealt with according to the relevant State provisions of security.

Where, due to special circumstances, it is necessary to extend the period of protection of a type of traditional Chinese medicine under first-class protection, the producing enterprise shall, 6 months before the expiration date of protection, submit an application for extension according to the procedures described in the Regulations. The extended period of protection shall be decided by the State examination and evaluation for the protection of types of traditional Chinese medicine, however, an extension approved each time shall not exceed the period of protection granted for the first time.

The period of protection of types of traditional Chinese medicine under second-class protection may be extended for 7 years upon expiration. If it is necessary to extend the period of protection of a type of traditional Chinese medicine under second-class protection, the producing enterprise shall, 6 months before the expiration date of protection, submit an application for extension according to the procedures described in the Regulations.

The production of protected types of traditional Chinese medicine within the period of protection shall be restricted to enterprises that have been granted the Certificate of Protection of Types of Traditional Chinese Medicine, unless otherwise provided for in the Regulations. Where more than one enterprise produces a type of traditional Chinese medicine under protection before the protection is granted by the administrative department of public health under the State Council, those enterprises who have not applied for the Certificate of Protection of Types of Traditional Chinese Medicine shall, within 6 months as of the date of announcement, report the case to the administrative department of public health under the State Council and provide relevant materials according to the provisions of the Regulations.

The administrative department of public health under the State Council shall designate a pharmaceutical inspection institution to inspect the quality of the reported type of medicine as has been done with the type under protection. Based on the inspection, the administrative department of public health under the State Council may take the following measures:

1. If it is up to the national pharmaceutical standards, the Certificate of Protection of Types of Traditional Chinese Medicine shall be issued through consultation with the State competent authority for the production and trading of traditional Chinese medicine.
2. If it is below the national pharmaceutical standards, the registered document of approval of this type of traditional Chinese medicine shall be revoked according to the laws and regulations governing pharmaceutical administration.

With regard to protected types of traditional Chinese medicine in short supply for clinical needs, the administrative departments of public health in provinces, autonomous regions, and municipalities directly under the Central Government shall, as proposed by the State administrative department for the production and trading of traditional Chinese medicine and with the approval of the administrative department of public health under the State Council, issue registered documents of approval to the enterprises that produce in their localities the same types of traditional Chinese medicine as the protected types for imitation.

The imitation enterprises shall pay reasonable use fees to the enterprises that hold the Certificate of Protection of Types of Traditional Chinese Medicine and transfer the prescriptions and pharmaceutical techniques of the protected types. The amounts of the use fees shall be decided by the two sides through consultation. If the two sides fail to reach an agreement, the administrative department of public health under the State Council shall make a ruling.

Enterprises producing protected types of traditional Chinese medicine and the administrative department for the production and trading of traditional Chinese medicine shall improve conditions of production and the qualities of the protected types as required by the administrative departments of public health in provinces, autonomous regions, and municipalities directly under the Central Government.

The registration of protected types of traditional Chinese medicine within the period of protection in any foreign country shall be subject to the approval of the administrative department of public health under the State Council.

If anyone divulges secrets in violation of the provisions of the Regulations, the unit to which he belongs or the higher authority shall impose upon him disciplinary sanctions. If a crime has been committed, criminal liability shall be investigated according to laws. If anyone, in violation of the provisions of the Regulations, imitates a protected type of traditional Chinese medicine without approval, the administrative departments of public health at or above the county level shall punish him as a producer of fake medicines. If anyone fabricates the Certificate of Protection of Types of Traditional Chinese Medicine and relevant certification documents to produce and sell medicines, the administrative departments of public health at or above the county level shall confiscate all medicines involved and illegal gains and may concurrently fine him not more than three times the price of the standard equivalents of medicines involved. If the aforesaid acts have constituted crimes, the judicial organs shall investigate for criminal liabilities. Anyone who refuses to accept the decision of punishment made by the administrative departments of public health may apply for administrative reconsideration or institute administrative proceedings according to the relevant provisions of laws and administrative regulations.

24.2.4 Current Regulations in the European Union (EU)

References Cited/Additional Reading Resources

The European Medicines Agency website: www.ema.europa.eu/ema/

The EU Directive on Traditional Herbal Medicinal Products (THMP)—Directive 2004/24/EC and Directive 2001/83/EC for medicinal products for human use—was established by the European Parliament and Council on March 31, 2004, to provide a simplified regulatory approval process for traditional herbal medicines in the European Community (EC). Prior to 2004, there was no formal EC-wide authorization procedure, so each EC member state regulated these types of products at the national level. Under this regulation, all herbal medicinal products are required to obtain an authorization to market within the EC.

Manufacturers of THMP marketed before this legislation came into force were permitted to continue to market their product until April 30, 2011, under the transitional measures defined in the Traditional Herbal Medicinal Products Directive. Once this deadline expired, all herbal medicinal products must have prior authorization before they can be marketed in the EC. For those herbal medicinal products that were not on the market before April 30, 2004, an authorization must be obtained prior to marketing.

The Directive includes the following definitions:

- **Traditional herbal medicinal product**: A herbal medicinal product that fulfils the conditions laid down in the Regulations.
- **Herbal medicinal product:** Any medicinal product, exclusively containing as active ingredients one or more herbal substances or one or more herbal preparations, or one or more such herbal substances in combination with one or more such herbal preparations.
- **Herbal substances:** All mainly whole, fragmented, or cut plants, plant parts, algae, fungi, lichen in an unprocessed, usually dried, form, but sometimes fresh. Certain exudates that have not been subjected to a specific treatment are also considered to be herbal substances. Herbal substances are precisely defined by the plant part used and the botanical name according to the binomial system (genus, species, variety, and author);
- **Herbal preparations:** Preparations obtained by subjecting herbal substances to treatments such as extraction, distillation, expression, fractionation, purification, concentration, or fermentation. These include comminuted or powdered herbal substances, tinctures, extracts, essential oils, expressed juices, and processed exudates.

The only herbal medicines that are exempted from the provisions of the Traditional Herbal Medicinal Products Directive are those unlicensed remedies that are made up for a patient following a consultation with a herbalist.

Herbal medicines must be now manufactured under Good Manufacturing Practices (GMP) to ensure the quality of the finished product and also demonstrate safety.

Under the Traditional Herbal Medicinal Products Directive, a company needs to demonstrate that the herbal medicine has been in use within the EU for at least 30 or 15 years within the EU and 30 years outside the EU. There was concern that some herbal remedies used over 30 years ago, which are no longer in widespread use, could still be sold but that valid new herbal remedies, which cannot meet the 30-year rule, may no longer be allowed to be sold. The rule also means that it may not be possible to license some traditional herbal medicines that were in common use more than 30 years ago, but have since fallen into disuse. Such matters are reviewed on a case-by-case basis.

There are specific eligibility criteria for a herbal medicine to qualify under this legislation:

- Only herbal medicines that are administered orally, externally, or by inhalation are suitable. Any medication that requires intravenous administration will not be authorized.
- Only herbal medicines that are intended to be used without supervision by a medical doctor will be authorized by this scheme.
- The intended use of herbal medicines will only be authorized on the basis of its traditional history and/or the recognized pharmacological properties of the herbal ingredient(s).
- Vitamins and minerals may be added to the herbal medicine provided that their use is ancillary to the herbal ingredient(s).

If a competent EC member state judges that the herbal medicine fulfills the criteria for a marketing authorization, then an authorization under Traditional Herbal Medicines Product Directive should be granted. Herbal medicine products manufactured using isolated active ingredients from plants will not be regarded as herbal medicines and will not receive an authorization under this scheme. The Traditional Herbal Medicines Product Directive does allow medicinal claims to be made on the label of the final product, although restrictions do apply on the final wording.

The European Traditional Herbal Medicinal Products Directive has established a regulatory approval process for herbal medicines in the European Community (EC). This Directive requires each EC Member State to set up a traditional herbal registration scheme for manufactured traditional herbal medicines that are suitable for use without medical supervision. The EU Medicinal Products Directive 2001/83/EC (4) requires that applications for authorization to place a medicinal product on the market be accompanied by a dossier containing particulars and documents relating in particular to the results of physicochemical, biological, or microbiological tests as well as pharmacological and toxicological tests and clinical trials carried out on the product and thus proving its quality, safety, and efficacy. Where the applicant can demonstrate by detailed references to published scientific literature that the constituent or the constituents of the medicinal product has or have a well-established medicinal use with recognized efficacy and an acceptable level of safety within the meaning of Directive 2001/83/EC, he/she should not be required to provide the results of pre-clinical tests or the results of clinical trials.

A significant number of medicinal products, despite their long tradition, do not fulfil the requirements of a well-established medicinal use with recognized efficacy and an acceptable level of safety and are not eligible for a marketing authorization. To maintain these products on the market, some of the Member States have enacted differing procedures and provisions. The differences that currently exist between the provisions laid down in the Member States may hinder trade in traditional medicinal products within the EC and lead to discrimination and distortion of competition between the manufacturers of these products. These differences may also have an impact on the protection of public health since the necessary guarantees of quality, safety, and efficacy are not always provided by the various laws.

Having regard to the particular characteristics of these medicinal products, especially their long tradition, it is desirable to provide a special, simplified registration procedure for certain traditional medicinal products. However, this simplified procedure should be used only where no marketing authorization can be obtained pursuant to Directive 2001/83/EC, in particular because of a lack of sufficient scientific literature demonstrating a well-established medicinal use with recognized efficacy and an acceptable level of safety.

The long tradition of the medicinal product makes it possible to reduce the need for clinical trials, in so far as the efficacy of the medicinal product is plausible on the basis of long-standing use and experience. Pre-clinical tests do not seem necessary, where the medicinal product on the basis of the information on its traditional use, proves not to be harmful in specified conditions of use. However, even a long tradition does not exclude the possibility that there may be concerns with regard to the product's safety and therefore the competent authorities are entitled to ask for all data necessary for assessing the safety.

The quality aspect of the medicinal product is independent of its traditional use so that no exemption should be granted with regard to the necessary physicochemical, biological, and microbiological tests. Products should comply with quality standards in relevant European Pharmacopoeia monographs or those in the pharmacopoeia of a Member State.

The vast majority of medicinal products with a sufficiently long and coherent tradition are based on herbal substances. It therefore seems appropriate to limit the scope of the simplified registration as a first step to traditional herbal medicinal products. The simplified registration should be acceptable only where the herbal medicinal product may rely on a sufficiently long medicinal use in the EC. Medicinal use outside the EC should be taken into account only if the medicinal product has been used within the EC for a certain time. Where there is limited evidence of use within the Community, it is necessary to assess carefully the validity and relevance of use outside the Community.

With the objective of further facilitating the registration of certain traditional herbal medicinal products and of further enhancing harmonization, there should be the possibility of establishing a Community list of herbal substances that fulfil certain criteria, such as having been in medicinal use for a sufficiently long time and hence being considered not to be harmful under normal conditions of use.

Having regard to the particularities of herbal medicinal products, a Committee for Herbal Medicinal Products should be established within the European Agency for the Evaluation of Medicinal Products (hereinafter "the Agency") set up by Council Regulation (EEC) No 2309/93(5). The Committee carries out the tasks necessary for the simplified registration and authorization of medicinal products as provided for in this Directive. Its tasks relate in particular to establishing Community herbal monographs relevant for the registration as well as the authorization of herbal medicinal products. The Committee is composed of experts in the field of herbal medicinal products. It is important to ensure full consistency between the new Committee and the Committee for Human Medicinal Products already existing within the Agency.

In order to promote harmonization, Member States should recognize registrations of traditional herbal medicinal products granted by another Member State based on Community herbal monographs or consisting of substances, preparations, or combinations thereof contained in a list to be established. For other products, Member States should take due account of such registrations. This Directive allows non-medicinal herbal products, fulfilling the criteria of food legislation, to be regulated under food legislation in the Community.

The Commission should present a report on the application of the chapter on traditional herbal medicinal products to the European Parliament and to the Council including an assessment on the possible extension of traditional-use registration to other categories of medicinal products.

Specific provisions applicable to traditional herbal medicinal products: A simplified registration procedure (hereinafter "traditional-use registration") is hereby established for herbal medicinal products that fulfil all of the following criteria:

(a) They have indications exclusively appropriate to traditional herbal medicinal products that, by virtue of their composition and purpose, are intended and designed for use without the supervision of a medical practitioner for diagnostic purposes or for prescription or monitoring of treatment.
(b) They are exclusively for administration in accordance with a specified strength and posology.
(c) They are an oral, external, and/or inhalation preparation.
(d) The period of traditional use as laid down in the regulations has elapsed.
(e) The data on the traditional use of the medicinal product are sufficient; in particular, the product proves not to be harmful in the specified conditions of use and the pharmacological effects or efficacy of the medicinal product are plausible on the basis of long-standing use and experience.

Notwithstanding the regulations, the presence in the herbal medicinal product of vitamins or minerals for the safety of which there is well-documented evidence shall not prevent the product from being eligible for registration, provided that the action of the vitamins or minerals is ancillary to that of the herbal active ingredients regarding the specified claimed indications.

Additional Requirements:
The applicant and registration holder shall be established in the Community.

1. In order to obtain traditional-use registration, the applicant shall submit an application to the competent authority of the Member State concerned.
2. The application shall include:

   (a) The particulars and documents required by the regulations;
   (b) The results of the pharmaceutical tests required by the regulations;
   (c) The summary of product characteristics required by the regulations;
   (d) In case of combinations, the information required by the regulations relating to the ingredients in the combination, if the individual active ingredients are not sufficiently known, the data shall also relate to the individual active ingredients;
   (e) Any authorization or registration obtained by the applicant in another Member State, or in a third-party country, to place the medicinal product on the market and details of any decision to refuse to grant an authorization or registration, whether in the Community or a third country and the reasons for any such decision;

3. Bibliographical or expert evidence to the effect that the medicinal product in question, or a corresponding product has been in medicinal use throughout a period of at least 30 years preceding the date of the application, including at least 15 years within the Community.At the request of the Member State where the application for traditional-use registration has been submitted, the Committee for Herbal Medicinal Products shall draw up an opinion on the adequacy of the evidence of the long-standing use of the product, or of the corresponding product. The Member State shall submit relevant documentation supporting the referral.
   (a) A bibliographic review of safety data together with an expert report and where required by the competent authority, upon additional request, data necessary for assessing the safety of the medicinal product. A corresponding product, as referred to earlier, is characterized by having the same active ingredients, irrespective of the excipients used, the same or similar intended purpose, equivalent strength and posology, and the same or similar route of administration as the medicinal product applied for. The requirement to show medicinal use throughout the period of 30 years, referred to earlier, is satisfied even where the marketing of the product has not been based on a specific authorization. It is likewise satisfied

if the number or quantity of ingredients of the medicinal product has been reduced during that period. Where the product has been used in the Community for less than 15 years, but is otherwise eligible for simplified registration, the Member State where the application for traditional-use registration has been submitted shall refer the product to the Committee for Herbal Medicinal Products. The Member State shall submit relevant documentation supporting the referral. The Committee shall consider whether the other criteria for a simplified registration as referred to in the regulations are fully complied with. If the Committee considers it possible, it shall establish a Community herbal monograph that shall be taken into account by the Member State when taking its final decision.

Simplified Herbal Medicinal Product Registrations shall be granted in accordance with the regulations, provided that:

(a) A Community herbal monograph has been established in accordance with the regulations.
(b) The herbal medicinal product consists of herbal substances, preparations, or combinations thereof contained in the approved list required by the regulations.

For other herbal medicinal products as referred to in the regulations, each Member State shall, when evaluating an application for traditional-use registration, take due account of registrations granted by another Member State in accordance with these regulations. Traditional-use registration shall be refused if the application does not comply with the regulations cited earlier, or if at least one of the following conditions is fulfilled:

(a) The qualitative and/or quantitative composition is not as declared.
(b) The indications do not comply with the conditions laid down in the regulations.
(c) The product could be harmful under normal conditions of use.
(d) The data on traditional use are insufficient, especially if pharmacological effects or efficacy are not plausible on the basis of long-standing use and experience.
(e) The pharmaceutical quality is not satisfactorily demonstrated.

The competent authorities of the Member States shall notify the applicant, the Commission, and any competent authority that requests it, of any decision they take to refuse traditional-use registration and the reasons for the refusal.

A list of herbal substances, preparations, and combinations for use in traditional herbal medicinal products shall be established in accordance with the procedure referred to in the regulations. The list shall contain, with regard to each herbal substance, the indication, the specified strength and the posology, the route of administration, and any other information necessary for the safe use of the herbal substance as a traditional medicinal product. If a herbal substance, preparation, or a combination ceases to be included in the list referred to earlier, registrations pursuant to the regulation of herbal medicinal products containing this substance shall be revoked unless the particulars and documents referred to in the regulations are submitted within 3 months.

In addition, any labeling and user package leaflet shall contain a statement to the effect that:

(a) The product is a traditional herbal medicinal product for use in specified indication(s) exclusively based upon long-standing use.
(b) The user should consult a doctor or a qualified healthcare practitioner if the symptoms persist during the use of the medicinal product or if adverse effects not mentioned in the package leaflet occur.

A Member State may require that the labeling and the user package leaflet shall also state the nature of the tradition in question. Additional regulations regarding any advertisement for a medicinal product registered under this chapter shall contain the following statement: "Traditional herbal medicinal product for use in specified indication(s) exclusively based upon longstanding use."

Under these regulations, a Committee for Herbal Medicinal Products has been established. That Committee is now part of the Health Agency and has the following competence:

(a) As regards to simplified registrations, to:
    - Perform the tasks arising under the regulations.
    - Prepare a draft list of herbal substances, preparations, and combinations.
    - Establish Community monographs for traditional herbal medicinal products, as referred to in the regulations.
(b) As regards to authorizations of herbal medicinal products, to establish Community herbal monographs for herbal medicinal products, as defined in the regulations.
(c) As regards to where other medicinal products containing herbal substances are referred to the Agency, to give an opinion on the herbal substance where appropriate.

Finally, the Committee for Herbal Medicinal Products shall perform any other tasks conferred upon it by Community law. The appropriate coordination with the Committee for Human Medicinal Products shall be ensured by a procedure to be determined by the Executive Director of the Agency in accordance with Article 57(2) of Regulation (EEC) No 2309/93.

The Committee for Herbal Medicinal Products shall establish Community herbal monographs for herbal medicinal products, as well as traditional herbal medicinal products. This Committee shall fulfil further responsibilities conferred upon it by provisions of the regulations and other Community law. See the following website for the current monographs: http://www.ema.europa.eu/ema/index.jsp?curl=/pages/medicines/landing/herbal_search.jsp&mid=WC0b01ac058001fa1d.

These Community herbal monographs shall be taken into account by the Member States when examining an application. Where no such Community herbal monograph has yet been established, other appropriate monographs, publications, or data may be referred to. As new Community herbal monographs are established, the registration holder shall consider whether it is necessary to modify the registration dossier accordingly. The registration holder shall notify any such modification to the competent authority of the Member State concerned.

The general provisions of Regulation (EEC) No 2309/93 relating to the Committee for Human Medicinal Products shall apply by analogy to the Committee for Herbal Medicinal Products. For the traditional herbal medicinal products as referred to in the regulations that are already on the market on the entry into force of this Directive, the competent authorities shall apply the provisions of this Directive within 7 years after its entry into force.

24.2.5 Current Regulations in India

References Cited/Additional Reading Resources

The Central Drugs Standard Control Organization website: www.cdsco.nic.in

India has a number of rich traditional medicine systems including Ayurveda, Siddha, and Unani Tibb practices and the government regulates these practices through the several different agencies.

24.2.5.1 India's Central Drugs Standard Control Organization Website: http://www.cdsco.nic.in

The Central Drugs Standard Control Organization (CDSCO) is India's main regulatory body for pharmaceuticals and medical devices. Within the CDSCO, the Drug Controller General of India (DCGI) is responsible for the regulation of pharmaceuticals and medical devices. The DCGI is advised by the Drug Technical Advisory Board (DTAB) and the Drug Consultative Committee (DCC).

The CDSCO establishes safety, efficacy, and quality standards for pharmaceuticals and medical devices. It publishes and updates the Indian Pharmacopeia, a list of regulated pharmaceuticals and devices. For all drug and device applications, the CDSCO appoints notified bodies to perform conformity assessment procedures, including testing, in order to ensure compliance with their standards. The CDSCO is also divided into several zonal offices that do pre-licensing and post-licensing inspections, post-market surveillance, and recalls when necessary. In addition to its regulatory functions, the CDSCO offers technical guidance, trains regulatory officials and analysts, and monitors adverse events. The CDSCO works with the World Health Organization to promote Good Manufacturing Practices (GMP) and international regulatory harmony.

24.2.5.2 Central Council of Indian Medicine Website: http://www.ccimindia.org

The Central Council of Indian Medicine (CCIM) is a statutory body under the Department of Ayurveda, Yoga and Naturopathy, Unani, Siddha and Homoeopathy (AYUSH), in the Ministry of Health and Family Welfare. This Council was established in 1971 under the Indian Medicine Central Council Act (Act 48), which was passed in 1970. It is one of the professional councils under the University Grants Commission

(UGC) to monitor higher education in Indian Systems of Medicine (ISM), including Ayurveda, Siddha, and Unani medical practices.

The main objects of the Central Council are

1. To prescribe minimum standards of education in Indian Systems of Medicine regarding Ayurveda, Siddha, and Unani Tibb practices.
2. To advise the Central Government in matters relating to recognition (inclusion/withdrawal) of medical qualification in/from second schedule to Indian Medicine Council Act, 1970.
3. To maintain a Central Register on Indian Medicine and revise the register from time to time.
4. To prescribe Standards of Professional Conduct, Etiquette, and Code of Ethics to be observed by the practitioners.

The CCIM has translated the complete teaching syllabus of Ayurveda, Unani, and Siddha medical practices into the English language to promote the worldwide appreciation of the Indian Systems of Medicine. Prior to this translation, the teaching syllabus for both undergraduate and post-graduate courses of Ayurveda, Unani, and Siddha were in the Sanskrit, Urdu, and Tamil languages, respectively. This language barrier was hindering the path of success and popularity of these medical systems inside and outside of India.

The Council also has instituted a new post-graduate diploma course to provide specialized services of ISM systems and to enhance the worldwide benefits of these ancient medical systems. Other improvements in the teaching and practical training of ISM practitioners adopted by the Council concern Minimum Teaching Standards mandated by the CCIM. Additional CCIM actions have included:

- Preparing a database of all ISM teachers. The aim of preparing the database was to keep a record of all ISM teachers and to assess their qualifications and ISM eligibility.
- Appointment of teaching staff in Ayurveda, Unani, and Siddha colleges. More than 4000 ISM teachers have been appointed in these colleges.
- Improving review and teaching methods so qualified ISM doctors may become more skilled practitioners, researchers, and scientists and can provide the best service to the community.

The Drugs and Cosmetics Act of 1940 specifies aspects of control of Ayurvedic, Siddha, and Unani drugs as follows:

- No person shall manufacture for sale or for distribution any Ayurvedic, Siddha, or Unani drug except in accordance with such standards, as may be prescribed in relation to that drug.
- From such date as the State Government may, by notification in the Official Gazette, specify in this behalf, no person, either by himself or by any other person on his behalf, shall

(a) Manufacture for sale or for distribution
   (i) Any misbranded, adulterated, or spurious Ayurvedic, Siddha, or Unani drugs
   (ii) Any patent or proprietary medicine, unless there is displayed in the prescribed manner on the label or container thereof the true list of all the ingredients contained in it; and Drugs and Cosmetics Act, 1940
   (iii) Any Ayurvedic, Siddha, or Unani drug in contravention of any of the provisions of this chapter or any rule made thereunder

(b) Sell, stock, or exhibit or offer for sale or distribution, any Ayurvedic, Siddha, or Unani drug that has been manufactured in contravention of any of the provisions of this Act, or any rule made thereunder

(c) Manufacture for sale or for distribution, any Ayurvedic, Siddha, or Unani drug, except under and in accordance with the conditions of a license issued for such purpose under this chapter by the prescribed authority
   - Provided that nothing in these regulations apply to physicians (Vaidyas and Hakims) who manufacture Ayurvedic, Siddha, or Unani drug for the use of their own patients
   - Provided further that nothing in this section shall apply to the manufacture, subject to the prescribed conditions, of small quantities of any Ayurvedic, Siddha, or Unani drug for the purpose of examination, test, or analysis

Without prejudice to any other provision contained in these regulations, if the Central Government is satisfied on the basis of any evidence or other material available before it that the use of any Ayurvedic, Siddha, or Unani drug is likely to involve any risk to human beings or animals or that any such drug does not have the therapeutic value claimed or purported to be claimed for it and that in the public interest it is necessary or expedient to do so, then that Government may, by notification in the Official Gazette, prohibit the manufacture, sale, or distribution of such drug. The Central Government or a State Government may, by notification in the Official Gazette, appoint such persons as it thinks fit, having the prescribed qualifications, to be Government Analysts for such areas as may be assigned to them by the Central Government or the State Government, as the case may be. Notwithstanding anything contained in these regulations, neither the Central Government nor a State Government shall appoint as a Government Analyst any official not serving under it without the previous consent of the Government under which he is serving.

The Central Government or a State Government may, by notification in the Official Gazette, appoint such persons as it thinks fit, having the prescribed qualifications, to be Inspectors for such areas as may be assigned to them by the Central Government or the State Government, as the case may be. The powers that may be exercised by an inspector and the duties that may be performed by him and the conditions, limitations, or restrictions subject to which such powers and duties may be exercised or performed shall be such as may be prescribed. Every inspector shall be deemed to be a public servant within the meaning of the Indian Penal Code and shall be officially subordinate to such authority as the Government appointing him may specify in this behalf.

Penalties for the manufacture, sale, etc., of Ayurvedic, Siddha, or Unani drug in contravention of these regulations are as follows:

Whoever himself or by any other person on his behalf

1. Manufactures for sale or for distribution
    - (a) Any Ayurvedic, Siddha, or Unani drug
        - (i) Deemed to be adulterated under these regulations
        - (ii) Without a valid license as required under these regulations, shall be punishable with imprisonment for a term that may extend to 1 year and with fine that shall not be less than ₹2000
2. Any Ayurvedic, Siddha, or Unani drug deemed to be spurious under these regulations shall be punishable with imprisonment for a term that shall not be less than 1 year but that may extend to 3 years and with fine that shall not be less than ₹5000
    - (a) Provided that the Court may, for any adequate and special reasons to be mentioned in the judgment, impose a sentence of imprisonment for a term of less than 1 year and of fine of less than ₹5000
3. Contravenes any other provisions of these regulations, shall be punishable with imprisonment for a term that may extend to 3 months and with fine that shall not be less than ₹500.
    - (a) Where any person has been convicted under these regulations, the stock of the Ayurvedic, Siddha, or Unani drug, in respect of which the contravention has been made, shall be liable to confiscation.

24.2.6 Current Regulations in Japan

References Cited/Additional Reading Resources

The Ministry of Health, Labour and Welfare website: http://www.mhlw.go.jp/english

In Japan, traditional medicines are classified into two broad groups: kampo medicine and traditional medicine indigenous to Japan. Traditional Chinese medicine, introduced to Japan between the third and eighth centuries, was modified to meet local needs and became known as kampo medicine.

Under the Medical Practitioners Law 201 of 1948, only allopathic physicians may practice medicine, including kampo medicine. However, there are no restrictions on the types of medical procedures allopathic physicians may use in their practice. According to the Pharmacists Law 146 of 1960, a person must be qualified as a pharmacist in order to engage in services related to traditional medicines.

The Subcommittee on Kampo Medicines and Products of Animal and Plant Origin of the Central Pharmaceutical Affairs Council has developed regulations governing kampo medicines as proprietary medicines. These regulations also apply, with necessary modifications, to prescription medicines.

The Pharmaceutical Affairs Law in Japan does not distinguish between traditional and allopathic medicines; both types of preparations are subject to the same regulations.

Kampo medicines are products prepared for use in accordance with kampo medicine formulae, which, according to the principles set out by the Central Pharmaceutical Affairs Council, are formulae described in established books on kampo medicine currently and frequently used in Japan. The formulae include standard formulae, added or subtracted formulae, and combined formulae. They include formulae containing vitamins B1, B2, and/or C for nutritional supplementation. The extracts prepared from kampo medicine formulae should be limited to those that have previously been used as decoctions. Any ingredient, efficacy, or indication that is not appropriate for proprietary medicines is not accepted.

Standards for medicinal plant materials are included in Japanese Pharmacopoeia, the Japanese Herbal Medicine Codex, and Japanese Standards for Herbal Medicines.

When the Pharmaceutical Affairs Law was amended in 1993, the Regulations for Manufacturing Control and Quality Control of Drugs were changed from manufacturing requirements for drug companies to a prerequisite for licenses to manufacture drugs. The Regulations, including new validation requirements, came into effect in 1996. Moreover, good manufacturing practices for investigational products were adopted via a notice issued by the Director-General of the Pharmaceutical Affairs Bureau of the Health Ministry in 1997.

In 1990, the Society of Japanese Oriental Medicine started a registration system of allopathic physicians specializing in kampo medicine. This system requires all registered specialists to attend authorized meetings of the Society and to present relevant scientific papers and medical journals at the meetings. This registration system requires registration as a specialist in kampo medicine to be renewed every 5 years, in accordance with the rules set out by the Society.

Herbal medicines have been used clinically in Japan and traditional Japanese herbal (Kampo) formulas are approved by the government as ethical drugs. Kampo formulas are often mixtures of the crude extracts of several herbs, each of which contains multiple components.

Each Kampo drug is typically a formula consisting of a mixture of 5–10 different herbs. New features have been introduced into the practice of Kampo medicines. Most of the modern ready-to-use forms of the original formulae are produced in industrialized granular, powdered, or other forms based on the classical decoction.

24.2.6.1 Evaluation of Japanese and Chinese Medicines The difference between Japanese herbal medicine and Chinese medicine lies in the evaluation methods. Therefore, the applicable approval processes are quite different. The effect of a herbal medicine depends entirely on the sum of the pharmacological actions of the effective ingredients contained in the raw herb. There is no significant difference between the methods of evaluation applicable to herbal medicines and those with chemical substances.

On the other hand, local traditional usage alone is not sufficient for approval as a drug. The claims and rules of combinations are determined on the basis of the pharmacological actions of the ingredients that may be contained in the raw herbs. Usually, the scope of claims is clearly specified in the corresponding monograph. In cases where the monograph is not yet completed, the claims shown in the Japanese Pharmacopoeia are used as a guide.

In the evaluation of efficacy of a Chinese medicine, importance is given to the "empirical facts or experience" such as the reference data, clinical test reports, etc., rather than the pharmacological action of each ingredient. In many cases, Chinese medicine is administered in the form of an infusion and/or powders. However, for reasons of convenience or industrialization, raw materials are transformed into extracts and produced and marketed as tablets, granules, and powders.

Safety and efficacy have been estimated based on general methods employed by modern medical science. In 1972, the Health Ministry designated 210 formulas as over-the-counter drugs. This selection was based mainly on the experience of doctors actually practicing Chinese galenical medicine. In 1976, the Ministry specified 146 prescriptions as national health insurance applicable prescription drugs. In the case of an application for approval of a prescription other than those mentioned, specified data on safety, stability, comparison with other drugs, clinical test results, etc., are required to be submitted.

New Kampo drugs are regulated in essentially the same way as Western drugs in Japan. They are regarded as a form of combined drug and the same data required for new Western drugs are required for new Kampo drugs in a New Drug Application. The time-consuming and expensive chronic toxicity tests and special toxicity tests such as for mutagenicity, carcinogenicity, and teratogenicity depend on the possible length of treatment and indications that apply to them. Data for three-phase clinical trials are also required. For generic Kampo drugs, bioequivalence data are required, which may discourage development, because pharmacokinetic studies of Kampo drugs are difficult to conduct and bioassay methods are quite limited.

24.2.6.2 Guidelines on Quality of Kampo Drugs The Advisory Committee for Kampo Drugs was established in 1982 in close association with the Pharmaceutical Affairs Bureau of the Health Ministry. A Working Group on the Quality of Kampo Drugs was established and, 3 years later, a new regulation was issued by the Pharmaceutical Affairs Bureau setting standards for the manufacture and quality control of Kampo drugs. This ensures that the quality of herbs used in each original formula meets precise standards. The regulations also call for quality monitoring of specific ingredients, using at least two different chemical or physical methods to test them.

Since October 1986, Good Manufacturing Practices (GMP), a standard required of pharmaceutical drugs issued by the Health Ministry in 1976, applies also to Kampo drugs. In addition, in 1988, the Japan Kampo Medicine Manufacturers' Association drew up self-imposed guidelines that take into consideration the unique nature of Kampo drugs. In 1985, guidelines for ethical extract products in oriental medicine formulations were developed, according to which the data from a comparative study of the extract and a standard decoction have to be provided by the manufacturer of an ethical extract product. Besides data on the crude drug and on the standard decoction prepared in accordance with the Chinese traditional medicine prescription, a comparative study has to describe the content of an indicator ingredient in the finished product, which is required to be more than 70% of the content of the indicator ingredient in the standard decoction.

24.2.6.3 Post-Marketing Surveillance The Health Ministry employs three major systems for the collection of domestic adverse reaction data. The first is the Adverse Drug Reaction Monitoring System under which 2915 monitoring hospitals have been designated and requested to report cases of adverse drug reactions to the Ministry. This is a "voluntary" monitoring system.

The second data collection system is the Pharmacy Monitoring System formed by 2733 pharmacies. This system mainly collects data on cases of adverse reactions to over-the-counter drugs. In recent years, about 400 cases have been reported annually. Among these, reactions caused by Kampo drugs are the most common, though most of these adverse reactions are minor, involving symptoms such as gastric discomfort and skin problems. The third system is Adverse Reaction Reporting from Manufacturers. Since 1988, a Good Post-Marketing Surveillance Practice (GPMSP) has been used on Western drugs dispensed in Japan. As new Kampo drugs are approved and appear on the market, this guideline also applies to them.

24.2.7 Current Regulations: United Kingdom

References Cited/Additional Reading Resources

The Medicines and Healthcare products Regulatory Agency (MHRA) website: www.mhra.gov.uk

The Medicines and Healthcare products Regulatory Agency (MHRA) is responsible for the regulation of medicines and medical devices and equipment used in healthcare and the investigation of harmful incidents. The MHRA also looks after blood and blood products, working with UK blood services, healthcare providers, and other relevant organizations to improve blood quality and safety. As with other governmental reviews of drugs, the MHRA rules and regulations are designed to ensure that all medicines meet acceptable standards of safety, quality, and efficacy.

In the United Kingdom, the law defines a medicine as something used in connection with a disease, whether it is used to prevent, treat, or diagnose it, in anesthesia, investigating conditions or interfering with the normal operation of the body. It does not include such things as contact lens fluids, food supplements, and cosmetics. Many factors are considered in deciding whether a product is actually a medicine such as what it contains, what it is advertised or used for, the way it will be used, any particular targeting of the marketing information, and what the promotional literature says. Claims that a product "supports" health or a healthy lifestyle is not usually considered as medicinal. Control of medicines starts as soon as they are first discovered and tested in healthy volunteers, all the way through to when a company wants to change the conditions its products are approved for, such as changing the color of the tablet or what it is used for.

With respect to phytotherapies, the United Kingdom has adopted the European Traditional Herbal Medicinal Products Directive (THMPD) 2004/24/EC, which came into effect on April 30, 2011. This Directive establishes a regulatory approval process for herbal medicines in the European Union (EU). It requires each EU Member State to set up a traditional herbal registration scheme for manufactured traditional herbal medicines that are suitable for use without medical supervision.

Under this directive, companies in the United Kingdom are no longer able to sell manufactured herbal medicines unless they have an appropriate product license, either:

1. A full marketing authorization (MA) based on the safety, quality, and efficacy of the product, as with any regular medicine
2. A traditional herbal registration (THR) based on the safety, quality, and evidence of traditional use of the product.

The MHRA has adopted a certification trademark for use by sellers of approved herbal medicines. The following questions (Q) and answers (A) are provided by MHRA regarding the proper use of this certification mark in the United Kingdom:

Q. *What is a Certification Mark?*
A. A Certification Mark is a specific type of trade mark. It provides a guarantee that the goods bearing the mark meet certain defined standards or possess particular characteristics as defined by the owner of the mark.
Q. *What does THR Certification Mark mean?*
A. The Certification Mark indicates that the herbal medicine has been registered with the Medicines and Healthcare products Regulatory Agency (MHRA) under the Traditional Herbal Registration (THR) scheme and meets the required standards relating to its quality, safety, evidence of traditional use, and other criteria as set out under the Traditional Herbal Medicinal Products Directive (THMPD) 2004/24/EC.
Q. *Who owns the THR Certification Mark?*
A. The Certification Mark is owned by the MHRA. As the owner of the Mark, the MHRA can give permission to companies to use the certification mark on products registered under the THR scheme.
Q. *Who can use the Certification Mark?*
A. Only those companies granted a THR for their traditional herbal medicinal products will be authorized to use the THR Certification Mark for those registered products.

As the right to use the Certification Mark is linked to a specific product, retailers or other third parties who do not hold the actual THR cannot use the mark in their own right, that is, in relation to the promotion of their business. They may, however, use the Mark as part of promotional advertising provided by the THR holder or as part of general advice to customers to help them identify herbal medicines made to assured standards under the THR scheme.

Q. *Will all products registered under the THR scheme have to use the Certification Mark?*
A. No. Use of the Certification Mark is not compulsory. Companies can choose whether to include the THR Certification Mark on their registered products; however, MHRA recommends its use in order to help consumers identify those herbal medicines made to assured standards under the THR scheme.

Q. *What permission is needed from MHRA to use the THR Certification Mark?*

A. Permission to use the THR Certification Mark would normally be granted as part of the THR registration process.

Q. *What happens if I already have a THR product?*

A. The MHRA will be contacting all companies who were granted a THR before the introduction of the Certification Mark to discuss/coordinate changes to their product packs and leaflets. Adding the Certification Mark to your labels will not at this stage incur a fee. Nevertheless you will need to submit the proposed packaging with the Certification Mark to the MHRA for formal approval prior to use.

Q. *What happens if I initially choose not to use the Certification Mark, will I be able to make use of the Mark at a later date?*

A. Yes. But you will need to notify the MHRA of your plans and submit updated documentation before you place the Mark on labelling or the patient information leaflet.

Q. *I already have a THR product but I have not yet had the product registration varied to incorporate the certification mark, can I use the certification mark in my other advertising and publicity material?*

A. Yes. But you will need explicit authorization from the MHRA to do so. Once you have notified MHRA of your wish to use the Certification Mark, we will confirm in writing that your product is eligible to use the mark. You can then use the mark in advertising and MHRA will advise you of the steps you need to take to enable the Mark to be placed on particular packaging components.

It should be noted that, it is a condition of use of the Certification Mark that you indicate that it is a Certification Mark. You can do this by including the text "Certification Mark" adjacent to the Mark itself. You can also find specific guidance on consumer advertising for registered traditional herbal medicines at the following website: http://www.mhra.gov.uk/Publications/Regulatoryguidance/Medicines/Othermedicinesregulatoryguidance/CON043820

Q. *Are there any rules about how the Certification Mark is displayed on packs and leaflets?*

A. There are two versions of the Mark, one with a border, one without. You can use either version. But the Mark may only be displayed once on the pack or the patient information leaflet, or both. It should be subordinate in placement and prominence to the statutory information. It is also a condition of use that the Certification Mark shall not be used without indicating that it is a Certification Mark. You can do this by including the text "Certification Mark" adjacent to the Mark itself.

Q. *Can I change the color of the Certification Mark?*

A. No. The text, motif, and outer border (where used) must always be in black. The background can be in white or the same color as the product packaging.

Q. *What happens if the Certification Mark is used inappropriately in advertising by someone who is entitled to use it?*

A. The advertising of medicines is controlled by a combination of statutory measures (with both criminal and civil sanctions) enforced by the MHRA, self-regulation through Codes of Practice administered by trade associations, and the Advertising Standards Authority. Misuse of the mark will be dealt with in the same manner as other breaches of the rules relating to the advertising of medicinal products.

Q. *What happens if the Certification Mark is used by someone not entitled to use it?*

A. As the owner of the Certification Mark, the MHRA will take appropriate regulatory or legal action against any company or individual who makes unauthorized use or misuses the Certification Mark in any way.

Q. *I want to complain about inappropriate use of the Certification Mark, what do I do?*

A. Complaints about inappropriate use of the THR Certification Mark may be made in writing to the MHRA, enclosing where possible a copy of the advertisement or promotional material. These should be sent to: MHRA Herbal Policy Unit, 16-1, MHRA, Market Towers, 1 Nine Elms Lane, London SW8 5NQ; or emailed to thmrsqueries@mhra.gsi.gov.uk.

Q. *If I want to explain what the Certification Mark means, for example, on a company website, what should I say?*

A. The MHRA strongly suggests that in order to avoid inappropriate use of the Certification Mark, you use the following statement:

> The THR Certification Mark shows that a product has been registered by the Medicines and Healthcare products Regulatory Agency under the UK Traditional Herbal Registration Scheme. A product bearing this Mark meets the required standards for safety, quality and patient information.

Under this scheme, the permitted indications for the medicine are based on traditional usage and not on evidence of effectiveness of the product. More information about the THR scheme can be found on the Traditional Herbal Medicines Registration Scheme Internet site at www.mhra.gov.uk/thr

24.2.8 Current Regulations in the United States

References Cited/Additional Reading Resources

The Food and Drug Administration website: www.fda.gov

Unlike the other countries discussed in this chapter, the United States does not treat natural health products as being drugs. Instead, it treats these products as dietary supplements—or food. Every product of this type sold in the United States, which includes any form of health claims, carries the following (or a similar) disclaimer:

> These statements have not been evaluated by the Food and Drug Administration.
> This product is not intended to diagnose, treat, cure or prevent any disease.

A dietary supplement is a product taken by mouth that is intended to supplement the diet and that contains one or more "dietary ingredients." The "dietary ingredients" in these products may include vitamins, minerals, herbs or other botanicals, amino acids, and other substances found in the human diet, such as enzymes.

Dietary supplements must be labeled as such and must not be represented for use as a conventional food or as the sole item of a meal or the diet.

One way to distinguish dietary supplements from conventional foods is by looking at the nutrition information on the label of the product. Conventional foods must have a "Nutrition Facts" panel on their labels, but dietary supplements must have a "Supplement Facts" panel. Dietary supplement manufacturers and distributors are not required to obtain approval from FDA before marketing dietary supplements. Before a firm markets a dietary supplement, the firm is responsible for ensuring that

1. The products it manufactures or distributes are safe.
2. Any claims made about the products are not false or misleading.
3. The products comply with the Federal Food, Drug and Cosmetic Act and FDA regulations in all other respects, including compliance with cGMP manufacturing requirements.

The Dietary Supplement Health and Education Act (DSHEA) of 1994, which amended the Federal Food, Drug and Cosmetic Act, created a new regulatory framework for the safety and labeling of dietary supplements. The DSHEA defines dietary supplements as a category of food. However, there is one exception: if a dietary supplement meets the definition of a drug, it is regulated as a drug. Because dietary supplements are under the "umbrella" of foods, FDA's Center for Food Safety and Applied Nutrition (CFSAN) is responsible for the agency's oversight of these products.

In 2007, the FDA implemented new regulations to ensure quality throughout the manufacturing, packaging, labeling, and storing of dietary supplements. The rule includes GMP requirements for establishing quality control procedures, designing and constructing manufacturing plants, and testing ingredients and the finished product. It also includes requirements for recordkeeping and handling consumer product complaints. The 2007 rule was established to help ensure that dietary supplements are manufactured with controls that result in a consistent product free of contamination, with accurate labeling. Under the rule, manufacturers are required to evaluate the identity, purity, strength, and composition of their dietary supplements. Any dietary supplements that contain contaminants or do not contain the dietary ingredient they are represented to contain, the FDA would consider those products to be adulterated or misbranded. Sellers of adulterated or misbranded products are typically required to issue a recall of the product and face other penalties including significant fines. The aim of the 2007 rule is to prevent inclusion of the wrong ingredients, too much, or too little, of a dietary ingredient, contamination by substances such as natural toxins, bacteria, pesticides, glass, lead, and other heavy metals, as well as improper packaging and labeling. The 2007 rule includes flexible requirements that can evolve with improvements in scientific methods used for verifying identity, purity, strength, and composition of dietary supplements.

Generally, manufacturers do not need to register their products with FDA or get FDA approval before producing or selling dietary supplements. Manufacturers must make sure that product label information is truthful and not misleading. Under FDA regulations, all domestic and foreign companies that manufacture, package, label, or hold dietary supplement, including those involved with testing, quality control, and dietary supplement distribution in the United States, must comply with the Dietary Supplement Current Good Manufacturing Practices (cGMPs) for quality control. In addition, the manufacturer, packer, or distributor whose name appears on the label of a dietary supplement marketed in the United States is required to submit to FDA all serious adverse event reports associated with use of the dietary supplement in the United States. FDA's responsibilities include product information, such as labeling, claims, package inserts, and accompanying literature. The Federal Trade Commission (FTC) regulates dietary supplement advertising.

The following questions (Q) and answers (A) regarding dietary supplements are from the FDA website:

Q. *What is a dietary supplement?*

A. Congress defined the term "dietary supplement" in the Dietary Supplement Health and Education Act (DSHEA) of 1994. A dietary supplement is a product taken by mouth that contains a "dietary ingredient" intended to supplement the diet. The "dietary ingredients" in these products may include vitamins, minerals, herbs or other botanicals, amino acids, and substances such as enzymes, organ tissues, glandulars, and metabolites. Dietary supplements can also be extracts or concentrates and may be found in many forms such as tablets, capsules, softgels, gelcaps, liquids, or powders. They can also be in other forms, such as a bar, but if they are, information on their label must not represent the product as a conventional food or a sole item of a meal or diet. Whatever their form may be, DSHEA places dietary supplements in a special category under the general umbrella of "foods," not drugs, and requires that every supplement be labeled a dietary supplement.

Q *What is a "new dietary ingredient" in a dietary supplement?*

A. The Dietary Supplement Health and Education Act (DSHEA) of 1994 defined both of the terms "dietary ingredient" and "new dietary ingredient" as components of dietary supplements. In order for an ingredient of a dietary supplement to be a "dietary ingredient," it must be one or any combination of the following substances: a vitamin, a mineral, a herb or other botanical, an amino acid, a dietary substance for use by man to supplement the diet by increasing the total dietary intake (e.g., enzymes or tissues from organs or glands), or a concentrate, metabolite, constituent, or extract.

A "new dietary ingredient" is one that meets the preceding definition for a "dietary ingredient" and was not sold in the United States in a dietary supplement before October 15, 1994.

Q *What is FDA's role in regulating dietary supplements versus the manufacturer's responsibility for marketing them?*

A. In October 1994, the Dietary Supplement Health and Education Act (DSHEA) was signed into law by President Clinton. Before this time, dietary supplements were subject to the same regulatory requirements as were other foods. This new law, which amended the Federal Food, Drug and Cosmetic Act, created a new regulatory framework for the safety and labeling of dietary supplements. Under DSHEA, a firm is responsible for determining that the dietary supplements it manufactures or distributes are safe and that any representations or claims made about them are substantiated by adequate evidence to show that they are not false or misleading. This means that dietary supplements do not need approval from FDA before they are marketed. Except in the case of a new dietary ingredient, where pre-market review for safety data and other information is required by law, a firm does not have to provide FDA with the evidence it relies on to substantiate safety or effectiveness before or after it markets its products. Also, manufacturers need to register themselves pursuant to the Bioterrorism Act with FDA before producing or selling supplements.

In June, 2007, FDA published comprehensive regulations for Current Good Manufacturing Practices for those who manufacture, package, or hold dietary supplement products. These regulations focus on practices that ensure the identity, purity, quality, strength, and composition of dietary supplements.

Q *When must a manufacturer or distributor notify the FDA about a dietary supplement it intends to market in the United States?*

A. The Dietary Supplement Health and Education Act (DSHEA) requires that a manufacturer or distributor notify the FDA if it intends to market a dietary supplement in the United States that contains a "new dietary ingredient." The manufacturer (and distributor) must demonstrate to the FDA why the ingredient is reasonably expected to be safe for use in a dietary supplement, unless it has been recognized as a food substance and is present in the food supply.

There is no authoritative list of dietary ingredients that were marketed before October 15, 1994. Therefore, manufacturers and distributors are responsible for determining if a dietary ingredient is "new" and if it is not, for documenting that the dietary supplements its sells, containing the dietary ingredient, were marketed before October 15, 1994.

Q. *What information must the manufacturer disclose on the label of a dietary supplement?*

A. FDA regulations require that certain information appear on dietary supplement labels. Information that must be on a dietary supplement label includes a descriptive name of the product stating that it is a "supplement"; the name and place of business of the manufacturer, packer, or distributor; a complete list of ingredients; and the net contents of the product.

In addition, each dietary supplement (except for some small volume products or those produced by eligible small businesses) must have nutrition labeling in the form of a "Supplement Facts" panel. This label must identify each dietary ingredient contained in the product.

Q. *Must all ingredients be declared on the label of a dietary supplement?*

A. Yes, ingredients not listed on the "Supplement Facts" panel must be listed in the "other ingredient" statement beneath the panel. The types of ingredients listed there could include the source of dietary ingredients, if not identified in the "Supplement Facts" panel (e.g., rose hips as the source of vitamin C), other food ingredients (e.g., water and sugar), and technical additives or processing aids (e.g., gelatin, starch, colors, stabilizers, preservatives, and flavors). For more details, see Federal Register Final Rule—62 FR 49826 September 23, 1997.

Q. *Are dietary supplement serving sizes standardized or are there restrictions on the amount of a nutrient that can be in one serving?*

A. Other than the manufacturer's responsibility to ensure safety, there are no rules that limit a serving size or the amount of a nutrient in any form of dietary supplements. This decision is made by the manufacturer and does not require FDA review or approval.

Q. *Where can I get information about a specific dietary supplement?*

A. Manufacturers and distributors do not need FDA approval to sell their dietary supplements. This means that FDA does not keep a list of manufacturers, distributors, or the dietary supplement products they sell. If you want more detailed information than the label tells you about a specific product, you may contact the manufacturer of that brand directly. The name and address of the manufacturer or distributor can be found on the label of the dietary supplement.

Q. *Who has the responsibility for ensuring that a dietary supplement is safe?*

A. By law (DSHEA), the manufacturer is responsible for ensuring that its dietary supplement products are safe before they are marketed. Unlike drug products that must be proven safe and effective for their intended use before marketing, there are no provisions in the law for FDA to "approve" dietary supplements for safety or effectiveness before they reach the consumer.

Under DSHEA, once the product is marketed, FDA has the responsibility for showing that a dietary supplement is "unsafe," before it can take action to restrict the product's use or removal from the marketplace. However, manufacturers and distributors of dietary supplements must record, investigate, and forward to FDA any reports they receive of serious adverse events associated with the use of their products that are reported to them directly. FDA is able to evaluate these reports and any other adverse event information reported directly to us by healthcare providers or consumers to identify early signals that a product may present safety risks to consumers.

Q. *Do manufacturers or distributors of dietary supplements have to tell FDA or consumers what evidence they have about their product's safety or what evidence they have to back up the claims they are making for them?*

A. No, except for rules described earlier that govern "new dietary ingredients," there is no provision under any law or regulation that FDA enforces that requires a firm to disclose to FDA or consumers the information they have about the safety or purported benefits of their dietary supplement products.

Likewise, there is no prohibition against them making this information available either to FDA or to their customers. It is up to each firm to set its own policy on disclosure of such information.

Q. *What is FDA's oversight responsibility for dietary supplements?*

A. Because dietary supplements are under the "umbrella" of foods, FDA's Center for Food Safety and Applied Nutrition (CFSAN) is responsible for the agency's oversight of these products. FDA's efforts to monitor the marketplace for potential illegal products (i.e., products that may be unsafe or make false or misleading claims) include obtaining information from inspections of dietary supplement manufacturers and distributors, the Internet, consumer and trade complaints, occasional laboratory analyses of selected products, and adverse events associated with the use of supplements that are reported to the agency.

Q. *Does FDA routinely analyze the content of dietary supplements?*

A. In that FDA has limited resources to analyze the composition of food products, including dietary supplements, it focuses these resources first on public health emergencies and products that may have caused injury or illness. Enforcement priorities then go to products thought to be unsafe or fraudulent or in violation of the law. The remaining funds are used for routine monitoring of products pulled from store shelves or collected during inspections of manufacturing firms.

The agency does not analyze dietary supplements before they are sold to consumers.

The manufacturer is responsible for ensuring that the "Supplement Facts" label and ingredient list are accurate, that the dietary ingredients are safe, and that the content matches the amount declared on the label. FDA does not have resources to analyze dietary supplements sent to the agency by consumers who want to know their content. Instead, consumers may contact the manufacturer or a commercial laboratory for an analysis of the content.

Q. *Is it legal to market a dietary supplement product as a treatment or cure for a specific disease or condition?*

A. No, a product sold as a dietary supplement and promoted on its label or in labeling as a treatment, prevention, or cure for a specific disease or condition would be considered an unapproved, and thus illegal, drug. Labeling refers to the label as well as accompanying material that is used by a manufacturer to promote and market a specific product.

To maintain the product's status as a dietary supplement, the label and labeling must be consistent with the provisions in the Dietary Supplement Health and Education Act (DSHEA) of 1994.

Q. *Who validates claims and what kinds of claims can be made on dietary supplement labels?*

A. FDA receives many consumer inquiries about the validity of claims for dietary supplements, including product labels, advertisements, media, and printed materials. The responsibility for ensuring the validity of these claims rests with the manufacturer, FDA, and, in the case of advertising, with the Federal Trade Commission.

By law, manufacturers may make three types of claims for their dietary supplement products: health claims, structure/function claims, and nutrient content claims. Some of these claims describe the link between a food substance and disease or a health-related condition; the intended benefits of using the product; or the amount of a nutrient or dietary substance in a product.

Q. *Why do some supplements have wording (a disclaimer) that says: "This statement has not been evaluated by the FDA. This product is not intended to diagnose, treat, cure, or prevent any disease"?*

A. This statement or "disclaimer" is required by law (DSHEA) when a manufacturer makes a structure/function claim on a dietary supplement label. In general, these claims describe the role of a nutrient or dietary ingredient intended to affect the structure or function of the body.

The manufacturer is responsible for ensuring the accuracy and truthfulness of these claims; they are not approved by FDA. For this reason, the law says that if a dietary supplement label includes such a claim, it must state in a "disclaimer" that FDA has not evaluated this claim.

The disclaimer must also state that this product is not intended to "diagnose, treat, cure or prevent any disease," because only a drug can legally make such a claim.

Q. *How are advertisements for dietary supplements regulated?*

A. The Federal Trade Commission (FTC) regulates advertising, including infomercials, for dietary supplements and most other products sold to consumers. FDA works closely with FTC in this area, but FTC's work is directed by different laws.

24.3 FUTURE OF PHYTOTHERAPIES: WORLD HEALTH ORGANIZATION (WHO)

References Cited/Additional Reading Resources

The WHO International Regulatory Cooperation for Herbal Medicines (IRCH) website: www.who.int/medicines/areas/traditional/irch/en/

In 2001, the World Health Organization produced a document entitled Legal Status of Traditional Medicine and Complementary/Alternative Medicine: A Worldwide

Review, which includes information on the regulation and registration of herbal medicines as well as of non-medication therapies and traditional and complementary or alternative medical practitioners in 123 countries. The paper required 10 years of data compilation by the authors and when completed it was acknowledged that some information was likely obsolete. However, as a general reference source it is a very useful publication. For Internet access to this publication, see http://apps.who.int/medicinedocs/en/d/Jh2943e/10.html.

The WHO's International Regulatory Cooperation for Herbal Medicines (IRCH) is a global network of regulatory authorities responsible for the regulation of herbal medicines. Its mission is to protect and promote public health and safety through improved regulation for herbal medicines. Membership in the IRCH is open to any national regulatory authority responsible for the regulation of herbal medicines and regional/sub-regional bodies responsible for the regulation of herbal medicines. An interested country must apply for membership through a national regulatory authority.

WHO conducts the administrative admission procedure that starts with a candidate country showing interest in joining IRCH. WHO will then screen and review the application according to the admission criteria set according to the terms of reference of the IRCH. For more information, visit the IRCH website at http://www.who.int/medicines/areas/traditional/irch/en/

The first WHO working group meeting on international regulatory cooperation for herbal medicines took place in 2005 in Ottawa, Canada, hosted by and with the financial support of the Natural Health Products Directorate, Health Products and Food Branch of Health Canada. Twenty-nine participants from 16 countries attended the meeting. Representatives of regional bodies and groups on herbal medicines were also invited to attend the meeting. After intensive discussion, the participants agreed by consensus to establish a network for International Regulatory Cooperation for Herbal Medicines (IRCH).

The second WHO working group meeting on international regulatory cooperation for herbal medicines took place in 2006 in Beijing, China. Delegations from 11 of 13 Members of IRCH attended the meeting together with observers from four national regulatory authorities. These terms of reference for IRCH were drafted based on the discussion at the first WHO working group meeting and reviewed and agreed by consensus at the second WHO working group meeting in late 2006. Yearly meetings have been held since then, with more participants at each meeting.

The 2009 World Health Assembly, in Resolution 62.13, urged its Member States, in accordance with national capacities, priorities, relevant legislation, and circumstances as follows:

1. To consider adopting and implementing the Beijing Declaration on Traditional Medicine (see http://www.who.int/medicines/areas/traditional/congress/en/) in accordance with national capacities, priorities, relevant legislation, and circumstances
2. To respect, preserve, and widely communicate, as appropriate, the knowledge of traditional medicine, treatments, and practices, appropriately based

on the circumstances in each country and on evidence of safety, efficacy, and quality

3. To formulate national policies, regulations, and standards, as part of comprehensive national health systems, to promote appropriate, safe, and effective use of traditional medicine
4. To consider, where appropriate, including traditional medicine into their national health systems based on national capacities, priorities, relevant legislation, and circumstances and on evidence of safety, efficacy, and quality
5. To further develop traditional medicine based on research and innovation, giving due consideration to the specific actions related to traditional medicine in the implementation of the global strategy and plan of action on public health, innovation, and intellectual property
6. To consider, where appropriate, establishing systems for the qualification, accreditation, or licensing of traditional medicine practitioners and to assist traditional medicine practitioners to upgrade their knowledge and skill in collaboration with relevant health providers, on the basis of traditions and customs of indigenous peoples and communities
7. To consider strengthening communication between conventional and traditional medicine providers and, where appropriate, establishing appropriate training programs with content related to traditional medicine for health professionals, medical students, and relevant researchers
8. To cooperate with each other in sharing knowledge and practices of traditional medicine and exchanging training programs on traditional medicine, consistent with national legislation and relevant international obligations

In 2013 the World Health Organization published a book entitled The WHO Traditional Medicine Strategy 2014–2023. This book was developed and launched in response to the World Health Assembly resolution on traditional medicine (WHA 62.13—2009). This strategy aims to support WHO Member States in developing proactive policies and implementing action plans that will strengthen the role traditional medicine plays in keeping populations healthy. For Internet access to this publication, see www.who.int/medicines/publications/traditional/trm_strategy14_23/en/.

Clearly the World Health Organization encourages the medical use of phytotherapies as part of traditional medicine programs. However, there is a trend in most industrialized countries to more, not less, government control of the manufacture and use of phytotherapies. Safety concerns are the usual reason given for more oversight of the manufacture and use of natural health products. Government regulations are commonly used to protect the public health when phytotherapies are made, sold, and used. As more countries adopt the WHO guidelines, it is likely that restrictions on the manufacture, sale, and use of phytotherapies will be expanded.

FURTHER READING

EU Directives on Traditional Herbal Medicinal Products (THMP)—Directive 2004/24/EC and Directive 2001/83/EC.

FDA's Dietary Supplement Health and Education Act (DSHEA) of 1994.

Japanese Herbal Medicine Codex.

Japanese Pharmacopoeia.

Japanese Standards for Herbal Medicines.

The Ayurvedic Pharmacopoeia of India.

The Pharmacopoeia of the People's Republic of China (PPRC).

WHO's 2001 Report—Legal Status of Traditional Medicine and Complementary/Alternative Medicine: A Worldwide Review.

WHO Traditional Medicine Strategy Report 2014–2023.

INDEX

Note: Page numbers in *italics* refer to Figures; those in **bold** to Tables.

Phytotherapies: Efficacy, Safety, and Regulation, First Edition. Edited by Iqbal Ramzan.

Printed and bound by CPI Group (UK) Ltd, Croydon, CR0 4YY
16/09/2026
14963421-0008